Selected Titles in This Series

24 **B. Brent Gordon, James D. Lewis, Stefan Müller-Stach, Shuji Saito, and Noriko Yui, Editors,** The arithmetic and geometry of algebraic cycles, 2000

23 **Pierre Hansen and Odile Marcotte, Editors,** Graph colouring and applications, 1999

22 **Jan Felipe van Diejen and Luc Vinet, Editors,** Algebraic methods and q-special functions, 1999

21 **Michel Fortin, Editor,** Plates and shells, 1999

20 **Katie Coughlin, Editor,** Semi-analytic methods for the Navier-Stokes equations, 1999

19 **Rajiv Gupta and Kenneth S. Williams, Editors,** Number theory, 1999

18 **Serge Dubuc and Gilles Deslauriers, Editors,** Spline functions and the theory of wavelets, 1999

17 **Olga Kharlampovich, Editor,** Summer school in group theory in Banff, 1996, 1998

16 **Alain Vincent, Editor,** Numerical methods in fluid mechanics, 1998

15 **François Lalonde, Editor,** Geometry, topology, and dynamics, 1998

14 **John Harnad and Alex Kasman, Editors,** The bispectral problem, 1998

13 **Michel C. Delfour, Editor,** Boundaries, interfaces, and transitions, 1998

12 **Peter C. Greiner, Victor Ivrii, Luis A. Seco, and Catherine Sulem, Editors,** Partial differential equations and their applications, 1997

11 **Luc Vinet, Editor,** Advances in mathematical sciences: CRM's 25 years, 1997

10 **Donald E. Knuth,** Stable marriage and its relation to other combinatorial problems: An introduction to the mathematical analysis of algorithms, 1997

9 **D. Levi, L. Vinet, and P. Winternitz, Editors,** Symmetries and integrability of difference equations, 1996

8 **J. Feldman, R. Froese, and L. M. Rosen, Editors,** Mathematical quantum theory II: Schrödinger operators, 1995

7 **J. Feldman, R. Froese, and L. M. Rosen, Editors,** Mathematical quantum theory I: Field theory and many-body theory, 1994

6 **Guido Mislin, Editor,** The Hilton Symposium 1993: Topics in topology and group theory, 1994

5 **D. A. Dawson, Editor,** Measure-valued processes, stochastic partial differential equations, and interacting systems, 1994

4 **Hershy Kisilevsky and M. Ram Murty, Editors,** Elliptic curves and related topics, 1994

3 **Rémi Vaillancourt and Andrei L. Smirnov, Editors,** Asymptotic methods in mechanics, 1993

2 **Philip D. Loewen,** Optimal control via nonsmooth analysis, 1993

1 **M. Ram Murty, Editor,** Theta functions: From the classical to the modern, 1993

Volume 24

CRM PROCEEDINGS & LECTURE NOTES

Centre de Recherches Mathématiques
Université de Montréal

The Arithmetic and Geometry of Algebraic Cycles

Proceedings of the CRM Summer School
June 7–19, 1998
Banff, Alberta, Canada

B. Brent Gordon
James D. Lewis
Stefan Müller-Stach
Shuji Saito
Noriko Yui
Editors

The Centre de Recherches Mathématiques (CRM) of the Université de Montréal was created in 1968 to promote research in pure and applied mathematics and related disciplines. Among its activities are special theme years, summer schools, workshops, postdoctoral programs, and publishing. The CRM is supported by the Université de Montréal, the Province of Québec (FCAR), and the Natural Sciences and Engineering Research Council of Canada. It is affiliated with the Institut des Sciences Mathématiques (ISM) of Montréal, whose constituent members are Concordia University, McGill University, the Université de Montréal, the Université du Québec à Montréal, and the Ecole Polytechnique. The CRM may be reached on the Web at www.crm.umontreal.ca.

American Mathematical Society
Providence, Rhode Island USA

The production of this volume was supported in part by the Fonds pour la Formation de Chercheurs et l'Aide à la Recherche (Fonds FCAR) and the Natural Sciences and Engineering Research Council of Canada (NSERC).

1991 *Mathematics Subject Classification.* Primary 11Gxx, 14C30, 14C35, 14C25, 14C17, 14F05, 14F10, 14F42, 14F43, 14F20, 14G35, 19Dxx, 19Fxx.

Library of Congress Cataloging-in-Publication Data

The arithmetic and geometry of algebraic cycles : proceedings of the CRM summer school, June 7–19, 1998, Banff, Alberta, Canada / B. Brent Gordon... [et al.], editors.
p. cm. — (CRM proceedings & lecture notes, ISSN 1065-8580 ; v. 24)
Includes bibliographical references.
ISBN 0-8218-1954-2 (alk. paper)
1. Algebraic cycles—Congresses. I. Gordon, B. Brent, 1953– II. Series.
QA564.A72 2000
516.3′5—dc21 99-057916

∞ The paper used in this book is acid-free and falls within the guidelines
established to ensure permanence and durability.
This volume was typeset using $\mathcal{A}\mathcal{M}\mathcal{S}$-TEX,
the American Mathematical Society's TEX macro system,
and submitted to the American Mathematical Society in camera ready
form by the Centre de Recherches Mathématiques.
Visit the AMS home page at URL: `http://www.ams.org/`

10 9 8 7 6 5 4 3 2 1 05 04 03 02 01 00

Contents

Preface vii

Contents of the Proceedings of the NATO Advance Study Institute: Arithmetic and Geometry of Algebraic Cycles xv

Conference Programme xvii

Conference Picture xxi

List of Participants xxiii

Authors' Addresses xxvii

Cohomological Methods 1

Filtrations on the Cohomology of Abelian Varieties 3
Abdulali, Salman

Building Mixed Hodge Structures 13
Arapura, Donu

The Atiyah-Chern Character Yields the Semiregularity Map as well as the Infinitesimal Abel-Jacobi Map 33
Buchweitz, Ragnar-Olaf, and Flenner, Hubert

Regulators and Characteristic Classes of Flat Bundles 47
Dupont, Johan, Hain, Richard, and Zucker, Steven

Height Pairings Asymptotics and Bott-Chern Forms 93
Harris, Bruno and Wang, Bin

Logarithmic Hodge Structures and Classifying Spaces 115
Kato, Kazuya and Usui, Sampei

Chow Groups and Motives 131

Motives and Algebraic de Rham Cohomology 133
Asakura, Masanori

Hermitian Vector Bundles and Characteristic Classes 155
Burgos Gil, José I.

The Mixed Motive of a Projective Variety 183
Hanamura, Masaki

Bloch's Conjecture and the K-Theory of Projective Surfaces 195
Pedrini, Claudio

From Jacobians to One-Motives: Exposition of a Conjecture of Deligne 215
Ramachandran, Niranjan

Motives, algebraic cycles and Hodge theory 235
Saito, Shuji

Arithmetic Methods 255

Picard-Fuchs Uniformization: Modularity of the Mirror Map and Mirror-Moonshine 257
Doran, Charles F.

Hilbert Modular Varieties in Positive Characteristic 283
Goren, Eyal Z.

On the Néron-Severi groups of some $K3$ surfaces 305
Goto, Yasuhiro

Torsion Zero-Cycles and the Abel-Jacobi Map over the Real Numbers 329
van Hamel, Joost

A Remark on the Griffiths Group of Certain Product Varieties 361
Kimura, Kenichiro

p-adic Abel-Jacobi Maps and p-adic Heights 367
Nekovář, Jan

Crystalline Fundamental Groups and p-adic Hodge Theory 381
Shiho, Atsushi

Thompson Series, and the Mirror Maps of Certain Pencils of $K3$ Surfaces 399
Verrill, Helena and Yui, Noriko

Preface

The Centre de recherches mathématiques (CRM), Montréal, Summer School on "The Arithmetic and Geometry of Algebraic Cycles" was held at the Banff Centre for Conferences in Banff (Alberta, Canada) from June 7 until June 19, 1998. This conference, organized simultaneously as a NATO Advanced Study Institute, served as the kick-off activity of the 1998 CRM theme year on Number Theory and Arithmetic Geometry.

There were 109 participants who came from 17 countries: Belgium, Canada, China, France, Germany, Greece, India, Italy, Japan, Mexico, the Netherlands, Romania, Russia, Spain, Switzerland, the United States and the United Kingdom.

During a period of two weeks, 41 invited lectures and 20 contributed lectures were presented. Four lectures by invited speakers were delivered every day, followed by two sessions of contributed talks. Many informal discussions and working sessions involving small groups were organized by individual participants. In addition, participants' reprints and preprints were displayed throughout in a lounge next to the auditorium, which further enhanced opportunities for communication and interaction.

As a subfield of algebraic geometry, the subject of algebraic cycles has gone through various interactions with algebraic K-theory, Hodge theory, arithmetic algebraic geometry, number theory, and topology. These interactions have led to such developments as: a description of Chow groups in terms of algebraic K-theory; the application of the Mercurjev-Suslin theorem to the arithmetic Abel-Jacobi mapping, with corresponding torsion results for Chow groups; progress on the celebrated conjectures of Hodge, and of Tate, which compute cycle class groups respectively in terms of Hodge theory or as the invariants of a Galois group action on étale cohomology; the conjectures of Bloch and Beilinson, which explain the zero or pole of a variety's L-function, as well as to interpret the leading non-zero coefficient of its Taylor expansion at a critical point, in terms of arithmetic and geometric invariants of the variety and its cycle class groups. The immense recent progress in algebraic cycles, based on so many interactions with so many other areas of mathematics, has contributed to a considerable degree of inaccessibility, especially for graduate students and non-specialists. Even specialists in one approach to algebraic cycles may not understand other approaches well.

To our knowledge, this was the first conference to focus at the same time on both arithmetic and geometric aspects of algebraic cycles, bringing together in one place experts who study algebraic cycles from several different perspectives. Our goal was to have leading experts speak on their various points of view, aiming their talks towards graduate students and recent Ph.D.'s interested in algebraic cycles, as well as to researchers who work on the arithmetic or geometry of algebraic cycles from other directions. The interaction among these schools was strikingly successful, each being beneficiaries of each other. For a student learning the subject of algebraic cycles, this meeting offered the unique opportunity to view the depth and the breadth of this subject.

We are quite confident that this goal was at least partially achieved, we even hope that the conference will continue to have a positive impact on the future relations among various schools of thought about algebraic cycles. The meeting was also a success from the point of view of its organization and the way that the beautiful location at the Banff Centre provided such a stimulating atmosphere.

The present volume is the second volume of the Proceedings of the meeting, and it contains 20 articles by some of the participants. The first volume, to be published by the Kluwer Academic Publishers in the NATO Advanced Study Institute Series, contains an additional 21 articles. All papers in both volumes were rigorously refereed. In addition, some participants are publishing extended papers in the CRM monograph series.

The papers in this volume have been organized into three categories:

1. Cohomological Methods;
2. Chow groups and Motives;
3. Arithmetic Methods.

Cohomological Methods

Let us describe the contents of the section *Cohomological Methods.*

Salman Abdulali's article is a nice discussion of some of the filtrations that have been defined on the rational cohomology of a smooth complex projective variety X, and what can be said about those filtrations in the case of Abelian varieties. First there is the rational Hodge filtration, defined as the largest rational Hodge structure contained in a step of the (usual) Hodge filtration. Next, there is Grothendieck's arithmetic, or coniveau, filtration, generated by the Gysin images of the cohomology of smooth resolutions of all subvarieties of X; Grothendieck's formulation of the General Hodge Conjecture is that these two filtrations coincide. More recently, Friedlander and Mazur defined the correspondence filtration, generated by the images of of all correspondence homomorphisms $Z_*\colon H^*(Y,\mathbb{Q}) \to H^*(X,\mathbb{Q})$ for all smooth Y and all $Z \subset Y \times X$ of appropriate dimension; and further, they conjectured that their correspondence filtration coincides with Grothendieck's coniveau filtration. Then Abdulali defines the group theoretic filtration as the sum of all rational Hodge substructures of $H^*(X,\mathbb{Q})$ that are isomorphic, up to Tate twist, to some rational Hodge substructure of $H^*(Y,\mathbb{Q})$ for some Y. The main results are that in general the correspondence filtration is contained in the group theoretic filtration, which in turn is contained in the rational Hodge filtration, and moreover, the correspondence filtration does indeed coincide with the coniveau filtration for any Abelian variety for which the General Hodge Conjecture is presently known to hold.

Donu Arapura's paper gives an informative and elementary account of how to build mixed Hodge structures. The basic idea, which goes back to Deligne, is to introduce a notion of cohomological mixed Hodge complexes on which an integral structure can be imposed. The second main idea is to use a cohomological descent argument to limit the problem of constructing such complexes to some particular spaces and coefficient sheaves. Arapura carefully explains these ideas, and includes a number of examples, such as some recent work due to Navarro Aznar on complexes of real analytic differential forms, and of Burgos Gil on a C^∞ analogue.

The article by Ragnar-Olaf Buchweitz and Hubert Flenner concerns the deformations of a closed analytic subset on an arbitrary analytic space, or a scheme over some field, X, and the corresponding semiregular map, whose existence is known *a priori* for certain special cases. This map was first introduced by Severi for a curve on a surface, and later generalized by Kodaira and Spencer to arbitrary divisors on a complex manifold. Bloch then extended the notion to arbitrary cycles on smooth complex projective schemes by constructing a semiregular map for cycles that are locally complete intersections, and applied it to Grothendieck's variational Hodge conjecture for semiregular locally complete intersections. The point of this paper is to show that a semiregular map *always* exists, by constructing a more general semiregular map for coherent sheaves on X from the classical Atiyah-Chern character. They then deduce a description of the infinitesimal Abel-Jacobi map as the leading term of this generalized semiregulator map.

Johan Dupont, Richard Hain and Steven Zucker contributed an article about regulators and characteristic classes of flat bundles. Let $X_{/\mathbb{C}}$ be an algebraic manifold, and $E \to X$ an algebraic vector bundle. In Deligne-Beilinson cohomology, Beilinson and Gillet have (independently) constructed Chern classes which agree with the usual Chern classes under the map into singular cohomology. On the other hand, if E is flat with a flat connection ∇, then there is a unique algebraic structure on E such that ∇ has only regular singularities. The first theorem is that for such (E, ∇) the characteristic class defined by Cheeger and Simon coincides with the Chern class constructed by Gillet and Beilinson; more generally, this result is shown to hold whenever the two classes can be interpreted as belonging to the same cohomology group. In the universal case, where the base is $\mathrm{BLG}(\mathbb{C})^\delta$, they show that the universal Cheeger-Simons class is one-half of the Borel regulator element. They also conjecture that more generally these classes agree in the universal case.

Bruno Harris and Bin Wang investigate the asymptotic behaviour of the Archimedian height pairing on a compact Kähler manifold, say X. More specifically, when A and B are disjoint subvarieties of X with $\dim A + \dim B = \dim X - 1$, then their height pairing may be computed as the integral over A of the Green's current G_B of B—but this may not converge if $A \cap B = \emptyset$. However, if one takes a family A_t which is analytic near $A \cap B$, with $A_t \cap B = \emptyset$, then, building on work of Hain, Wang showed that asymptotically the height pairing of A_t with B has the form $e \log |t|^2 + k + o(1)$, where the integer e is a certain Chern number. In this paper Harris and Wang give an interpretation of the constant k as a sum of three integrals, one of which is related to the Bott-Chern secondary classes of vector bundle exact sequences. The constant depends on the deformation.

The article of Kazuya Kato and Sampei Usui looks into logarithmic Hodge structures and classifying spaces in their attempt to add *points at infinity* to the classifying space of polarized Hodge structures. If D is a Hermitian symmetric domain and Γ is an arithmetic group, then it is known by Mumford et. al. that $\Gamma\backslash D$ has toroidal compactification. The authors try to generalize this to the case where D is the classifying space of polarized Hodge structures of arbitrary Hodge type. The enlarged space $\Gamma\backslash D_\Sigma \subset \Gamma\backslash D$ has the structure of a generalized logarithmic analytic space which is a fine moduli space of polarized logarithmic Hodge structures of certain type.

Chow Groups and Motives

In many other papers in this (and the NATO) volume, motivic cohomology occurs in the form of higher Chow groups which one may view as extension groups in the category of mixed motives. This point of view leads us to the papers which fall into the category *Chow Groups and Motives.*

Masanori Asakura investigates motives from the point of view of algebraic de Rham cohomology. If X is a complex projective algebraic manifold, there are two well known cycle class map constructions, namely the fundamental class of an algebraic cycle into singular cohomology, and the Abel-Jacobi map defined on its kernel. The images of both maps lie inside certain groups that can be described in terms of extension data associated to the category of mixed Hodge structures (MHS). While the Hodge conjecture says something about the image of the fundamental class map, in general the image and the kernel of the Abel-Jacobi map are very complicated, as follows from work of Mumford and Griffiths. Bloch first, and Beilinson later, conjectured that this situation is explained by a descending filtration on the Chow group of X, whose graded pieces can be described in terms of the Grothendieck motive of X. Beilinson conjectured another description of the graded pieces in terms of a hyper-Ext group in the conjectural category of mixed motives. Unfortunately, the Hodge realizations of these Ext groups vanish for higher extensions, i.e., $\mathrm{Ext}^{\ell}_{\mathrm{MHS}} = 0$ for $\ell \geq 2$, due to the fact that there are "too many" mixed Hodge structures over $\mathbb{Q}$. What Asakura does instead is to consider arithmetic Hodge structures, where the higher extension groups do not vanish in general.

The article of José Burgos Gil describes the Beilinson regulator for higher K-groups in terms of Chern-Weil theory, generalizing some work of Gillet and Soulé that described the regulator for K_1 using Bott-Chern forms. There are a number of different ways to construct Chern classes of complex vector bundles in cohomology. One approach, due to Grothendieck, is via the cohomology of projective bundles. Although a weight argument shows that on the higher K-groups of smooth complex projective varieties the Chern classes in ordinary cohomology are at most torsion, the Chern classes into Deligne cohomology are expected to be non-trivial. Gillet has generalized Grothendieck's method to construct characteristic classes from the higher K-groups to any "reasonable" cohomology theory. In particular, when the characteristic class is the Chern character, and the cohomology is real Deligne-Beilinson cohomology, Burgos Gil shows that the Chern class map can be obtained from Beilinson's regulator map.

In his paper in this volume, Masaki Hanamura adds to his previous work on mixed motives and algebraic cycles by describing how to define the cohomological mixed motive of a projective variety which is not necessarily smooth, assuming strong resolution of singularities. The general approach here is to consider complexes of cycles, similarly as Bloch used originally to define higher Chow groups. In order to deal with the possible presence of singularities, a so-called cubical hyperresolution of a simplicial scheme built from smooth projective schemes and having the same cohomology as the original variety is constructed. Then the issue is to show that the resulting object in the derived category of mixed motives, which Hanamura had defined earlier, is independent of the choice of cubical hyperresolution, and that the induced Chow cohomology coincides with Bloch's higher Chow groups when the variety is actually smooth.

Claudio Pedrini proves a sort of decomposability result for the K-theory of projective surfaces X whose Chow group of codimension two cycles satisfies $\mathrm{CH}^2(X) = \mathbb{Z}$. The Bloch-Quillen isomorphism from the Chow groups of X to the Zariski cohomology of the sheaf of K-groups on X, together with the Milnor pairing on the K-groups of X, induces maps τ from the tensor product of a Chow group of X with a K-group of its field of definition to a certain Zariski cohomology of sheaf of K-groups. The goal of this paper is to give conditions which will guarantee that τ is an isomorphism, up to torsion. From work of Collino, Müller-Stach, Gordon and Lewis (see for example the paper by Gordon and Lewis in the companion Proceedings of the NATO Advanced Study Institute volume), and others, it is by no means the case that τ is in general either surjective or injective. Indeed one would expect to have to impose fairly stringent conditions on X in order to obtain an isomorphism (up to torsion). Here Pedrini is able to give definitive results when $L = \mathbb{C}$ and X is smooth, and $\mathrm{CH}^2(X) \simeq \mathbb{Z}$. Examples include surfaces of general type for which the genus $p_g(X) = 0$ and Bloch's conjecture holds.

Niranjan Ramachandran's paper is a motivated expository review of the current state of affairs in the theory of one-motives. The basic idea is that a one-motive should be something that can be constructed algebraically, analogously to the Jacobian of a smooth projective curve or the Albenese or Picard varieties, or, in some cases, the intermediate Jacobian, of a smooth projective variety, but with the difference that the cohomological realization of a one-motive should be the first cohomology (or homology) of an arbitrary algebraic variety. In particular, its Hodge realization would be a mixed Hodge structure. Despite the results that have been achieved with so-called generalized Jacobians and generalized Albenese varieties, Ramachandran makes a good case that there is still far to go towards developing a good theory of one-motives.

Shuji Saito's paper is a summary of the talks he gave at the conference. In the companion volume, the Proceedings of the NATO Advanced Study Institute, Saito discusses some formal consequences of the existence of a Bloch-Beilinson filtration, and his candidate prescription for such a filtration; in this volume Saito addresses the very interesting problem of finding candidate higher order Abel-Jacobi maps whose kernels may yield the Bloch-Beilinson filtration. Raskind had previously considered an arithmetical approach, involving ℓ-adic Abel-Jacobi maps into continuous cohomology and using the Hochschild-Serre spectral sequence, while Saito's approach is more Hodge theoretic. Here he introduces a version of higher Abel-Jacobi maps involving families of varieties over a base variety, together with the associated variation of Hodge structure. For instance, the 'usual' Abel-Jacobi map is now replaced by a Poincaré normal function, and his higher Abel-Jacobi maps are perhaps better described as higher normal functions. Then he applies this variational Abel-Jacobi map to his candidate Bloch-Beilinson filtration, as well as to zero-cycles on general complete intersections, and to the problem of constructing non-torsion elements in higher Griffiths groups.

Arithmetic Methods

Finally let us explain the contents of the contributions which use *Arithmetic Methods*:

The article of Charles Doran is concerned with the *Mirror-Moonshine Conjecture* of B. Lian and S.-T. Yau that the mirror maps arising in string theory are *modular* in the sense that their q-expansions are related to Thompson series (or Hauptmoduln of some genus zero subgroups of $\mathrm{PSL}(2, \mathbb{R})$). In this paper, Doran determines when the mirror maps of certain families of elliptic curves and $K3$ surfaces are Hauptmoduln. This is done by giving geometric criteria for modularity which characterize orbifold uniformization properties of the Picard-Fuchs differential equations. The main ingredients in his proof are classical theory of ordinary differential equations.

In his paper, Eyal Goren gives a survey on some recent developments concerning Hilbert modular varieties in positive characteristic. This includes discussions on stratifications of Hilbert modular varieties, and those of a class of Shimura varieties, the structure of the p-torsion and Hasse invariants. There are several methods to stratify moduli spaces of Abelian varieties in characteristic p. The paper describes the Ekedahl-Oort stratification, the Newton polygon stratification, as well as several examples.

Yasuhiro Goto's paper considers certain classes of $K3$ surfaces which are obtained from resolving singularities of some one-parameter deformation of weighted Delsarte surfaces. These $K3$ surfaces come with a lot of exceptional divisors. In this article, Goto looks at these K3 surfaces over fields of positive characteristic. For one-parameter deformation of a weighted quasi-diagonal $K3$ surface, he computes the Artin invariants, and then calculates the discriminant of the Néron-Severi groups over $\mathbb{Z}$ numerically,, and also determines algebraic cycles that form a $\mathbb{Z}$-basis for the Néron-Severi groups.

Joost van Hamel's paper is devoted to studying torsion in the Chow group of zero-cycles on a variety over the real numbers, in particular, the restriction of the Abel-Jacobi map to torsion subgroups. Using Roitman's theorem over the complex numbers and a version of Bloch's cohomological Abel-Jacobi map over the real numbers, he shows that this map can be completely described in terms of étale cohomology. He then illustrates his theory by explicitly determining the torsion in the kernel and cokernel of the Abel-Jacobi maps for products of curves, for Abelian varieties, and for certain fiber bundles.

Ken-Ichiro Kimura looks at the product variety $W = S \times_{\mathbb{Q}} C$ where S is a smooth projective surface over $\mathbb{Q}$ and C is a curve defined as the normalization of the projective completion of an affine curve $y^m = f(u)$, for a positive integer m and some $f(u) \in \mathbb{Q}[u]$. He addresses the relative version of the conjecture of Belinson and Bloch where W is defined over a Galois extension $K/\mathbb{Q}$. His result is that the order of the L-function of W over K at $s = 2$ is equal to the relative rank of the Griffith group of W over K.

In his article, Jan Nekovář gives a survey of foundational results on p-adic Abel-Jacobi maps and p-adic height pairings for algebraic cycles on varieties over number fields. Some new results are presented here, though proofs will appear elsewhere. After giving a brief discussion on p-adic cycle class maps, a special case of Tate's conjecture and conjectures of Block-Kato about p-adic Abel-Jacobi map are stated. Then the author sets a goal of establishing the validity of these conjectures for varieties over number fields. first reducing the problem to a local question at each non-archimedean prime. Since at primes not dividing p, the situation is more or less under control, the author concentrates on primes dividing p, especially, primes with good, or potentially good reductions. Then using the key ingredient of *syntomic cohomology* of Fontaine-Messing, a special case of conjectures of Bloch-Kato is partially proved for varieties over p-adic fields at primes with potentially good reduction. The second subject is the p-adic height pairings. Under certain conditions, one has the existence of p-adic height pairings on Chow groups, factoring through p-adic Abel-Jacobi maps. Finally p-adic height pairings in families are discussed, introducing Selmer complexes to deal with duality theorems.

In his paper, Atsushi Shiho surveys the various definitions he has given, using the theory of Tannakian categories, for the rational crystalline fundamental group for a smooth variety over a perfect field of positive characteristic. Analogously with the more well-known comparison isomorphisms that exist among the étale, Betti, de Rham and crystalline cohomology of a smooth projective variety, a number of authors have found comparison theorems relating the ℓ-adic, Betti and de Rham rational fundamental groups. Here Shiho proves two comparison theorems, one that links the rational de Rham fundamental group with the rational crystalline fundamental group, and another that connects a rational p-adic fundamental group (as in p-adic Hodge theory) with the rational crystalline fundamental group.

The article of Helena Verrill and Noriko Yui is concerned with the mirror maps of certain pencils of $K3$ surfaces and their (possible) relationship to Thompson series in the context of *Mirror-Moonshine Conjecture* of B. Lian and S.-T. Yau. The pencils of $K3$ surfaces discussed in the paper are one-(or multi)-parameter deformations of weighted orbifold Calabi-Yau hypersurfaces of dimension 2, and those obtained via toric construction from root lattices such as A_1^3, A_3 (and possibly C_3). The mirror maps are computed for these $K3$ surfaces, and the mirror-moonshine phenomena are illustrated with a number of explicit examples. Possible generalizations and open problems are also discussed. The main ingredients employed in this article are of geometric nature. This paper may be considered a complement to the article by Doran in this volume, which is coated with analytic flavour.

Acknowledgments

The editors wish to express their appreciation to all the contributors for their effort in preparing their manuscripts. We are deeply grateful to all our referees for their immense efforts and discipline in evaluating the papers.

The joint 1998 CRM Summer School and NATO Advanced Study Institute at Banff was supported in addition by the U.S. National Science Foundation and National Security Agency, as well as the University of Alberta. We thank them for their support. We also wish to thank: Louis Pelletier (CRM) for his valuable assistance with the planning and organization of the conference; the staff at the Banff Centre for Conferences for their helpful on-site organization of services and facilities; and André Montpetit (CRM) for his technical assistance during the final preparation of these proceedings.

On a lighter note, we also give honorable mention to: the University of Chicago, for the institution with the most participants; V. Srinivas, for traveling the farthest; Tony Scholl, for the most colourful slides; Chad Schoen, as the most persistent lecturer; and Henri Gillet, for the closest encounter with an elk and her calf.

The Editors: B. Brent Gordon
James D. Lewis
Stefan Müller-Stach
Shuji Saito
Noriko Yui

Virtual Cycle City, August, 1999

Contents of the Proceedings of the NATO Advance Study Institute: Arithmetic and Geometry of Algebraic Cycles

Cohomological Methods

Lectures on Algebro-Geometric Chen-Weil and Cheeger-Chern-Simons Theory for Vector Bundles
Bloch, Spencer and Esnault, Hélène

Deligne Cohomology and Geometric (Co-)Bar Constructions
Gajer, Pawel

Kuga-Satake Varieties and the Hodge Conjecture
van Geemen, Bert

Hodge and Weil Classes on Abelian Varieties
Murty, V. Kumar

Bloch-Kato Conjecture and Motivic Cohomology with Finite Coefficients
Suslin, Andrei and Voevodsky, Vladimir

Chow Groups and Motives

Indecomposable Higher Chow Cycles
Gordon, B. Brent and Lewis, James D.

Equivalence Relations on Algebraic Cycles
Jannsen, Uwe

Letter to Dick Gross on Higher Abel-Jacobi Maps
Jannsen, Uwe

Finiteness of Torsion in the Codimension-Two Chow Group: An Axiomatization
Langer, Andreas

Algebraic Cycle Complexes
Müller-Stach, Stefan

Algebraic Cycles on Abelian Varieties
Murre, Jacob P.

Motives and Filtrations on Chow Groups. II
Saito, Shuji

Zero Cycles on Singular Varieties
Srinivas, V.

Arithmetic Methods

Prepotentials of Yukawa Couplings of Certain Calabi-Yau 3-folds and Mirror Symmetry
Saito, Masahiko

Weight-Monodromy Conjecture for ℓ-adic Representations Associated to Modular Forms: A Supplement to the Paper [**S1**]
Saito, Takeshi

Cohomology Computations Related to the ℓ-adic Abel-Jacobi Map Modulo ℓ
Schoen, Chad

Integral Elements in K-Theory and Products of Modular Curves
Scholl, Anthony J.

Appendix to Scholl's Article: A Counterexample to a Conjecture of Beilinson
de Jeu, Rob

Reduction of Abelian Varieties
Silverberg, Alice and Zahrin, Yuri

The Arithmetic of Certain Calabi-Yau Varieties over Number Fields
Yui, Noriko

Classical and Elliptic Polylogarithms and Special Values of L-Series
Zagier, Don, and Gangl, Herbert

Titles in the CRM Monograph Series

Higher Regulators, Algebraic K-Theory, and Zeta Functions of Elliptic Curves
Bloch, Spencer

Conference Programme

Sunday June 7, 1998

Arrival

Monday June 8, 1998

9:00–9:20	Opening ceremonies and information	
9:30–10:20	P. Gajer	*Geometry of Deligne cohomology*
10:45–11.35	M. Green	*Algebraic cycles and Hodge theory.* I
12:00–13:00	Lunch	
13:30–14:20	V. K. Murty	*Algebraic cycles on Abelian varieties*
14:45–15:35	J. D. Lewis	*Remarks on indecomposable motives*
15:45–16:00	Coffee and tea	
16:15–16:45	R.-O. Buchweitz	*The Atiyah-Chern character yields the general semiregularity as well as the infinitesimal Abel-Jacobi map*
17:00–17:30	N. Ramachandran	*A conjecture of Deligne on one-motives*
18:30–20:00	Dinner	
20:00–	Poster sessions	

Tuesday June 9, 1998

9:30–10:20	D. Ramakrishnan	*Zero cycles on surfaces.* I
10:45–11:35	W. Raskind	*Higher Abel-Jacobi mappings and filtrations on Chow groups*
12:00–13:00	Lunch	
13:30–14:20	V. Voevodsky	*Motivic cohomology.* I
14:45–15:35	H. Gillet	*Motives for arithmetic varieties*
15:45–16:00	Coffee and tea	
16:15–16:45	S. Usui	*Logarithmic Hodge structures and classifying spaces*
17:00–17:30	K. Künnemann	*Height pairings for algebraic cycles on Abelian varieties*
18:30–20:00	Dinner	
20:00–	Poster sessions	

Wednesday June 10, 1998

9:30–10:20	A. Langer	*Zero-cycles on Hilbert-Blumenthal surfaces*
10:45–11:35	C. Schoen	*Cycle class maps for nullhomologous cycles on a variety over a finite field.* I

12:00–13:00	Lunch	
13:30–14:20	B. Gordon	*Indecomposable Bloch-Quillen cycles*
14:45–15:35	H. Esnault	*Some remarks on Boliruch's theorem*
15:45–16:00	Coffee and tea	
16:15–16:45	S. Abdulali	*Cycles on Abelian varieties of type* III
17:00–17:30	E. Goren	*Hilbert modular varieties in positive characteristic*
18:30–20:00	Dinner	
20:00–	Poster sessions	

Thursday June 11, 1998

9:30–10:20	D. Zagier	*Polylogarithms, elliptic polylogarithms, periods and L-values.* I
10:45–11:35	Y. Zarhin	*Semistable reduction of Abelian varieties and inertia groups*
12:00–13:00	Lunch	
13:30–14:20	V. Voevodsky	*Motivic cohomology.* II
14:45–15:35	M. Kolster	*The Lichtenbaum conjectures revisited*
15:45–16:00	Coffee and tea	
16:15–16:45	J. Scholten	*Non-selfdual* 3 *and* 4 *dimensional Galois representations*
17:00–17:30	K. Sato	*Finiteness of a certain motivic cohomology group of varieties over local or global fields*
18:30–20:00	Dinner	
20:00–	Poster sessions	

Friday June 12, 1998

9:30–10:20	S. Saito	*Filtrations on Chow groups and higher Abel-Jacobi maps.* I
10:45–11:35	C. Soulé	*Perfect forms and the Vandiver conjecture*
12:00–13:00	Lunch	
13:30–14:20	M. Green	*Algebraic cycles and Hodge theory.* II
14:45–15:35	J. P. Murre	*On cycles on Abelian varieties*
15:45–16:00	Coffee and tea	
16:15–16:45	K. Kimura	*On the Griffiths groups of certain product varieties*
17:00–17:30	E. Izadi	*The intermediate jacobian of the theta divisor of an Abelian fourfold*
18:30–20:00	Dinner	
20:00–	Poster sessions	

Saturday June 13, 1998

9:30–10:20	T. Scholl	*Algebraic cycles and modular forms*
10:45–11:35	D. Ramakrishnan	*Special value problems of L-series.* II
12:00–13:00	Lunch	
13:30–14:20	M. Saito	*Prepotentials of Yukawa couplings of certain Calabi-Yau 2-filds and lattice Theta functions*

14:45–15:35	N. Yui	*The L-series, periods, and special values of L-series of certain Calabi-Yau varieties*
15:45–16:00	Coffee and tea	
16:00–18:30	Free time. Informal discussion	
18:30–20:00	Dinner	
20:00–	Poster sessions	

Sunday June 14, 1998

Free day

Monday June 15, 1998

9:30–10:20	D. Ramakrishnan	*Zero cycles on surfaces.* III
10:45–11:35	T. Saito	*Modular forms and p-adic Hodge theory*
12:00–13:00	Lunch	
13:30–14:20	S. Bloch	*Algebraic cycles and differential forms.* I
14:45–15:35	U. Jannsen	*Equivalence relations on algebraic cycles, I*
15:45–16:00	Coffee and tea	
16:15–16:45	B. Wang	*Incidence structure on the Chow variety*
17:00–17:30	M. Asakura	*Motives and algebraic de Rham cohomology*
18:30–20:00	Dinner	
20:00–	Poster sessions	

Tuesday June 16, 1998

9:30–10:20	M. Green	*Algebraic cycles and Hodge theory.* III
10:45–11:35	V. Srinivas	*Zero cycles on singular varieties*
12:00–13:00	Lunch	
13:30–14:20	D. Zagier	*Polylogarithms, elliptic polylogarithms, periods and L-values.* II
14:45–15:35	B. van Geemen	*Cycles on Abelian fourfolds*
15:45–16:00	Coffee and tea	
16:15–16:45	J. van Hamel	*Algebraic cycles and cohomology of varieties over the real numbers*
17:00–17:30	Y. Goto	*Algebraic cyces on some weighted Delsarte K3 surfaces and on their deformations*
18:30–20:00	Dinner	
20:00–	Poster sessions	

Wednesday June 17, 1998

9:30–10:20	C. Schoen	*Cycle class maps for nullhomologous cycles on a variety over a finite field.* II*: Application*
10:45–11:35	Colliot-Thélène	*Local-global principle for zero-cycles on conic bundles*
12:00–13:00	Lunch	
13:30–14:20	S. Saito	*Filtrations on Chow groups and higher Abel-Jacobi maps.* II
14:45–15:35	U. Jannsen	*Equivalence relations on algebraic cycles, II*
15:45–16:00	Coffee and tea	

16:15–16:45	A. Shiho	*Crystalline fundamental groups and p-adic Hodge theory*
17:00–17:30	T. Geisser	*Motivic cohomology of smooth varieties over fields*
18:30–20:00	Dinner	
20:00–	Poster sessions	

Thursday June 18, 1998

9:30–10:20	S. Müller-Stach	*Algebraic cycle complexes*
10:45–11:35	S. Bloch	*Algebraic cycles and differential forms.* II
12:00–13:00	Lunch	
13:30–14:20	J. Nekovář	*p-adic Abel-Jacobi maps and p-adic heights*
14:45–15:35	S. Saito	*Filtrations on Chow groups and higher Abel-Jacobi maps.* III
15:45–16:00	Coffee and tea	
16:15–16:45	H. Gangl	*Polylogarithmic identities in cubical higher Chow groups*
17:00–17:30	I. Burgos Gil	*Higher Bott-Chern forms and Beilinson's regulator*
18:30–20:00	Dinner	
20:00–	Poster sessions	

Friday June 19, 1998

9:30–10:20	D. Zagier	*Polylogarithms, elliptic polylogarithms, periods and L-values.* III
10:30–12:00	Free time. Informal discussion	
12:00–13:00	Lunch	

Departure

NATO ASI/CRM Summer School
"The Arithmetic and Geometry of Algebraic Cycles"

List of Participants

Abdulali, Salman, East Carolina University (U.S.A.)
Achter, Jeffrey A., University of Pennsylvania (U.S.A.)
Arapura, Donu, Purdue University (U.S.A.)
Arcara, Daniele, University of Georgia (U.S.A.)
Archava, Sviatoslav, University of California at Los Angeles (U.S.A.)
Asakura, Masanori, Osaka University (Japan)
Baba, Srinath, City University of New York (U.S.A.)
Bell, Robert, The Cooper Union School of Engineering (U.S.A.)
Bloch, Spencer, University of Chicago (U.S.A.)
Brosnan, Patrick, University of Chicago (U.S.A.)
Buchweitz, Ragnar-Olaf, University of Toronto (U.S.A.)
Burgos Gil, Jose Ignacio, Universidad de Barcelona (Spain)
Castravet, Ana-Maria, Massachussetts Institute of Technology (U.S.A.)
Ching, Avery, University of California at Los Angeles (U.S.A.)
Coffey, Joseph L., State Univeristy of New York at Stony Brook (U.S.A.)
Colliot-Thélène, Jean-Louis, Université de Paris-Sud (France)
Colombo, Elisabetta, Universitá di Milano (Italy)
Colwell, Jason, University of Alberta (Canada)
Consani, Katia, University of Cambridge (United Kingdom)
Cushman, Matthew, The University of Chicago (U.S.A.)
Dee, Jonathan, University of Cambridge (United Kingdom)
Doran, Charles Francis Jr., Harvard University (U.S.A.)
Durand, Philippe, Conservatoire National des Arts et Metiers (France)
Elbaz-Vincent, Philippe, Université de Montpellier II (France)
Elizondo, Javier E., Ciudad Universidad, Universidad Nacional Aotonoma de Mexico (Mexico)
Esnault, Helene, Universität-GH-Essen (Germany)
Fernandez de Bobadilla de Olazabal, Javier, University of Nijmegen (The Netherlands)
Fimmel, Torsten, Universität zu Köln (Germany)
Flach, Matthias, California Institute of Technology (U.S.A.)
Gajer, Pawel, The Johns Hopkins University (U.S.A.)
Galluzzi, Federica, Universitá di Torino (Italy)
Gangl, Herbert, Institut für Experimentelle Mathematik, Essen (Germany)
van Geemen, Bert, Universitá di Pavia (Italy)
Geisser, Thomas, University of Illinois at Urbana-Champaign (U.S.A.)
Gillet, Henri, University of Illinois at Chicago (U.S.A.)
Gordon, B. Brent, University of Oklahoma (U.S.A.)

Goren, Eyal, McGill University (Canada)
Goto, Yasuhiro, Kyoto University (Japan)
Green, Mark, University of California at Los Angeles (U.S.A.)
van Hamel, Joost, Universiteit Utrecht (The Netherlands)
Hanamura, Masaki, Kyushu University (Japan)
Harris, Bruno, Brown University (U.S.A.)
Hoobler, Raymond, City College of New York (U.S.A.)
Howson, Susan, Corpus Christi College, Cambridge (United Kingdom)
Hulays, Hayssam, University of Toronto (Canada)
Izadi, Elham, University of Georgia (U.S.A.)
Jannsen, Uwe, Universität zu Köln (Germany)
de Jeu, Rob, California Institute of Technology (U.S.A.)
Joshi, Kirti, University of Arizona (U.S.A.)
Joshua, Roy, Ohio State University (U.S.A.)
Khazanov, Aleksandr, Pennsylvania State University (U.S.A.)
Kimura, Kenichiro, University of Tsukuba (Japan)
Künnemann, Klaus, Newton Institute for Mathematical Sciences (United Kingdom)
Langer, Andreas, Universität Münster (Germany)
Laterveer, Robert, Institut de Recherches Mathématiques Applicées, Strasbourg (France)
Levin, Norman, University of Minnesota (U.S.A.)
Lewis, James D., University of Alberta (Canada)
Lima-Filho, Paulo, Texas A&M University (U.S.A.)
Mouroukos, Evangelos, University of Chicago (U.S.A.)
Müller-Stach, Stefan, Universität-GH-Essen (Germany)
Murre, Jacob P., Rijksuniversiteit Leiden (The Netherlands)
Murty, V. Kumar, University of Toronto (Canada)
Nagel, Jan, Universität-GH-Essen (Germany)
Nekovář, Jan, University of Cambridge (United Kingdom)
Otsubo, Noriyuki, University of Tokyo (Japan)
Pedrini, Claudio, Universitá di Genova (Italy)
Portelli, Dario, Universitá di Trieste (Italy)
Ramachandran, Niranjan, University of Michigan (U.S.A.)
Ramakrishnan, Dinakar, California Institute of Technology (U.S.A.)
Raskind, Wayne, University of Southern California (U.S.A.)
Rosenschon, Andreas, University of Maryland (U.S.A.)
Saito, Masahiko, Kobe University (Japan)
Saito, Shuji, Tokyo Insitute of Technology (Japan)
Saito, Takeshi, University of Tokyo (Japan)
Santos, Pedro, State University of New York at Stony Brook (U.S.A.)
Sato, Kanetomo, Tokyo Institute of Technology (Japan)
Schoen, Chad, Duke University (U.S.A.)
Scholl, Anthony, University of Durham (United Kingdom)
Scholten, Jasper, Rijksuniversiteit Groningen (The Netherlands)
Shiho, Atsushi, Tohoku University (Japan)
Shimizu, Yuji, Kyoto University (Japan)
Somodi, Marius, Louisiana State University (U.S.A.)
Soulé, Christopher, Institut des Hautes Études Scientifiques (France)

Sreekantan, Ramesh, University of Chicago (U.S.A.)
Srinivas, Vasudevan, Tata Institute of Fundamental Research (India)
Stahlke, Colin, Universität Düsseldorf (Germany)
Stopple, Jeffrey, University of California at Santa Barbara (U.S.A.)
Stroila, Matei, University of Southern California (U.S.A.)
Toen, Bertrand, Université Paul Sabatier (France)
Usui, Sampei, Osaka University (Japan)
Verrill, Helena, Max-Planck-Institut für Mathematik (Germany)
Veys, Willem, Katholieke Universiteit Leuven (Belgium)
Voevodsky, Vladimir, Northwestern University (U.S.A.)
Vologodski, Vitali, University of Georgia (U.S.A.)
Walker, Mark, University of Nebraska (U.S.A.)
Wang, Bin, Yale University (U.S.A.)
Wang, Song, California Institute of Technology (U.S.A.)
Weibel, Charles, Rutgers University (U.S.A.)
Wortmann, Sigrid, Universität zu Köln (Germany)
Xarles, Xavier, Universidad Autonoma de Barcelona (Spain)
Yui, Noriko, Queen's University (Canada)
Zagier, Don, Max-Planck-Institut für Mathematik (Germany)
Zarhin, Yuri, Pennsylvannia State University (U.S.A.)
Zintl, Jörg, Universität Kaiserslautern (Germany)
Zucker, Steven, The Johns Hopkins University (U.S.A.)

Authors' Addresses

Abdulali, Salman, Department of Mathematics, East Carolina University, Greenville, NC 27858, USA; salman@math.ecu.edu

Arapura, Donu, Department of Mathematics, Purdue University, West Lafayette, IN 47907, USA; dvb@math.purdue.edu

Asakura, Masanori, Department of Mathematics, Graduate School of, Science, Osaka University, Toyonaka, Osaka 560, Japan; asakura@klingon.math.wani.osaka-u.ac.jp

Current address: Research Institute for Mathematical Sciences, Kyoto University, Sakyo-ku, Kyoto, 606-8502 Japan; asakura@kurims.kyoto-u.ac.jp

Buchweitz, Ragnar-Olaf, Department of Mathematics, University of Toronto, Toronto, Ontario M5S 3G3, Canada; ragnar@math.toronto.edu

Burgos Gil, Jose Ignacio, Departmento de Algebra y Geometriá, Universidad de Barcelona, Gran Via 585, 08007 Barcelona, Spain; burgos@cerber.mat.ub.es

Doran, Charles, Department of Mathematics, Harvard University, Cambridge, MA 02138, USA; doran@math.harvard.edu

Current address: Department of Mathematics, Pennsylvania, State University, College Park, PA 16802, USA; doran@math.psu.edu

Dupont, Johan, Matematisk Institut, Aarhus Universitet, DK-8000 Aarhus C, Denmark; dupont@imf.au.dk

Flenner, Hubert, Fakultät für Mathematik, Ruhr-Universität Bochum, Geb. NA 2/72, D-44781 Bochum, Germany; Hubert.Flenner@ruhr-uni-bochum.de

Goren, Eyal, Department of Mathematics, McGill University, 805 Sherbrooke Street, Montreal, Quebec H3A 2K6, Canada ; goren@math.mcgill.ca

Goto, Yasuhiro, Research Institute for Mathematical Sciences, Kyoto, University, Sakyo-ku, Kyoto, 606-8502 Japan; got@kurims.kyoto-u.ac.jp

Current address: Department of Mathematics, Hokkaido University of Education, 1-2 Hachiman-cho, Hakodate, Hokkaido, 040-8567 Japan; ygoto@cc.hokkyodai.ac.jp

Hain, Richard, Department of Mathemataics, Duke University, Durham, NC 27708-0320, USA; hain@math.duke.edu

van Hamel, Joost, Mathematisch Instituut, Universiteit Utrecht, Budapestlaan 6, 3584 CD Utrecht, The Netherlands; vanhamel@math.uu.nl

Hanamura, Masaki, Graduate School of Mathematics, Kyushu University, Fukuoka, 812 Japan; hanamura@math.kyushu-u.ac.jp

Harris, Bruno, Department of Mathematics, Brown University, Providence, RI 02912, USA; Bruno_Harris@Brown.edu

Kato, Kazuya, Department of Mathematical Sciences, University of Tokyo, Komaba, Tokyo 153, Japan; kkato@ms.u-tokyo.ac.jp

Kimura, Ken-Ichiro, Department of Mathematics, University of Tsukuba, Tsukuba, Ibaraki, 305 Japan; kimurak@math.tsukuba.ac.jp

Nekovàř, Jan, Department of Pure Mathematics, University of Cambridge, 16 Mill Lane, Cambridge CB2 1SB, The United Kingdom; nekovar@pmms.cam.ac.uk

Pedirni, Claudio, Dipartmento di Matematica, Universitá di Genova, Via Dedecaneso, 35 16146, Genova, Italy; pedrini@dima.unige.it

Ramachandran, Niranjan, Department of Mathematics, University of Michigan, Ann Arbor, MI 48109-1109, USA; atma@math.lsa.umich.edu

Saito, Shuji, Department of Mathematics, Tokyo Institute of Technology, Oh-Okayama, Meguro-ku, Tokyo, 152 Japan; sshuji@math.titech.ac.jp

Shiho, Atsushi, Mathematical Institute, Tohoku University, Aoba, Sendai, 980-8578 Japan; shiho@math.tohoku.ac.jp

Usui, Sampei, Graduate School of Science, Osaka University, Toyonaka, Osaka, 550 Japan; usui@math.wani.osaka-u.ac.jp

Verrill, Helena, Max-Planck-Insitut für Mathematik, Vivatsgasse 7, D-53111 Bonn, Germany; verrill@mpim-bonn.mpg.de

Current address: Matematisk Institut Københavns Universitet, Universitetsparken 5, DK-2100 København Ø, Denmark; verrill@math.ku.dk

Wang, Bin, Department of Mathematics, Brown University, Provicence, RI 02912 USA; binwang@math.brown.edu

Current address: Department of Mathematics, Harvard University, Cambridge, MA 02138, USA

Yui, Noriko, Department of Mathematics and Statics, Queen's University, Kingston, Ontario K7L 3N6, Canada ; yui@mast.queensu.ca

Zucker, Steven, Department of Mathematics, The Johns Hopkins University, Baltimore, MD 21218, USA; sz@math.jhu.edu

Cohomological Methods

Centre de Recherches Mathématiques
CRM Proceedings and Lecture Notes
Volume **24**, 2000

Filtrations on the Cohomology of Abelian Varieties

Salman Abdulali

1. Introduction

The arithmetic filtration $F_{\mathrm{a}}^{\bullet}$ on the cohomology $H^n(X)$ of a smooth projective variety X is obtained by defining $F_{\mathrm{a}}^r H^n(X)$ to be the linear span of the set of cohomology classes supported on algebraic subvarieties of codimension at least r. Though this filtration is defined geometrically, Grothendieck [**19**] called it the *arithmetic* filtration, "as it embodies deep arithmetic properties of the scheme X." There are several conjectural descriptions of the arithmetic filtration. Over the field of complex numbers, the general Hodge conjecture [**22**], as corrected by Grothendieck [**19**], gives a conjectural description of the arithmetic filtration in terms of the Hodge filtration, i.e., in terms of harmonic analysis. For a variety over a finite field, Grothendieck [**17**, §10, pp. 164–170] gives a conjectural description of the arithmetic filtration in terms of eigenvalues of the Frobenius automorphism, and discusses its relations with the Weil and Tate conjectures.

Friedlander and Mazur [**11**] use operations in Lawson homology to define a filtration, called the topological filtration, which they conjecture equals the arithmetic filtration. Friedlander [**7**] shows that this conjecture is implied by Grothendieck's standard conjectures [**18**]. Though this provides strong reason to believe the conjecture, there are very few varieties for which it has been proved. In an attempt to understand the topological filtration, I defined a Hodge-theoretic analog called the group theoretic filtration in [**1**], and used it to prove the general Hodge conjecture for some Abelian varieties.

The aim of this paper is to examine the relationship between the arithmetic, topological, group theoretic, and, Hodge filtrations, with special reference to Abelian varieties. In Sections 2 to 5 we recall the definitions of the filtrations, and discuss some general results. In Sections 6 to 8 we specialize to Abelian varieties. Section 7 is a list of all Abelian varieties for which the general Hodge conjecture is known, including a few new examples. In §8 we prove the conjecture of Friedlander and Mazur for all of these Abelian varieties.

1991 *Mathematics Subject Classification.* Primary 14C30, 14K99; Secondary 14F45.
Received by the editors August 30, 1998; Accepted November 2, 1998.
This is the final form of the paper.

I would like to take this opportunity to thank Eric Friedlander for patiently answering many questions and, in particular, for suggesting the use of the moving lemma for families of cycles to prove Proposition 3.2.

2. The General Hodge Conjecture

Let X be a smooth projective variety over $\mathbb{C}$, Y an irreducible subvariety of X, and $\widetilde{Y} \to Y$ a resolution of singularities. Let j be the composition $\widetilde{Y} \to Y \hookrightarrow X$. Then j induces a Gysin homomorphism $j_\star \colon H^{n-2r}(\widetilde{Y}, \mathbb{Q}) \to H^n(X, \mathbb{Q})$, where $r = \operatorname{codim} Y$. A cohomology class is said to be supported on Y if it lies in the image of $j_\star$. The *arithmetic* (or *coniveau*) *filtration* $F_{\mathrm{a}}^\bullet$ on $H^n(X, \mathbb{Q})$ is defined by taking $F_{\mathrm{a}}^r H^n(X, \mathbb{Q})$ to be the set of cohomology classes supported on algebraic subvarieties of codimension at least r.

The *Hodge filtration* is given by

$$F^r H^n(X, \mathbb{C}) := \bigoplus_{\substack{p+q=n \\ p \geq r}} H^{p,q}(X).$$

The *rational Hodge filtration* is obtained by defining $F_{\mathbb{Q}}^r H^n(X, \mathbb{Q})$ to be the largest $\mathbb{Q}$-Hodge substructure of $F^r H^n(X, \mathbb{C}) \cap H^n(X, \mathbb{Q})$.

We have $F_{\mathrm{a}}^r H^n(X, \mathbb{Q}) \subset F_{\mathbb{Q}}^r H^n(X, \mathbb{Q})$. The *general Hodge conjecture* [**22**], as amended by Grothendieck [**19**], states that this inclusion is an equality.

The analogue of the arithmetic filtration on homology is called the *niveau* (or *geometric*) filtration. It is defined by letting $F_{\mathrm{N}}^r H_n(X, \mathbb{Q})$ be the subspace of $H_n(X, \mathbb{Q})$ spanned by the images of maps $H_n(\widetilde{Y}, \mathbb{Q}) \to H_n(X, \mathbb{Q})$, where Y runs over all irreducible subvarieties of X of dimension $n - r$, and, $\widetilde{Y}$ denotes a resolution of singularities of Y.

The Poincaré duality isomorphism $H^n(X, \mathbb{Q}) \cong H_{2d-n}(X, \mathbb{Q})$, where $d := \dim X$, induces an isomorphism

$$F_{\mathrm{a}}^r H^n(X, \mathbb{Q}) \cong F_{\mathrm{N}}^{r+d-n} H_{2d-n}(X, \mathbb{Q}). \tag{2.1}$$

REMARK 2.1. The niveau filtration defined above is the same as the geometric filtration defined in [**11**, §7.2, p. 67]. It is related to the niveau filtration $F_{\mathrm{BO}}^r H_n(X, \mathbb{Q})$ of Bloch and Ogus [**5**] by $F_{\mathrm{N}}^r H_n(X, \mathbb{Q}) = F_{\mathrm{BO}}^{n-r} H_n(X, \mathbb{Q})$.

3. The Correspondence Filtration

In [**11**, §§7.3–7.4, p. 68] Friedlander and Mazur introduce two filtrations, called the topological filtration and the correspondence filtration, on the homology of a complex projective variety X. These two filtrations are in fact equal (see [**11**, Theorem 7.3, p. 70] for the smooth case, and [**12**, Theorem 4.3, p. 724] for the general case). We recall the definition of the correspondence filtration below. I recommend Lawson's article [**29**, Chapter IV, §9, pp. 192–195] as a starting point for the reader who wishes to learn more about these filtrations.

Let X and Y be projective varieties, with Y smooth of dimension m. Let $Z \subset Y \times X$ be an irreducible subvariety of dimension $m + r$ ($r \geq 0$). Then the class of Z is a homology class ζ in $H_{2m+2r}(Y \times X, \mathbb{Q})$. Using the Künneth formula and

Poincaré duality, we have

$$H_{2m+2r}(Y \times X, \mathbb{Q}) \cong \bigoplus_i H_{2m-i}(Y, \mathbb{Q}) \otimes H_{i+2r}(X, \mathbb{Q})$$
$$\cong \bigoplus_i \hom\big(H_i(Y, \mathbb{Q}), H_{i+2r}(X, \mathbb{Q})\big).$$

Thus Z determines a homomorphism

$$\varphi_Z \colon H_i(Y, \mathbb{Q}) \to H_{i+2r}(X, \mathbb{Q})$$

for each i. These homomorphisms are called *correspondence homomorphisms* [**11**, p. 39].

DEFINITION 3.1. The *correspondence filtration* on homology is defined by letting $F_{\mathrm{C}}^r H_n(X, \mathbb{Q})$ be the subspace of $H_n(X, \mathbb{Q})$ spanned by the images of correspondence homomorphisms

$$\varphi_Z \colon H_{n-2r}(Y, \mathbb{Q}) \to H_n(X, \mathbb{Q})$$

as Y ranges over all smooth varieties of dimension $n-2r$, and Z ranges over all cycles on $Y \times X$ equidimensional over Y of relative dimension r.

The restrictions on Y and Z in the definition of the correspondence filtration appear to be rather severe. We shall now show that they can be somewhat relaxed, at least when X is smooth. The following proposition should be thought of as a generalization of [**7**, Proposition 4.1, p. 338].

PROPOSITION 3.2. *Let X be smooth. Then $F_{\mathrm{C}}^r H_n(X, \mathbb{Q})$ is the subspace of $H_n(X, \mathbb{Q})$ spanned by the images of correspondence homomorphisms*

$$\varphi_Z \colon H_{n-2r}(Y, \mathbb{Q}) \to H_n(X, \mathbb{Q})$$

as Y ranges over all smooth varieties of dimension $\geq n-2r$, and Z ranges over all irreducible subvarieties of $Y \times X$ of dimension $\dim Y + r$.

PROOF. Suppose $\dim Y > n - 2r$. Let W be a smooth hyperplane section of Y. Then the weak Lefschetz theorem says that the inclusion $W \hookrightarrow Y$ induces a surjection

$$H_{n-2r}(W, \mathbb{Q}) \to H_{n-2r}(Y, \mathbb{Q})$$

The composition $H_{n-2r}(W, \mathbb{Q}) \to H_{n-2r}(Y, \mathbb{Q}) \to H_n(X, \mathbb{Q})$ is then a correspondence homomorphism induced by an algebraic cycle $Z_0 \subset W \times X$ of dimension $\dim W + r$. Thus, applying the weak Lefschetz theorem $\dim Y - (n-2r)$ times we may replace Y by a smooth variety of dimension $n - 2r$.

Next, we claim that Z is rationally equivalent to a cycle which is equidimensional over Y. This follows by applying a moving lemma of Friedlander and Lawson [**10**] to the family of cycles $\big\{\{y\} \times X \bigm| y \in Y\big\}$ (see [**10**, Remark, p. 115] and [**9**, Theorem B1, p. 561]), to obtain a cycle Z' which meets each $\{y\} \times X$ properly, and is therefore equidimensional over Y. □

Friedlander and Mazur show that $F_{\mathrm{C}}^r H_n(X, \mathbb{Q}) \subset F_{\mathrm{N}}^r H_n(X, \mathbb{Q})$ [**11**, Corollary 3, p. 71], and conjecture that equality holds.

CONJECTURE 3.3 (Friedlander and Mazur [**11**, p. 71]). *The geometric and topological filtrations coincide on the homology of any smooth, projective variety over $\mathbb{C}$.*

We define the correspondence filtration on cohomology by letting $F_{\mathrm{C}}^{r}H^{n}(X,\mathbb{Q})$ be the Poincaré dual of $F_{\mathrm{C}}^{r+d-n}H_{2d-n}(X,\mathbb{Q})$, where $d=\dim X$. Then (2.1) shows that Conjecture 3.3 is equivalent to the equality of the correspondence and arithmetic filtrations on cohomology.

Friedlander and Lawson also define a topological filtration $F_{\mathrm{T}}^{\bullet}$ on cohomology ([**8**, Definition 5.3, p. 399], see also [**29**, Chapter V, §6, pp. 203–205]). This is an increasing filtration, related to the topological filtration on homology via Poincaré duality by $F_{\mathrm{T}}^{r}H^{n}(X,\mathbb{Q}) \cong F_{\mathrm{C}}^{d-r}H_{2d-n}(X,\mathbb{Q})$ [**9**, Theorem 7.9, p. 556]. Thus the correspondence and topological filtrations on cohomology are related by $F_{\mathrm{C}}^{r}H^{n}(X,\mathbb{Q}) = F_{\mathrm{T}}^{n-r}H^{n}(X,\mathbb{Q})$.

4. Standard Conjecture B

Let X be a smooth projective variety over $\mathbb{C}$ of dimension n, and $\zeta \in H^{2}(X,\mathbb{Q})$ the class of a smooth hyperplane section. Let $L\colon H^{i}(X,\mathbb{Q}) \to H^{i+2}(X,\mathbb{Q})$ be the map $\omega \mapsto \omega \wedge \zeta$. The strong Lefschetz theorem states that for $0 \le i \le n$, the map $L^{n-i}\colon H^{i}(X,\mathbb{Q}) \to H^{2n-i}(X,\mathbb{Q})$ is an isomorphism. Grothendieck's standard conjecture B [**18**] states that the inverse Λ^{n-i} of L^{n-i} is an algebraic correspondence, i.e., induced by an algebraic cycle on $X \times X$. Since Λ^{n-i} is a morphism of Hodge structures, this conjecture is implied by the usual Hodge conjecture for $X \times X$. For further details on the standard conjectures we refer the reader to the articles of Grothendieck [**18**] and Kleiman [**23, 25**].

Lieberman [**30**] proved standard conjecture B for all complex Abelian varieties. A simpler proof, due to Kleiman and Lieberman, may be found in [**24**, p. 86]. Alternatively, we may note that standard conjecture B holds for generic Abelian varieties since the usual Hodge conjecture is known for all powers of such Abelian varieties. Since standard conjecture B is stable under specialization, it must hold for all Abelian varieties.

PROPOSITION 4.1 (Friedlander [**7**, Proposition 4.2, p. 339]). *If X is a smooth projective variety, and standard conjecture B is true for a resolution of singularities of each irreducible subvariety of X, then Conjecture* 3.3 *is true for X.*

5. The Group Theoretic Filtration

Let X be a smooth complex projective variety. The *group theoretic filtration* [**1**] on $H^{n}(X,\mathbb{Q})$ is defined by letting $F_{\mathrm{g}}^{r}H^{n}(X,\mathbb{Q})$ be the sum of those Hodge substructures of $H^{n}(X,\mathbb{Q})$ which are isomorphic to Tate twists of Hodge substructures of $H^{s}(Y,\mathbb{Q})$ for some smooth projective variety Y, and some $s \le n-2r$. We say that a class $\mathcal{A}$ of smooth projective varieties *dominates* X if each irreducible Hodge substructure of $F_{\mathrm{g}}^{r}H^{n}(X,\mathbb{Q})$ is isomorphic to a Tate twist of a Hodge substructure of $H^{s}(Y,\mathbb{Q})$ for some $Y \in \mathcal{A}$, and some $s \le n-2r$.

The group theoretic filtration is so named because for Abelian varieties it can frequently be explicitly described in terms of representations of the Hodge group; see [**1**] for evidence of this.

PROPOSITION 5.1. *Let X be a smooth projective variety over $\mathbb{C}$. For any r, n, we have*

$$F_{\mathrm{C}}^{r}H^{n}(X,\mathbb{Q}) \subset F_{\mathrm{g}}^{r}H^{n}(X,\mathbb{Q}) \subset F_{\mathbb{Q}}^{r}H^{n}(X,\mathbb{Q}).$$

If the usual Hodge conjecture is true for $Y \times X$ for all Y in a class $\mathcal{A}$ of algebraic varieties which dominates X, then

$$F^r_{\mathrm{C}} H^n(X,\mathbb{Q}) \subset F^r_{\mathrm{g}} H^n(X,\mathbb{Q}) \subset F^r_{\mathrm{a}} H^n(X,\mathbb{Q}) \subset F^r_{\mathbb{Q}} H^n(X,\mathbb{Q}).$$

PROOF. The inclusion $F^r_{\mathrm{g}} H^n(X,\mathbb{Q}) \subset F^r_{\mathbb{Q}} H^n(X,\mathbb{Q})$ is Proposition 3.1 of [**1**, p. 344]. To prove the inclusion $F^r_{\mathrm{C}} H^n(X,\mathbb{Q}) \subset F^r_{\mathrm{g}} H^n(X,\mathbb{Q})$, let V be an irreducible Hodge structure contained in $F^r_{\mathrm{C}} H^n(X,\mathbb{Q})$. Then V lies in the image of a correspondence homomorphism

$$\varphi_Z \colon H^{n-2r}(Y,\mathbb{Q}) \cong H_{n-2r}(Y,\mathbb{Q}) \to H_{2d-n}(X,\mathbb{Q}) \cong H^n(X,\mathbb{Q})$$

where $\dim Y = n-2r$ and $d := \dim X$. Therefore $H^{n-2r}(Y,\mathbb{Q})$ contains a Hodge substructure isomorphic to $V(-r)$. This shows that $F^r_{\mathrm{C}} H^n(X,\mathbb{Q}) \subset F^r_{\mathrm{g}} H^n(X,\mathbb{Q})$.

The second part of the theorem is implied by Proposition 3.2 of [**1**, p. 344]. □

THEOREM 5.2. *Let X be a smooth complex projective variety dominated by a class $\mathcal{A}$ of smooth varieties. Assume that for each $Y \in \mathcal{A}$, the usual Hodge conjecture is true for $Y \times X$, and standard conjecture B is true for Y. Then the group theoretic filtration equals the correspondence filtration.*

PROOF. Let V be an irreducible Hodge structure contained in $F^r_{\mathrm{g}} H^n(X,\mathbb{Q})$. Then there exist $Y \in \mathcal{A}$, a positive integer $s \le n-2r$, and a Hodge substructure U of $H^s(Y,\mathbb{Q})$, such that U is a Tate twist of V. Since the group theoretic and correspondence filtrations are both decreasing, we may assume without loss of generality that $s = n-2r$. Let $m := \dim Y$. Then there exists a Hodge cycle $\zeta \in H^{2m+2r}(Y \times X,\mathbb{Q})$ which induces an isomorphism of U with V. The usual Hodge conjecture for $Y \times X$ implies that ζ is the class of an algebraic cycle $Z \subset Y \times X$, with $\operatorname{codim} Z = m + r$. Then Z induces a correspondence homomorphism

$$\varphi_Z \colon H_{2m-n+2r}(Y,\mathbb{Q}) \to H_{2d-n}(X,\mathbb{Q})$$

whose image contains $V^\star$, the Poincaré dual of V.

Suppose first that $\dim Y \le n-2r$. Then φ_Z raises degrees by $2d-2m-2r$, which is at least $2(r+d-n)$. Proposition 3.2 now shows that $V^\star \subset F^{r+d-n}_{\mathrm{C}} H_{2d-n}(X,\mathbb{Q})$, and hence $V \subset F^r_{\mathrm{C}} H^n(X,\mathbb{Q})$.

Next suppose $\dim Y > n-2r$. Then standard conjecture B for Y implies that there is an isomorphism $H_{n-2r}(Y,\mathbb{Q}) \to H_{2m-n+2r}(Y,\mathbb{Q})$ induced by an algebraic cycle in $Y \times Y$. The composition $H_{n-2r}(Y,\mathbb{Q}) \to H_{2d-n}(X,\mathbb{Q})$ is thus a correspondence homomorphism. It follows that $V^\star \subset F^{r+d-n}_{\mathrm{g}} H_{2d-n}(X,\mathbb{Q})$, and hence $V \subset F^r_{\mathrm{C}} H^n(X,\mathbb{Q})$. □

6. Abelian Varieties

6.1. Polarizations. Let A be an Abelian variety over $\mathbb{C}$. Let $V := H_1(A,\mathbb{Q})$. We view V as a left vector space over the endomorphism algebra $D := \operatorname{End}(A) \otimes \mathbb{Q}$. As a complex manifold, A is a torus $V_{\mathbb{R}}/V_{\mathbb{Z}}$ with a complex structure J on $V_{\mathbb{R}}$, where $V_{\mathbb{Z}}$ is the lattice $H_1(A,\mathbb{Z})$. Recall that a polarization of A induces an involution ρ on D, and a Riemann form β; i.e., an alternating $\mathbb{Q}$-bilinear form on V such that $\beta(x, Jy)$ is symmetric and positive definite, $\beta(V_{\mathbb{Z}}, V_{\mathbb{Z}}) \subset \mathbb{Z}$, and such that $\beta(ax, y) = \beta(x, a^\rho y)$ for $a \in D$. There is a unique skew-Hermitian form T on V such that $\beta(x,y) = \operatorname{Tr}_{D/\mathbb{Q}} T(x,y)$ [**40**, Lemma 1.2, p. 162]. If D is a CM-field (in which case ρ is necessarily complex conjugation), and $\alpha \in E^\times$ is such that $\alpha^\rho = -\alpha$, then there exists a unique Hermitian form H on V such that

$\beta(x,y) = \mathrm{Tr}_{D/\mathbb{Q}}(\alpha H(x,y))$ [**6**, Lemma 4.6, p. 55]. We shall freely refer to the Riemann form β, or the skew-Hermitian form T, or (when D is a CM-field) the Hermitian form H as a polarization of A.

6.2. The Hodge and Lefschetz Groups. The Hodge group and Lefschetz group of an Abelian variety A are reductive algebraic groups over $\mathbb{Q}$ which act on the cohomology of any power of A. The Hodge group is defined by Mumford and Tate in [**33**]; its invariants are precisely the Hodge cycles. The Lefschetz group L is the reductive algebraic group over $\mathbb{Q}$, such that $L(\mathbb{Q}) = \mathrm{Aut}_D(V,T)$, where V and T are as in §6.1; its invariants form the subring of the Hodge ring which is generated by the classes of divisors ([**34**, Lemma 3.6, p. 204] and [**32**, Lemma 11(iii), p. 87]). Note that the group called $L(A)$ by Murty in [**34**] is the connected component of the group we are calling the Lefschetz group. The Hodge group is always connected, and contained in the Lefschetz group.

6.3. Abelian Varieties of PEL-Type. Let G denote either the Hodge group or the Lefschetz group of A, and, G' the derived group (or semisimple part) of G. Then the connected component $G'(\mathbb{R})^0$ is of Hermitian type, i.e., the quotient $X = G'(\mathbb{R})^0/K$ is a bounded symmetric domain (possibly a point), where K is a maximal compact subgroup of $G'(\mathbb{R})^0$. If Γ is a torsion-free arithmetic subgroup of G', then, the quotient $V = \Gamma\backslash X$ has the structure of a smooth arithmetic variety, which parametrizes families of Abelian varieties which play a crucial role in understanding cycles on Abelian varieties. The families of Abelian varieties obtained from the Lefschetz group are the PEL-families of Shimura [**41**]. The families of Abelian varieties obtained from the Hodge group are Mumford's Hodge families [**33**]. In general, the Hodge family is a subfamily of the PEL-family. They are equal if and only if the derived groups of the Hodge group and the connected component of the Lefschetz group are the same, in which case we say that A is of *PEL-type*. The class of Abelian varieties of PEL-type includes all Abelian varieties of CM-type, all Abelian varieties of dimension less than 4, and all Abelian varieties for which either the general Hodge conjecture or Conjecture 3.3 is known at present.

7. Abelian Varieties And The General Hodge Conjecture

7.1. A List. The following is a list of Abelian varieties for which the general Hodge conjecture is known to be true (see also [**16**]). For an Abelian variety A, we denote by $D(A)$ the endomorphism algebra $\mathrm{End}(A) \otimes \mathbb{Q}$. Recall that A is said to be of type I if $D(A)$ is a totally real number field, and of type II if $D(A)$ is a totally indefinite quaternion algebra over a totally real number field.

1. An Abelian variety A of PEL-type such that the Hodge group of A is semisimple, and, each simple factor of A is one of the following:
   (a) Any simple Abelian variety A_0 of PEL-type of type I or type II. The usual Hodge conjecture for all powers of A_0 is due to Kuga [**28**, pp. 79–80]. The general Hodge conjecture was first proved by Hazama [**21**] and Tankeev [**43**]. Special cases of this have been proved by Mattuck [**31**], Knight [**26, 27**], Gordon [**14, 15**], and, Tankeev [**44, 45**].
   (b) A simple $2m$-dimensional Abelian variety A_0 of PEL-type such that $D(A_0) = \mathbb{Q}\left(\sqrt{-d}\right)$, and polarization given by a Hermitian form H of signature (m,m), in the following three cases:
      (i) $d = 3$ and $m = 2$.

(ii) $d = 3$, $m = 3$, and H has discriminant -1.
(iii) $d = 1$, $m = 2$, and H has discriminant 1.
The usual Hodge conjecture is proved by Schoen [**39**] for the first two cases, and by van Geemen [**13**] for the last case. This implies the usual Hodge conjecture for all powers of A_0 by [**2**, Theorem 3.1].

(c) A 6-dimensional Abelian variety A_0 of PEL-type such that $D(A_0)$ is a definite quaternion algebra over $\mathbb{Q}$ which contains $\mathbb{Q}\left(\sqrt{-3}\right)$, and polarization given by a skew-Hermitian form of discriminant -1 [**2**, Example 5.1].

By Theorem 5.1 of [**1**, p. 348], the general Hodge conjecture for A follows from the usual Hodge conjecture for all powers of A, which in turn follows from the usual Hodge conjecture for all powers of each simple factor of A.

2. Any Abelian variety B of CM-type such that the Hodge ring of each power of B is generated by divisors [**3**]. A special case of this is proved by Tankeev [**43**, Theorem 2, p. 180].
3. An Abelian variety isogenous to a product $A \times B$, where A is as in Case 1, and B is an Abelian variety of CM-type belonging to Case 2. Since the Hodge group of $A \times B$ is the product of the Hodge groups of A and B (Hazama [**20**, Lemma 3.2, p. 306]), this follows from [**1**, Proposition 4.4.1, p. 347].
4. An Abelian m-fold A of PEL-type with $D(A) = \mathbb{Q}\left(\sqrt{-d}\right)$, and polarization given by a Hermitian form of signature $(m-1, 1)$ in the following five cases:
   (a) $d = 3$ and $m = 3$ (Schoen [**37**, Remark 3.13, p. 31]; see also Pirola [**35**]).
   (b) $d = 3$ and $m = 4$ (see the proof of Theorem 8.1).
   (c) $d = 3$ and $m = 6$ (Schoen [**38**, p. 153]).
   (d) $d = 1$ and $m = 3$ (see the proof of Theorem 8.1).
   (e) $d = 1$ and $m = 4$ (Schoen [**38**, p. 152]).

7.2. Abelian Varieties of Small Dimension.

7.2.1. The general Hodge conjecture is true for any product A of elliptic curves [**1**, Theorem 6.1, p. 353]. This is included in Cases 1, 2, and 3 above, depending upon whether A has no, all, or, some factors of CM-type. Special cases were proved earlier by Shioda [**42**, p. 63] and Bardelli [**4**, (3.5), (3.6), and (3.7), p. 115].

7.2.2. An Abelian surface is either a product of two elliptic curves, or, of CM-type (Case 2), or, its Hodge group is simple and it belongs to Case 1(a). Thus the general Hodge conjecture is true for any power of an Abelian surface.

7.2.3. Bardelli proves the general Hodge conjecture for various Abelian 3-folds in [**4**]. All of these are listed in §7.1. Of the seven cases he lists on p. 115, (3.1), (3.2), and (3.4) are included in Case 1. As already noted, (3.5), (3.6), and (3.7) are products of elliptic curves. The Abelian variety in (3.3) is isogenous to the product of a general Abelian surface and an elliptic curve. If the elliptic curve is of CM-type, then it belongs to Case 3 of §7.1; otherwise, it belongs to Case 1.

8. Abelian Varieties and the Friedlander-Mazur Conjecture

We shall now prove Conjecture 3.3 for all Abelian varieties for which the general Hodge conjecture is known. This had previously been proved for generic Abelian varieties by Hain [**7**, Proposition 4.4, p. 340] and for products of elliptic curves by

Lawson [**29**, p. 195]. Lest the reader be misled into thinking that Conjecture 3.3 is equivalent to the general Hodge conjecture, let us note that Proposition 4.1 implies Conjecture 3.3 for all Abelian 3-folds, even though the general Hodge conjecture is not known for an Abelian 3-fold, A, with $D(A) = \mathbb{Q}\left(\sqrt{-d}\right)$ unless $d = 1$ or 3.

THEOREM 8.1. *Conjecture* 3.3 *is true for all Abelian varieties listed in* §7.1.

PROOF. Let A be one of these Abelian varieties. We claim that A is dominated by a class $\mathcal{A}$ of Abelian varieties, and the group theoretic filtration equals the rational Hodge filtration. If we can also prove the usual Hodge conjecture for $B \times A$ for all $B \in \mathcal{A}$, then, by Proposition 5.1 and Theorem 5.2, the correspondence, arithmetic, group theoretic, and, rational Hodge filtrations are all equal, and we are done.

Suppose A belongs to Case 1, 2, or 3. Then A is dominated by the set of powers of itself, and the group theoretic filtration equals the rational Hodge filtration. For Cases 1 and 2, this is shown in [**1**, Theorem 5.1, p. 348] and [**3**] respectively. For Case 3, this follows from [**1**, Proposition 4.4.1, p. 347]. Since the usual Hodge conjecture is true for all powers of A, our claim is proved.

Now let A be as in Case 4, with $K := D(A) = \mathbb{Q}\left(\sqrt{-d}\right)$. Ribet [**36**, Theorem 3, p. 526] has proved the usual Hodge conjecture for all powers of A. However, in this case, A is not dominated by the set of its powers. The obstacle is a Hodge structure $U \subset H^m(A, \mathbb{Q})$ of Hodge type $\{(m-1, 1), (1, m-1)\}$. The Tate twist $U(-1)$ is not isomorphic to any Hodge structure in the cohomology of a power of A, but it does arise in $H^{m-2}(E^{m-2}, \mathbb{Q})$, where E is an elliptic curve with complex multiplications by K. Thus A is dominated by the set of powers of A, together with E^{m-2}, and to prove both the general Hodge conjecture and Conjecture 3.3 for A it is sufficient to prove the usual Hodge conjecture for $A \times E^{m-2}$.

The Abelian variety $A \times E^{m-2}$ is a (non-generic) member of a Weil family [**46**] of $(2m-2)$-dimensional Abelian varieties with endomorphisms by K, and polarization given by a Hermitian form of discriminant $(-1)^{m-1}$. The Hodge ring of $A \times E^{m-2}$ is generated by divisors, together with the extraordinary" cycles of Weil. The algebraicity of the Weil cycles for Cases 4(a), (b), and (d) follow from the usual Hodge conjecture for the Abelian varieties in Case 1(b).

In Cases 4(c) and (e), the Weil cycles are not known to be algebraic for a generic fiber of the Weil family. In proving the general Hodge conjecture for these cases, Schoen [**38**] finds a variety Y of dimension $m-2$ and an algebraic cycle Z on $Y \times A$, such that U lies in the image of the correspondence homomorphism $H^{m-2}(Y, \mathbb{Q}) \to H^m(A, \mathbb{Q})$ induced by Z. Proposition 3.2 now implies Conjecture 3.3 in these cases. □

References

1. S. Abdulali, *Abelian varieties and the general Hodge conjecture*, Compositio Math. **109** (1997), 341–355.
2. ———, *Abelian varieties of type* III *and the Hodge conjecture*, Preprint available from `http://www.math.ecu.edu/~salman/publications.html`.
3. ———, *Hodge structures on Abelian varieties of CM-type* (in preparation).
4. F. Bardelli, *A footnote to a paper by A. Grothendieck (the Grothendieck generalized Hodge conjecture for some geometric families of Abelian threefolds)*, Proceedings of the Geometry Conference (Milan and Gargnano, 1987), Rend. Sem. Mat. Fis. Milano, vol. 57, 1987, pp. 109–124.

5. S. Bloch and A. Ogus, *Gersten's conjecture and the homology of schemes*, Ann. Sci. École Norm. Sup. (4) **7** (1974), 181–202.
6. P. Deligne , *Hodge cycles on Abelian varieties*, Hodge Cycles, Motives, and Shimura Varieties, Lecture Notes in Math. , vol. 900, Springer-Verlag, Berlin-Heidelberg-New York, 1982, pp. 9–100 (notes by J. S. Milne).
7. E. M. Friedlander, *Filtrations on algebraic cycles and homology*, Ann. Sci. École Norm. Sup. (4) **28** (1995), 317–343.
8. E. M. Friedlander and H. B. Lawson, Jr., *A theory of algebraic cocycles*, Ann. of Math. (2) **136** (1992), 361–428.
9. ———, *Duality relating spaces of algebraic cocycles and cycles*, Topology **36** (1997), 533–565.
10. ———, *Moving algebraic cycles of bounded degree*, Invent. Math. **132** (1998), 91–119.
11. E. M. Friedlander and B. Mazur, *Filtrations on the homology of algebraic varieties*, Mem. Amer. Math. Soc. **529** (1994).
12. ———, *Correspondence homomorphisms for singular varieties*, Ann. Inst. Fourier (Grenoble) **44** (1994), 703–727.
13. B. van Geemen, *Theta functions and cycles on some Abelian varieties*, Math. Z. **221** (1996), 617–631.
14. B. B. Gordon, *Topological and algebraic cycles in Kuga-Shimura varieties*, Math. Ann. **279** (1988), 395–402.
15. ———, *Algebraic cycles and the Hodge structure of a Kuga fiber variety*, Trans. Amer. Math. Soc. **336** (1993), 933–947.
16. ———, *A survey of the Hodge conjecture for Abelian varieties*, Preprint, math.AG/9709030.
17. A. Grothendieck, *Le groupe de Brauer* III: *Exemples et compléments*, Dix Exposés sur la Cohomologie des Schémas, Advanced Studies in Pure Mathematics, vol. 3, North-Holland, Amsterdam, 1968, pp. 88–188.
18. ———, *Standard conjectures on algebraic cycles*, Algebraic Geometry (Bombay, 1968), Tata Inst. Fund. Res. Studies in Math., vol. 4, Oxford University Press, Bombay-London, 1969, pp. 193–199.
19. ———, *Hodge's general conjecture is false for trivial reasons*, Topology **8** (1969), 299–303.
20. F. Hazama, *Algebraic cycles on Abelian varieties with many real endomorphisms*, Tôhoku Math. J. (2) **35** (1983), 303–308.
21. ———, *The generalized Hodge conjecture for stably nondegenerate Abelian varieties*, Compositio Math. **93** (1994), 129–137.
22. W. V. D. Hodge, *The topological invariants of algebraic varieties*, Proc. Internat. Congr. Math. (Cambridge, MA, 1950), vol. 1, Amer. Math. Soc., Providence, RI, 1952, pp. 182–192.
23. S. L. Kleiman, *Algebraic cycles and the Weil conjectures*, Dix Exposés sur la Cohomologie des Schémas, Advanced Studies in Pure Mathematics, vol. 3, North-Holland, Amsterdam, 1968, pp. 359–386.
24. ———, *Finiteness theorems for algebraic cycles* , Algebraic Geometry (Oslo 1970) (F. Oort, ed.), Wolters-Noordhoff, Groningen, 1972, pp. 83–88 (Appendix 1 to Motives).
25. ———, *The standard conjectures*, Motives (Seattle, WA, 1991) (U. Jannsen, S. L. Kleiman, and J.-P. Serre, eds.), Proc. Sympos. Pure Math., vol. 55, Part 1, Amer. Math. Soc., Providence, RI, 1994, pp. 3–20.
26. A. J. Knight, *A classification of the cycles on an Abelian variety*, Proc. London Math. Soc. (3) **13** (1963), 184–192.
27. ———, *On general Riemann matrices and their attached Abelian varieties*, Proc. London Math. Soc. (3) **20** (1970), 573–596.
28. M. Kuga, *Fibred variety over symmetric space whose fibres are Abelian varieties*, Proceedings of the U.S.-Japan Seminar in Differential Geometry (Kyoto, 1965), Nippon Hyoronsha, Tokyo, 1966, pp. 72–81.
29. H. B. Lawson, Jr., *Spaces of algebraic cycles*, Surveys in Differential Geometry (C.-C. Hsiung and S.-T. Yau, eds.), vol. 2, International Press, Cambridge, MA, 1995, pp. 137–213.
30. D. Lieberman, *Numerical and homological equivalence of algebraic cycles on Hodge manifolds*, Amer. J. Math. **90** (1968), 366–374.
31. A. Mattuck, *Cycles on Abelian varieties*, Proc. Amer. Math. Soc. **9** (1958), 88–98.
32. B. J. J. Moonen and Yu. G. Zarhin, *Weil classes on Abelian varieties*, J. Reine Angew. Math. **496** (1998), 83–92.

33. D. Mumford, *Families of Abelian varieties*, Algebraic Groups and Discontinuous Subgroups (Boulder, CO, 1965) (A. Borel, and G. D. Mostow, eds.), Proc. Sympos. Pure Math., vol. 9, Amer. Math. Soc., Providence, RI, 1966, pp. 347–351.
34. V. K. Murty, *Exceptional Hodge classes on certain Abelian varieties*, Math. Ann. **268** (1984), 197–206.
35. G. P. Pirola, *On a conjecture of Xiao*, J. Reine Angew. Math. **431** (1992), 75–89.
36. K. A. Ribet, *Hodge classes on certain types of Abelian varieties*, Amer. J. Math. **105** (1983), 523–538.
37. C. Schoen, *Hodge classes on self-products of a variety with an automorphism*, Compositio Math. **65** (1988), 3–32.
38. ———, *Cyclic covers of* $\mathbf{P}^v$ *branched along* $v+2$ *hyperplanes and the generalized Hodge conjecture for certain Abelian varieties*, Arithmetic of Complex Manifolds) (Erlangen, 1988) (W.-P. Barth and H. Lange, eds.), Lecture Notes in Math., vol. 1399, Springer-Verlag, Berlin-Heidelberg-New York, 1989, pp. 137–154.
39. ———, *Addendum to: Hodge classes on self-products of a variety with an automorphism* (preprint).
40. G. Shimura, *On the field of definition for a field of automorphic functions*, Ann. of Math. (2) **80** (1964), 160–189.
41. ———, *Moduli of Abelian varieties and number theory*, Algebraic Groups and Discontinuous Subgroups (Boulder, CO, 1965) (A. Borel and G. D. Mostow, eds.), Proc. Sympos. Pure Math., vol. 9, Amer. Math. Soc., Providence, RI, 1966, pp. 312–332.
42. T. Shioda, *What is known about the Hodge conjecture?*, in Algebraic Varieties and Analytic Varieties (S. Iitaka, ed.), Adv. Stud. Pure Math., vol. 1, Kinokuniya, Tokyo, 1983, pp. 55–68.
43. S. G. Tankeev, *Abelian varieties and the general Hodge conjecture*, Izv. Ross. Akad. Nauk Ser. Mat. **57** (1993), 192–206 (Russian); English transl., Russian Acad. Sci. Izv. Math. **43** (1994), 179–191.
44. ———, *Algebraic cycles on an Abelian variety without complex multiplication*, in Izv. Ross. Akad. Nauk Ser. Mat. **58** (1994), 103–126 (Russian); Russian Acad. Sci. Izv. Math. **44** (1995), 531–553.
45. ———, *Cycles on Abelian varieties and exceptional numbers*, Izv. Ross. Akad. Nauk Ser. Mat. **60** (1996), 159–194 (Russian); English transl., Izv. Math. **60** (1996), 391–424.
46. A. Weil, *Abelian varieties and the Hodge ring*, Œuvres Scientifiques Collected Papers, Vol. III (1964–1978), Springer-Verlag, New York, 1979; Corrected second printing, 1980, pp. 421–429.

Department of Mathematics, East Carolina University, Greenville, NC 27858, USA
E-mail address: salman@math.ecu.edu

Centre de Recherches Mathématiques
CRM Proceedings and Lecture Notes
Volume **24**, 2000

Building Mixed Hodge Structures

Donu Arapura

This article is aimed at people who already know what mixed Hodge structures are and what they are good for, but who are not sure how to construct them. Even if one wants to treat mixed Hodge theory primarily as a black box, there will be times when one needs to open it up, look inside, and perhaps reassemble it in a different way. I would like to discuss some of the mechanics involved along the lines of Deligne's original construction [**D2**]. I should mention that a somewhat different and far reaching approach to constructing mixed Hodge structures has been given by M. Saito [**Sa3**]; I will say a few words about it at the end. However, I believe that there is still some value in this old-fashioned approach. Some of the techniques involved are quite useful in other situations, especially descent theory, which has a number of applications beyond those discussed here (e.g., [**B**], [**GS**], [**GNPP**], [**GN**], [**Ha**], [**N**]).

The first basic principle of the construction is that in order to get mixed Hodge structures on cohomology of a space, the necessary filtrations should be defined on the level of complexes that give rise to the cohomology. Appropriate homological conditions need to be imposed in order to get things to work. In this article, these conditions will be embodied in the notion of a $\mathbb{C}$-Hodge complex which is a fairly minimalistic reworking of Deligne's original axioms for a cohomological mixed Hodge complex. $\mathbb{C}$-Hodge complexes need not have any underlying integral or real structure, but they do possess the necessary structure to carry out the key degeneration of spectral sequence arguments. Once these technical results are established, it is fairly straightforward to impose integral structures on the $\mathbb{C}$-Hodge complexes so as to obtain mixed Hodge structures on cohomology.

The second basic principle is that in order to construct mixed Hodge structures on the cohomology of a whole class of spaces and coefficient sheaves, it is only necessary to construct explicit Hodge complexes on a subset of these pairs, provided that the general pairs can be "approximated" by the special ones. This reduction is ultimately handled by cohomological descent. However, in order to clarify the structure of the argument, the condition for guaranteeing the existence of approximations, called density, will be introduced first; the actual descent theory

1991 *Mathematics Subject Classification.* 14C30.
Received by the editors October 13, 1998; Accepted June 23, 1999.
Partially supported by NSF.
This is the final form of the paper.

will be done afterwards. Verification of density in practice requires resolution of singularities.

The last section contains a rather elaborate illustration of these ideas in construction of mixed Hodge structures on the cohomology of certain constructible sheaves.

My heartfelt thanks goes out to the referees for their remarkable thoroughness. I have learned, after the fact, that one of them was Steve Zucker.

1. First Attempt

Let me begin by trying to make a bare hands construction of a functorial mixed Hodge structure on the cohomology of a singular projective variety. This attempt will fail, but it will be instructive nevertheless.

For the record, I should recall what a mixed Hodge structure is. I will use the standard convention that filtrations indexed by superscripts are decreasing while those with subscripts are increasing. These can be interchanged by the rule $W^{-n} = W_n$, and we do this without further comment. A mixed Hodge structure is usually taken to be a finitely generated Abelian group $H_{\mathbb{Z}}$, along with an ascending filtration $W_\bullet$ on $H_{\mathbb{Q}} = H_{\mathbb{Z}} \otimes \mathbb{Q}$, and a decreasing filtration $F^\bullet$ on $H = H_{\mathbb{Z}} \otimes \mathbb{C}$ subject to certain axioms [**D2**]. After jettisoning $H_{\mathbb{Z}}$ and $H_{\mathbb{Q}}$, one arrives at the notion of the $\mathbb{C}$-*mixed Hodge structure* (or simply just a $\mathbb{C}$-Hodge structure) which is a finite dimensional vector space H together with three filtrations $(F^\bullet, \widehat{F}^\bullet, W_\bullet)$ which are opposed in the sense of Deligne [**D2**], i.e.,

$$\mathrm{Gr}_n^W H = F^p \,\mathrm{Gr}_n^W H \oplus \widehat{F}^q \,\mathrm{Gr}_n^W H$$

whenever $p + q = n$, where $\mathrm{Gr}_n^W H = W_n/W_{n-1}$. By tradition W is taken to be increasing, but it will be frequently convenient to regard a $\mathbb{C}$-Hodge structure as given by three decreasing filtrations $(F^\bullet, \widehat{F}^\bullet, W^\bullet)$. The third filtration $\widehat{F}^\bullet$ plays the role of the conjugate filtration to $F^\bullet$. Morphisms of $\mathbb{C}$-Hodge structures are simply linear maps preserving the filtrations. In the presence of several $\mathbb{C}$-Hodge structures H, H', ... , I will indicate the corresponding filtrations by $F^\bullet H$, $F^\bullet H'$, ... $\mathbb{C}$-Hodge structures possess most of the homological properties of mixed Hodge structures. For example the category of $\mathbb{C}$-Hodge structures is Abelian, and the functors Gr_F, $\mathrm{Gr}_{\widehat{F}}$, Gr_W are exact [**D2**, 1.2.5]. A $\mathbb{C}$-Hodge structure is pure of weight n if $\mathrm{Gr}_n^W H = H$. The filtration W will usually be suppressed for pure Hodge structures. There are no nonzero morphisms between pure $\mathbb{C}$-Hodge structures of different weights. An integral lattice on a $\mathbb{C}$-Hodge structure $(H, F^\bullet, \widehat{F}^\bullet, W_\bullet)$ is a finitely generated Abelian group $H_{\mathbb{Z}}$ with an isomorphism $H_{\mathbb{Z}} \otimes \mathbb{C} \cong H$ such that $W_\bullet$ is rational, that is spanned by $W_\bullet \cap (H_{\mathbb{Z}} \otimes \mathbb{Q})$, and $\widehat{F}^\bullet = \overline{F}^\bullet$ the conjugate of F with respect to the induced real structure on H. A mixed Hodge structure in the usual sense is the same thing as a $\mathbb{C}$-Hodge structure together with a lattice. A morphism of mixed Hodge structures is simply a morphism of $\mathbb{C}$-Hodge structures and a compatible map of lattices.

I will generally use the term complex variety to mean a reduced separated scheme of finite type over $\mathbb{C}$ with its classical topology. Suppose that X is complex proper variety. By Hironaka's theorem [**Hi**], there exists a commutative diagram

$$\begin{array}{ccc} E & \stackrel{i}{\hookrightarrow} & \tilde{X} \\ \downarrow p & & \downarrow p \\ S & \stackrel{j}{\hookrightarrow} & X \end{array}$$

where π is a resolution of singularities with exceptional divisor E, and $S = \pi(E)$. Then

PROPOSITION 1.1. *For any sheaf of Abelian groups $\mathcal{F}$ on X, the "Mayer-Vietoris" sequence*

$$\cdots \to H^i(X, \mathcal{F}) \xrightarrow{\pi^* + i^*} H^i(\tilde{X}, \pi^* F) \oplus H^i(S, j^*\mathcal{F}) \xrightarrow{i^* - p^*} H^i(E, (\pi \circ i)^*\mathcal{F}) \to \cdots$$

is exact.

Although this fact is well known, I will give the proof since it foreshadows the results of Section 5.

PROOF. I will use some basic notions from derived categories [**V**] for succinctness, but the argument can be rewritten without them.

Let $g = \pi \circ i$. It is enough to show that the sequence

$$\mathcal{F} \to i_*\mathcal{F} \oplus \mathbb{R}\pi_*\mathcal{F} \to \mathbb{R}g_*\mathcal{F}$$

is part of a distinguished triangle. This can be checked at the stalks. If $x \notin S$, then the stalk of the sequence at x reduces to

$$\mathcal{F}_x \xrightarrow{=} 0 \oplus \mathcal{F}_x \to 0$$

which is certainly part of a distinguished triangle. If $x \in S$, let E_x denote the fiber $g^{-1}(x) = \pi^{-1}(x)$. Then the stalk of the sequence decomposes as sum of sequences

$$\mathcal{F}_x \xrightarrow{=} \mathcal{F}_x \to 0$$

and

$$0 \to \mathbb{R}\Gamma(E_x, \mathcal{F}|_{E_x}) \xrightarrow{=} \mathbb{R}\Gamma(E_x, \mathcal{F}|_{E_x})$$

which both extend to distinguished triangles. □

This proposition suggests a plan for putting a mixed Hodge structure on $H^*(X, \mathbb{Z})$. First use classical Hodge theory to put pure Hodge structures on the cohomology of smooth projective varieties. Then assume by induction that the cohomology of all proper varieties of dimension less than $dimX$ has been equipped with a mixed Hodge structure. Choose a diagram as above, and try to impose a mixed Hodge structure on it so that the Mayer-Vietoris sequence is an exact sequence of mixed Hodge structures. Well that's the idea. The problem is that the category of mixed Hodge structures has nontrivial extensions, so compatibility with Mayer-Vietoris will not uniquely determine the mixed Hodge structure on $H^*(X)$. Even if one worked in the category of $\mathbb{C}$-mixed Hodge structures, which is semisimple, one would only be able to determine the $\mathbb{C}$-mixed Hodge structure on $H^*(X)$ up to noncanonical isomorphism. To circumvent these problems, it will be necessary to work on the level of complexes. Once this is done, it is possible to fix up this inductive approach. However it will turn out to be simpler to do the construction in one step using descent theory (following Deligne's idea). An elementary construction along lines similar to the above sketch has been given by El Zein [**E**]. A rather different construction has been given by Saito [**Sa1, Sa3**].

2. $\mathbb{C}$-Hodge Complexes

Recall that a filtration on a complex $A^\bullet$ is a sequence of filtrations $F^\bullet A^n \subset A^n$ compatible with the differentials. The filtration is biregular if for each n, there exist a and b such that $F^a A^n = A^n$ and $F^b A^n = 0$. In the sequel, I will try to conserve notation by writing $F^\bullet$ or simply F instead of $F^\bullet A^\bullet$ when no confusion is likely.

The basic device for constructing $\mathbb{C}$-Hodge structures is a $\mathbb{C}$-*Hodge complex* [**A**] It is a biregularly trifiltered complex of $\mathbb{C}$-vector spaces $(A^\bullet, F^\bullet A^\bullet, \widehat{F}^\bullet A^\bullet, W^\bullet A^\bullet)$ bounded below and satisfying:

HC1. $A^\bullet$ has finite dimensional cohomology.

HC2. For each p, the filtrations on $\mathrm{Gr}^p_W A$ induced by $F^\bullet$ and $\widehat{F}^\bullet$ are strictly preserved by the differentials (i.e., $dF^p = \mathrm{image}(d) \cap F^p$ for all p, and likewise for $\widehat{F}^\bullet$).

HC3. $H^i(\mathrm{Gr}^p_W A)$, when equipped with the filtrations

$$\mathrm{image}[H^i(F^\bullet \mathrm{Gr}^p_W A) \to H^i(\mathrm{Gr}^p_W A)]$$
$$\mathrm{image}[H^i(\widehat{F}^\bullet \mathrm{Gr}^p_W A) \to H^i(\mathrm{Gr}^p_W A)],$$

becomes a pure $\mathbb{C}$-Hodge structure of weight $i - p$ for all i, p.

This is related to Deligne's notion of a mixed Hodge complex [**D2**]; a more detailed comparison will be given below.

Example 2.1. Consider a complex $A^\bullet$ of $\mathbb{C}$-mixed Hodge structures with its filtrations $F^\bullet$, $\widehat{F}^\bullet$, $W^\bullet$. As it stands, this is not a $\mathbb{C}$-Hodge complex. But if one defines

$$\widetilde{W}^i A^n = W^{i-n} A^n$$

then $(A^\bullet, F^\bullet, \widehat{F}^\bullet, \widetilde{W}^\bullet)$ is.

Example 2.2. Given a $\mathbb{C}$-Hodge complex $(A^\bullet, F^\bullet, \widehat{F}^\bullet, W^\bullet)$, let $(A[i]^\bullet, d_{A[i]}) = (A^{\bullet+i}, (-1)^i d_A)$ be the translated complex. Then this becomes a $\mathbb{C}$-Hodge complex when endowed with filtrations F, $\widehat{F}$ and $W^\bullet[i] = W^{\bullet+i}$. For any pair of integers the twist $A(a,b)$ which is A with the filtrations $F^\bullet[a]$, $\widehat{F}^\bullet[b]$ and $W^\bullet[-a-b]$ is a $\mathbb{C}$-Hodge complex. $A(1,1)$ is the Tate twist.

Example 2.3. Let X be a compact Kähler manifold. The de Rham complex $\mathcal{E}^\bullet(X)$ with filtrations

$$F^p \mathcal{E}^\bullet(X) = \bigoplus_{p' \geq p} \mathcal{E}^{p'q}(X)$$
$$\widehat{F}^q \mathcal{E}^\bullet(X) = \bigoplus_{q' \geq q} \mathcal{E}^{pq'}(X)$$
$$W_0 \mathcal{E}^\bullet(X) = \mathcal{E}^\bullet(X); \quad W_{-1}\mathcal{E}^\bullet(X) = 0$$

is a $\mathbb{C}$-Hodge complex by classical Hodge theory.

Example 2.4. Let X be as above. A polarized complex variation of Hodge structure is a C^∞ vector bundle V with an indefinite C^∞ Hermitian metric $\langle\ ,\ \rangle$, a decomposition $V = \bigoplus_{p+q=w} V^{pq}$, and a flat connection ∇ on V subject to the following axioms:

- The metric is positive definite on V^{pq} for p even and negative definite when p is odd, and the decomposition is orthogonal.

- The connection is compatible with the metric (i.e., ∇ satisfies the Leibniz rule with respect to $\langle\ ,\ \rangle$) and satisfies the following form of Griffith's transversality:

$$\nabla(V^{pq}) \subseteq \mathcal{E}^{01}(V^{p+1,q-1}) \oplus \mathcal{E}^{10}(V^{p,q}) \oplus \mathcal{E}^{01}(V^{p,q}) \oplus \mathcal{E}^{10}(V^{p-1,q+1})$$

where $\mathcal{E}^{rs}(U)$ is the sheaf C^∞ U-valued rs-forms.

The de Rham complex $(\mathcal{E}^\bullet(X,V),\nabla)$ becomes a $\mathbb{C}$-Hodge complex when it is equipped with the filtrations

$$F^p\mathcal{E}^\bullet(X,V) = \bigoplus_{p'+p''\geq p} \mathcal{E}^{p'q'}(X, V^{p''q''})$$

$$\widehat{F}^q\mathcal{E}^\bullet(X) = \bigoplus_{q'+q''\geq q} \mathcal{E}^{p'q'}(X, V^{p''q''})$$

$$W_w\mathcal{E}^\bullet(X) = \mathcal{E}^\bullet(X,V);\ W_{w-1}\mathcal{E}^\bullet(X) = 0.$$

This follows from the generalized Kähler identities of Deligne and Simpson [**S**], [**Z**]. A lattice is a locally constant sheaf $\mathcal{V}$ of Abelian groups with finitely generated stalks and an isomorphism $\mathcal{V}\otimes\mathbb{C}\cong\ker(\nabla)$ such that V^{pq} is conjugate to V^{qp} with respect to resulting $\mathbb{R}$-structure. The usual notion of a polarized variation of Hodge structure is equivalent to a complex variation of Hodge structure and a lattice. See [**D3**] [**GS**], [**S**] for more information.

Complex (usually nonintegral) variations of Hodge structure, occur naturally in Simpson's work [**S**] as fixed points in the moduli space of Higgs bundles under the natural $\mathbb{C}^*$-action. Using this, he has shown that any representation of the fundamental group of a smooth projective variety can be deformed to the monodromy of a complex variation of Hodge structure.

In the previous examples, the complexes arose by taking global sections of complexes of sheaves. This can be generalized as follows. A $\mathbb{C}$-Hodge complex on a space X is a quadruple $(A^\bullet, F^\bullet, \widehat{F}^\bullet, W^\bullet)$ consisting of a bounded below complex of sheaves of $\mathbb{C}$-vector spaces $A^\bullet$ with finite dimensional cohomology, together with three biregular filtrations on it such that:

$\mathbb{HC}1$. $A^\bullet$ has finite dimensional hypercohomology.

$\mathbb{HC}2$. For each p, the filtrations on $\mathbb{R}\Gamma\operatorname{Gr}^p_W A$ induced by F and $\widehat{F}$ are strictly preserved by the differentials.

$\mathbb{HC}3$. For every i,p, F and $\widehat{F}$ induce a pure $\mathbb{C}$-Hodge structure of weight $i-p$ on $\mathbb{H}^i(\operatorname{Gr}^p_W A)$.

A map of trifiltered complexes

$$(A^\bullet, F^\bullet A^\bullet, \widehat{F}^\bullet A^\bullet, W^\bullet A^\bullet) \to (B^\bullet, F^\bullet B^\bullet, \widehat{F}^\bullet B^\bullet, W^\bullet B^\bullet)$$

will be called an *acyclic resolution* if the associated maps

$$\operatorname{Gr}^\bullet_{FA}\operatorname{Gr}^\bullet_{WA} A^\bullet \to \operatorname{Gr}^\bullet_{FB}\operatorname{Gr}^\bullet_{WB} B^\bullet$$

and

$$\operatorname{Gr}^\bullet_{\hat{F}A}\operatorname{Gr}^\bullet_{WA} A^\bullet \to \operatorname{Gr}^\bullet_{\hat{F}B}\operatorname{Gr}^\bullet_{WB} B^\bullet$$

are acyclic resolutions in the usual sense. Canonical acyclic resolutions can be constructed using Godement's flasque resolutions. The axioms imply that if the quadruple $(A^\bullet, F^\bullet, \widehat{F}^\bullet, W^\bullet)$ is a $\mathbb{C}$-Hodge complex, then after replacing it by an acyclic resolution, $\Gamma(A^\bullet)$ with its induced filtrations is a $\mathbb{C}$-Hodge complex in the earlier sense.

EXAMPLE 2.5. Let $D \subset X$ be a divisor with normal crossings in a compact Kähler manifold and let $j : X - D \hookrightarrow X$ be the inclusion. Let $A^\bullet_X$ be the sheaf of $\mathbb{C}$-valued real analytic differential forms on X. Following Navarro Aznar [**N**], define $A^\bullet_X(\log D)$ to be the A^0_X-subalgebra of $j_*A^\bullet_{X-D}$ generated by $A^\bullet_X$ and $\log|f|$, $\mathrm{Re}\, df/f$, and $\mathrm{IM}\, df/f$ for any (local) holomorphic function f vanishing along a component of D. This is a subcomplex of $j_*A^\bullet_{X-D}$, which is an incarnation of $\mathbb{R}j_*\mathbb{C}$. Furthermore it has a bigrading

$$A^n_X(\log D) = \bigoplus_{p+q=n} A^{p,q}_X(\log D)$$

by (p,q) type. With filtrations

$$F^p A^\bullet_X(\log D) = \bigoplus_{p'\geq p} A^{p',q}_X(\log D)$$

$$\widehat{F}^q A^\bullet_X(\log D) = \bigoplus_{q'\geq q} A^{p,q'}_X(\log D)$$

$$W_k A^m_X(\log D) = image(A^k_X(\log D) \wedge A^{m-k}_X \to A^m_X(\log D))$$

$A_X(\log D)$ becomes a $\mathbb{C}$-Hodge complex. The complex $A^\bullet_X(\log D)$ can be replaced by the C^∞ analogue constructed by Burgos [**Bu**] to obtain an explicit acyclic resolution of this $\mathbb{C}$-Hodge complex.

EXAMPLE 2.6. Let X and D be as above. Let (V^o, ∇^o) be a holomorphic bundle on $X - D$ with a flat connection which is unitary in the sense that $\mathcal{V} = ker(\nabla^o)$ corresponds to a unitary representation of $\pi_1(X - D)$. Then there exists a unique extension of V^o to a vector bundle V on X such that ∇^o extends to a logarithmic connection ∇ with residues having eigenvalues in $[0,1)$ [**D1**]. The complex $(A^\bullet_X(\log D) \otimes V, \nabla)$ is quasi-isomorphic to $\mathbb{R}j_*\mathcal{V}$. There is a filtration W on $(A^\bullet_X(\log D) \otimes V, \nabla)$, such that together with the filtrations

$$F^p A^\bullet_X(\log D) \otimes V = \bigoplus_{p'\geq p} A^{p',q}_X(\log D) \otimes V$$

$$\widehat{F}^q A^\bullet_X(\log D) \otimes V = \bigoplus_{q'\geq q} A^{p,q'}_X(\log D) \otimes V$$

this becomes a $\mathbb{C}$-Hodge complex [**A**, IV.1].

EXAMPLE 2.7. A morphism of $\mathbb{C}$-Hodge complexes is a map of complexes which is compatible with the filtrations. Let $f : (A^\bullet, F^\bullet A^\bullet, \ldots) \to (B^\bullet, F^\bullet B^\bullet, \ldots)$ be a morphism. The mapping cone is a new complex $\mathrm{Cone}(f)^\bullet = C^\bullet$ with

$$C^n = A^{n+1} \oplus B^n$$

and $d(\alpha, \beta) = (-d\alpha, d\beta - f(\alpha))$. With the filtrations

$$F^p C^n = F^p A^{n+1} \oplus F^p B^n$$

$$\widehat{F}^p C^n = \widehat{F}^p A^{n+1} \oplus \widehat{F}^p B^n$$

and

$$W_k C^n = W_{k-1} A^{n+1} \oplus W_k B^n$$

$C^\bullet$ becomes a $\mathbb{C}$-Hodge complex. By construction there are morphisms $B \to \mathrm{Cone}(f)$ and $\mathrm{Cone}(f) \to A[1]$ where $A[1]$ is defined in Example 2.2

The previous example can be generalized.

EXAMPLE 2.8. A differential graded $\mathbb{C}$-Hodge complex consists of a collection $(A^{n\bullet}, FA^{n\bullet}, \dots)$ of $\mathbb{C}$-Hodge complexes, and morphisms $\delta_n \colon A^{n,\bullet} \to A^{n+1,\bullet}$ satisfying $\delta_{n+1}\delta_n = 0$. Define the total (or associated single) complex $T^\bullet = tot^\bullet(A)$ by

$$T^n = \bigoplus_{i+j=n} A^{ij}$$

with differential $d + (-1)^j\delta_j$. With filtrations

$$F^pT^n = \bigoplus_{i+j=n} F^pA^{ij}$$
$$\widehat{F}^pT^n = \bigoplus_{i+j=n} \widehat{F}^pA^{ij}$$

and

$$W_kT^n = \bigoplus_{i+j=n} W_{k+i}A^{ij}$$

$T^\bullet$ becomes a $\mathbb{C}$-Hodge complex.

The above constructions will be used extensively later. Here are a few concrete instances.

EXAMPLE 2.9. Let X be a compact Kähler manifold, $i\colon Y \hookrightarrow X$ a closed submanifold and $j\colon U = X - Y \hookrightarrow X$ the the complement. Let $C^\bullet$ be the mapping cone of the restriction map $\mathcal{E}_X^\bullet \to \mathcal{E}_Y^\bullet$. This is a $\mathbb{C}$-Hodge complex quasi-isomorphic to $j_!\mathbb{C}_U$, and $H^i(j_!\mathbb{C}_U) \cong H^i(X, Y)$.

The above example is very special case of the constructions of the final section.

EXAMPLE 2.10. Let X be a compact Kähler manifold, and let $\theta \in H^0(X, \Omega_X^1)$ and N be some positive integer. Let θ' denote the operator $\phi \mapsto (-1)^{\deg\phi}\theta \wedge \phi$. Then $(\theta')^2 = 0$, we get a differential graded $\mathbb{C}$-Hodge complex

$$\mathcal{E}_X^\bullet \xrightarrow{\theta'} \mathcal{E}_X^\bullet(1,0)[1] \xrightarrow{\theta'} \cdots \xrightarrow{\theta'} \mathcal{E}_X^\bullet(N,0)[N]$$

where $\mathcal{E}_X(i,0)[i]$ is defined in Examples 2.2 and 2.3.

This example is from [**A**]. Related constructions occur in the work of Steenbrink [**St**].

In Example 2.1, the cohomology groups of $A^\bullet$ carry $\mathbb{C}$-Hodge structures with filtrations induced by F, $\widehat{F}$ and W, because the category of $\mathbb{C}$-Hodge structures is Abelian. This generalizes to all $\mathbb{C}$-Hodge complexes. The following, which was announced in [**A**] and is an analogue of [**D2**, scholium 8.1.9], is the basic result.

THEOREM 2.11. *Let* $(A^\bullet, F^\bullet, \widehat{F}^\bullet, W^\bullet)$ *be a* $\mathbb{C}$*-Hodge complex then*

(a) *The spectral sequence induced by* W *degenerates at* E_2.

(b) *$H^i(A^\bullet)$ carries a $\mathbb{C}$-Hodge structure with filtrations given by $F^\bullet$, $\widehat{F}^\bullet$ and $W^\bullet$ shifted by i, or explicitly by:*

$$F^pH^i(A^\bullet) = \text{image}[H^i(F^pA^\bullet) \to H^i(A^\bullet)]$$

$$\widehat{F}^pH^i(A^\bullet) = \text{image}[H^i(\widehat{F}^pA^\bullet) \to H^i(A^\bullet)]$$

$$W^kH^i(A^\bullet) = \text{image}[H^i(W^{k+i}A^\bullet) \to H^i(A^\bullet)]$$

(c) *The spectral sequences associated to F and $\widehat{F}$ degenerate at E_1.*

The proof of the above theorem will be given in Section 3. A variant of part (a), proved by a similar argument, is:

PROPOSITION 2.12. *The spectral sequence induced by $W^\bullet$ on $\text{Gr}^i_F A^\bullet$ degenerates at E_2.*

Analogues of (c) also hold, for F and $\widehat{F}$, namely the spectral sequences induced by $F^\bullet$ and $\widehat{F}^\bullet$ on $\text{Gr}^i_W A^\bullet$ degenerate at E_1. This is nothing but a restatement of axiom HC1. As a corollary, one gets similar statements for $\mathbb{C}$-Hodge complexes on topological space with cohomology replaced by hypercohomology.

The degeneration statements have a number of important applications. When part (c) of the theorem is applied to Example 2.6, along with the filtered quasi-isomorphism $(\Omega^\bullet_X(\log D) \otimes V, F^\bullet) \hookrightarrow (A^\bullet_X(\log D) \otimes V, F^\bullet)$, one obtains the degeneration of the spectral sequence

$$H^q(\Omega^p_X(\log D) \otimes V) \implies H^{p+q}(X - D, \mathcal{V})$$

This fact, and its generalizations, can be used to prove vanishing theorems (see [**EV**]). For example, if $X - D$ is Stein (e.g., if X is projective and D ample) then it has the homotopy type of a CW-complex of dimension at most $n = \dim X$; this implies the following version of Kodaira's vanishing theorem:

$$H^i(X, \omega_X(D) \otimes V) \subseteq H^{n+i}(X - D, \mathcal{V}) = 0$$

for $i > 0$. When Proposition 2.12 is applied to the complex of Example 2.10, one recovers a theorem of Green and Lazarsfeld [**GL**] that the spectral sequence

$$E_1 = H^q(X, \Omega^p_X) \implies \mathbb{H}^{p+q}(X; O_X \xrightarrow{\theta} \Omega^1_X \xrightarrow{\theta} \cdots)$$

degenerates at E_2. This kind of result, while technical, is crucial for establishing their generic vanishing theorems. The argument is adapted from [**A**] which contains further details and refinements.

With the basic homological apparatus in place, the integral structure can now be added. An integral structure for a $\mathbb{C}$-Hodge complex $(A^\bullet, W, F, \widehat{F})$ on a topological space consists of a bounded below complex of sheaves of Abelian groups $A^\bullet_{\mathbb{Z}}$ with finitely generated cohomology and quasi-isomorphisms

$$A^\bullet_{\mathbb{Z}} \otimes \mathbb{Q} \xleftarrow{\sim} A'^\bullet \xrightarrow{\sim} A^\bullet_{\mathbb{C}}$$

inducing lattices on cohomology. Note that the intermediate complex $A'^\bullet$ is part of the data (or to put it another way: fix an explicit representative for an isomorphism $A^\bullet_{\mathbb{Z}} \otimes \mathbb{Q} \cong A^\bullet_{\mathbb{Q}}$ in the derived category). The easist way to ensure that $W_\bullet H^\bullet(A^\bullet)$ is rational is to construct a filtered complex of $\mathbb{Q}$-vector spaces $(A^\bullet_{\mathbb{Q}}, W_{\mathbb{Q}})$ and isomorphisms

$$A^\bullet_{\mathbb{Z}} \otimes \mathbb{Q} \cong A^\bullet_{\mathbb{Q}}$$

$$(A^\bullet_{\mathbb{Q}}, W_{\mathbb{Q}}) \otimes \mathbb{C} \cong (A^\bullet, W)$$

in the derived and filtered derived categories. A Hodge complex will consist of a $\mathbb{C}$-Hodge complex together with a choice of integral structure. Hodge complexes carry mixed Hodge structures on cohomology. A trivial example of an integral structure is the inclusion of the constant sheaf $\mathbb{Z}_X \subset \mathcal{E}_X^\bullet$ into the De Rham complex of a compact Kähler manifold. More generally for variations of Hodge structure, the inclusion of the lattice in $\mathcal{E}_X^\bullet(V)$ supplies an integral structure. In Example 2.5, $A_\mathbb{Z}$ can be taken to be the direct image under j_* of the complex of sheaves of singular C^∞ cochains, $A' = A = A_X^\bullet(\log D)$, and $A' \to A_\mathbb{Z} \otimes \mathbb{C}$ the map which assigns to a form α, the cochain $\gamma \mapsto \int_\gamma \alpha$ (the rationality of W is a consequence of a filtered quasi-isomorphism $(A_\mathbb{Z}, \tau_{\le \bullet}) \otimes \mathbb{C} \cong (A, W_\bullet)$, [**D2, N**]).

A few remarks ought to be made about the relationship between Hodge complexes and (cohomological) mixed Hodge complexes [**D2**, 8.1]. The set up for the latter is similar, one has a collection of complexes with filtrations and compatability isomorphisms subject to appropriate conditions. Although I do not expect a precise equivalence, it does seem to be case in practice that whenever one has a natural Hodge complex, one has a corresponding cohomological mixed Hodge and conversely. However, there are interesting examples of $\mathbb{C}$-Hodge complexes (or structures) which do not possess any integral or even real structure. The $\mathbb{C}$-Hodge complexes of Examples 2.4, 2.6 and 2.10 will not generally possess natural integral structures. For certain applications (such as those discussed above), the degeneration of the above spectral sequences is precisely what is required; the integral structure is irrelevant even when it exists. For these reasons, the notion of a $\mathbb{C}$-Hodge complex is useful on its own.

For any space X, the collection of $\mathbb{C}$-Hodge complexes forms a category, denoted $C_{\mathbb{C}\text{-Hodge}}(X)$. More generally, let $C_{\mathbb{C}\text{-Hodge}}$ be the category whose objects are pairs consisting of a topological space and a $\mathbb{C}$-Hodge complex on it. A morphism from $(X, (A, \dots))$ to $(Y, (B, \dots))$ is pair consisting of a continuous map $f \colon X \to Y$ and a map of trifiltered complexes $(f^*B, \dots) \to (A, \dots)$. This category is fibered over the category of topological spaces by the functor $|\cdots|$, and $C_{\mathbb{C}\text{-Hodge}}(X)$ is the fiber over X. The category of Hodge complexes C_{Hodge} is defined similarly.

3. Proof of Theorem 2.11

Let $A^\bullet$ be a bounded below complex of vector spaces with two biregular filtrations $W^\bullet$, $F^\bullet$. There is a spectral sequence corresponding to the first filtration with

$$E_1^{pq}(A, W) = H^{p+q}(\operatorname{Gr}_W^p A^\bullet)$$

and differential

$$d_1 \colon E_1^{pq}(A, W) \to E_1^{p+1,q}(A, W)$$

given by the connecting map associated to

$$0 \to \operatorname{Gr}_W^{p+1} A \to W^p A / W^{p+2} A \to \operatorname{Gr}_W^p A \to 0.$$

This can be filtered by F in many (usually) different ways. The *first direct filtration* is

$$F_{\text{dir}}^i E_r^{pq}(A, W) = \operatorname{image}[E_r^{pq}(F^i, F^i \cap W) \to E_r^{pq}(A, W)].$$

The *recursive filtration* is defined inductively by $F_{\text{rec}}^i E_1^{pq} = F_{\text{dir}}^i E_1^{pq}$ and

$$F_{\text{rec}}^i E_{r+1}^{pq}(A, W) = \operatorname{image}[F_{\text{rec}}^i \cap \ker_r^{pq}(A, W)]$$

where

$$\ker_r^{pq}(A,W) = \ker[E_r^{pq}(A,W) \to E_r^{p+r,q-r+1}(A,W)]$$

and the image is taken in $E_{r+1}^{pq}(A,W)$. E_∞ is a subquotient of $H^*(A)$ and it inherits a filtration induced by

$$F^i H^*(A) = \text{image}[H^*(F^i A) \to H^*(A)]$$

which will be called the *final filtration*. The first direct filtration is useful because it is preserved by differentials $d_r\colon E_r^{pq}(A,W) \to E_r^{p+r,q-r+1}(A,W)$, while the recursive filtration is what one is seeing if one follows the spectral sequence from E_1 to E_2 to $\dots$. Fortunately in situations arising Hodge theory, agreement is guaranteed by:

THEOREM 3.1 ([**D2**, 1.3.16,1.3.17, 7.2.5]). *If d_i strictly preserve F_{dir} for $i = 0, 1, \dots, r$, then the first direct and recursive filtrations on E_{r+1} coincide. If $r = \infty$ then these filtrations coincide with the final filtration on E_∞.*

REMARK 1. The above condition for $i = 0$ just says that the filtration on $\mathrm{Gr}_W^\bullet A$ induced by F is strict.

PROOF OF THEOREM 2.11. The argument goes along the lines presented in [**D2**]. Axiom HC3 and the exactness of Gr_F and $\mathrm{Gr}_{\widehat{F}}$ on the category of $\mathbb{C}$-Hodge structures implies that d_1 strictly preserves F_{dir} and $\widehat{F}_{\text{dir}}$. The same goes for d_0 by HC2. Thus the first direct and recursive filtrations for F and $\widehat{F}$ on $E_2(A,W)$ coincide. Therefore $E_2^{pq}(A,W)$, when equipped with these filtrations, becomes a pure $\mathbb{C}$-Hodge structure of weight q because it is a subquotient of $E_1^{pq}(A,W)$. Consequently $d_2\colon E_2^{pq}(A,W) \to E_2^{p+2,q-1}(A,W)$ is a morphism of Hodge structures of different weights and so it must vanish. As zero strictly preserves everything, the first direct and recursive filtrations on E_3 agree, then by similar reasoning $d_3 = 0$. Continuing in this fashion proves (a).

It follows from the previous paragraph that F and $\widehat{F}$ induce weight $i-k$ $\mathbb{C}$-Hodge structures on $\mathrm{Gr}_W^k H^i(A^\bullet)$. This implies (b).

The proof of the last part is based on the fact that given a spectral sequence of finite dimensional vector spaces (lying in a translate of the first quadrant), there is an inequality

$$\sum_{p+q=n} \dim E_r^{pq} \geq \dim H^n(A^\bullet)$$

and equality holds if and only if $E_r^{pq} = E_\infty^{pq}$ for all p,q. It suffices to prove the opposite inequality for $E_r = E_1(A,F)$. Applying the above inequality for a different spectral sequence gives

$$\sum_q \dim E_2^{q,n-q}(\mathrm{Gr}_F^p A, W) \geq \dim H^n(\mathrm{Gr}_F^p A) = \dim E_1^{p,n-p}(A,F).$$

There is a commutative diagram

$$\begin{array}{ccccc}
\cdots \to & E_1^{q,n-q}(\mathrm{Gr}_F^p A, W) & \to & E_1^{q+1,n-q}(\mathrm{Gr}_F^p A, W) \\
& \| & & \| \\
\cdots \longrightarrow & H^n(\mathrm{Gr}_F^p \mathrm{Gr}_W^q) & \longrightarrow & H^{n+1}(\mathrm{Gr}_F^p \mathrm{Gr}_W^{q+1}) \\
& \| & & \| \\
\cdots \to & \mathrm{Gr}_F^p E_1^{q,n-q}(A,W) & \to & \mathrm{Gr}_F^p E_1^{q+1,n-q}(A,W)
\end{array}$$

where the top isomorphisms follows from Zassenhaus' lemma [**D2**, 1.2.1], and the bottom from axiom HC1. As $E_1(A, W)$ carries a $\mathbb{C}$-Hodge structure and Gr_F is exact for these, it follows that

$$E_2^{q,n-q}(\mathrm{Gr}_F^p A, W) \cong \mathrm{Gr}_F^p E_2^{q,n-q}(A, W).$$

By (a),

$$\sum_q \dim E_2^{q,n-q}(A, W) = \dim H^n(A^\bullet)$$

and hence

$$\sum_p \dim E_1^{p,n-p}(A, F) \leq \sum_{p,q} \dim \mathrm{Gr}_F^p E_2^{q,n-q}(A, W) = \dim H^n(A^\bullet)$$

as required. The argument for $\widehat{F}$ is identical. □

4. Functorial Hodge Structures

Now I will address the basic problem of constructing functorial mixed Hodge structures on the cohomology of a given category of spaces and sheaves. The obvious strategy is to produce a functorial Hodge complex for each pair under consideration, but this is usually very hard to implement. A more useful approach is to construct explicit complexes for special pairs, and then try and approximate the general pairs by the special ones.

Here is the precise set up. Suppose that V is a category with pairwise fibered products, a terminal object, and a "forgetful" functor (denoted by $|\cdots|$) to the category Top of topological spaces which preserves these operations. Assume furthermore that for each $X \in \mathrm{Ob}(V)$ there is a category $\mathcal{CF}(X)$ equipped with a forgetful functor (also denoted by $|\cdots|$) to the category $C^+(|X|)$ of bounded below complexes of sheaves of Abelian groups on $|X|$. For each morphism $f : X \to Y$, there is a pullback functor $f^*colon\mathcal{CF}(Y) \to \mathcal{CF}(X)$ compatible with f^* for sheaves. One should think of an element of $\mathcal{CF}(X)$ as a complex of sheaves on X enriched with additional data. $|X|$ will be denoted by X if there is no danger of confusion. Let $\mathcal{CF}$ be the category of pairs $X \in \mathrm{Ob}(V), \mathcal{F} \in \mathcal{CF}(X)$, a morphism of pairs is a pair of morphisms $(f\colon X_1 \to X_2, f^*\mathcal{F}_2 \to \mathcal{F}_1)$. There is an obvious functor from $\mathcal{CF} \to V$ and $\mathcal{CF}(X)$ is just the fiber over $X \in \mathrm{Ob}(V)$. $\mathcal{CF}$ will be called a *family of coefficients*. I will say that $\mathcal{CF}$ is an *Abelian family of coefficients* if each $\mathcal{CF}(X)$ is an Abelian category and the functors f^* and $|\cdots|$ are exact.

Define a Hodge complex on the family of coefficients $\mathcal{CF}$ to be a functor $A\colon \mathcal{CF} \to C_{\mathrm{Hodge}}$ such that

$$\begin{array}{ccc} \mathcal{CF} & \longrightarrow & C_{\mathrm{Hodge}} \\ \downarrow & & \downarrow \\ V & \longrightarrow & \mathrm{Top} \end{array}$$

commutes, and a natural quasi-isomorphism between the composite of A with $C_{\mathrm{Hodge}} \to C^+(-)$ and $|\cdots|$. In other words, it is just a functorial choice of a Hodge complex on $|X|$ naturally quasi-ismorphic to $\mathcal{F}$, for each pair $(X, \mathcal{F})$ with $X \in V$ and $\mathcal{F} \in \mathcal{CF}(X)$. The definition of a $\mathbb{C}$-Hodge complex on $\mathcal{CF}$ is similar.

LEMMA 4.1. *Given a* ($\mathbb{C}$-)*Hodge complex on* $\mathcal{CF}$, $H^i(|X|, |\mathcal{F}|)$ *carries a* ($\mathbb{C}$-) *mixed structure which is functorial in both variables.*

This lemma is somewhat limited in applicability, because it's hard to construct Hodge complexes on all of $\mathcal{CF}$. In order to formulate a more useful result, let me call a full subcategory Δ of V *dense* provided that:

1. Given any object $Y \in ObV$, there is a morphism $X \to Y$ with $X \in \Delta$ such that the underlying map on spaces is proper and surjective. Call this a *resolution* of Y in Δ.
2. Given a morphism $X \to Y$ and a proper surjective morphism $Y' \to Y$, there is a resolution $X' \to X$ which fits into a commutative diagram:

$$\begin{array}{ccc} X' & \to & Y' \\ \downarrow & & \downarrow \\ X & \to & Y \end{array}$$

(It's convenient to phrase the definition this way, even though the second condition implies the first.) The notion of denseness is extended to coefficients in the most straightforward possible way. The preimage of Δ under $\mathcal{CF} \to V$ will be called a subfamily of coefficients; it is called dense if Δ is dense. The reason for introducing these notions comes from:

THEOREM 4.2. *Given a ($\mathbb{C}$-)Hodge complex defined on a dense subfamily of coefficients, $H^\bullet(|X|, |\mathcal{F}|)$ carries a bifunctorial ($\mathbb{C}$-)mixed Hodge structure for all pairs $(X, \mathcal{F}) \in \mathrm{Ob}\,\mathcal{CF}$. If $\mathcal{CF}$ is an Abelian family of coefficients then any short exact sequence of coeffients induces a long exact sequence of mixed Hodge structures on cohomology.*

An outline of the proof of Theorem 4.2 will be given in Section 5.

EXAMPLE 4.3. Let V be the category of proper complex varieties. Let Δ be the full subcategory of nonsingular projective varieties. Density is deduced from either Hironaka's theorem [**Hi**] (and Chow's lemma), or the easier theorem of de Jong [**dJ**] on the existence of nonsingular alterations. For $\mathcal{CF}(X)$, one can take the category consisting of the constant sheaf $\mathbb{Z}_X$ and the identity. A Hodge complex on Δ is given by Example 2.3. This yields the Deligne mixed Hodge structure on the cohomology of proper varieties (and indeed this is pretty much Deligne's original construction).

EXAMPLE 4.4. Let V and Δ be as above. Let $\mathcal{CF}(X)$ be the category pairs $(f\colon X \to Y, \mathcal{F})$ where f is a map to a smooth variety and $\mathcal{F}$ a polarized complex variation of Hodge structure on Y. Define $|(f, \mathcal{F})| = f^*\mathcal{F}$. The $\mathbb{C}$-Hodge complex of Example 2.4 yields $\mathbb{C}$-mixed Hodge structures on cohomologies of proper varieties with these coefficients.

The previous example is essentially due to Lasell [**L**] who used it in proving his generalization of Nori's Lefschetz theorem of π_1. One of the key steps amounts to verification that if $f : Z \to X$ is a morphism of smooth projective varieties (with X and $Y = \mathrm{image}(f)$ connected) then

$$\ker[H^1(X, \mathcal{V}) \to H^1(Y, \mathcal{V})] = \ker[H^1(X, \mathcal{V}) \to H^1(Z, f^*\mathcal{V})]$$

for any polarized complex variation of Hodge structure $\mathcal{V}$ on X. This follows rather formally from the existence and basic properties of the $\mathbb{C}$-Hodge structure on these cohomology groups.

EXAMPLE 4.5. Let V be the category of pairs (X, Z) consisting of a proper complex variety X and a Zariski closed subset Z. A morphism $(X_1, Z_1) \to (X_2, Z_2)$ is a morphism of the underlying schemes such that Z_1 is the preimage of Z_2. Δ is the full subcategory generated by pairs consisting of smooth projective varieties and divisors with normal crossings. Once again density follows from [**Hi**] or [**dJ**]. The Hodge complex described in Example 2.5 (with the integral structure described in the discussion following Theorem 2.11) yields the Deligne mixed Hodge structures on the cohomology of varieties with constant coefficients.

EXAMPLE 4.6. With the same V and Δ as in the previous example, let $\mathcal{CF}(X,Z)$ be the category of unitary local systems on $X - Z$. This is an Abelian family of coefficients. Then the $\mathbb{C}$-Hodge complex of Example 2.6 gives a $\mathbb{C}$-Hodge structure on the cohomology of varieties with coeffients in unitary local systems. This is compatible with the cohomology long exact sequence by the theorem, but it can be also deduced directly from the semisimplicity of $\mathcal{CF}(X, Z)$.

A more elaborate example will be given in Section 6.

5. Cohomological Descent

I will outline the proof of Theorem 4.2. The name of the game is cohomological descent, which is a way of computing the cohomology of space in terms of the cohomology of a (simplicial) diagram of spaces lying over it. It is a generalization of Čech's procedure for computing cohomology, where the diagram in question is just the nerve of the open cover. Recall that a strict simplicial object in some category C is a collection of objects X_n indexed by the natural numbers, and "face" morphisms $\delta_i \colon X_n \to X_{n-1}$ $i = 0,1, \ldots, n$ such that $\delta_i\delta_j = \delta_{j-1}\delta_i$ for $i < j$. I will usually drop the adjective "strict", although this is technically incorrect[1]. An augmentation to X is a collection of morphisms $\epsilon_n \colon X_n \to X$ which commute with the face maps. A cosimplicial object is the dual notion where arrows go the other way, or equivalently a simplicial object in the opposite category. Suppose that $A^0 \rightrightarrows A^1 \cdots$ is a cosimplicial object in an additive category, then defining $d = \sum_i (-1)^i \delta_i$ results in a complex $A^\bullet$.

Suppose that $\epsilon_\bullet \colon X_\bullet \to X$ is an augmented simplicial space, i.e., an augmented simplicial object in the category of topological spaces. A sheaf on $X_\bullet$ is a collection of sheaves F^n on X_n and morphisms $\delta_i \colon \delta_i^* F^{n-1} \to F^n$ satisfying the simplicial identities. For example, given an augmentation $\epsilon_\bullet \colon X_\bullet \to X$ and a sheaf F on X, the pullback $\epsilon_\bullet^* F$ with $\delta_i^* = \mathrm{id}$ is a sheaf on $X_\bullet$. Suppose that F is a sheaf of Abelian groups. Then the Godement flasque resolution $\mathcal{G}^\bullet(F^\bullet)$ forms a complex of sheaves on $X_\bullet$. Taking direct images yields a cosimplicial object $\epsilon_*(\mathcal{G}^\bullet(F^\bullet))$ in the category of complexes of sheaves on X. This in turn determines a double complex, and let $s\epsilon_*(\mathcal{G}^\bullet(F^\bullet))$ denote the total or associated single complex (I am being a bit cavalier about signs; the important thing is that they be chosen in a consistent way so as to yielda complex). The map $\epsilon \colon X_\bullet \to X$ satisfies *cohomological descent* if the natural map

$$F \to s\epsilon_*(\mathcal{G}^\bullet(\epsilon^* F))$$

[1]Strictly speaking, a simplicial object has additional degeneracy maps $X_{n-1} \to X_n$. In all our examples, C will have finite coproducts, so we can turn a strict simplicial object $X_\bullet$ in a true simplicial object X_n^+ by inductively replacing X_n by $X_n \amalg X_{n-1} \amalg \cdots$, one X_{n-1} for each degeneracy map ....

is quasi-isomorphism for any sheaf F on X. In less prosaic terms, this says that the natural map $F \to \mathbb{R}\epsilon_*\epsilon^*F$ is an isomorphism in $D^+(X)$.

This phenomenon has already been encountered earlier in this article. Let

$$\begin{array}{ccc} Ei & \stackrel{l}{\hookrightarrow} & \widetilde{X} \\ \downarrow & & \downarrow \\ Si & \stackrel{l}{\hookrightarrow} & X \end{array}$$

be the diagram given in Proposition 1.1. Then the proof of that proposition actually showed:

LEMMA 5.1.

$$E \rightrightarrows \widetilde{X} \amalg S \to X$$

satisfies cohomological descent.

The next step is to find a general effective criterion for cohomological descent. Suppose that $X_\bullet$ is a simplicial set, and $x \in X_n$. If $x_i = \delta_i x$, then these satisfy $\delta_j x_i = \delta_i x_{j+1}$ for all $i \leq j$. Let me call such an $(n+1)$-tuple in X_{n-1} an $n-1$ cycle, and let $Z_{n-1}(X_\bullet)$ denote the set of all of these. Note that $Z_{n-1}(X_\bullet)$ depends only on the maps $\delta_i : X_{n-1} \to X_{n-2}$. This observation will be useful for the inductive constructions later on. If $X_\bullet \to X$ is an augmented simplicial set, then define $Z_{n-1}(X_\bullet \to X)$ by imposing the additional constraint that the x_i take the same value in X. If $X_\bullet \to X$ is an augmented simplicial space then $Z_{n-1}(X_\bullet \to X)$ is also a topological space with a continuous map to X. Say that an augmented simplicial space $X_\bullet \to X$ is a *hypercover*[2] if $X_0 \to X$ and the obvious continuous maps $X_n \to Z_{n-1}(X_\bullet \to X)$ are proper and surjective for each n.

THEOREM 5.2. *A hypercover satisfies cohomological descent.*

This result will be sufficient for our needs, although it is far from optimal: it does not include Lemma 5.1 for example. For a more general statement and proofs, see [**SGA4**, V bis] or [**GNPP**] (this last reference takes a slightly different point of view into which Lemma 5.1 fits quite naturally). Note that these references use the stronger form of simplicial object having also *degeneracy maps*, but this will not affect the proofs since (in the notation of footnote 1) we can pass from $X_\bullet \to X$ to $X_\bullet^+ \to X$. Also, in order to convert our definition to those in the references, $Z_{n-1}(X_\bullet)$ is often written as $(\cos k\, sk_{n-1} X_\bullet)_n$.

Let V be a category satisfying the conditions of the previous section and let Δ be a dense subcategory. Let Res be the set of resolutions in the sense of Section 4, i.e., morphisms with domain in Δ such that the underlying continuous map is proper and surjective. Given an augmented simplicial object $X_\bullet \to X$ in the category V, $Z_{n-1}(X_\bullet \to X)$ can be defined as the object that represents the functor

$$T \mapsto \{(f_0, \ldots, f_n) \in \mathrm{Hom}(T, X_{n-1})^{n+1} \mid \forall i \leq j, \delta_j f_i = \delta_i f_{j+1}\}$$

in the category of objects over X (the representability of this functor follows from our assumptions about V). I will treat $Z_{n-1}(X_\bullet \to X)$ as an object of V by forgetting the map to X. An augmented simplicial object $X_\bullet \to X$ in V, will be called a *simplicial resolution* of X if $X_0 \to X$ and each $X_n \to Z_{n-1}(X_\bullet \to X)$ lies in Res. A simplicial resolution is a hypercover.

---

[2]It would be more accurate to call this a hypercover for the Grothendieck topology of proper maps.

THEOREM 5.3. *Every object possesses a simplicial resolution. Given a hypercover $Y_\bullet \to Y$ and a morphism $X \to Y$, a simplicial resolution $X_\bullet \to X$ can be chosen so as to fit into a commutative diagram*

$$\begin{array}{ccc} X_\bullet & \to & Y_\bullet \\ \downarrow & & \downarrow \\ X & \to & Y \end{array}$$

Any two simplicial resolutions are dominated by a third simplicial resolution.

In a nutshell, the proof is by induction. The key point is that once X_i and the associated maps are defined for all $i < n$, then $Z_{n-1}(X_\bullet \to X)$ exists. And by its very construction, it carries morphisms to X_{n-1} satisfying the appropriate face relations. Now take X_n to be a resolution of $Z_{n-1}(X_\bullet \to X)$ (which fits into a commutative square

$$\begin{array}{ccc} X_n & \longrightarrow & Y_n \\ \downarrow & & \downarrow \\ Z_{n-1}(X_\bullet \to X) & \to & Z_{n-1}(Y_\bullet \to Y) \end{array}$$

for the second part). To start the whole process, set X_0 equal to any resolution of X (compatible with $Y_0 \to Y$). For the last statement, apply the previous part with $Y = X$ and $Y_\bullet$ equal to the fiber product of two given simplicial resolutions of X.

Now, I can outline the proof of Theorem 4.2. To simplify the discussion, let me dispense with the $\mathbb{Z}$-structure. Suppose that $(X, \mathcal{F}) \in Ob\mathcal{CF}$, then choose a simplicial resolution $X_\bullet \to X$. The assumptions, guarantees that there are $\mathbb{C}$-Hodge complexes $(A^{\bullet n}(\mathcal{F}), W, F, \widehat{F})$ quasi-isomorphic to $\epsilon_n^* \mathcal{F}$ on each X_n, and face morphisms δ_i between them. Replace these complexes by their Godement resolutions. This results in a differential graded $\mathbb{C}$-Hodge complex

$$\Gamma(A^{0\bullet}(\mathcal{F})) \xrightarrow{\delta_0 - \delta_1} \Gamma(A^{1\bullet}(\mathcal{F})) \to \cdots .$$

The total complex $S^\bullet(\mathcal{F})$ is a $\mathbb{C}$-Hodge complex (Example 2.8) which is naturally quasi-isomorphic to $\mathbb{R}\Gamma(\mathcal{F})$. This yields a $\mathbb{C}$-Hodge structure on $H^\bullet(X, \mathcal{F})$ which depends, a priori, on the choice of the simplicial resolution $X_\bullet \to X$. However given a second resolution $X'_\bullet \to X$ dominating $X_\bullet$, one gets a morphism of the corresponding $\mathbb{C}$-Hodge structures which is an isomorphism of vector spaces and hence of Hodge structures, because the forgetful functor to vector spaces is exact and faithful. As any two simplicial resolutions are dominated by a third, this shows independence. Similarly since any morphism can be lifted to a morphism of simplicial resolutions, this proves functoriallity. Finally, suppose that $\mathcal{CF}$ is Abelian, and let

$$0 \to \mathcal{F}_1 \xrightarrow{f} \mathcal{F}_2 \xrightarrow{g} \mathcal{F}_3 \to 0$$

be exact. Then there exists morphisms $S(f)\colon S^\bullet(\mathcal{F}_1) \to S^\bullet(\mathcal{F}_2)$ and $S(g)\colon S^\bullet(\mathcal{F}_2) \to S^\bullet(\mathcal{F}_3)$ of Hodge complexes. As $S(g) \circ S(f) = 0$, there is a well defined map $\mathrm{Cone}(S^\bullet(f))^\bullet \to S^\bullet(\mathcal{F}_3)$ which is a (nontrilftered) quasi-isomorphism. Therefore the $\mathbb{C}$-Hodge structures on cohomology coincide. The maps in the triangle

$$S^\bullet(\mathcal{F}_1) \to S^\bullet(\mathcal{F}_2) \to \mathrm{Cone}(f)^\bullet \to S^\bullet(\mathcal{F}_1)[1]$$

yields maps of Hodge structures in the long exact sequence for cohomology.

The following construction is implicit in the above argument. Let $D_{\mathbb{C}\text{-Hodge}}(pt)$ be obtained by localizing the category of $\mathbb{C}$-Hodge complexes $C_{\mathbb{C}\text{-Hodge}}(pt)$ with respect to (not necessarily filtered) quasi-isomorphisms (see [**V**] for the definition). Then this has the structure of a triangulated category with translation given in Example 2.2 and distinguished triangles represented by

$$A^\bullet \xrightarrow{f} B^\bullet \to \mathrm{Cone}(f)^\bullet \to A^\bullet[1]$$

where $\mathrm{Cone}(f)$ is defined in Example 2.7. There is a Δ-functor from $D_{\mathbb{C}\text{-Hodge}}(pt)$ to the category of $\mathbb{C}$-Hodge structures which are compatible with the usual cohomology functor $H^\bullet \colon D^+(\mathbb{C}) \to \mathbb{C}\text{-Mod}$. Similar remarks apply to the category of Hodge complexes. See [**B**] or [**E**] for further details about the construction.

6. Constant Constructible Sheaves

Recall that a sheaf of Abelian groups $\mathcal{F}$ on a complex variety X is *constructible* if it has finitely generated stalks and there exists a partition of X into Zariski locally closed subsets such that $\mathcal{F}|_U$ is locally constant for each member U of the partition. I will say that $\mathcal{F}$ is *constant constructible* if each restriction is in fact constant. As an illustration of the previous ideas, I want to prove[3]:

THEOREM 6.1. *Let $\mathcal{F}$ be a constant constructible sheaf on a proper complex variety X, then $H^i(X, \mathcal{F})$ carries a mixed Hodge structure which is functorial in both variables. A short exact sequence of sheaves induces a long exact sequence of mixed Hodge structures on cohomology.*

Let X be a variety over $\mathbb{C}$. A *stratification* is a finite Boolean algebra $\mathcal{B}$ of subsets of X such that the Zariski closure of any element of $\mathcal{B}$ is also in $\mathcal{B}$. This somewhat unorthodox usage of terminology requires some explanation. Fix a stratification $\mathcal{B}$, and let $A(\mathcal{B})$ be the set of atoms or nonempty minimal elements. As $\mathcal{B}$ is a finite Boolean algebra, any nonempty element is a union of atoms. Let me define a partial order on A, by $Z_1 \leq Z_2$ if and only if $\overline{Z}_1 \subseteq \overline{Z}_2$. Then, one easily obtains

LEMMA 6.2. *If $Z \in A(\mathcal{B})$, then $\overline{Z} = \bigcup_{Z' \leq Z} Z'$ and $Z = \overline{Z} - \bigcup_{Z' < Z} \overline{Z}'$.*

Consequently, $A(\mathcal{B})$ is a partition of X into Zariski locally closed sets, and the elements of $\mathcal{B}$ are constructible. Conversely, given such a partition, the set of unions of elements of the partition yield a stratification.

Let $\mathcal{SV}$ be the set of all pairs $(X, \mathcal{B})$ consisting of a variety with a stratification. This can be made into a category by taking as morphisms $f \colon (X_1, \mathcal{B}_1) \to (X_2, \mathcal{B}_2)$, morphisms $f \colon X_1 \to X_2$ of varieties such that $f^{-1}(Z) \in \mathcal{B}_1$ for all $Z \in \mathcal{B}_2$. A morphism corresponding to the identity on the underlying variety is just a refinement of the stratification. A sheaf $\mathcal{F}$ on X with finitely generated stalks will be called constant constructible with respect to a stratification $\mathcal{B}$ provided that $\mathcal{F}|_U$ is constant for each $U \in A(\mathcal{B})$.

Given a (Weil) divisor E on an irreducible variety X, let $\mathcal{B}(E)$ be the smallest stratification containing the irreducible components of E. If X is smooth and E a divisor with normal crossings, then $\mathcal{B}(E)$ will be refered to as a good stratification. Let Δ be the full subcategory of $\mathcal{SV}$ whose objects are disjoint unions of smooth projective varieties with good stratifications.

PROPOSITION 6.3. *$\mathcal{SV}$ satisfies the assumptions of Section 4, and Δ is dense.*

[3]This argument was worked at MSRI during 1988–1989; my thanks for their support.

PROOF. We verify that $\mathcal{SV}$ satisfies the axioms of Section 4. $\mathcal{SV}$ certainly contains a terminal object. Given two morphisms $f_i\colon (X_i, \mathcal{B}_i) \to (Y, \mathcal{B})$ with $i = 1, 2$, let $P = (X_1 \times_Y X_2)_{\mathrm{red}}$. Let $\mathcal{B}_P$ be the Boolean algebra generated by fibered products $(Z_1 \times_Y Z_2)_{\mathrm{red}}$ with $Z_i \in \mathcal{B}_i$. Then $(P, \mathcal{B}_P)$ is the fibered product of f_1 and f_2 in $\mathcal{SV}$.

The next step is to show that every object possesses a resolution with respect to Δ. If $(X, \mathcal{B})$ is a stratified variety, let X' be the disjoint union of the irreducible components. We can pull $\mathcal{B}$ back to X' to obtain a surjective proper morphism $(X', \mathcal{B}') \to (X, \mathcal{B})$. Since we can work with one component at a time, we may assume that X is irreducible. The proof will proceed by induction on the cardinality of $\mathcal{B}$. If $\mathcal{B}$ has less than or equal to 2 elements, in other words if the stratification is trivial, then the existence of a resolution follows immediately from [**Hi**] or [**dJ**]. In general, let A be the set of atoms of $\mathcal{B}$. Choose an element $Z \in A$, which is minimal with respect to $\leq$. Then Z is necessarily closed. Let $\mathcal{B}' \subset \mathcal{B}$ be the Boolean subalgebra generated by the closures of elements of $A - \{Z\}$. $\mathcal{B}'$ is easily seen to be a stratification not containing Z. Therefore, by induction, there is a resolution $(X_1, \mathcal{B}(E_1)) \to (X, \mathcal{B}')$. Let Z_1 be the preimage of Z. Then there exists a proper surjective morphism $X_2 \to X_1$ from a second nonsingular variety such that the preimage E_2 of $Z_1 \cup E_1$ is a divisor with normal crossings [loc. cit.]. Then $(X_2, \mathcal{B}(E_2))$ is a resolution of $(X, \mathcal{B})$.

It remains to check the last condition for denseness. Given a morphism $(Y_1, \mathcal{B}_{Y_1}) \to (Y_2, \mathcal{B}_{Y_2})$ and a proper surjective morphism $(X_2, \mathcal{B}_{X_2}) \to (Y_2, \mathcal{B}_{Y_2})$, let $(X_1, \mathcal{B}_{X_1})$ be a resolution of the fibered product. $\square$

Given an object $(X, \mathcal{B})$ of $\mathcal{SV}$, let $S(X, \mathcal{B}) = \coprod_{U \in A(\mathcal{B})} U$. Let $J(X, \mathcal{B})\colon S(X, \mathcal{B})) \to X$ denote the canonical map. Set

$$T = T(X, \mathcal{B}) = J(X, \mathcal{B})_* \circ J(X, \mathcal{B})^*\colon \mathrm{Sh}(X) \to \mathrm{Sh}(X),$$

where $\mathrm{Sh}(X)$ is the category of sheaves on X. From the adjointness of J_* and J^*, there are natural transformations:

$$\eta\colon I \to T, \quad \mu\colon T^2 \to T$$

satisfying the usual identities of a "triple" or monad [**M**]. Define the natural transformations:

$$\delta_i = T^i \eta T^{n-i}\colon T^n \to T^{n+1}$$

For any sheaf $\mathcal{F} \in \mathrm{Sh}(X)$, these maps determine a cosimplicial sheaf augmented over $\mathcal{F}$:

$$\mathcal{F} \to T\mathcal{F} \rightrightarrows T^2\mathcal{F} \cdots$$

and hence also a complex (augmented over $\mathcal{F}$):

$$\mathcal{F} \to T\mathcal{F} \xrightarrow{d} T^2\mathcal{F} \to \cdots$$

where $d = \sum (-1)^i \delta_i$. Either of these will be denoted by $T^\bullet\mathcal{F}$ or $T(X, \mathcal{B})^\bullet\mathcal{F}$. This can be made more explicit as follows. Let $j_U\colon U \to X$ denote the inclusion for $U \in A(\mathcal{B})$. Since $j_V^* j_{U*} = 0$ unless $V \leq U$, it follows that:

$$T^n\mathcal{F} = \bigoplus_{U_0 \leq \cdots \leq U_n} j_{U_0*}^* \cdots j_{U_n*}^* \mathcal{F}$$

where I abbreviate $j_{U*}j_U^*$ by j_{U*}^*. The ith face map δ_i is the sum of the adjunction maps:

$$j_{U_0*}^* \cdots \hat{j}_{U_i*}^* \cdots j_{U_n*}^* \mathcal{F} \to j_{U_0*}^* \cdots j_{U_n*}^* \mathcal{F}$$

LEMMA 6.4. *$T^\bullet \mathcal{F}$is quasi-isomorphic to $\mathcal{F}$.*

PROOF. Fix $x \in X$. It suffices to construct a functorial retraction $h\colon (T\mathcal{F})_x \to \mathcal{F}_x$ to the map $\mathcal{F}_x \to (T\mathcal{F})_x$, because then the augmented complex $\mathcal{F}_x \to (T^\bullet\mathcal{F})_x$ carries a contracting homotopy defined by

$$h\colon (T^{n+1}\mathcal{F})_x = (TT^n\mathcal{F})_x \to (T^n\mathcal{F})_x.$$

See [**G**, Appendix 5] for details. We have

$$(T\mathcal{F})_x = \bigoplus_{U \in A(\mathcal{B})} (j_{U*}^*\mathcal{F})_x$$

If $x \in U$ then $(j_{U*}^*\mathcal{F})_x = \mathcal{F}_x$, define h to act as the identity on each summand $(j_{U*}^*\mathcal{F})_x$ with $x \in U$, and zero on the other summands. □

The above constructions are functorial. Given a morphism $f\colon (X, \mathcal{B}) \to (X', \mathcal{B}')$ and a sheaf $\mathcal{F}$ on X', there is a morphism of cosimplicial sheaves:

$$f^*T(X', \mathcal{B}')^\bullet\mathcal{F} \to T(X, \mathcal{B})^\bullet f^*\mathcal{F}.$$

PROOF OF THEOREM 6.1. Let $\mathcal{PSV}$ be the full subcategory of $\mathcal{SV}$ where the underlying variety is proper. The intersection $P\Delta = \Delta \cap \operatorname{Ob}\mathcal{PSV}$ is dense in $\mathcal{PSV}$. Let $\mathcal{CF}(X, \mathcal{B})$ denote the category of constant constructible sheaves on $(X, \mathcal{B}) \in \operatorname{Ob}\mathcal{PSV}$. This determines an Abelian family of coefficients.

Let $(X, \mathcal{B})$ be an object of $P\Delta$, and let $\mathcal{F}$ be a constant constructible sheaf on it. There is no loss of generality in assuming that X is connected, so that $\mathcal{B}$ is the Boolean algebra $\mathcal{B}(E)$ generated by the components of a divisor with normal crossings. In this case, the components of $T^\bullet(\mathcal{F})$ are fairly easy to describe: $j_{\overline{U}_0}^* j_{U_1*}^* \cdots j_{U_n*}^*\mathcal{F} = j_{\overline{U}_0}^* j_{U_n*}^*\mathcal{F}$ is a constant sheaf. Let

$$A^{n\bullet}(\mathcal{F}) = \bigoplus_{U_0 \le \cdots \le U_n} j_{\overline{U}_0*}(\mathcal{E}_{\overline{U}_0}^\bullet \otimes j_{\overline{U}_0}^* j_{U_n*}^*\mathcal{F}, \dots)$$

be the associated Hodge complex. The differentials of $T^\bullet(\mathcal{F})$ extend to give a differential graded Hodge complex $A^{\bullet\bullet}(\mathcal{F})$. Then $\mathcal{F} \mapsto tot^\bullet(A(\mathcal{F}))$ is a Hodge complex on $\mathcal{CF}|_{P\Delta}$. The theorem now follows from Theorem 4.2. □

There are a couple of variations on this construction. Call a sheaf $\mathcal{F}$ of $\mathbb{C}$-vector spaces *unitary constructible* if it has finite dimension stalks and there is stratification such that $\mathcal{F}|_U$ is isomorphic to a unitary local system for each $U \in A(\mathcal{B})$. In this case, the sheaves $j_{\overline{U}_0}^* j_{U_1*}^* \cdots j_{U_n*}^*\mathcal{F}$ are all unitary local systems, let $(V_{(U_0,\dots,U_n)}, \nabla)$ denote the extension of the associated flat vector bundle to $\overline{U}_0$ (see Example 2.6). Then a $\mathbb{C}$-Hodge complex:

$$A^{n\bullet}(\mathcal{F}) = \bigoplus_{U_0 \le \cdots \le U_n} j_{\overline{U}_0*}(W_0 A_{\overline{U}_0}^\bullet(\log(\overline{U}_0 - U_0)) \otimes V_{(U_0,\dots,U_n)}, \dots)$$

can be constructed on $P\Delta$, where $A_{\overline{U}_0}^\bullet(\log(\overline{U}1_0 - U_0)) \otimes V_{(U_0,\dots,U_n)}$ is described (Example 2.6). This produces a natural $\mathbb{C}$-Hodge structure on cohomology with coefficients in $\mathcal{F}$. Similar arguments yield ($\mathbb{C}$-)mixed Hodge structures on the cohomology of arbitrary varieties with constant (or unitary) constructible coefficients.

The bookkeeping is a bit more involved in this case. $\mathcal{SV}$ would be replaced by the category of triples $(X, \mathcal{B}, \overline{X})$ where $\mathcal{B}$ is a stratification of X and $\overline{X}$ a compactification. The objects of Δ would be triples where $\overline{X}$ is disjoint union of smooth projective varieties, $\mathcal{B}$ a good stratification of the form $\mathcal{B}(E)$ and $(\overline{X} - X) \cup E$ a divisor with normal crossings.

I want to conclude with a few words about how all of this ought to relate to Saito's work on mixed Hodge modules [**Sa1, Sa2, Sa3**]. I will treat mixed Hodge module theory entirely as black box, and just summarize some of the main properties. For a gentle introduction to the subject see [**BZ**]. To every variety X over $\mathbb{C}$, Saito constructs the Abelian category $\mathrm{MHM}(X)$ of mixed Hodge modules. For a point, $\mathrm{MHM}(pt)$ is equivalent to the category of $\mathbb{Q}$-mixed Hodge structures H which are polarizable in the sense that $\mathrm{Gr}^W_\bullet H$ admits polarizations. There is an additive functor $\mathrm{rat}\colon \mathrm{MHM}(X) \to D^b_c(X, \mathbb{Q})$ to the derived category of sheaves of $\mathbb{Q}_X$-modules with bounded constructible cohomology; in fact, rat is an exact and faithful functor to the Abelian subcategory of perverse sheaves [**BBD**]. When X is smooth, the category of perverse sheaves over $\mathbb{C}$ is exactly the image of the category of regular holonomic D_X-modules under the De Rham functor DR (by the Riemann-Hilbert correspondence). In this case, $\mathrm{MHM}(X)$ is constructed as a subcategory of the category whose objects are pairs consisting of a filtered regular holonomic D_X-module (M, F) and a filtered perverse sheaf over $\mathbb{Q}$ (K, W) such that $\mathrm{DR}(M) \cong K \otimes \mathbb{C}$. The functor rat extends to a morphism of triangulated categories denoted by the same symbol $\mathrm{rat}\colon D^b\,\mathrm{MHM}(X) \to D^b_c(X, \mathbb{Q})$. The basic operations of sheaf theory extend to $D^b\,\mathrm{MHM}$. In particular, given a morphism $f\colon X \to Y$, there exists morphisms of triangulated categories $\mathbb{R}f_*\colon D^b\,\mathrm{MHM}(X) \to D^b\,\mathrm{MHM}(Y)$ and $\mathbb{L}f^*\colon D^b\,\mathrm{MHM}(Y) \to D^b\,\mathrm{MHM}(X)$ which are compatible with the corresponding morphisms of D^b_c (I have added the "$\mathbb{R}$" and "$\mathbb{L}$" for consistency; Saito leaves these out by convention). These facts together imply that $H^i(X, \mathrm{rat}(M))$ is assigned a mixed Hodge structure for each $M \in D^b\,\mathrm{MHM}(X)$.

Given a constant constructible sheaf $\mathcal{F}$ on X, I expect that there should be a natural choice of $M \in D^b\,\mathrm{MHM}(X)$ with $\mathrm{rat}(M) = \mathcal{F} \otimes \mathbb{Q}$ such that $H^i(X, M)$ gives the same $\mathbb{Q}$-mixed Hodge structure as above. When $\mathcal{F} = A_X$ is constant, the obvious choice for M is $\mathbb{L}f^*(A \otimes \mathbb{Q}(0))$, and when X is smooth the mixed Hodge structures on $H^i(X, A)$ do coincide by [**Sa2**, 3.3]. I have a candidate for M for general $\mathcal{F}$, but I don't yet have a proof that it works. The main problem is that the derived category is the wrong place for doing descent, a more "rigid" description of M is needed first. Mixed Hodge modules do not work on simplicial schemes for these reasons; some of these issues are touched upon in a recent preprint of Saito [**Sa4**], but it appears that more work is needed to resolve them.

References

[A] D. Arapura, *Geometry of cohomology support loci for local systems.* I, J. Algebraic Geom. **6** (1997), 563–597

[SGA4] M. Artin, A. Grothendieck, and J. Verdier (eds.), *Théorie des topos et cohomologie étale de schémas.* I, Lecture Notes in Math., vol. 269, Springer-Verlag, Berlin-New York, 1972; II, vol. 270, 1972; III, vol. 305, 1973.

[B] A. A. Beilinson, *Notes on absolute Hodge cohomology,* Applications of algebraic K-theory to algebraic geometry and number theory. Part I (Boulder, 1986) (S. J. Bloch, R. K. Dennis, E. M. Friedlander, and M. R. Stein, eds.), Contemp. Math., vol 55, Amer. Math. Soc., Providence, RI, 1986, pp. 35–68

[BBD] A. A. Beilinson, J. Bernstein and P. Deligne, *Faisceaux pervers*, Analysis and Topology on singular spaces. I (Luminy, 1981), Astérisque, vol. 100, Soc. Math. France, Paris, 1982, pp. 5–171.

[BZ] J-L, Brylinski, S. Zucker, *An overview of recent advances in Hodge theory*, Several Complex Variables. VI, Encyclopaedia Math. Sci., vol. 69, Springer, Berlin, 1990, pp. 39–142.

[Bu] J. I. Burgos, *A C^∞ logarithmic Dolbeault complex*, Compositio Math **92** (1994), 61–86.

[D1] P. Deligne, *Équations différentielles à points singuliers réguliers* , Lecture Notes in Math., vol. 163, Springer-Verlag, Berlin-New York, 1969.

[D2] ______, *Théorie de Hodge.* II, Inst. Hautes Études Sci. Publ. Math. **40** (1972), 5–57; III, **44** (1974), 5–77.

[D3] ______, *Un théorème de finitude pour la monodromie*, Discrete Groups in Geometry and Analysis (New Haven, CT, 1984), Progr. Math., vol. 67, Birkhäuser, Boston, 1987, pp. 1–19.

[E] F. El Zein, *Mixed Hodge structures*, Trans. Amer. Math. Soc. **275** (1983), 71–106.

[EV] H. Esnault and E. Viehweg, *Logarithmic de Rham complexes and vanishing theorems*, Invent. Math. **86** (1986), 161–194 (with appendix by K. Timmerscheidt).

[GS] H. Gillet and C. Soulé, *Descent, motives and K-theory*, J. Reine Angew. Math. **478** (1996), 127–176

[G] R. Godement, *Topologie algébrique et théorie des faisceaux*, Publ. Math. Univ. Strasbourg, no. 13, Actualités Sci. Indust., vol. 1252, Hermann, Paris, 1958

[GL] M. Green and R. Lazarsfeld, *Deformation theory, generic vanishing theorems, and some conjectures of Enriques* Invent. Math. **90** (1987), 389–407.

[GS] P. Griffiths and W. Schmid, *Recent developments in Hodge theory: a discussion of techniques and results*, Discrete Subgroups of Lie groups and Applicatons to Moduli (Bombay, 1973), Oxford Univ. Press, Bombay, 1975. pp. 31–127.

[GN] F. Guillén and V. Navarro Aznar, *Un critère d'extension d'un foncteurr défini sur les schémas lisses*, alg-geom/9505008, 1995.

[GNPP] F. Guillén, V. Navarro Aznar, P. Pascual-Gainza, and P. Puerta, *Hyperrésolutions cubiques et descente cohomologique* Lecture Notes in Math., vol. 1335, Springer, Berlin-New York, 1988.

[Ha] R. Hain, *The de Rham homotopy theory of complex algebraic varieties.* I, K-theory **1** (1987), 271–324; II, 481–497.

[Hi] H. Hironaka, *Resolution of singularities of an algebraic variety of characteristic zero.* I, Ann. of Math. (2) **79** (1964), 109–203; II, 205-326.

[dJ] A. J. de Jong, *Smoothness, semi-stability and alterations*, Inst. Hautes Études Sci. Publ. Math. **83** (1996), 51–93.

[L] B. Lasell, *Complex local systems and morphisms of varieties*, Compositio Math. **98** (1995), 141–166.

[M] S. Maclane, *Categories for the working mathematician*, Grad. Texts in Math., vol. 5, Springer-Verlag, New York-Berlin, 1971.

[N] V. Navarro Aznar, *Sur la théorie de Hodge-Deligne*, Invent Math. **90** (1987), 11–76.

[Sa1] M. Saito, *Modules de Hodge polarisables*, Publ. Res. Inst. Math. Sci. **24** (1988), 849–995

[Sa2] ______, *Mixed Hodge modules and admissible variations*, C. R. Acad. Sci. Paris Sér. I Math. **309** (1989), 351–356.

[Sa3] ______, *Mixed Hodge modules*, Publ. Res. Inst. Math. Sci. 26 (1990), 221–333.

[Sa4] ______, *Mixed Hodge complexes on algebraic varieties*, alg-geom/9906088, 1999.

[S] C. T. Simpson, *Higgs bundles and local systems*, Inst. Hautes Études Sci. Publ. Math. **75** (1992), 5–95

[St] J. Steenbrink, *Limits of Hodge structures*, Invent. Math. **31** (1975/76), 229–257.

[V] J.-L. Verdier, *Des catégories dérivées des catégories abéliennes*, Astérisque, vol. 239, Soc. Math. France, Paris, 1996.

[Z] S. Zucker, *Hodge theory with degenerating coefficients. L_2 cohomology in the Poincaré metric*, Ann. of Math. (2) **109** (1979), 415–476.

Department of Mathematics, Purdue University, West Lafayette IN 47907, USA
E-mail address: dvb@math.purdue.edu

Centre de Recherches Mathématiques
CRM Proceedings and Lecture Notes
Volume **24**, 2000

The Atiyah-Chern Character yields the Semiregularity Map as well as the Infinitesimal Abel-Jacobi Map

Ragnar-Olaf Buchweitz and Hubert Flenner

Introduction

The purpose of this work is to construct a general semiregularity map for cycles on a complex analytic or algebraic manifold and to show that such semiregularity map can be obtained from the classical tool of the Atiyah-Chern character. The first part of the paper is fairly detailed, deducing the existence and explicit form of a generalized semiregularity map from known results and constructions. As a corollary we obtain in the second part as well a description of the infinitesimal Abel-Jacobi map for smooth cycles as the leading term of this generalized semiregularity map, indicate why for locally complete intersections the appropriate component of our semiregularity map coincides with the one constructed by Bloch [**Blo**], and give applications to embedded deformations and deformations of coherent modules.

At the request of several participants at this conference, we restrict ourselves mainly to the case of a smooth ambient space, avoiding thus the formidable technical machinery of resolvents, powers of cotangent complexes, traces and so on. The underlying "metaresult" in its utmost generality is only stated at the end with details to be given in [**BFl**].

We wish finally to thank the editorial committee of these proceedings for their inexhaustible patience awaiting completion of this paper.

1. Chern Character and Griffiths' Transversality

1.1. Let $Z \subseteq X$ be a codimension q cycle on a compact Kähler manifold or a smooth quasi-projective variety X defined over a field of characteristic zero. With Ω^p_X the sheaf of (holomorphic or algebraic) differential forms of order p on X, the (cohomological) cycle or fundamental class of Z in $\mathrm{H}^q(X, \Omega^q_X)$ has the following classical description due to Grothendieck [**Gr1**, p. 151], see also [**Mur**, 2.18] for a recent treatment.

1991 *Mathematics Subject Classification.* 14C25, 32G05, 13D03.

Received by the editors April 22, 1999; Accepted July 29, 1999.

R.O.B. was partly supported by NSERC grant 3-642-114-80.

Both authors were partly supported by a grant from the Volkswagen Foundation under the Research in Pairs program at Math. Forschungsinstitut Oberwolfach.

This is the final form of the paper.

Let $c_p(\mathcal{O}_Z)$, $\mathrm{ch}_p(\mathcal{O}_Z) \in \mathrm{H}^p(X, \Omega^p_X)$ denote the pth *Chern class* of Z and the pth component of the *Chern character* of Z, respectively.

The *fundamental (Hodge) class of Z* is

$$[Z] = \frac{(-1)^{q-1} c_q(\mathcal{O}_Z)}{(q-1)!} = \mathrm{ch}_q(\mathcal{O}_Z) \in \mathrm{H}^q(X, \Omega^q_X).$$

As the components $\mathrm{ch}_p(\mathcal{O}_Z)$ vanish for p below the codimension, the fundamental class is thus the leading term of the Chern character, $\mathrm{ch}(\mathcal{O}_Z) = \sum_{p \geq q} \mathrm{ch}_p(\mathcal{O}_Z) \in \prod_p \mathrm{H}^p(X, \Omega^p_X)$.

1.2. Let X be an analytic space or a scheme over a field. If X is *smooth*, its first order (flat) deformations are parametrized by $\mathrm{H}^1(X, \Theta_X)$, the first cohomology group of the locally free sheaf of (holomorphic or algebraic) vector fields $\Theta_X = \mathcal{H}om_X(\Omega^1_X, \mathcal{O}_X)$.

The group $\mathrm{H}^1(X, \Theta_X) \cong \mathrm{Ext}^1_X\left(\Omega^1_X, \mathcal{O}_X\right)$ acts through graded derivations with respect to the total degree $p+q$ on the algebra $\prod_{p,q} \mathrm{H}^q(X, \Omega^p_X)$, sending a class of type (p, q) to one of type $(p-1, q+1)$. Denoting the action of the class of a first order deformation $*$ on $c \in \mathrm{H}^q\left(X, \Omega^p_X\right)$ by $* \lrcorner\, c$, as in [**ALG**, III.164, Prop. 10] or [**Blo**, 6.1], a codimension q cycle Z on X defines thus a commutative diagram

$$\begin{array}{ccc} \mathrm{H}^1(X, \Theta_X) & \xrightarrow{\;* \lrcorner \mathrm{ch}(\mathcal{O}_Z)\;} & \prod_p \mathrm{H}^{p+1}(X, \Omega^{p-1}_X) \\ & {\scriptstyle * \lrcorner [Z]} \searrow \quad \swarrow {\scriptstyle \text{projection}} & \\ & \mathrm{H}^{q+1}(X, \Omega^{q-1}_X). & \end{array}$$

The geometric meaning of the action of first order deformations on cycle classes is explained by Bloch's interpretation of Griffiths' Transversality Theorem:

THEOREM 1.1 ([**Blo**, (4.2)]; see [**Voi**, 2.3] for the analytic case). *Given $\xi \in \mathrm{H}^1(X, \Theta_X)$, the unique horizontal lift of $[Z]$ relative to the Gauß-Manin connection defined by the infinitesimal deformation ξ stays of type (q, q) if and only if $\xi \lrcorner [Z] = 0$.* □

One way of stating Grothendieck's (infinitesimal) *Variational Hodge Conjecture* [**Gr2**], see also [**Ste**, §3], is to ask whether the obstruction expressed by Griffiths' Transversality is algebraic: If the unique horizontal lift of $[Z]$ along a first order deformation ξ remains of Hodge type (q, q), is there then some cycle representing the cycle class of $[Z]$ that deforms along ξ in X? To begin with one may ask, as Bloch did in [**Blo**], what is the relationship between the obstruction to deform the given Z along ξ and the cohomological obstruction $\xi \lrcorner [Z]$?

This leads to the notion of *semiregularity* that we now recall.

2. The Structure of the Hilbert Scheme and Semiregularity

Let X be an arbitrary analytic space or scheme over some field and let $Z \subseteq X$ be any closed subspace defined by a sheaf of ideals $\mathcal{I} \subseteq \mathcal{O}_X$. The embedded deformations of $Z \subseteq X$ are parametrized in the analytic case by the *Douady space* H_X at the point corresponding to $Z \subseteq X$, and in the algebraic case by the *Hilbert scheme* Hilb_X at Z.

2.1. The tangent space to H_X or Hilb_X at the point corresponding to Z is

$$T^1_{Z/X} = \mathrm{Hom}_Z(\mathcal{I}/\mathcal{I}^2, \mathcal{O}_Z) \cong \mathrm{H}^0(Z, \mathcal{N}_{Z/X})$$

where $\mathcal{N}_{Z/X}$ is the normal sheaf to Z in X. The obstruction space is contained in $T^2_{Z/X}$, the second cohomology group of the relative cotangent complex of Z in X, see, e.g., [**Ill**],[**Rim**],[**Fle**], [**Pal**]. One has the following inclusions and equalities that approximate this group and describe it in many situations:

$$\begin{array}{ccccc} \mathrm{H}^1(Z, \mathcal{N}_{Z/X}) & \subseteq & \mathrm{Ext}^1_Z(\mathcal{I}/\mathcal{I}^2, \mathcal{O}_Z) & \subseteq & T^2_{Z/X} \\ & \uparrow & & \uparrow & \\ & \begin{array}{l} = \text{if } Z \text{ is locally} \\ \quad \text{a complete} \\ \quad \text{intersection} \end{array} & & \begin{array}{l} = \text{if } Z \text{ is generically a} \\ \quad \text{complete intersection} \\ \quad \text{and has no embedded} \\ \quad \text{components.} \end{array} & \end{array}$$

2.2. With X still an arbitrary space, denote T^1_X its space of first order deformations. There is then a natural obstruction map $\mathrm{ob}_{Z/X}\colon T^1_X \to T^2_{Z/X}$ so that Z can be deformed along a given (first order) deformation ξ of X iff $\mathrm{ob}_{Z/X}(\xi) = 0$.

If X is smooth, then $T^1_X \cong \mathrm{H}^1(X, \Theta_X)$ and the obstruction map factors through the restriction map $\mathrm{H}^1(X, \Theta_X) \to \mathrm{H}^1(Z, \Theta_X \otimes \mathcal{O}_Z) \cong \mathrm{H}^1(Z, \mathcal{H}om_Z(\Omega^1_X \otimes \mathcal{O}_Z, \mathcal{O}_Z))$ followed by $\mathrm{H}^1(Z, \ )$ of the $\mathcal{O}_Z$-dual of the Jacobi map $j\colon \mathcal{I}/\mathcal{I}^2 \to \Omega^1_X \otimes \mathcal{O}_Z$,

$$\mathrm{ob}_{Z/X}\colon T^1_X \cong \mathrm{H}^1(X, \Theta_X) \to \mathrm{H}^1(Z, \Theta_X \otimes \mathcal{O}_Z) \xrightarrow{\mathrm{H}^1(Z, j^*)} \mathrm{H}^1(Z, \mathcal{N}_{Z/X}) \subseteq T^2_{Z/X}.$$

2.3. The question referred to above is now: Is there a natural *semiregularity map*

$$\sigma\colon T^2_{Z/X} \to \mathrm{H}^{q+1}(X, \Omega^{q-1}_X)$$

such that the diagram

$$\begin{array}{ccc} \mathrm{H}^1(X, \Theta_X) & \xrightarrow{\quad \mathrm{ob}_{Z/X}(*) \quad} & T^2_{Z/X} \\ & {\scriptstyle * \lrcorner [Z]} \searrow \qquad \swarrow {\scriptstyle \sigma} & \\ & \mathrm{H}^{q+1}(X, \Omega^{q-1}_X) & \end{array}$$

commutes?

2.4. If a semiregularity map exists, and if it is furthermore *injective*, then Z is called *semiregular* in X.

Such a semiregularity map was first introduced by Severi [**Sev**] for a curve on a surface and he coined the term semiregular in that context. Kodaira-Spencer [**KSp**] later generalized Severi's construction and results to arbitrary divisors on a complex manifold using the tools from the deformation theory they had just developed. Bloch [**Blo**] extended the notion to arbitrary cycles on smooth complex projective schemes, constructed a semiregularity map for cycles that are locally complete intersection, showed that the Hilbert scheme Hilb_X is smooth at the point corresponding to a semiregular locally complete intersection $Z \subseteq X$, and deduced then the variational Hodge conjecture for the corresponding cycle class $[Z]$.

We will show here that a semiregularity map always exists, and that its construction could have been achieved more than 25 years ago, by anyone reading [**Ill**] and [**Blo**] side by side!

The key ingredient is to forget part of the structure and to construct a more general semiregularity map for coherent sheaves on X. The underlying idea is simple: The fundamental class is not only defined for a cycle in X but for any element of $K^0(X)$, the Grothendieck group of the smooth space X. Thus one may ask for a semiregularity map for any representative of a class in that group, that is for any coherent sheaf on X, or, if one allows X to become singular, for any *perfect complex* of sheaves on X.

3. Deformations of Modules and the Atiyah-Chern Character

If $Z \subseteq X$ is a closed subspace of X, we may consider instead of embedded deformations of Z the deformations of $\mathcal{O}_Z$ as $\mathcal{O}_X$-module. This is a forgetful functor: a deformation of $\mathcal{O}_Z$ as module will only be an invertible sheaf on its support, not necessarily its structure sheaf.

3.1. For any coherent sheaf $\mathcal{F}$ on X, the space of first order deformations is given by $\operatorname{Ext}^1_X(\mathcal{F},\mathcal{F})$ with obstructions in $\operatorname{Ext}^2_X(\mathcal{F},\mathcal{F})$, and so there are accordingly natural forgetful transformations

$$f^i : T^i_{Z/X} \to \operatorname{Ext}^i_X(\mathcal{O}_Z,\mathcal{O}_Z). \tag{1}$$

These maps can be explained quite simply within the theory of resolvents and cotangent complexes, as developed in [**Fle**], [**Ill**], [**Pal**] based on [**Qui**]: a resolvent for Z over X is a differential graded free $\mathcal{O}_X$-algebra that resolves $\mathcal{O}_Z$, the groups $T^i_{Z/X}$ constitute the hypercohomology groups of the $\mathcal{O}_X$-derivations of said resolvent, whereas the groups $\operatorname{Ext}^i_X(\mathcal{O}_Z,\mathcal{O}_Z)$ represent the hypercohomology of its $\mathcal{O}_X$-linear endomorphisms. Now any $\mathcal{O}_X$-derivation is $\mathcal{O}_X$-linear by definition, whence the comparison maps; see [**BFl**] for further details.

3.2. Note that $\operatorname{Ext}^i_X(\mathcal{O}_Z,\mathcal{O}_Z)$ always contains $\mathrm{H}^i(X,\mathcal{O}_Z) = \mathrm{H}^i(Z,\mathcal{O}_Z)$ naturally as a direct summand and that f^i always maps into the complement. For $i=1$ one has indeed a split exact sequence

$$0 \to T^1_{Z/X} = \mathrm{H}^0(X,\mathcal{N}_{Z/X}) \xrightarrow{f^1} \operatorname{Ext}^1_X(\mathcal{O}_Z,\mathcal{O}_Z) \to \mathrm{H}^1(Z,\mathcal{O}_Z) \to 0$$

and $\mathrm{H}^1(Z,\mathcal{O}_Z)$ is the tangent space to the Picard scheme of Z at $\mathcal{O}_Z$. This formalizes that a first order deformation of $\mathcal{O}_Z$ as $\mathcal{O}_X$-module splits into an embedded deformation of Z and a deformation of the invertible sheaf $\mathcal{O}_Z$ on Z.

When restricted to $\operatorname{Ext}^{i-1}_Z(\mathcal{I}/\mathcal{I}^2,\mathcal{O}_Z) \subseteq T^i_{Z/X}$, the forgetful maps f^i become particularly explicit: The extension defined by the first fundamental neighbourhood of Z in X,

$$\mathcal{Z}^{(2)} \quad \equiv \quad 0 \to \mathcal{I}/\mathcal{I}^2 \to \mathcal{O}_X/\mathcal{I}^2 \to \mathcal{O}_Z \to 0,$$

defines a connecting homomorphism $\operatorname{Ext}^{i-1}_X(\mathcal{I}/\mathcal{I}^2,\mathcal{O}_Z) \to \operatorname{Ext}^i_X(\mathcal{O}_Z,\mathcal{O}_Z)$ which composed with the forgetful map $\operatorname{Ext}^{i-1}_Z(\mathcal{I}/\mathcal{I}^2,\mathcal{O}_Z) \to \operatorname{Ext}^{i-1}_X(\mathcal{I}/\mathcal{I}^2,\mathcal{O}_Z)$ yields the restriction of f^i.

3.3. Returning to the context of coherent $\mathcal{O}_X$-modules on a smooth space X we want thus to construct a general semiregularity map for coherent $\mathcal{O}_X$-modules $\mathcal{F}$,

$$\tau^2 = \left(\tau^{2,p}\right)_{p\geq 0} : \operatorname{Ext}^2_X(\mathcal{F},\mathcal{F}) \to \prod_{p\geq 0} \mathrm{H}^{p+2}\left(X,\Omega^p_X\right),$$

such that the desired semiregularity map for a cycle Z of codimension q will be

$$\sigma = \tau^{2,q-1} f^2 \colon T^2_{Z/X} \to \mathrm{Ext}^2_X(\mathcal{O}_Z, \mathcal{O}_Z) \to \mathrm{H}^{q+1}(X, \Omega^{q-1}_X).$$

A natural such map τ has been known for a long time!

3.4. To describe it, recall the definition of the *Atiyah class* of an $\mathcal{O}_X$-module $\mathcal{F}$: Let $X^{(2)} \subseteq X \times X$ denote the first fundamental neighbourhood of the diagonal $\Delta\colon X \hookrightarrow X \times X$ and denote $p_i\colon X^{(2)} \to X$ the projection onto the ith factor. The sheaf of 1-jets or principal parts of first order associated to $\mathcal{F}$ is defined to be $\mathcal{P}^1(F) = p_{2*}p_1^*\mathcal{F}$. Restricting $\mathcal{P}^1(F)$ to the diagonal itself defines a canonical exact sequence of (right) $\mathcal{O}_X$-modules, the *Atiyah sequence* of $\mathcal{F}$,

$$\mathrm{At}_X(\mathcal{F}) \quad \equiv \quad 0 \to \mathcal{F} \otimes \Omega^1_X \to \mathcal{P}^1(\mathcal{F}) \to \mathcal{F} \to 0$$

whose class

$$\mathrm{at}_X(\mathcal{F}) \in \mathrm{Ext}^1_X(\mathcal{F}, \mathcal{F} \otimes \Omega^1_X)$$

is the (first) *Atiyah class* of $\mathcal{F}$.

3.5. Combining the composition or Yoneda product with the exterior product on $\Omega^{\cdot}_X$ defines on the vector space $\prod_{i,j} \mathrm{Ext}^i_X(\mathcal{F}, \mathcal{F} \otimes \Omega^j_X)$ an algebra structure with multiplication

$$\smile\colon \mathrm{Ext}^a_X(\mathcal{F}, \mathcal{F} \otimes \Omega^b_X) \times \mathrm{Ext}^c_X(\mathcal{F}, \mathcal{F} \otimes \Omega^d_X) \to \mathrm{Ext}^{a+c}_X(\mathcal{F}, \mathcal{F} \otimes \Omega^{b+d}_X).$$

One can thus form powers of the Atiyah class,

$$\mathrm{at}^p_X(\mathcal{F}) = \mathrm{at}_X(\mathcal{F})^{\smile p} \in \mathrm{Ext}^p_X(\mathcal{F}, \mathcal{F} \otimes \Omega^p_X),$$

and then, in characteristic zero, the *Atiyah-Chern character*

$$e^{\mathrm{at}_X(\mathcal{F})} = \sum_{p \geq 0} \frac{\mathrm{at}^p_X(\mathcal{F})}{p!} \in \prod_p \mathrm{Ext}^p_X(\mathcal{F}, \mathcal{F} \otimes \Omega^p_X).$$

The (powers of the) Atiyah class and the Atiyah-Chern character are functorial in the appropriate sense; in particular these notions localize to yield local sections in $\prod_p \mathcal{E}xt^p_X(\mathcal{F}, \mathcal{F} \otimes \Omega^p_X)$. Various explicit representatives of these extension classes are worked out in [**ALJ**] in the algebraic case.

3.6. As X is supposed to be smooth, any coherent $\mathcal{O}_X$-module $\mathcal{F}$ admits locally a finite resolution by locally free sheaves, that is, it represents a *perfect complex*, and thus there exist natural trace maps

$$\mathrm{tr}^p\colon \mathrm{Ext}^p_X(\mathcal{F}, \mathcal{F} \otimes \Omega^p_X) \to \mathrm{H}^p(X, \Omega^p_X)$$

that uniquely extend the usual trace tr^0 of an endomorphism of $\mathcal{F}$. The quite intricate technicalities in the general setting were overcome in the algebraic case by Illusie, first in [**SGA6**, Exp. I], then in [**Ill**], and the analytic case was dealt with explicitly by O'Brian, Toledo, Tong in [**OTT**].

It is a classical result that the trace of (the inverse of) the Atiyah-Chern character yields the Chern character:

THEOREM 3.1 (Atiyah [**At**]; Illusie [**Ill**]; O'Brian-Toledo-Tong [**OTT**]). *If $\mathcal{F}$ is a coherent sheaf on the smooth space X, then*

$$\mathrm{ch}(\mathcal{F}) = \mathrm{tr}\big(e^{-\mathrm{at}_X(\mathcal{F})}\big) \in \prod_p \mathrm{H}^p(X, \Omega^p_X).$$

In particular, for a closed subspace $Z \subseteq X$ of codimension q, one has

$$[Z] = \operatorname{ch}_q(\mathcal{O}_Z) = \frac{(-1)^q \operatorname{tr}(\operatorname{at}_X^q(\mathcal{O}_Z))}{q!}.$$

The minus sign appears here as traditionally the first Chern class is the trace of the opposite of the Atiyah class, see [**At**, Prop. 12], [**Ill**, V.5.4.1, 5.9.4]. Note also that extension to the algebraic case removed the factor $2\pi i$ originally found in the exponent.

The link between this theorem and the existence of a semiregularity map is the following result, due to Illusie [**Ill**, IV.3.1.8] in much vaster generality.

PROPOSITION 3.2. *If $\mathcal{F}$ is a coherent sheaf on the smooth space X, then the obstruction $\operatorname{ob}_{\mathcal{F}}\colon T_X^1 \to \operatorname{Ext}_X^2(\mathcal{F},\mathcal{F})$ to deform $\mathcal{F}$ (flatly) along a given first order deformation of X can be realized as*

$$\operatorname{ob}_{\mathcal{F}}\colon T_X^1 \cong \operatorname{Ext}_X^1(\Omega_X^1, \mathcal{O}_X) \xrightarrow{\mathcal{F}\otimes(\ )} \operatorname{Ext}_X^1(\mathcal{F}\otimes\Omega_X^1,\mathcal{F}) \xrightarrow{(\ )\circ(-\operatorname{at}_X(\mathcal{F}))} \operatorname{Ext}_X^2(\mathcal{F},\mathcal{F}).$$

In comparison to Illusie's original treatment we changed here the sign in front of the Atiyah class to incur fewer signs later on.

3.7. The obstructions for embedded deformations of a subspace Z versus those for deformations of the module $\mathcal{O}_Z$ are related explicitly by means of the following commutative diagram whose rows are exact sequences of $\mathcal{O}_X$-modules, namely the Atiyah sequence of the $\mathcal{O}_X$-module $\mathcal{O}_Z$ and the exact sequence of the first infinitesimal neighbourhood of Z in X, respectively:

$$\begin{array}{ccccccccccc}
\operatorname{At}_X(\mathcal{O}_Z) \equiv & 0 & \longrightarrow & \mathcal{O}_Z\otimes\Omega_X^1 & \longrightarrow & \mathcal{P}^1(\mathcal{O}_Z) & \longrightarrow & \mathcal{O}_Z & \longrightarrow & 0 \\
\uparrow & & & \uparrow{\scriptstyle j} & & \uparrow{\scriptstyle j'} & & \| & & \\
\mathcal{Z}^{(2)} \equiv & 0 & \longrightarrow & \mathcal{I}/\mathcal{I}^2 & \longrightarrow & \mathcal{O}_X/\mathcal{I}^2 & \longrightarrow & \mathcal{O}_Z & \longrightarrow & 0.
\end{array}$$

The map j' is given by $j'(x) = 1\otimes p_2^*(x) \in \mathcal{P}^1(\mathcal{O}_Z) = p_1^*\mathcal{O}_Z \otimes (\mathcal{O}_{X\times X}/\mathcal{I}_\Delta^2)$, where x is a local section of $\mathcal{O}_X/\mathcal{I}^2$ and $\mathcal{I}_\Delta$ denotes the ideal of the diagonal in $X\times X$. Indeed, for a local section $f \in \mathcal{I}$ one has that $j(f)$ lifts to

$$\begin{aligned}
\tilde{\jmath}(f) = 1\otimes df \in \mathcal{O}_X\otimes\Omega_X^1 &\mapsto (p_2^*(f) - p_1^*(f)) = 1\otimes f - f\otimes 1 \in \mathcal{O}_{X\times X}/\mathcal{I}_\Delta^2 \\
&\mapsto 1\otimes p_2^*(f) \in p_1^*\mathcal{O}_Z \cong \mathcal{O}_{Z\times X}/\mathcal{O}_{Z\times X}\mathcal{I}_\Delta^2 \\
&\mapsto j'(f) \in \mathcal{P}^1(\mathcal{O}_Z).
\end{aligned}$$

In view of (3.2) this means that indeed $\operatorname{ob}_{\mathcal{O}_Z} = f^2 \circ \operatorname{ob}_{Z/X}$ relates the obstruction maps of the two deformation problems.

Now we are ready to define

4. The Generalized Semiregularity Map

4.1. The homomorphism of $\mathcal{O}_X$-algebras $\mathcal{O}_X \to \Omega_X^{\cdot}$ induces an algebra homomorphism

$$\prod_i \operatorname{Ext}_X^i(\mathcal{F},\mathcal{F}) \to \prod_{i,j} \operatorname{Ext}_X^i(\mathcal{F},\mathcal{F}\otimes\Omega_X^j)$$

which extends the natural right action of the Yoneda Ext-algebra $\operatorname{Ext}_X(\mathcal{F},\mathcal{F})$ on the target. Contraction combined with Yoneda product defines further a left action

$$\operatorname{Ext}_X^r(\Omega_X^s,\mathcal{O}_X) \times \operatorname{Ext}_X^i(\mathcal{F},\mathcal{F}\otimes\Omega_X^j) \xrightarrow{\lrcorner} \operatorname{Ext}_X^{r+i}(\mathcal{F},\mathcal{F}\otimes\Omega_X^{j-s})$$

and its restriction to $\operatorname{Ext}^1_X(\Omega^1_X, \mathcal{O}_X)$ acts through graded algebra derivations with respect to total degree,

$$\xi \lrcorner (\omega \smile \omega') = (\xi \lrcorner \omega) \smile \omega' + (-1)^{i+j} \omega \smile (\xi \lrcorner \omega'),$$

for $\xi \in \operatorname{Ext}^1_X(\Omega^1_X, \mathcal{O}_X)$, $\omega \in \operatorname{Ext}^i_X(\mathcal{F}, \mathcal{F} \otimes \Omega^j_X)$.

4.2. Combining these actions, one obtains for any coherent sheaf $\mathcal{F}$ a family of *generalized semiregularity maps*,

$$\tau^i = \left(\tau^{i,p}\right)_{p \geq 0} = \operatorname{tr}\left(e^{-\operatorname{at}_X(\mathcal{F})} \smile *\right) \colon \operatorname{Ext}^i_X(\mathcal{F}, \mathcal{F}) \to \prod_p \mathrm{H}^{i+p}(X, \Omega^p_X)$$

that is natural in $\mathcal{F}$ and whose value on $\mathrm{id}_{\mathcal{F}} \in \operatorname{Ext}^0_X(\mathcal{F}, \mathcal{F})$ is the Chern character of $\mathcal{F}$.

4.3. If $Z \subseteq X$ contains the support of $\mathcal{F}$, then the trace map factors through the local cohomology with support in Z, see [**Ill**, V.6] or [**OTT**, 4.6], whence one may define Chern character and generalized semiregularity maps with support,

$$\operatorname{ch}_Z(\mathcal{F}) = \operatorname{tr}_Z\left(e^{-\operatorname{at}_X(\mathcal{F})}\right) \in \prod_p \mathrm{H}^p_Z(X, \Omega^p_X),$$

$$\tau_Z = \operatorname{tr}_Z\left(e^{-\operatorname{at}_X(\mathcal{F})} \smile *\right) \colon \prod_i \operatorname{Ext}^i_X(\mathcal{F}, \mathcal{F}) \to \prod_{i,p} \mathrm{H}^{i+p}_Z(X, \Omega^p_X).$$

The key result is now the following.

THEOREM 4.1. *A coherent $\mathcal{O}_X$-module $\mathcal{F}$ on a smooth space X whose support is contained in $Z \subseteq X$ gives rise to a commutative diagram*

$$\begin{array}{ccc}
\mathrm{H}^1(X, \Theta_X) & \xrightarrow{\ \operatorname{ob}_{\mathcal{F}}\ } & \operatorname{Ext}^2_X(\mathcal{F}, \mathcal{F}) \\
{\scriptstyle * \lrcorner \operatorname{ch}(\mathcal{F})} \big\downarrow & \swarrow {\scriptstyle \tau^2} & \big\downarrow {\scriptstyle \tau^2_Z} \\
\prod_{p \geq 1} \mathrm{H}^{p+1}(X, \Omega^{p-1}_X) & \xleftarrow[\ \textit{canonical}\ ]{} & \prod_{p \geq 1} \mathrm{H}^{p+1}_Z(X, \Omega^{p-1}_X).
\end{array}$$

PROOF. Illusie's result (3.2) shows that $\operatorname{ob}_{\mathcal{F}}(\xi) = \xi \lrcorner - \operatorname{at}_X(\mathcal{F})$ for any first order deformation $\xi \in \mathrm{H}^1(X, \Theta_X)$. The class $\operatorname{at}_X(\mathcal{F})$ is of total degree two, and so

$$\begin{aligned}
\xi \lrcorner \operatorname{ch}(\mathcal{F}) &= \xi \lrcorner \operatorname{tr}\left(e^{-\operatorname{at}_X(\mathcal{F})}\right) \\
&= \operatorname{tr}\left(\xi \lrcorner e^{-\operatorname{at}_X(\mathcal{F})}\right) && \text{by naturality of the trace,} \\
&= \operatorname{tr}\left(e^{-\operatorname{at}_X(\mathcal{F})} \smile (\xi \lrcorner - \operatorname{at}_X(\mathcal{F}))\right)
\end{aligned}$$

as $\operatorname{tr}(ab) = \operatorname{tr}(ba)$ for even classes a and b, whence $\operatorname{tr}(D(e^{-y})) = \operatorname{tr}(e^{-y} D(-y))$ for any (graded) derivation D and any (even) element y,

$$\begin{aligned}
&= \operatorname{tr}\left(e^{-\operatorname{at}_X(\mathcal{F})} \smile \operatorname{ob}_{\mathcal{F}}(\xi)\right) && \text{by (3.2),} \\
&= \tau^2\left(\operatorname{ob}_{\mathcal{F}}(\xi)\right).
\end{aligned}$$

□

Factoring the obstruction map $\operatorname{ob}_{\mathcal{O}_Z}$ through f^2 gives now the desired generalized semiregularity map for cycles.

THEOREM 4.2. *If $Z \subseteq X$ is a closed (analytic or algebraic) subspace of the smooth space X, then the diagram*

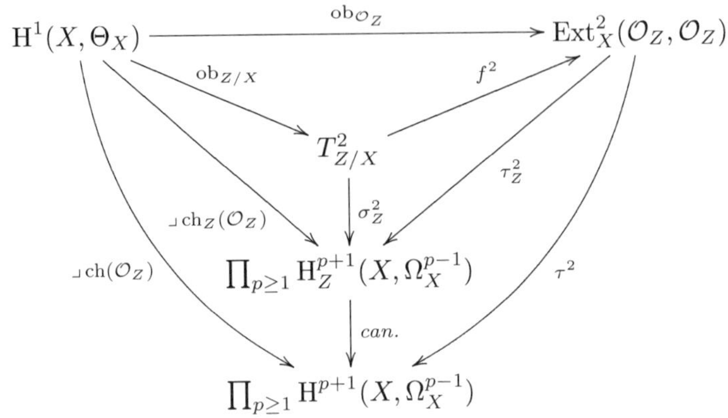

commutes, where we define $\sigma^2_Z = \tau^2_Z \cdot f^2$.

If Z is of codimension q, then the component $\sigma^{2,q-1}_Z = \tau^{2,q-1}_Z \cdot f^2$ is a semiregularity map with support for Z, whereas its composition with the canonical map yields the desired (absolute) semiregularity map

$$\sigma^{2,q-1} = \tau^2 \cdot f^2 \colon T^2_{Z/X} \to \mathrm{H}^{q+1}(X,\Omega^{q-1}_X).$$

4.4. More generally, we can define as well a family of generalized semiregularity maps for Z (with support) through

$$\sigma^i_{(Z)} = \tau^i_{(Z)} \cdot f^i \colon T^i_{Z/X} \to \prod_p \mathrm{H}^{i+p}_{(Z)}(X,\Omega^p_X).$$

5. Duality and the Infinitesimal Abel-Jacobi Map

5.1. Due to their naturality, the trace maps with support as well as the (powers of the) Atiyah class localize to yield generalized semiregularity maps on the level of sheaves

$$\tilde{\tau}^i \colon \mathcal{E}xt^i_X(\mathcal{F},\mathcal{F}) \to \prod_p \mathcal{H}^{i+p}_Z(X,\Omega^p_X)$$

and these morphisms of sheaves fit together to morphisms of the associated local-global spectral sequences

$$\begin{array}{ccc}
\mathrm{H}^i\left(X,\mathcal{E}xt^j_X(\mathcal{F},\mathcal{F})\right) & \Longrightarrow & \mathrm{Ext}^{i+j}_X(\mathcal{F},\mathcal{F}) \\
{\scriptstyle \mathrm{H}^i(X,\tilde{\tau}^j)}\Big\downarrow & & \Big\downarrow{\scriptstyle \tau^{i+j}_Z} \\
\prod_p \mathrm{H}^i\left(X,\mathcal{H}^{p+j}_Z(X,\Omega^p_X)\right) & \Longrightarrow & \prod_p \mathrm{H}^{i+j+p}_Z(X,\Omega^p_X).
\end{array}$$

As X is assumed to be smooth, each Ω^p_X is locally free and $\mathcal{H}^k_Z(X,\Omega^p_X) = 0$ for $k < \operatorname{depth}_Z \mathcal{O}_X$. It follows that the components $\tau^{i,p}_Z$, and then a fortiori $\tau^{i,p}$ and $\sigma^{i,p}_Z$ or $\sigma^{i,p}$ for Z, vanish whenever $i + p < \operatorname{depth}_Z \mathcal{O}_X$.

5.2. If $Z \subseteq X$ is a closed subspace of codimension q that is locally Cohen-Macaulay, then the local cohomology sheaves $\mathcal{H}^{q'}_Z(X, \Omega^p_X)$ vanish for $q' < q$ and any p, and so there are natural inclusions $\mathrm{H}^i(X, \mathcal{H}^q_Z(X, \Omega^p_X)) \hookrightarrow \mathrm{H}^{i+q}_Z(X, \Omega^p_X)$.

Now assume that the subspace Z satisfies furthermore $T^2_{Z/X} = \mathrm{H}^1(X, \mathcal{N}_{Z/X})$. The isomorphisms of $\mathcal{O}_X$-modules

$$\mathcal{N}_{Z/X} \cong \mathcal{H}om_Z(\mathcal{I}/\mathcal{I}^2, \mathcal{O}_Z) \cong \mathcal{H}om_X(\mathcal{I}, \mathcal{O}_Z) \cong \mathcal{E}xt^1_X(\mathcal{O}_Z, \mathcal{O}_Z)$$

show that the semiregularity maps with support, $\sigma^{2,q-1}_Z \colon T^2_{Z/X} \to \mathrm{H}^{q+1}_Z(X, \Omega^{q-1}_X)$, factor through $\mathrm{H}^1(X, \tilde{\sigma}^{1,q-1}_Z)$. Thus the semiregularity map $\sigma^{2,q-1}$ itself fits into the following commutative diagram

$$\begin{array}{ccc} T^2_{Z/X} = \mathrm{H}^1(X, \mathcal{N}_{Z/X}) & \xrightarrow{\mathrm{H}^1(X, \tilde{\sigma}^{1,q-1}_Z)} & \mathrm{H}^1\left(X, \mathcal{H}^q_Z(X, \Omega^{q-1}_X)\right) \\ \Big\downarrow {\scriptstyle \sigma^{2,q-1}} & & \Big\downarrow {\scriptstyle \text{inclusion}} \\ \mathrm{H}^{q+1}(X, \Omega^{q-1}_X) & \xleftarrow{\text{canonical}} & \mathrm{H}^{q+1}_Z(X, \Omega^{q-1}_X) \end{array}$$

5.3. If Z is locally a complete intersection of codimension q in X, it is in particular locally Cohen-Macaulay and satisfies $T^2_{Z/X} = \mathrm{H}^1(X, \mathcal{N}_{Z/X})$. With $d = \dim Z$ and $\mathcal{I}$ the defining ideal of Z, take exterior powers of the Jacobi map $j \colon \mathcal{I}/\mathcal{I}^2 \to \Omega^1_X \otimes \mathcal{O}_Z$ to define for $0 \le i \le q$ natural pairings

$$\Omega^{d+i}_X \otimes \Lambda^{q-i}\left(\mathcal{I}/\mathcal{I}^2\right) \xrightarrow{1 \otimes \Lambda^{q-i} j} \Omega^{d+i}_X \otimes \Omega^{q-i}_X \otimes \mathcal{O}_Z \xrightarrow{(\ \wedge\ ) \otimes 1} \omega_X \otimes \mathcal{O}_Z.$$

As $\mathcal{I}/\mathcal{I}^2$ is locally free, these give rise, equivalently, to $\mathcal{O}_X$-homomorphisms

$$\gamma_i \colon \Omega^{d+i}_X \to \omega_X \otimes \Lambda^{q-i}\mathcal{N}_{Z/X} \cong \omega_Z \otimes \Lambda^i\left(\mathcal{I}/\mathcal{I}^2\right)$$

where we have used the adjunction formula $\omega_Z \cong \omega_X \otimes \det \mathcal{N}_{Z/X}$ for the dualizing module ω_Z, and the isomorphism $\Lambda^{q-i}\mathcal{N}_{Z/X} \cong \det \mathcal{N}_{Z/X} \otimes \Lambda^i\left(\mathcal{I}/\mathcal{I}^2\right)$.

The duality theorem for finite morphisms says that

$$\mathrm{Ext}^p_X(\omega_Z \otimes \Lambda^i\left(\mathcal{I}/\mathcal{I}^2\right), \omega_X) \cong \mathrm{Ext}^{p-q}_Z(\Lambda^i\left(\mathcal{I}/\mathcal{I}^2\right), \mathcal{O}_Z) \cong \mathrm{H}^{p-q}(Z, \Lambda^i\mathcal{N}_{Z/X})$$

and clearly

$$\mathrm{Ext}^p_X(\Omega^{d+i}_X, \omega_X) \cong \mathrm{H}^p(X, \Omega^{q-i}_X)$$

for any p or i, whence $\check{\gamma}^{i,p} = \mathrm{Ext}^p_X(\gamma_i, \omega_X)$ can be identified as a map between cohomology groups,

$$\check{\gamma}^{i,p} \colon \mathrm{H}^{p-q}(Z, \Lambda^i\mathcal{N}_{Z/X}) \to \mathrm{H}^p(X, \Omega^{q-i}_X).$$

Equivalently, $\check{\gamma}^{i,p}$ is the Serre dual of the map $\mathrm{H}^{d+q-p}_c(X, \Omega^{d+i}_X) \to \mathrm{H}^{d+q-p}_c(Z, \omega_Z \otimes \Lambda^i\mathcal{I}/\mathcal{I}^2)$ induced by γ_i.

5.4. Some of these maps are classical objects:

1. For $i = 0$, $p = q$, and with $f \colon Z \hookrightarrow X$ the closed embedding, one finds by [**FGA**, 149-20, Thm. 4] the leading term of the direct image map f_* in cohomology,

$$\check{\gamma}^{0,q} = f_*|_{\mathrm{H}^0(Z, \mathcal{O}_Z)} \colon \mathrm{H}^0(Z, \mathcal{O}_Z) \to \mathrm{H}^q(X, \Omega^q_X).$$

By definition, $\check{\gamma}^{0,q}(1_Z) = [Z]$, the fundamental (Hodge) class of Z, and a simple application of the Grothendieck-Riemann-Roch Theorem, see [**Mur**,

2.11], shows $\check{\gamma}^{0,q}(1_Z) = \mathrm{ch}_q(\mathcal{O}_Z)$ which is the result quoted at the beginning (1.1).

2. For $i = 1, p = q$ and Z smooth on a compact Kähler manifold, the map $\check{\gamma}^{1,q}$ is the *infinitesimal Abel-Jacobi map* at Z as shown in [**Gre**, Lect.2, p.29],
$$\check{\gamma}^{1,q} = AJ_{X,Z}\colon \mathrm{H}^0(X, \mathcal{N}_{Z/X}) \to \mathrm{H}^q(X, \Omega_X^{q-1}).$$
3. For $i = 1, p = q - 1$ and Z again locally a complete intersection, one finds Bloch's semiregularity map
$$\check{\gamma}^{1,q+1} = \sigma_{\mathrm{Bloch}}\colon \mathrm{H}^1(Z, \mathcal{N}_{Z/X}) \to \mathrm{H}^{q+1}(X, \Omega_X^{q-1}).$$

These observations lead to the following duality theorem for the leading terms of the generalized semiregularity maps.

THEOREM 5.1. *Let $Z \subseteq X$ be a closed subspace of codimension q that is locally a complete intersection.*

(1) *One has $T^p_{Z/X} \cong \mathrm{H}^{p-1}(Z, \mathcal{N}_{Z/X})$ and*
$$\sigma^{p,q-1} \cong \check{\gamma}^{1,p+q-1}\colon T^p_{Z/X} \to \mathrm{H}^{p+q-1}(X, \Omega_X^{q-1}).$$
Each of these maps fits into a commutative diagram
$$\begin{array}{ccc}
\mathrm{H}^p(X, \mathcal{N}_{Z/X}) & \xrightarrow{\mathrm{H}^p(X, \tilde{\sigma}^{1,q-1})} & \mathrm{H}^p\left(X, \mathcal{H}^q_Z(X, \Omega_X^{q-1})\right) \\
{\scriptstyle \sigma^{p,q-1}}\downarrow & & \downarrow{\scriptstyle \mathit{inclusion}} \\
\mathrm{H}^{p+q}(X, \Omega_X^{q-1}) & \xleftarrow{\mathit{canonical}} & \mathrm{H}^{p+q}_Z(X, \Omega_X^{q-1}).
\end{array}$$
(2) *Bloch's semiregularity map satisfies $\sigma_{\mathrm{Bloch}} = \sigma^{2,q-1}$.*
(3) *If Z is smooth in a compact Kähler manifold X, then the infinitesimal Abel-Jacobi map at Z is given by the component $\sigma^{1,q-1}$ of the generalized semiregularity map.*

Clearly (1) $\Longrightarrow$ (2), (3) in view of the explicit construction of both σ_{Bloch} and $AJ_{X,Z}$. Verification of (1) is achieved through a local calculation that identifies $\tilde{\sigma}^{1,q-1}$. The patient reader is referred to the details in [**BFl**] whereas the impatient one may wish to extract the calculation from the explicit formalism in [**ALJ**]. $\square$

5.5. Of course, this result raises several questions, in particular whether for any algebraic cycle Z on a smooth space the diagram
$$\begin{array}{ccc}
\mathrm{H}^0(X, \mathcal{N}_{Z/X}) = T^1_{Z/X} & \xrightarrow{f^1} & \mathrm{Ext}^1_X(\mathcal{O}_Z, \mathcal{O}_Z) \\
{\scriptstyle \text{Infinitesimal Abel-Jacobi map}}\downarrow & & \downarrow{\scriptstyle \tau_Z^{1,q-1} = \mathrm{tr}_Z(e^{-\mathrm{at}_X(\mathcal{O}_Z)} \smile *)^{q-1}} \\
\mathrm{H}^q(X, \Omega_X^{q-1}) & \xleftarrow{\text{canonical}} & \mathrm{H}^q_Z(X, \Omega_X^{q-1})
\end{array}$$
commutes. Note that f^1 is always a split monomorphism by (3.2) so that the general infinitesimal Abel-Jacobi map should be a retract of $\tau^{1,q-1}$.

Even more generally, there should be an Abel-Jacobi map for deformations of arbitrary coherent sheaves on a compact algebraic or Kähler manifold such that its differential is essentially given by the trace applied to multiplication with powers of the Atiyah class.

6. Applications to Deformation Theory

6.1. Ideally, one would like to interpret the family of semiregularity maps σ or τ as maps between cohomology groups of cotangent complexes induced by morphisms of deformation theories. Whereas the domain of these maps is of the desired form, it is not clear what kind of deformation theory, if any, would allow the groups $\mathbb{H}^q := \prod_p \mathrm{H}^{p+q}(X, \Omega_X^p)$ as its tangent cohomology. Hodge theory shows that if such a deformation theory existed, it were necessarily non obstructed for any complex manifold X that is bimeromorphically equivalent to a Kähler manifold: According to Deligne [**Del**, (5.5)], each of the functors

$$\mathcal{N} \mapsto \mathrm{H}^q(X_T, \Omega^p_{X_T/T} \otimes_{\mathcal{O}_T} \mathcal{N})$$

on the coherent modules $\mathcal{N}$ over an artinian complex germ T is then *exact*, and this applied to $\mathbb{H}^1$ would show that every "infinitesimal deformation" could be lifted. This line of reasoning applies indeed directly in some special cases, see [**Ran4**].

In general, even though there might not be an actual deformation theory underlying the groups $\mathbb{H}^q$, it was first pointed out by Ran, [**Ran3, Ran2, Ran1**], and then made more precise in [**Kaw1, Kaw2**] and [**FMa**], that Deligne's theorem can be used to obtain results on embedded deformations or deformations of coherent modules. Complete proofs and further generalizations of the following results will be contained in [**BFl**].

THEOREM 6.1. *Assume that $f\colon X \to \Sigma$ is a smooth morphism of complex spaces with $X_0 := f^{-1}(0)$ a compact manifold bimeromorphically equivalent to a Kähler manifold. Let $\mathcal{F}_0$ be a coherent module on X_0 and*

$$\tau^2 = \mathrm{tr}(e^{-\,\mathrm{at}(\mathcal{F}_0)} \smile *)\colon \mathrm{Ext}^2_{X_0}(\mathcal{F}_0, \mathcal{F}_0) \to \prod_{p \geq 0} H^{p+2}(X_0, \Omega^p_{X_0})$$

the corresponding generalized semiregularity map. The dimension of the basis S of a semi-universal deformation of $\mathcal{F}_0$ satisfies

$$\dim S \geq \dim_{\mathbb{C}} \mathrm{Ext}^1_{X_0}(\mathcal{F}_0, \mathcal{F}_0) - \dim_{\mathbb{C}} \ker \tau^2.$$

If τ^2 is injective then S is smooth over a closed subspace of Σ.

6.2. This result generalizes in characteristic zero the *Artamkin-Mukai criterion* [**Art, Muk, Kaw1**] which is the above statement but restricted to the component $\tau^{2,0}\colon \mathrm{Ext}^2_X(\mathcal{F}, \mathcal{F}) \to H^2(X, \mathcal{O}_X)$ in the absolute case where $X = X_0$ is a nonsingular projective variety. The map $\tau^{2,0}$ is just the ordinary trace and this component of the generalized semiregularity map has indeed a description as a transformation between tangent cohomologies induced by a morphism of deformation theories: As X is smooth, every coherent sheaf $\mathcal{F}$ admits a well defined determinant line bundle $\det \mathcal{F}$, and associating to $\mathcal{F}$ its determinant is a morphism of deformation functors that induces $\tau^{i,0}\colon \mathrm{Ext}^i_X(\mathcal{F}, \mathcal{F}) \to H^i(X, \mathcal{O}_X)$ on the corresponding tangent cohomology groups.

As the target, the Picard functor of X, is smooth, it follows that all obstructions to deform $\mathcal{F}$ are already contained in the kernel of the trace map $\tau^{2,0}$. This deformation theoretic interpretation allows Artamkin to obtain as well results in positive characteristic that are beyond reach here.

The analogous application to embedded deformations of subspaces generalizes Bloch's original result [**Blo**, Thm. 7.3] relating semiregularity and smoothness of the Hilbert scheme. It encompasses the previous generalizations of that result in [**Ran2, Kaw1**].

THEOREM 6.2. *Assume that $f\colon X \to \Sigma$ is a smooth morphism of complex spaces and that $X_0 := f^{-1}(0)$ is a compact manifold bimeromorphically equivalent to a Kähler manifold. Let $Z \subseteq X_0$ be a closed subspace and*

$$\sigma^2\colon T^2_{Z/X_0}(\mathcal{O}_Z) \to \prod_{p\geq 0} H^{p+2}(X_0, \Omega^p_{X_0}).$$

the generalized semiregularity map.

The dimension of the Douady space $H_{X/\Sigma}$ at $[Z]$ satisfies

$$\dim_{[Z]} H_{X/\Sigma} \geq \dim_{\mathbb{C}} T^1_{Z/X_0}(\mathcal{O}_Z) - \dim_{\mathbb{C}} \ker \sigma^2.$$

If σ^2 is injective, then $H_{X/\Sigma}$ is smooth over a closed subspace of Σ in a neighbourhood of $[Z]$.

7. The Singular Case

In this announcement of results we restricted ourselves to smooth spaces X, but all the necessary ingredients

- Description of the obstruction for embedded deformations or for deformations of complexes of modules in terms of the Atiyah class,
- Construction of a trace for perfect complexes,
- Construction of the Atiyah-Chern character for perfect complexes,

have been established by L. Illusie in arbitrary generality in the algebraic case, and the analytic case can be handled through the methods of [**Fle, OTT**].

Using this machinery, one constructs first the cotangent complex $\mathbb{L}_{X/\Sigma}$ for an arbitrary morphism $X \to \Sigma$ of analytic spaces, schemes or even ringed topoi, secondly one defines traces $\mathrm{Ext}^q(\mathcal{F}, \mathcal{F} \otimes^{\mathbb{L}} \Lambda^p\mathbb{L}_{X/\Sigma}) \to \prod_p \mathrm{H}^{p+q}(X, \Lambda^p\mathbb{L}_{X/\Sigma})$ for a perfect complex $\mathcal{F}$, which needs special care as $\Lambda^p\mathbb{L}_{X/\Sigma}$, the derived exterior powers of the cotangent complex, will in general not be perfect, and finally one defines the Atiyah class and, in characteristic zero, an Atiyah-Chern-Illusie character in $\prod_p \mathrm{Ext}^p(\mathcal{F}, \mathcal{F} \otimes^{\mathbb{L}} \Lambda^p\mathbb{L}_{X/\Sigma})$ to obtain a commutative diagram extending (4.1) to something resembling a metatheorem:

THEOREM 7.1. *Let $X \to \Sigma$ be a morphism of complex spaces or of schemes over a field of characteristic zero. If $\mathcal{F}$ is a perfect complex on X and $\mathcal{N}$ is a coherent $\mathcal{O}_X$-module then there is a commutative diagram*

$$\begin{array}{ccc}
T^1_{X/\Sigma}(\mathcal{N}) = \mathrm{Ext}^1_X(\mathbb{L}_{X/\Sigma}, \mathcal{N}) & \xrightarrow{\mathrm{ob}_{\mathcal{F}} = * \lrcorner - \mathrm{at}_{X/\Sigma}(\mathcal{F})} & \mathrm{Ext}^2_X(\mathcal{F}, \mathcal{F} \otimes^{\mathbb{L}} \mathcal{N}) \\
\downarrow {\scriptstyle * \lrcorner \mathrm{ch}_{X/\Sigma}(\mathcal{F})} & \swarrow {\scriptstyle \mathrm{tr}(e^{-\mathrm{at}_{X/\Sigma}(\mathcal{F})} \otimes^{\mathbb{L}} \mathcal{N} \smile *)} & \\
\prod_p H^{p+2}(X, \Lambda^p(\mathbb{L}_{X/Y}) \otimes^{\mathbb{L}} \mathcal{N}) & &
\end{array}$$

Moreover, if $\mathcal{F} = \mathcal{O}_Z$ for $Z \subseteq X$ a closed subspace, then the morphism on top, that is, contracting against the opposite of the Atiyah class, factors naturally through $T^2_{Z/X}(\mathcal{N}|_Z) = \mathrm{Ext}^2_Z(\mathbb{L}_{Z/X}, \mathcal{O}_Z \otimes^{\mathbb{L}} \mathcal{N})$.

References

[ALJ] B. Angéniol and M. Lejeune-Jalabert, *Calcul différentiel et classes caractéristiques en géométrie algébrique.* Travaux en Cours **38**, Hermann, Paris, 1989.

[Art] I. V. Artamkin, *On the deformation of sheaves*, Izv. Akad. Nauk SSSR Ser. Mat. **52** (1988), no. 3, 660–665, 672 (Russian); English transl., Math. USSR-Izv. **32** (1989), no. 3, 663–668.

[At] M. F. Atiyah, *Complex analytic connections in fibre bundles*, Trans. Amer. Math. Soc. **85** (1957), 181–207.

[Blo] S. Bloch, *Semi-regularity and de Rham cohomology.* Invent. Math. **17** (1972), 51–66.

[ALG] N. Bourbaki, *Éléments de mathématique. Algèbre. Chapitres* 1 *à* 3 Hermann, Paris 1970

[BFl] R.-O. Buchweitz and H. Flenner, *A semiregularity map for modules and applications to deformations* (preprint).

[Del] P. Deligne, *Théorème de Lefschetz et critères de dégénérescence de suites spectrales*, Inst. Hautes Études Sci. Publ. Math. **35** (1968), 107–126.

[FMa] B. Fantechi and M. Manetti, *On the T^1-lifting theorem*, J. Algebraic Geom. **8** (1999), no. 1, 31–39.

[Fle] H. Flenner, *Deformationen holomorpher Abbildungen*, Habilitationsschrift, Osnabrück, 1978.

[Gre] M. L. Green, *Infinitesimal methods in Hodge theory*, Algebraic Cycles and Hodge Theory (Torino, 1993), (A. Albano and F. Bardelli, eds.), Lecture Notes in Math., vol 1594, Springer-Verlag, Berlin-New York, 1994, pp. 1–92.

[Gr1] A. Grothendieck, *La théorie des classes de Chern.* Bull. Soc. Math. France **86** (1958), 137–154.

[FGA] ———, *Fondements de la géométrie algébrique*, Secrétariat mathématique, Paris, 1962.

[Gr2] ———, *On the de Rham cohomology of algebraic varieties.* Inst. Hautes Études Sci. Publ. Math. **29** (1966), 95–103.

[Ill] L. Illusie, *Complexe cotangent et déformations.* I, Lecture Notes in Math., vol. 239, Springer Verlag, Berlin-Heidelberg-New York 1971.

[Kaw1] Y. Kawamata, *Unobstructed deformations. A remark on a paper of Z. Ran: "Deformations of manifolds with torsion or negative canonical bundle"* [J. Algebraic Geom. **1** (1992), no. 2, 279–291] J. Algebraic Geom. **1** (1992), 183–190; Erratum, **6** (1997), 803–804.

[Kaw2] ———, *Unobstructed deformations.* II, J. Algebraic Geom. **4** (1995), 277–279.

[KSp] K. Kodaira and D. C. Spencer, *A theorem of completeness of characteristic systems of complete continuous systems.* Amer. J. Math. **81** (1959), 477–500.

[Muk] S. Mukai, *Symplectic structure of the moduli space of sheaves on an Abelian or $K3$ surface.* Invent. Math. **77** (1984), 101–116.

[Mur] J. -P.Murre, *Algebraic cycles and algebraic aspects of cohomology and K-theory*, Algebraic Cycles and Hodge Theory (Torino, 1993) (A. Albano and F. Bardelli, eds.), Lecture Notes in Math., vol. 1594, Springer-Verlag, Berlin-New York, 1994, pp. 93–152.

[OTT] N. R. O'Brian, D. Toledo, D., and Y. L. L. Tong, *The trace map and characteristic classes for coherent sheaves.* Amer. J. Math. **103** (1981), no. 2, 225–252.

[Pal] V. P. Palamodov, *Deformations of complex spaces*, Several Complex Variables. IV. Algebraic Aspects of Complex Analysis, Encyclopaedia Math. Sci., vol. 10, Springer-Verlag, Berlin-New York, 1990, pp. 105–194.

[Qui] D. Quillen, *On the (co-)homology of rings.* Proc. Symp. in Pure Math., vol. 17, Amer. Math. Soc. Providence, RI, 1970, pp. 65–87.

[Ran1] Z. Ran, *Deformations of manifolds with torsion or negative canonical bundle*, J. Algebraic Geom. **1** (1992), 183–190.

[Ran2] ———, *Hodge theory and the Hilbert scheme*, J. Differential Geom. **37** (1993), no. 1, 191–198.

[Ran3] ———, *Hodge theory and deformations of maps*, Compositio Math. **97** (1995), no. 3, 309–328.

[Ran4] ———, *Semiregularity, obstructions and deformations of Hodge classes*, 1998 (preprint).

[Rim] D. S. Rim, *Formal deformation theory*, Groupes de monodromie en géométrie algébrique. I, Lecture Notes in Math., vol. 288, Springer-Verlag, Berlin-New York, 1972, pp. 32–132.

[Sev] F. Severi, *Sul teorema fondamentale dei sistemi continui di curve sopra una superficie algebrica*, Ann. Mat. Pura Appl. **23** (1944), no. 4, 149–181.

[SGA6] P. Berthelot, A. Grothendieck, and L. Illusie (eds.), *Théorie des intersections et théorème de Riemann-Roch*, Lecture Notes in Math. vol. 225, Springer-Verlag, Berlin-New York, 1971.

[Ste] J. H. M. Steenbrink, *Some remarks about the Hodge conjecture*, Hodge Theory (Sant Cugat, 1985), Lecture Notes in Math., vol. 1246, Springer-Verlag, Berlin-New York, 1987, pp. 165–175.

[Voi] C. Voisin, *Transcendental methods in the study of algebraic cycles*, Algebraic Cycles and Hodge Theory (Torino, 1993) (A. Albano and F. Bardelli, eds.), Lecture Notes in Math., vol. 1594, Springer, Berlin, 1994, pp. 153–222.

Department of Mathematics, University of Toronto, Toronto, ON M5S 3G3, Canada
E-mail address: ragnar@math.utoronto.ca

Fakultät für Mathematik, Ruhr-Universität Bochum, Geb. NA 2/72, 44781 Bochum, Germany
E-mail address: Hubert.Flenner@ruhr-uni-bochum.de

Centre de Recherches Mathématiques
CRM Proceedings and Lecture Notes
Volume **24**, 2000

Regulators and Characteristic Classes of Flat Bundles

Johan Dupont, Richard Hain, and Steven Zucker

ABSTRACT. In this paper, we prove that on any non-singular algebraic variety, the characteristic classes of Cheeger-Simons and Beilinson agree whenever they can be interpreted as elements of the same group (e.g., for flat bundles). In the universal case, where the base is $B\,\mathrm{GL}(\mathbb{C})^{\delta}$, we show that the universal Cheeger-Simons class is half the Borel regulator element. We were unable to prove that the universal Beilinson class and the universal Cheeger-Simons classes agree in this universal case, but conjecture they do agree.

CONTENTS

1. Introduction 48
2. Universal Connections 52
3. The Cheeger-Simons Chern Class 55
4. Deligne-Beilinson Cohomology 60
5. Proof of Theorem 1.3 70
6. Towards a Proof of Conjecture 1.4 76
Appendix A. Fiber Integration and the Homotopy Formula 79
Appendix B. Universal Weak F^1-Connections 80
Appendix C. Simplicial de Rham Theory 81
Appendix D. Additional Techniques 89
References 91

1991 *Mathematics Subject Classification.* Primary 14C30, 14F05.

Received by the editors November 9, 1998; Accepted May 23, 1999.

First-named author supported in part by grants from the Aarhus Universitets Forskningsfond, Statens Naturvidenskabelige Forskningsraad, and the Paul and Gabriella Rosenbaum Foundation.

Second-named author supported in part by the National Science Foundation under grants DMS-8601530 and DMS-8901608.

Third-named author supported in part by the National Science Foundation under grants DMS-8800355 and DMS-9102233.

This is the final form of the paper.

1. Introduction

For each complex algebraic variety X, Deligne and Beilinson [**2**] have defined cohomology groups

$$H^k_{\mathcal{D}}(X, \mathbb{Z}(p)) \quad k, p \in \mathbb{N}$$

which map to the usual singular cohomology groups:

$$H^k_{\mathcal{D}}(X, \mathbb{Z}(p)) \to H^k(X, \mathbb{Z}(p)). \tag{1.1}$$

(Here, $\mathbb{Z}(p)$ denotes the subgroup $(2\pi i)^p\mathbb{Z}$ of $\mathbb{C}$.) Beilinson [**2**] and Gillet [**20**] have constructed Chern classes for algebraic vector bundles $E \to X$

$$c_p^B(E) \in H^{2p}_{\mathcal{D}}(X, \mathbb{Z}(p))$$

whose images under (1.1) are the usual Chern classes.

In a different direction, Cheeger and Simons [**9**] have defined characteristic classes

$$\hat{c}_p(E, \nabla) \in H^{2p-1}(M, \mathbb{C}/\mathbb{Z}(p))$$

for vector bundles $E \to M$ endowed with a flat connection ∇ over a smooth manifold. Flat complex vector bundles over an algebraic manifold X can be made algebraic; we always want to take, as algebraic structure on the bundle, the unique one with respect to which the connection has regular singularities [**12**, p. 98].[1] We can then associate to such bundles the pair of characteristic classes

$$\hat{c}_p(E, \nabla) \in H^{2p-1}(X, \mathbb{C}/\mathbb{Z}(p)) \quad \text{and} \quad c_p^B \in H^{2p}_{\mathcal{D}}(X, \mathbb{Z}(p)).$$

There is a natural homomorphism

$$H^{2p-1}(X, \mathbb{C}/\mathbb{Z}(p)) \to H^{2p}_{\mathcal{D}}(X, \mathbb{Z}(p)). \tag{1.2}$$

Our first result is:

THEOREM 1.1. *If $E \to X$ is a flat algebraic vector bundle with regular singularities over the algebraic manifold X, then $c_p^B(E)$ is the image of $\hat{c}_p(E, \nabla)$ under* (1.2) *for all $p \geq 1$.*

Previously known cases of this result are due to Bloch [**3**] (flat bundles with unitary monodromy over a smooth projective base) and Soulé [**33**] (arbitrary flat bundles over a smooth projective base). In [**5**], there is a proof of the quasi-projective case of Theorem 1.2 below that is similar to ours (in Section 4.5) and which invokes our Proposition 6.4.

Since both characteristic classes are functorial, the proof of Theorem 1.1 would be rather straightforward if, given $E \to X$ as in the theorem, one could find a morphism of X into some Grassmannian $f\colon X \to G(n, \mathbb{C}^m)$ and a connection ∇ on the universal n-plane bundle over $G(n, \mathbb{C}^m)$ that classified *both* the bundle E and its flat connection. However, it is easy to see that this is impossible in general; the context of flat vector bundles is just too rigid. For example, if X were compact and E had nontrivial monodromy, then f could not be constant, implying that $c_1(E)$ is non-zero in $H^2(X, \mathbb{C})$, which contradicts the flatness of E. As we shall see, there are ways to evade this problem.

We first state a generalization of Theorem 1.1 that goes beyond the case of flat vector bundles, providing a sort of Chern-Weil theory for DB-cohomolgy for

[1] If E is given *a priori* as an algebraic bundle on X with flat connection, it is possible that the two algebraic structures fail to coincide. See the paragraphs following Remark 4.13.

algebraic manifolds. This applies to algebraic vector bundles with what we call an F^1-connection—see Definition 4.12; flat connections with regular singularities at infinity are examples of F^1-connections. In order to state the result precisely, one has to first write $X = \overline{X} - D$, where $\overline{X}$ is compact and D is a normal crossings divisor. Then one has to define appropriate subgroups

$$\widehat{H}^{2p}(\overline{X} \log D, \mathbb{Z}(p))$$

of $\widehat{H}^{2p}(X, \mathbb{Z}(p))$ and show that there is a natural map

$$\widehat{H}^{2p}(\overline{X} \log D, \mathbb{Z}(p)) \to H^{2p}_{\mathcal{D}}(X, \mathbb{Z}(p)) \tag{1.3}$$

and that the Cheeger-Simons Chern classes bundles E over X with an F^1-connection lie in this subgroup. This is done in Section 4.4.

THEOREM 1.2. *If $E \to X$ is an algebraic vector bundle with F^1-connection over the algebraic manifold X, then $c^B_p(E)$ is the image of $\hat{c}_p(E, \nabla)$ under* (1.3) *for all $p \geq 1$.*

We first observe that a necessary condition for Theorem 1.2 is that the image of $\hat{c}_p(E, \nabla)$ in $H^{2p}_{\mathcal{D}}(X, \mathbb{Z}(p))$ is the same for all F^1-connections ∇ on E. This is verified *a priori*,—though it appears later in our exposition (as Proposition 6.4)—and it greatly simplifies the task.

The proof of Theorem 1.2 in the quasi-projective case, given in Section 4.5, goes as follows. For any algebraic extension $\overline{E}$ of E to a compactification $\overline{X}$ of X, there exists an ample line bundle L on $\overline{X}$ such that both $\overline{E} \otimes L$ and L are generated by global sections, hence, *are* pullbacks of universal bundles by regular maps. Since the conclusion of Theorem 1.2 is a tautology for the universal bundles, we get the result by functoriality for $\overline{E} \otimes L$ and L; this implies the same for $\overline{E}$ itself, hence for E.[2] The reduction from algebraic manifolds to the quasi-projective case is fairly standard; see our Section 4.6.

We also formulate and prove the analogous result in the Kähler case in Section 4.7. In order to effect this, one has to make distinctions among the various meromorphic equivalence classes of compactifications, and then of vector bundle extensions. In the case that X is algbraic, we have been tacitly using the obvious choices, viz. the algebraic ones (cf. Remark 4.11).

From a different point of view, a trick [**16**] (see Proposition 6.5) can be used to construct a model (depending on $E \to X$ and its connection) of the universal n-plane bundle $E_n \to B\,\mathrm{GL}_n(\mathbb{C})$ in the category of *simplicial* varieties, a connection on this bundle, and a morphism of simplicial varieties $X \to B\,\mathrm{GL}_n(\mathbb{C})$ that simultaneously classifies the bundle and the connection. For this reason, it is both natural and tempting to work in the category of simplicial varieties; Cheeger-Simons classes, DB-cohomology and Beilinson Chern classes all extend directly to the simplicial setting (cf. Appendix C). We would then want to view Theorem 1.2 as a special case of its simplicial analogue. Unfortunately, there remain difficulties in trying to carry out this approach; see Section 6. An abbreviated account of the above is given in [**39**].

[2] This is, in essence, the argument in [**33**], where $\overline{X} = X$; see also our crucial Proposition 6.4. Note that this argument cannot be used in the complex analytic setting, not even for compact manifolds.

Another reason for working within the simplicial category is that we can work with the "universal case" of a flat bundle. Denote the general linear group of complex $n \times n$ matrices, *endowed with the discrete topology*, by $\mathrm{GL}_n(\mathbb{C})^\delta$. Its classifying space $B\,\mathrm{GL}_n(\mathbb{C})^\delta$ classifies flat complex vector bundles of rank n. The Cheeger-Simons classes of the universal flat bundle

$$\hat{c}_p \in H^{2p-1}(B\,\mathrm{GL}_n(\mathbb{C})^\delta, \mathbb{C}/\mathbb{Z}(p))$$

are the universal Cheeger-Simons classes.

To generalize a classical theorem of Dirichlet, Borel [**4**] defined canonical cohomology classes

$$b_p \in H^{2p-1}(B\,\mathrm{GL}(\mathbb{C})^\delta, \mathbb{C}/\mathbb{R}(p))$$

which he used to define "regulators" $r_p\colon K_{2p-1}(\mathbb{C}) \to \mathbb{C}/\mathbb{R}(p)$ from Quillen's algebraic K-theory of $\mathbb{C}$ into $\mathbb{C}/\mathbb{R}(p) \cong H^1_{\mathcal{D}}(\operatorname{Spec}\mathbb{C}, \mathbb{R}(p))$. Denote by x_p the class in $H^{2p-1}(\mathrm{GL}_n(\mathbb{C}), \mathbb{Z}(p))$ that transgresses to the universal Chern class c_p which belongs to $H^{2p}(B\,\mathrm{GL}_n(\mathbb{C}), \mathbb{Z}(p))$. For each n, the cohomology class b_p restricts to the image in

$$H^{2p-1}(\mathrm{GL}_n(\mathbb{C})^\delta, \mathbb{C}/\mathbb{R}(p)) \cong H^{2p-1}(B\,\mathrm{GL}_n(\mathbb{C})^\delta, \mathbb{C}/\mathbb{R}(p))$$

of the element ξ_p of $H^{2p-1}_{\mathrm{cts}}(\mathrm{GL}_n(\mathbb{C}), \mathbb{C}/\mathbb{R}(p))$ that corresponds to the element x_p of $H^{2p-1}(\mathrm{GL}_n(\mathbb{C}), \mathbb{R}(p))$, under the canonical isomorphism

$$H^{2p-1}(\mathrm{GL}_n(\mathbb{C}), \mathbb{R}(p)) \xrightarrow{\cong} H^{2p-1}_{\mathrm{cts}}(\mathrm{GL}_n(\mathbb{C}), \mathbb{C}/\mathbb{R}(p)).$$

This construction and the relevant background is reviewed in Sections 5.1 and 5.2. Our second main result, proved in Section 5 is:

THEOREM 1.3. *The image of* $\hat{c}_p$ *in* $H^{2p-1}(B\,\mathrm{GL}(\mathbb{C})^\delta, \mathbb{C}/\mathbb{R}(p))$ *is* $b_p/2$.

Since $B\,\mathrm{GL}_n(\mathbb{C})^\delta$ is a simplicial variety, we also have the Beilinson Chern classes of the universal flat bundle:

$$c_p^B \in H^{2p}_{\mathcal{D}}(B\,\mathrm{GL}_n(\mathbb{C})^\delta, \mathbb{Z}(p)) \cong H^{2p-1}(B\,\mathrm{GL}_n(\mathbb{C})^\delta, \mathbb{C}/\mathbb{Z}(p)).$$

The homomorphism $\mathbb{C}/\mathbb{Z}(p) \to \mathbb{C}/\mathbb{R}(p)$ gives rise to a commutative square:

$$\begin{array}{ccc} H^{2p}_{\mathcal{D}}(B\,\mathrm{GL}(\mathbb{C})^\delta, \mathbb{Z}(p)) & \xrightarrow{\cong} & H^{2p-1}(B\,\mathrm{GL}(\mathbb{C})^\delta, \mathbb{C}/\mathbb{Z}(p)) \\ \downarrow & & \downarrow \\ H^{2p}_{\mathcal{D}}(B\,\mathrm{GL}(\mathbb{C})^\delta, \mathbb{R}(p)) & \xrightarrow{\cong} & H^{2p-1}(B\,\mathrm{GL}(\mathbb{C})^\delta, \mathbb{C}/\mathbb{R}(p)) \end{array}$$

We had hoped to prove that

$$c_p^B = \hat{c}_p \text{ in } H^{2p-1}(B\,\mathrm{GL}_n(\mathbb{C})^\delta, \mathbb{C}/\mathbb{Z}(p)).$$

as a case of the simplicial version of Theorem 1.2, which is stated as Statement 6.1. Our approach in trying to prove that, and some of the difficulties we encountered in our attempt, are discussed in Section 6. Thus, we are obliged to state the preceding as a conjecture:

CONJECTURE 1.4. *For all* $p \geq 1$ *and* $n \in \mathbb{N} \cup \{\infty\}$,

$$c_p^B = \hat{c}_p \in H^{2p-1}(B\,\mathrm{GL}_n(\mathbb{C})^\delta, \mathbb{C}/\mathbb{Z}(p)).$$

This conjecture, if true, would imply that the universal Beilinson Chern class for flat bundles is represented by half the Borel regulator element. Combining Theorem 1.3 with the conjecture yields the following conjectural refinement of Beilinson's result [**2**, A5.3] (see also the article by Rapoport in [**29**]) which asserts that there is a non-zero rational constant λ such that $b_p \equiv \lambda c_p^B$ modulo the decomposable elements

$$\left[H^{+}(\mathrm{GL}_n(\mathbb{C})^{\delta}, \mathbb{R}) \cdot H^{+}(\mathrm{GL}_n(\mathbb{C})^{\delta}, \mathbb{R})\right] \otimes \mathbb{C}/\mathbb{R}(p) \subseteq H^{\bullet}(\mathrm{GL}(\mathbb{C})^{\delta}, \mathbb{C}/\mathbb{R}(p)).$$

He used this to prove that his regulators agreed with those of Borel up to a non-zero rational constant. Theorem 1.3 yields as a corollary the following strengthening of Beilinson's result .

PROPOSITION 1.5. *If Conjecture* 1.4 *is true, then the Borel regulator is two times the Beilinson regulator* $K_{2p-1}(\mathbb{C}) \to \mathbb{C}/\mathbb{R}(p)$. *Consequently, for all number fields* F*, the Borel regulator and twice the Beilinson regulator*

$$K_{2p-1}(F) \to H^1_{\mathcal{D}}(\operatorname{Spec} F, \mathbb{R}(p)) \cong [\mathbb{C}/\mathbb{R}(p)]^{d_p}$$

are equal. Here

$$d_p = \begin{cases} r_1 + r_2 & p \text{ odd}, \\ r_2 & p \text{ even}, \end{cases}$$

where r_1 *is the number of real embeddings of* F*, and* r_2 *is the number of conjugate pairs of complex embeddings.*

We wish to stress that even if we considered only flat bundles in our theorems, it would be necessary to consider Cheeger-Simons classes of bundles with arbitrary connections [**9**], as we make essential use of the universal n-plane bundle which does not admit a flat connection. By definition, these classes take values in the ring of differential characters. We give a detailed account of Cheeger-Simons classes and differential characters in Section 3, as we could not find in the literature an exposition suitable for our purposes. One of the original approaches to Cheeger-Simons classes uses the construction of a universal connection by Narasimhan-Ramanan [**28**]. In Section 2 we give an alternative and more canonical approach to universal connections. In Section 4, we review Deligne-Beilinson cohomology, and prove Theorem 1.2. Next, Theorem 1.3 is proved in Section 5. Finally, we treat the issues and traps involved in our attempt to prove Conjecture 1.4 in Section 6.

We would like to thank the numerous mathematicians with whom we have had useful discussions related to this paper, especially Pierre Deligne, Hélène Esnault, Bernard Shiffman, Vyacheslav Shokurov and Christophe Soulé. We are particularly grateful to the referee for his or her patience over our reluctance to recognize the relevance of the distinctions now made in Sections 4.3 and 4.4, even though this was known to the authors.

Hain and Zucker would also like to thank the Max-Planck-Institut für Mathematik for its hospitality and support during the fall of 1987. Hain would also like to thank the Mathematics Institute at Århus for its support during several visits.

1.1. Conventions. To make all Chern classes compatible with those used in algebraic geometry, we introduce the algebraic geometers' Tate twist. Denote by $\mathbb{Z}(p)$ the subgroup of $\mathbb{C}$ generated by $(2\pi i)^p$. For each subgroup Λ of $\mathbb{C}$, set

$$\Lambda(p) = \Lambda \otimes_{\mathbb{Z}} \mathbb{Z}(p).$$

The isomorphism $\mathbb{Z} \to \mathbb{Z}(p)$ that takes 1 to $(2\pi i)^p$ induces a canonical isomorphism

$$H^\bullet(X, \mathbb{Z}) \to H^\bullet(X, \mathbb{Z}(p)). \tag{1.4}$$

In this paper, the pth Chern class of a $\mathrm{GL}_n(\mathbb{C})$ (equivalently, a complex n-plane) bundle over X is the element of $H^{2p}(X, \mathbb{Z}(p))$ which is the image under (1.4) of the usual topological Chern class as defined, for example, in [**27**].

Let

$$C_k \colon \mathfrak{gl}_n(\mathbb{C}) \to \mathbb{C} \quad k = 0, \dots, n$$

be the $\mathrm{GL}_n(\mathbb{C})$-invariant polynomials uniquely determined by

$$\det(tI - A) = \sum_{k=0}^{n} C_k(A) t^{n-k}.$$

Explicitly,

$$C_k(A) = (-1)^k \operatorname{Tr} \wedge^k A.$$

If $E \to M$ is a complex vector bundle with connection ∇ and curvature Θ, then it is well known (see [**21**, p. 403]) that $C_k(\Theta)$ is a closed form that represents $c_k(E)$ in $H^{2k}(M, \mathbb{C})$.

All varieties in this paper are defined over the field $\mathbb{C}$ of complex numbers. We will work in the complex topology unless we explicitly say otherwise. The complex of smooth, $\mathbb{C}$-valued forms on a manifold X will be denoted by $A^\bullet(X)$.

2. Universal Connections

2.1. Classical Formulation. A construction of universal connections was first given in [**28**], where the universal bundle U on a Grassmannian $G_{\mathbb{C}}(n)$ (or equivalently, its frame bundle) is endowed with a connection ∇^U, such that any vector bundle with a unitary connection (E, ∇) on the manifold M is isomorphic to a pullback of (U, ∇^U) via a classifying mapping $g \colon M \to G_{\mathbb{C}}(n)$. More precisely:

THEOREM 2.1 ([**28**, I§4]). *Fix positive integers m and n. Then there is a positive integer ℓ such that any complex n-plane bundle with a unitary connection (E, ∇) on an m-dimensional manifold M is the pullback of the universal one (U, ∇^U) on the Grassmannian $G(n, \mathbb{C}^\ell)$ of n-planes in $\mathbb{C}^\ell$. Moreover, the universal connections are compatible with the inclusions $G(n, \mathbb{C}^\ell) \hookrightarrow G(n, \mathbb{C}^{\ell+1})$.*

REMARK 2.2. The integer ℓ is given explicitly in [**28**]. Note that it is necessarily larger than would be needed for classifying bundles without connection. Universal connections are constructed for principal bundles with arbitrary connected structure group in [**28**].

2.2. Alternate Formulation. For our own purposes, it is better to have an alternate formulation of universal connections. Let P be a principal bundle, with structure group G (not assumed to be connected), over the manifold Y. A connection on P is, by definition, a G-equivariant lifting of the tangent bundle TY of Y to TP (the tangent bundle of P). Consider, then, the diagram:

$$\begin{array}{ccccc} TP & \longrightarrow & TP/G & \xrightarrow{p} & TY \\ \downarrow & & \downarrow{\scriptstyle \pi} & & \downarrow \\ P & \longrightarrow & P/G & = & Y \end{array}$$

In fact, π is a vector bundle projection, and the left-hand square is Cartesian. Put

$$\widetilde{Y} = \{\alpha \in \mathrm{Hom}(TY, TP/G) : p \circ \alpha = \mathrm{id}_{TY}\} \tag{2.1}$$

Let $q\colon \widetilde{Y} \to Y$ denote the natural projection. The following is evident:

PROPOSITION 2.3. *The connections on P are in one-to-one correspondence with the cross-sections of q.*

Let $\mathfrak{g}$ denote the Lie algebra of G. Associated to any connection is its *connection 1-form*:

$$\omega \in A^1(P, \mathrm{Ad}(\mathfrak{g})).$$

In terms of the preceding, ω can be described as follows. Let $\widetilde{\alpha}\colon TY \to TP$ be the horizontal lift associated to α, and let $\xi \in TP$. Then one has the simple formula:

$$\omega(\xi) = \xi - \widetilde{\alpha}(p_*\xi). \tag{2.2}$$

It is convenient to describe $\widetilde{Y}$ in terms of a local trivialization of P. Thus, we replace Y by a sufficiently small open subset, *which we still call Y*. Then

$$P \cong Y \times G, \tag{2.3}$$

$$TP/G \cong TY \times (TG/G) \cong TY \times \mathfrak{g}, \tag{2.4}$$

so any α in (2.1) is determined by an element $\bar{\alpha} \in \mathrm{Hom}(TY, \mathfrak{g})$. Thus we have:

PROPOSITION 2.4. *$\widetilde{Y}$ is an affine-space bundle over Y, with fiber* $\mathrm{Hom}(T_yY, \mathfrak{g})$ *(for any $y \in Y$). In particular, $\widetilde{Y}$ is a manifold of the same homotopy type as Y.*

Given $f\colon M \to Y$, the pullback of P is, in terms of (2.3),

$$f^*P \cong M \times G,$$

so

$$T(f^*P)/G \cong TM \times \mathfrak{g}. \tag{2.5}$$

The pullback of the connection is represented by $\bar{\alpha} \circ Tf \in \mathrm{Hom}(TM, \mathfrak{g})$.

Consider next the diagram

$$\begin{array}{ccccccccc}
 & & q^*TP/G & = & q^*TP/G & = & q^*TP/G & & \\
 & & \uparrow & & & & \downarrow & & \\
T\widetilde{P} & \longrightarrow & T\widetilde{P}/G & \longrightarrow & T\widetilde{Y} & \longrightarrow & q^*TY & \longrightarrow & TY \\
\downarrow & & \downarrow & & \downarrow & & \downarrow & & \downarrow \\
\widetilde{P} & \longrightarrow & \widetilde{P}/G & = & \widetilde{Y} & = & \widetilde{Y} & \xrightarrow{\ q\ } & Y
\end{array}$$

where $\widetilde{P} = q^*P$, the pullback of P along q. By construction, $\widetilde{P}$ has a tautological connection $\widetilde{\nabla}$, given by the pullback of

$$\hat{\beta} \in \{\beta \in \mathrm{Hom}(q^*TY, q^*TP/G) : \beta \text{ projects to } \mathrm{id}_{q^*TY}\}, \tag{2.6}$$

where $\hat{\beta}$ is defined by: if $\tilde{y} \in q^{-1}(y)$, $\hat{\beta}(\tilde{y})$ is $\tilde{y}\colon T_yY \to (TP/G)_y$. When $P \cong Y \times G$, (2.6) is determined by $\bar{\beta} \in \mathrm{Hom}(q^*TY, \mathfrak{g})$ (recall (2.3) and (2.4)).

COROLLARY 2.5. *For any immersion of manifolds $f\colon M \to Y$ and connection ∇ on f^*P, there is a lifting $\tilde{f}\colon M \to \widetilde{Y}$ such that $\tilde{f}^*\widetilde{\nabla} = \nabla$.*

PROOF. By Proposition 2.3, the connection ∇ corresponds to a cross-section σ of

$$\begin{array}{ccl} \widetilde{M} & = & \{\mu \in \operatorname{Hom}(TM, f^*TP/G) : f^*p \circ \mu = \mathrm{id}_{TM}\} \\ \downarrow & & \\ M & & \end{array}$$

Since f is, by hypothesis, an immersion, the bundle mapping

$$Tf\colon TM \to f^*TY$$

is an injection. It follows that the natural mapping

$$\begin{array}{ccl} f^*\widetilde{Y} & = & \widetilde{Y} \times_Y M \subset \operatorname{Hom}(f^*TY, f^*TP/G) \\ \downarrow{\scriptstyle r} & & \\ \widetilde{M} & & \end{array}$$

is surjective (it is induced by the dual of Tf). Any splitting of Tf determines a section s of r. Pick one, and take $\tilde{f} = j \circ s \circ \sigma$ (see diagram below).

$$\begin{array}{ccccc} \widetilde{M} & \xleftarrow{\;r\;} & f^*\widetilde{Y} & \xrightarrow{\;j\;} & \widetilde{Y} \\ \downarrow & & \uparrow{\scriptstyle \tilde{f}} & & \downarrow \\ M & = & M & \xrightarrow[f]{} & Y \end{array}$$

$\square$

Taking Y to be $G_{\mathbb{C}}(n)$, and $G = \mathrm{GL}_n(\mathbb{C})$, we obtain:

PROPOSITION 2.6. *The universal bundle $\widetilde{U} = q^*U$ on $\widetilde{G}_{\mathbb{C}}(n)$, with its tautological connection $\widetilde{\nabla}$, is universal for connections on $\mathrm{GL}_n(\mathbb{C})$-bundles.*

PROOF. Any vector bundle of rank n on a manifold can be classified by an immersion into $G_{\mathbb{C}}(n)$. Now apply Corollary 2.5. $\square$

From the above interpretation of universal connections, one gets, for free, the following useful fact.

COROLLARY 2.7. *If $\tilde{g}_0, \tilde{g}_1\colon M \to \widetilde{G}_{\mathbb{C}}(n)$ are two immersions which classify the bundle with connection (E, ∇). Then there is a piecewise-smooth homotopy*

$$\hat{h}\colon I \times M \to \widetilde{G}_{\mathbb{C}}(n).$$

from g_0 to g_1 such that $\hat{h}^*\widetilde{\nabla}$ is the "constant" connection $d_t \oplus \nabla = p^*\nabla$ on p^*E ($p\colon I \times M \to M$ being the projection).

PROOF. Let $g_j\colon M \to G_{\mathbb{C}}(n)$ be the projection $q \circ \tilde{g}_j$ of $\tilde{g}_j$. As g_0 and g_1 are necessarily homotopic, pick any homotopy k from g_0 to g_1; we may take $k\colon I \times M \to G_{\mathbb{C}}(n)$ to be an immersion. Let $\tilde{k}\colon I \times M \to \widetilde{G}_{\mathbb{C}}(n)$ be a lifting that classifies $p^*(E, \nabla)$. This is, of course, a homotopy between its ends, $\tilde{k}_0$ and $\tilde{k}_1$. By construction,

$$q \circ \tilde{k}_j = q \circ \tilde{g}_j \quad (j = 0, 1),$$

so $\tilde{k}_j$ and $\tilde{g}_j$ can be connected linearly in the affine-space bundle $\widetilde{G}_{\mathbb{C}}(n)$. These homotopies also classify the constant connection. By combining them with $\tilde{k}$, we obtain $\hat{h}$ as desired. □

3. The Cheeger-Simons Chern Class

In this section we recall and elaborate on the definition and properties of the Cheeger-Simons invariant, which can be interpreted as a Chern classes in the ring of differential characters, associated to a vector bundle with connection on a manifold [**9**].

3.1. Generalities. Let M be a C^∞ manifold, $S_\bullet(M)$ the complex of C^∞ singular chains on M with integer coefficients, and $Z_\bullet(M)$ the subgroup of cycles. For any Abelian group Λ, one has

$$S^\bullet(M,\Lambda) = \operatorname{Hom}_{\mathbb{Z}}(S_\bullet(M),\Lambda), \tag{3.1}$$

the corresponding complex of Λ-valued smooth singular cochains, whose coboundary operator we shall denote by δ.

Let $A^\bullet(M)$ denote the complex of C^∞ differential forms on M with complex coefficients (one could use real coefficients here as well). An element $w \in A^k(M)$ determines a cochain

$$c_w \in S^k(M,\mathbb{C})$$

by the formula

$$c_w(\sigma) = \int_{\Delta^k} \sigma^* w \tag{3.2}$$

whenever $\sigma\colon \Delta^k \to M$ is a C^∞ singular simplex. This embeds $A^\bullet(M)$ as a subcomplex of $S^\bullet(M,\mathbb{C})$. For any subgroup Λ of $\mathbb{C}$, we then get a morphism of complexes.

$$\iota_\Lambda\colon A^\bullet(M) \to S^\bullet(M,\mathbb{C}/\Lambda), \tag{3.3}$$

which is an injection whenever Λ is totally disconnected.

3.2. Differential Characters. Following [**9**], one makes the following:

DEFINITION 3.1. The group of mod Λ *differential characters* of degree k on M is the group

$$\widehat{H}^k(M,\Lambda) = \{f,\alpha) \in \operatorname{Hom}_{\mathbb{Z}}(Z_{k-1}(M),\mathbb{C}/\Lambda) \oplus A^k(M) : \delta f = \iota_\Lambda \alpha, \text{ and } d\alpha = 0\}.$$

One sees that $\widehat{H}^\bullet(M,\Lambda)$ is a contravariant functor of M; when Λ is a ring there is a functorial product [**9**, (1.11)] that imparts a ring structure to the differential characters.

REMARK 3.2. (i) Though [**9**] would have us writing $\widehat{H}^k(M,\mathbb{C}/\Lambda)$ in Definition 3.1, we have introduced the above change of notation for the sake of consistency with that to be used in §5. We have also taken the liberty of shifting by one the degree of a differential character from that of [**9**], so that it becomes compatible with other Chern classes. This also makes the ring structure a graded one.

(ii) If Λ is totally disconnected, then α above is uniquely determined, (3.3), and the second condition of Definition 3.1 is a consequence of the first.

It is useful to understand Definition 3.1 in terms of conventional homological algebra. We have:

3.2.1.

(i) If $\bar{f} \in S^{k-1}(M, \mathbb{C}/\Lambda)$ is any $\mathbb{Z}$-linear extension of f to $S_{k-1}(M)$, then $\delta\bar{f} = \iota_\Lambda \alpha$; in particular, if $\iota(\Lambda)\alpha = 0$, then $\bar{f}$ is a cocycle.

(ii) If $\bar{f}_1$ and $\bar{f}_2$ are two extensions as in (i), then $\bar{f}_2 - \bar{f}_1$ is a $(\mathbb{C}/\Lambda)$-cocycle vanishing on $Z_{k-1}(M)$ which, by a simple argument using the fact that $C_{k-1}(M) = Z_{k-1}(M) \oplus F$, where F is free, implies that the class of $\bar{f}_2 - \bar{f}_1$ vanishes in $H^{k-1}(M, \mathbb{C}/\Lambda)$. We can therefore view $\widehat{H}^k(M, \Lambda)$ as a subgroup of

$$S^{k-1}(M, \mathbb{C}/\Lambda)/\delta S^{k-2}(M, \mathbb{C}/\Lambda)$$

when Λ is totally disconnected.

(iii) The form α, representing zero in $H^k(M, \mathbb{C}/\Lambda)$, has its periods (on $Z_k(M)$) in Λ.

From this, one obtains:

PROPOSITION 3.3. (i) *There is a canonical and functorial exact sequence*

$$0 \to H^{k-1}(M, \mathbb{C}/\Lambda) \to \widehat{H}^k(M, \Lambda) \to A^k_{cl}(M, \Lambda) \to 0,$$

where

$$A^k_{cl}(M, \Lambda) = \{\varphi \in A^k(M) : d\varphi = 0, \text{ and all periods of } \varphi \text{ on } Z_k(M) \text{ lie in } \Lambda\}.$$

(ii) (cf. [**17**, §4])

$$\begin{aligned}\widehat{H}^k(M, \Lambda) &\cong H^k(\operatorname{cone}\{A^{\geq k}(M) \xrightarrow{\iota_\Lambda} S^\bullet(M, \mathbb{C}/\Lambda)\}[-1]) \\ &= \mathbb{H}^{k-1}(M, \operatorname{cone}\{\mathcal{A}^{\geq k}_M \to \mathcal{S}^\bullet_M(\mathbb{C}/\Lambda)\});\end{aligned}$$

here, $\mathcal{A}^\bullet_M$ and $\mathcal{S}^\bullet_M(\mathbb{C}/\Lambda)$ denote the sheaves of C^∞ forms and singular $(\mathbb{C}/\Lambda)$-cochains respectively.

COROLLARY 3.4. *If $H^{k-1}(M, \mathbb{C}/\Lambda) = 0$, then*

$$\widehat{H}^k(M, \Lambda) \cong A^k_{cl}(M, \Lambda).$$

3.3. Cheeger-Simons Classes. Let (E, ∇) be a C^∞ complex vector bundle with connection on M. As in Section 1.1, the image of the Chern class

$$c_p(E) \in H^{2p}(M, \mathbb{Z}(p))$$

in $H^{2p}(M, \mathbb{C})$ is represented in de Rham cohomology by the polynomial $C_p(\Theta)$ in the curvature Θ of ∇:

$$C_p(\Theta) \in A^{2p}_{cl}(M, \mathbb{Z}(p)). \tag{3.4}$$

As such, (3.4) is functorial for pullbacks of bundles with connection. The *Cheeger-Simons invariant* of (E, ∇) is a functorial lifting of $c_p(E, \nabla)$ to

$$\hat{c}_p(E, \nabla) \in \widehat{H}^{2p}(M, \mathbb{Z}(p))$$

in Proposition 3.3(i). In terms of Proposition 3.3(ii), it become a pth Chern class with values in the group of differential characters with $\mathbb{Z}(p)$ coefficients.

REMARK 3.5. If $c_p(E, \nabla) = 0$ (e.g., if E is a flat bundle), then $\hat{c}_p(E, \nabla)$ is an element of $H^{2p-1}(M, \mathbb{C}/\mathbb{Z}(p))$.

3.4. Existence and Uniqueness of $\hat{c}_p$. The existence and uniqueness of the Cheeger-Simon invariants can be deduced from the existence of universal connections (§2). By functoriality, the invariant is completely determined by its value on the universal connection $(\widetilde{U}, \widetilde{\nabla})$ of Proposition 2.6. On the other hand, because the odd-dimensional cohomology of a Grassmannian is trivial, there is a unique lifting $\hat{c}_p(\widetilde{U}, \widetilde{\nabla})$ of its Chern form to a differential character (see Proposition 3.3(i)). Thus, at most one Cheeger-Simons invariant can be defined: if $\tilde{g}\colon M \to \widetilde{G}_{\mathbb{C}}(n)$ is a smooth mapping that classifies (E, ∇), one *must* take

$$\hat{c}_p(E, \nabla) = \tilde{g}^* \hat{c}(\widetilde{U}, \widetilde{\nabla}). \tag{3.5}$$

One sees that (3.5) actually gives a definition of $\hat{c}_p$ by checking that it is, in fact, independent of the choice of g. For this, let g_0 and g_1 be two classifying mappings, and apply Corollary 2.7 to produce a "nice" homotopy $\hat{h}$ between $\tilde{g}_0$ and $\tilde{g}_1$. Then the homotopy formula (see (A.13) in Appendix A) yields

$$\begin{aligned}(3.6)\quad \tilde{g}_1^* \hat{c}_p(\widetilde{U}, \widetilde{\nabla}) - \tilde{g}_0^* \widetilde{c}_p(\widetilde{U}, \widetilde{\nabla}) - \delta(B\hat{h}^* \hat{c}_p(\widetilde{U}, \widetilde{\nabla}))& \\ = B(\hat{h}^* \delta \hat{c}_p(\widetilde{U}, \widetilde{\nabla})) = B(\hat{h}^* c_p(\widetilde{U}, \widetilde{\nabla})) &= 0,\end{aligned}$$

for $\hat{h}^* c_p(\widetilde{U}, \widetilde{\nabla}) \in A^{2p}(I \times M)$ is pulled back from M, so is annihilated by the fiber integration B.

3.5. An Intrinsic Construction of $\hat{c}_p$. There is a more intrinsic definition of $\hat{c}_p(E, \nabla)$, one that uses only functorial constructions on bundles and connections [**9**, §4]. (See also [**11**, §3].) A truly intrinsic construction is given in Proposition 6.5; that construction uses simplicial methods.

Given integers $n \geq p \geq 1$, let V_n^p denote the Stiefel manifold of linearly independent $(n-p+1)$-tuples in $\mathbb{C}^n$. Then V_n^p has the homotopy type of its submanifold $U(n)/U(p-1)$ of unitary $(n-p+1)$-frames. From this, one sees:

PROPOSITION 3.6 (see [**35**, (25.7)]). (i)

$$\pi_i(V_n^p) = \begin{cases} 0 & \textit{for } i < 2p-1, \\ \mathbb{Z} & \textit{for } i = 2p-1. \end{cases}$$

(ii) $H_{2p-1}(V_n^p, \mathbb{Z}) \cong \mathbb{Z}$, *and is generated by the homology class of* $S^{2p-1} \cong U(p)/U(p-1)$.

From a complex vector bundle E rank n on M, one can form the associated *Stiefel bundle*

$$\pi\colon V^p(E) \to M,$$

with fiber V_n^p. (For instance, $V^1(E)$ is just the frame bundle of E.) By construction, $\pi^* E$ has $n-p+1$ tautological sections, i.e., contains a (canonically) trivial bundle of rank $n-p+1$. Therefore, the Chern classes of $\pi^* E$ are those of any complementary $(p-1)$-plane bundle W. In particular,

$$\pi^* c_p(E) = c_p(\pi^* E) = 0 \text{ in } H^{2p}(V^p(E), \mathbb{Z}(p)) \tag{3.7}$$

Given any connection ∇ on E, the Chern form $\pi^* c_p(E, \nabla)$ is thus exact. The first goal is to achieve the exactness in a functorial fashion.

3.5.1. For any W as above, let ∇^W denote the unique connection on π^*E which satisfies both:

(i) ∇^W and $\pi^*\nabla$ agree on W,
(ii) the tautological sections are flat with respect to ∇^W.

There is a functorial formula expressing the fact that the Chern forms of two connections represent the same cohomology class (see [**10**, p. 48]):

$$c_p(E,\nabla_1) - c_p(E,\nabla_0) = d\eta_p, \tag{3.8}$$

where

$$\eta_p = p\int_0^1 \mu(\omega,\Theta_t,\dots,\Theta_t)dt; \tag{3.9}$$

here, μ is the p-linear symmetric form with

$$\mu(\Theta,\dots,\Theta) = C_p(\Theta), \tag{3.10}$$

$\omega = \nabla_1 - \nabla_0$, and Θ_t is the curvature of

$$\nabla_t = \nabla_0 + t\omega, \quad t\in[0,1].$$

OBSERVATION 3.7. Let σ_0, $\sigma_1\colon M \to \widetilde{M}$ be the sections corresponding to ∇_0, ∇_1 (recall Proposition 2.3); let $h\colon I\times M \to \widetilde{M}$ classify the connection

$$\overline{\nabla}\colon d_t + (1-t)\nabla_0 + t\nabla_1.$$

Then formula (3.8) is just what comes out of the homotopy formula (A.8) when it is applied to the Chern form of the tautological connection $\widetilde{\nabla}$ on $\widetilde{M}$, or equivalently, when (A.7) is applied to the Chern form of $\overline{\nabla}$. To see this, one need only observe that the curvature Θ of $\overline{\nabla}$ is given by

$$\Theta = \Theta_t + dt\wedge\omega.$$

For $\nabla_0 = \nabla^W$ and $\nabla_1 = \pi^*\nabla$, (3.9) gives

$$\pi^*c_p(E,\nabla) - c_p(\pi^*E,\nabla^W) = d\eta^W. \tag{3.11}$$

Since

$$c_p(\pi^*E,\nabla^W) = 0 \quad \text{in } A^{2p}(V^p(E)),$$

we can rewrite (3.11) as:

$$\pi^*c_p(E,\nabla) = d\eta^W. \tag{3.12}$$

It remains to study the effect of changing W.

PROPOSITION 3.8. *There is a unique natural choice of*

$$\bar\eta(E,\nabla) \in A^{2p-1}(V^p(E))/dA^{2p-2}(V^p(E))$$

such that $d\bar\eta(E,\nabla) = \pi^*c_p(E,\nabla)$.

PROOF. Let $F \cong V^p_n$ denote a fiber of π. From Proposition 3.6 and the Leray spectral sequence for π, one deduces the exactness of:

$$0 \to H^{2p-1}(M,\mathbb{Z}) \xrightarrow{\pi^*} H^{2p-1}(V^p(E),\mathbb{Z}) \to H^{2p-1}(F,\mathbb{Z}) \xrightarrow{\delta} H^{2p}(M,\mathbb{Z}).$$

The image of one of the generators of $H^{2p-1}(F,\mathbb{Z}) \cong \mathbb{Z}$ under δ is, in fact, $c_p(E)$, which can be seen easily by considering the Serre spectral sequence of the fibration. In particular, for the universal bundle U on $G_{\mathbb{C}}(n)$

(3.13) $$H^{2p-1}(V^p(U),\mathbb{Z}) \cong H^{2p-1}(G_{\mathbb{C}}(n),\mathbb{Z}) = 0.$$

Given φ, the condition $d\eta = \varphi$ determines η up to a *closed* form. From (3.13), a closed $(2p-1)$-form on $V^p(U)$ is exact. It follows that $\bar{\eta}(\widetilde{U},\widetilde{\nabla})$ is uniquely determined. Therefore, there is at most one natural construction of $\bar{\eta}(E,\nabla)$: it must be $\tilde{g}^*\bar{\eta}(\widetilde{U},\widetilde{\nabla})$ whenever $g\colon M \to G_{\mathbb{C}}(n)$ classifies E. The proof that this is well-defined goes as in (3.6). □

As the preceding is contrary to the spirit with which we began Section 3.5, we will show directly, after all, that η^W (from (3.8)) modulo exact forms, independent of W. Let W_0 and W_1 be two such. Consider the following connection on the pullback of E to $I \times I \times M$:

(3.14) $$d_s + d_t + (1-t)[(1-s)\nabla^{W_0} + s\nabla^{W_1}] + t(\pi^*\nabla);$$

it corresponds to a mapping

$$I \times I \times M \to \widetilde{M}.$$

Let

(3.15) $$B\colon A^{2p}(I \times I \times M) \to A^{2p-2}(M)$$

denote the fiber integration, which is just the iteration of (A.1) in Appendix A; as such, we write $B = B_2 \circ B_1$. From (A.7), it follows that

(3.16) $$d(B\varphi) = (B_2\varphi_{1,t} - B_2\varphi_{0,t}) - (B_1\varphi_{1,s} - B_1\varphi_{0,s})$$

where $\varphi_{0,t}$ is the restriction of φ to $\{0\} \times I \times M$, $\varphi_{1,s}$ to $I \times \{1\} \times M$, etc. Let φ be the pth Chern form of (3.14). Then, by Observation 3.7

$$B_1\varphi_{1,s} = 0, \quad B_2\varphi_{1,t} = \eta^{W_1}, \quad B_2\varphi_{0,t} = \eta^{W_0}$$

This gives

$$d(B\varphi) = \eta^{W_1} - \eta^{W_0} + B_1\varphi_{0,s}.$$

The desired conclusion follows from the realization that $B_1\varphi_{0,s}$ is the zero form. The reason for this is that when $t = 0$, (3.14) gives the linear interpolation between ∇^{W_0} and ∇^{W_1}, so $\nabla^{W_1} - \nabla^{W_0}$ is an endomorphism of π^*E that is zero on the tautological sub-bundle (so is of rank at most $p-1$). The same holds for the curvature Θ_s. Since μ in (3.10) is induced by polarization from trace $\wedge^p$, it follows that $B_1\varphi_{0,s} = 0$.

We can now complete the direct construction of $\hat{c}_p(E,\nabla)$.

PROPOSITION 3.9. *There is a unique element* $\widehat{\eta} \in \widehat{H}^{2p}(M,\mathbb{Z}(p))$ *with* $\pi^*\widehat{\eta} = [\eta^W, \pi^*c_p(E,\nabla)]$ *(and* $\delta\widehat{\eta} = \iota_{\mathbb{Z}(p)}c_p(E,\nabla)$*).*

PROOF. The uniqueness of $\widehat{\eta}$ is easy, for it follows from the diagram

$$\begin{array}{ccccccccc}
0 & \longrightarrow & H^{2p-1}(F,\mathbb{C}/\mathbb{Z}(p)) & \longrightarrow & \widehat{H}^{2p}(F,\mathbb{Z}(p)) & & & & \\
 & & \uparrow & & \uparrow & & & & \\
0 & \rightarrow & H^{2p-1}(V^p(E),\mathbb{C}/\mathbb{Z}(p)) & \rightarrow & \widehat{H}^{2p}(V^p(E),\mathbb{Z}(p)) & \rightarrow & A^{2p}_{cl}(V^p(E),\mathbb{Z}(p)) & \rightarrow & 0 \\
 & & \uparrow & & \uparrow{\scriptstyle \pi^*} & & \uparrow & & \\
0 & \longrightarrow & H^{2p-1}(M,\mathbb{C}/\mathbb{Z}(p)) & \longrightarrow & \widehat{H}^{2p}(M,\mathbb{Z}(p)) & \longrightarrow & A^{2p}_{cl}(M,\mathbb{Z}(p)) & \longrightarrow & 0 \\
 & & \uparrow & & & & \uparrow & & \\
 & & 0 & & & & 0 & &
\end{array}$$

that π^* is injective on differential characters. Likewise, we argue that $[\eta^W]$ is in the image of π^* by showing that the restriction of η^W to the fiber F is the zero differential character, or equivalently,

$$\int_{S^{2p-1}} \eta^W \in \mathbb{Z}(p).$$

(Recall that $d\eta^W = \pi^* c_p(\nabla, E)$ vanishes on F.) A little bizarrely, we pass over what might have been a direct calculation on the space V^p_n, the partial frames in the trivial n-bundle over a point, and check it in the universal situation. There, S^{2p-1} bounds in $V^p(U)$, so we write $S^{2p-1} = \partial\Gamma$. Then $Z = \pi_*\Gamma$ is a cycle on the Grassmannian, with

$$\int_{S^{2p-1}} \eta^W = \int_\Gamma \pi^* c_p(\nabla^U) = \int_Z c_p(\nabla^U) \in \mathbb{Z}(p)$$

This completes the proof of Proposition 3.9. □

With the preceding accomplished, we now make the:

DEFINITION 3.10. The differential character $\widehat{\eta}$ above is the *Cheeger-Simons Chern class* of (E, ∇). We write $\widehat{\eta} = \hat{c}_p(E, \nabla)$.

REMARK 3.11. By Section 3.2.1(ii), the Cheeger-Simons class $\hat{c}_p$ is the choice of an element

$$y \in S^{2p-1}(M, \mathbb{C}/\mathbb{Z}(p))/\delta S^{2p-2}(M, \mathbb{C}/\mathbb{Z}(p))$$

with $\delta y = \iota_{\mathbb{Z}(p)} c_p(E, \nabla)$. In terms of the cone on $\iota_{\mathbb{Z}(p)}$ (recall Proposition 3.3(ii)), it is represented by

$$(c_p(E,\nabla), -y) \in A^{2p}(M) \oplus S^{2p-1}(M, \mathbb{C}/\mathbb{Z}(p))/\delta S^{2p-2}(M, \mathbb{C}/\mathbb{Z}(p)),$$

which is a cocycle by virtue of the calculation

$$D(c_p(E,\nabla), -y) = (-dc_p(E,\nabla), -\delta y + \iota_{\mathbb{Z}(p)} c_p(E,\nabla)) = (0,0).$$

4. Deligne-Beilinson Cohomology

In this section we recall the definition of Deligne-Beilinson cohomology, and establish the precise relation with differential characters. References on Deligne-Beilinson cohomology include [**2**] and [**18**].

4.1. Deligne-Beilinson (DB-) Cohomology. Suppose that Y is a smooth algebraic variety. By resolution of singularities [**24**], we can find a smooth completion $\overline{Y}$ of Y such that $\overline{Y} - Y$ is a normal crossings divisor D. Denote by I_p the composite

$$F^p A^\bullet(\overline{Y} \log D) \to A^\bullet Y \xrightarrow{I} S^\bullet(Y),$$

where F denotes the Hodge filtration of the de Rham complex.

DEFINITION 4.1. Suppose that Λ is a subring of $\mathbb{R}$ and that $p \in \mathbb{N}$. The *Deligne-Beilinson* (or DB-) *cohomology* $H^\bullet_{\mathcal{D}}(Y, \Lambda(p))$ of the variety Y is the cohomology of the complex

$$D^\bullet(Y, \overline{Y}; \Lambda(p)) = \text{cone}\{F^p A^\bullet(\overline{Y} \log D) \xrightarrow{I_p} S^\bullet(Y, \mathbb{C}/\Lambda(p))\}[-1].$$

As the notation suggests, the DB-cohomology of a variety Y is independent of the compactification $\overline{Y}$ chosen. This follows from standard arguments (cf. [**13**, (8.3.2)]). As DB-cohomology is constructed from a cone, we have:

PROPOSITION 4.2. *For each smooth algebraic variety Y, there is natural long exact sequence*

$$\text{(4.1)} \quad \cdots \to H^{k-1}(Y, \mathbb{C}/\Lambda(p)) \to H^k_{\mathcal{D}}(Y, \Lambda(p)) \\ \to F^p H^k(Y, \mathbb{C}) \to H^k(Y, \mathbb{C}/\Lambda(p)) \to \cdots.$$

COROLLARY 4.3. *For each variety*

(i) *For each variety Y, there is a natural homomorphism*

$$H^{k-1}(Y, \mathbb{C}/\Lambda(p)) \to H^k_{\mathcal{D}}(Y, \Lambda(p)).$$

(ii) *If $H^{2p-1}(Y, \mathbb{C}/\Lambda(p)) = 0$, there is a natural isomorphism*

$$H^{2p}_{\mathcal{D}}(Y, \Lambda(p)) \cong H^{p,p}(Y) \cap \text{image}\{H^{2p}(Y, \Lambda(p)) \to H^{2p}(Y, \mathbb{C})\}.$$

There are products defined in Deligne cohomology:

$$\text{(4.2)} \qquad H^k_{\mathcal{D}}(Y, \Lambda(p)) \otimes H^l_{\mathcal{D}}(Y, \Lambda(q)) \to H^{k+l}_{\mathcal{D}}(Y, \Lambda(p+q))$$

(see [**2**, 1.5.1]). In particular, $\bigoplus_p H^{2p}_{\mathcal{D}}(Y, \Lambda(p))$ is a ring.

Beilinson [**2**] and Gillet [**20**] have defined Chern classes

$$c^B_p(E) \in H^{2p}_{\mathcal{D}}(Y, \mathbb{Z}(p))$$

for vector bundles $E \to Y$ over a variety. The construction in [**2**, 1.7], is in the manner of [**22**]: since the splitting principle holds for DB-cohomology, the existence of Chern classes in Deligne cohomology reduces to the existence of first Chern classes

$$c_1(L) \in H^2_{\mathcal{D}}(Y, \mathbb{Z}(1))$$

for line bundles $L \to Y$.

4.2. Differential Characters and DB-Cohomology. Recall (Section 3.2) that the mod Λ differential characters of degree k on a C^∞ manifold M, $\widehat{H}^k(M,\Lambda)$, are represented by the cocycles of degree $k-1$ in cone$\{A^{\geq k}(M) \xrightarrow{\iota_\Lambda} S^\bullet(M,\mathbb{C}/\Lambda)\}$.

When X is a *complex* manifold, of the form $X = \overline{X} - D$, with D a divisor with normal crossings $\overline{X}$, we can incorporate the Hodge filtration and growth conditions on $A^\bullet(X)$ to define subgroups of $\widehat{H}^k(X,\Lambda)$. Specifically, take

$$F^p\widehat{H}^k(\overline{X}\log D,\Lambda) = H^{k-1}(\text{cone}\{F^pA^{\geq k}(\overline{X}\log D) \xrightarrow{\iota_\Lambda} S^\bullet(X,\mathbb{C}/\Lambda)\}) \tag{4.3}$$

and

$$\widehat{H}^k(\overline{X}\log D,\Lambda) = F^0\widehat{H}^k(\overline{X}\log D,\Lambda).$$

It is a subgroup of $\widehat{H}^k(X,\Lambda)$. When $\overline{X}$ is compact, there is an exact sequence:

$$0 \longrightarrow H^k(X,\mathbb{C}/\Lambda) \longrightarrow \widehat{H}^k(\overline{X}\log D,\Lambda) \longrightarrow F^pA^k_{cl}(\overline{X}\log D,\Lambda) \longrightarrow 0 \tag{4.4}$$

where $A^k_{cl}(\overline{X}\log D,\Lambda)$ is the set of closed elements of $A^\bullet(\overline{X}\log D)$ whose periods lie in Λ.

The reason for introducing (4.3) comes from its similarity to Definition 4.1. The cone in (4.3) is a subset of the cone in Definition (4.1); for $k=2p$, one obtains, for X algebraic and for any Λ, a functorial exact sequence:

$$F^pA^{2p-1}(\overline{X}\log D) \to F^p\widehat{H}^{2p}(\overline{X}\log D,\Lambda) \to H^{2p}_{\mathcal{D}}(X,\Lambda) \to 0. \tag{4.5}$$

4.3. Meromorphic Equivalence. Though it may seem premature, it is helpful to leave the setting of algebraic varieties and algebraic vector bundle for awhile.

Let X be a compactifiable complex manifold, in the sense that X admits a compactification $\overline{X}$ that is a compact manifold, and for which $D = \overline{X} - X$ is an analytic subvariety. We may then modify $\overline{X}$ by blow-ups with smooth center [**25**] to make D a divisor with normal crossings.

It becomes necessary to divide the compactifications of X into meromorphic equivalence classes.

DEFINITION 4.4. Two compactifications $\overline{X}_1$ and $\overline{X}_2$ of X are said to be *meromorphically equivalent* if there exists a compactification $\overline{X}_3$ of X and morphisms of compactifications, i.e., extensions of the identity map of X, $\overline{X}_3 \to \overline{X}_1$ and $\overline{X}_3 \to \overline{X}_2$.

The above is easily seen to be an equivalence relation.

We point out that (the underlying complex manifold of) a smooth algebraic variety can admit compactifications that are not meromorphically equivalent to the algebraic ones. A simple example of this is provided by [**6**]:

EXAMPLE 4.5. Let C be an elliptic curve. Then $X = \mathbb{C}^{p+q} \times C$ admits non-algebraic compactifications $\overline{X}_{p,q}$ ($p>0$, $0 \leq q \leq p$) with the following properties:

(i) There is a principal C-fibration $\pi\colon \overline{X}_{p,q} \to \mathbb{P}^p(\mathbb{C}) \times \mathbb{P}^q(\mathbb{C})$.
(ii) Every meromorphic function on $\overline{X}_{p,q}$ is a pullback from $\mathbb{P}^p(\mathbb{C}) \times \mathbb{P}^q(\mathbb{C})$; the algebraic dimension of $\overline{X}_{p,q}$ is less than its complex dimension, so it is not algebraic.
(iii) $\overline{X}_{p,q}$ is diffeomorphic to the product of spheres $S^{2p+1} \times S^{2q+1}$. It follows that $H^2(\overline{X}_{p,q},\mathbb{C}) = 0$, and thus $\overline{X}_{p,q}$ is not Kähler.

Furthermore, it is possible for a complex manifold to admit more than one algebraic structure; it can have inequivalent algebraic compactifications:

EXAMPLE 4.6. Let C again be an elliptic curve. The *universal vector extension* of C, as a complex manifold, can be given Hodge-theoretically as follows. Put $H_{\mathbb{Z}} = H^1(C, \mathbb{Z})$, $H = H^1(C, \mathbb{C})$, and $F = F^1H^1(C, \mathbb{C})$. There is a short exact sequence of Abelian groups:

$$0 \to F \to H/H_{\mathbb{Z}} \to H/(H_{\mathbb{Z}} + F) \to 0,$$

which is isomorphic to

$$0 \to \mathbb{C} \to (\mathbb{C}^*)^2 \to C \to 0.$$

The associated $\mathbb{P}^1$-bundle:

$$\begin{array}{ccc} \mathbb{P}^1 & \longrightarrow & \overline{X} \\ & & \downarrow \\ & & C \end{array}$$

then gives an algebraic variety $\overline{X}$ that is a compactification of $X = (\mathbb{C}^*)^2$. However, $\overline{X}$ and $\mathbb{P}^1 \times \mathbb{P}^1$ are not meromorphically (birationally) equivalent, for they have non-isomorphic function fields; only the latter one is an algebraic completion of the *algebraic* variety X.

We also recall the notion of a meromorphic mapping, from complex geometry. Let X and Y be complex analytic varieties, and $\overline{X}$, $\overline{Y}$, respective partial compactifications.

DEFINITION 4.7. A map $f \colon X \to Y$ is said to be *meromorphic* with respect to $\overline{X}$ and $\overline{Y}$ (and one writes $f \colon \overline{X} \dashrightarrow \overline{Y}$) if the closure of the graph of f in $\overline{X} \times \overline{Y}$ is a subvariety of $\overline{X} \times \overline{Y}$.

Because of the existence of resolutions of singularities for complex analytic varieties ([**25**] again), two compactifications of X are meromorphically equivalent if and only if the identity map of X defines a meromorphic map between them.

There is an obvious notion of the *extendability* of a holomorphic vector bundle E on X to $\overline{X}$. We next recall the notion of meromorphic equivalence of bundle extensions; it is presented in a somewhat different, though equivalent, manner in [**12**, p. 65ff]).

DEFINITION 4.8. Let $\overline{X}$ be a complex manifold, D a divisor on $\overline{X}$, and put $X = \overline{X} - D$. Let E be a vector bundle on X. Two extensions $\overline{E}$ and $\overline{E}'$ of E to a vector bundle on $\overline{X}$ are said to be *meromorphically equivalent* if, whenever U is a nice open neighborhood of a point of D in $\overline{X}$, on which both bundles are trivial, the change-of-frame matrix from one to the other (*a priori* holomorphic on $U \cap X$) is meromorphic (i.e., does not have essential singularities) along $U \cap D$.

REMARK 4.9. If $\mathfrak{M}$ is a meromorphic equivalence class of compactifications of X, there are corresponding notions of the $\mathfrak{M}$-extendability of E and $\mathfrak{M}$-meromorphic equivalence classes of extensions of E.

One can always regard a vector bundle as a complex manifold, forgetting the linear structure. The following is almost a tautology:

LEMMA 4.10. *With notation as above,*

(i) *Two bundle extensions $\overline{E}$ and $\overline{E}'$ are meromorphically equivalent if and only if $\overline{E}$ and $\overline{E}'$ are equivalent partial compactifications of E.*

(ii) *If* $\overline{E}$ *and* $\overline{E}'$ *are meromorphically equivalent, then the meromorphic map* $\overline{E} \dashrightarrow \overline{E}'$ *induces a meromorphic map* $\mathbb{P}(\overline{E}) \dashrightarrow \mathbb{P}(\overline{E}')$.

REMARK 4.11. When X is an algebraic manifold, there is a canonical equivalence class of compactifications of X, namely the smooth algebraic completions. Algebraic vector bundles admit algebraic extensions to suitable completions, and this class of extensions is likewise canonical. We understand algebraic extensions of algebraic vector bundles on an algebraic variety if it is not specified otherwise.

4.4. F^1-Connections. Throughout, we let $X \subseteq \overline{X}$ be as in Section 4.2, and E a vector bundle on X.

DEFINITION 4.12. An F^1*-connection* on E (relative to $\overline{X}$) is a connection for which there exists a vector bundle $\overline{E}$ over $\overline{X}$ such that

(i) the restriction of $\overline{E}$ to X is E;
(ii) the local connection forms (with respect to holomorphic frames of $\overline{E}$), and therefore also the curvature, lie in $F^1 A^\bullet(\overline{X} \log D, \mathrm{End}(\overline{E}))$.

We define a sort of "universal F^1-connection" in Appendix B. Note that Definition 4.12 coincides with Definition B.1 if and only if $X = \overline{X}$.

REMARK 4.13. This notion occurs in a well-known fact about Hermitian geometry (see [**21**, p. 73]): If E is an Hermitian vector bundle over the *compact* complex manifold X, there exists a unique F^1-connection on E with respect to which the Hermitian metric on E is horizontal. In particular, an extendable vector bundle always admits an F^1-connection (without singularities).

The following is evident:

PROPOSITION 4.14. *If* ∇ *is an* F^1*-connection on* E *(relative to* $\overline{X}$*), then the Cheeger-Simons class* $\hat{c}_p(E, \nabla)$ *is in the subgroup* $F^p \widehat{H}^{2p}(\overline{X} \log D, \Lambda)$ *of* $\widehat{H}^{2p}(X, \Lambda)$.

To be able to talk sensibly about F^1-connections, one needs to know the following:

PROPOSITION 4.15. *Let* $\overline{X}$ *be a compact complex manifold,* D *a divisor with normal crossings on* $\overline{X}$*, and put* $X = \overline{X} - D$*. Let* (E, ∇) *be a vector bundle with connection on* X*. Suppose that* $\overline{E}$ *and* $\overline{E}'$ *are vector bundles on* $\overline{X}$ *with respect to which* ∇ *is an* F^1*-connection. Then* $\overline{E}$ *and* $\overline{E}'$ *are meromorphically equivalent.*

PROOF. The issue is fairly elementary. Let U be open in $\overline{X}$. If ω is the connection form with respect to a frame of $\overline{E}$ on $U \subset \overline{X}$, and ω' is the connection form with respect to a frame of $\overline{E}'$ on U, suppose that ω and ω' have first order-poles[3] on $U \cap D$. Let A be the matrix expressing the frame for $\overline{E}'$ in terms of the one for $\overline{E}$ (thus $\omega' = A\omega A^{-1} + (dA)A^{-1}$). We wish to conclude that the entries of A must be meromorphic along $U \cap D$.

Like holomorphy, meromorphy can be detected by a curve test, i.e., a function of several variables is meromorphic if and only if its restriction to sufficiently many curves is—e.g., if it is meromorphic in each variable separately—(see [**32**]; cf. [**12**, II, 4.1.1]). Since connections are functorial, we may assume that U is the unit disc in $\mathbb{C}$, and $D = \{0\}$. In this case, the assertion is proved in [**12**, II, 1.19]. □

[3]Since we will be reducing to curves in U, there is actually no need to assume that the poles are logarithmic, for the two notions coincide on a curve.

REMARK 4.16. Note that Proposition 4.15 does not say that there exists $\overline{E}$ for which the connection form is logarithmic; nor does it say that if such $\overline{E}$ exists, the connection form has logarithmic poles for all $\overline{E}'$ meromorphically equivalent to $\overline{E}$ (counterexamples abound). Also, there is no contradiction between Remark 4.13 and Proposition 4.15.

According to [**12**, II, (5.2)], given *any* compactification $\overline{X}$ of X, a flat bundle E on X admits a vector bundle extension $\overline{E}$ to $\overline{X}$ with respect to which the flat connection can be seen to be an F^1-connection. In fact, according to Proposition 4.15 any two such $\overline{E}$ are meromorphically equivalent. We point out that $\overline{E}$ can be determined from constructions that are local on $\overline{X}$ along D. Local connection forms are then to be computed on $\overline{X}$ in terms of local frames of $\overline{E}$. In particular, there is no analogue of Proposition 4.15 for equivalence classes of compactifications.

On the other hand, if one starts with an F^1-connection on an *algebraic* vector bundle on X, relative to an algebraic compactification $\overline{X}$, it does *not* follow that the meromorphic equivalence class of extensions of E distinguished by Proposition 4.15 are the algebraic vector bundles on $\overline{X}$. However, the two are known to coincide in an important class of examples, namely the flat bundles "coming from algebraic geometry." By that, one means that the fibers of E are cohomology groups for a family of algebraic varieties over X, and ∇ is the Gauss-Manin connection. This result is commonly called *the regularity theorem in algebraic geometry.* (See [**26**] and its generalizations, e.g., [**34**, §5].)

The following should shed some light on the issue.

EXAMPLE 4.17. Let $j\colon X \hookrightarrow \overline{X}$ denote the inclusion. Given the holomorphic vector bundle E, the specification of $\overline{E}$ is equivalent to selecting a locally-free subsheaf $(\mathcal{O}_{\overline{X}}(\overline{E}))$ of $j_*\mathcal{O}_X(E)$ on $\overline{X}$.

Suppose that $\dim X = 1$ (so $\overline{X}$ is automatically algebraic). Let $\Delta \cong U \subset \overline{X}$ be a disc, with coordinate t, for which the restriction $j|_U$ is $\Delta^* \hookrightarrow \Delta$. Also, take E to be a line bundle. Restricting to U, we consider the two inequivalent extensions of E, $\overline{E} = \mathcal{O}_\Delta$ and $\overline{E}' = \mathcal{O}_\Delta \cdot e^{t^{-1}}$. Let ∇ be $\frac{d}{dt}$ (connection matrix $\omega = 0$) and $\nabla' = \nabla + \omega'$, where $\omega' = -t^{-2}dt$. Clearly, ∇ is F^1 with respect to $\overline{E}$ (indeed with respect to $t^k\overline{E}$ for any $k \in \mathbb{Z}$); whereas ∇', not F^1 with respect to $\overline{E}$, *is* F^1 with respect to $\overline{E}'$. (We leave it to the reader to squelch the possible misconception that the preceding contradicts GAGA.) Note that the preceding discussion is global when $X = \mathbb{C}^*$.

Fixing a meromorphic equivalence class $\mathfrak{M}$ of compactifications of X, we use (4.3) to define

$$F^p\widehat{H}^{2p}(X,\Lambda)_{\mathfrak{M},\log} = \varinjlim F^p\widehat{H}^{2p}(\overline{X}\log D,\Lambda), \tag{4.6}$$

where the limit is taken over $\overline{X} \in \mathfrak{M}$; and $F^pA^{2p-1}(X,\Lambda)_{\mathfrak{M},\log}$ is defined analogously.

The following variant of Definition 4.12 is inevitable.

DEFINITION 4.18. An F^1*-connection* on E *(relative to* $\mathfrak{M}$*)* is a connection that is F^1 relative to some member of $\mathfrak{M}$.

Then Proposition 4.14 yields immediately:

PROPOSITION 4.19. *If* ∇ *is an* F^1*-connection on* E *relative to* $\mathfrak{M}$*, then the Cheeger-Simons class* $\hat{c}_p(E,\nabla)$ *lies in the subgroup* $F^p\widehat{H}^{2p}(X,\Lambda)_{\mathfrak{M},\log}$ *of* $\widehat{H}^{2p}(X,\Lambda)$.

When X is an algebraic manifold and $\mathfrak{M}$ is the [class generated by] algebraic completions, we drop the subscript "$\mathfrak{M}$" in equation (4.6) to Proposition 4.19; cf. Remark 4.11.

4.5. Proof of Theorem 1.2 (Quasi-Projective Case). We return to algebraic manifolds. First let X be a smooth projective variety (*a fortiori* compact), and E an algebraic vector bundle on X. There exists an ample line bundle L on X such that both L and $E' = E \otimes L$ are generated by global sections, so that both vector bundles are pullbacks of universal bundles on Grassmannians under holomorphic maps. We get a pullback diagram:

$$\begin{array}{ccc} E & \longrightarrow & p_1^* U_n \otimes p_2^* U_1^{-1} \\ \downarrow & & \downarrow \\ X & \xrightarrow{g} & G_{\mathbb{C}}(n) \times G_{\mathbb{C}}(1) \end{array}$$

in which n denotes the rank of E, $p_1 \circ g$ classifies E', and $p_2 \circ g$ classifies L.

Also, choose any F^1-connection ∇^L on L. The latter, together with any given F^1-connection ∇^E on E, induces the tensor product F^1-connection:[4]

$$\nabla^{E'}(e \otimes \ell) = \nabla^E e \otimes \ell + e \otimes \nabla^L \ell.$$

Likewise, the tensor product of $\nabla^{E'}$ and the dual of ∇^L is just ∇^E again (under the isomorphism $E \cong E' \otimes L^{-1}$).

As the image in Deligne cohomology is independent of the F^1-connection (see Proposition 6.4), we may assume that the connections are pullbacks of F^1-connections on $G_{\mathbb{C}}(n)$ and $G_{\mathbb{C}}(1)$, which exist by Remark 4.13. Since $G_{\mathbb{C}}(n) \times G_{\mathbb{C}}(1)$ satisfies the hypothesis of Corollaries 3.4 and 4.3(ii), the assertion of Theorem 1.2 holds for $p_1^* U_n \otimes p_2^* U_1^{-1}$, so by functoriality for E. This completes the proof when X is compact.

We also use functoriality to get the assertion for non-compact X. Specifically, let $\overline{X}$ be an algebraic completion of the sort considered in Section 4.2 for which E extends to an algebraic vector bundle $\overline{E}$ on $\overline{X}$. Then there is a commutative diagram:

$$\begin{array}{ccc} F^p \widehat{H}^{2p}(\overline{X}, \mathbb{Z}(p)) & \longrightarrow & H^{2p}_{\mathcal{D}}(\overline{X}, \mathbb{Z}(p)) \\ \downarrow & & \downarrow \\ F^p \widehat{H}^{2p}(\overline{X} \log D, \mathbb{Z}(p)) & \longrightarrow & H^{2p}_{\mathcal{D}}(X, \mathbb{Z}(p)) \end{array} \tag{4.7}$$

Let ∇' be any F^1-connection on $\overline{E}$. By functoriality, $\hat{c}_p(\overline{E}, \nabla')$ restricts to $\hat{c}_p(E, \nabla')$ and $c_p^B(\overline{E})$ maps to $c_p^B(E)$. By Proposition 6.4 again, $\hat{c}_p(E, \nabla')$ has the same image in $H^{2p}_{\mathcal{D}}(X, \mathbb{Z}(p))$ as $\hat{c}_p(E, \nabla)$, and we are done.

4.6. General Algebraic Manifolds. We wish to generalize Theorem 1.2 to the context of general algebraic manifolds, i.e., to ones that are not quasi-projective, so need not even be Kähler. This is done by reducing to the quasi-projective case as we next describe.

---

[4]In terms of (B.2), the tensor product construction corresponds to $\mathrm{Hom}(T^{1,0}Y, \mathfrak{g}) \times \mathrm{Hom}(T^{1,0}Y, \mathbb{C}) \to \mathrm{Hom}(T^{1,0}Y, \mathfrak{g})$.

PROPOSITION 4.20. *Suppose that* $f\colon Y \to X$ *is a morphism of algebraic varieties satisfying:*

(i) $f^*\colon H^{2p-1}(X,\mathbb{Z})/\mathrm{tors} \to H^{2p-1}(Y,\mathbb{Z})/\mathrm{tors}$ *is injective with image a direct summand in the sense of mixed Hodge structures over* $\mathbb{Z}$*;*

(ii) $f^*\colon H^{2p}(X,\mathbb{Z})_{\mathrm{tors}} \to H^{2p}(Y,\mathbb{Z}))_{\mathrm{tors}}$ *is injective.*

Then $f^*\colon H^{2p}_{\mathcal{D}}(X,\mathbb{Z}(p)) \to H^{2p}_{\mathcal{D}}(Y,\mathbb{Z}(p))$ *is injective.*

PROOF. The exact sequence (4.1) gives a commutative diagram with exact rows:

$$
\begin{array}{ccccccc}
0 & \longrightarrow & J'^p(X) & \longrightarrow & H^{2p}_{\mathcal{D}}(X,\mathbb{Z}(p)) & \longrightarrow & F^pH^{2p}(X,\mathbb{C}) \\
 & & \downarrow & & \downarrow & & \downarrow \\
0 & \longrightarrow & J'^p(Y) & \longrightarrow & H^{2p}_{\mathcal{D}}(Y,\mathbb{Z}(p)) & \longrightarrow & F^pH^{2p}(Y,\mathbb{C})
\end{array}
\tag{4.8}
$$

The following is standard:

LEMMA 4.21. *Given the commutative diagram with exact rows:*

$$
\begin{array}{ccccccc}
0 & \longrightarrow & A & \longrightarrow & B & \longrightarrow & C \\
 & & \downarrow{\scriptstyle\alpha} & & \downarrow{\scriptstyle\beta} & & \downarrow{\scriptstyle\gamma} \\
0 & \longrightarrow & A' & \longrightarrow & B' & \longrightarrow & C'
\end{array}
$$

a sufficient condition for the injectivity of β *is that* α *and* γ *be injective.*

We apply the lemma twice. The hypotheses imply that the rightmost vertical arrow in (4.8) is an injection. Thus, to reach the conclusion, it suffices by Lemma 4.21 to show that leftmost arrow is injective. Recall that J'^p in the above is an extension by the intermediate Jacobian J^p, having the functorial exact sequence

$$0 \to J^p(X) \to J'^p(X) \to H^{2p}(X;\mathbb{Z}(p))_{\mathrm{tors}} \to 0.$$

The hypothesis implies the injectivity of $J^p(X) \to J^p(Y)$ and of $H^{2p}(X;\mathbb{Z}(p))_{\mathrm{tors}} \to H^{2p}(Y;\mathbb{Z}(p))_{\mathrm{tors}}$. By Lemma 4.21, we are done. □

COROLLARY 4.22. *Under the hypotheses of Proposition 4.20, if the conclusion of Theorem 1.2 holds for* Y*, then it also holds for* X*.*

PROOF. Consider the diagram that one gets by functoriality:

$$
\begin{array}{ccc}
F^p\widehat{H}^{2p}(X) & \longrightarrow & H^{2p}_{\mathcal{D}}(X,\mathbb{Z}(p)) \\
\downarrow & & \downarrow \\
F^p\widehat{H}^{2p}(Y) & \longrightarrow & H^{2p}_{\mathcal{D}}(Y,\mathbb{Z}(p))
\end{array}
\tag{4.9}
$$

Then $\hat{c}_p(f^*E,\nabla) \in F^p\widehat{H}^{2p}(X)$ maps to $c_p^B(f^*E) = f^*(c_p^B(E))$. Since the right vertical map is injective by Corollary 4.22, $\hat{c}_p(E,\nabla)$ can only map to $c_p^B(E)$ in $H^{2p}_{\mathcal{D}}(X,\mathbb{Z}(p))$. □

We next show that such Y exist:

PROPOSITION 4.23. (i) *Given a smooth algebraic variety* X*, there is a smooth quasi-projective variety* Y *with surjective birational map* $Y \to X$*.*

(ii) *Under the conditions of* (i), *there is a direct sum decomposition in the derived category of* X:

$$Rf_*\mathbb{Z}_Y \cong \mathbb{Z}_X \oplus \ker \operatorname{Tr}(f).$$

(iii) *In particular,* $H^\bullet(Y,\mathbb{Z}) \cong H^\bullet(X,\mathbb{Z}) \oplus \ker f_*$, *and the corresponding mixed Hodge structures are isomorphic over* $\mathbb{Z}$.

PROOF. The argument for (i) was used already in [**13**, II:(3.2)] for extending Hodge theory to general algebraic varieties, and we sketch it here. By a theorem of Nagata and the Chow Lemma, X fits into a Cartesian diagram

$$\begin{array}{ccc} Y & \hookrightarrow & \overline{Y} \\ {\scriptstyle f}\downarrow & & \downarrow{\scriptstyle \bar f} \\ X & \hookrightarrow & \overline{X} \end{array} \tag{4.10}$$

with $\overline{X}$ a completion of X, $\overline{Y}$ projective, and $\bar f$ birational. (By resolution of singularities [**24**], we may assume that $\overline{X}$ and $\overline{Y}$ are smooth, and $\overline{X}-X$ and $\overline{Y}-Y$ are divisors with normal crossings.)

Statement (ii) is a topological assertion that follows from the identity $\operatorname{Tr}(f)\circ f^* = \mathbf{1}_X$. Then (iii) follows immediately. □

Combining Propositions 4.20 and 4.23, and keeping in mind our convention (cf. Remark 4.11), we obtain:

THEOREM 4.24. *If* X *is a non-singular complex algebraic variety,* E *an algebraic vector bundle on* X *that extends to an algebraic vector bundle on some completion of* X, *and* ∇ *an* F^1*-connection on* E *(relative to said extension of* E*), then* $c_p^B(E)$ *is the image of* $\hat c_p(E,\nabla)$ *for all* $p \geq 1$.

COROLLARY 4.25. *The conclusion of Theorem* 1.1 *holds for all non-singular algebraic varieties.*

4.7. Chern Classes in $\mathfrak{M}$-DB-Cohomology. We wish to define the analogue of $c_p^B(E)$ in the complex analytic setting. Fix the meromorphic equivalence class $\mathfrak{M}$ of compactifications $\overline{X}$ of X. We can use the formula in Definition 4.1 to define $H^\bullet_{\mathcal{D}}(X,\overline{X};\Lambda(p))$. When $\overline{X}$ is Kähler, this is seen to be independent of $\overline{X}\in\mathfrak{M}$, for the same reason it was true in Section 4.1 (see paragraph containing (4.16)), so we rename it:

DEFINITION 4.26. The $\mathfrak{M}$-DB-cohomology $H^\bullet_{\mathcal{D}}(X,\Lambda(p))_{\mathfrak{M}}$ of X with coefficients in $\Lambda(p)$ is the common value of $H^\bullet_{\mathcal{D}}(X,\overline{X};\Lambda(p))$ for $\overline{X}\in\mathfrak{M}$.

Then (4.5) gives rise to the exact sequence:

$$F^pA^{2p-1}(X,\Lambda)_{\mathfrak{M},\log} \to F^p\widehat{H}^{2p}(X,\Lambda)_{\mathfrak{M},\log} \to H^{2p}_{\mathcal{D}}(X,\Lambda)_{\mathfrak{M}} \to 0. \tag{4.11}$$

Next, fix a meromorphic equivalence class $\mathfrak{E}$ of extensions of E to the members of $\mathfrak{M}$ (note that $\mathfrak{E}$ subsumes $\mathfrak{M}$). We call these the $\mathfrak{E}$-extensions of the vector bundle E on X. We will construct Chern classes

$$c_p^B(E;\mathfrak{E}) \in H^\bullet_{\mathcal{D}}(X,\mathbb{Z}(p))_{\mathfrak{M}} \tag{4.12}$$

in more or less the same way as in the algebraic setting. Choose $\overline{X}\in\mathfrak{M}$ to which E extends to $\overline{E}\in\mathfrak{E}$, and let $j\colon X\hookrightarrow\overline{X}$ denote the inclusion.

As usual, we consider first the case where E is a line bundle. Then there is a short exact sequence of complexes of sheaves on $\overline{X}$:

$$(4.13)\quad 0 \to \operatorname{cone}\{\mathbb{Z}_{\overline{X}}(1) \to \mathcal{O}_{\overline{X}}\}[-1] \to \operatorname{cone}\{Rj_*\mathbb{Z}_X(1) \to \mathcal{O}_{\overline{X}}\}[-1] \\ \to \operatorname{cone}\{\mathbb{Z}_{\overline{X}}(1) \to Rj_*\mathbb{Z}_X(1)\} \to 0.$$

From this, we extract the following commutative diagram:

$$\begin{array}{ccccc} H^2_{\mathcal{D}}(\overline{X}, \mathbb{Z}(1)) & \longrightarrow & H^2_{\mathcal{D}}(X, \mathbb{Z}(1))_{\mathfrak{M}} & & \\ \simeq \Big\uparrow d\log & & \Big\uparrow c_1^B(\mathfrak{E}) & & \\ H^1(\overline{X}, \mathcal{O}^*_{\overline{X}}) & \longrightarrow & H^1(X, \mathcal{O}^*_X)_{\mathfrak{E}} & \hookrightarrow & H^1(X, \mathcal{O}^*_X), \end{array}$$

where $H^1(X, \mathcal{O}^*_X)_{\mathfrak{E}}$ denotes the $\mathfrak{E}$-extendable line bundles on X. Since any two extensions in $\mathfrak{E}$ differ by a divisor class supported on D, we see that the Chern class $c_1^B(E; \mathfrak{E}) \in H^2_{\mathcal{D}}(X, \mathbb{Z}(1))_{\mathfrak{M}}$ of an $\mathfrak{E}$-extendable line bundle E is well-defined.

To obtain the same for higher-rank E, one invokes the splitting principle. In the formulation of [**22**, p. 140, A1], one must know Proposition 4.28 below. Because we will invoke a little Hodge theory in the argument, we must impose a condition on X.

Definition 4.27. A complex manifold X is said to be *quasi-c-Kähler* if it admits a compactification $\overline{X}$ that is a compact Kähler manifold, and for which $D = \overline{X} - X$ is an analytic subvariety.

When X is quasi-c-Kähler, one can then modify such $\overline{X}$ by a sequence of blow-ups with smooth center, a process that preserves the Kähler condition, to make D a divisor with normal crossings [**25**].

Let $\psi\colon \mathbb{P}(E) \to X$ be the projectivization of E, L the tautological line subbundle $\mathcal{O}(1)$ of $\psi^*(E)$, and $\widetilde{\mathfrak{M}}$ the meromorphic equivalence class of $\{\mathbb{P}(\overline{E}) : \overline{E}$ an $\mathfrak{E}$-extension of $E\}$.

Proposition 4.28. *Let X be quasi-c-Kähler, and $\mathfrak{M}$ a meromorphic equivalence class of Kähler compactifications. Let $\xi = c_1(L; \widetilde{\mathfrak{M}})$. Then there is an isomorphism of additive groups*

$$H^\bullet_{\mathcal{D}}(\mathbb{P}(E), \mathbb{Z}(p))_{\widetilde{\mathfrak{M}}} \cong \bigoplus_{0 \le j < r} H^\bullet_{\mathcal{D}}(X, \mathbb{Z}(p))_{\mathfrak{M}} \cdot \xi^j,$$

where r is the rank of E; in other words, there is an injective map

$$H^\bullet_{\mathcal{D}}(X, \mathbb{Z}(p))_{\mathfrak{M}} \to H^\bullet_{\mathcal{D}}(\mathbb{P}(E), \mathbb{Z}(p))_{\widetilde{\mathfrak{M}}},$$

and the right-hand side of the above is freely generated, as a module over the left-hand side, by $\{\xi^j : 0 \le j < r\}$.

Proof. We want to reduce the assertion to more elementary cohomology for which one already knows the splitting principle. As in Proposition 4.2, there is a long exact sequence:

$$(4.14)\quad \cdots \to H^{2p-1}(F^pA^\bullet(\overline{X}\log D)) \to H^{2p-1}(X, \mathbb{C}/\mathbb{Z}(p)) \\ \to H^{2p}_{\mathcal{D}}(X, \mathbb{Z}(p))_{\mathfrak{M}} \to H^{2p}(F^pA^\bullet(\overline{X}\log D)) \to \cdots.$$

Next, understanding ξ to denote generically the Chern class of L in any cohomology group, multiply (4.14) by ξ^j $(0 \leq j < r)$ and take the direct sum over j yields

(4.15)
$$\begin{array}{ccccc} \bigoplus H^{2p-1-2j}(X,\mathbb{C}/\mathbb{Z}(p-j))\cdot\xi^j & \to & \bigoplus H_{\mathcal{D}}^{2p-2j}(X,\mathbb{Z}(p-j))_{\mathfrak{M}}\cdot\xi^j & \to & \bigoplus H^{2p-2j}(F^{p-j}A^\bullet(\overline{X}\log D))\xi^j \\ \downarrow & & \downarrow & & \downarrow \\ H^{2p-1}(\mathbb{P}(E),\mathbb{C}/\mathbb{Z}(p)) & \longrightarrow & H_{\mathcal{D}}^{2p}(\mathbb{P}(E),\mathbb{Z}(p))_{\widetilde{\mathfrak{M}}} & \longrightarrow & H^{2p}(F^pA^\bullet(\mathbb{P}(\overline{E})\log\psi^*D)) \end{array}$$

Now, we know the splitting principle for $\mathbb{Z}$-coefficients, hence also for $\mathbb{Z}(p)$- and $\mathbb{C}$-coefficients, and then also for $(\mathbb{C}/\mathbb{Z}(p))$-coefficients. In other words, the left vertical arrow in the display (4.15) is an isomorphism. The easiest way to deal with the one on the right is to use Hodge theory: there is, in general, a canonical surjection

$$H^{2k}(F^kA^\bullet(\overline{X}\log D)) \to F^kH^{2k}(X,\mathbb{C}), \tag{4.16}$$

and this is an isomorphism when $\overline{X}$ is Kähler. Thus, the right vertical arrow is the Hodge filtered version of the one for $\mathbb{C}$-coefficients, so it is an likewise an isomorphism. We conclude by the five-lemma that our assertion holds. □

The above determines (4.12). We recall that $c_p(E;\mathfrak{E})$ is, up to a sign, the coefficient in $\mathbb{H}_{\mathcal{D}}^{2p}(X,\mathbb{Z}(p))_{\mathfrak{M}}$ of ξ^{r-p} in the formula for ξ^r in terms of the additive decomposition given in Proposition 4.28.

Because of Proposition 6.4, we obtain the generalization of Theorem 1.2 to the Kähler case:

THEOREM 4.29. *Let X be a quasi-c-Kähler manifold, $\mathfrak{M}$ an equivalence class of Kähler compactifications of X, E an $\mathfrak{M}$-extendable vector bundle on X, and ∇ an F^1-connection on E. Then $c_p^B(E;\mathfrak{E})$ is the image of $\hat{c}_p(E,\nabla)$ for all $p \geq 1$. Here, $\mathfrak{E}$ is the unique equivalence class of $\mathfrak{M}$-extensions of E implied by Proposition* 4.15.

COROLLARY 4.30. *The conclusion of Theorem* 1.1 *holds when X is a quasi-c-Kähler manifolds for the $\mathfrak{M}$-DB Chern classes of flat vector bundles that are regular with respect to $\mathfrak{E}$.*

5. Proof of Theorem 1.3

Throughout this section, the complexification $\mathfrak{g}\otimes\mathbb{C}$ of a real Lie algebra $\mathfrak{g}$ will be denoted $\mathfrak{g}_{\mathbb{C}}$. In addition, we make use of simplicial methods; all relevant definitions can be found in Appendix C.

5.1. Continuous Cohomology. The continuous cohomology $H^\bullet_{\mathrm{cts}}(G,V)$ of a topological group G with coefficients in a real topological vector space V, on which G acts, is defined to be the cohomology of the complex $C^\bullet_{\mathrm{cts}}(G,V)$, whose elements in degree m are G-equivariant continuous maps $f\colon G^{m+1} \to V$ (with the usual coboundary map).

Now let G be the real points of a reductive algebraic group defined over $\mathbb{R}$, and K a maximal compact subgroup. We recall that the space G/K is contractible, and thus G and K are of the same homotopy type. Let $\mathfrak{g} = \mathfrak{k}\oplus\mathfrak{p}$ be the corresponding Cartan decomposition of the Lie algebra of G. By van Est's Theorem [**19**], whenever V is finite-dimensional with trivial G-action, there is a canonical isomorphism

$$H^\bullet_{\mathrm{cts}}(G,V) \cong H^\bullet(\mathfrak{g},K)\otimes V. \tag{5.1}$$

Here, the right-hand side is the relative Lie algebra cohomology, and is canonically isomorphic to $(\wedge^\bullet \mathfrak{p}^*)^K \otimes V$, the complex of G-invariant V-valued differentials forms on G/K. The isomorphism can be realized as follows. Fix a point e of G/K, and endow the latter with a G-invariant Riemannian metric. For $(g_0, \dots, g_m) \in G^{m+1}$, define $\Delta_e(g_0, \dots, g_m)$ to be the "geodesic simplex" in G/K with vertices $g_0 e$, ... , $g_m e$; it is constructed inductively as the cone swept out by all geodesics from $g_0 e$ to $\Delta_e(g_1, \dots, g_m)$. (Note that, in general, the order of the vertices makes a difference, since the curvature may not be constant.) Define a homomorphism

$$\wedge^\bullet \mathfrak{p}^* \to C^\bullet_{\mathrm{cts}}(G, \mathbb{R})$$

by taking an m-form ω to the function

$$(g_0, \dots, g_m) \mapsto \int_{\Delta_e(g_0,\dots,g_m)} \omega. \tag{5.2}$$

This induces (5.1).

We now recall the standard trick for computing the continuous cohomology of a reductive group. The compact real form of $\mathfrak{g} \otimes \mathbb{C}$ is the Lie algebra $\mathfrak{u} = \mathfrak{k} \oplus i\mathfrak{p} \subseteq \mathfrak{g}_{\mathbb{C}}$. Let U be the corresponding subgroup of $G(\mathbb{C})$. Since U is compact, there are canonical isomorphisms

$$H^\bullet(U/K, V) \cong H^\bullet(\mathfrak{u}, K; V) \cong (\wedge^\bullet (i\mathfrak{p})^*)^K \otimes V.$$

Composing these isomorphisms with the isomorphism

$$\wedge^\bullet (i\mathfrak{p})^* \to \wedge^\bullet \mathfrak{p}^*$$

induced by multiplication by i, we obtain an isomorphism

$$H^\bullet(U/K, \mathbb{C}) \xrightarrow{\sim} H^\bullet_{\mathrm{cts}}(G, \mathbb{C}), \tag{5.3}$$

which carries $H^m(U/K, \mathbb{R}(p))$ onto the subspace $i^m H^m_{\mathrm{cts}}(G, \mathbb{R}(p))$ of $H^m_{\mathrm{cts}}(G, \mathbb{C})$.

5.2. The Borel Regulator Elements. We now take G to be $\mathrm{GL}_n(\mathbb{C})$, viewed as a real group, and we take K to be $\mathrm{TU}(n)$. In this case, one identifies the complexification of G as $\mathrm{GL}_n(\mathbb{C}) \times \mathrm{GL}_n(\mathbb{C})$; the natural map $G \to G_{\mathbb{C}}$ is the homomorphism

$$\iota\colon \mathrm{GL}_n(\mathbb{C}) \to \mathrm{GL}_n(\mathbb{C}) \times \mathrm{GL}_n(\mathbb{C}),$$

with $\iota(g) = (g, \bar{g})$. The corresponding map on Lie algebras

$$\mathfrak{gl}_n(\mathbb{C}) \to \mathfrak{gl}_n(\mathbb{C}) \oplus \mathfrak{gl}_n(\mathbb{C}) \tag{5.4}$$

takes X to $(X, \overline{X})$; indeed, it is not hard to see that the identification

$$\mathfrak{gl}_n(\mathbb{C}) \otimes_{\mathbb{R}} \mathbb{C} \cong \mathfrak{gl}_n(\mathbb{C}) \oplus \mathfrak{gl}_n(\mathbb{C}) \tag{5.5}$$

is given by: for $X, Y \in \mathfrak{gl}_n(\mathbb{C})$,

$$X \otimes 1 + Y \otimes i \mapsto \left(X + iY, \overline{X} + i\overline{Y}\right). \tag{5.6}$$

It follows that the Lie algebra $\mathfrak{u}_n$ of $\mathrm{TU}(n)$ appears in (5.4) as the "anti-diagonal"

$$\{(X, -{}^tX) \mid X \in \mathfrak{u}_n\}.$$

It is clear that the compact real form U of $G_{\mathbb{C}}$ is $\mathrm{TU}(n) \times \mathrm{TU}(n)$. The quotient space U/K is therefore $(\mathrm{TU}(n) \times \mathrm{TU}(n)) / \mathrm{TU}(n)$, which we identify with $\mathrm{TU}(n)$

via the map induced by inclusion as the first factor of the product. Since $\mathrm{GL}_n(\mathbb{C})$ and $\mathrm{TU}(n)$ are homotopy-equivalent, we obtain from (5.3) an isomorphism

$$H^\bullet(\mathrm{GL}_n(\mathbb{C}),\mathbb{C}) \cong H^\bullet_{\mathrm{cts}}(\mathrm{GL}_n(\mathbb{C}),\mathbb{C})$$

which takes $H^m(\mathrm{GL}_n(\mathbb{C}),\mathbb{R}(p))$ onto $i^m H^m_{\mathrm{cts}}(\mathrm{GL}_n(\mathbb{C}),\mathbb{R}(p))$. This yields a canonical isomorphism

$$H^{2p-1}(\mathrm{GL}_n(\mathbb{C}),\mathbb{R}(p)) \xrightarrow{\sim} H^{2p-1}_{\mathrm{cts}}(\mathrm{GL}_n(\mathbb{C}),\mathbb{C}/\mathbb{R}(p)). \tag{5.7}$$

The real cohomology of $\mathrm{GL}_n(\mathbb{C})$ is an exterior algebra on generators $x_1, \ldots,$ x_n, where x_p is in degree $2p-1$. We can choose x_p canonically by insisting that it be the unique (necessarily primitive) class which transgresses, via

$$d_{2p}\colon H^0(B\,\mathrm{GL}_n(\mathbb{C}), H^{2p-1}(\mathrm{GL}_n(\mathbb{C}))) \to H^{2p}(B\,\mathrm{GL}_n(\mathbb{C})),$$

to the universal Chern class $c_p \in H^{2p}(B\,\mathrm{GL}_n(\mathbb{C}))$ in the Leray spectral sequence associated to the universal $\mathrm{GL}_n(\mathbb{C})$-bundle; it lies in

$$W_0 H^{2p-1}(\mathrm{GL}_n(\mathbb{C}),\mathbb{Z}(p)),$$

as $\mathbb{Z}(p)$ is of weight $-2p$. Define $b_p \in H^{2p-1}(B\,\mathrm{GL}_n(\mathbb{C})^\delta,\mathbb{C}/\mathbb{R}(p))$ to be the image of x_p under the composite of (5.7) and the natural map

$$H^\bullet_{\mathrm{cts}}(\mathrm{GL}_n(\mathbb{C}),\mathbb{C}/\mathbb{R}(p)) \to H^\bullet_{\mathrm{cts}}(\mathrm{GL}_n(\mathbb{C})^\delta,\mathbb{C}/\mathbb{R}(p)) \cong H^\bullet(B\,\mathrm{GL}_n(\mathbb{C})^\delta,\mathbb{C}/\mathbb{R}(p)).$$

This is called the p-th *Borel regulator element.*[5]

5.3. The Weil Algebra. Let G be a real reductive group and K any compact subgroup. Denote the corresponding Lie algebras by $\mathfrak{g}$ and $\mathfrak{k}$. Consider the (*real*) *Weil algebra* [**7**]

$$W(\mathfrak{g}) = \wedge^\bullet\mathfrak{g}^* \otimes S^\bullet(\mathfrak{g}^*), \tag{5.8}$$

a d.g.a. that is a $\mathfrak{g}$-module via the coadjoint action (here S denotes the symmetric algebra over $\mathbb{R}$). Denote the set of invariant polynomials $S^\bullet(\mathfrak{g}^*)^G$, which is a subalgebra of $W(\mathfrak{g})$, by $I^\bullet(G)$.

Let $\omega \in A^1(P,\mathfrak{g})$ be a connection on a principal bundle $P \to M$ over a (simplicial) manifold, which will be viewed as a map $\mathfrak{g}^* \to A^1(P)$. Then ω extends uniquely to a d.g.a. homomorphism

$$k_P(\omega)\colon W(\mathfrak{g}) \to A^\bullet(P), \tag{5.9}$$

such that for $\Phi \in S^p(\mathfrak{g}^*)$,

$$k_P(\omega)(\Phi) = \Phi(\Theta^p),$$

where Θ denotes the curvature of the connection.

The *relative Weil algebra* $W(\mathfrak{g},K)$ [**8**] is the subspace of "K-basic" elements

$$\left[\wedge^\bullet(\mathfrak{g}/\mathfrak{k})^* \otimes S^\bullet(\mathfrak{g}^*)\right]^K \tag{5.10}$$

---

[5]Some prefer to define the Borel regulator element to be the class in $H^{2p-1}(\mathrm{GL}_n(\mathbb{C})(\mathbb{C})^\delta,\mathbb{C}/\mathbb{R}(p))$ that corresponds to the element y_p of

$$W_0H^{2p-1}(\mathrm{GL}_n(\mathbb{C}),\mathbb{Q}(p)) \cong W_0H^{2p-1}(GL_p(\mathbb{C}),\mathbb{Q}(p)) \cong H^{2p-1}(\mathbb{C}^p - \{0\},\mathbb{Q}(p)) \cong \mathbb{Q}$$

that takes the value 1 on the generator of $\pi_{2p-1}(GL_n(\mathbb{C})(\mathbb{C}))$. Since $x_p = \pm(p-1)!y_p$ [**30**, (24.5.2)], it follows that y_p corresponds, up to a sign, to twice the degree p part of the Chern character

$$ch_p\colon K_{2p-1}(F) \to H^1_{\mathcal{D}}(\operatorname{Spec} F,\mathbb{R}(p)).$$

for F a number field or $\mathbb{R}$ or $\mathbb{C}$.

of $W(\mathfrak{g})$. One recovers (5.8) when K is trivial; $W(\mathfrak{g}, K)$ is a contravariant functor of pairs $(\mathfrak{g}, K)$, and it contains $I^\bullet(G)$ for all K. A connection $\omega \in A^1(P, \mathfrak{g})$ on a principal bundle $P \to M$ induces a d.g.a. homomorphism (*Chern-Weil homomorphism*)

$$k_P(\omega)\colon W(\mathfrak{g}, K) \to A^\bullet(P/K) \tag{5.11}$$

by restriction of (5.9). The natural map $W(\mathfrak{g}, K) \to I^\bullet(K)$ (which factors through $I^\bullet(G)$) is a quasi-isomorphism [**8**], hence

$$H^\bullet(W(\mathfrak{g}, K)) \cong H^\bullet(K, \mathbb{R}).$$

A consequence of this is that if $\Phi \in I^\bullet(G)$ has the property that its image in $I^\bullet(K)$ is zero, then there exists $T \in W(\mathfrak{g}, K)$ such that $dT = \Phi$. Note that when K is *maximal* compact, $K \hookrightarrow G$ is a homotopy equivalence, so $P/K \to P/G = M$ is a homotopy equivalence, and (5.11) induces a map

$$H^\bullet(G, \mathbb{R}) \to H^\bullet(M, \mathbb{R}).$$

For a real vector space V, denote the Weil algebra $W(\mathfrak{g}) \otimes V$ with coefficients in V by $W_V(\mathfrak{g})$, and the corresponding Weil algebra in the relative case by $W_V(\mathfrak{g}, K)$. In particular, we have the Weil algebras $W_{\mathbb{C}}(\mathfrak{g})$ (quasi-isomorphic to $A^\bullet(EG) \cong \mathbb{C}$) and $W_{\mathbb{C}}(\mathfrak{g}, K)$, as well as $W_{\mathbb{R}(p)}(\mathfrak{g}, K)$. When $\mathfrak{g}$ and $\mathfrak{k}$ are *complex* Lie algebras, one can also define the complex Weil algebra $\mathfrak{W}(\mathfrak{g}, K)$ by doing the linear algebra in (5.10) over $\mathbb{C}$ instead of $\mathbb{R}$. We observe that when $\mathfrak{g}$ is a real Lie algebra, $W_{\mathbb{C}}(\mathfrak{g}, K) \cong \mathfrak{W}(\mathfrak{g}_{\mathbb{C}}, K_{\mathbb{C}})$ as $\mathbb{C}$-algebras.

5.4. The Class τ_p and Cheeger-Simons Classes. In this section, $\mathfrak{g}$ is $\mathfrak{gl}_n(\mathbb{C})$ viewed as a real Lie algebra, and $\mathfrak{k} = \mathfrak{u}_n$. Let

$$C_p(X) = (-1)^p \operatorname{Tr}(\wedge^p X),$$

the invariant polynomial (of degree p) on $\mathfrak{gl}_n(\mathbb{C})$ which determines the pth Chern class. We can express it as

$$C_p(X) = P_p(X) + Q_p(X),$$

where

$$P_p(X) = \frac{1}{2}\left[C_p(X) + (-1)^p C_p(\overline{X})\right] \text{ and } Q_p(X) = \frac{1}{2}\left[C_p(X) - (-1)^p C_p(\overline{X})\right]$$

are elements of $S^p(\mathfrak{g}^*)$. Note that P_p is $\mathbb{R}(p)$-valued, and Q_p is $\mathbb{R}(p-1)$-valued. If $X \in \mathfrak{u}_n$, then $\overline{X} = -{}^tX$, so $Q_p(X) = 0$. It follows from the discussion in the previous section that there exists $T_p \in W_{\mathbb{R}(p)}(\mathfrak{gl}_n(\mathbb{C}), \mathrm{TU}(n))$ such that $dT_p = Q_p$.

The connection on the universal flat bundle $P \to B\,\mathrm{GL}_n(\mathbb{C})^\delta$ induces a Chern-Weil homomorphism

$$k\colon W_{\mathbb{C}}(\mathfrak{gl}_n(\mathbb{C}), \mathrm{TU}(n)) \to A^\bullet\left(P/\,\mathrm{TU}(n)\right).$$

Denote $k(T_p) \in A^{2p-1}(P/\,\mathrm{TU}(n))$ by τ_p; the curvature Θ of P is zero. We have

$$d\tau_p = Q_p(\Theta) = 0,$$

and thus τ_p defines a class in

$$H^{2p-1}(P/\,\mathrm{TU}(n), \mathbb{C}) \cong H^{2p-1}(\mathrm{GL}_n(\mathbb{C})^\delta, \mathbb{C}).$$

The first task is to show:

PROPOSITION 5.1. *The class of τ_p in $H^{2p-1}(\mathrm{GL}_n(\mathbb{C})^\delta, \mathbb{C}/\mathbb{R}(p))$ is the negative of the universal real Cheeger-Simons class $\hat{c}_p$.*

PROOF. Let $E \to B$ be the model of the universal $\mathrm{GL}_n(\mathbb{C})$-bundle constructed from $P \to B\,\mathrm{GL}_n(\mathbb{C})^\delta$ (over the standard model of $B\,\mathrm{GL}_n(\mathbb{C})^\delta$), as in Proposition 6.5. Recall that E has a canonical connection, induced by that of P, and there is a tautological map $B\,\mathrm{GL}_n(\mathbb{C})^\delta \to B$ which classifies both the universal flat bundle and its connection. Consider the Chern-Weil homomorphism

$$k_E\colon W_{\mathbb{C}}(\mathfrak{gl}_n(\mathbb{C}), \mathrm{TU}(n)) \to A^\bullet\,(E/\,\mathrm{TU}(n))$$

associated to E and its connection, writing τ_p^E for $k_E(T_p)$. Since the polynomial P_p is $\mathbb{R}(p)$-valued,

$$C_p(\Theta_E) \equiv Q_p(\Theta_E) \mod \mathbb{R}(p));$$

and since $dT_p = Q_p$, it follows that $d\tau_p^E = Q_p(\Theta_E)$, so $(Q_p(\Theta_E), -\tau_p^E)$ represents the "universal Cheeger-Simons class"

$$\hat{c}_p \in \widehat{H}^{2p}(E/\,\mathrm{TU}(n), \mathbb{R}(p)),$$

i.e., that of the pullback of $E \to B$ to $E/\,\mathrm{TU}(n) \simeq B$. The naturality of Chern-Weil construction implies that the diagram

$$\begin{array}{ccc} W_{\mathbb{C}}(\mathfrak{gl}_n(\mathbb{C}), \mathrm{TU}(n))k_E & \longrightarrow & A^\bullet(E/\,\mathrm{TU}(n)) \\ & \searrow^{k} & \downarrow \\ & & A^\bullet\,(P/\,\mathrm{TU}(n)) \end{array}$$

commutes. It follows that the Cheeger-Simons class of the pullback of $P \to B\,\mathrm{GL}_n(\mathbb{C})^\delta$ to $P/\,\mathrm{TU}(n)$ is represented by $(Q_p(\Theta), -\tau_p) = (0, -\tau_p)$, and this gives the desired assertion. □

5.5. τ_p and T_p. The next step is to obtain a formula for the cocycle given by τ_p in terms of T_p. For this we will need an explicit section of the bundle $P/\,\mathrm{TU}(n) \to B\,\mathrm{GL}_n(\mathbb{C})^\delta$. It is more convenient to write down a section of $E\,\mathrm{GL}_n(\mathbb{C})/\,\mathrm{TU}(n) \to B\,\mathrm{GL}_n(\mathbb{C})$ and pull it back to the universal flat bundle. Actually, we will write down a section of the bundle $\bar{\pi}\colon EG/K \to BG$, where K is a maximal compact subgroup of an arbitrary real reductive group G, following [**14**].

The m-simplex G^{m+1} of the usual model of EG (see Proposition 6.5, and take $X_\bullet$ to be a point) is identified with G^{m+1} in the *inhomogeneous*, or *reduced*, model via the map

$$(g_0, g_1, \ldots, g_m) \mapsto (g_0, g_1 g_0^{-1}, \ldots, g_m g_{m-1}^{-1}).$$

This map is G-equivariant with respect to the diagonal right G-action on the left-hand side, and the right action of G on the first factor of the right-hand side. It follows that in the reduced model, the space of m-simplices of EG/K is $G/K \times G^m$. In particular, when $G = K$, we have that the space of m-simplices of the reduced model of BG is G^m.

Suppose now that G is reductive and that K is a maximal compact subgroup. For each m, define a map (which is a homotopy equivalence)

$$s_m\colon G^m \times \Delta^m \to G/K \times G^m \times \Delta^m \tag{5.12}$$

by

$$((g_1,\dots,g_m),(t_0,\dots,t_m))$$
$$\mapsto \Big((\tilde{\Delta}_e((g_1,g_2g_1,\dots,g_mg_{m-1}\cdots g_1),(t_0,\dots,t_m)),(g_1,\dots,g_m),(t_0,\dots,t_m)\Big),$$

where $\tilde{\Delta}_e\colon G^m\times\Delta^m\to G/K$ parametrizes the geodesic simplices defined in (5.2).

PROPOSITION 5.2. [**14**, p. 241] *The maps* (5.12) *are compatible with the face maps and therefore induce a section* $s\colon BG\to(EG)/K$ *of* $\bar{\pi}$.

We now show that the cohomology class on $\mathrm{GL}_n(\mathbb{C})^\delta$ defined by the continuous cohomology class corresponding to $T_p\in\wedge^{2p-1}\mathfrak{p}^*$ equals the cohomology class defined by $s^*\tau_p$. With respect to the inhomogeneous model (see (5.12)), the connection form of the universal flat bundle is given locally by the Maurer-Cartan form in the vertical direction, and is zero in the horizontal directions. Consequently, with respect to the inhomogeneous model, the restiction of τ_p to $\mathrm{GL}_n(\mathbb{C})/\mathrm{TU}(n)\times G^m\times\Delta^m$ satisfies

$$\tau_p\in A^{2p-1}(\mathrm{GL}_n(\mathbb{C})/\mathrm{TU}(n))\subseteq A^{2p-1}(\mathrm{GL}_n(\mathbb{C})/\mathrm{TU}(n)\times G^m\times\Delta^m),$$

and that it is the image of T_p under the natural inclusion

$$\wedge^{2p-1}\mathfrak{p}^*\hookrightarrow A^{2p-1}(\mathrm{GL}_n(\mathbb{C})/\mathrm{TU}(n)).$$

It follows immediately that the value of the cocycle

$$s^*\tau_p\in A^{2p-1}(B\,\mathrm{GL}_n(\mathbb{C})^\delta)$$

on the simplex of $B\,\mathrm{GL}_n(\mathbb{C})$ with vertices $(g_0,\dots,g_m)$ is given by the integral

$$\int_{\Delta_e(g_0,\dots,g_m)} T_p.$$

5.6. T_p and the Borel Regulator Element. In the final step, the idea is to use the complexification to show that T_p can be interpreted as a class that transgresses to the pth Chern class of $E\to B$.

An invariant polynomial on $\mathfrak{gl}_n(\mathbb{C})$ defines, in an obvious way, an invariant function on $\mathfrak{g}_\mathbb{C}$. Using (5.5) to write the elements of $\mathfrak{g}_\mathbb{C}$ as pairs (X,Y), with $X,Y\in\mathfrak{gl}_n(\mathbb{C})$, we have for Q_p of Section 5.4:

$$Q_p(X,Y)=\frac{1}{2}\left[C_p(X)-(-1)^pC_p(Y)\right]. \tag{5.13}$$

From this, we see that for all $X\in\mathfrak{gl}_n(\mathbb{C})$,

$$C_p(X)=2\,Q_p(X,0). \tag{5.14}$$

Likewise, we can view T_p (also from Section 5.4) as an element of the complex Chern-Weil algebra $\mathfrak{W}(\mathfrak{g}_\mathbb{C},\mathrm{TU}(n)_\mathbb{C})$, which then determines elements of $W_\mathbb{C}(\mathfrak{u}_n\oplus\mathfrak{u}_n,\mathrm{TU}(n))$ and

$$W_\mathbb{C}(\mathfrak{u}_n\oplus\{0\})\cong W_\mathbb{C}(\mathfrak{u}_n),$$

which we continue to denote T_p.

From (5.14), it follows that, in $W_\mathbb{C}(\mathfrak{u}_n)$, the relation $2\,dT_p=C_p$ holds. In other words, the invariant form on $\mathrm{TU}(n)$ defined by $2\,T_p$ represents a class which transgresses to the pth Chern class. The corresponding class in

$$\wedge^\bullet(\mathfrak{u}_n)^*\xrightarrow{\sim}_{pr_1{}^*}\wedge^\bullet((\mathfrak{u}_n\oplus\mathfrak{u}_n)/\mathfrak{u}_n)^*\cong\wedge^\bullet(i\mathfrak{p}^*)$$

is the image of $2T_p$ in $\wedge^{2p-1}(i\mathfrak{p}^*)$. By the definition given in Section 5.2, the Borel element is represented by the differential form

$$i^{2p-1}\left(2T_p|_{\wedge^{2p-1}i\mathfrak{p}}\right) = (-1)^{2p-1}\left(2T_p|_{\wedge^{2p-1}\mathfrak{p}}\right) = -2T_p \in \wedge^{2p-1}\mathfrak{p}^*.$$

Combining the above with Sections 5.4 and 5.5, we have shown that the Borel element is twice the Cheeger-Simons class, and Theorem 1.3 is proved.

6. Towards a Proof of Conjecture 1.4

In this section we make use of simplicial methods. Again, see Appendix C for definitions and constructions.

6.1. The Goal. We would have preferred to have Theorem 1.2 and Conjecture 1.4 as special cases of the simplicial analogue of Theorem 1.2, which we have not been able to prove:

STATEMENT 6.1. *Suppose that* $\pi\colon \overline{P}_\bullet \to \overline{X}_\bullet$ *is a morphism in the category of smooth simplicial varieties, with* $\overline{X}_\bullet$ *complete, and that* $D_\bullet$ *and* $Q_\bullet = \pi^{-1}(D_\bullet)$ *and are divisors with normal crossings in* $\overline{X}_\bullet$ *and* $\overline{P}_\bullet$*, respectively. Let* $P_\bullet = \overline{P}_\bullet - Q_\bullet$ *and* $X_\bullet = \overline{X}_\bullet - D_\bullet$*. If the restriction* $P_\bullet \to X_\bullet$ *of* π *is a principal* $\mathrm{GL}_n(\mathbb{C})$*-bundle with* F^1*-connection—i.e., the connection form* ω *satisfies*

$$\omega \in F^1E^1(|\overline{P}_\bullet|\log Q_\bullet, \mathrm{End}(\overline{E}_\bullet)),$$

then the image of its Cheeger-Simons class $\hat{c}_p$ *in* $H^{2p}_{\mathcal{D}}(X_\bullet, \mathbb{Z}(p))$ *under the natural homomorphism*

$$F^p\widehat{H}^{2p}(\overline{X}_\bullet \log D, \mathbb{Z}(p)) \longrightarrow H^{2p}_{\mathcal{D}}(X_\bullet, \mathbb{Z}(p))$$

is the Beilinson Chern class c_p^B.

REMARK 6.2. Note that we are *not* assuming in the above that $\overline{P}_\bullet$ is proper over $\overline{X}_\bullet$, but only that it is so "in the horizontal direction". In practice, $P_\bullet$ would be the frame bundle of a vector bundle $E_\bullet$ on $X_\bullet$, and $\overline{P}_\bullet$ the frame bundle of an extension $\overline{E}_\bullet$ of $E_\bullet$ to $\overline{X}_\bullet$.

The following lemma, whose proof is quite direct, provides good evidence for the conjecture. (We retain the previous notation.)

LEMMA 6.3. *Suppose that one has two connections* ∇_0 *and* ∇_1 *on* $E_\bullet$*, and denote by* Θ_0 *and* Θ_1 *the respective curvature forms. Write* $\omega = \nabla_1 - \nabla_0$ *and* η_p *for the solution of*

$$d\eta_p = c_p(E_\bullet, \nabla_1) - c_p(E_\bullet, \nabla_0)$$

given by (3.9). *If*

$$\Theta_0, \Theta_1 \in F^1A^2(|\overline{X}_\bullet|\log D_\bullet, \mathrm{End}(\overline{E}_\bullet)) \quad \textit{and} \quad \omega \in F^1(A^1(|\overline{X}_\bullet|\log D_\bullet, \mathrm{End}(\overline{E}_\bullet))),$$

then

(i) $\eta_p \in F^pA^{2p-1}(|\overline{X}_\bullet|\log D_\bullet)$;
(ii) $c_p(E_\bullet, \nabla_0)$ *and* $c_p(E_\bullet, \nabla_1)$ *represent the same cohomology class in*

$$H^{2p}(F^pA^\bullet(|\overline{X}_\bullet|\log D_\bullet)) = F^pH^{2p}(|X_\bullet|, \mathbb{C}).$$

PROOF. This is a direct consequence of (3.8) and (3.9), as the assumptions imply that Θ_t is in F^1. □

The following "refinement" of Lemma 6.3, which we will soon prove, is a necessary condition for Statement 6.1:

PROPOSITION 6.4. *Under the conditions of Lemma 6.3, these two Cheeger-Simons classes have the same image in* $H^{2p}_{\mathcal{D}}(X_\bullet, \mathbb{Z}(p))$.

The basic theme in any proof of Satement 6.1 is to reduce to the universal case (where the assertion is a tautology; see Proposition 3.3, and Proposition 6.8 below), invoking functoriality. We wanted to make use of the following device:

PROPOSITION 6.5. *If* $X_\bullet$ *is a simplicial manifold and* $P_\bullet \to X_\bullet$ *is a principal* $\mathrm{GL}_n(\mathbb{C})$*-bundle with connection given by*

$$\omega \in A^1(|P_\bullet|, \mathrm{Ad}(\mathfrak{gl}_n(\mathbb{C}))),$$

then there exists a smooth bisimplicial variety $B_{\bullet\bullet}$ *which has the homotopy type of* $B\,\mathrm{GL}_n(\mathbb{C})$ *and a* $\mathrm{GL}_n(\mathbb{C})$*-principal bundle* $U_{\bullet\bullet} \to B_{\bullet\bullet}$ *with connection*

$$\omega_U \in A^1(|U_{\bullet\bullet}|, \mathrm{Ad}(\mathfrak{gl}_n(\mathbb{C})))$$

and a morphism of $\mathrm{GL}_n(\mathbb{C})$*-bundles*

$$\begin{array}{ccc} P_\bullet & \xrightarrow{G} & U_{\bullet\bullet} \\ \downarrow & & \downarrow \\ X_\bullet & \xrightarrow{g} & B_{\bullet\bullet} \end{array}$$

such that $G^*\omega_U = \omega$. *Moreover, if* $X_\bullet$ *is a smooth simplicial variety such that*

$$\omega \in F^1A^1(|P_\bullet|, \mathrm{Ad}(\mathfrak{gl}_n(\mathbb{C}))),$$

then we can choose ω_U *to satisfy*

$$\omega_U \in A^{1,0}(|U_{\bullet\bullet}|, \mathrm{Ad}(\mathfrak{gl}_n(\mathbb{C}))).$$

PROOF. Consider the bisimplicial variety $U_{\bullet\bullet}$ whose simplicial variety of m simplices is the simplicial variety $(P_\bullet)^{m+1}$ with face maps

$$\begin{array}{llll} d_i\colon & (P_\bullet)^{m+1} \longrightarrow (P_\bullet)^m & & 0 \le i \le m \\ & (u_0, \dots, u_m) \longmapsto (u_0, \dots, \hat{u}_i, \dots, u_m). & & \end{array}$$

The geometric realization of this simplicial variety is contractible as it is the coskeleton of the trivial covering (cf. [**1**] or [**31**, p. 107]).[6] The free $\mathrm{GL}_n(\mathbb{C})$ action on $P_\bullet$ induces a free $\mathrm{GL}_n(\mathbb{C})$ action on $U_{\bullet\bullet}$. Let $B_{\bullet\bullet}$ be the quotient of $U_{\bullet\bullet}$. The resulting principal bundle $U_{\bullet\bullet} \to B_{\bullet\bullet}$ is a model of the universal $\mathrm{GL}_n(\mathbb{C})$-bundle.

Define a connection $\omega_U \in A^1(|U_{\bullet\bullet}|, \mathfrak{gl}_n(\mathbb{C}))$ on the bundle $|U_{\bullet\bullet}| \to |B_{\bullet\bullet}|$ by the compatible family of 1-forms

$$\sum_{j=0}^{m} t_j \pi_j^* \omega \in A^1(|P_\bullet|^{m+1} \times \Delta^m),$$

where $(t_0, \dots, t_m)$ are the barycentric coordinates of Δ^m, and $\pi_j\colon |P_\bullet|^{m+1} \to |P_\bullet|$ denote the canonical projections, $0 \le j \le m$.

[6] This may be proved directly as follows. First show that the geometric realization of the simplicial space is simply connected. This is not difficult, as the fundamental group depends only on the 2-skeleton. One then shows that this space has trivial integral homology, and is therefore contractible, by standard arguments.

The canonical isomorphism $P_\bullet \to U_{\bullet 0}$ induces a $\mathrm{GL}_n(\mathbb{C})$-equivariant map $G\colon P_\bullet \to U_{\bullet\bullet}$, and therefore a map $g\colon X_\bullet \to B_{\bullet\bullet}$ such that the pair of maps (g, G) classifies the bundle and the connection. The remaining assertion is easy to verify. □

REMARK 6.6. In other words, given a principal $\mathrm{GL}_n(\mathbb{C})$-bundle with connection over a complete smooth simplicial variety , we can build a model of the universal $\mathrm{GL}_n(\mathbb{C})$-bundle, and put a connection on this bundle, such that a *fixed* mapping of our variety into this model of $B\,\mathrm{GL}_n(\mathbb{C})$ simultaneously classifies the bundle and its connection. Unfortunately the above does not yield a proof of Statement 6.1, for the Maurer-Cartan forms of $\mathrm{GL}_n(\mathbb{C})$, and hence the connection constructed in Proposition 6.5, are not logarithmic at infinity for $n > 1$. This was our fundamental obstacle.

However, Proposition 6.5 does yield at once the following useful observation:

PROPOSITION 6.7. *Suppose that ∇_1 is a second connection on $E_\bullet$ and that $\hat{c}_p(E_\bullet, \nabla_0)$ is represented by $(c_p(E_\bullet, \nabla_0), -y)$, as in Remark* 3.11. *Let η_p be as in* (3.9). *Then $\hat{c}_p(E_\bullet, \nabla_1)$ is represented by*

$$(c_p(E_\bullet, \nabla_1), -(y + \iota_{\mathbb{Z}(p)}(\eta_p))) = (c_p(E_\bullet, \nabla_0), -y) + (d\eta_p, -\iota_{\mathbb{Z}(p)}(\eta_p))).$$

PROOF. By the functoriality of c_p, η_p and $\hat{c}_p$, it suffices to check this "universally", on the model of $B\,\mathrm{GL}_n(\mathbb{C})$ from Proposition 6.5. But there, the Cheeger-Simons class is completely determined by the Chern form (see Section 3.4), so it comes down to the calculation:

$$\delta(y + \iota_{\mathbb{Z}(p)}(\eta_p)) = \iota_{\mathbb{Z}(p)}(c_p(E_\bullet, \nabla_0)) + \iota_{\mathbb{Z}(p)}(d\eta_p) = \iota_{\mathbb{Z}(p)}(c_p(E_\bullet, \nabla_1)).$$

□

Using Proposition 6.7, we can now give:

PROOF OF PROPOSITION 6.4. In the DB-complex of $X_\bullet$, we have that:

$$(c_p(E, \nabla_0), -y) - (c_p(E, \nabla_1), -(y + \iota_{\mathbb{Z}(p)}(\eta_p))) = (-d\eta_p, \iota_{\mathbb{Z}(p)}(\eta_p)) = D(\eta_p, 0),$$

so Proposition 6.4 follows. □

The simplicial version of (4.3) is

(6.1)

$$F^p\widehat{H}^k(|\overline{X}_\bullet|\log D_\bullet, \Lambda) = H^{k-1}(\mathrm{cone}\{F^pA^{\geq k}(|\overline{X}_\bullet|\log \overline{D}_\bullet) \to S^\bullet(|X_\bullet|, \mathbb{C}/\Lambda)\}).$$

Taking $k = 2p$, and $\Lambda = \mathbb{Z}(p)$, and recalling the definitions in Section 4, we obtain the following diagram, in which the rows are exact:

(6.2)

$$\begin{array}{ccccccc} 0 \to H^{2p-1}(|X_\bullet|, \mathbb{C}/\mathbb{Z}(p)) & \to & F^p\widehat{H}^{2p}(\overline{X}_\bullet \log D_\bullet, \mathbb{Z}(p)) & \to & F^pA^{2p}(|\overline{X}_\bullet|\log D_\bullet, \mathbb{Z}(p))_{cl} \to 0 \\ \| & & \downarrow & & \downarrow \\ H^{2p-1}(|X_\bullet|, \mathbb{C}/\mathbb{Z}(p)) & \longrightarrow & H^{2p}_{\mathcal{D}}(X_\bullet, \mathbb{Z}(p)) & \longrightarrow & F^pH^{2p}(|X_\bullet|, \mathbb{C}) \end{array}$$

We note that the following is easily read from (6.2):

PROPOSITION 6.8. *If $H^{2p-1}(|X_\bullet|, \mathbb{C}/\mathbb{Z}(p)) = 0$, then the assertion in Statement* 6.1 *holds for every holomorphic vector bundle on $|X_\bullet|$.*

Next, consider the diagram

$$\begin{array}{ccc} H^{2p}_{\mathcal{D}}(\overline{X}_\bullet,\mathbb{Z}(p)) & \longrightarrow & H^{2p}_{\mathcal{D}}(X_\bullet,\mathbb{Z}(p)) \\ \uparrow & & \uparrow \\ F^p\widehat{H}^{2p}(\overline{X}_\bullet,\mathbb{Z}(p)) & \longrightarrow & F^p\widehat{H}^{2p}(\overline{X}_\bullet \log D_\bullet,\mathbb{Z}(p)) \end{array}$$

Here, we would like to use Proposition 6.4 to replace the given connection by one that extends without singularity to the compactification, without changing the image in Deligne-Beilinson cohomology; the desired conclusion then follows by functoriality. This can always be done when $X_\bullet$ is just a single algebraic manifold $X = \overline{X} - D$. Thus, the proof of Theorem 1.2 (the non-simplicial version of Statement 6.1) reduced to the compact case (see Section 4.5).

REMARK 6.9. It should be apparent that the preceding development goes through verbatim to the setting of simplicial complex manifolds $X_\bullet$ that are of the form $\overline{X}_\bullet - D_\bullet$, with $\overline{X}_\bullet$ compact Kähler and $D_\bullet$ a divisor with normal crossings in $\overline{X}_\bullet$. In fact, even the condition that $\overline{X}_\bullet$ be Kähler can be dispensed with; one simply makes the distinction between the two sides (isomorphic when $\overline{X}_\bullet$ is Kähler) of

$$F^pH^\bullet(X_\bullet,\mathbb{C}) \longleftarrow H^\bullet(F^pA^\bullet(|\overline{X}_\bullet|\log D_\bullet))$$

and

$$H^\bullet(X_\bullet,\mathbb{C})/F^pH^\bullet(X_\bullet,\mathbb{C}) \hookrightarrow H^\bullet(A^\bullet(|\overline{X}_\bullet|\log D_\bullet)/F^p),$$

and takes the right-hand member wherever we have written the left.

We present in Appendix D a couple of plausible techniques we tried in our unsuccessful attempt to prove Statement 6.1.

Appendix A. Fiber Integration and the Homotopy Formula

In this appendix, we present some basic facts needed in Section 3. Let $p\colon I \times M \to M$ denote the canonical projection of the product of the unit interval with an m-manifold M onto M. For all i, there is a linear mapping

$$B\colon A^{i+1}(I\times M) \to A^i(M) \tag{A.1}$$

characterized by the formula

$$T(B\varphi) = ([I]\times T) \tag{A.2}$$

for all compactly supported i-currents T on M. Whenever T is the current T_ψ defined by

$$T_\psi(\varphi) = \int_M \varphi\wedge\psi \quad (\psi \in A_c^{m-i}(M)), \tag{A.3}$$

for which

(A.4) (i) $\partial T_\psi = (-1)^{i+1}T_{d\psi}$ and (ii) $[I]\times T_\psi = T_{p^*\psi}$,

(A.2) becomes $\int_M B\varphi\wedge\psi = \int_{I\times M}\varphi\wedge p^*\psi$. In fact, in general, $B\varphi$ is just the integral over I of $(\partial_t\rfloor\varphi)$. The mapping B is functorial, in the sense that if $g\colon N\to M$ is smooth, inducing $\tilde{g}\colon I\times N \to I\times M$, then

$$B(\tilde{g}^*\varphi) = g^*B\varphi. \tag{A.5}$$

The identity

$$\partial([I] \times T) = \partial[I] \times T - [I] \times \partial T \tag{A.6}$$

yields at once

$$B(d\varphi) + d(B\varphi) = \varphi_1 - \varphi_0, \tag{A.7}$$

where $\varphi_t = \varphi|_{\{t\}\times M}$. Applied to $\varphi = h^*\omega$, where $h\colon M \times I \to N$ is a smooth homotopy, and $\omega \in A^i(N)$, (A.7) gives the *homotopy formula*:

$$(B \circ h^*)d\omega + d(B \circ h^*)\omega = h_1^*\omega - h_0^*\omega. \tag{A.8}$$

The dual version for currents is

$$\partial(h_*([I] \times T)) + h_*([I] \times \partial T) = (h)_*T - (h_0)_*T. \tag{A.9}$$

There is a parallel treatment for singular cochains with any coefficients, as follows. A smooth singular simplex $\sigma\colon \Delta^i \to M$ defines a current:

$$[\sigma] = \sigma_*[\Delta^i], \tag{A.10}$$

such that $[\partial\sigma] = \partial[\sigma]$. Using the standard triangulation of $I \times \Delta^i$, a smooth mapping

$$\tau\colon I \times \Delta^i \to N$$

defines a singular chain $[\tau]$ on N. Taking τ of the form

$$h_*(1 \times \sigma)\colon I \times \Delta^i \to I \times M \to N,$$

we have

$$\partial h_*(1 \times \sigma) + h_*(1 \times \partial\sigma) = (h_1)_*\sigma - (h_0)_*\sigma, \tag{A.11}$$

which under (A.10) recovers (A.9). Define, for a smooth singular cochain f on $I \times M$,

$$(Bf)(\sigma) = f(1 \times \sigma) \tag{A.12}$$

(cf. (A.2)); putting $f = h^*\tau$, we get from (A.11)

$$B(h^*\delta\tau) + \delta(Bh^*\tau) = h_1^*\tau - h_0^*\tau. \tag{A.13}$$

When τ is a differential form (see Section 3.1), one recovers (A.8), i.e.:

PROPOSITION A.1. *The homotopy formula for smooth singular cochains, when restricted to differential forms, coincides with the homotopy formula for differential forms.*

Appendix B. Universal Weak F^1-Connections

We show that if one ceases to impose conditions at infinity, it is possible to construct a universal "F^1-connection." We start with a definition:

DEFINITION B.1. A *weak F^1-connection* on E is a connection for which the local connection forms and therefore also the curvature lie in $F^1A^\bullet(X, \mathrm{End}(E))$.

Imposing the growth condition along D, as in (4.3), in (B.1) gives Section 4.4.

We introduce the notion of a *universal weak F^1-connection*. The discussion proceeds as in that of Section 2.2. In the setting of (2.1), suppose that Y is a *complex* manifold, and take $G = \mathrm{GL}_n(\mathbb{C})$. Let $P \to Y$ be a holomorphic principal $\mathrm{GL}_n(\mathbb{C})$-bundle, and

$$p\colon T^{1,0}P \to T^{1,0}Y$$

be the induced map of holomorphic tangent bundles. We put

$$\widetilde{Y}^{(1)} = \{\widehat{\alpha} \in \mathrm{Hom}(T^{1,0}Y, T^{1,0}P/G) : \widehat{p} \circ \widehat{\alpha} = \mathrm{id}_{T^{1,0}Y}\}, \tag{B.1}$$

where $\widehat{p}\colon T^{1,0}P/G \to T^{1,0}Y$ is induced by p. Then $\widetilde{Y}^{(1)}$ can be identified with a submanifold of the space $\widetilde{Y}$ in (2.1) in the following way. One makes use of the canonical splitting $(TY)_{\mathbb{C}} = T^{1,0}Y \oplus T^{0,1}Y$. The projection $q^{(1)}\colon \widetilde{Y}^{(1)} \to Y$ then becomes a complex affine space sub-bundle of $q\colon \widetilde{Y} \to Y$ (via $\alpha = \widehat{\alpha} \oplus 0$); a local holomorphic trivialization of P gives rise to the local description (cf. Proposition 2.4):

$$\widetilde{Y}^{(1)} \cong \mathrm{Hom}(T^{1,0}Y, \mathfrak{g}). \tag{B.2}$$

There is a tautological *holomorphic* connection on $q^{(1)*}P$.

PROPOSITION B.2. *The weak F^1-connections on P are in one-to-one correspondence with the C^∞ sections of $q^{(1)}$; the holomorphic sections of $q^{(1)}$ correspond to the holomorphic connections on P.*

One must remember that the notion of weak F^1-connection is respected under pullback only by *holomorphic* maps. For a holomorphic map $f\colon X \to Y$, the pullback connection on f^*P is induced by

$$\mathrm{Hom}(T^{1,0}_{f(x)}Y, \mathfrak{g}) \to \mathrm{Hom}(T^{1,0}_x X, \mathfrak{g}).$$

Taking $Y = G_{\mathbb{C}}(n)$, we deduce (cf. Corollary 2.5, Proposition 2.6):

COROLLARY B.3. *If a holomorphic vector bundle E is classified by a holomorphic immersion $f\colon X \to G_{\mathbb{C}}(n)$, then every weak F^1-connection on E is the pullback of the tautological weak F^1-connection on $\widetilde{G}_{\mathbb{C}}(n)^{(1)}$ by a C^∞ lifting $\tilde{f}\colon X \to \widetilde{G}_{\mathbb{C}}(n)^{(1)}$ of f.*

Appendix C. Simplicial de Rham Theory

In this appendix we give a resume of simplicial manifolds, tailored to our present needs. The principal reference is [**15**].

C.1. Simplicial Manifolds. Let $\mathbf{\Delta}$ be the category whose objects are the finite ordinals

$$[n] = \{0 < 1 < \cdots < n\} \quad n \in \mathbb{N}$$

and whose morphisms are order-preserving *monomorphisms*. Of special importance are the face maps $d_j\colon [n-1] \to [n]$, the unique morphism that omits the value j.

View $[n]$ as the set of the set of vertices of the standard n-simplex Δ^n. Then, each morphism $f\colon [n] \to [m]$ of $\mathbf{\Delta}$ induces a simplicial map

$$|f|\colon \Delta^n \to \Delta^m.$$

Definition C.1. A (*strict*) *simplicial* (resp. *cosimplicial*) object of a category $\mathcal{C}$ is a contravariant (resp. covariant) functor $\mathbf{\Delta} \to \mathcal{C}$. Morphisms between strict (co)simplicial objects in $\mathcal{C}$ are natural transformations of functors.

Denote the category of smooth manifolds, each of whose connected components are second countable, and smooth maps by $\mathcal{M}$.

Definition C.2. A *simplicial manifold* $M_\bullet$ is a strict simplicial object of $\mathcal{M}$. The object of $\mathcal{M}$ corresponding to the ordinal $[n]$ will be denoted by M_n. The smooth map induced by the morphism $f\colon [n] \to [m]$ will be denoted by

$$f_M\colon M_m \to M_n.$$

The *geometric realization* $|M_\bullet|$ of the simplicial manifold $M_\bullet$ is defined to be the topological space

$$\coprod_{n\geq 0} (M_n \times \Delta^n)/\sim$$

endowed with the quotient topology, where $\sim$ is the equivalence relation which identifies the points

$$(x, |f|\xi) \in M_m \times \Delta^m \quad \text{and} \quad (f_M(x), \xi) \in M_n \times \Delta^n$$

whenever $f\colon [n] \to [m]$ is a morphism.

Observe that simplicial sets can be viewed as simplicial manifolds. In particular, the standard simplicial model of the classifying space of $\mathrm{GL}_n(\mathbb{C})^\delta$, the general linear group *with the discrete topology*, is a simplicial manifold.

There is a fully faithful imbedding of $\mathcal{M}$ into $\mathcal{M}^{\mathbf{\Delta}}$, the category of simplicial manifolds: The manifold M is taken to the simplicial manifold $M_\bullet$ with

$$M_n = \begin{cases} M & n = 0 \\ \emptyset & n > 0. \end{cases}$$

C.2. The de Rham Functor. Suppose that $C^\bullet$ is a strict cosimplicial cochain complex of $\mathbb{C}$-vector spaces. Denote the cochain complex corresponding to $[n]$ by $C^\bullet[n]$.

Definition C.3. [**23**, (5.3)] The *de Rham complex* $DC^\bullet$ of $C^\bullet$ is the total complex of the double complex $D^{\bullet\bullet}C^\bullet$, where $D^{s,t}C^\bullet$ consists of those elements (w_n) of

$$\prod_{n\geq 0} A^s(\Delta^n) \otimes C^t[n]$$

that satisfy the compatibility condition:

$$(|f|^* \otimes \mathrm{id})w_n = (id \otimes f_*)w_m \in A^s(\Delta^m) \otimes C^t[n],$$

whenever $f\colon [m] \to [n]$ is a morphism. The differentials in the double complex arise from exterior differentiation in the simplicial direction s and from the differential of $C^\bullet$ in the t direction.

Now suppose that $M_\bullet$ is a simplicial manifold. Applying the usual de Rham complex functor to $M_\bullet$ we obtain a cosimplicial cochain complex $A^\bullet(M_\bullet)$ whose value on the ordinal $[n]$ is $A^\bullet(M_n)$.

DEFINITION C.4. The *de Rham complex* $A^\bullet|M_\bullet|$ *of the simplicial manifold* $M_\bullet$ is the de Rham complex $DA^\bullet(M_\bullet)$ of $A^\bullet(M_\bullet)$. It is naturally bigraded:

$$A^k|M_\bullet| = \bigoplus_{s+t=k} A^{s,t}|M_\bullet|,$$

where

$$A^{s,t}|M_\bullet| = D^s A^t(M_\bullet).$$

If $M_\bullet$ is a simplicial set, then $A^\bullet|M_\bullet|$ is just the Thom-Whitney de Rham complex of $\mathbb{C}$-valued forms on $M_\bullet$ (cf. [**37**], [**38**]).

PROPOSITION C.5. *The association of* $A^\bullet|M_\bullet|$ *to the simplicial manifold* $M_\bullet$ *defines a contravariant functor from the category of simplicial manifolds into the category of commutative differential graded algebras.*

C.3. The de Rham Theorem. In this section we sketch the de Rham theorem for simplicial manifolds. It is first convenient to introduce the functor $C^\bullet$ from the category of (strict) cosimplicial Abelian groups into the category of cochain complexes: The cosimplicial Abelian group A is taken to the cochain complex $C^\bullet A$, where

$$C^n A = A[n]$$

and the differential $\delta\colon C^n A \to C^{n+1} A$ is the alternating sum

$$\delta = \sum_{j=0}^{n+1} (-1)^j d^j$$

of the coface maps $d^i\colon A[n] \to A[n+1]$. When $A^\bullet$ is a strict cosimplicial cochain complex, $C^\bullet A^\bullet$ is a double complex.

The classical de Rham theorem states that if M is a smooth manifold, the integration map

$$\int : A^\bullet(M) \to S^\bullet(M, \mathbb{C})$$

is a quasi-isomorphism which induces an algebra isomorphism on cohomology.

Now, to each simplicial manifold $M_\bullet$, we can associate the strict cosimplicial cochain complex $S^\bullet(M_\bullet)$ whose value on the ordinal $[n]$ is $S^\bullet(M_n)$. Applying the cochain functor C above, we obtain a double complex $C^\bullet S^\bullet(M_\bullet)$. Denote the associated single complex (total complex) by $S^\bullet(M_\bullet)$. Note that if $M_\bullet$ is a strict simplicial set, then $S^\bullet(M_\bullet)$ denotes the usual complex of simplicial cochains associated to $M_\bullet$.

For each simplicial manifold $M_\bullet$, define a $\mathbb{C}$-linear function

$$A^{s,t}|M_\bullet| \to S^t(M_s, \mathbb{C}) \tag{C.1}$$

by taking the compatible family

$$(w_n) \in \prod_{n\geq 0} A^s(\Delta^n) \otimes A^t(M_n)$$

to the t-cochain

$$\sigma \mapsto \int_{\Delta^s \times \Delta^t} w_s, \quad \sigma : \Delta^t \to M_s.$$

The following result is a direct consequence of Stokes' Theorem and the compatibility conditions.

PROPOSITION C.6. *The mapping* (C.1) *is a morphism of double complexes.*

COROLLARY C.7. *The mapping* (C.1) *induces a morphism of complexes*

$$\int : A^\bullet|M_\bullet| \to S^\bullet(M_\bullet, \mathbb{C}).$$

The following de Rham theorem is a modest generalization of those of Thom (see [**37**]) and Sullivan [**36**]. Proofs of statements similar to the following appear in [**14, 15**], [**23**, (5.4)].

THEOREM C.8. *For each simplicial manifold $M_\bullet$, the integration mapping*

$$\int : A^\bullet|M_\bullet| \to S^\bullet(M_\bullet, \mathbb{C})$$

is a quasi-isomorphism that induces an algebra isomorphism

$$H^\bullet(A^\bullet|M_\bullet|) \to H^\bullet(S^\bullet(M_\bullet, \mathbb{C})) \cong H^\bullet(|M_\bullet|, \mathbb{C}).$$

C.4. Bundles over Simplicial Manifolds. As is well-known, we may talk equivalently about vector bundles or about principal bundles, and here we choose the latter. Let G be a Lie group. A principal G-bundle over the simplicial manifold $M_\bullet$ is a simplicial manifold $P_\bullet$ and a morphism $\pi\colon P_\bullet \to M_\bullet$ of simplicial manifolds satisfying:

(i) for each n, $\pi\colon P_n \to M_n$ is a principal G-bundle;
(ii) for each morphism $f\colon [m] \to [n]$ of $\mathbf{\Delta}$, f_P is a morphism of G-bundles:

$$\begin{array}{ccc} P_n & \xrightarrow{f_P} & P_m \\ \downarrow & & \downarrow \\ M_n & \xrightarrow{f_M} & M_m \end{array}$$

The second condition guarantees that the principal G-action on $P_\bullet$ passes to its geometric realization:

PROPOSITION C.9. *If $\pi\colon P_\bullet \to M_\bullet$ is a principal G-bundle over a simplicial manifold, then $|\pi|\colon |P_\bullet| \to |M_\bullet|$ is a principal G-bundle, with G-action induced by*

$$\Delta^n \times P_n \times G \to \Delta^n \times P_n$$
$$(t, x, g) \mapsto (t, xg).$$

One can define vector bundles over simplicial manifolds similarly.

DEFINITION C.10. A connection ∇ on the principal G-bundle $P_\bullet \to M_\bullet$ over a simplicial manifold $M_\bullet$ is a G-invariant 1-form

$$\omega \in A^1(|P_\bullet|, \operatorname{Ad}\mathfrak{g})$$

taking values in $\mathfrak{g}$ (cf. Definition C.3), the Lie algebra of G, on which G acts via the adjoint representation.

As in the classical case, the $\mathfrak{g}$-valued 2-form

$$d\omega + \frac{1}{2}[\omega, \omega] \in A^2(|P_\bullet|, \operatorname{Ad}\mathfrak{g})$$

descends to a $\mathfrak{g}$-valued 2-form $\Theta \in A^2(|M_\bullet|, \operatorname{Ad}\mathfrak{g})$. The 2-form Θ is necessarily unique and is called the *curvature* of ∇.

We conclude this subsection with a review of the extension of Chern-Weil theory to simplicial manifolds (cf. [**14, 15**]). We restrict our attention to Chern classes.

DEFINITION C.11. The kth *Chern form* $c_k(\nabla)$ of a connection ∇ on the principal $\mathrm{GL}_n(\mathbb{C})$-bundle $P_\bullet \to M_\bullet$ over a simplicial manifold is:

$$c_k(\nabla) := C_k(\Theta) \in A^{2k}|M_\bullet|,$$

where Θ is the curvature of ∇ and C_k is the invariant polynomial defined in Section 1.1.

The standard arguments can be used to prove the following result (see [**15**]).

THEOREM C.12. *If $c_0(\nabla), \dots, c_n(\nabla)$ are the Chern forms of a connection on a principal $\mathrm{GL}_n(\mathbb{C})$ bundle over a simplicial manifold, then*

(i) *each $c_k(\nabla)$ is closed;*

(ii) *the cohomology class of $c_k(\nabla)$ is the image under the canonical map*

$$H^{2k}(|M_\bullet|, \mathbb{Z}(k)) \to H^{2k}(|M_\bullet|, \mathbb{C})$$

of the kth Chern class of $|P_\bullet| \to |M_\bullet|$.

C.5. Cheeger-Simons Classes. Suppose that M is a simplicial manifold and Λ an Abelian group. Denote by $C^\bullet(M_\bullet, \Lambda)$ the cochain complex of compatible cochains on $|M_\bullet|$. This is, $C^k(M_\bullet, \Lambda)$ consists of those elements (c_n) of

$$\prod_{n \geq 0} S^k(\Delta^n \times M_n)$$

that satisfy the compatibility condition

$$(|f| \times \mathrm{id})^* c_n = (\mathrm{id} \times f_M)^* c_m \in S^k(\Delta^m \times M_n)$$

for each morphism $f\colon [m] \to [n]$ of $\mathbf{\Delta}$.

The following fact is standard:

PROPOSITION C.13. *The natural inclusion*

$$C^\bullet(M_\bullet, \Lambda) \hookrightarrow S^\bullet(|M_\bullet|, \Lambda)$$

is a quasi-isomorphism.

A direct consequence of the compatibility condition of Definition C.3 is that integration induces a chain map

$$I\colon A^\bullet|M_\bullet| \to C^\bullet(M_\bullet, \mathbb{C}).$$

This is easily seen to be a quasi-isomorphism.

DEFINITION C.14. Suppose that Λ is a subgroup of $\mathbb{C}$. The group of mod Λ *differential characters* of degree k of a simplicial manifold $M_\bullet$ is defined by

$$\widehat{H}^k(M_\bullet, \Lambda) = H^{k-1}(\operatorname{cone}\{A^{\geq k}|M_\bullet| \xrightarrow{\iota_\Lambda} C^\bullet(M_\bullet, \mathbb{C}/\Lambda)\}),$$

where ι_Λ is the composite of I with the natural map $C^\bullet(M_\bullet, \mathbb{C}) \to C^\bullet(M_\bullet, \mathbb{C}/\Lambda)$.

Since the differential characters are constructed from a cone, we have the short exact sequence

$$0 \to H^{k-1}(|M_\bullet|, \mathbb{C}/\Lambda) \to \widehat{H}^k(M_\bullet, \Lambda) \to A^k_{cl}(|M_\bullet|, \Lambda) \to 0, \tag{C.2}$$

where

$$A^k_{cl}(|M_\bullet|, \Lambda) = \{\varphi \in A^k|M_\bullet| \colon d\varphi = 0 \text{ and the periods of } \varphi \text{ on } H_k(|M_\bullet|) \text{ lie in } \Lambda\}.$$

The construction of the Cheeger-Simons invariant $\hat{c}_p$ given in Section 3.5 works equally well in the simplicial setting. We use it to define Cheeger-Simons classes

$$\hat{c}_p(E_\bullet, \nabla) \in \widehat{H}^{2p}(M_\bullet, \mathbb{Z}(p))$$

for complex vector bundles $E_\bullet \to M_\bullet$ with connection over simplicial manifolds.

C.6. Simplicial Varieties. Let *Alg* be the category whose objects are arbitrary disjoint unions of complex algebraic manifolds and whose morphisms are morphisms of varieties on each component.

DEFINITION C.15. A *smooth simplicial variety* is a strict[7] simplicial object of *Alg*. The category of simplicial varieties will be denoted by Alg^{Δ}.

For later use we record the following result whose proof follows from standard results in algebraic geometry.

PROPOSITION C.16. (a) *If $Y_\bullet$ is a simplicial variety, then there is a simplicial variety $\overline{Y}_\bullet$ and an open immersion $Y_\bullet \to \overline{Y}_\bullet$ such that each $\overline{Y}_n$ is complete and such that $\overline{Y}_n - Y_n$ is a normal crossing divisor in $\overline{Y}_n$. Moreover, if $f\colon Y_\bullet \to Z_\bullet$ is a morphism of simplicial varieties, then there exist smooth completions $\overline{Y}_\bullet$ of $Y_\bullet$ and $\overline{Z}_\bullet$ of $Z_\bullet$ as above and a morphism $\bar{f}\colon \overline{Y}_\bullet \to \overline{Z}_\bullet$ whose restriction to $Y_\bullet$ is f.*

(b) *If $Y'_\bullet$ and $Y''_\bullet$ are two smooth completions as in (a) of a simplicial variety $Y_\bullet$, then there exists a third completion $\overline{Y}_\bullet$ of $Y_\bullet$ and morphisms $\overline{Y}_\bullet \to Y'_\bullet$, $\overline{Y}_\bullet \to Y''_\bullet$ such that the diagram*

$$\begin{array}{ccccc} & & Y_\bullet & & \\ & \swarrow & \downarrow & \searrow & \\ Y'_\bullet & \longleftarrow & \overline{Y}_\bullet & \longrightarrow & Y''_\bullet \end{array}$$

commutes.

C.7. Complements. Because the cohomology of $B\,\mathrm{GL}_n(\mathbb{C})^\delta$ and other simplicial varieties that occur naturally in the proof of Theorems 1.1 and 1.2 are not of finite type, we need to say a few words about generalizations of Hodge theory to that case.

We will say that a simplicial manifold is of *finite type* if the cohomology of its n-simplices is finite dimensional for each n. Every simplicial algebraic variety can be written as the direct limit of simplicial varieties which are of finite type. It follows that the cohomology, with characteristic zero field coefficients, of every simplicial variety is the inverse limit of an inverse system of mixed Hodge structures. We view the cohomology of a simplicial variety as a topological vector space, where the neighbourhoods of 0 are the kernels of restriction maps to the cohomology of

[7]The modifier *strict* means that this is the category whose objects come *without* degeneracy maps.

simplicial varieties of finite type. It is complete in this topology. We extend the notion of mixed Hodge structures to such topological spaces.

Suppose that Λ is a subring of $\mathbb{R}$ with quotient field $\mathbb{L}$. Denote the category of Λ-mixed Hodge structures by $\mathcal{H}_\Lambda$. Consider the category of *completed Λ-mixed Hodge structures* $\widehat{\mathcal{H}}_\Lambda$ whose objects consist of triples

$$(H_\Lambda, (H_\mathbb{L}, W_\bullet), (H_\mathbb{C}, F^\bullet)),$$

where H_Λ is a Λ-module; $H_\mathbb{L} = H_\Lambda \otimes_\Lambda \mathbb{L}$ a complete topological $\mathbb{L}$-vectorspace with an increasing filtration $W_\bullet$ by closed subspaces; $H_\mathbb{C}$ the complexification of $H_\mathbb{L}$ with the corresponding topology and $F^\bullet$ a decreasing filtration by closed subspaces. In addition we suppose that there is a base of neighborhoods (N_α) of 0 in $H_\mathbb{L}$ such that for each α;

$$(H_\Lambda / (H_\Lambda \cap N_\alpha), (H_\mathbb{L}/N_\alpha, W_\bullet), (H_\mathbb{C}/N_\alpha \otimes \mathbb{C}, W_\bullet \otimes \mathbb{C}, F^\bullet))$$

is a Λ-mixed Hodge structure. A morphism

$$(H_\Lambda, (H_\mathbb{L}, W_\bullet), (H_\mathbb{C}, F^\bullet)) \to (H'_\Lambda, (H'_\mathbb{L}, W_\bullet), (H'_\mathbb{C}, F^\bullet))$$

consists of a Λ-module homomorphism $H_\Lambda \to H'_\Lambda$ that induces continuous, filtration preserving homomorphisms $H_\mathbb{L} \to H'_\mathbb{L}$ and $H_\mathbb{C} \to H'_\mathbb{C}$.

It is straightforward to show that the cohomology of every simplicial variety inherits a functorial mixed Hodge structure, in the above sense, from its de Rham complex (cf. [**13**, (8.1.19)]).

C.8. Deligne-Beilinson (DB) Cohomology. Suppose that $Y_\bullet$ is a smooth simplicial variety. By Proposition C.16 we can find a smooth completion $\overline{Y}_\bullet$ of $Y_\bullet$ such that each $\overline{Y}_n - Y_n$ is a normal crossings divisor D_n. Denote by I_p the composite

$$F^p A^\bullet(|\overline{Y}_\bullet| \log D_\bullet) \to A^\bullet|Y_\bullet| \xrightarrow{I} S^\bullet(Y_\bullet),$$

where F denotes the Hodge filtration of the de Rham complex.

Definition C.17. Suppose that Λ is a subring of $\mathbb{R}$ and that $p \in \mathbb{N}$. The *Deligne-Beilinson* (or DB) *cohomology* $H^\bullet_\mathcal{D}(Y_\bullet, \Lambda(p))$ of the simplicial variety $Y_\bullet$ is the cohomology of the complex

$$D^\bullet(Y_\bullet, \overline{Y}_\bullet; \Lambda(p)) = \operatorname{cone}\{F^p A^\bullet(|\overline{Y}_\bullet| \log D_\bullet) \xrightarrow{I_p} S^\bullet(Y_\bullet, \mathbb{C}/\Lambda(p))\}[-1].$$

As DB-cohomology is constructed from a cone, we have:

Proposition C.18. *For each simplicial variety $Y_\bullet$, there is natural long exact sequence*

$$\text{(C.3)} \quad \cdots \to H^{k-1}(|Y_\bullet|, \mathbb{C}/\Lambda(p)) \to H^k_\mathcal{D}(Y_\bullet, \Lambda(p)) \to F^p H^k(|Y_\bullet|, \mathbb{C}) \to H^k(|Y_\bullet|, \mathbb{C}/\Lambda(p)) \to \cdots.$$

Corollary C.19. *For each simplicial variety $Y_\bullet$, there is a natural homomorphism*

$$H^{k-1}(|Y_\bullet|, \mathbb{C}/\Lambda(p)) \to H^k_\mathcal{D}(Y_\bullet, \Lambda(p)).$$

Since a point is a smooth projective variety of dimension 0, every simplicial set can be regarded as a simplicial variety. Since the Hodge filtration of the de Rham complex of such a simplicial variety satisfies $F^p = 0$ when $p > 0$ we have:

COROLLARY C.20. *If $K_\bullet$ is a simplicial set and $p > 0$, then the natural homomorphism*

$$H^{k-1}(|K_\bullet|, \mathbb{C}/\Lambda(p)) \to H^k_{\mathcal{D}}(K_\bullet, \Lambda(p))$$

is an isomorphism.

COROLLARY C.21. *There is a natural isomorphism*

$$H^{2p}_{\mathcal{D}}(B\,\mathrm{GL}_n(\mathbb{C})^\delta, \Lambda(p)) \cong H^{2p-1}(B\,\mathrm{GL}_n(\mathbb{C})^\delta, \mathbb{C}/\Lambda(p)),$$

where $B\,\mathrm{GL}_n(\mathbb{C})^\delta$ denotes the classifying space of the general linear group endowed with the discrete topology.

C.9. Chern Classes in DB Cohomology. A vector bundle of rank r over a simplicial variety $Y_\bullet$ is a morphism of simplicial varieties $E_\bullet \to Y_\bullet$ where, for each n, $E_n \to Y_n$ is an algebraic vector bundle of rank r. Beilinson [**2**] and Gillet [**20**] have defined Chern classes

$$c_p^B(E_\bullet) \in H^{2p}_{\mathcal{D}}(Y_\bullet, \mathbb{Z}(p))$$

for vector bundles $E_\bullet \to Y_\bullet$ over a simplicial varieties: Since the splitting principle holds for DB-cohomology, the existence of Chern classes in Deligne cohomology reduces to the existence of first Chern classes

$$c_1(L_\bullet) \in H^2_{\mathcal{D}}(Y_\bullet, \mathbb{Z}(1))$$

for line bundles $L_\bullet \to Y_\bullet$. These are constructed in [**2**, (1.7)], in the manner of [**22**].

By a connection on a vector bundle $E_\bullet \to Y_\bullet$ we mean a connection on the associated $\mathrm{GL}_n(\mathbb{C})$ bundle $P(E_\bullet) \to Y_\bullet$ over the underlying simplicial manifold.

PROPOSITION C.22. *Suppose that $Y_\bullet$ is a smooth simplicial variety and that*

$$H^{2p-1}(|Y_\bullet|, \mathbb{C}/\mathbb{Z}(p)) = 0.$$

If ∇ is a connection on the vector bundle $E_\bullet \to Y_\bullet$ whose pth Chern form satisfies

$$c_p(\nabla) \in F^p A^{2p}(|\overline{Y}_\bullet| \log D_\bullet),$$

for some smooth completion $\overline{Y}_\bullet$ of $Y_\bullet$ where $\overline{Y}_\bullet - Y_\bullet$ is a divisor with normal crossings, then

$$(c_p(\nabla), -\eta) \in D^{2p}(Y_\bullet, \overline{Y}_\bullet; \mathbb{Z}(p))$$

represents $c_p^B(E_\bullet)$, where η is any element of $S^{2p-1}(Y_\bullet, \mathbb{C}/\mathbb{Z}(p))$ such that $\delta\eta = c_p(\nabla)$.

PROOF. The result follows from Proposition C.18 and the vanishing of $H^{2p-1}(|Y_\bullet|, \mathbb{C}/\mathbb{Z}(p))$ which imply that

$$H^{2p}_{\mathcal{D}}(Y_\bullet, \mathbb{Z}(p)) \to F^p H^{2p}(|Y_\bullet|, \mathbb{C})$$

is injective. □

C.10. Multisimplicial Manifolds and Varieties. In some constructions that one would like to do, such as Proposition 6.5, one needs to work with *multisimplicial varieties.* Define the category of 0-simplicial manifolds to be $\mathcal{M}$, the category of manifolds. For each $n \geq 1$, we define inductively an *n-simplicial manifold* to be a strict simplicial object in the category of $(n-1)$-simplicial manifolds; morphisms between two n-simplicial manifolds are natural transformations. The category of n-simplicial manifolds and morphisms will be denoted by $\mathcal{M}^{\mathbf{\Delta}^n}$. Just as every manifold can be regarded as a simplicial manifold, every n-simplicial manifold can be regarded as an $(n+1)$ simplicial manifold by placing it in degree 0 of the $(n+1)$-simplicial manifold, and the empty set in positive degrees. Objects of one of the categories $\mathcal{M}^{\mathbf{\Delta}^n}$ are called a *multisimplicial manifolds.*

The geometric realization $|M_\bullet|$ of an n-simplicial space $M_\bullet$ is defined inductively to be the geometric realization of the simplicial topological space $|M|_\bullet$ whose space of k-simplices is the geometric realization $|M_k|$ of the $(n-1)$-simplicial manifold M_k.

One can extend the definition of de Rham complex, the de Rham theorem, Chern-Weil theory and Cheeger-Simons Chern classes to multisimplicial manifolds, and the definition of Deligne-Beilinson cohomology to multisimplicial varieties, all in a straightforward way.

Appendix D. Additional Techniques

D.1. Killing the Unipotent Directions. Let B be a Borel subgroup of $G = \mathrm{GL}_n(\mathbb{C})$. Then G is a principal B-bundle over the Grassmannian G/B. If one decomposes B as $T \times U$, where T is the maximal torus and U is the unipotent radical, one sees that the trouble in Remark 6.6 is that the Maurer-Cartan forms have second-order poles in the direction of U. But U is contractible. One might try to replace G by G/U, an algebraic variety of the same homotopy type. We didn't see that this was helpful.

D.2. The Q-Filtration. Because of Remark 6.6, we were led to introduce another filtration on the de Rham complex that is coarser than the Hodge filtration, yet induces the same filtration on cohomology.

We begin by letting X be the disc Δ, and $D = \{0\}$. The logarithmic (holomorphic) de Rham complex

$$\Omega_X^\bullet(\log D) = \{\mathcal{O}_X \to \Omega_X^1(\log D)\}$$

comes with its usual Hodge filtration F. Let $\Omega_X^\bullet(*D)$ denote the larger complex of forms that are meromorphic along D. We define a decreasing filtration Q on $\Omega_X^\bullet(*D)$ by:

$$Q^0\Omega_X^\bullet(*D) = \Omega_X^\bullet(*D), \quad Q^1\Omega_X^\bullet(*D) = F^1\Omega_X^\bullet(\log D) = \Omega_X^1(\log D)[-1], \tag{D.1}$$

and $Q^2\Omega_X^\bullet(*D) = 0$. In other words, we "discount" a 1-form that has worse than logarithmic singularities. It is standard that the inclusion $\Omega_X^\bullet(\log D) \subset \Omega_X^\bullet(*D)$ is a quasi-isomorphism, and it respects the filtrations.

PROPOSITION D.1. *The inclusion of filtered complexes*

$$(\Omega_X^\bullet(\log D), F) \hookrightarrow (\Omega_X^\bullet(*D), Q)$$

is a filtered quasi-isomorphism.

PROOF. The real content of the above assertion is that

$$Gr_F^0 \Omega_X^\bullet(\log D) \to Gr_Q^0 \Omega_X^\bullet(*D)$$

is a quasi-isomorphism. This is equivalent to the exactness of

$$0 \to \mathcal{O}_X \to \mathcal{O}_X(*D) \xrightarrow{d} \Omega_X^1(*D)/\Omega_X^1(\log D) \to 0,$$

which is elementary. □

Next, we consider products of the preceding. For $X = \Delta^k$, and D the union of the coordinate axes, so that $X - D = (\Delta^*)^k$, we use the tensor product of the Q-filtrations of the factors to define the Q-filtration on $\Omega_X^\bullet(*D)$; and for the product of the this pair with a polydisc Δ^l, we take the tensor product of the preceding with the Hodge filtration F of Δ^l. It is an easy exercise to check:

LEMMA D.2. *Let X and D be as above, so $X - D \cong (\Delta^*)^k \times \Delta^l$. Then the filtration Q is independent of the choice of coordinates. The inclusion*

$$(\Omega_X^\bullet(\log D), F) \hookrightarrow (\Omega_X^\bullet(*D), Q)$$

is a filtered quasi-isomorphism.

This yields at once the following globalization:

PROPOSITION D.3. *Let X be a complex manifold, and D a divisor with normal crossings on X. Then there is a uniquely-defined filtration Q of $\Omega_X^\bullet(*D)$, given by the above in any system of local coordinates adapted to D. Moreover, the inclusion*

$$(\Omega_X^\bullet(\log D), F) \hookrightarrow (\Omega_X^\bullet(*D), Q)$$

is a filtered quasi-isomorphism.

REMARK D.4. (i) In contrast to F, the filtration Q is *not* functorial: e.g., in two variables $dz/w^2 \in Q^1$, whereas its restriction to the diagonal is $dz/z^2 \notin Q^1$. However, if X and D are as in the proposition, and $f\colon X' \to X$ is a morphism of complex manifolds, such that $f^*(D)$ is a *reduced* divisor with normal crossings, then pulling back by f respects the Q-filtrations.

(ii) Q is not multiplicative: both dw and dz/w^2 are in Q^1, but $dw \wedge dz/w^2 \notin Q^2$. However, it does induce a functorial, multiplicative filtration on $H^\bullet(X - D, \mathbb{C})$, via the quasi-isomorphism of Proposition D.3.

Finally, let $X_\bullet$ be a smooth simplicial variety with smooth completion $\overline{X}_\bullet$, such that $D_\bullet = \overline{X}_\bullet - X_\bullet$ has normal crossings, and where the face maps preserve the Q filtration. One defines in the obvious way a subcomplex $A^\bullet(|\overline{X}_\bullet| * D_\bullet)$ of the simplicial de Rham complex $A^\bullet(|X_\bullet|)$, that contains $A^\bullet(|\overline{X}_\bullet| \log D_\bullet)$. One places the Q-filtration on $A^\bullet(|\overline{X}_\bullet| * D_\bullet)$ by taking the Q-filtration on each $A^\bullet(|\overline{X}_n| * D_n)$. Since the Hodge filtration F of $A^\bullet(|\overline{X}_\bullet| \log D_\bullet)$ is defined similarly, it follows from Proposition D.3 that

PROPOSITION D.5. *Let $\overline{X}_\bullet$ and $D_\bullet$ be as above. Then the inclusion*

$$(A^\bullet(|\overline{X}_\bullet| \log D_\bullet), F) \hookrightarrow (A^\bullet(|\overline{X}_\bullet| * D_\bullet, Q)$$

is a filtered quasi-isomorphism.

One can determine, at least, that the Maurer-Cartan connection forms, and then the Chern forms for the connection constructed in Proposition 6.5, are in Q^1 along the zero and polar loci of the determinant function (recall Remark 6.6). However, for this fact to be useful, one needs a serviceable compactification of the simplicial model of BG.

References

1. M. Artin and B. Mazur, *Etale homotopy*, Lecture Notes in Math., vol. 100, Springer-Verlag, Berlin-New York, 1969.
2. A. A. Beilinson, *Higher regulators and values of L-functions*, Current Problems in Mathematics, vol. 24, Itogi Nauki i Tekhniki, Akad. Nauk SSSR, Vsesoyuz. Inst. Nauchn. i Tekhn. Inform., Moscow, 1984, pp. 181–238 (Russian); English transl., J. Soviet Math. **30** (1985), 2036–2070.
3. S. Bloch, *Applications of the dilogarithm function in algebraic K-theory and algebraic geometry*, International Symposium on Algebraic Geometry (Kyoto, 1977) (M. Nagata, ed.), Kinokuniya Book Store, Tokyo, 1978, pp. 103–114.
4. A. Borel, *Cohomologie de* SL_n *et valeurs de fonctions de zeta aux points entiers*, Ann. Scuola Norm. Sup. Pisa Cl. Sci. (4) **4** (1977), 613–636.
5. J.-L. Brylinski, *Comparison of the Beilinson-Chern classes with the Chern-Cheeger-Simons classes*, Advances in Geometry, Progr. Math., vol. 172, Birkhäuser, Boston, 1999, pp. 95–105.
6. E. Calabi and B. Eckmann, *A class of compact, complex manifolds which are not algebraic*, Ann. of Math. (2) **58** (1953), 494–500.
7. H. Cartan, *Notions d'algèbre différentielle; application aux groupes de Lie et aux variétés où opère un groupe de Lie*, Colloque de topologie (espaces fibrés) (Bruxelles, 195), G. Thone, Liège, 1950, pp. 15–27.
8. ———, *La transgression dans un groupe de Lie et dans un espace fibré principal*, Colloque de topologie (espaces fibrés), G. Thone, Liège, 1950, pp. 57–71.
9. J. Cheeger and J. Simons, *Differential characters and geometric invariants*, Geometry and Topology, (College Park, MD, 1983/84) (J. Alexander and J. Harer, eds.), Lecture Notes in Math., vol. 1167, Springer, Berlin-New York, 1985, pp. 50–80.
10. S. S. Chern *Complex manifolds without potential theory*, 2nd edition, Universitext, Springer-Verlag, New York-Heidelberg, 1979.
11. S. S. Chern and J. Simons, *Characteristic forms and geometric invariants*, Ann. of Math. (2) **99** (1974), 48–69.
12. P. Deligne,*Équations différentielles à points singuliers réguliers*, Lecture Notes in Math., vol. 163, Springer-Verlag, Berlin- New York, 1970.
13. ———, *Théorie de Hodge.* II, Inst. Hautes Études Sci. Publ. Math. **40** (1971), 5–57; III, **44** (1974), 5–77.
14. J. L. Dupont, *Simplicial de Rham cohomology and characteristic classes of flat bundles*, Topology 15 (1976), 233–245
15. ———, *Curvature and characteristic classes*, Lecture Notes in Math., vol. 640, Springer-Verlag, Berlin-New York, 1978.
16. J. L. Dupont and F. W. Kamber, *On a generalization of Cheeger-Chern-Simons classes*, Illinois J. Math. **34** (1990), 221–254.
17. H. Esnault, *Characteristic classes of flat bundles*, Topology 27 (1988), 323–352.
18. H. Esnault and E. Viehweg, *Deligne-Beilinson cohomology*, Beilinson's Conjectures on Special Values of L-Functions (M. Rapoport, N. Schappacher, P. Schneider, eds.), Perspect. Math., vol. 4, Academic Press, Boston, 1988, pp. 43–91
19. W. T. van Est, *On algebraic cohomology concepts in Lie groups.* II, Nederl. Acad. Wetensch. Proc. Ser. A **58** (1955), 286–294.
20. H. Gillet, *Riemann-Roch theorems for higher algebraic K-theory*, Adv. in Math. **40** (1981), 203–289.
21. P. Griffiths and J. Harris, *Principles of algebraic geometry*, Pure Appl. Math., Wiley-Interscience, New York, 1978.
22. A. Grothendieck, *La théorie des classes de Chern*, Bull. Soc. Math. France **86** (1958), 137–154
23. R. Hain, *The de Rham homotopy theory of complex algebraic varieties.* I, K-Theory 1 (1987), 271–324.

24. H. Hironaka, *Resolution of singularities of an algebraic variety over a field of characteristic zero.* I, Ann. Math. **79** (1964), 109–203,; II, 205–326.
25. ———, *Bimeromorphic smoothing of a complex-analytic space*, Acta Math. Vietnam. 2 (1977), 103–168.
26. N. Katz, *The regularity theorem in algebraic geometry*, Actes du Congrs International des Mathmaticiens (Nice, 1970), vol. 1, Gauthier-Villars, Paris, 1971, pp. 437–443.
27. J. Milnor and J. Stasheff, *Characteristic classes.* Ann. of Math. Stud., vol. 76, Princeton Univ. Press, Princeton, NJ, 1974.
28. R. Narasimhan and S. Ramanan, *Existence of universal connections*, Amer. J. Math. **83** (1961), 563–572; II, **85** (1963), 223–231.
29. M. Rapoport, N. Schappacher, and P. Schneider (eds.) *Beilinson's conjectures on special values of L-functions*, Perspect. Math., vol. 4, Academic Press, Boston, 1988.
30. R. L. E. Schwarzenberger, Appendix 1, Topological Methods in Algebraic Geometry, 3rd edition, Grundlehren Math. Wiss., vol. 131, Springer-Verlag, New York, 1966.
31. G. Segal, *Classifying spaces and spectral sequences*, Inst. Hautes Études Sci. Publ. Math. **34** (1968), 105–112.
32. B. Shiffman, *Separately meromorphic functions and separately holomorphic mappings*, Several Complex Variables and Complex Geometry (Santa Cruz, 1989), Proc. Sympos. Pure Math., vol. 52, Part I, Amer. Math. Soc., Providence, RI, 1991, pp. 191–198.
33. Ch. Soulé, *Connexions et classes caractéristiques de Beilinson*, Algebraic K-Theory and Algebraic Number Theory (Honolulu, 1987) (M. R. Stein and R. K. Dennis, eds.) Contemp. Math., vol. 83, Amer. Math. Soc., Providence, RI, 1989, pp. 349–376.
34. J. Steenbrink and S. Zucker, *Variation of mixed Hodge structure.* I, Invent. Math. **80** (1985), 489–542.
35. N. Steenrod, *The topology of fibre bundles*, Princeton Math. Ser., vol. 14, Princeton Univ. Press, Princeton, NJ, 1951.
36. D. Sullivan, *Infinitesimal computations in topology*, Inst. Hautes Études Sci. Publ. Math. **47** (1977), 269–331.
37. R. G. Swan, *Thom's theory of differential forms on simplicial sets*, Topology **14** (1974), 271–273.
38. H. Whitney, *Geometric integration theory*, Princeton Univ. Press, Princeton, NJ, 1957.
39. S. Zucker, *The Cheeger-Simons invariant as a Chern class*, Algebraic Analysis, Geometry and Number Theory (Baltimore, MD,1988), Johns Hopkins Univ. Press, Baltimore, MD, 1989, pp. 397–417.

Matematisk Institut, Aarhus Universitet, DK-80000 Århus C, Denmark
E-mail address: dupont@imf.au.dk

Department of Mathematics, Duke University, Durham, NC 27708-0320, USA
E-mail address: hain@math.duke.edu

Department of Mathematics, Johns Hopkins University, Baltimore, MD 21218, USA
E-mail address: sz@math.jhu.edu

Centre de Recherches Mathématiques
CRM Proceedings and Lecture Notes
Volume **24**, 2000

Height Pairings Asymptotics and Bott-Chern Forms

Bruno Harris and Bin Wang

Introduction

Let X be a compact Kähler n-dimensional manifold and A, B complex subvarieties of dimensions p, q with $p + q = n - 1$. If A and B are disjoint then their Archimedean height pairing (A, B) is a real number, defined by integrating over A a differential form G_B smooth on $X \backslash B$ but singular on B (called a Green's current for B, see Section 1 below). This pairing has an analogue for real Riemannian manifolds X, which reduces to the topological linking number if A, B are homologous to 0. If X, A, B are algebraic over $\operatorname{Spec} \mathbb{Z}$ there are also non-Archimedean pairings $(A, B)_p$. If however the complex subvarieties A, B are allowed to intersect then the definition of (A, B) as $\int_A G_B$ no longer makes sense: one has to consider whether the integral converges since G_B becomes infinite on $A \cap B$ (this can be overcome if $A \neq B$ and G_B has slow growth), and more importantly whether the result depends on the choice of G_B. We attack these problems, as in the topological real case, by moving A away from B in a family A_t (t varying in a small disk in $\mathbb{C}$, $A_0 = A$, $A_t \cap B = \emptyset$ for $t \neq 0$) and studying the asymptotics of (A_t, B) as function of t for $t \to 0$. Hain considered this problem in [**6**]: assuming A_t is a holomorphic family, he applied the theory of mixed Hodge structures and their variations to show that $(A_t, B) = e \log |t|^2 + \mathcal{O}(1)$ for some integer e, but did not specify e.

The next step was taken by Bin Wang [**8**]: see also [**10**]. He assumed that A, B intersect nicely—"clean intersection" along a submanifold M of dimension m, and that the family A_t is complex analytic only near $M = A \cap B$ (and just differentiable elsewhere): this greatly extends the generality since the whole of A may not have a motion, but rules out the Hodge theory. He then proved the existence of the asymptotics $(A_t, B) = e \log |t|^2 + k + o(1)$ where he gave a simple description of the integer e as a Chern number: $e = \int_M c_m$ where c_m is the mth Chern form of the vector bundle $TX/(TA + TB)$ over M, so that e is independent of the deformation. The next term of the asymptotics is a constant k which is the subject of this paper: it depends on the deformation and involves Bott-Chern forms.

We calculate k as a sum of 3 terms: the first is $\int_{A \backslash M} G_B$ for a suitably chosen Green's current G_B, the second is $\int_M \log |T|^2 c_m(TX/TA + TB + \mathbb{C}T)$ where T is the

1991 *Mathematics Subject Classification.* 14C30, 14F05.
Received by the editors October 13, 1998; Accepted December 8, 1998.
This is the final form of the paper.

image in $TX/TA+TB$ of the vector field, and the third is the integral over $\mathbf{P}(TA+TB/TB)$ of the Bott-Chern secondary class of a vector bundle exact sequence (see Theorem 5.2). The specification of the choice of G_B is made completely explicit in the special case that B is given as the locus of zeroes of a section of a metrized vector bundle on X of rank equal to the codimension of B.

The results were announced in [**10**]. We must acknowledge our debt to the papers [**2**], [**4**], and [**5**] of Gillet, Soulé, and Bost which gave constructions of Green's currents and proved some of their basic properties as well as those of Bott- Chern forms. There is some overlap between [**10**] and the final version of [**3**], which were written at about the same time. For instance the Chern number formula for e (due to Bin Wang in his Ph.D. thesis) and the fact that it can be positive or negative, gives a comprehensive explanation for the negative answer to the general problem (which we learned from Soulé) of the existence of positive Green's forms. [**3**] gives some special examples of deformations and counterexamples to positivity. Another idea discussed in both [**10**] and [**3**] is the case of subvarieties which are transverse zeros of vector bundle sections.

Finally we want to describe another point of view on the height pairing which will give further explanation of the results here: Barry Mazur has proposed to us the project of constructing a metrized line bundle and section on a parameter space for pairs $[A, B]$ of subvarieties of X such that the height pairing is given by

$$(A, B) = \log \|s[A, B]\|^2$$

and $s[A, B] = 0$ exactly when A, B intersect. The leading term e and the constant k should then be interpreted in terms of the behavior of the section at this vanishing locus. This problem has been solved by Bin Wang [**9**].

The paper proceeds as follows: Section 1 recalls some basic definitions and properties of Green's currents. We use a construction for a Green's current G_B due to Gillet and Soulé [**4**], which uses the blow up $\widetilde{X}$ of X along B (assuming B smooth): first one constructs a current $\tilde{G}$ on $\widetilde{X}$ and takes for G_B the push forward of $\tilde{G}$ to X. $\tilde{G}$ has the form $\log \|s\|^2 \alpha + \gamma$ where s is the line bundle section defining the exceptional divisor $\widetilde{B}$ and α, γ are smooth forms on $\widetilde{X}$. One has to study $\int_{A_t} G_B = \int_{\tilde{A}_t} \tilde{G}$ ($\tilde{A}_t$ = inverse image of A_t in $\widetilde{X}$ for $t \neq 0$, isomorphic to A_t). As $t \to 0$, $\tilde{A}_t$ regarded as current approaches $\tilde{A} + \mathbf{P}(\mu)$ where $\tilde{A}$ is the proper transform of A and $\mathbf{P}(\mu)$ is a projective space bundle over M, contained in $\tilde{B}$. Furthermore the term $\log \|s\|^2$ in $\tilde{G}$ approaches infinity on approaching $\mathbf{P}(\mu)$. However $\log \|s\|^2 - \log |t|^2$ (t the deformation parameter) approaches a C^∞ function on a Zariski-open subset of $\mathbf{P}(\mu)$. In Section 2 (following the basic definitions in Section 1), we describe the geometry of these spaces and calculate the limit of $\log \|s\|^2 - \log |t|^2$ on $\mathbf{P}(\mu)$ when $t \to 0$, in Proposition 2.2. The limit is a current and dd^c applied to it is given in Proposition 2.3. In Section 3, Proposition 3.1, we calculate the integral over $\mathbf{P}(\mu)$ of this limiting function multiplied by the form α (which is a pth Chern form). In Section 4 we complete the construction of the form $\tilde{G} = \log \|s\|^2 \alpha + \gamma$ by choosing γ in Proposition 4.1 so as to simplify the final formula: this is easiest in the case when B is a zero set of a vector bundle section. Section 5 puts together these ingredients to give our main result in Theorem 5.2. Section 6 shows that our asymptotic formula does not change if B is moved instead of A, provided the vector field along M remains essentially the same. Section 7 deals with the simplest example, linear subspaces of projective space and shows clearly the role of the vector field.

1.

We give here the main technical definitions of Green's currents [**2, 4**].

Recall that X has dimension n and Kähler metric and $p+q=n-1$. Let $d=\partial+\bar{\partial}$, $d^c=i(\bar{\partial}-\partial)/4/pi$ be the usual operators on differential forms and currents on X (X is compact). For a q-dimensional subvariety B of X, δ_B denotes the Dirac current of integration over B, $\delta_B(\alpha)=\int_B \alpha$. If G is a smooth (p,p) form on $X \backslash B$ which is L^1 on X, i.e., for all continuous $2(n-p)$-forms α on X, the integral $\int_{X\backslash B} \alpha \wedge G$ is absolutely convergent then G defines, by this integral, a current on X of bidegree (p,p).

We will call G a Green's current for B and write $G=G_B$ if as before G is smooth on $X\backslash B$ and L^1 on X, of bidegree (p,p) where $\dim B=q=n-p-1$ and further satisfies two conditions:

$$dd^c G_B=\delta_B-\omega_B \tag{1.1}$$

as currents where ω_B is (the current represented by) the harmonic form on X Poincaré dual to the homology class of B. (Note that our Green's currents are thus the negatives of those of Gillet-Soulé, but we avoid many negative signs in formulas.)

If η is any harmonic $2(n-p)$ form on X then

$$\int_X \eta \wedge G_B=0. \tag{1.2}$$

We will give some explicit Green's currents G_B below. If G'_B is another Green's current for B then it is known that $G_B-G'_B=d\eta_1+d^c\eta_2$, where η_i are currents on X and smooth on $X\backslash B$.

To define the height pairing (A,B) for A any p-dimensional complex subvariety disjoint from B, we choose any Green's current G_B, recall that it is smooth on $X\backslash B\supset A$ and define

$$(A,B)=\int_A G_B. \tag{1.3}$$

Since $\int_A d\eta_1+d^c\eta_2=0$ by Stokes, the integral is independent of the choice of G_B. If instead of subvarieties A,B we use cycles with disjoint support and *homologous to zero* then it is not hard to see that the harmonic forms occurring in (1.1), (1.2) become irrelevant and the pairing (A,B) is independent of the Kähler metric.

Finally, "reduction to the diagonal" is valid just as in ordinary intersection theory: that is if we replaced X by $X\times X$ with product metric and consider the two subvarieties $A\times B$ and $\Delta=$ diagonal, then (Bost [**2**])

$$(A,B)_X=(A\times B,\Delta)_{X\times X}. \tag{1.4}$$

One can use this to show $(A,B)=(B,A)$. (1.4) will be very useful for us since we need only construct G_Δ independently of A, B and use the fact that Δ is smooth. However it will be slightly simpler and more general to continue to use A, B in X and *assume henceforth B is smooth.*

If now A, B happen to intersect then we cannot use the definition (1.3) any more since $\int_A G_B$ may not converge. One can sometimes achieve convergence by using Green's currents which are not just L^1 but grow slowly—"log growth" (see below)—near B, but then the problem of independence of the integral (1.3) of the choice of G_B arises. It will turn out that by adding to (1.3) some "correction terms" reflecting the geometry of how A, B intersect, we can achieve a satisfactory answer.

2.

We assume B is a smooth q-dimensional subvariety of X, an n-dimensional compact Kähler manifold, and A is a p-dimensional subvariety of X, with $p+q = n-1$. Let $M = A \cap B$. We do not assume A is smooth, merely that *it is smooth in a neighborhood* of M, and in fact only what happens in this neighborhood will be used. M is assumed *smooth* but not necessarily connected.

We assume A, B intersect *cleanly*, i.e., for $x \in M$ the tangent spaces at x satisfy:

$$T_x(A) \cap T_x(B) = T_x(M). \tag{2.1}$$

(Note that this is to be distinguished from transversal intersection which in our case would be just empty intersection since $p+q = n-1$)

Let τ' be a vector field along M which is nowhere 0 modulo $TA+TB$ (restricted to M). Denote by τ the residue class of τ' modulo TB (restricted to M), so that τ is a section of the normal bundle $N_B(X)$ (restricted to M). Let T be residue class of τ' modulo $TA+TB$, a section of $TX/TA+TB$. We will use the following notation for certain vector bundles over M:

$$\begin{aligned}
&\nu = TX/TB \quad \text{(restricted to } M).\\
&\tau \text{ is a nowhere 0 section of } \nu.\\
&\mu_0 = (TA+TB)/TB \subset \nu\\
&\mu = \mu_0 + \mathbb{C}\tau \text{ so } \mu_0 \subset \mu \subset \nu.\\
&T \text{ is the image of } \tau \text{ in } \nu/\mu_0 = TX/TA+TB.
\end{aligned} \tag{2.2}$$

μ_0, μ, ν have fiber dimensions $p-m$, $p-m+1$, $p+1$ over a component of M of dimension m. For simplicity we write M as if it is connected of dimension m, but in the final formula we sum over connected components.

The clean intersection assumption shows that μ_0 is isomorphic to the normal bundle of M in A. However the choice of metrics requires more care: in general, if V is a vector space with Hermitian metric and W is a subspace, we metrize V/W to be isometric to $W^\perp \subset V$ by the orthogonal projection $P\colon V \to W^\perp$ which induces $V/W \to W^\perp$. Note that if W' is another subspace of V, this convention gives a metric to $W'/W\cap W'$ by use of the orthogonal projection $Q\colon V \to (W\cap W')^\perp$ so that Q induces an isometry $W'/W\cap W' \to W' \cap (W\cap W')^\perp$. However the natural isomorphism $W'/W\cap W' \to W+W'/W$ is *not* an isometry.Instead, we have a commutative diagram (since $PQ = P$):

$$\begin{array}{ccc}
W'/W\cap W' & \xrightarrow{\;Q\;} & (W\cap W')^\perp \cap W' \\
\big\downarrow & & \big\downarrow{\scriptstyle P} \\
(W+W')/W & \xrightarrow{\;P\;} & W^\perp \cap (W+W')
\end{array}$$

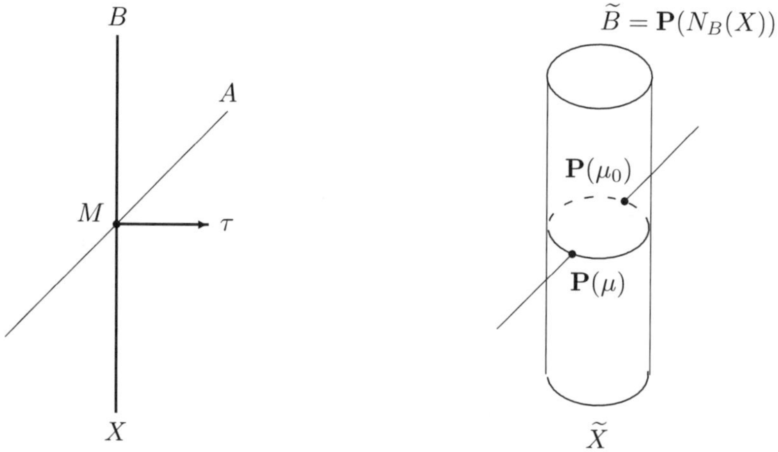

FIGURE 1. Geometry of X, A, B, M

where the vertical $P\colon (W\cap W')^{\perp}\cap W' \to W^{\perp}\cap(W+W')$ (induced by $P\colon V\to W^{\perp}$) is not an isometry but only an isomorphism of vector spaces, and the horizontal P, Q are isometries by definition.

Coming back to the geometry of X, A, B, M sketched in the Figure 1, we blow up X along B getting $\pi\colon \widetilde{X}\to X$ with exceptional divisor $\widetilde{B}=\mathbf{P}(N_B(X))\to B$. Over $M\subset B$ we have the projective space bundles $\mathbf{P}(\nu)\supset\mathbf{P}(\mu)\supset\mathbf{P}(\mu_0)$ whose total spaces have dimensions $p+m$, p, $p-1$ respectively. The figure shows the case $A\neq M$, i.e. $m<p$. If $A=M\subset B$, so $m=p$, then μ_0 is just the rank 0 bundle and $\mathbf{P}(\mu_0)$ is empty, while μ is of rank 1 and $\mathbf{P}(\mu)$ is identified with M by the nowhere 0 section τ.

The normal bundle $N_{\widetilde{B}}(\widetilde{X})$ can be identified with $\mathcal{O}_{\mathbf{P}(N_B(X))}(-1)$ (a line bundle over $\widetilde{B}=\mathbf{P}(N_B(X))$). A vector in this normal bundle is a pair $(z_x,\tilde{x})$, $x\in B$, $\tilde{x}$ a 1-dimensional subspace of the fiber $N_B(X)_x$ over x, and $z_x\in\tilde{x}$. We take $x\in M, \tilde{x}\subset\nu_x$.

$$\text{Under } \pi\colon \widetilde{X}\to X$$
$$d\pi\colon N_{\widetilde{B}}(\widetilde{X})\to N_B(X)$$
$$d\pi(z_x,\tilde{x})=z_x.$$

Note that $d\pi$ is injective on each fiber of the line bundle.

DEFINITION 2.1. If τ is the given section of μ over M, let $\tilde{\tau}$ be the unique section of the restriction of $N_{\widetilde{B}}(\widetilde{X})$ to $\mathbf{P}(\mu)\backslash\mathbf{P}(\mu_0)$ such that $d\pi(\tilde{\tau})-\tau$ is a section of μ_0.

Thus if $\tilde{x}\in\mathbf{P}(\mu)\backslash\mathbf{P}(\mu_0)$ is a 1-dimensional subspace of μ_x not contained in $\mu_{0,x}$ then $\mu_x=\mu_{0,x}\oplus\mathbb{C}\tau_x\supset\tilde{x}$ and $\tilde{\tau}(\tilde{x})$ is the unique vector in $\tilde{x}$ whose projection is τ_x (and "approaches ∞" as $\tilde{x}$ approaches $\mathbf{P}(\mu_0)$). $\tilde{\tau}$ is nowhere 0 on $\mathbf{P}(\mu)\backslash\mathbf{P}(\mu_0)$. If $A=M$ then $\mu_x=\mathbb{C}\tau_x$ and $\tilde{\tau}=\tau$.

Next let L be the line bundle over $\widetilde{X}$ and s the section of L with $s=0$ defining the exceptional divisor $\widetilde{B}$. Here $\widetilde{B}$ was assumed smooth so the graph of s

is transverse to $\widetilde{B}$ and we have the canonical isomorphism

$$ds\colon N_{\widetilde{B}}(\widetilde{X}) \to L|_{\widetilde{B}}. \tag{2.3}$$

Thus $ds(\tilde{\tau})$ is a nowhere 0 section of L over $\mathbf{P}(\mu)\backslash\mathbf{P}(\mu_0)$.

Let $\|v\|^2 = \langle v, v\rangle$ be a Hermitian metric on L. Then $\log\|ds(\tilde{\tau})\|^2$ is a C^∞-function on $\mathbf{P}(\mu)\backslash\mathbf{P}(\mu_0)$.

We will next *assume* given a locally closed complex submanifold $\mathcal{A} \subset X$, i.e., a closed submanifold of a neighborhood of M in X, containing A (intersected with this neighborhood) as a closed codimension 1 submanifold. We assume $\mathcal{A}$ has clean intersection with B, and $\mathcal{A}\cap B = M$, and further for $x \in M$, $T_x(\mathcal{A}) = T_x(A) + \mathbb{C}\tau'_x$ (τ' being the original vector field along M and τ its residue class modulo TB). Thus

$$N_M(\mathcal{A}) = N_M(A) \oplus \mathbb{C}\tau = \mu_0 \oplus \mathbb{C}\tau = \mu. \tag{2.4}$$

Suppose $M \neq A$, i.e., $m < p$. Then we can regard $\mathbf{P}(\mu)$ as the exceptional divisor for the blow up of $\mathcal{A}$ along M, denoted $\tilde{\mathcal{A}}$. $\tilde{\mathcal{A}}$ is also the strict transform of $\mathcal{A}$ in $\widetilde{X}$ as $N_M(A) \subset N_B(X)|_M$. In the same way, the blow up of A along M, denoted $\tilde{A}$, is also the strict transform of A in $\widetilde{X}$. In $\widetilde{X}$, $\widetilde{B}\cap\tilde{\mathcal{A}} = \mathbf{P}(\mu)$. Therefore if we restrict L and s to $\tilde{\mathcal{A}} \subset \widetilde{X}$, then $(s = 0)$ in $\tilde{\mathcal{A}}$ defines $\mathbf{P}(\mu)$.

If $A = M$ then $\mathbf{P}(\mu)$ is identified with M and $\tilde{\mathcal{A}}$ with $\mathcal{A}$, while $\mathbf{P}(\mu_0)$ and $\tilde{A}$ are both empty.

Let now (Γ, g) be a line bundle and section over $\mathcal{A}$ with $g = 0$ defining A. Their pull backs to $\tilde{\mathcal{A}}$ are $(\pi^*\Gamma, g\circ\pi)$ and $(g\circ\pi) = 0$ defines $\mathbf{P}(\mu)\cup\tilde{A}$, and $\mathbf{P}(\mu)\cap\tilde{A} = \mathbf{P}(\mu_0)$. We will consider $g\circ\pi$ in a neighborhood of $\mathbf{P}(\mu)\backslash\mathbf{P}(\mu_0)$ in $\tilde{\mathcal{A}}$ disjoint from $\tilde{A}$ (e.g., $\tilde{\mathcal{A}}\backslash\tilde{A}$). We will choose a metric on Γ so that under the isomorphism over the zero set A of g,

$$dg\colon N_A(\mathcal{A}) \to \Gamma|_A$$

we have $\|dg(\tau)\|^2 = 1$.

We can now describe the behavior of the sections s and $g\circ\pi$ at points $\tilde{y} \in \tilde{\mathcal{A}}\backslash\tilde{A}$ as $\tilde{y}$ approaches $\tilde{x} \in \mathbf{P}(\mu)\backslash\mathbf{P}(\mu_0)$:

PROPOSITION 2.2. *Given the nowhere 0 section τ of $\nu = N_B(X)|_M$ (τ being also non-zero mod $\mu_0 = N_M(A) \subset \nu$), given also $\mathcal{A} \supset A$ with codimension 1 with $N_A(\mathcal{A})|_M = \mathbb{C}\tau$, and given the line bundles L over $\widetilde{X}$ with section s and $\widetilde{B} = (s = 0)$ and Γ over $\mathcal{A}$ with section g and $A = (g = 0)$ with metrics $\|\ \|^2$ and $\|\ \|^2_\Gamma$ with $\|dg(\tau)\|^2_\Gamma = 1$, then we have the following limit as $\tilde{y} \in \tilde{\mathcal{A}}\backslash A$ approaches $\tilde{x} \in \mathbf{P}(\mu)\backslash\mathbf{P}(\mu_0)$:*

$$\lim_{\tilde{y}\to\tilde{x}}[\log\|s(\tilde{y})\|^2_L - \log\|g\circ\pi(\tilde{y})\|^2_\Gamma] = \log\|ds(\tilde{\tau}(\tilde{x}))\|^2_L \tag{2.5}$$

(recall $\tilde{\tau}$ was defined in Definition 2.1).

PROOF. On $\mathcal{A}$ we had the isomorphism $dg|_A : N_A(\mathcal{A}) \to \Gamma|_A$. Lifting to $\tilde{\mathcal{A}}$ by $\pi : \tilde{\mathcal{A}} \to \mathcal{A}$ and then restricting to $\mathbf{P}(\mu)\backslash\mathbf{P}(\mu_0)$, $d(g\circ\pi)$ is an isomorphism of $N_{\tilde{A}}(\tilde{\mathcal{A}})|_{\mathbf{P}(\mu)\backslash\mathbf{P}(\mu_0)} = \mathcal{O}_{\mathbf{P}(\mu)}(-1)|_{\mathbf{P}(\mu)\backslash\mathbf{P}(\mu_0)}$ with $\pi^*\Gamma|_{\mathbf{P}(\mu)\backslash\mathbf{P}(\mu_0)}$.

The bundle $\mathcal{O}_{\mathbf{P}(\mu)}(-1)|_{\mathbf{P}(\mu)\backslash\mathbf{P}(\mu_0)}$ has the nowhere 0 section $\tilde{\tau}$ and for $x \in M$ and $\tilde{x} \in \mathbf{P}(\mu)\backslash\mathbf{P}(\mu_0)$ with $\pi(\tilde{x}) = x$, we have

$$d(g\circ\pi)(\tilde{\tau}(\tilde{x})) = dg(d\pi(\tilde{\tau}(\tilde{x}))) = dg(\tau(x)) \tag{2.6}$$

since $\tilde{\tau} - \tau \in \mu_{0,x} \subset \mu_x$ and $\mu_{0,x} = N_M(A)_x$ is annihilated by dg (since $g = 0$ on A). Since $\tau_x \notin N_M(A)$, $dg(\tau(x)) \neq 0$ and so $d(g \circ \pi)(\tilde{\tau}(\tilde{x})) \neq 0$. Thus in the above neighborhood of $\mathbf{P}(\mu)\backslash\mathbf{P}(\mu_0)$ in $\tilde{\mathcal{A}}$, $(g \circ \pi) = 0$ defines $\mathbf{P}(\mu)\backslash\mathbf{P}(\mu_0)$, as also does $s = 0$. We can therefore write $s = (g \circ \pi) \otimes u$ where u is a nowhere 0 section of $(\pi^*\Gamma)^{-1} \otimes L$. Since (writing U for the neighborhood above)

$$ds\colon N_{\mathbf{P}(\mu)\backslash\mathbf{P}(\mu_0)}(U) \to L|_{\mathbf{P}(\mu)\backslash\mathbf{P}(\mu_0)} = \mathbb{C}\tilde{\tau}$$

is an isomorphism, and similarly for $d(g \circ \pi)$ and du,

$$\begin{aligned} ds(\tilde{\tau}) &= d(g \circ \pi)(\tilde{\tau}) \otimes u(\tilde{x}) + (g \circ \pi)(\tilde{x}) \otimes du(\tilde{\tau}) \\ &= d(g \circ \pi)(\tilde{\tau}) \otimes u(\tilde{x}) \qquad\qquad \text{as } g \circ \pi = 0 \end{aligned}$$

on $\mathbf{P}(\mu)\backslash\mathbf{P}(\mu_0)$. At $\tilde{x}$, $x = \pi(\tilde{x})$, by (2.6) we get

$$ds(\tilde{\tau}(\tilde{x})) = dg(\tau(x)) \otimes u(\tilde{x}).$$

Now suppose we have metrics on L, Γ, $L \otimes \pi^*\Gamma^{-1}$ (over U) such that

$$\| \; \|_{L \otimes \pi^*\Gamma^{-1}} = \| \; \|_L \bullet \| \; \|_{\pi^*\Gamma^{-1}}.$$

Then for $\tilde{y} \in U$,

$$\log \|s(\tilde{y})\|_L^2 - \log \|g(\pi(\tilde{y}))\|_\Gamma^2 = \log \|u(\tilde{y})\|_{L \otimes \Gamma^{-1}}^2.$$

For $\tilde{x} \in \mathbf{P}(\mu)\backslash\mathbf{P}(\mu_0)$

$$\log \|ds(\tilde{\tau}(\tilde{x}))\|_L^2 - \log \|d(g \circ \pi)(\tilde{\tau}(\tilde{x}))\|_\Gamma^2 = \log \|u(\tilde{x})\|^2.$$

To simplify, note that we can change the metric on Γ so that $\|dg(\tau(x))\|_\Gamma^2 = 1$, all $x \in M$ (since $\|dg(\tau(x))\|_\Gamma^2$ is a positive C^∞ function on M which can be extended to a positive C^∞ function on $\mathcal{A}$, and the metric can then be divided by this function). With this choice of metric on Γ, $\log \|d(g \circ \pi)(\tilde{\tau}(\tilde{x}))\| = 0$, so the two formulas above give, for $\tilde{y} \to \tilde{x}$,

$$\begin{aligned} \log \|ds(\tilde{\tau}(\tilde{x}))\|_L^2 &= \log \|u(\tilde{x})\|^2 \\ &= \lim_{\tilde{y} \to \tilde{x}} [\log \|s(\tilde{y})\|_L^2 - \log \|g \circ \pi(\tilde{y})\|_\Gamma^2] \end{aligned}$$

proving (2.5) and Proposition 2.2. □

PROPOSITION 2.3. *The C^∞ function $\log \|ds(\tilde{\tau})\|^2$ on $\mathbf{P}(\mu)\backslash\mathbf{P}(\mu_0)$, when regarded as a current on $\mathbf{P}(\mu)$, satisfies the equation*

$$dd^c \log \|ds(\tilde{\tau})\|_L^2 = -\delta_{\mathbf{P}(\mu_0)} - c_1(L, \| \; \|_L)$$

where c_1 denotes the first Chern form of a line bundle with metric. (Here we assume $m < p$ so $\mathbf{P}(\mu_0)$ is non-empty.)

PROOF. Restricting the isomorphism (2.3) to $\mathbf{P}(\mu)$, we get the isomorphism

$$ds\colon \mathcal{O}_{\mathbf{P}(\mu)}(-1) = N_{\widetilde{B}}(\widetilde{X})|_{\mathbf{P}(\mu)} \to L|_{\mathbf{P}(\mu)}.$$

$\tilde{\tau}$ is a section of $\mathcal{O}_{\mathbf{P}(\mu)}(-1)$, denoted $\mathcal{O}(-1)$ now, only over $\mathbf{P}(\mu)\backslash\mathbf{P}(\mu_0)$. However we look at the dual line bundle $\mathcal{O}(1)$ over $\mathbf{P}(\mu)$ and the dual section $\tilde{\mathrm{l}}$. To define $\tilde{\mathrm{l}}$ we must give for each $x \in M$ a linear function $\mathrm{l}_x\colon \mu_x \to \mathbb{C}$ whose restriction to the one-dimensional subspace $\tilde{x} \subset \mu_x$, $\tilde{x}$ not in $\mu_{0,x}$, satisfies $\mathrm{l}_x(\tilde{\tau}(\tilde{x})) = 1$. Recalling that $\mu_x = \mu_{0,x} \oplus \mathbb{C}\tau_x$ and $\tilde{\tau} \in \tilde{x}$ projects to τ_x, the unique way to define l_x is:

$$\mathrm{l}_x\colon \mu_x \to \mathbb{C}\tau_x \to \mathbb{C}$$

the first arrow being projection to the factor $\mathbb{C}\tau_x$ and the second arrow taking τ_x to 1. Thus $\mathrm{l}_x(\mu_{0,x}) = 0$ and $\mathrm{l}_x(\tau_x) = 1$, so that $\tilde{\mathrm{l}}$ as section of $\mathcal{O}_{\mathbf{P}(\mu)}(1)$ has zero scheme $\mathbf{P}(\mu_0) \subset \mathbf{P}(\mu)$, and $\tilde{\mathrm{l}}(\tilde{\tau}) = 1$. Let $(ds)^T \colon L^* \to \mathcal{O}(1)$ be the transposed isomorphism to ds and $(ds^T)^{-1}$ its inverse. Then $(ds^T)^{-1}(\tilde{\mathrm{l}})(ds(\tilde{\tau})) = 1$, and $(ds^T)^{-1}(\tilde{\mathrm{l}})$ is a section of the dual bundle L^* with zero scheme $\mathbf{P}(\mu_0)$.

Given a metric on L and dual metric on L^*, i.e., for $\lambda \in L^*$, $\mu \in L$ we have $|\lambda(\mu)| = \|\lambda\|\,\|\mu\|$, then

$$\log \|ds(\tilde{\tau})\|^2 = -\log \|(ds^T)^{-1}(\tilde{\mathrm{l}})\|^2.$$

Finally we apply to L^* and the section $(ds^T)^{-1}(\tilde{\mathrm{l}})$ the Poincaré-Lelong formula: if σ is a section of a line bundle with metric, with zero scheme Z, then $dd^c \log \|\sigma\|^2 = \delta_Z - c_1$. This proves Proposition 2.3. In the extreme case $M = A$, $\mathbf{P}(\mu_0)$ is empty and $\mathbf{P}(\mu)$ is identified with M using τ, while $\tilde{\tau} = \tau$. Under this identification the line bundle on M with nowhere 0 section τ is isomorphic to the restriction of L to $\mathbf{P}(\mu)$, under ds. If we choose the metrics so that ds is an isometry then $\log \|ds(\tilde{\tau})\|^2 = \log \|\tau\|^2$. □

To simplify the notation we will henceforth assume $N_B(X)$ has the orthogonal projection of the metric on X, thus metrizing the line bundle $\mathcal{O}(-1)$ over $\widetilde{B} = \mathbf{P}(N_B(X)$, which is a subbundle of $\pi^* N_B(X)$. Let $\mathcal{C}$ denote the exact sequence, *whose terms we metrize also from* $N_B(X)$, of bundles on $\widetilde{B}$:

$$\mathcal{C} : 0 \to \mathcal{O}(-1) \to \pi^* N_B(X) \to Q \to 0. \tag{2.7}$$

Metrize L so that $ds \colon \mathcal{O}(-1) \to L|_{\tilde{B}}$ is an isometry. Then we can write

$$\log \|ds(\tilde{\tau})\|^2 = \log \|\tilde{\tau}\|^2 \tag{2.8}$$

on $\mathbf{P}(\mu)\backslash\mathbf{P}(\mu_0)$, and Proposition 2.3 becomes

$$dd^c \log \|\tilde{\tau}\|^2_{\mathcal{O}(-1)} = -\delta_{\mathbf{P}(\mu_0)} - c_1(\mathcal{O}(-1)) \tag{2.9}$$

on $\mathbf{P}(\mu)$.

The function $\log \|\tilde{\tau}\|^2$ when restricted to the fiber $\mathbf{P}(\mu_x)$ of $\mathbf{P}(\mu)$ coincides, up to addition of a constant, with the Levine function Λ (see [**7**] and [**5**, Section 5]), namely, consider any finite dimensional vector space V with Hermitian inner product (e.g., μ_x), and let V_0 be a codimension 1 subspace (e.g., $\mu_{0,x}$). Denote by P_bot the orthogonal projection to the orthogonal complement of V_0. The function Λ on $\mathbf{P}(V)\backslash\mathbf{P}(V_0)$ is defined as $\log(|v|^2) - \log |P_bot(v)|^2$ for $v \in V\backslash V_0$. If $V = \mu_x$ then for any $\tilde{x} \in \mathbf{P}(\mu_x)\backslash\mathbf{P}(\mu_{0,x})$, $\tilde{\tau}(\tilde{x})$ and $\tau(x)$ are both in μ_x and $\tilde{\tau}(\tilde{x}) - \tau(x) \in \mu_{0,x}$ so that $P_bot(\tilde{\tau}) = P_bot(\tau) = T$ (see line following (2.2)).

Finally,

$$\Lambda(\tilde{x}) = \log \|\tilde{\tau}(\tilde{x})\|^2 - \log \|P_bot(\tau(x))\|^2$$

(note that the last term $\log \|P_bot(\tau(x))\|^2$ is constant on $\mathbf{P}(\mu_x)$ for fixed x).

The Hermitian metric on V induces a metric on the product bundle $\mathbf{P}(V) \times V$ and hence on the subbundle $\mathcal{O}(-1)$ and the quotient bundle Q, and these bundles are acted on by the unitary group of V so as to preserve the metrics. It follows that their Chern forms and secondary Bott-Chern forms (see below) are invariant under this unitary group and hence closed and harmonic on the symmetric space $\mathbf{P}(V)$. The first Chern form $c_1(\mathcal{O}(1))$ defines the invariant Kähler metric on $\mathbf{P}(V)$ of total volume 1, i.e., if $\dim V = k+1$ then $\int_{\mathbf{P}(V)} c_1(\mathcal{O}(1))^k = 1$.

Further the usual exact sequence on $\mathbf{P}(V)$

$$0 \to \mathcal{O}(-1) \to \mathbf{P}(V) \times V \to Q \to 0$$

leads to the relation on Chern *cohomology* classes:

$$[1 + c_1(\mathcal{O}(-1))][1 + c_1(Q) + \cdots + c_k(Q)] = 1$$

which is here valid for the invariant Chern *forms* as well (since the ring of invariant forms maps isomorphically to the cohomology) and since $c_1(\mathcal{O}(-1)) = -c_1(\mathcal{O}(1))$ we get $c_k(Q) = c_1(\mathcal{O}(1))^k$.

For any function f on $\mathbf{P}(V)$ we will use the notation

$$H(f) = \int_{\mathbf{P}(V)} f c_k(Q). \tag{2.10}$$

3.

In this section we will evaluate an integral

$$\int_{\mathbf{P}(\mu)} (\log \|\tilde{\tau}\|^2)\alpha \tag{3.1}$$

where α is the top Chern form $c_p(Q_\nu)$ restricted to $\mathbf{P}(\mu) \subset \mathbf{P}(\nu)$ (recall that the restriction of $N_B(X)$ to M was denoted ν in (2.2)) and all the vector bundles inherit the metric of $N_B(X)$, either as subbundles or quotients with orthogonal projection metric. These vector bundles over $\mathbf{P}(\mu)$ are shown in the following diagram with exact rows denoted by $\mathcal{L}_1$, $\mathcal{L}_2$, $\mathcal{L}_3$ and exact columns denoted by $\mathcal{C}_1$, $\mathcal{C}_2$ and $\mathcal{C}_3$ in the following computations:

$$\begin{array}{cccccccc}
 & & \mathcal{C}_1 & & \mathcal{C}_2 & & \mathcal{C}_3 & \\
 & & 0 & & 0 & & 0 & \\
 & & \downarrow & & \downarrow & & \downarrow & \\
\mathcal{L}_1 & 0 \longrightarrow & \mathcal{O}(-1) & \xrightarrow{=} & \mathcal{O}(-1) & \longrightarrow & 0 & \longrightarrow 0 \\
 & & \downarrow & & \downarrow & & \downarrow & \\
\mathcal{L}_2 & 0 \longrightarrow & \pi^*\mu & \longrightarrow & \pi^*\nu & \longrightarrow & \pi^*(\nu/\mu) & \longrightarrow 0 \\
 & & \downarrow & & \downarrow & & \downarrow & \\
\mathcal{L}_3 & 0 \longrightarrow & Q_\mu & \longrightarrow & Q_\nu & \longrightarrow & \pi^*(\nu/\mu) & \longrightarrow 0 \\
 & & \downarrow & & \downarrow & & \downarrow & \\
 & & 0 & & 0 & & 0 &
\end{array} \tag{3.2}$$

Recall that for any exact sequence of vector bundles, each of which has a metric,

$$\mathcal{E} : 0 \to E' \to E \to E'' \to 0$$

the top Chern forms (we will only need these) satisfy (if E has rank r say)

$$c_{\text{top}}(E')c_{\text{top}}(E'') - c_{\text{top}}(E) = dd^c \tilde{c}_r(\mathcal{E})$$

where the Bott-Chern form $\tilde{c}_r(\mathcal{E})$ is of type $(r-1, r-1)$ and is defined by the exact sequence $\mathcal{E}$ (including the metrics) modulo forms $d\alpha + d^c\beta$ (see [**5**]).

PROPOSITION 3.1. *With the above notation, assuming* $p > m$,

$$\int_{\mathbf{P}(\mu)} (\log \|\tilde{\tau}\|^2)\alpha = \int_M (\log \|P_b ot(\tau)\|^2) c_m(\nu/\mu) - \int_{\mathbf{P}(\mu)} \tilde{c}_{p+1}(\mathcal{C}_2) + \int_{\mathbf{P}(\mu_0)} \tilde{c}_p(\mathcal{L}_3).$$

We can also write $\|P_b ot(\tau)\|^2 = \|T\|^2$ *(*T *defined just after* (2.2). *Further, if* $m = 0$ *then* $\tilde{c}_p(\mathcal{L}_3) = 0$ *and so the last integral is* 0, *and in the first integral* $c_0(\nu/\mu) = 1$.

PROOF. π will always denote $\mathbf{P}(\mu) \to M$ and all bundles will be restricted to $\mathbf{P}(\mu) \subset \widetilde{X}$, or to $M \subset X$, respectively. The exact sequence $\mathcal{L}_3$ in (3.2), namely

$$\mathcal{L}_3 : 0 \to Q_\mu \to Q_\nu|_{\mathbf{P}(\mu)} \to \pi^*(\nu/\mu) \to 0$$

(where the ranks are $p - m, p, m$) leads to the equation

$$\alpha = c_p(Q_\nu|_{\mathbf{P}(\mu)}) = c_{p-m}(Q_\mu) c_m(\pi^*(\nu/\mu)) - dd^c \tilde{c}_p(\mathcal{L}_3)$$

with $\tilde{c}_p(\mathcal{L}_3) \in A^{p-1,p-1}(M)/(\operatorname{Im} d + \operatorname{Im} d^c)$.

If M is a single point x then $\mathcal{L}_3$ splits (both with regard to vector bundles and to metrics): over a 1-dimensional subspace $\tilde{x}$ of ν_x, the fibers are

$$Q_{\nu,\tilde{x}} = \nu_x/\tilde{x} = \mu_x/\tilde{x} \oplus (\nu_x/\mu_x) \quad (\text{orthogonal } \oplus)$$

so $Q_\nu|_{\mathbf{P}(\mu)} = Q_\mu \oplus [(\mathbf{P}(\mu) \times (\nu_x/\mu_x))]$.

It follows that $\tilde{c}_p(\mathcal{L}_3) = 0$ if $\dim M = 0$.

In the general case the above equation for α gives

$$(3.3) \quad \int_{\mathbf{P}(\mu)} (\log \|\tilde{\tau}\|^2)\alpha)$$
$$= \int_{\mathbf{P}(\mu)} (\log \|\tilde{\tau}\|^2 c_{p-m}(Q_\nu) \pi^* c_m(\nu/\mu)) - \int_{\mathbf{P}(\mu)} (\log \|\tilde{\tau}\|^2) dd^c \tilde{c}_p(\mathcal{L}_3)$$
$$= \int_M \pi_*[\log \|\tilde{\tau}\|^2 c_{p-m}(Q_\nu)] c_m(\nu/\mu) - \int_{\mathbf{P}(\mu)} \log \|\tilde{\tau}\|^2 dd^c \tilde{c}_p(\mathcal{L}_3)$$

where π_* denotes integration over the fibers $\mathbf{P}(\mu_x)$ of π. Since these fibers have the same real dimension as the degree of the form $\log \|\tilde{\tau}\|^2 c_{p-m}(Q_\mu)$ this fiber integration is just restriction to $\mathbf{P}(\mu_x)$ followed by integration over $\mathbf{P}(\mu_x)$.

As discussed towards the end of Section 2, on $\mathbf{P}(\mu_x)$ we have

$$\log \|\tilde{\tau}\|^2 = \log \|P_b ot(\tau(x))\|^2 + \Lambda$$

where Λ is the Levine function, and further

$$c_{p-m}(Q_\nu) = c_1(\mathcal{O}(1))^{p-m}$$

so that using the notation (2.10)

$$\int_{\mathbf{P}(\mu_x)} [\log \|\tilde{\tau}\|^2 c_{p-m}(Q_\mu)] = \log \|P_b ot(\tau(x)\|^2 + \int_{\mathbf{P}(\mu_x)} \Lambda c_{p-m}(Q_\mu)$$
$$= \log \|P_b ot(\tau(x)\|^2 + H(\Lambda)$$

[$H(\Lambda)$ is an explicitly known constant independent of x].

The first integral on the right hand side of (3.3) thus is found to be

$$(3.4) \qquad \int_M \log \|P_b ot(\tau)\|^2 c_m(\nu/\mu) + H(\Lambda) \int_M c_m(\nu/\mu).$$

Next we look at the second integral in (3.3) and will calculate that (notation as in (3.2))

$$(3.5)\qquad -\int_{\mathbf{P}(\mu)}(\log\|\tilde{\tau}\|^2)dd^c\tilde{c}_p(\mathcal{L}_3)$$
$$= -\int_{\mathbf{P}(\mu)}\tilde{c}_{p+1}(\mathcal{C}_2)+\int_{\mathbf{P}(\mu_0)}\tilde{c}_p(\mathcal{L}_3)-H(\Lambda)\int_M c_m(\nu/\mu)$$

so that adding (3.4) and (3.5) will give us Proposition 3.1.

To prove (3.5) we start with the integral involving $\log\|\tilde{\tau}\|^2$ which is continuous only on $\mathbf{P}(\mu)\backslash\mathbf{P}(\mu_0)$ but can be regarded as a current, with $dd^c\log\|\tilde{\tau}\|^2 = -[\delta_{\mathbf{P}(\mu_0)}+c_1\mathcal{O}(-1)]$ (see (2.9)). Since $\tilde{c}_p(\mathcal{L}_3)$ is smooth everywhere on $\mathbf{P}(\mu)$ (but is only defined $\mod\{d\alpha+d^c\beta\}$),

$$(3.6)\qquad -\int_{\mathbf{P}(\mu)}(\log\|\tilde{\tau}\|^2)dd^c\tilde{c}_p(\mathcal{L}_3) = -\int_{\mathbf{P}(\mu)}(dd^c\log\|\tilde{\tau}\|^2)\tilde{c}_p(\mathcal{L}_3)$$
$$= \int_{\mathbf{P}(\mu_0)}\tilde{c}_p(\mathcal{L}_3)+\int_{\mathbf{P}(\mu)}c_1(\mathcal{O}(-1))\tilde{c}_p(\mathcal{L}_3).$$

We need to transform these last two terms further using the diagram (3.2) and some calculations of Gillet and Soulé [**5**, 1.3.4] about such rectangular diagrams. First,

$$(3.7)\qquad \tilde{c}_{p+1}(\mathcal{L}_1\oplus\mathcal{L}_3)-\tilde{c}_{p+1}(\mathcal{L}_2)=\tilde{c}_{p+1}(\mathcal{C}_1\oplus\mathcal{C}_3)-\tilde{c}_{p+1}(\mathcal{C}_2).$$

In the notation of this reference (and the usual symmetric function notation), c_{p+1} corresponds to

$$\sigma_{p+1}(T_1,\dots,T_{p+1})=\sigma_1(T_1)\sigma_p(T_2,\dots,T_{p+1}).$$

From [**5**, 1.3.2.1] we conclude

$$(3.8)\qquad \tilde{c}_{p+1}(\mathcal{L}_1\oplus\mathcal{L}_3)=\tilde{c}_1(\mathcal{L}_1)c_p(Q_\nu)+c_1(\mathcal{O}(-1))\tilde{c}_p(\mathcal{L}_3)$$
$$=c_1(\mathcal{O}(-1))\tilde{c}_p(\mathcal{L}_3)$$

since $\tilde{c}_1(\mathcal{L}_1)=0$ (in fact $\tilde{c}_j(\mathcal{L}_1)$ and $\tilde{c}_j(\mathcal{C}_3)$ are 0 for all j).

Similarly, from

$$\sigma_{p+1}(T_1,\dots,T_{p-m+1},T_{p-m+2},\dots,T_{p+1})$$
$$=\sigma_{p-m+1}(T_1,\dots,T_{p-m+1})\sigma_m(T_{p-m+2},\dots,T_{p+1})$$

one concludes

$$(3.9)\qquad \tilde{c}_{p+1}(\mathcal{C}_1\oplus\mathcal{C}_3)=\tilde{c}_{p+1-m}(\mathcal{C}_1)c_m\pi^*(\nu/\mu).$$

Substituting (3.8), (3.9) into (3.7) we find

$$c_1(\mathcal{O}(-1))\tilde{c}_p(\mathcal{L}_3)=-\tilde{c}_{p+1}(\mathcal{C}_2)+\tilde{c}_{p+1}(\mathcal{L}_2)+\tilde{c}_{p+1-m}(\mathcal{C}_1)\pi^*c_m(\nu/\mu)$$

so on integrating over $\mathbf{P}(\mu)$

$$(3.10)\qquad \int_{\mathbf{P}(\mu)}c_1(\mathcal{O}(-1))\tilde{c}_p(\mathcal{L}_3)$$
$$=-\int_{\mathbf{P}(\mu)}\tilde{c}_{p+1}(\mathcal{C}_2)+\int_{\mathbf{P}(\mu)}\tilde{c}_{p+1}(\mathcal{L}_2)+\int_M\pi_*(\tilde{c}_{p+1-m}(\mathcal{C}_1))c_m(\nu/\mu).$$

Since $\mathcal{L}_2$ is the lift $\pi^* L_2$ of an exact sequence on M, namely

$$L_2 : 0 \to \mu \to \nu \to \nu/\mu \to 0, \quad \tilde{c}_{p+1}(\mathcal{L}_2) = \pi^* \tilde{c}_{p+1}(L_2).$$

If now $p - m > 0$, as we assumed, then the bundle $\pi : \mathbf{P}(\mu) \to M$ has fibers of positive dimension and so automatically

$$\int_{\mathbf{P}(\mu)} \pi^* \tilde{c}_{p+1}(L_2) = 0$$

in (3.10).

The third integral in (3.10) is

$$\int_{x \in M} \left[\int_{\mathbf{P}(\mu_x)} \tilde{c}_{p+1-m}(\mathcal{C}_1) \right] c_m(\nu/\mu)$$

since degree$(\tilde{c}_{p+1-m}) = 2(p-m) = \operatorname{real}\dim(\mathbf{P}(\mu_x))$. But by [**5**, Proposition 5.3] on the projective space $\mathbf{P}(\mu_x)$ the standard sequence $\mathcal{C}_1$, has

$$\tilde{c}_{p+1-m}(\mathcal{C}_1|_{\mathbf{P}(\mu_x)}) = -H(\Lambda) c_1(\mathcal{O}_{\mathbf{P}(\mu_x)}(1))^{p-m}$$

so $\int_{\mathbf{P}(\mu_x)} \tilde{c}_{p+1-m}(\mathcal{C}_1) = -H(\Lambda)$. Finally we have proved (3.5) and hence also Proposition 3.1 □

4.

In this section we construct a current $\tilde{G}$ on $\widetilde{X}$ whose push-forward to X is a Green's current for B in X: $\pi_* \tilde{G} = G_B$. We then calculate in the next section (for $t \neq 0$) $\int_{A_t} G_B$ for a family of subvarieties A_t of X which are the fibers of a map $\mathcal{A} \to D$ (the disk $|t| < \epsilon$), with $A_0 = A$ and A_t disjoint from B for $t \neq 0$.

As before we have the line bundle L and section s on $\widetilde{X}$ with the exceptional divisor $\widetilde{B} = \mathbf{P}(N_B(X))$ defined by the vanishing of s. Along $\widetilde{B}$ we identify L with $\mathcal{O}_{\mathbf{P}}(-1)$ via ds and metrize L over $\widetilde{X}$ so ds is an isometry over $\widetilde{B}$ (where the metric on $\mathcal{O}_{\mathbf{P}}(-1)$ is inherited from X via $N_B(X)$).

We denote by $\mathcal{C}$ the standard exact sequence on $\widetilde{B}$:

$$\mathcal{C} : 0 \to \mathcal{O}_{\mathbf{P}(N_B(X))}(-1) \to \pi^* N_B(X) \to Q \to 0. \tag{4.1}$$

We make the assumption that the cohomology map $H^{2p}(\widetilde{X}) \to H^{2p}(\widetilde{B})$ is surjective. This will hold if $H^{2p}(X) \to H^{2p}(B)$ is surjective and in particular holds if X is replaced by $X \times X$ and B by the diagonal. Finally we assume we have chosen a Kähler metric on $\widetilde{X}$. Then $H^{p,p}(\widetilde{X}) \to H^{p,p}(\widetilde{B})$ is also surjective and so the (p,p) closed form $c_p(Q)$ on $\widetilde{B}$ is the restriction of some (p,p) closed form, denoted by α, on $\widetilde{X}$.

PROPOSITION 4.1. *Let $\tilde{G} = (\log \|s\|^2)\alpha + \gamma$ where α is any closed (p,p) form on $\widetilde{X}$ restricting to $c_p(Q)$ on $\widetilde{B}$, and γ is a (p,p) form $\widetilde{X}$ satisfying the two conditions*

a) *$dd^c \gamma = c_1(L) \wedge \alpha - \pi^* \omega_B$ where ω_B is the harmonic Poincaré dual form to B on X,*

b) *If P_H denotes the projection to harmonic forms on X, then*

$$P_H(\pi_* \gamma) = -P_H(\pi_*(\log \|s\|^2 \alpha)).$$

Then $G_B = \pi_*(\tilde{G})$ is a Green's current for B on X, i.e., satisfies (1.1), (1.2). *Such forms γ exist.*

PROOF. See B. Wang [8, Theorem 2.5]. The conditions a, b correspond exactly to (1.1), (1.2). □

PROPOSITION 4.2. *The (p,p) form γ on $\widetilde{X}$ can be chosen to satisfy the conditions* a, b *in Proposition* 4.1 *and the following further condition: There is a (p,p) form $\overline{\gamma}$ on X such that the restrictions of γ, $\tilde{c}_{p+1}(\mathcal{C})$ and $\pi^*\overline{\gamma}$ to either $\mathbf{P}(\mu)$ or $\mathbf{P}(\nu)$ satisfy*

$$\gamma = \tilde{c}_{p+1}(\mathcal{C}) + \pi^*\overline{\gamma} \pmod{\operatorname{Im} d + \operatorname{Im} d^c} \tag{4.2}$$

on either $\mathbf{P}(\mu)$ or $\mathbf{P}(\nu)$.

If further $m < p$ (for a given component of M) then

$$\gamma = \tilde{c}_{p+1}(\mathcal{C}) \tag{4.3}$$

on $\mathbf{P}(\mu)$ or $\mathbf{P}(\nu)$ always $\operatorname{mod} \operatorname{Im} d + \operatorname{Im} d^c$.

PROOF. Denote by γ' the (p,p) form γ asserted to exist and satisfy a, b in Proposition 4.1. Since on $\widetilde{B}$

$$dd^c\tilde{c}_{p+1}(\mathcal{C}) = c_1(\mathcal{O}(-1))c_p(Q) - \pi^*c_{p+1}(N_B(X)) \tag{4.4}$$

on subtracting (4.4) from a, we get

$$dd^c[\gamma'|_{\widetilde{B}} - \tilde{c}_{p+1}(\mathcal{C})] = \pi^*[c_{p+1}(N_B(X)) - \omega_B|_B].$$

But $c_{p+1}(N_B(X))$, $\omega_B|_B$ are cohomologous on B, so

$$c_{p+1}(N_B(X)) - \omega_B|_B = dd^c\overline{\gamma}'$$

for some form $\overline{\gamma}'$, on the Kähler manifold B, which can be extended arbitrarily to a (p,p) form on X. (If $\dim B < p+1$ we can choose $\overline{\gamma}' = 0$). So

$$dd^c[\gamma'|_{\widetilde{B}} - \tilde{c}_{p+1}(\mathcal{C}) - \pi^*\overline{\gamma}'] = 0$$

on $\widetilde{B}$, which is Kähler, so there is a closed form β' and forms η_1, η_2 on $\widetilde{B}$ so that

$$\gamma'|_{\widetilde{B}} - \tilde{c}_{p+1}(\mathcal{C}) - \pi^*\overline{\gamma}' = \beta' + d\eta_1 + d^c\eta_2$$

on $\widetilde{B}$.

By the cohomology assumption $H^{2p}(\widetilde{X}) \to H^{2p}(\widetilde{B})$ surjective, $\exists$ a closed (p,p) form β on $\widetilde{X}$ restricting to β' $(\operatorname{mod} \operatorname{Im} d + \operatorname{Im} d^c)$ on $\widetilde{B}$.

Let $\gamma = \gamma' - \beta + \pi^*(P_H(\pi_*\beta))$ on $\widetilde{X}$ (P_H = harmonic projection, in X). Then $P_H(\pi_*\gamma) = P_H(\pi_*\gamma')$.

The restriction of γ to $\widetilde{B}$ satisfies

$$\gamma|_{\widetilde{B}} = \tilde{c}_{p+1}(\mathcal{C}) + \pi^*[P_H(\pi_*\beta) + \overline{\gamma}'] \pmod{\operatorname{Im} d + \operatorname{Im} d^c}.$$

Let $\overline{\gamma} = P_H(\pi_*\beta) + \overline{\gamma}'$ on X. Then γ satisfies (4.2). Further if $m < p$ then $\pi^*\overline{\gamma} = 0$ on the total space $\mathbf{P}(\nu)$ or $\mathbf{P}(\mu)$ over M since $\overline{\gamma}$ restricts to 0 on M ($\overline{\gamma}$ being a (p,p) form). Note that if B is replaced by the diagonal Δ then $\dim[(A \times B) \cap \Delta] < \dim(A \times B)$ except in the case $\dim A = 0 = \dim B$. Finally, γ satisfies 1a since γ' does and β, $\pi^*(P_H(\pi_*\beta))$ are closed, and γ satisfies 1b since γ' does and $P_H(\pi_*\gamma) = P_H(\pi_*\gamma')$. This concludes the proof. □

The choices of the various forms become canonical in the special case we consider next: suppose

$$\begin{array}{ll} & E \text{ is a vector bundle on } X \text{ of rank } p+1 = \operatorname{codim} B,\ \sigma \text{ is a section of } E \\ (4.5) & \text{whose vanishing defines } B \text{ (so } \sigma \text{ is transverse to the zero section along } B), \\ & \|\ \|_E \text{ is a metric on } E \text{ such that } d\sigma\colon N_B(X) \to \mathrm{E}|_B \text{ is an isometry.} \end{array}$$

In this situation (4.5) let $\mathbf{P}(E)$ be the projective space bundle over X. σ defines a section $[\sigma]$ of $\mathbf{P}(E)$ over $X\backslash B$ and the closure of $[\sigma](X\backslash B)$ in $\mathbf{P}(E)$ can be identified with $\widetilde{X}$, the blow up of X along B. (All this is quite standard: we are embedding $\widetilde{X}$ in $\mathbf{P}(E)$ by using the section σ). The lift $\pi^*\sigma$ is a section of π^*E over $\mathbf{P}(E)$ whose vanishing defines $\widetilde{B} = \mathbf{P}(E|_B)$. If we restrict π^*E to $\widetilde{X}$ (here $\widetilde{X} = \{(x, \mathrm{l}_x) : x \in X, \mathrm{l}_x = 1-\dim \text{subspace of } E_x, \sigma(x) \in \mathrm{l}_x\}$), we see that $\pi^*\sigma|_{\widetilde{X}}$ is a section of $\mathcal{O}_{\mathbf{P}(E)}(-1)$ over $\widetilde{X}$.

Consider the usual exact sequence restricted from $\mathbf{P}(E)$ to $\widetilde{X}$:

$$\mathcal{C} : 0 \to \mathcal{O}(-1) \to \pi^*E \to Q_E \to 0. \tag{4.6}$$

PROPOSITION 4.3. *Assuming* (4.5), *let* ω_B *be the harmonic Poincaré dual form of* B *in* X, *then*

$$c_{p+1}(E) = \omega_B + dd^c\overline{\gamma} \tag{4.7}$$

where $\overline{\gamma}$ *is a* (p,p) *form on* X, *uniquely determined mod forms* $d\rho_1 + d^c\rho_2$ *when we specify its harmonic part* $P_H(\overline{\gamma})$, *which we do by requiring*

$$P_H[\overline{\gamma} + \pi_*(\log\|\pi^*\sigma\|^2 c_p(Q_E) + \tilde{c}_{p+1}(\mathcal{C})] = 0. \tag{4.8}$$

Further define

$$\tilde{G} = (\log\|\pi^*\sigma\|^2)c_p(Q_E) + \tilde{c}_{p+1}(\mathcal{C}) + \pi^*\overline{\gamma} \tag{4.9}$$

on $\widetilde{X}$.

Then $\pi_*\tilde{G} = G_B$ *is a Green's current for* B *in* X. (*Here we define* $\|\pi^*\sigma\|^2$ *as* $\|\sigma\|^2$).

PROOF. This is a special case of Proposition 4.1 together with Proposition 4.2 where the forms α, γ have been canonically specified under assumption (4.5). $\square$

As a further example of (4.5), we can replace X by $\mathbf{P}(E)$, E by Q_E and σ by the image of $\pi^*\sigma$ in Q_E, with zero locus $\widetilde{X}$.

5.

We can now put together the ingredients for calculating $\int_{A_t} G_B$. Much of the work was already done in [**8**] by examining local coordinates around points in $\widetilde{X}$ lying over M (see Figure 1), so we need only give a short geometric description of what happens to $\tilde{A}_t = \pi^{-1}(A_t)$ in $\widetilde{X}$ as $t \to 0$. In the notation of Section 2, we are looking at a map of $\mathcal{A}$ onto an ϵ-disk $D : |t| < \epsilon$ in $\mathbb{C}$. If we identify the line bundle Γ over $\mathcal{A}$ with the pull-back to $\mathcal{A}$ of the trivial bundle $D \times \mathbb{C}$ over D and the section g of Γ with the pull-back of the "diagonal" section $t \to (t,t)$ of $D \times \mathbb{C}$ then we can identify g with the map $\mathcal{A} \to D$, i.e., write $A_t = g^{-1}(t)$. We will also write $\log|t|^2$ for $\log\|g\|^2$ (i.e., regard t as a function on $\mathcal{A}$ via pull-back from D to $\mathcal{A}$). We may write: $dg(\tau) = \partial/\partial t$ at $t = 0$, More precisely (see Section 2) τ' is the unique vector field tangent to $\mathcal{A}$ along A_0 such that the map to D takes it to d/dt at $t = 0$, τ is the residue class of τ' modulo TB, along M.

Although $\mathcal{A}$ represents a holomorphic deformation of an open subset of A containing M, and not necessarily of all of A, we may assume that this deformation extends differentiably to all of A, denoting it again by A_t, so that for each t in the disk, A_t is a differentiable chain which approaches $A = A_0$ as a current as t approaches 0. We have to see that for any two Green's currents G_B, G'_B for B the two functions of t given by integrating over A_t have the same asymptotics, i.e., the same leading term $e \log |t|^2$ and the same constant term. This needs proof because, although $G_B - G'_B = d\rho_1 + d^c\rho_2$ where ρ_i are C^∞ outside B, Stokes' theorem does not immediately give $\int_{A_t} d^c\rho_i = 0$ since A_t is not everywhere complex analytic. However this difficulty is overcome in [**8**] where the following is proved:

PROPOSITION 5.1. *If G_B, G'_B are Green's currents for B then $\int_{A_t}(G_B - G'_B) \to 0$ as $t \to 0$.*

PROOF. The main point is to reduce this to the holomorphic case, and is done as follows. Write A_t as a union of a *holomorphic* manifold A'_t with boundary and a complementary chain A'' with the same boundary (oriented in the opposite way), where, as $t \to 0$, A'_t approaches holomorphically the open neighborhood of M in A_0 denoted A'_0, and A''_t approaches the complement A''_0. Then

$$\int_{A''_t} d^c\rho_2 = \left(\int_{A''_0} d^c\rho_2\right) + o(1)$$

(since $d^c\rho_2$ is singular only on B and here we are away from B), and $\int_{A''_0} d^c\rho_2$ reduces to an integral over $\partial A''_0$ using the complex analyticity of A''_0 and Stokes. The other integral $\int_{A'_t} d^c\rho_2$ in exactly the same way evaluates as an integral over $\partial A'_t$ which equals $o(1)$ plus an integral over $\partial A'_0$ that cancels the previous integral over $\partial A''_0$. □

We will now state our main result: the geometry of A, B, M, τ is as described in Section 2 and the deformation A_t as described above in this section.

THEOREM 5.2. *Let A, B intersect cleanly along M and let M have connected components M_i of dimensions m_i. We assume A, B connected, A smooth near M, and B, M smooth.*

Let A_t be a deformation of A as described above, which is holomorphic in a neighborhood of M and is a differentiable chain elsewhere and which approaches $A = A_0$ as a current as $t \to 0$. Let the vector field τ' be tangent to the deformation as before, let τ be $\tau' \mod TB$, and $\mu_0 = TA + TB/TB$, $\mu = \mu_0 + \mathbb{C}\tau$ (vector bundles on M). Then $\int_{A_t} G_B$ is defined for $t \neq 0$ and satisfies the following asymptotics independently of the choice of Green's current G_B:

$$\int_{A_t} G_B = e \log |t|^2 + k + o(1)$$

where e is an integer independent of the deformation A_t, of τ, and of the choice of G_B and is given by

$$e = \sum_i \int_{M_i} c_{m_i}(TX/TA + TB) \tag{5.1}$$

(here we define $c_0 = 1$ and the integral as 1 if M_i is a point), and the constant term k depends only on A, B and τ along M but not on the deformation A_t, or

the choice of G_B. It can be calculated by choosing G_B as in Proposition 4.2 by the formulas

$$(5.2)\quad k = \int_{A\setminus M} G_B + \sum_i \left[\int_{M_i} \log |P_b ot(\tau)|^2 c_{m_i}(TX/TA + TB + \mathbb{C}\tau) + \int_{\mathbf{P}(\mu_0)} \tilde{c}_p(\mathcal{L}) \right]$$

provided all components M_i of M have dimensions $m_i < p$ ($p = \dim A$), where $P_b ot(\tau)$ i.e. T is the orthogonal projection of τ to $\mu_0^\perp$, and $\mathcal{L}$ is the exact sequence of vector bundles on $\mathbf{P}(\mu)$ with metrics from $N_B(X)$ (see (3.2)): $(\mathcal{L})\ 0 \to Q_\mu \to Q_\nu \to \pi^*(\nu/\mu) \to 0$. If in adition (4.5) holds then G_B can be chosen canonically as in Proposition 4.3 and (denoting π^{-1} the isomorphism $X\setminus B \to \widetilde{X}\setminus\widetilde{B}$)

$$(5.3)\qquad \int_{A\setminus M} G_B = \int_{A\setminus M} [\log \|\sigma\|^2 (\pi^{-1})^* c_p(Q_E) + (\pi^{-1})^* \tilde{c}_{p+1}(\mathcal{C}) + \overline{\gamma}].$$

In the special case $M = A$, instead of (5.2) we have

$$(5.4)\qquad k = \int_M \log |\tau|^2 c_m(TX/TA + TB + \mathbb{C}\tau) + \tilde{c}_{p+1}(\mathcal{C}) + \overline{\gamma}$$

where as in Proposition 4.2, (4.2), $\gamma = \tilde{c}_{p+1}(\mathcal{C}) + \pi^*\overline{\gamma}$ and in this formula $\tilde{c}_{p+1}(\mathcal{C})$ on $\mathbf{P}(\mu)$ is identified with a form on M by the isomorphism of M with $\mathbf{P}(\mu)$ given by τ. ($\mathcal{C}$ is the exact sequence in (4.1).)

Proof of Theorem 5.2. As noted above, we are free to choose G_B as

$$\pi_*(\log \|s\|^2 \alpha + \gamma).$$

Since A_t does not intersect B for $t \neq 0$ and π is an isomorphism of $\widetilde{X}\setminus\widetilde{B}$ with $X\setminus B$, we can define $\tilde{A}_t = \pi^{-1}(A_t)$ for $t \neq 0$ and will have $\int_{A_t} G_B = \int_{\tilde{A}_t} \tilde{G}$, where $\tilde{G} = \log \|s\|^2 \alpha + \gamma$ on $\widetilde{X}$. To see what happens as $t \to 0$, we can look separately at i) neighborhoods in $\widetilde{X}$ of points in $\tilde{A}\setminus\mathbf{P}(\mu_0)$, where $\tilde{A} \cap \mathbf{P}(\mu) = \mathbf{P}(\mu_0)$ (see Section 2) and ii) neighborhoods of points in $\mathbf{P}(\mu)$.

At points of type i), $\tilde{G}$ is continuous and the integral over the intersection of A_t with such a neighborhood converges to the integral over an open set in A_0. Finally i) leads to the convergent integral $\int_{A\setminus M} G_B$ in (5.2).

In case ii) the term $\log \|s\|^2$ becomes infinite as we approach $\mathbf{P}(\mu)$ since $s = 0$ defines $\widetilde{B}$ which contains $\mathbf{P}(\mu)$. The main point, as in [**8**], is to write:

$$\log \|s\|^2 = \log |t|^2 + [\log \|s\|^2 - \log |t|^2].$$

Next one has to see (e.g., by using local coordinates as in [**8**]) that near type ii) points, the integral of $(\log \|s\|^2)\alpha + \gamma$ over such a neighborhood intersected with A_t, is the sum of

a) $\log |t|^2 \int \alpha$ and
b) $\int [\log \|s\|^2 - \log |t|^2]\alpha + \gamma$.

The term a) then gives the leading term $e \log |t|^2$.

The analysis of $\log \|s\|^2 - \log |t|^2$ in Section 2 shows that this function extends smoothly to $\mathbf{P}(\mu)$, i.e. $s = 0, t = 0$ (Proposition 2.2). Then Propositions 3.1 and 4.2 complete the calculation of the integral b) and the proof of (5.2).

The case $M = A$, that is (5.4), is actually easier since the integral over $A \backslash M$ is replaced by 0. Furthermore the exact sequence $\mathcal{C}$ (from (4.1)) when restricted to $\mathbf{P}(\mu)$, which we identify with M by the map $M \to \mathbf{P}(\mu) = \mathbf{P}(\mathbb{C}\tau)$ given by τ, becomes the following exact sequence over M(here $\nu = (TX/TB)|_M = TX/TA + TB$)

$$0 \to \mathbb{C}\tau \to TX/TA + TB \to TX/TA + TB + \mathbb{C}\tau \to 0.$$

However we also need the form $\overline{\gamma}$ here, unlike in (5.2) where its restriction to M vanishes.

In both (5.2), (5.4) the forms $\gamma, \overline{\gamma}$ have not been canonically specified. Under the further assumption that the vector bundle E and section σ are given (see Assumption 4.5) these forms as well as α and s are completely specified, as in (5.3). Thus Theorem 5.2 is completely proved. □

Furthermore we can make the following observation:

5.1. Under the assumption (4.5) about the existence of a vector bundle E, section σ, and metric on E, with $B = (\sigma = 0)$, the formulas (5.1) for the integer e and (5.2), (5.3) for the real number k ($e = e(A, B)$, $k = k(A, B, \tau)$), allow us to attach to (A, B, τ) the function of a complex variable t:

$$e \log |t|^2 + k. \tag{5.5}$$

Note that if t is changed to ct, c a non-zero constant, then (5.5) is unchanged.

If further a deformation A_t exists then (5.5) gives the first two terms of the height pairing asymptotics, but (5.5) is defined (assuming (4.5)) even without such a deformation. If A, B are disjoint and we interpret e as 0, then $k = (A, B)$. The first term $e \log |t|^2$ is analogous to the arithmetic term $n_p \log p$ in the global height pairing formula for arithmetic varieties. It is not clear whether one can make a *sufficiently simple* choice of the forms γ, α (and the section s) in general, i.e., without assuming (4.5), to allow us to make a definition (5.5) without this extra assumption.

6.

We discuss next symmetry of the height pairing asymptotics under interchange of which A or B is moved. It is most convenient to work in $X \times X$ and use $A \times B$ and the diagonal Δ to play the roles of A, B.

As before let A_t be a deformation of A in X which is holomorphic only near M, with tangent vector field τ' along M (τ denoted τ' modulo TB). Let G_B in X and G_Δ in $X \times X$ be Green's currents constructed as in Proposition 4.1. As discussed by Bost in [**2**, Theorem 2.1 i] G_Δ can be restricted to $X \times B$ and integrated in the second variable over B (i.e., we look at $\mathrm{pr}\colon X \times B \to X$ and take $\mathrm{pr}_*(G_\Delta)$) and this gives a Green's current G'_B for B on X. As discussed just before (Theorem 5.2), $\int_{A_t} G_B$ and $\int_{A_t} G'_B$ have the same asymptotics (mod $o(1)$) as $t \to 0$. Since

$$\int_{A_t} G'_B = \int_{A_t \times B} G_\Delta \tag{6.1}$$

we can study the asymptotics of $\int_{A_t} G_B$ by using $\int_{A_t \times B} G_\Delta$ instead. This has a further advantage: *it is not necessary* to assume B smooth everywhere: just near $A \cap B = M$, since Δ is smooth.

THEOREM 6.1. *Let A, B be smooth in a neighborhood of $M = A \cap B$ and intersect cleanly along M. Let A_t, B_t be deformations as described in Sections 2 and 5, with respective tangent vector fields τ' and ρ' along M (see (2.2) and following discussion). Suppose that*

$$\text{(6.2)}\quad \begin{array}{l}\textit{at each } x \in M, \textit{ the projections } T = p(\tau') \textit{ and } R = p(\rho') \textit{ orthogonal to} \\ T_xA + T_xB \textit{ span the same one-dimensional subspace } \mathbb{C}T = \mathbb{C}R \textit{ and have} \\ \textit{the same (non-zero) length.}\end{array}$$

Then $\int_{A_t \times B} G_\Delta$ and $\int_{A \times B_t} G_\Delta$ have the same asymptotics: $e \log |t|^2 + k + o(1)$ (i.e., the same e and k, calculable in $X \times X$ by (5.1), (5.2).

We assume the dimensions p, q are not both 0.

PROOF. In $X \times X$, $(A \times B) \cap \Delta = \Delta_M$, the diagonal in $M \times M$. Furthermore we have an isomorphism

$$T(X \times X)/T(\Delta) \to TX$$
$$(\mathrm{v}_1, \mathrm{v}_2) + T(\Delta) \to (\mathrm{v}_1 - \mathrm{v}_2)/\sqrt{2}$$

which is even an isometry if the quotient on the left is metrized as the orthogonal complement of $T\Delta$. (We have been metrizing all quotients this way, as a convenient choice.) This isometry along Δ gives us an isometry of vector bundles on Δ_M, respectively M,:

$$T(X \times X)/T(A \times B) + T(\Delta) \to TX/T(A) + T(B)$$

(these bundles were denoted ν/μ_0). The integral over M of the Chern class c_m gave the integer e, independently of the deformation or vector field, so e is the same for $A_t \times B$ and $A \times B_t$.

Next we look at the constant term k as given in (5.2) where we work in $X \times X$ instead of X. The first term in (5.2) is now (noting $\dim \Delta_M < \dim A \times B$ unless A, B are zero dimensional) $\int_{A \times B} G_\Delta$ which is the same for $A_t \times B$ and $A \times B_t$ as it does not use the deformation.

In the second term, if we are working in X then $T = P_b ot(\tau)$ is just the projection of τ' to $(TA + TB)^\perp$, while in $X \times X$ the vector field for $A_t \times B$ is $(\tau', 0)$ and we have to project it to $(T(A \times B) + T(\Delta))^\perp$ and take its length there which is $|T|/\sqrt{2}$ by (6.2). Now the assumption (6.2) shows that the second term on the right in (5.2) is the same for τ' as for ρ'.

As for the third term in (5.2), the space is $\mathbf{P}(\mu_0) = \mathbf{P}(T(A \times B) + T(\Delta)/T(\Delta))$ and the exact sequence is

$$\mathcal{L} : 0 \to \pi^*\mu/\mathcal{O}(-1) \to \pi^*\nu/\mathcal{O}(-1) \to \pi^*(\nu/\mu) \to 0$$

where

$$\mu = T(A \times B) + T(\Delta) + \mathbb{C}(\tau', 0)/T(\Delta)$$
$$\nu = T(X \times X)/T(\Delta)$$

and these are the same for $(\tau', 0)$ and $(0, \mu')$ by the assumption (6.2). Thus the constant terms k agree, concluding the proof. □

7.

An example.

We let $X = \mathbf{P}^n$ and A, B be linear subspaces of dimensions p, q with $p+q = n-1$, with $M = A \cap B$ of dimension $m \geq 0$. A_t will be a holomorphic family which we can take as $g(t)A$ where $t \to g(t)$ is a holomorphic map of an ϵ-disk to $GL(n+1)$ with $g(0) =$ Identity. Let α, β be the linear subspaces of $\mathbb{C}^{n+1}$ corresponding to A, B and write $A = [\alpha]$ for the corresponding point in the Grassmanian $G_{p,n}$ and $\alpha(t) = g(t)\alpha$.

In $\mathbf{P}^n$ we have the tangent vector field τ' along $M = A \cap B$ and its residue class T which is a nowhere 0 section of $T(\mathbf{P}^n)/TA + TB$. We can identify T with an *isomorphism* of linear spaces $\alpha \cap \beta \to \mathbb{C}^{n+1}/(\alpha+\beta)$ and composing with the isometry $\mathbb{C}^{n+1}/(\alpha+\beta) \to (\alpha+\beta)^{\perp}$ we obtain a linear isomorphism, which we also denote T,

$$T : \alpha \cap \beta \to (\alpha + \beta)^{\perp} \tag{7.1}$$

of vector spaces with metrics (*not* claimed to be an isometry).

Then $|\det T|$ is a well defined positive real number. (The top exterior powers are one-dimensional with metrics.)

One can see that the isomorphism $\alpha \cap \beta \to \mathbb{C}^{n+1}/\alpha+\beta$ is determined by the tangent vector $\xi \in T_{[\alpha]}(\mathbf{G}_{p,n}) = \mathrm{Hom}(\alpha, \mathbb{C}^{n+1}/\alpha)$ which is the image under $t \to [\alpha(t)]$ of d/dt at $t=0$, i.e., T is the linear transformation induced by $\xi \colon \alpha \to \mathbb{C}^{n+1}/\alpha$.

PROPOSITION 7.1. *Let $P(\alpha,\beta) = P$ be the linear isomorphism $\alpha \cap (\alpha\cap\beta)^{\perp} \to (\alpha+\beta)\cap\beta^{\perp}$ of subspaces of $\mathbb{C}^{n+1}$ induced by the orthogonal projection $\mathbb{C}^{n+1} \to \beta^{\perp}$, and $T\colon \alpha\cap\beta \to (\alpha+\beta)^{\perp}$ the isomorphism* (7.1). *Then we have the height pairing asymptotics*

$$\begin{aligned}(A_t, B) &= \log(|\det(tT)|^2 |\det P|^2) + C + \mathcal{O}(|t|)\\ &= (m+1)\log|t|^2 + \log|\det T|^2 + \log|\det P|^2 + C + \mathcal{O}(|t|)\end{aligned}$$

where C is a constant independent of A, B (or T), (see (7.3)*), T is determined by the tangent vector to the deformation, and P depends on A, B but not on the deformation. The integer e is $m+1$ and $k = \log(|\det T|^2) + \log(|\det P|^2) + C$.*

PROOF. We make use of the incidence correspondence I with

$$\begin{aligned}\pi_1 &: I \to G_{p,n}\\ \pi_2 &: I \to \mathbf{P}^n\end{aligned}$$

and construct a metrized line bundle L on $G_{p,n}$ and section s with zero locus $D = \pi_1(\pi_2^{-1}B)$ such that for any A' disjoint from B, e.g., $A' = A_t$,

$$\log\|s(A')\|^2 + C = (A', B). \tag{7.2}$$

First of all we have the vector bundle on $G_{p,n}$ with fiber over the point $[\alpha]$ given by $\mathrm{Hom}(\alpha, \beta^{\perp})$. L is the determinant line bundle with fiber $\mathrm{Hom}(\wedge^{p+1}\alpha, \wedge^{p+1}(\beta^{\perp}))$. s is the composition of the inclusion $\wedge^{p+1}\alpha \to \wedge^{p+1}\mathbb{C}^{n+1}$ with the orthogonal projection. The metric is given as follows: if $f\colon \wedge^{p+1}(\alpha) \to \wedge^{p+1}(\beta^{\perp})$ is linear and λ is non-zero in $\wedge^{p+1}(\alpha)$ then $\|f\|^2 = \|f(\lambda)\|^2/\|\lambda\|^2$ (using the metrics in $\alpha, \beta^{\perp}$). L is the pull-back of $\mathcal{O}(1)$ under the Plücker embedding and s is the section given by one Plücker coordinate. The divisor D of s consists of all $[\alpha]$ with $\alpha\cap\beta \neq 0$ and is the union of manifolds D_m (strata) defined by: $\dim(\alpha \cap \beta) = m + 1$. (A normal vector to D_m at $[\alpha]$ is given by a homomorphism $\alpha \cap \beta \to \mathbb{C}^{n+1}/(\alpha + \beta)$ which is an isomorphism if this normal vector is not in the limit of tangent spaces to D_i for $i < m$.)

We show first that (7.2) holds. The correspondence I between $\mathbf{P}^n$ and $G_{p,n}$ is such that both projections π_1, π_2 are fiber bundle maps and are equivariant with respect to the compact group $\mathrm{U}(n+1)$ of isometries of $\mathbb{C}^{n+1}$. We then have maps $\pi_{1*}\pi_2^*$ of forms on $\mathbf{P}^n$ to forms on $G_{p,n}$, commuting with the $\mathrm{U}(n+1)$ action, so harmonic forms, being the same thing as invariant forms, are sent to harmonic forms. Since π_i^* and π_{i*} can be regarded as transposes of each other, π_i^* is defined on currents and extends its definition on forms. We have $\pi_{1*}\pi_2^*\delta_B = \delta_{\pi_1\pi_2^{-1}(B)} = \delta_D$ (note that $\pi_2^{-1}(B) \to D$ is an isomorphism over the open set D_1 of D of $[\alpha]$ with $\alpha \cap \beta$ one-dimensional). Finally we have $\pi_{1*}\pi_2^* G_B = G_D$.

Since we have the section s of L with divisor D on $G_{p,n}$ and since $\mathrm{U}(n+1)$ acts on L preserving the metric, the Chern form $c_1(L)$ is invariant and hence harmonic: $c_1(L) = \omega_D$.

$$G_D = \log \|s\|^2 + C$$

where C is chosen so that the harmonic projection of G_D is 0. This gives (7.2). To evaluate C we take $A' = B^\perp$, i.e., $\alpha' = \beta^\perp$ in $\mathbb{C}^{n+1}$. Then $\|s(B^\perp)\| = 1$ and $C = (B^\perp, B)$. Finally, $(B^\perp, B)$ is evaluated in [[**3**, p. 997, iii and (1.4.3)], recalling that the Green's forms of [**3**] are the negatives of ours, we get

$$C = 2(\sigma_n - \sigma_q - \sigma_p) \tag{7.3}$$

where $\sigma_r = \frac{1}{2}\sum_{k=1}^{r}\sum_{m=1}^{k} 1/m$.

To prove Proposition 7.1 we now have to calculate $\log\|s(\alpha(t))\|^2$. Choose an orthonormal basis $e_1, \ldots, e_{m+1}$ for $(\alpha+\beta)^\perp$ in $\mathbb{C}^{n+1}$, and extend it to an orthonormal basis $e_1, \ldots, e_{p+1}$ for $\beta^\perp$. Further, choose an orthonormal basis $e_{p+2}, \ldots, e_{p+m+2}$ for $\alpha \cap \beta$ and extend it to an orthonormal basis $e_{p+2}, \ldots, e_{n+1}$ for β. Then $e_{m+2}, \ldots, e_{p+1}$ is an orthonormal basis for $(\alpha+\beta)\cap\beta^\perp$ (since $\beta^\perp = (\alpha+\beta)\cap\beta^\perp \oplus (\alpha+\beta)^\perp$) and $e_{m+2}, \ldots, e_{n+1}$ is a basis for $\alpha+\beta$. To obtain an orthonormal basis for α, denoted $\mathrm{v}_1, \ldots, \mathrm{v}_{p+1}$ we let $\mathrm{v}_1 = e_{p+2}, \ldots, \mathrm{v}_{m+1} = e_{p+m+2}$ be the basis chosen for $\alpha\cap\beta$, and further vectors $\mathrm{v}_{m+2}, \ldots, \mathrm{v}_{p+1}$ which are an orthonormal basis for $\alpha \cap (\alpha\cap\beta)^\perp$.

Recall that $\alpha(t) = g(t)\alpha$, $\alpha = \alpha(0)$ ($g(t) \in \mathrm{GL}(n+1)$) and write $\mathrm{v}_i(0) = \mathrm{v}_i$, $\mathrm{v}_i(t) = g(t)\mathrm{v}_i$: this gives a basis for $\alpha(t)$ which is orthonormal only for $t = 0$. Write $\mathrm{v}_i(t) = a_{i1}(t)e_1 + \cdots + a_{in+1}(t)e_{n+1}$ for $i = 1, \ldots, p+1$.

The orthogonal projection $\mathbb{C}^{n+1} \to \beta^\perp$ (i.e., to the subspace spanned by $e_1, \ldots, e_{p+1}$) when restricted to $\alpha(t)$ can be written

$$F(t)\colon \mathrm{v}_i(t) \to \sum_{j=1}^{p+1} a_{ij}(t)e_j \quad i,j = 1,\ldots,p+1. \tag{7.4}$$

The matrix $a_{ij}(t)$ of this map $F(t)$ for the indicated bases can be written in block form with four blocks: $A_{11}(t)$ consisting of $a_{ij}(t)$ with $i, j = 1, \ldots, m+1$, $A_{12}(t)$ for $i = 1, \ldots, m+1$, $j = m+2, \ldots, p+1$, $A_{21}(t)$, $A_{22}(t)$. These block matrices have the following structure: $A_{11}(0) = 0$ (since $\mathrm{v}_i(0) = \mathrm{v}_i = e_{p+1+i}$ for $i \le m+1$)

$$A'_{11}(0) = \frac{d}{dt}A_{11}(t), \quad \text{at } t = 0$$

is the matrix of the map T of (7.1) with respect to the bases v_i, e_j (for $i, j \le m+1$), see remarks just before (7.1).

$A_{12}(0) = 0$ just as $A_{11}(0) = 0$. The matrices $A_{21}(0)$, $A_{22}(0)$ represent the map from $\alpha \cap (\alpha \cap \beta)^\perp$ (basis $\mathrm{v}_{m+2}, \dots, \mathrm{v}_{p+1}$) to $\beta^\perp$ induced by orthogonal projection $\mathbb{C}^{n+1} \to \beta^\perp$. This map has image $(\alpha + \beta) \cap \beta^\perp$ (see commutative diagram a little after (2.2)) which has basis $e_{m+2}, \dots, e_{p+1}$. Thus $A_{21}(0) = 0$ and $A_{22}(0)$ represents the isomorphism P of Proposition 7.1.

Finally we find for the determinant of this $(p+1) \times (p+1)$ matrix representing $F(t)\colon \alpha(t) \to \beta^\perp$ (in (7.4)),

$$\det[a_{ij}(t)] = \det(tT)\det(P) \mod t^{m+2}$$

where $\det(tT) = t^{m+1}\det(T)$. The section s had norm at $\alpha(t)$ given by

$$\|s(\alpha(t))\|^2 = \frac{\|F(t)\mathrm{v}_1(t) \wedge \cdots \wedge F(t)\mathrm{v}_{p+1}(t)\|^2}{\|\mathrm{v}_1(t) \wedge \cdots \wedge \mathrm{v}_{p+1}(t)\|^2}.$$

Since $F(t)\mathrm{v}_i(t) = \sum a_{ij}(t)e_j$ and the e_j are orthonormal, the numerator is just $|\det a_{ij}(t)|^2$.

In the denominator, $\mathrm{v}_i(t) = \mathrm{v}_i(0) \mod t$ and the $\mathrm{v}_i(0)$ are orthonormal so the denominator is $1 + \mathcal{O}(|t|)$. Finally $\log\|s(\alpha(t))\|^2 = \log|\det a_{ij}(t)|^2 + \mathcal{O}(|t|)$.

This concludes the proof of Proposition 7.1. □

References

1. A. Blanchard, *Sur les variétés analytiques complexes*, Ann. Sci. École Norm. Sup. (3), (2) **73** (1956), 157–202.
2. J.-B. Bost, *Green's currents and height pairing on complex tori*, Duke Math. J. **61** (1990), 899–912.
3. J.-B. Bost, H. Gillet, and C. Soulé, *Heights of projective varieties and positive Green's forms*, J Amer. Math. Soc. **7** (1994), 903–1027.
4. H. Gillet and C. Soulé, *Arithmetic intersection theory*, Inst. Hautes Études Scient. Publ. Math. **72** (1990), 93–174.
5. ———, *Characteristic classes for algebraic vector bundles with Hermitian metric.* I, Ann. of Math. (2) **131** (1990), 169–203; II, 205–238.
6. R. Hain, *Biextensions and heights associated to curves of odd genus*, Duke Math. J. **61** (1990), 859–898.
7. H. Levine, *A theorem on holomorphic mappings into complex projective space*, Ann. of Math. (2) **71** (1960), 529–535.
8. B. Wang, *Asymptotics of the Archimedean height pairing*, Amer. J. Math. **120** (1998), 229–249.
9. ———, *Mazur's incidence structure for projective varieties* I, Compositio Math. **115** (1999), 303–327.
10. B. Wang and B. Harris, *Archimedean height pairing of intersecting cycles*, Internat. Math. Res. Notices (1993), 107–111.

Department of Mathematics, Brown University, Providence, RI 02912, USA

E-mail address, B. Harris: `Bruno_Harris@Brown.edu`

E-mail address, B. Wang: `binwang@math.brown.edu`

Centre de Recherches Mathématiques
CRM Proceedings and Lecture Notes
Volume **24**, 2000

Logarithmic Hodge Structures and Classifying Spaces

Kazuya Kato and Sampei Usui

ABSTRACT. If D is a Hermitian symmetric domain and Γ is an arithmetic group, then $\Gamma\backslash D$ has toroidal compactifications by Mumford et al. We generalize this to the case where D is the classifying space of polarized Hodge structures of arbitrary Hodge type. Our enlarged space $\Gamma\backslash D_\Sigma \supset \Gamma\backslash D$ has a structure of generalized fs logarithmic analytic space which is a fine moduli space of polarized logarithmic Hodge structures of certain type. This realizes one of the dreams of Griffiths in [**7**]

Contents

Introduction 115
Acknowledgments 117
1. Preliminaries 117
2. Spaces of Nilpotent Orbits 120
3. Spaces of Nilpotent i-Orbits 121
4. Strong Topology, Generalized fs Logarithmic Analytic Spaces 123
5. Polarized Logarithmic Hodge Structures 124
6. Main Result 126
7. Examples 127
References 129

Introduction

This work is an attempt to add 'points at infinity' to the classifying space of polarized Hodge structures (PH, for short) of arbitrary weight. It is based on the following two ideas:

1991 *Mathematics Subject Classification.* Primary 14C30; Secondary 14D07, 32G20.

Received by the editors September 9, 1998; Accepted March 31, 1999.

Partly supported by the Grants-in-Aid for Scientific Research (1) (B): 08304002, the Ministry of Education, Science and Culture, Japan.

This is the final form of the paper.

0.1. We introduce the notion of *polarized logarithmic Hodge structures* (*PLH*, for short) (§5). Polarized logarithmic Hodge structures are defined by using the theory of logarithmic structures of Fontaine-Illusie (cf. [**9**], [**10**]), and they work well in the analysis of degenerations of polarized Hodge structures. Our principle is that we enlarge the classifying space of polarized Hodge structures in the following way (for precise definitions, see Definitions 2.1, 5.5):

$$\begin{aligned} D &:= (\text{Classifying space of PH}) \\ &\quad \subset D_\Sigma := (\text{Space of nilpotent orbits in directions in } \Sigma), \\ \Gamma\backslash D_\Sigma &= (\text{Classifying space of PLH with local monodromies in directions in } \\ &\quad \Sigma \text{ and with global monodromy } \Gamma). \end{aligned}$$

Here Γ is a discrete subgroup of $\mathrm{Aut}(D)$ satisfying certain conditions.

0.2. We introduce the notions of *generalized analytic spaces* and *generalized fs logarithmic analytic spaces* (see Definition 4.4). Our extended space $\Gamma\backslash D_\Sigma \supset \Gamma\backslash D$ is not in general an analytic space. For instance, it can be something like

$$S := \{(x, y) \in \mathbb{C}^2; x \neq 0\} \cup \{(0, 0)\}.$$

The space $\Gamma\backslash D_\Sigma$ and the above space S admit the structure of generalized analytic space or even generalized fs logarithmic analytic space. In order to give S the structure of generalized analytic space, the essential point is to endow it with the so-called *strong topology* (see Definition 4.1), which is strictly stronger than the topology as a subspace of $\mathbb{C}^2$.

Our main result says that the generalized fs logarithmic analytic space $\Gamma\backslash D_\Sigma$ is a fine moduli space of polarized logarithmic Hodge structures of certain type (§6). In a way, this realizes one of the dreams of Griffiths in [**7**].

We would like to add a remark. D is always a homogeneous complex manifold. Assume that the horizontal tangent bundle of D is not trivial. Then, it is a Hermitian symmetric domain if and only if the set of Hodge numbers is one of the following three cases:

(i) $w = 2t + 1$, $h^{p,q} = 0$ unless $p = t + 1$, t.
(ii) $w = 2t$, $h^{p,q} = 1$ for $p = t + 1$, $t - 1$, $h^{t,t}$ is arbitrary, $h^{p,q} = 0$ otherwise.
(iii) $w = 2t$, $h^{p,q} = 1$ for $p = t + a$, $t + a - 1$, $t - a + 1$, $t - a$ for some $a \geq 2$, $h^{p,q} = 0$ otherwise.

Familiar (analytic) compactifications of $\Gamma\backslash D$ such as the Satake-Baily-Borel, the Borel-Serre, and toroidal are constructed only for those D which are Hermitian symmetric domains. The only partially successful attempts to go beyond the symmetric case are the work [**5**] of Cattani-Kaplan and the work [**19**]. By the construction of the present article, we can talk about, for example, the extended period mapping associated to a degeneration of surfaces of general type, or of Calabi-Yau manifolds. One of the remaining open problems is to find 'big fans' Σ so that the spaces $\Gamma\backslash D_\Sigma$ contain all possible boundary points. This work is a continuation of the study [**19**], which treats the part of $\Gamma\backslash D_\Sigma$, and is also a continuation of the study [**12**] on logarithmic complex geometry.

This article is an abbreviated account. A paper [**13**] with complete proofs will appear elsewhere.

Acknowledgments

The authors are grateful to Professor Kazuhiro Fujiwara for stimulating discussions.

They are grateful to Professor Steven Zucker for his careful reading of the earlier versions of this paper, and for his valuable suggestions and comments on presentations. Without his help, this paper has never reached the present form.

They are also grateful to the organizers of NATO ASI/1998 CRM Summer School at Banff, Canada.

1. Preliminaries

Classifying Spaces of Hodge Structures. Let $(h^{p,q})_{p+q=w}$, with $h^{p,q} = h^{q,p}$, be a set of Hodge numbers of weight w. Let H_0 be a free $\mathbb{Z}$-module of rank $\sum h^{p,q}$ and let $\langle\ ,\ \rangle_0$ be a non-degenerate $\mathbb{Q}$-bilinear form on $H_{0,\mathbb{Q}} := \mathbb{Q} \otimes H_0$ which is symmetric if w is even and anti-symmetric if w is odd. We denote by D the set of all decreasing filtrations F on $H_{0,\mathbb{C}} := \mathbb{C} \otimes H_0$ such that $(H_0, \langle\ ,\ \rangle_0, F)$ is a polarized Hodge structure of Hodge type $(h^{p,q})_{p+q=w}$. We denote by $\check{D}$ the set of all decreasing filtrations F on $H_{0,\mathbb{C}}$ such that $\dim F^p = \sum_{r\geq p} h^{r,w-r}$ and $\langle F^p, F^q\rangle_0 = 0$ for $p+q > w$.

Let $G_{\mathbb{Z}} := \mathrm{Aut}(H_0, \langle\ ,\ \rangle_0)$. For $A = \mathbb{Q}, \mathbb{R}, \mathbb{C}$, we denote $G_A := \mathrm{Aut}(H_{0,A}, \langle\ ,\ \rangle_0)$ and $\mathfrak{g}_A := \mathrm{Lie}\, G_A$. Then $\check{D}$ is a compact homogeneous space under $G_{\mathbb{C}}$, D is an open subset of $\check{D}$ that is homogeneous under $G_{\mathbb{R}}$. Let Γ be a subgroup of $G_{\mathbb{Z}}$. Then $\Gamma\backslash D$ is the classifying space of polarized Hodge structures on H_0 with the given Hodge numbers $(h^{p,q})_{p+q=w}$ and with the given global monodromy Γ.

Fans.

DEFINITION 1.1. (1) A subset $\sigma \subset \mathfrak{g}_{\mathbb{Q}}$ is a *nilpotent cone* if it satisfies the following two conditions:
- (i) σ is a finitely generated cone in $\mathfrak{g}_{\mathbb{Q}}$ consisting of mutually commuting nilpotent elements.
- (ii) $\sigma \cap (-\sigma) = \{0\}$.

(2) A *fan in* $\mathfrak{g}_{\mathbb{Q}}$ is a non-empty set Σ of nilpotent cones in $\mathfrak{g}_{\mathbb{Q}}$ satisfying the following two conditions:
- (i) If $\sigma \in \Sigma$ then any face of σ belongs to Σ.
- (ii) If $\sigma, \sigma' \in \Sigma$ then $\sigma \cap \sigma'$ is a face of σ and of σ'.

(3) A subgroup $\Gamma \subset G_{\mathbb{Z}}$ is *strongly compatible* with a fan Σ in $\mathfrak{g}_{\mathbb{Q}}$ if it satisfies the following two conditions:
- (i) If $\gamma \in \Gamma$ and $\sigma \in \Sigma$ then $\gamma\sigma\gamma^{-1} \in \Sigma$.
- (ii) If $\sigma \in \Sigma$ and if we define

$$\Gamma(\sigma) := \{\gamma \in \Gamma; \gamma : \text{unipotent}, \log\gamma \in \sigma\},$$

then $\log\Gamma(\sigma)$ generates the cone σ.

EXAMPLE 1.2. Let

$$\Xi := \{\sigma; \sigma \text{ is a nilpotent cone in } \mathfrak{g}_{\mathbb{Q}}, \mathrm{rank}(\sigma) \leq 1\}.$$

Then Ξ is a fan. For any subgroup Γ of $G_{\mathbb{Z}}$ with finite index, Γ is automatically strongly compatible with Ξ. Whereas, $\Gamma = \{1\}$ is not strongly compatible with Ξ, i.e., the condition (ii) in Definition 1.1(3) is not fulfilled.

Nilpotent Orbits and Nilpotent i-Orbits.

DEFINITION 1.3. Let σ be a nilpotent cone in $\mathfrak{g}_{\mathbb{Q}}$. For $A = \mathbb{R}$, $\mathbb{C}$, we denote by σ_A the A-linear span of σ in $\mathfrak{g}_A$. A subset Z of $\check{D}$ is a *σ-nilpotent orbit* (resp. *σ-nilpotent i-orbit*) if it satisfies the following three conditions for some $F \in Z$.

(i) $Z = \exp(\sigma_{\mathbb{C}})F$ (resp. $Z = \exp(i\sigma_{\mathbb{R}})F$).
(ii) $NF^p \subset F^{p-1}$ ($\forall p$, $\forall N \in \sigma$), (Griffiths transversality).
(iii) $\exp(\sum_{j=1}^{r} iy_j N_j)F \in D$ ($\forall y_j \gg 0$), (positivity). Here $(N_j)_{1\le j\le r}$ is a system of generators of the cone σ.

Logarithmic Structures. Let X be a ringed space over $\mathbb{C}$ with structure sheaf $\mathcal{O}_X$. We recall briefly the notion of logarithmic structures on X of Fontaine-Illusie (for details, cf. [**9**], [**10**]).

A *pre-logarithmic structure* on X is a sheaf of commutative monoids M_X together with a homomorphism $\alpha\colon M_X \to \mathcal{O}_X$ of monoids, where $\mathcal{O}_X$ is regarded as a sheaf of monoids by multiplication. A *logarithmic structure* on X is a pre-logarithmic structure (M_X, α) on X which satisfies $\alpha^{-1}(\mathcal{O}_X^\times) \xrightarrow{\sim} \mathcal{O}_X^\times$.

Let (M_X, α) be a pre-logarithmic structure on X. The *associated logarithmic structure* $(\widetilde{M}_X, \tilde{\alpha})$ is defined as the push-forward $\widetilde{M}_X$ of $\mathcal{O}_X^\times \leftarrow \alpha^{-1}(\mathcal{O}_X^\times) \to M_X$, namely the diagram

$$\begin{array}{ccc} \alpha^{-1}(\mathcal{O}_X^\times) & \longrightarrow & M_X \\ \downarrow & & \downarrow \\ \mathcal{O}_X^\times & \longrightarrow & \widetilde{M}_X. \end{array}$$

is co-Cartesian, together with the homomorphism $\tilde{\alpha}\colon \widetilde{M}_X \to \mathcal{O}_X$ induced by $\alpha\colon M_X \to \mathcal{O}_X$ and the inclusion $\mathcal{O}_X^\times \hookrightarrow \mathcal{O}_X$.

Let $f\colon (X, \mathcal{O}_X) \to (Y, \mathcal{O}_Y)$ be a morphism of ringed spaces and (M, α) be a logarithmic structure on Y. Then the sheaf-theoretic inverse image $f^{-1}M$ together with the composite morphism $f^{-1}M \to f^{-1}\mathcal{O}_Y \to \mathcal{O}_X$ form a pre-logarithmic structure on X. The *inverse image* $f^*(M, \alpha)$ of (M, α) is defined as the logarithmic structure on X associated to the above pre-logarithmic structure.

A monoid $\mathcal{S}$ is *fs* (= finitely-generated, integral and saturated) if it satisfies the following three conditions.

(i) $\mathcal{S}$ is finitely generated.
(ii) If a, b, $c \in \mathcal{S}$ and $ab = ac$, then $b = c$.
(iii) If $a \in \mathcal{S}^{\mathrm{gp}}$ and $a^n \in \mathcal{S}$ for some positive integer n, then $a \in \mathcal{S}$.

Here $\mathcal{S}^{\mathrm{gp}}$ is the group generated by $\mathcal{S}$. A logarithmic structure (M_X, α) on X is *fs* if there exists an open covering $(U_\lambda)_\lambda$ of X such that each restriction $(M_X, \alpha)|_{U_\lambda}$ is the logarithmic structure associated to a pre-logarithmic structure of the form $(\mathcal{S}_{U_\lambda}, \alpha_\lambda)$ where $\mathcal{S}_{U_\lambda}$ is a constant sheaf of monoids on U_λ induced by an fs monoid $\mathcal{S}$. In this case, $(\mathcal{S}_{U_\lambda}, \alpha_\lambda)$ is called a *chart* of (M_X, α) on U_λ.

The Ringed Space $(X^{\log}, \mathcal{O}_X^{\log})$. Let X be an fs logarithmic local ringed space (= a local ringed space endowed with an fs logarithmic structure) over $\mathbb{C}$ which satisfies the following two conditions:

(V1) $\mathcal{O}_{X,x}/\mathfrak{m}_x = \mathbb{C}$.
(V2) For any open set $U \subset X$ and any $f \in \mathcal{O}_X(U)$, the evaluation map $U \to \mathbb{C}, x \mapsto f(x)$, is continuous.

We recall briefly the associated ringed space $(X^{\log}, \mathcal{O}_X^{\log})$ introduced in [**12**] (cf. also [**14**]).

As a set, $X^{\log}$ is the set of all pairs (x, h), consisting of a point $x \in X$ and a homomorphism $h\colon M_{X,x} \to \mathbb{S}^1$ whose restriction to $\mathcal{O}_X^\times$ is $u \mapsto u(x)/|u(x)|$. Here $\mathbb{S}^1 := \{t \in \mathbb{C}; |t| = 1\}$.

The topology of $X^{\log}$ is defined as follows. We work locally on X. Take a chart $\mathcal{S} \to M_X$, then we have an injective map

$$X^{\log} \hookrightarrow X \times \operatorname{Hom}(\mathcal{S}^{\mathrm{gp}}, \mathbb{S}^1), \quad (x, h) \mapsto (x, h_{\mathcal{S}}),$$

where $h_{\mathcal{S}}$ denotes the composite map $\mathcal{S}^{\mathrm{gp}} \to M_{X,x}^{\mathrm{gp}} \to \mathbb{S}^1$. We endow $X^{\log}$ with the topology as a subset of $X \times \operatorname{Hom}(\mathcal{S}^{\mathrm{gp}}, \mathbb{S}^1)$. It can be seen that this topology is independent of the choice of chart and hence is globally well-defined, and that the canonical map $\tau\colon X^{\log} \to X$, $(x, h) \mapsto x$ is surjective, continuous and proper.

The sheaf of $\mathbb{C}$-algebras $\mathcal{O}_X^{\log}$ is defined as follows. Let $\mathcal{L}$ be the fiber product of

$$\begin{array}{ccc} & & \tau^{-1}(M_X^{\mathrm{gp}}) \\ & & \downarrow \\ \operatorname{Cont}(\ , i\mathbb{R}) & \xrightarrow{\exp} & \operatorname{Cont}(\ , \mathbb{S}^1), \end{array}$$

where $\operatorname{Cont}(\ , T)$, for a topological space T, denotes the sheaf on $X^{\log}$ of continuous maps to T, and $\tau^{-1}(M_X^{\mathrm{gp}}) \to \operatorname{Cont}(\ , \mathbb{S}^1)$ comes from the definition of $X^{\log}$. We define

$$\mathcal{O}_X^{\log} := (\tau^{-1}(\mathcal{O}_X) \otimes_{\mathbb{Z}} \operatorname{Sym}_{\mathbb{Z}}(\mathcal{L}))/\mathfrak{a},$$

where $\operatorname{Sym}_{\mathbb{Z}}(\mathcal{L})$ denotes the symmetric algebra of $\mathcal{L}$ over $\mathbb{Z}$, and $\mathfrak{a}$ is the ideal of $\tau^{-1}(\mathcal{O}_X) \otimes_{\mathbb{Z}} \operatorname{Sym}_{\mathbb{Z}}(\mathcal{L})$ generated by the image of

$$\tau^{-1}(\mathcal{O}_X) \to \tau^{-1}(\mathcal{O}_X) \otimes_{\mathbb{Z}} \operatorname{Sym}_{\mathbb{Z}}(\mathcal{L}), \quad f \mapsto f \otimes 1 - 1 \otimes \iota(f).$$

Here the map $\iota\colon \tau^{-1}(\mathcal{O}_X) \to \mathcal{L}$ is the one induced by

$$\tau^{-1}(\mathcal{O}_X) \to \operatorname{Cont}(\ \ , i\mathbb{R}), \quad f \mapsto \tfrac{1}{2}(f - \bar{f}),$$

and

$$\tau^{-1}(\mathcal{O}_X) \xrightarrow{\exp} \tau^{-1}(\mathcal{O}_X^\times) \subset \tau^{-1}(M_X^{\mathrm{gp}}).$$

Let $y \in X^{\log}$ and $x := \tau(y) \in X$. There exist a family $(\ell_j)_{1 \le j \le n}$ of elements of $\mathcal{L}_y$, whose images in $(M_X^{\mathrm{gp}}/\mathcal{O}_X^\times)_x$ form a system of free generators, and which are algebraically independent over $\mathcal{O}_{X,x}$ and generate the stalk $\mathcal{O}_{X,y}^{\log} = \tau^{-1}(\mathcal{O}_{X,x})[\ell_1, \ldots, \ell_n]$. Note that this is *not* a local ring if $(M_X^{\mathrm{gp}}/\mathcal{O}_X^\times)_x \neq \{1\}$.

EXAMPLE 1.4. Let $X := \mathbb{C}$ be the complex plane and let z be the coordinate of X. The fs logarithmic structure on X associated to the divisor $\{0\}$ is defined by

$$M_X := \{f \in \mathcal{O}_X; f \text{ is invertible outside } \{0\}\} \hookrightarrow \mathcal{O}_X.$$

We can take a chart $\mathbb{N} \to M_X = \bigsqcup_{n \ge 0} z^n \mathcal{O}_X^\times$, $n \mapsto z^n$. We have an isomorphism

$$X^{\log} \xrightarrow{\sim} \mathbb{R}_{\ge 0} \times \mathbb{S}^1, \quad (x, h) \mapsto (|z(x)|, h(z)).$$

Let $\ell := \log z$, then we have $\mathcal{O}_X^{\log} = \tau^{-1}(\mathcal{O}_X)[\ell]$.

2. Spaces of Nilpotent Orbits

Let Σ be a fan in $\mathfrak{g}_{\mathbb{Q}}$ and let Γ be a subgroup of $G_{\mathbb{Z}}$ which is strongly compatible with Σ.

The Set D_Σ.

DEFINITION 2.1. As a set, we define the *space* D_Σ *of nilpotent orbits in the directions in* Σ by

$$D_\Sigma := \{(\sigma, Z); \sigma \in \Sigma, Z \subset \check{D} \text{ is a } \sigma\text{-nilpotent orbit}\}.$$

REMARK 2.2. Note that the set D is contained in D_Σ by $F \mapsto (\{0\}, \{F\})$.

Let σ be a nilpotent cone in $\mathfrak{g}_{\mathbb{Q}}$. Then, {faces of σ} is a fan, and we denote

$$D_\sigma := D_{\{\text{faces of } \sigma\}}.$$

EXAMPLE 2.3. Let $w = 1$, $h^{1,0} = h^{0,1} = 1$ and $h^{p,q} = 0$ otherwise. Let e_1, e_2 be a $\mathbb{Z}$-free basis of H_0 such that $\langle e_2, e_1 \rangle_0 = 1$. Then

$$D = \mathfrak{h} \subset \check{D} = \mathbb{P}^1(\mathbb{C}).$$

A point $t \in \mathbb{P}^1(\mathbb{C})$ corresponds to the filtration $F_t \in \check{D}$ defined by

$$F_t^p := H_{0,\mathbb{C}} \qquad \text{for } p \leq 0,$$
$$F_t^p := \{0\} \qquad \text{for } p \geq 2,$$
$$F_t^1 := \begin{cases} \mathbb{C}(te_1 + e_2) & \text{if } t \in \mathbb{C}, \text{ and} \\ \mathbb{C}e_1 & \text{if } t = \infty. \end{cases}$$

Let Ξ be as in Example 1.2. Then

$$D_\Xi = \mathfrak{h} \cup \mathbb{P}^1(\mathbb{Q}).$$

The point $(\sigma, Z) \in D_\Xi$ corresponding to $\infty \in \mathbb{P}^1(\mathbb{Q})$ is given by

$$\sigma = \mathbb{Q}_{\geq 0} N, \quad N = \begin{pmatrix} 0 & 1 \\ 0 & 0 \end{pmatrix}, \quad Z = \mathbb{C} \subset \mathbb{P}^1(\mathbb{C}) = \check{D},$$

and we have $D_\sigma = \mathfrak{h} \cup \{\infty\}$.

The Toric Variety toric($\Gamma(\sigma)$). In order to fix notations, we review here toric varieties (cf. [**1**]). For $\sigma \in \Sigma$, we see that $\Gamma(\sigma)$ is an fs monoid with $\Gamma(\sigma)^\times = \{1\}$. The associated toric variety is defined by

$$\operatorname{toric}(\Gamma(\sigma)) := \operatorname{Spec}(\mathbb{C}[\Gamma(\sigma)^\vee])_{\mathrm{an}}.$$

Here $\Gamma(\sigma)^\vee := \operatorname{Hom}(\Gamma(\sigma), \mathbb{N})$. Note that the set of complex points of toric($\Gamma(\sigma)$) can be identified with $\operatorname{Hom}(\Gamma(\sigma)^\vee, \mathbb{C}^{\mathrm{mult}})$. Here $\mathbb{C}^{mathrmmult}$ means the set $\mathbb{C}$ of complex numbers with multiplication.

We have bijections

$$(2.1)\quad \operatorname{toric}(\Gamma(\sigma)) \xleftarrow[\sim]{\alpha} \bigsqcup_{\tau \prec \sigma} (\Gamma(\sigma)^{\mathrm{gp}}/\Gamma(\tau)^{\mathrm{gp}}) \otimes \mathbb{C}^\times \xrightarrow[\sim]{\beta} \bigsqcup_{\tau \prec \sigma} \exp(\sigma_{\mathbb{C}})/(\exp(\tau_{\mathbb{C}})\Gamma(\sigma)^{\mathrm{gp}})$$

Here $\tau \prec \sigma$ means that τ is a face of σ, and α is defined as follows: For $a \in (\Gamma(\sigma)^{\mathrm{gp}}/\Gamma(\tau)^{\mathrm{gp}}) \otimes \mathbb{C}^\times$ and $h \in \Gamma(\sigma)^\vee$,

$$\alpha(\tau, a)(h) := \begin{cases} 0 & \text{if } h(\Gamma(\tau)) \neq 0, \\ h(a) & \text{if } h(\Gamma(\tau)) = 0. \end{cases}$$

In the second case in the above, we regard h as the induced homomorphism $h\colon (\Gamma(\sigma)^{\mathrm{gp}}/\Gamma(\tau)^{\mathrm{gp}}) \otimes \mathbb{C}^\times \to \mathbb{Z} \otimes \mathbb{C}^\times = \mathbb{C}^\times$. And β is defined by

$$\beta(\gamma \otimes \exp(2\pi i z)) := \exp(z \log \gamma) \quad (\gamma \in \Gamma(\sigma)^{\mathrm{gp}}, z \in \mathbb{C}).$$

The subset $E_\sigma \subset \mathrm{toric}(\Gamma(\sigma)) \times \check{D}$.

DEFINITION 2.4. For $\sigma \in \Sigma$, we define the set E_σ by

$$E_\sigma := \{(\theta, F) \in \mathrm{toric}(\Gamma(\sigma)) \times \check{D}; \exp(\tau_{\mathbb{C}})gF \text{ is a } \tau\text{-nilpotent orbit}\}$$

Here $(\tau, g \bmod \exp(\tau_{\mathbb{C}})\Gamma(\sigma)^{\mathrm{gp}})$ denotes the pair corresponding to θ under the composite (2.1).

Define the map $E_\sigma \to \Gamma(\sigma)^{\mathrm{gp}}\backslash D_\sigma$ by $(\theta, F) \mapsto (\tau, \exp(\tau_{\mathbb{C}})gF)$.

Define the action of $\sigma_{\mathbb{C}}$ on E_σ over $\Gamma(\sigma)^{\mathrm{gp}}\backslash D_\sigma$ by

$$a \cdot (\theta, F) := (\exp(2\pi i a)\theta, \exp(-a)F) \quad (a \in \sigma_{\mathbb{C}}).$$

Here, for $b \in \sigma_{\mathbb{C}}$, $\exp(b)\theta$ is the element of $\mathrm{Hom}_{\mathbb{C}\text{-alg}}(\mathbb{C}[\Gamma(\sigma)^\vee], \mathbb{C})$ defined by

$$(\exp(b)\theta)(h) := \exp(h(\exp(b)))\theta(h) \quad (h \in \Gamma(\sigma)^\vee).$$

Then, $E_\sigma \to \Gamma(\sigma)^{\mathrm{gp}}\backslash D_\sigma$ is surjective (more precisely, a $\sigma_{\mathbb{C}}$-torsor, see Proposition 6.1(1) below).

EXAMPLE 2.5. In the case $\mathrm{rank}(\sigma) = 1$, we have $\mathrm{toric}(\Gamma(\sigma)) \simeq \mathbb{C}$. Let γ be the generator of $\Gamma(\sigma)$ and let $N := \log(\gamma)$. Then we have

$$E_\sigma := \left\{ \begin{array}{ll} (\theta, F) \in \mathbb{C} \times \check{D}; & \text{If } \theta \neq 0,\ \exp((\log(\theta)/2\pi i)N)F \in D \\ & \text{If } \theta = 0,\ \exp(\sigma_{\mathbb{C}})F \text{ is a } \sigma\text{-nilpotent orbit} \end{array} \right\}.$$

Here the cases $\theta \neq 0$, $\theta = 0$ correspond to $\tau = 0$, $\tau = \sigma$, respectively. An element $a \in \sigma_{\mathbb{C}} \simeq \mathbb{C}$ acts on E_σ over $\Gamma(\sigma)^{\mathrm{gp}}\backslash D_\sigma$ by

$$a \cdot (\theta, F) = (\exp(2\pi i a)\theta, \exp(-aN)F).$$

3. Spaces of Nilpotent i-Orbits

Let Σ be a fan in $\mathfrak{g}_{\mathbb{Q}}$ and let Γ be a subgroup of $G_{\mathbb{Z}}$ which is strongly compatible with Σ as before.

The set $D_\Sigma^\sharp$.

DEFINITION 3.1. As a set, we define the *space $D_\Sigma^\sharp$ of nilpotent i-orbits in the directions in Σ* by

$$D_\Sigma^\sharp := \{(\sigma, Z); \sigma \in \Sigma, Z \subset \check{D} \text{ is a } \sigma\text{-nilpotent } i\text{-orbit}\}.$$

Define the map

$$D_\Sigma^\sharp \to D_\Sigma, \quad (\sigma, Z) \mapsto (\sigma, \exp(\sigma_{\mathbb{C}})Z).$$

REMARK 3.2. Note that the set D is contained in $D_\Sigma^\sharp$ by $F \mapsto (\{0\}, \{F\})$.

For a nilpotent cone σ in $\mathfrak{g}_{\mathbb{Q}}$, we denote

$$D_\sigma^\sharp := D_{\{\text{faces of } \sigma\}}^\sharp.$$

The Real Toric Variety $|\mathrm{toric}|(\sigma)$. In order to fix notations, we review here real toric varieties (for details, see [**1**, I §1], [**8**, 2.1]). For $\sigma \in \Sigma$, we define

$$|\mathrm{toric}|(\sigma) := \mathrm{Hom}(\sigma^\vee, (\mathbb{R} \cup \{\infty\})^{\mathrm{add}}) \supset \sigma_{\mathbb{R}} = \mathrm{Hom}(\sigma^\vee, \mathbb{R}^{\mathrm{add}}).$$

Here $\sigma^\vee := \mathrm{Hom}(\sigma, \mathbb{Q}^{\mathrm{add}}_{\geq 0})$. We have the bijection

$$(3.1) \qquad |\mathrm{toric}|(\sigma) \xleftarrow[\sim]{\alpha} \bigsqcup_{\tau \prec \sigma} \sigma_{\mathbb{R}}/\tau_{\mathbb{R}},$$

defined by

$$\alpha(\tau, H \bmod \tau_{\mathbb{R}})(h) := \begin{cases} \infty & \text{if } h(\tau) \neq 0, \\ h(H) & \text{if } h(\tau) = 0, \end{cases} \qquad (h \in \sigma^\vee).$$

By using the isomorphism

$$(\mathbb{R} \cup \{\infty\})^{\mathrm{add}} \xrightarrow{\sim} \mathbb{R}^{\mathrm{mult}}_{\geq 0}, \quad y \mapsto e^{-2\pi y},$$

we have

$$|\mathrm{toric}|(\sigma) \simeq \mathrm{Hom}(\sigma^\vee, \mathbb{R}^{\mathrm{mult}}_{\geq 0}) \simeq \mathrm{Hom}(\Gamma(\sigma)^\vee, \mathbb{R}^{\mathrm{mult}}_{\geq 0}).$$

If we identify $\mathrm{toric}(\Gamma(\sigma))$ with $\mathrm{Hom}(\Gamma(\sigma)^\vee, \mathbb{C}^{\mathrm{mult}})$, we see that the maps

$$\mathbb{R}^{\mathrm{mult}}_{\geq 0} \to \mathbb{C}^{\mathrm{mult}}, \ r \mapsto r, \quad \text{and} \quad \mathbb{C}^{\mathrm{mult}} \to \mathbb{R}^{\mathrm{mult}}_{\geq 0}, \ q \mapsto |q|,$$

induce the maps

$$(3.2) \qquad |\mathrm{toric}|(\sigma) \to \mathrm{toric}(\Gamma(\sigma)) \to |\mathrm{toric}|(\sigma),$$

whose composite is the identity.

The Subset $E^\sharp_\Sigma \subset |\mathrm{toric}|(\sigma) \times \check{D}$.

DEFINITION 3.3. For $\sigma \in \Sigma$, we define the set $E^\sharp_\sigma$ by

$$E^\sharp_\sigma := \{(b, F) \in |\mathrm{toric}|(\sigma) \times \check{D}; \exp(i\tau_{\mathbb{R}})\exp(iH)F \text{ is a } \tau\text{-nilpotent } i\text{-orbit}\}.$$

Here $(\tau, H \bmod \tau_{\mathbb{R}})$ is the point in $\bigsqcup_{\tau \prec \sigma} \sigma_{\mathbb{R}}/\tau_{\mathbb{R}}$ corresponding to b in $|\mathrm{toric}|(\sigma)$ under the bijection (3.1).

Define the map $E^\sharp_\sigma \to D^\sharp_\sigma$ by $(b, F) \mapsto \exp(i\tau_{\mathbb{R}})\exp(iH)F$.

Define the action of $\sigma_{\mathbb{R}}$ on $E^\sharp_\sigma$ over $D^\sharp_\sigma$ by

$$a \cdot \big((\tau, H \bmod \tau_{\mathbb{R}}), F\big) := \big((\tau, (H + a) \bmod \tau_{\mathbb{R}}), \exp(-ia)F\big) \quad (a \in \sigma_{\mathbb{R}}).$$

Then, $E^\sharp_\sigma \to D^\sharp_\sigma$ is surjective (more precisely, a $\sigma_{\mathbb{R}}$-torsor, see Proposition 6.1(1) below). The maps in (3.2) induce the maps

$$(3.3) \qquad E^\sharp_\sigma \to E_\sigma \to E^\sharp_\sigma,$$

whose composite is the identity.

4. Strong Topology, Generalized fs Logarithmic Analytic Spaces

The Strong Topology.

DEFINITION 4.1. Let X be an analytic space and S be a subset on X. The *strong topology* on S is the strongest topology on S for which the map $\lambda\colon Y \to S$ is continuous for any analytic space Y and for any analytic morphism $\lambda\colon Y \to X$ with $\lambda(Y) \subset S$.

REMARK 4.2. In the notation of the above definition, if S is an analytic subspace of X then the strong topology on S coincides with the usual topology, i.e., the topology as a subspace of X.

EXAMPLE 4.3. Let $X := \mathbb{C}^2$ be the complex two-dimensional plane and let $S := \mathbb{C}^2 - \{0\} \times \mathbb{C} + \{(0,0)\} \subset X$. For the strong topology on S and the usual topology on S as a subspace of X, we can observe the following:

If $k \in \mathbb{Z}_{\geq 1}$ and $z \in \mathbb{C}$ converges to 0 then (z^k, z) converges to $(0,0)$ in the strong topology on S.

If $a \in \mathbb{R}_{>1}$ and $s \in \mathbb{R}$, $s \to \infty$, then $(1/a^s, 1/s)$ does not converge to $(0,0)$ in the strong topology on S, whereas $(1/a^s, 1/s)$ converges to $(0,0)$ in the usual topology on S as a subspace of X.

Generalized (fs Logarithmic) Analytic Spaces.

DEFINITION 4.4. (1) Let $\mathcal{C}$ be a category and let $\mathcal{A}$ be a full subcategory of $\mathcal{C}$. For an object S of $\mathcal{C}$, let $h^S_{\mathcal{A}}$ be the contravariant functor

$$\mathcal{A} \to \mathit{Sets}, \quad Y \mapsto h^S_{\mathcal{A}}(Y) := \mathrm{Mor}(Y, S).$$

Denote by $\bar{\mathcal{A}} \supset \mathcal{A}$ the full subcategory of $\mathcal{C}$ consisting of all objects S satisfying

$$\mathrm{Mor}(S, Z) \xrightarrow{\sim} \mathrm{Mor}(h^S_{\mathcal{A}}, h^Z_{\mathcal{A}}) \quad \text{for any object } Z \text{ of } \mathcal{C}.$$

(2) For

$$\mathcal{C} := (\text{local ringed spaces over } \mathbb{C}\ ) \supset \mathcal{A} := (\text{analytic spaces}),$$

we call an object of $\bar{\mathcal{A}}$ a *generalized analytic space.*

(3) For

$$\mathcal{C} := (\text{fs logarithmic local ringed spaces over } \mathbb{C})$$
$$\supset \mathcal{A} := (\text{fs logarithmic analytic spaces}),$$

we call an object of $\bar{\mathcal{A}}$ a *generalized fs logarithmic analytic space.*

It can be proved that, if we forget logarithmic structures, generalized fs logarithmic analytic spaces become generalized analytic spaces. Generalized analytic spaces satisfy the conditions (V1), (V2) in the definition of the ringed space $(X^{\log}, \mathcal{O}^{\log}_X)$ in §1.

REMARK 4.5. In Example 4.3, S, endowed with the strong topology and the pull-back sheaf of rings $\mathcal{O}$ and the pull-back fs logarithmic structure M from those on X, is a generalized fs logarithmic analytic space. Here the fs logarithmic structure on X is the one associated to the divisor $\{0\} \times \mathbb{C} \subset X$.

5. Polarized Logarithmic Hodge Structures

Polarized Logarithmic Hodge Structures.

DEFINITION 5.1. Let X be a generalized fs logarithmic analytic space. A *pre-polarized logarithmic Hodge structure* (*pre-PLH*, for short) on X, of weight w and of Hodge type $(h^{p,q})_{p+q=w}$ is a triple $(H_{\mathbb{Z}}, \langle\ ,\ \rangle, F)$ consisting of a locally constant sheaf of free $\mathbb{Z}$-modules $H_{\mathbb{Z}}$ on $X^{\log}$, of a non-degenerate bilinear form $\langle\ ,\ \rangle\colon H_{\mathbb{Q}} \times H_{\mathbb{Q}} \to \mathbb{Q}$, symmetric for even w and anti-symmetric for odd w and of a decreasing filtration F of the $\mathcal{O}_X^{\log}$-module $\mathcal{O}_X^{\log} \otimes H_{\mathbb{Z}}$, which satisfy the following three conditions.

(i) Each F^p is the pull-back under $\tau : (X^{\log}, \mathcal{O}_X^{\log}) \to (X, \mathcal{O}_X)$ of a locally free $\mathcal{O}_X$-module of rank $\sum_{r \geq p} h^{r,w-r}$.

(ii) Each F^p is, locally on $X^{\log}$, an $\mathcal{O}_X^{\log}$-direct summand of $\mathcal{O}_X^{\log} \otimes H_{\mathbb{Z}}$.

(iii) $\langle F^p, F^q \rangle = 0$ if $p + q > w$.

Let X be an fs logarithmic analytic space. Let Ω_X^1 be the sheaf of Kähler differentials on X. The sheaf of logarithmic differential forms on X is defined by

$$\omega_X^1 := (\Omega_X^1 \oplus M_X^{\mathrm{gp}})/N,$$

where N is the $\mathcal{O}_X$-submodule generated by $\{(-d\alpha(x), \alpha(x) \otimes x); x \in M_X\}$. The derivation $d\colon \mathcal{O}_X \to \Omega_X^1$ induces the one $d\colon \mathcal{O}_X \to \omega_X^1$ via the map $\Omega_X \to \omega_X^1$. The sheaf $\omega_X^{1,\log}$ on $X^{\log}$ is defined by

$$\omega_X^{1,\log} := \tau^*(\omega_X^1) = \mathcal{O}_X^{\log} \otimes_{\tau^{-1}(\mathcal{O}_X)} \tau^{-1}(\omega_X^1).$$

The derivation

$$d\colon \mathcal{O}_X^{\log} \to \omega_X^{1,\log}$$

is induced by $d\colon \mathcal{O}_X \to \omega_X^1$ and by $\mathcal{L} \xrightarrow{\exp} \tau^{-1}(M_X^{\mathrm{gp}}) \xrightarrow{d\log} \tau^{-1}(\omega_X^1)$, where $d\log\colon x \mapsto (0, 1 \otimes x)$.

Let now x be an fs logarithmic analytic space whose underlying analytic space is $\mathrm{Spec}(\mathbb{C})$ and let $r := \mathrm{rank}_{\mathbb{Z}}(M_x^{\mathrm{gp}}/\mathcal{O}_x^\times)$. Then $x^{\log} = (\mathbb{S}^1)^r$. Let $y \in x^{\log}$. We define

$$\mathrm{sp}(y) := \mathrm{Hom}_{\mathbb{C}\text{-alg}}(\mathcal{O}_{x,y}^{\log}, \mathbb{C}),$$
$$\mathrm{sp}(x) := \{t \in \mathrm{Hom}(M_x, \mathbb{C}^{\mathrm{mult}}); t|_{\mathbb{C}^\times} = \mathrm{id}\},$$

and

$$\mathrm{sp}(y) \to \mathrm{sp}(x), \quad s \mapsto t := \big(M_x \ni a \mapsto \exp(s(\log(a))) \in \mathbb{C}^\times \subset \mathbb{C}\big).$$

Note that the structure homomorphism $\alpha\colon M_x \to \mathbb{C}$ of the logarithmic structure of x is an element of $\mathrm{sp}(x)$, which does not belong to the image of the canonical map $\mathrm{sp}(y) \to \mathrm{sp}(x)$ if $r > 0$. Note also that $\mathrm{sp}(x) \simeq \mathbb{C}^r$, whereas $\mathrm{sp}(y)$ ($y \in x^{\log}$) form a $\mathbb{C} \otimes_{\mathbb{Z}} \pi_1(x^{\log})$-torsor over $x^{\log}$ (cf. the statement just before Proposition 5.4).

For a pre-PLH $(H_{\mathbb{Z}}, \langle\ ,\ \rangle, F)$ on x, for $y \in x^{\log}$ and for $s \in \mathrm{sp}(y)$, we have a decreasing filtration

$$F(s) := \big(\mathbb{C} \otimes_{\mathcal{O}_{x,y}^{\log}} F_y^p\big)_{p \in \mathbb{Z}}, \quad \text{where } s\colon \mathcal{O}_{x,y}^{\log} \to \mathbb{C},$$

on the $\mathbb{C}$-vector space $H_{\mathbb{C},y} := \mathbb{C} \otimes_{\mathbb{Z}} H_{\mathbb{Z},y}$.

DEFINITION 5.2. Let x be an fs logarithmic analytic space whose underlying analytic space is $\mathrm{Spec}(\mathbb{C})$. A pre-PLH $(H_{\mathbb{Z}}, \langle\ ,\ \rangle, F)$ on x is called a *polarized logarithmic Hodge structure* (*PLH*, for short) on x if it satisfies the following two conditions.

(i) $(d\otimes 1)(F^p) \subset \omega_x^{1,\log} \otimes_{\mathcal{O}_x^{\log}} F^{p-1}$ for all p. Here $d\otimes 1$ means $d\otimes 1_{H_{\mathbb{Z}}} \colon \mathcal{O}_x^{\log} \otimes_{\mathbb{Z}} H_{\mathbb{Z}} \to \omega_x^{1,\log} \otimes_{\mathbb{Z}} H_{\mathbb{Z}}$.

(ii) For $y \in x^{\log}$ and for $s \in \mathrm{sp}(y)$ whose image in $\mathrm{sp}(x)$ is sufficiently near to the structure homomorphism $\alpha \in \mathrm{sp}(x)$, $(H_{\mathbb{Z},y}, \langle\ ,\ \rangle_y, F(s))$ is a polarized Hodge structure in the usual sense.

Note that the validity of the condition 5.2(ii) is independent of the choice of point $y \in x^{\log}$.

DEFINITION 5.3. Let X be a generalized fs logarithmic analytic space. A pre-PLH $(H_{\mathbb{Z}}, \langle\ ,\ \rangle, F)$ on X is called a *polarized logarithmic Hodge structure* (*PLH*, for short) on X if, for any $x \in X$, the inverse image of $(H_{\mathbb{Z}}, \langle\ ,\ \rangle, F)$ on x is a PLH. Here we regard x as an fs logarithmic analytic space with ring $\mathbb{C}$ and with the inverse image of the logarithmic structure of X.

Polarized Logarithmic Hodge Structures of Type Φ. Let x be an fs logarithmic analytic space whose underlying analytic space is $\mathrm{Spec}(\mathbb{C})$. Then $x^{\log} \simeq (\mathbb{S}^1)^r$, where $r = \mathrm{rank}(M_x^{\mathrm{gp}}/\mathcal{O}_x^{\times})$, as before. Let $y \in x^{\log}$. We define an action of $\mathbb{C} \otimes \pi_1(x^{\log})$ on $\mathrm{sp}(y)$ as follows. Let $\mathcal{L}$ be the sheaf of logarithms of $\tau^{-1}(M_x^{\mathrm{gp}})$ constructed in §1. Then the exact sequence

$$0 \to \mathbb{Z} \xrightarrow{2\pi i} \mathcal{L} \xrightarrow{\exp} \tau^{-1}(M_x^{\mathrm{gp}}) \to 1$$

on $x^{\log}$ induces the exact sequence of cohomology groups

$$\mathcal{O}_x \xrightarrow{\exp} M_x^{\mathrm{gp}} \to H^1(x^{\log}, \mathbb{Z}) \simeq \mathrm{Hom}(\pi_1(x^{\log}), \mathbb{Z}) \to 0.$$

Hence we have

$$\pi_1(x^{\log}) \simeq \mathrm{Hom}(M_x^{\mathrm{gp}}/\mathcal{O}_x^{\times}, \mathbb{Z}) \xrightarrow{\sim} \mathrm{Hom}(\mathcal{L}_y/\mathcal{O}_x^{\mathrm{add}}, \mathbb{Z}),$$

where the last isomorphism is induced by exp. On the other hand, the restriction of $s \in \mathrm{sp}(y)$ to $\mathcal{L}_y$ yields a bijection

$$\mathrm{sp}(y) \xrightarrow{\sim} (\text{splittings of the group homomorphism } \iota\colon \mathbb{C}^{\mathrm{add}} = \mathcal{O}_x^{\mathrm{add}} \to \mathcal{L}_y).$$

Now we define the action of $\varphi \in \mathbb{C}\otimes_{\mathbb{Z}}\pi_1(x^{\log}) \underset{\sim}{\xrightarrow{\nu}} \mathrm{Hom}(\mathcal{L}_y/\mathcal{O}_x^{\mathrm{add}}, \mathbb{C}^{\mathrm{add}})$ on $s \in \mathrm{sp}(y)$ by

$$(\varphi s)|_{\mathcal{L}_y} := (\varphi - \mathrm{id}) + s|_{\mathcal{L}_y}.$$

Note that $(\varphi - \mathrm{id})\colon \mathcal{L}_y \to \mathbb{C}^{\mathrm{add}}$ is the lifting of $\nu(\varphi)$. Then, the $\mathrm{sp}(y)$ $(y \in x^{\log})$ form a $\mathbb{C} \otimes_{\mathbb{Z}} \pi_1(x^{\log})$-torsor over $x^{\log} \simeq (\mathbb{S}^1)^r$ under this action.

The following proposition gives a connection between a PLH and nilpotent orbits.

PROPOSITION 5.4. *For $x \in X$, $y \in x^{\log}$, $s \in \mathrm{sp}(y)$ and $\varphi = \sum_{j=1}^n z_j \otimes \gamma_j \in \mathbb{C} \otimes \pi_1(x^{\log})$, we have*

$$F(\varphi s) = \exp\Big(-\sum_{j=1}^{n} z_j \log(\gamma_j)\Big) F(s).$$

Let

$$\Phi := (w, (h^{p,q})_{p+q=w}, H_0, \langle\ ,\ \rangle_0, \Gamma, \Sigma) \tag{5.1}$$

be a 6-tuple consisting of a weight, a set of Hodge numbers, a free $\mathbb{Z}$-module H_0 of rank $\sum h^{p,q}$, a $\mathbb{Q}$-bilinear form $\langle\ ,\ \rangle_0$ on $H_{0,\mathbb{Q}}$ as in the beginning of §1, a subgroup of $G_{\mathbb{Z}}$ and a fan with which the subgroup Γ is strongly compatible.

DEFINITION 5.5. Let X be a generalized fs logarithmic analytic space. A *polarized logarithmic Hodge structure of type* Φ (*PLH of type* Φ, for short) on X is a 4-tuple $(H_{\mathbb{Z}}, \langle\ ,\ \rangle, F, \mu)$ consisting of a pre-PLH $(H_{\mathbb{Z}}, \langle\ ,\ \rangle, F)$ on X of weight w and of Hodge type $(h^{p,q})_{p,q}$, and of a global section μ of the sheaf

$$\Gamma\backslash \underline{\mathrm{Isom}}((H_{\mathbb{Z}}, \langle\ ,\ \rangle), (H_0, \langle\ ,\ \rangle_0))$$

on $X^{\log}$ which satisfy condition (i) below. Here H_0 with $\langle\ ,\ \rangle_0$ is regarded as a constant sheaf on $X^{\log}$, $\underline{\mathrm{Isom}}$ is the sheaf of isomorphisms, and $\gamma \in \Gamma$ acts on $\underline{\mathrm{Isom}}(\cdots)$ by $h \mapsto \gamma \circ h$.

(i) For any $x \in X$ and any $y \in x^{\log} = \tau^{-1}(x)$, if

$$\tilde{\mu}_y\colon (H_{\mathbb{Z},y}, \langle\ ,\ \rangle_y) \xrightarrow{\sim} (H_0, \langle\ ,\ \rangle_0)$$

denotes a representative of the stalk of μ at y, then there exists $\sigma \in \Sigma$ such that the image of the composite map

$$\mathrm{Hom}(M_{X,x}/\mathcal{O}_{X,x}^{\times}, \mathbb{Z}_{\le 0}) \hookrightarrow \pi_1(x^{\log}) \to \mathrm{Aut}_{\mathbb{Z}}(H_{\mathbb{Z},y}, \langle\ ,\ \rangle_y) \overset{\text{by } \tilde{\mu}_y}{\simeq} \mathrm{Aut}_{\mathbb{Z}}(H_0, \langle\ ,\ \rangle_0)$$

is contained in $\Gamma(\sigma)$ and that the filtrations $\tilde{\mu}_y(F|_{x^{\log}}(s))$ ($s \in \mathrm{sp}(y)$) on $H_{0,\mathbb{C}}$ form a σ-nilpotent orbit (cf. Definition 1.3).

REMARK 5.6. Using Proposition 5.4, we can prove that the conditions (i), (ii) in Definition 5.2 follow from the conditions (ii) (Griffiths transversality), (iii) (positivity) in Definition 1.3, respectively. Hence if $(H_{\mathbb{Z}}, \langle\ ,\ \rangle, F, \mu)$ is a PLH on X of type Φ in the above definition, then $(H_{\mathbb{Z}}, \langle\ ,\ \rangle, F)$ is actually a PLH on X in the sense of Definition 5.2.

6. Main Result

Let $\Phi := (w, (h^{p,q})_{p+q=w}, H_0, \langle\ ,\ \rangle_0, \Gamma, \Sigma)$ be as in (5.1) and let $\sigma \in \Sigma$.

We endow $E_\sigma \subset \check{E}_\sigma := \mathrm{toric}(\Gamma(\sigma)) \times \check{D}$ with the strong topology, the pull-back sheaf of rings $\mathcal{O}$ and the pull-back fs logarithmic structure M from those on $\check{E}_\sigma$. Here the fs logarithmic structure on $\check{E}_\sigma$ is the pull-back of the one on $\mathrm{toric}(\Gamma(\sigma))$ associated to the normal crossing divisor $\mathrm{toric}(\Gamma(\sigma)) - \Gamma(\sigma)^{\mathrm{gp}} \otimes \mathbb{C}^{\times}$.

Via $\pi_\sigma\colon E_\sigma \to \Gamma(\sigma)^{\mathrm{gp}}\backslash D_\sigma \to \Gamma\backslash D_\Sigma$, we endow $\Gamma(\sigma)^{\mathrm{gp}}\backslash D_\sigma$ with the quotient topology, and $\Gamma\backslash D_\Sigma$ with the strongest topology for which π_σ is continuous for all $\sigma \in \Sigma$. For any open set U of $\Gamma\backslash D_\Sigma$, let $U_\sigma := \pi_\sigma^{-1}(U)$ and define

$$\begin{aligned}&\mathcal{O}(U)\ (\text{resp. } M(U))\\ &\qquad := \{\text{map } f\colon U \to \mathbb{C};\, f \circ \pi_\sigma \in \mathcal{O}(U_\sigma)\ (\text{resp. } \in M(U_\sigma))\ (\forall \sigma \in \Sigma)\}.\end{aligned}$$

Here we regard $M(U_\sigma)$ as a subset of $\mathcal{O}(U_\sigma)$ via $\alpha\colon M(U_\sigma) \to \mathcal{O}(U_\sigma)$ which is injective.

We introduce the quotient topology on $E_\sigma^\sharp$ via the surjection $E_\sigma \to E_\sigma^\sharp$ in (3.3). Via $\pi_\sigma^\sharp\colon E_\sigma^\sharp \to D_\sigma^\sharp \to \Gamma\backslash D_\Sigma^\sharp$, we endow $D_\sigma^\sharp$ with the quotient topology, and $D_\Sigma^\sharp$ with the strongest topology for which $\pi_\sigma^\sharp$ is continuous for all $\sigma \in \Sigma$.

A subgroup Γ of $G_{\mathbb{Z}}$ is called *neat* if the subgroup of $\mathbb{C}^\times$, generated by all the eigenvalues of all $\gamma \in \Gamma$, is torsion free. It is known that there exist neat subgroups of $G_{\mathbb{Z}}$ with finite index.

PROPOSITION 6.1. *Assume Γ is neat. Then, in the above notation, we have*

(1) $E_\sigma \to \Gamma(\sigma)^{\mathrm{gp}}\backslash D_\sigma$ *(resp. $E_\sigma^\sharp \to D_\sigma^\sharp$) is a $\sigma_{\mathbb{C}}$-torsor (resp. $\sigma_{\mathbb{R}}$-torsor) in the category of topological spaces.*
(2) *The maps $\Gamma(\sigma)^{\mathrm{gp}}\backslash D_\sigma \to \Gamma\backslash D_\Sigma$ and $D_\sigma^\sharp \to \Gamma\backslash D_\Sigma^\sharp$ are open.*

Our main result is

THEOREM 6.2. *Let $\Phi := (w, (h^{p,q})_{p+q=w}, H_0, \langle\ ,\ \rangle_0, \Gamma, \Sigma)$ be as in (5.1). We assume Γ is neat. Then we have*

(1) *$\Gamma\backslash D_\Sigma$ is a generalized fs logarithmic analytic space, which is Hausdorff, but not locally compact in general. There is a homeomorphism of the topological spaces $(\Gamma\backslash D_\Sigma)^{\log}$ and $\Gamma\backslash D_\Sigma^\sharp$, which is compatible with the canonical map $\tau\colon (\Gamma\backslash D_\Sigma)^{\log} \to \Gamma\backslash D_\Sigma$ and the map $\Gamma\backslash D_\Sigma^\sharp \to \Gamma\backslash D_\Sigma$ induced by $D_\Sigma^\sharp \to D_\Sigma$ in Definition 3.1.*
(2) *Let $\bar{\mathcal{A}}$:= (generalized fs logarithmic analytic spaces) and define the contravariant functor $\underline{\mathrm{PLH}}_\Phi\colon \bar{\mathcal{A}} \to \mathit{Sets}$ by*

$$\underline{\mathrm{PLH}}_\Phi(X) := (\textit{isomorphism classes of PLH on } X \textit{ of type } \Phi).$$

Then this functor $\underline{\mathrm{PLH}}_\Phi$ is represented by $\Gamma\backslash D_\Sigma$.

REMARK 6.3. In the case where D is a Hermitian symmetric domain, the strong topology on $E_\sigma \subset \check{E}_\sigma$ coincides with the usual topology on E_σ as a subspace of $\check{E}_\sigma$, and hence they coincide on $\Gamma\backslash D_\Sigma$. Thus our space $\Gamma\backslash D_\Sigma$ becomes an fs logarithmic analytic space whose underlying analytic space is nothing but a toroidal compactification of Mumford et al. for a 'big fan' Σ. Even in this case, our construction has the advantage that it carries the universal family of PLH.

7. Examples

Example of Weight 3. In the case of Hodge structures of weight 2, Cattani and Kaplan constructed a partial compactification $\Gamma\backslash D_{CK}$ ($\Gamma\backslash\bar{D}$ in their notation in [**5**]) as a topological space, which is a generalization of the Satake type. For the example of weight 3 below, their argument is applicable to obtain $\Gamma\backslash D_{CK}$. Consider the case of weight 2, or weight 3 of the type below. Let Ξ be the fan in Example 1.2 and let Γ be a subgroup of $G_{\mathbb{Z}}$ with finite index. Then, there is the standard map $\Gamma\backslash D_\Xi \to \Gamma\backslash D_{\mathrm{CK}}$ from our space to the Cattani-Kaplan, which is continuous.

In the following, we give an example which shows that the strong topology is needed. Let $w = 3$ and $h^{p,q} = 1$ ($p + q = 3$, p, $q \geq 0$), $h^{p,q} = 0$ otherwise. Let e_1, e_2, f_1, f_2 be a free $\mathbb{Z}$-basis of H_0 and let $\langle\ ,\ \rangle_0$ be an anti-symmetric bilinear form on H_0 defined by

$$\langle e_1, e_2\rangle_0 = \langle f_1, f_2\rangle_0 = 1, \quad \langle e_j, f_k\rangle_0 = 0 \quad (\forall j, \forall k).$$

Let N be a nilpotent endomorphism of H_0 defined by

$$Nf_1 = f_2, \quad Nf_2 = Ne_j = 0 \quad (\forall j).$$

For $s \in \mathbb{C}$, we define a family of Hodge filtrations $F_s \in \check{D}$ by

$$F_s^3 := \{sf_1 + e_1 - ie_2\}_{\mathbb{C}} \subset F_s^2 := \{f_1, e_1 - ie_2\}_{\mathbb{C}} \\ \subset F_s^1 := \{f_1, e_1 - ie_2, se_2 - f_2\}_{\mathbb{C}} \subset F_s^0 := H_{0,\mathbb{C}},$$

where $\{\dots\}_{\mathbb{C}}$ means the $\mathbb{C}$-linear span. Using this family, we can prove

PROPOSITION 7.1. (1) *The following are equivalent:*
(a) (N, F_s) *generates a nilpotent orbit.*
(b) (N, F_s) *satisfies Griffiths transversality.*
(c) $s = 0$.
(2) *We use the notation in* (1). *For* $y \in \mathbb{R}$, *we see* $e^{iyN}F_s \in D$ *if and only if* $0 < y < |s|^{-2}$. (*Here we understand* $|s|^{-2} = \infty$ *for* $s = 0$).
(3) *Let* Ξ *be the fan in Example* 1.2. *If we endow* E_σ ($\sigma \in \Xi$) *with the usual topology as a subset of* $\check{E}_\sigma$ *and endow* $\Gamma\backslash D_\Xi$ *with the quotient topology via* $\bigsqcup_{\sigma\in\Xi} E_\sigma \to \Gamma\backslash D_\Xi$, *then* $\Gamma\backslash D_\Xi$ *is not Hausdorff and the standard map* $\Gamma\backslash D_\Xi \to \Gamma\backslash D_{\mathrm{CK}}$ *is not continuous.*

The example in this subsection is a degeneration of the Hodge structures of weight 3 of the mirrors of quintic hypersurfaces in $\mathbb{P}^4$ (cf., for example, [**15**]).

Example of Weight 2. Let Ξ be the fan in Example 1.2. In this subsection, we give an example of $\Gamma\backslash D_\Xi$ which is locally isomorphic to $\mathbb{C} \times \big(\mathbb{C}^2 - \{0\} \times \mathbb{C} + \{(0,0)\}\big)$ (cf. Example 4.3).

Put

$$S := \mathbb{C} \times \mathbb{C} - \{0\} \times \mathbb{C} + \{(0,0)\},$$
$$\widetilde{S} := \mathbb{R}_{\geq 0} \times \mathbb{C} - \{0\} \times \mathbb{C} + \{(0,0)\}.$$

We define a topology $\mathcal{T}$ (resp. $\widetilde{\mathcal{T}}$) on S (resp. on $\widetilde{S}$) as follow. On $S - \{(0,0)\}$ (resp. on $\widetilde{S} - \{(0,0)\}$), $\mathcal{T}$ (resp. $\widetilde{\mathcal{T}}$) coincides with the usual topology, and, at the point $(0,0)$, a fundamental system of neighborhoods is given by

$$W(\varepsilon) := \bigcup_{n\geq 1} \{(v,z); |v| < \varepsilon_n, |z| < \varepsilon_n, |z|^n < |v|\}$$
$$\left(\text{resp. } \tilde{W}(\varepsilon) := \bigcup_{n\geq 1} \{(r,z); r < \varepsilon_n, |z| < \varepsilon_n, |z|^n < r\}.\right).$$

Here $\varepsilon = (\varepsilon_n)_{n\geq 1}$ runs over all families of positive real numbers. Then, $\mathcal{T}$ coincides with the strong topology of $S \subset \mathbb{C}^2$ considered in Example 4.3

Let $w = 2$, $h^{2,0} = h^{0,2} = 2$, $h^{1,1} = 1$ and $h^{p,q} = 0$ otherwise. Let e_j $(1 \leq j \leq 5)$ be a free $\mathbb{Z}$-basis of H_0 and let $\langle\ ,\ \rangle_0$ be a symmetric bilinear form on H_0 defined by

$$\langle e_j, e_j\rangle_0 = 1 \ (1 \leq j \leq 3), \quad \langle e_4, e_5\rangle_0 = \langle e_5, e_4\rangle_0 = 1, \quad \langle e_j, e_k\rangle_0 = 0 \text{ otherwise.}$$

Then $\dim_{\mathbb{C}} D = 3$ and we can prove

PROPOSITION 7.2. *In the above notation, we have the following.*

(i) D_Ξ *contains all nilpotent orbits.*

(ii) *Let* $p \in \Gamma\backslash D_\Xi - \Gamma\backslash D$. *Then, there exist an open neighborhood* U *of* p *in* $\Gamma\backslash D_\Xi$ *and an open set* $\tilde{U}$ *in* $D^\sharp_\Xi$, *in the topologies introduced at the beginning of* §1, *which fit in the following commutative diagram:*

$$\begin{array}{ccccc} \Gamma\backslash D_\Xi & \supset & U & \xhookrightarrow{\alpha} & \mathbb{C}\times S \\ \uparrow & & \uparrow & & \uparrow{\scriptstyle\beta} \\ D^\sharp_\Xi & \supset & \widetilde{U} & \hookrightarrow & \mathbb{C}\times\mathbb{R}\times\widetilde{S}. \end{array}$$

Here $\alpha\colon p \mapsto (0,0,0)$, $\beta\colon (w,a,r,z) \mapsto (w, re^{ai}, z)$. *If we endow* $\mathbb{C}\times S$ *(resp.* $\mathbb{C}\times\mathbb{R}\times\widetilde{S}$*) with the product topology of the usual topology of* $\mathbb{C}$ *(resp.* $\mathbb{C}\times\mathbb{R}$*) and* $\mathcal{T}$ *(resp.* $\widetilde{\mathcal{T}}$*), then both* $\hookrightarrow$ *are open immersions.*

References

1. A. Ash, D. Mumford, M. Rapoport, and Y. Tai, *Smooth compactification of locally symmetric varieties*, Lie Group: History, Frontiers and Applications, vol. 4, Math. Sci. Press, Brookline, MA 1975.
2. W. L. Baily and A. Borel, *Compactification of arithmetic quotients of bounded symmetric domains*, Ann. of Math. (2) **84** (1966), 442–528.
3. A. Borel and J.-P. Serre, *Corners and arithmetic groups*, Comment. Math. Helv. **48** (1973), 436–491.
4. J. A. Carlson, E. H. Cattani and A. G. Kaplan, *Mixed Hodge structures and compactifications of Siegel's space*, Journées de géométrie algébrique d'Angers 1979 (A. Beauville, ed.), Sijthohoff & Noordhoff, Alphen aan den Rijn-Germantown, MD, 1980. 1980, pp. 77–105.
5. E. H. Cattani and A. G. Kaplan, *Extension of period mappings for Hodge structures of weight* 2, Duke Math. J. **44** (1977), 1–43.
6. E. H. Cattani and A. G. Kaplan and W. Schmid, *Degenerations of Hodge structures*, Ann. of Math. (2) **123** (1986), 457–535.
7. P. A. Griffiths, *Periods of integrals on algebraic manifolds: Summary of main results and discussions of open problems*, Bull. Amer. Math. Soc. **76** (1970), 228–296.
8. M. Harris and S. Zucker, *Boundary cohomology of Shimura varieties.* I. *Coherent cohomology on toroidal compactifications*, Ann. Sci. École Norm. Sup. (4) **27** (1994), 249–344.
9. L. Illusie, *Logarithmic spaces* (*according to K. Kato*), Barsotti Symposium in Algebraic Geometry (Abano Terme, 1991) (V. Critstante and W. Messing, eds.), Perspect. Math., vol. 15, Academic Press, San Diego, CA, 1994, pp. 183–203.
10. K. Kato, *Logarithmic structures of Fontaine-Illusie*, Algebraic Analysis, Geometry, and Number Theory (Baltimore, MD, 1988) (J.-I. Igusa, ed.), Johns Hopkins Univ. Press, Baltimore, MD, 1989, pp. 191–224.
11. ———, *Toric singularities*, Amer. J. Math. **116** (1994), 1073–1099.
12. K. Kato and C. Nakayama, *Log Betti cohomology, log étale cohomology, and log de Rham cohomology of log schemes over* $\mathbb{C}$, Tech. Report no. UTMS 95-16, Univ. Tokyo, 1995.
13. K. Kato and S. Usui, *Logarithmic Hodge structures and classifying spaces* (in preparation).
14. Y. Kawamata and Y. Namikawa, *Logarithmic deformations of normal crossing varieties and smoothing of degenerate Calabi-Yau varieties*, Invent. Math. **118** (1994), 395–409.
15. D. R. Morrison, *Mirror symmetry and rational curves on quintic threefolds: A guide for mathematicians*, J. Amer. Math. Soc. **6** (1993), 223–247.
16. I. Satake, *On compactification of the quotient spaces for arithmetically defined discontinuous groups*, Ann. of Math. (2) **72** (1960), 555–580.
17. W. Schmid, *Variation of Hodge structure; the singularities of the period mappings*, Invent. Math. **22** (1973), 211–319.
18. S. Usui, *A numerical criterion for admissibility of semi-simple elements*, Tôhoku Math. J. (2) **45** (1993), 471–484.
19. ———, *Complex structures on partial compactifications of arithmetic quotients of classifying spaces of Hodge structures*, Tôhoku Math. J. (2) **47** (1995), 405–429.

GRADUATE SCHOOL OF MATHEMATICAL SCIENCE, UNIVERSITY OF TOKYO, KOMABA, TOKYO, 153, JAPAN
E-mail address: kkato@ms.u-tokyo.ac.jp

GRADUATE SCHOOL OF SCIENCE, OSAKA UNIVERSITY TOYONAKA, OSAKA, 560, JAPAN
E-mail address: usui@math.wani.osaka-u.ac.jp

Chow Groups and Motives

Centre de Recherches Mathématiques
CRM Proceedings and Lecture Notes
Volume **24**, 2000

Motives and Algebraic de Rham Cohomology

Masanori Asakura

Abstract. In this paper, we define a certain Hodge-theoretic structure for an arbitrary variety X over the complex number field by using the theory of mixed Hodge module due to Morihiko Saito. We call it an arithmetic Hodge structure of X. It is shown that extension groups of arithmetic Hodge structure do not vanish even for degree ≥ 2. Moreover, we define higher Abel-Jacobi maps from Bloch's higher Chow groups of X to these extension groups. These maps essentially involve the classical Abel-Jacobi maps by Weil and Griffiths, and Mumford's infinitesimal invariants of 0-cycles on surfaces.

1. Introduction

Let X be a nonsingular projective surface over $\mathbb{C}$. D. Mumford was the first to show that the Chow group $\mathrm{CH}_0(X)$ of 0-cycles on X is more complicated than what was earlier believed [**M**]; in fact it can be "enormous". This has led to a difficulty in understanding the precise nature of structure of the Chow group of zero-cycles. His theorem asserts that if the geometric genus of X is not zero, then the kernel $T(X)$ of the Albanese map

$$A_0(X) \to \mathrm{Alb}(X)$$

cannot be finite-dimensional. Namely $A_0(X)$ cannot be given a geometric structure such as a complex torus. Here $A_0(X)$ denotes the subgroup of 0-cycles of degree 0.

S. Bloch [**B1**] provided some insight in this situation by conjecturing that $T(X)$ is controlled by the transcendental part of the cohomology $H^2(X,\mathbb{Q})_{\mathrm{tr}} = H^2(X,\mathbb{Q})/\mathrm{NS}(X)$. This is now called the Bloch conjecture (see Conjecture 4.6 in §4). Combining this with the mixed motives as introduced by P. Deligne and A. Beilinson, his conjecture led to a conjectured existence of a filtration $F_{\mathcal{M}}$ on all of the higher Chow groups $\mathrm{CH}^r(X,m;\mathbb{Q}) = \mathrm{CH}^r(X,m) \otimes \mathbb{Q}$ [**B2**], called the motivic filtration. (In this paper we will simply write $\mathrm{CH}^r(X,m)$ instead of $\mathrm{CH}^r(X,m;\mathbb{Q})$, to be understood as Bloch's higher Chow groups tensored with $\mathbb{Q}$.) This is fortified by the following beautiful conjectural formula for any nonsingular projective

---

1991 *Mathematics Subject Classification.* 14F40, 14D07, 14C25, 14C30.
Received by the editors September 22, 1998; Accepted July 12, 1999.
The author is supported by JSPS Research Fellowships for Young Scientists.
This is the final form of the paper.

variety X:

$$\mathrm{Gr}^{\nu}_{F_{\mathcal{M}}} \mathrm{CH}^r(X,m) = \mathrm{Ext}^{\nu}_{\mathcal{M}}(\mathbb{Q}(0), H^{2r-m-\nu}(X)(r)). \tag{1.1}$$

Here $\mathcal{M}$ is the category of mixed motives over $\mathrm{Spec}(\mathbb{C})$ and $\mathbb{Q}(0)$ is the trivial motive. It is proved that the motivic filtration is determined uniquely if it exists (cf. [**J1**], [**SaS2**]). For example, it is conjectured that $F^1_{\mathcal{M}} \mathrm{CH}^r(X) = \mathrm{CH}^r(X)_{\mathrm{hom}}$ (the subgroup of homologically trivial cycles), $F^2_{\mathcal{M}} \mathrm{CH}_0(X) = T(X)$, and $F^{r+1}_{\mathcal{M}} \mathrm{CH}^r(X) = 0$. In spite of much effort by many people, nobody has found a suitable definition of $\mathcal{M}$ and $F^{\bullet}_{\mathcal{M}}$ (see [**J1**], [**Mur1**], [**SaS1**]).

On the other hand, the category of the mixed motives $\mathcal{M}$ is considered to possess an exact faithful functor $\mathcal{M} \to \mathrm{MHS}$ (called the *realization functor*) to the category of graded polarizable $\mathbb{Q}$-mixed Hodge structures. Therefore there should be the map

$$\mathrm{Gr}^{\nu}_{F_{\mathcal{M}}} \mathrm{CH}^r(X) \to \mathrm{Ext}^{\nu}_{\mathrm{MHS}}(\mathbb{Q}(0), H^{2r-\nu}(X)(r)).$$

Unfortunately, the higher extension group $\mathrm{Ext}^{\nu}_{\mathrm{MHS}}$ ($\nu \geq 2$) always vanishes in the category MHS. This means that $T(X) = \mathrm{Gr}^2_{F_{\mathcal{M}}} \mathrm{CH}_0(X)$ for a surface X cannot be captured by the extension of mixed Hodge structures.

The purpose of this paper is to define a certain Hodge-theoretic structure for an arbitrary variety X over the complex number field. We call it an arithmetic Hodge structure of X. We define the decreasing filtration $F^{\bullet} \mathrm{CH}^r(X,m)$ on higher Chow groups and the following map (which we call the *higher Abel-Jacobi map*):

$$\rho^{\nu}_X \colon \mathrm{Gr}^{\nu}_F \mathrm{CH}^r(X,m) \to \mathrm{Ext}^{\nu}_{\underline{\mathrm{M}}(\mathbb{C})}(\mathbb{Q}(0), H^{2r-m-\nu}(X)(r)).$$

Here $\underline{\mathrm{M}}(\mathbb{C})$ denotes the category of arithmetic Hodge structure which possesses the realization functor $r\colon \underline{\mathrm{M}}(\mathbb{C}) \to \mathrm{MHS}$. When $\nu = 1$, the map ρ^1_X and the functor r induces the usual Abel-Jacobi maps (cf. §2.5). When $\nu \geq 2$, the higher Abel-Jacobi map ρ^{ν}_X gives a new tool for Hodge theoretic study of algebraic cycles. For example, it defines the *second Albanese map*

$$T(X) \to \mathrm{Ext}^2_{\underline{\mathrm{M}}(\mathbb{C})}(\mathbb{Q}(0), H^{2n-2}(X)(n)),$$

for a projective nonsingular variety X of dimension n. If it is injective, the Bloch conjecture is true (Theorem 4.9). More generally, we conjecture that $F^{\bullet}$ is discrete, that is, $F^N = 0$ for $N \gg 0$. If our conjecture holds, our filtration $F^{\bullet}$ gives the motivic filtration (cf. Theorem 1.1 below).

Let us explain arithmetic Hodge structures. Let X be a quasi-projective nonsingular variety over $\mathbb{C}$. Then X is defined by finitely many equations which have finitely many coefficients. By considering the coefficients as parameters of a space S, we obtain the smooth family $f\colon X_S \to S$. Then we get the mixed Hodge module $H^{\bullet}(X_S/S)$ (cf. [**SaM2**]). The usual mixed Hodge structure $H^{\bullet}(X,\mathbb{Q})$ appears on the fiber over a point of S. The *arithmetic Hodge structure* is defined to be the inductive limit of the above mixed Hodge modules, S running over the embedding of the function field to $\mathbb{C}$. The most important ingredient of arithmetic Hodge structure is the arithmetic Gauss Manin connection on algebraic de Rham cohomology (see §3.2). Because of this connection, the higher extension group does not necessarily vanish.

Our main result is as follows:

THEOREM 1.1. *Let X be a nonsingular projective variety over $\mathbb{C}$. Let $\mathrm{CH}^r(X,m)$ be the higher Chow group tensored with $\mathbb{Q}$. There is a decreasing filtration $F^\bullet$ on $\mathrm{CH}^r(X,m)$ and a natural map*

$$\rho_X^\nu \colon \mathrm{Gr}_F^\nu \mathrm{CH}^r(X,m) \to \mathrm{Ext}^\nu_{\underline{\mathrm{M}}(\mathbb{C})}(\mathbb{Q}(0), H^{2r-m-\nu}(X)(r)), \tag{1.2}$$

which has the following properties:

(1) *If $F^{r+1}\mathrm{CH}^r(X,m) = 0$ for arbitrary X and r, then $F^\bullet$ coincides with the motivic filtration $F^\bullet_{\mathcal{M}}$. In particular, the vanishing implies the Bloch conjecture* 4.6.

(2) (a) $F^0\mathrm{CH}^r(X,m) = \mathrm{CH}^r(X,m)$. $F^1\mathrm{CH}^r(X) = \mathrm{CH}^r(X)_{\mathrm{hom}}$.
  (b) $F^0\mathrm{CH}^r(X,m) = F^1\mathrm{CH}^r(X,m)$ *for* $m \geq 1$.
  (c) $F^2\mathrm{CH}^r(X,m)$ *is contained in the kernel of the cycle map to the Deligne cohomology group. These coincide if the realization functor* $r\colon \underline{\mathrm{M}}(\mathbb{C}) \to \mathrm{MHS}$ *is fully faithful.*
  (d) $F^2\mathrm{CH}_0(X) = T(X)$.
  (e) $F^{r+1}\mathrm{CH}^r(X,m) = F^{r+2}\mathrm{CH}^r(X,m) = \cdots$.

(3) *There are the following natural maps:*

$$\mathrm{Ext}^p_{\underline{\mathrm{M}}(\mathbb{C})}(\mathbb{Q}(0), H^q(X)(r)) \to \Xi_X^{r-p,q-r+p}(p) \to \Lambda_X^{r-p,q-r+p}(p),$$

where $\Xi_X^{p,q}(r)$ (resp. $\Lambda_X^{p,q}(r)$) is defined as the cohomology of the following complex (see Definitions 3.2 *and* 3.5*):*

$$F^{p+1}H^{p+q}_{\mathrm{dR}}(X/\mathbb{C}) \otimes \Omega^{r-1}_{\mathbb{C}/\overline{\mathbb{Q}}} \xrightarrow{\nabla} F^pH^{p+q}_{\mathrm{dR}}(X/\mathbb{C}) \otimes \Omega^r_{\mathbb{C}/\overline{\mathbb{Q}}}$$
$$\xrightarrow{\nabla} F^{p-1}H^{p+q}_{\mathrm{dR}}(X/\mathbb{C}) \otimes \Omega^{r+1}_{\mathbb{C}/\overline{\mathbb{Q}}},$$
$$(\textit{resp. } H^{q-1}(\Omega^{p+1}_{X/\mathbb{C}}) \otimes \Omega^{r-1}_{\mathbb{C}/\overline{\mathbb{Q}}} \xrightarrow{\overline{\nabla}} H^q(\Omega^p_{X/\mathbb{C}}) \otimes \Omega^r_{\mathbb{C}/\overline{\mathbb{Q}}}$$
$$\xrightarrow{\overline{\nabla}} H^{q+1}(\Omega^{p-1}_{X/\mathbb{C}}) \otimes \Omega^{r+1}_{\mathbb{C}/\overline{\mathbb{Q}}}).$$

In particular, the map (1.2) *induces homomorphisms*

$$\xi_X^\nu \colon \mathrm{Gr}_F^\nu \mathrm{CH}^r(X,m) \to \Xi_X^{r-\nu,r-m}(\nu)$$

and

$$\delta_X^\nu \colon \mathrm{Gr}_F^\nu \mathrm{CH}^r(X,m) \to \Lambda_X^{r-\nu,r-m}(\nu).$$

In general, these maps have non-zero images, even when $\nu \geq 2$, (which follows, for example, from Theorem 4.3*).*

(4) *Assume that X is defined over a number field k, and $X = X_k \otimes_k \mathbb{C}$. If $F^{\dim X+1}\mathrm{CH}_0(X) = 0$, then $F^2\mathrm{CH}_0(X_k) = 0$ and the rank of $\mathrm{CH}_0(X_k)$ is finite.*

We call the above map (1.2) the *νth higher Abel-Jacobi map*. Also, we remark that philosophically speaking, the fact that $F^0\mathrm{CH}^r(X,m) = F^1\mathrm{CH}^r(X,m)$ for $m \geq 1$ should not come as a surprise. Indeed, for $m \geq 1$ and by a standard weight argument, the cycle class map from $\mathrm{CH}^r(X,m)$ to ordinary Betti cohomology is zero. Thus $\mathrm{CH}^r(X,m)$ is already the kernel of a "regulator" map.

By way of acknowledgment, a similar idea of arithmetic Hodge structures was obtained by M. Green and P. Griffiths [**G2**]. They defined the arithmetic Hodge structures independently of the author[1]. The terminology "*arithmetic Hodge structure*" was introduced by them. It is appropriate to adopt their terminology in this paper. However, our motivation for the above idea comes from the the work of Shuji Saito; that is, the higher normal functions [**AS**]. He defined the relative filtration $F_S^\bullet \mathrm{CH}^\bullet(X)$ and likewise $F_S^\bullet H_D^\bullet(X, \mathbb{Q}(\bullet))$ for a smooth projective family $X \to S$, and defined higher normal functions

$$\mathrm{Gr}^{\nu}_{F_S} \mathrm{CH}^r(X) \to \Gamma(S, DJ^{r,\nu}_{X/S}).$$

It was surprising for us that the above map has non-trivial image.

This paper is organized as follows.

In §2, we review Morihiko Saito's theory of mixed Hodge modules. Here we omit the definition, because what we need is not the precise definition but rather the formalism. In §3, we introduce the notion of arithmetic Hodge structure. Moreover we define the spaces of Mumford's infinitesimal invariants, and construct the natural maps from extension groups in $\underline{\mathrm{M}}(\mathbb{C})$ to those (Proposition 3.6). Finally in §4, we construct the higher Abel-Jacobi maps, and study algebraic cycles, in particular, the Bloch conjecture.

Acknowledgments

We are especially indebted to Shuji Saito for his many helpful suggestions. It is my great pleasure to have had the opportunity to work together with him.

We thank Shin-Ichi Mochizuki, who kindly provided a proof for the existence of the curve which appears in Theorem 4.5, and Morihiko Saito, who has read through this paper carefully and has come up with many helpful comments.

We also thank Sampei Usui and Takeshi Saito for fruitful discussions and constant encouragement.

Special thanks are due to the two referees who have read through the earlier versions of this paper pointing out many inaccuracies and suggesting numerous improvements.

At last but not least, we would like to thank Noriko Yui and James D. Lewis for their help for improving English and mathematical presentations putting the paper into a presentable form to the reader. Without their generous help and patience, this paper has never reached this final form.

It is great pleasure to thank the organizers of the Banff Conference on "Arithmetic and Geometry of Algebraic Cycles" for the excellent conference.

Notation and Conventions

1. A *variety* means a quasi-projective algebraic variety over a field. We mainly work with algebraically closed fields of characteristic 0 (e.g., $\mathbb{C}$, $\overline{\mathbb{Q}}$).
2. For a variety X over $\mathbb{C}$, X^{an} denotes the associated analytic space: $X^{\mathrm{an}} = X(\mathbb{C})$.
3. We always ignore torsion. We assume that all Abelian groups (for instance, $\mathrm{CH}^r(X, m)$ above) are tensored with $\mathbb{Q}$.
4. In this paper, we fix an embedding $\overline{\mathbb{Q}} \hookrightarrow \mathbb{C}$.

[1]There is a slight difference between their definition and ours. In their definition, the datum of the $\mathbb{Q}$-structure does not appear. Therefore the category of their arithmetic Hodge structures does not form an Abelian category, but only an exact category.

2. Morihiko Saito's Formalism of Mixed Hodge Modules

In this section, we review the theory of mixed Hodge modules.

2.1. Perverse Sheaf. A perverse sheaf is not a "sheaf" in the usual sense, but rather a complex of sheaves with algebraically constructible cohomology. In many ways it behaves like a sheaf [**BBD**].

Let $D^b_c(\mathbb{Q}_X)$ be the derived category of bounded complexes of $\mathbb{Q}_{X^{\mathrm{an}}}$-sheaves with algebraically constructible cohomology for a complex algebraic variety X. The *perverse t-structure* $(D^{\geq 0}, D^{\leq 0})$ of $D^b_c(\mathbb{Q}_X)$ is the pair of additive full subcategories defined as follows:

(1) $K^\bullet_{\mathbb{Q}} \in D^{\leq 0} \iff \dim \operatorname{Supp} H^i(K^\bullet_{\mathbb{Q}}) \leq -i \ (\forall i)$,

(2) $K^\bullet_{\mathbb{Q}} \in D^{\geq 0} \iff \mathbb{D}(K^\bullet_{\mathbb{Q}}) \in D^{\leq 0} \ (\forall i)$, where $\mathbb{D}$ is the Verdier dual functor on $D^b_c(\mathbb{Q}_X)$.

Here we define the dimension of empty set to be $-\infty$. We call an object of $D^{\geq 0} \cap D^{\leq 0}$ a *perverse sheaf*, and denote its category by $\operatorname{Perv}(\mathbb{Q}_X)$.

The category of perverse sheaves is Abelian. There are the truncation functors ${}^p\tau_{\leq 0}\colon D^b_c(\mathbb{Q}_X) \to D^{\leq 0}$ and ${}^p\tau_{\geq 0}\colon D^b_c(\mathbb{Q}_X) \to D^{\geq 0}$, which are characterized as the right and left adjoint of $D^{\leq 0} \hookrightarrow D^b_c(\mathbb{Q}_X)$ and $D^{\geq 0} \hookrightarrow D^b_c(\mathbb{Q}_X)$ respectively. We put ${}^pH^0 := {}^p\tau_{\leq 0} \circ {}^p\tau_{\geq 0} (\simeq {}^p\tau_{\geq 0} \circ {}^p\tau_{\leq 0})$, and call it the perverse cohomology functor.

There are the standard functors on $D^b_c(\mathbb{Q}_X)$:

$$f_*, \quad f_!, \quad f^*, \quad f^!, \quad \mathbb{D}, \quad \psi_g, \quad \phi_g, \quad \otimes, \quad \underline{\mathrm{Hom}}. \tag{2.1}$$

We write ${}^pH^0 f_*(-)[n]$ by ${}^pR^n f_*$ and so on.

The simplest example of perverse sheaf is a local system $L[\dim X]$ shifted by $\dim X$ when X is nonsingular. The category of local systems is not appropriate for the treatment of homological algebra. Indeed, it is not closed under the above standard functors. The category of constructible sheaves is closed under those, but the one of perverse sheaves has more advantages. Those properties of perverse sheaves are essentially used to construct the theory of mixed Hodge modules.

2.2. Regular Holonomic *D*-Modules. We recall holonomic D-modules[2] and regular singularities.

Let X be a nonsingular variety over a field of characteristic zero.

Let M be an algebraic D_X-module which is quasi-coherent over $\mathcal{O}_X$. We denote $F_iD_X \subset D_X$ the differential order filtration, that is, it is generated by differential operators $\sum_{|I| \leq i} f_I(x)\partial^I$. An increasing filtration $F_\bullet M$ of coherent $\mathcal{O}_X$-submodules is called a *good filtration* if

1. $\bigcup_i F_iM = M$, $F_{-k}M = 0$ for some integer $k \gg 0$,
2. $F_iD_X \cdot F_jM \subset F_{i+j}M$ for $\forall i, j$,
3. $\operatorname{gr}^F M := \bigoplus_i \operatorname{gr}^F_i M$ is coherent over $\operatorname{gr}^F D_X$.

Any coherent D_X-module has at least one good filtration. Note that $\operatorname{gr}^F D_X$ is isomorphic to $\mathcal{O}_{T^*X}$ the structure sheaf of the cotangent bundle T^*X over X. The support of $\operatorname{gr}^F M$ in T^*X is called the *characteristic variety* of M, which does not depend on the choice of a good filtration.

In general, the dimension of the characteristic variety of a coherent D_X-module is larger than or equal to $\dim X$ (Bernstein's inequality). If the equality holds, we call the coherent D_X-module *holonomic*

[2]We use left D-modules, though M. Saito prefers right D-modules to it.

Let us recall the regularity of holonomic D-module[3]. When $\dim X = 1$, a holonomic D_X-module M has *regular singularities* if and only if there is a dense open set $U \subset X$ such that $M|_U$ is a connection with regular singularities, that is, there is a locally free sheaf $\overline{M}$ of finite rank over $\mathcal{O}_{\overline{U}}$ ($\overline{U}$ is the smooth completion of U) and log connection $\overline{\nabla}\colon \overline{M} \to \Omega^1_{\overline{U}/\mathbb{C}}(\log D) \otimes \overline{M}$ ($D := \overline{U} - U$) such that $(\overline{M}, \overline{\nabla})|_U \simeq (M|_U, \nabla)$. For X of arbitrary dimension, a holonomic D_X-module M has *regular singularities* if and only if so does $H^0(i_C^* M)$ for any $i_C \colon C \to X$ with C a nonsingular curve over $\mathbb{C}$, where i_C^* is the derived inverse image functor of D-modules.

The following theorem is due to Kashiwara and Mebkhout:

THEOREM 2.1 (Riemann-Hilbert correspondence). *Let X be a nonsingular variety over $\mathbb{C}$. Then the de Rham functor* $\mathrm{DR} := \Omega^\bullet_{X^{\mathrm{an}}} \otimes (-)[\dim X]$ *induces the equivalence from the category of regular holonomic D-modules on X to the one of perverse sheaves with $\mathbb{C}$-coefficient:*

$$\mathrm{DR}\colon \mathrm{Mod}_{\mathrm{rh}}(X) \xrightarrow{\sim} \mathrm{Perv}(\mathbb{C}_X).$$

The above equivalence also induces the equivalence between local systems and locally free $\mathcal{O}_X$-sheaves with integrable connection with regular singularities [**D3**].

There are the standard operations as in (2.1) on the derived category of bounded complexes of regular holonomic D-modules. Those operations are compatible under the Riemann-Hilbert correspondence.

Again, let X be a nonsingular variety over any field of characteristic zero. $\mathrm{MF}_{\mathrm{rh}}(X)$ denotes the category of filtered regular holonomic D_X-modules (M, F), where M is a regular holonomic D_X-module, and F is a good filtration. Moreover we denote by $\mathrm{MFW}_{\mathrm{rh}}(X)$ the category of filtered regular holonomic D_X-modules with weight filtration (M, F, W), where W is a finite increasing filtration of D_X-submodules (= the weight filtration). These are not Abelian, but exact categories such that the kernel and cokernel objects exist for any morphisms. A complex in $\mathrm{MF}_{\mathrm{rh}}(X)$ (resp. $\mathrm{MFW}_{\mathrm{rh}}(X)$)

$$(M', F) \to (M, F) \to (M'', F)$$
$$(\text{resp. } (M', F, W) \to (M, F, W) \to (M'', F, W))$$

is exact if and only if

$$\mathrm{gr}^F M' \to \mathrm{gr}^F M \to \mathrm{gr}^F M''$$
$$(\text{resp. } \mathrm{gr}^F \mathrm{gr}^W M' \to \mathrm{gr}^F \mathrm{gr}^W M \to \mathrm{gr}^F \mathrm{gr}^W M'')$$

is an exact sequence of sheaves of $\mathcal{O}_X$-modules.

2.3. Mixed Hodge Modules. Morihoko Saito defined mixed Hodge modules, and proved the expected properties.

What is a mixed Hodge module? Roughly speaking, it is a $\mathbb{Q}$-coefficient perverse sheaf with mixed Hodge structure. Its notion contains *admissible variation of mixed Hodge structure.*

First we recall it:

---

[3]There are several ways to define regularities of holonomic D-modules. We follow the text of Z. Mebkhout [**Me**, p. 163, Proposition 5.4.2].

DEFINITION 2.2. Let X be a nonsingular variety over $\mathbb{C}$. Then an admissible variation of mixed Hodge structures on X is defined to be the data

$$(H_{\mathbb{Q}}, H_{\mathcal{O}}, W_\bullet, F^\bullet, \nabla, i)$$

where:

- $H_{\mathbb{Q}}$ is a local system of finite dimensional $\mathbb{Q}$-vector spaces on X^{an},
- $H_{\mathcal{O}}$ is a locally free (Zariski) sheaf of $\mathcal{O}_X$-module of the same rank as $H_{\mathbb{Q}}$,
- $W_\bullet$ is a finite increasing filtration of $H_{\mathbb{Q}}$, called the *weight filtration*,
- $F^\bullet$ is a finite decreasing filtration on $H_{\mathcal{O}}$ by locally free $\mathcal{O}_X$-submodules, called the *weight filtration* (*resp. Hodge filtration*),
- $\nabla\colon H_{\mathcal{O}} \to H_{\mathcal{O}} \otimes_{\mathcal{O}_X} \Omega^1_{X/\mathbb{C}}$ an integrable connection (called the (*algebraic*) *Gauss-Manin connection*),
- $i\colon H_{\mathbb{Q}} \otimes \mathbb{C} \xrightarrow{\sim} \ker \nabla^{\mathrm{an}}$ (called the *comparison isomorphism*), or equivalently, i induces an isomorphism $H_{\mathbb{Q}} \otimes \mathcal{O}_X^{\mathrm{an}} \xrightarrow{\sim} H_{\mathcal{O}}^{\mathrm{an}}$.

and these satisfy:

(1) For all points $s \in X^{\mathrm{an}}$, the fiber $H_{\mathbb{Q},s} \overset{i}{\hookrightarrow} H_{\mathcal{O}} \otimes \mathbb{C}(s)$ with the induced filtrations $W_{\bullet,s}$ on $H_{\mathbb{Q},s}$ and $F_s^\bullet$ on $H_{\mathcal{O}} \otimes \mathbb{C}(s)$ defines a mixed Hodge structure.

(2) (Griffiths transversality) $W_\bullet$ and $F^\bullet$ satisfy the following:

$$\nabla(W_\ell) \subset W_\ell \otimes \Omega^1_{X/\mathbb{C}}, \quad \nabla(F^p) \subset F^{p-1} \otimes \Omega^1_{X/\mathbb{C}} \quad \text{for } \forall \ell, p.$$

(3) (polarizability) For each ℓ, there are $\mathbb{Q}$-bilinear form

$$Q\colon \mathrm{Gr}_\ell^W(H_{\mathbb{Q}}) \otimes \mathrm{Gr}_\ell^W(H_{\mathbb{Q}}) \to \mathbb{Q}$$

and $\mathcal{O}_X$-bilinear form

$$Q_X\colon \mathrm{Gr}_\ell^W(H_{\mathcal{O}}) \otimes \mathrm{Gr}_\ell^W(H_{\mathcal{O}}) \to \mathcal{O}_X$$

satisfying:

(a) Q and Q_X are compatible under the comparison isomorphism i.

(b) For all $s \in X^{\mathrm{an}}$, Q defines a polarization form on the $\mathbb{Q}$-Hodge structure $(\mathrm{Gr}_\ell^W(H_{\mathbb{Q},s}), F_s^\bullet)$.

(c) $Q_X(\nabla(x), y) + Q_X(x, \nabla(y)) = dQ_X(x, y)$ for any local sections $x, y \in H_{\mathcal{O}}$.

(4) (admissibility) When the data $(H_{\mathbb{Q}}, H_{\mathcal{O}}, W_\bullet, F^\bullet, \nabla, i)$ is pulled back to a nonsingular complex algebraic curve C, it satisfies the admissibility [**K**, 1.9]:

(a) Any local monodromy around $\overline{C} - C$ is quasi-unipotent. Here $\overline{C}$ denotes the smooth completion of C.

(b) The logarithm N of the unipotent part of the local monodromy admits a weight filtration relative to $W_\bullet$ [**SZ**, §2].

(c) The Hodge filtration $F^\bullet$ can be extended to a locally free subsheaf $\overline{F}^\bullet$ of the Deligne's canonical extension $\overline{H}_C$ such that $\mathrm{Gr}_{\overline{F}}^\bullet \mathrm{Gr}_\bullet^W(\overline{H}_C)$ is locally free.

We denote the category of admissible variations of mixed Hodge structures on X by VMHS(X).

In order to define mixed Hodge module, we replace $H_{\mathbb{Q}}$ by perverse sheaf, $(H_{\mathcal{O}}, F^\bullet, \nabla)$ by filtered regular holonomic D_X-module, and i by the Riemann-Hilbert correspondence.

More precisely, let $\mathrm{PervW}(\mathbb{Q}_X)$ be the category of perverse sheaves $(K^\bullet_{\mathbb{Q}}, W_{\mathbb{Q}})$ with finite increasing filtration $W_{\mathbb{Q}}$ of sub perverse sheaves (= the weight filtration). There is a natural functor from $\mathrm{MFW}_{\mathrm{rh}}(X)$ to $\mathrm{PervW}(\mathbb{C}_X)$, which maps (M, F, W) to $(\mathrm{DR}(M), \mathrm{DR}(W))$. Denote the fiber product $\mathrm{PervW}(\mathbb{Q}_X) \times_{\mathrm{PervW}(\mathbb{C}_X)} \mathrm{MFW}_{\mathrm{rh}}(X)$ by $\mathrm{MFW}_{\mathrm{rh}}(X, \mathbb{Q})$, that is, it consists of the objects $(K^\bullet_{\mathbb{Q}}, W_{\mathbb{Q}}, M, F, W, i)$ where $i\colon (K^\bullet_{\mathbb{Q}} \otimes \mathbb{C}, W_{\mathbb{Q}} \otimes \mathbb{C}) \simeq \mathrm{DR}(M, W)$ is an isomorphism of filtered objects (cf. Theorem 2.1).

THEOREM 2.3 (Morihiko Saito [**SaM1, SaM2**]). *There is an Abelian full subcategory*

$$\mathrm{MHM}(X) \subset \mathrm{MFW}_{\mathrm{rh}}(X, \mathbb{Q})$$

for each nonsingular complex algebraic variety X, which satisfies the following properties. We call an object of $\mathrm{MHM}(X)$ *a mixed Hodge module on X. W (resp. F) is called the weight filtration (resp. Hodge filtration).*

(1) *The forgetful functor* rat: $\mathrm{MHM}(X) \to \mathrm{Perv}(\mathbb{Q}_X)$ *is exact and faithful.*
(2) *Any morphisms of mixed Hodge modules are strict with respect to Hodge and weight filtration.*
(3) *Any pure weight full subcategory of* $\mathrm{MHM}(X)$ *(called polarizable Hodge modules) is semi-simple.*
(4) *There are the standard operations on $D^b(\mathrm{MHM}(X))$ the derived category of bounded complex of mixed Hodge modules:*

$$f_*, \quad f_!, \quad f^*, \quad f^!, \quad \mathbb{D}, \quad \psi_g, \quad \phi_{g,1}, \quad \otimes, \quad \underline{\mathrm{Hom}}.$$

Those functors satisfies the adjointness, projection formulas, and are compatible with the ones on perverse sheaves under the forgetful functor rat.
(5) *Let* $\mathrm{MHM}(X)_{\mathrm{sm}}$ *be the full subcategory of* $\mathrm{MHM}(X)$ *whose object consists of those whose underlying perverse sheaf is smooth (i.e. local system). Then there is the equivalence between the category of admissible variations of mixed Hodge structure (Definition 2.2) and* $\mathrm{MHM}(X)_{\mathrm{sm}}$*:*

$$\mathrm{VMHS}(X) \xrightarrow{\sim} \mathrm{MHM}(X)_{\mathrm{sm}}.$$

Here an admissible variation of mixed Hodge structure $(H_{\mathbb{Q}}, H_{\mathcal{O}}, W_\bullet, F^\bullet, \nabla, i)$ *corresponds to the mixed Hodge module* $(H_{\mathbb{Q}}[\dim X], W_{\bullet - \dim X}, H_{\mathcal{O}}, F_\bullet)$*, where $F_p := F^{-p}$. An admissible variation of Hodge structure of pure weight n corresponds to the Hodge module of weight $n + \dim X$. In particular,* $\mathrm{MHM}(\mathrm{Spec}\,\mathbb{C})$ *is isomorphic to the category of graded polarizable $\mathbb{Q}$-mixed Hodge structures.*
(6) (decomposition theorem) *Let $f\colon X \to Y$ be a proper morphism between nonsingular varieties over $\mathbb{C}$. Then there is a noncanonical isomorphism*

$$f_*M \simeq \bigoplus_k H^k f_* M[-k]$$

in the derived category $D^b(\mathrm{MHM}(Y))$ for any pure weight Hodge module M.
(7) *If H is a mixed Hodge module of weight $\leq n$ (resp. $\geq n$), then $H^k f_!(H)$ and $H^k f^*(H)$ (resp. $H^k f_*(H)$, $H^k f^!(H)$, resp. $\mathbb{D}H$) are of weight $\leq k + n$ (resp. $\geq k + n$, resp. $\leq -n$).*

Morihiko Saito firstly defined the category of polarizable Hodge modules using vanishing cycle functor, and proved the stability of direct images of projective morphism, Verdier dual and so on, which is the main result in [**SaM1**]. Next he defined mixed Hodge modules as a successive extension of Hodge modules satisfying some stability conditions [**SaM2**]. The most important property is the existence of standard functors (loc. cit. Theorem 4.3. etc.). However, the precise construction of the category of mixed Hodge modules is long and complicated. Moreover it is quite difficult to write it down explicitly.

DEFINITION 2.4. $\mathbb{Q}_X(r)[\dim X]$ denotes the *Tate Hodge module*

$$(\mathbb{Q}_X(r)[\dim X], W_\bullet, \mathcal{O}_X, F_\bullet)$$

with $\mathrm{Gr}^W_{\dim X-2r}(\mathbb{Q}_X(r)[\dim X]) = \mathbb{Q}_X(r)[\dim X]$ and $\mathrm{Gr}^F_r(\mathcal{O}_X) = \mathcal{O}_X$, (which is actually a mixed Hodge module of pure weight $\dim X - 2r$ due to Theorem 2.3(5).)

We write the Tate Hodge module shifted by $-\dim X$ by $\mathbb{Q}_X(r)$:

$$\mathbb{Q}_X(r) := (\mathbb{Q}_X(r)[\dim X])[-\dim X] \in D^b(\mathrm{MHM}(X)).$$

2.4. Mixed Hodge Modules Over a Field of Characteristic 0. Morihiko Saito has already pointed out that there are several modifications of mixed Hodge modules. Here we recall one of those.

DEFINITION 2.5. Let $k \subset \mathbb{C}$ be a subfield. For a nonsingular variety X_k over k, we define $\mathrm{MHM}(X_k)$ to be the full subcategory of the fiber product

$$\mathrm{MFW}_{\mathrm{rh}}(X_k) \times_{\mathrm{MFW}_{\mathrm{rh}}(X_{\mathbb{C}})} \mathrm{MHM}(X_{\mathbb{C}}) \quad (X_{\mathbb{C}} := X_k \otimes \mathbb{C})$$

such that the polarization on each weight quotient is defined over k.

Since all operations such as (2.1) can be also defined on $\mathrm{MFW}_{\mathrm{rh}}(X_k)$, we have those induced on $\mathrm{MHM}(X_k)$. Each property of those operations can be reduced to those of $\mathrm{MHM}(X_{\mathbb{C}})$. In particular, Theorem 2.3 still holds ture, even after we replace $\mathrm{MHM}(X_{\mathbb{C}})$ by $\mathrm{MHM}(X_k)$, and $\mathrm{VMHS}(X_{\mathbb{C}})$ by the full subcategory of $\mathrm{MFW}_{\mathrm{rh}}(X_k)_{\mathrm{sm}} \times_{\mathrm{MFW}_{\mathrm{rh}}(X_{\mathbb{C}})_{\mathrm{sm}}} \mathrm{VMHS}(X_{\mathbb{C}})$ such that the polarization on each weight quotient is defined over k. ($\mathrm{MFW}_{\mathrm{rh}}(X_k)_{\mathrm{sm}}$ denotes the full subcategory of $\mathrm{MFW}_{\mathrm{rh}}(X_k)$ generated by objects whose underlying D-module is a locally free $\mathcal{O}_X$-module of finite rank.) For more details, see his exposition [**SaM3**, (1.8)(ii)].

COROLLARY 2.6. *Let $f\colon X \to Y$ be a proper morphism between nonsingular varieties over k. Then there is the Leray spectral sequence*

$$(2.2)\quad E_2^{pq} = \mathrm{Ext}^p_{\mathrm{MHM}(Y)}(\mathbb{Q}_Y(0), H^q f_*\mathbb{Q}_X(r)) \implies \mathrm{Ext}^{p+q}_{\mathrm{MHM}(X)}(\mathbb{Q}_X(0), \mathbb{Q}_X(r)),$$

which degenerates at E_2 terms.

PROOF. The existence of the spectral sequence (2.2) follows from the existence of the standard functors (Theorem 2.3(4)) and the fact $f^*\mathbb{Q}_Y(r) = \mathbb{Q}_X(r)$. The E_2-degeneration follows from the decomposition theorem (Theorem 2.3(6)). □

COROLLARY 2.7 ([**SaM3**, (8.3)]). *Let X be a nonsingular variety over k. Then there is a cycle map from the higher Chow group to the extension group of mixed Hodge modules:*

$$(2.3)\qquad \mathrm{CH}^r(X, m) \to \mathrm{Ext}^{2r-m}_{\mathrm{MHM}(X)}(\mathbb{Q}_X(0), \mathbb{Q}_X(r)).$$

PROOF. (Here we recall the construction of the cycle map (2.3) for the convenience of the reader.)

Let Y be a nonsingular variety over k of dimension d_Y, and $i_Z\colon Z \hookrightarrow Y$ a closed subscheme (not necessarily smooth nor irreducible) of pure dimension d_Z. Let $j_Z\colon U \hookrightarrow Y$ be the complement of Z. We define

$$i_{Z*}i_Z^!\mathbb{Q}_Y(r) = \mathrm{Cone}(\mathbb{Q}_Y(r) \to j_{Z*}j_Z^*\mathbb{Q}_Y(r))[-1],$$

and

$$i_{Z*}\mathbb{Q}_Z(r) = \mathrm{Cone}(j_{Z!}j_Z^*\mathbb{Q}_Y(r) \to \mathbb{Q}_Y(r)),$$

which are complexes of mixed Hodge modules on Y by Theorem 2.3(4). Note that $\mathrm{rat}(i_{Z*}\mathbb{Q}_Z(r)) = i_{Z*}\mathbb{Q}_{Z^{\mathrm{an}}}$ and $\mathrm{rat}(i_{Z*}i_Z^!\mathbb{Q}_Y(r)) = i_{Z*}i_Z^!\mathbb{Q}_{Y^{\mathrm{an}}}$.

Let Z_k be an irreducible component of Z, and $z = \sum_k n_k Z_k$ be a cycle on Y with $n_k \in \mathbb{Q}$. First we construct the natural morphism

$$c_z\colon i_{Z*}\mathbb{Q}_Z(0) \to \mathbb{Q}_Y(d_Y - d_Z)[2d_Y - 2d_Z] \tag{2.4}$$

associated to the cycle z in the derived category $D^b(\mathrm{MHM}(Y))$ as follows. Let $h_k\colon \widetilde{Z}_k \to Z_k$ be a resolution of singularity, and $h\colon \coprod \widetilde{Z}_k \to Y$ the composition with the inclusion i_Z. There are the pull-back morphisms $\alpha\colon i_{Z*}\mathbb{Q}_Z(0) \to i_{Z*}h_{k*}\mathbb{Q}_{\widetilde{Z}_k}(0)^\oplus = h_*\mathbb{Q}_{\widetilde{Z}_k}(0)^\oplus$ and $\beta\colon \mathbb{Q}_Y(0) \to h_*\mathbb{Q}_{\widetilde{Z}_k}(0)^\oplus$. Applying the Verdier dual $\mathbb{D}$ on the latter morphism β, we have $h_*\mathbb{Q}_{\widetilde{Z}_k}(d_Z)^\oplus[2d_Z] \to \mathbb{Q}_Y(d_Y)[2d_Y]$ because Y and Z_k are nonsingular. Composing this with $\oplus n_k\colon h_*\mathbb{Q}_{\widetilde{Z}_k}(0)^\oplus \to h_*\mathbb{Q}_{\widetilde{Z}_k}(0)^\oplus$ and the morphism α, we have (2.4). It is a routine work to show its independency of the choice of resolutions of Z_k.

Put $Y = X$, and compose the morphism (2.4) with $\mathbb{Q}_X(0) \to i_{Z*}\mathbb{Q}_Z(0)$. Then we have the cycle map (2.3) for $m = 0$.

Next we construct the cycle map (2.3) for $m \geq 1$. We write

$$\Delta^m = \mathrm{Spec}\, k[t_0, \dots, t_m]/\left(\sum_{i=0}^m t_i - 1\right)$$

and its faces by

$$\Delta_I\colon t_{i_0} = \cdots = t_{i_k} = 0$$

for a subset $I = \{i_0, \dots, i_k\} \subset \{0, \dots, m\}$. Put $Y = X \times \Delta^m$, and $i\colon X \times (\bigcup_i \Delta_i) \hookrightarrow Y$ the closed immersion and $j\colon X \times U \hookrightarrow Y$ its complement. Let $z = \sum n_k Z_k$ be a cycle on Y (where Z_k are reduced and irreducible subvarieties of dimension $d_Y - r$) which intersects with all faces $X \times \Delta_I$ properly, and

$$z \cdot [X \times \Delta_i] = 0 \quad \text{for } 0 \leq \forall i \leq m. \tag{2.5}$$

Put $Z = \bigcup_k Z_k$ the support of z.

We first claim:

LEMMA 2.8. *The morphism c_z (2.4) is uniquely lifted to the morphism*

$$i_{Z*}\mathbb{Q}_Z(0) \to j_!j^*\mathbb{Q}_Y(r)[2r].$$

In fact, using the distinguished triangle

$$j_!j^*\mathbb{Q}_Y(r) \to \mathbb{Q}_Y(r) \to i_*\mathbb{Q}_{X\times(\bigcup_i \Delta_i)}(r) \xrightarrow{+1},$$

it follows from

$$\mathrm{Ext}^\nu_{\mathrm{MHM}(Y)}(i_{Z*}\mathbb{Q}_Z(0), i_*\mathbb{Q}_{X\times(\bigcup_i \Delta_i)}(r)) = 0 \quad \text{for } \forall \nu < 2r, \tag{2.6}$$

and

(2.7) the composition of c_z with $\mathbb{Q}_Y(r) \to i_*\mathbb{Q}_{X\times(\bigcup_i \Delta_i)}(r)$ is zero.

Since $i_*\mathbb{Q}_{X\times(\bigcup_i \Delta_i)}(r)$ is isomorphic to

$$\bigoplus_{i=0}^{m} \mathbb{Q}_{X\times\Delta_i}(r) \to \bigoplus_{i<j} \mathbb{Q}_{X\times\Delta_{ij}}(r) \to \cdots \to \bigoplus_{|I|=m} \mathbb{Q}_{X\times\Delta_I}(r),$$

(2.6) and (2.7) follow from

(2.8) $\operatorname{Ext}^{\nu}_{\mathrm{MHM}(Y)}(i_{Z*}\mathbb{Q}_Z(0), i_I * \mathbb{Q}_{X\times\Delta_I}(r)) = 0$ for $\forall I$ and $\forall \nu < 2r$

where $i_I \colon X \times \Delta_I \hookrightarrow Y$, and

(2.9) the composition of c_z with $\mathbb{Q}_Y(r) \to i_*\mathbb{Q}_{X\times\Delta_i}(r)$ is zero for $0 \le \forall i \le m$.

(2.9) follows from (2.5). In order to see (2.8), observe that we have

$$\begin{aligned}\operatorname{Ext}^{\nu}_{\mathrm{MHM}(Y)}&(i_{Z*}\mathbb{Q}_Z(0), i_{I*}\mathbb{Q}_{X\times\Delta_I}(r))\\ &= \operatorname{Ext}^{\nu}_{\mathrm{MHM}(X\times\Delta_I)}(i_{Z_I*}\mathbb{Q}_{Z_I}(0), \mathbb{Q}_{X\times\Delta_I}(r))\\ &= \operatorname{Ext}^{\nu}_{\mathrm{MHM}(X\times\Delta_I)}(i_{Z_I*}\mathbb{Q}_{Z_I}(0), i_{Z_I*}i^!_{Z_I}\mathbb{Q}_{X\times\Delta_I}(r))\end{aligned}$$

by the adjunction, where we put $Z_I := Z \cap (X \times \Delta_I)$. Then (2.8) follows from the fact that

$$H^k((i_{Z_I*}\mathbb{Q}_{Z_I}(0)) = 0 \quad \text{for } k > d_{Z_I}$$

and

$$H^k(i_{Z_I*}i^!_{Z_I}\mathbb{Q}_{X\times\Delta_I}(r)) = 0 \quad \text{for } k < d_{X\times\Delta_I} + r,$$

which can be checked on the underlying perverse sheaves (cf. [**SaM3**, (7.15)]). Thus we have shown Lemma 2.8.

LEMMA 2.9. *Let $p \colon Y \to X$ be the projection. Then $p_*j_!j^*\mathbb{Q}_Y(r) = \mathbb{Q}_X(r)[-m]$.*

In fact, since $j = 1 \times j_0 \colon X \times (\Delta^m \setminus \bigcup_i \Delta_i) \hookrightarrow Y = X \times \Delta^m$, we can assume that $X = \operatorname{Spec} k$. Then the assertion follows from [**SaM3**, (7.6)] .

By Lemma 2.8 and Lemma 2.9, we have the morphism

$$\mathbb{Q}_X(0) = p_*\mathbb{Q}_Y(0) \to p_*i_{Z*}\mathbb{Q}_Z(0) \to p_*j_!j^*\mathbb{Q}_Y(r)[2r] = \mathbb{Q}_X(r)[2r-m],$$

which gives the cycle map (2.3). □

2.5. The Carlson Isomorphisms. The category MHS is not semi simple, and the Yoneda extension groups in MHS have the well known explicit description, due to Carlson [**Ca**].

Here we will review it. Since MHS carries the neutral Tannakian structure, we have:

$$\operatorname{Ext}^{\bullet}_{\mathrm{MHS}}(H_1, H_2) = \operatorname{Ext}^{\bullet}_{\mathrm{MHS}}(\mathbb{Q}(0), H_1^* \otimes H_2).$$

THEOREM 2.10 (Carlson). *Let $H = (H_{\mathbb{Q}}, W_\bullet, F^\bullet)$ be a graded polarizable $\mathbb{Q}$-mixed Hodge structure. We define*

$$C(H)^0 = \ker(F^0W_0H_{\mathbb{C}} \oplus W_0H_{\mathbb{Q}} \to \operatorname{Gr}^W_0 H_{\mathbb{C}}), \quad C(H)^1 = W_{-1}H_{\mathbb{C}}.$$

Then

(1) $\mathrm{Ext}^1_{\mathrm{MHS}}(\mathbb{Q}(0), H) = \mathrm{Ext}^1_{\mathrm{MHS}}(\mathbb{Q}(0), W_0 H) = \mathrm{Coker}(C(H)^0 \to C(H)^1)$. *Here an element $\xi \in W_{-1}H_{\mathbb{C}}$ corresponds to the following extension of graded polarizable mixed Hodge structures:*

$$0 \to W_0 H \to \widetilde{H} \to \mathbb{Q}(0) \to 0,$$

where $\widetilde{H} = (\widetilde{H}_{\mathbb{Q}}, W_\bullet, F^\bullet)$ is the mixed Hodge structure with $\widetilde{H}_{\mathbb{Q}} = W_0 H_{\mathbb{Q}} \oplus \mathbb{Q}(0)$, and the weight filtration $W_0 \widetilde{H}_{\mathbb{Q}} = \widetilde{H}_{\mathbb{Q}}$, $W_\ell \widetilde{H}_{\mathbb{Q}} = W_\ell H_{\mathbb{Q}}$ $(\ell \leq -1)$, and the Hodge filtration $F^p \widetilde{H}_{\mathbb{C}} = F^p W_0 H_{\mathbb{C}}$ $(p \geq 1)$, $F^q \widetilde{H}_{\mathbb{C}} = F^q W_0 H_{\mathbb{C}} + F^0 \widetilde{H}_{\mathbb{C}}$ $(q \leq -1)$ and

$$F^0 \widetilde{H}_{\mathbb{C}} = F^0 W_0 H_{\mathbb{C}} + \mathbb{C} \cdot (\xi \oplus 1).$$

(2) $\mathrm{Ext}^\nu_{\mathrm{MHS}}(\mathbb{Q}(0), H) = 0$ *for* $\nu \geq 2$. (*This is a formal consequence of the fact that the functor* $\mathrm{Ext}^1(\mathbb{Q}(0), -)$ *is right exact, which follows from the above explicit description.*)

REMARK 2.11. The author learned the above formula from Morihiko Saito, which is more complicated than the original one by J. Carlson. He calculated the Yoneda extension groups in the category of mixed Hodge structures which are not necessarily graded polarizable, and obtained a simple description [**Ca**]

$$\mathrm{Ext}^1(\mathbb{Q}(0), H) = W_0 H_{\mathbb{C}} / F^0 W_0 H_{\mathbb{C}} + W_0 H_{\mathbb{Q}}.$$

In particular, if $H = H^{2r-1}(X^{\mathrm{an}}, \mathbb{Q}(r))$ for a complex projective nonsingular variety X,

$$\mathrm{Ext}^1_{\mathrm{MHS}}(\mathbb{Q}(0), H) = J^r(X)_{\mathbb{Q}}, \tag{2.10}$$

where $J^r(X)_{\mathbb{Q}} = H_{\mathbb{C}} / F^0 H_{\mathbb{C}} + H_{\mathbb{Q}}$ is the r*th intermediate Jacobian* of X. Using the mixed Hodge modules, we can define the Abel-Jacobi map

$$\rho \colon \mathrm{CH}^r(X)_{\mathrm{hom}} \to J^r(X)_{\mathbb{Q}}, \tag{2.11}$$

where $\mathrm{CH}^r(X)_{hom}$ denotes the $\mathbb{Q}$-tensored Chow group of X generated by algebraic cycles of codimension r which are homologically equivalent to 0.

Let

$$\mathrm{CH}^r(X) \to \mathrm{Ext}^{2r}_{\mathrm{MHM}(X)}(\mathbb{Q}_X(0), \mathbb{Q}_X(r)). \tag{2.12}$$

be the cycle map as in Corollary 2.7. By the Leray spectral sequence for the structure morphism $f \colon X \to \mathrm{Spec}\,\mathbb{C}$ (Corollary 2.6) and the vanishing of $\mathrm{Ext}_{\mathrm{MHS}}$ of deree ≥ 2, we get the following commutative diagram:

$$\begin{array}{ccc}
0 & & 0 \\
\downarrow & & \downarrow \\
\mathrm{CH}^r(X)_{\mathrm{hom}} & \xrightarrow{\rho} & \mathrm{Ext}^1_{\mathrm{MHS}}(\mathbb{Q}(0), H^{2r-1}(X, \mathbb{Q}(r))) \\
\downarrow & & \downarrow \\
\mathrm{CH}^r(X) & \longrightarrow & \mathrm{Ext}^{2r}_{\mathrm{MHM}(X)}(\mathbb{Q}_X(0), \mathbb{Q}_X(r)) \\
\downarrow & & \downarrow \\
\mathrm{CH}^r(X)/\mathrm{CH}^r(X)_{\mathrm{hom}} & \longrightarrow & H^{2r}(X, \mathbb{Q}) \cap H^{r,r} \\
\downarrow & & \downarrow \\
0 & & 0.
\end{array} \tag{2.13}$$

Then the top horizontal arrow ρ is defined to be the Abel-Jacobi map.

The Abel-Jacobi maps were originally defined by A. Weil and P. Griffiths. Let us recall their definition.

Let $Z \in \mathrm{CH}^r(X)_{\mathrm{hom}}$ be an algebraic cycle. Z is the boundary of a topological $(2\dim X - 2r + 1)$-cycle Γ: $\partial\Gamma = Z$. Then the Abel-Jacobi class is defined as follows:

$$\rho(Z) = \sum_{j=1}^{g} \left(\int_\Gamma \omega_j \right) \omega_j^* \in J^r(X), \tag{2.14}$$

where $\omega_1, \ldots, \omega_g \in F^{n-r+1}H^{2n-2r+1}(X, \mathbb{C})$ is a basis, and $\omega_1^*, \ldots, \omega_g^*$ which belongs to $H^{2r-1}(X, \mathbb{C})/F^r$ denotes the Serre dual class of those: $\langle \omega_i, \omega_j^* \rangle = \delta_{ij}$.

It is known that the both definitions of Abel-Jacobi maps coincide (cf. [**EZ**], [**J3**]).

3. Arithmetic Hodge Structure

Let X be a nonsingular algebraic variety over $\mathbb{C}$. When we study the algebraic cycles of X (in particular, contained in the kernel of the Abel-Jacobi map), the many difficulties are often caused from the fact of the vanishing of extension groups of mixed Hodge structure of degree ≥ 2.

In this section, we introduce the notion of arithmetic Hodge structure (or more generally, arithmetic Hodge module), which has non-trivial higher extension groups.

3.1. Let X be a quasi-projective nonsingular variety over $\mathbb{C}$. Then X is defined by finitely many equations which possess finitely many coefficients. By considering the coefficients as parameters of a space S, we can obtain a model $f_S \colon X_S \to S$ and the Cartesian diagram:

$$\begin{array}{ccc} X_S & \longleftarrow & X \\ {\scriptstyle f_S}\downarrow & & \downarrow \\ S & \xleftarrow{\ a\ } & \operatorname{Spec}\mathbb{C}, \end{array} \tag{3.1}$$

where S is a nonsingular variety over $\overline{\mathbb{Q}}$, and the map a factors through the generic point $\operatorname{Spec}\overline{\mathbb{Q}}(S) \hookrightarrow S$.

Definition 3.1. We define the Abelian category

$$\underline{\mathrm{M}}(X) = \varinjlim_{X_S} \mathrm{MHM}(X_S).$$

Here $\mathrm{MHM}(X_S)$ is the category of mixed Hodge modules on the variety X_S over $\overline{\mathbb{Q}}$. In the above limit, X_S runs over all models (3.1), and for a morphism $j \colon X_{S'} \to X_S$ of the models, we take the pull-back $j^* \colon \mathrm{MHM}(X_S) \to \mathrm{MHM}(X_{S'})$. We call it the category of *arithmetic Hodge modules*. In particular, we call $\underline{\mathrm{M}}(\mathbb{C}) := \underline{\mathrm{M}}(\operatorname{Spec}\mathbb{C})$ the category of *arithmetic Hodge structures*.

$\underline{\mathrm{M}}(X)$ is an Abelian category. By Theorem 2.3(4) and the remark about Definition 2.5, the category of arithmetic Hodge modules carries the standard functors:

$$f_*, \quad f_!, \quad f^*, \quad f^!, \quad \mathbb{D}, \quad \otimes, \quad \underline{\mathrm{Hom}}.$$

The arithmetic Tate Hodge module $\mathbb{Q}_X(r)$ is defined as the equivalence class of $[\mathbb{Q}_{X_S}(r)]$ of the Tate Hodge module on a model X_S. It does not depend on the choice of a model X_S. It follows from the fact that for any two models X_S and $X_{S'}$, there is another model $X_{S''}$ with morphisms $X_{S''} \to X_S$ and $X_{S''} \to X_{S'}$ of models. We write $H^\bullet(X, \mathbb{Q}(r)) := R^\bullet f_* \mathbb{Q}_X(r)$ (or, simply $H^\bullet(X)(r)$) for the structure morphism $f\colon X \to \operatorname{Spec}\mathbb{C}$.

3.2. The Space of Mumford's Infinitesimal Invariants. Yoneda extension groups in the category of arithmetic Hodge structures are complicated. I do not know its explicit form in general. Here we construct a natural map from the extension groups to the spaces of Mumford's infinitesimal invariants, which have explicit forms.

Let X be a nonsingular projective variety over $\mathbb{C}$. Let us recall the *arithmetic Gauss-Manin connection* on the algebraic de Rham cohomology (cf. [**H**]):

$$\nabla\colon H^q_{\mathrm{dR}}(X/\mathbb{C}) \to H^q_{\mathrm{dR}}(X/\mathbb{C}) \otimes_{\mathbb{C}} \Omega^1_{\mathbb{C}/\overline{\mathbb{Q}}}. \tag{3.2}$$

Let $f\colon X \to \operatorname{Spec}\mathbb{C}$ be the structure morphism. The exact sequence

$$0 \to f^*\Omega^1_{\mathbb{C}/\overline{\mathbb{Q}}} \to \Omega^1_{X/\overline{\mathbb{Q}}} \to \Omega^1_{X/\mathbb{C}} \to 0$$

induces the following exact sequence

$$0 \to \Omega^{\bullet-1}_{X/\mathbb{C}} \otimes f^*\Omega^1_{\mathbb{C}/\overline{\mathbb{Q}}} \to \Omega^\bullet_{X/\overline{\mathbb{Q}}}/U^2 \to \Omega^\bullet_{X/\mathbb{C}} \to 0, \tag{3.3}$$

where $U^2 := \operatorname{Image}(\Omega^{\bullet-2}_{X/\overline{\mathbb{Q}}} \otimes f^*\Omega^2_{\mathbb{C}/\overline{\mathbb{Q}}} \to \Omega^\bullet_{X/\overline{\mathbb{Q}}})$. Applying Rf_* on (3.3), we obtain the coboundary map

$$H^q_{\mathrm{dR}}(X/\mathbb{C}) := R^q f_* \Omega^\bullet_{X/\mathbb{C}} \to R^q f_*(\Omega^{\bullet-1}_{X/\mathbb{C}} \otimes f^*\Omega_{\mathbb{C}/\overline{\mathbb{Q}}}) \simeq H^q_{\mathrm{dR}}(X/\mathbb{C}) \otimes \Omega^1_{\mathbb{C}/\overline{\mathbb{Q}}}$$

(where the last isomorphism is the projection formula), and hence the arithmetic Gauss-Manin connection (3.2).

By definition, the arithmetic Gauss-Manin connection (3.2) satisfies the Griffiths transversality

$$\nabla(F^p) \subset F^{p-1} \otimes \Omega^1_{\mathbb{C}/\overline{\mathbb{Q}}},$$

where $F^s = F^s H^q_{\mathrm{dR}}(X/\mathbb{C}) = H^q(X, \Omega^{\bullet \geq s}_{X/\mathbb{C}})$ denotes the Hodge filtration.

DEFINITION 3.2. Let p, q, r be integers. Then $\Xi^{p,q}_X(r)$ is defined as the cohomology at the middle term of the following complex induced from the arithmetic Gauss-Manin connection (3.2):

$$F^{p+1}H^{p+q}_{\mathrm{dR}}(X/\mathbb{C}) \otimes \Omega^{r-1}_{\mathbb{C}/\overline{\mathbb{Q}}} \xrightarrow{\nabla} F^p H^{p+q}_{\mathrm{dR}}(X/\mathbb{C}) \otimes \Omega^r_{\mathbb{C}/\overline{\mathbb{Q}}} \xrightarrow{\nabla} F^{p-1}H^{p+q}_{\mathrm{dR}}(X/\mathbb{C}) \otimes \Omega^{r+1}_{\mathbb{C}/\overline{\mathbb{Q}}}.$$

We construct a natural map from the extension groups in $\underline{\mathrm{M}}(\mathbb{C})$ to the space $\Xi^{p,q}_X(r)$.

Let $X_S \to S$ be the model of X (3.1). Recall the category $\mathrm{MHM}(S)$ is defined to be a full subcategory of the fiber product $\mathrm{MFW}_{\mathrm{rh}}(S) \times_{\mathrm{MFW}_{\mathrm{rh}}(S_{\mathbb{C}})} \mathrm{MHM}(S_{\mathbb{C}})$ such that the polarization on each weight quotient is defined over $\overline{\mathbb{Q}}$. There is the forgetful functor $\mathrm{MHM}(S) \to \mathrm{MF}_{\mathrm{rh}}(S)$, which is exact by Theorem 2.3(2):

$$\mathrm{MHM}(S) \to \mathrm{MF}_{\mathrm{rh}}(S), \quad (K^\bullet_{\mathbb{Q}}, M, F, W) \mapsto (M, F). \tag{3.4}$$

Therefore we have the well-defined map of Yoneda extension groups:

$$\operatorname{Ext}^p_{\mathrm{MHM}(S)}(\mathbb{Q}_S(0), R^q f_* \mathbb{Q}_{X_S}(r)) \to \operatorname{Ext}^p_{\mathrm{MF}_{\mathrm{rh}}(S)}(\mathcal{O}_S, H^q_{\mathrm{dR}}(X_S/S)(r)) \tag{3.5}$$

Note that $\mathrm{MF}_{\mathrm{rh}}(S)$ is an exact category (see. §2.2), and we can define its derived category $D^b(\mathrm{MF}_{\mathrm{rh}}(S))$ (cf. [**BBD**] 1.1). Then we can describe $\operatorname{Ext}^p_{\mathrm{MF}_{\mathrm{rh}}(S)}(A, B) =$

$\mathrm{Hom}_{D^b(\mathrm{MF}_{\mathrm{rh}}(S))}(A, B[p])$. We will take the inductive limit over S of the right hand side in (3.5). So we may assume $S = \operatorname{Spec} \mathcal{O}$ where $\mathcal{O}$ is a regular local ring over $\overline{\mathbb{Q}}$ with regular parameter $x_1, \dots, x_n$. Then we can see $D_S = \mathcal{O}[\partial_1, \dots, \partial_n]$ where ∂_i is the derivation on $\mathcal{O}$ over $\overline{\mathbb{Q}}$ such that $[\partial_i, x_j] = \delta_{ij}$. Let $D_S(m)$ be the filtered D-module $(D_S, F_{\bullet+m})$ with the differential order filtration shifted by m.

LEMMA 3.3. $\mathrm{Ext}^p_{\mathrm{MF}_{\mathrm{rh}}(S)}(D_S(m), M) = 0 \quad \forall p \geq 1.$

PROOF. Standard. □

There is the Koszul resolution of $\mathcal{O}_S$

$$0 \to D_S(-n) \otimes_{\overline{\mathbb{Q}}} \overset{n}{\wedge} V \to \cdots \to D_S(-1) \otimes_{\overline{\mathbb{Q}}} V \to D_S \to \mathcal{O}_S \to 0, \tag{3.6}$$

where V is a n-dimensional $\overline{\mathbb{Q}}$-vector space with a basis $e_1, \dots, e_n$, and the boundary map

$$D_S(-p) \otimes_{\overline{\mathbb{Q}}} \overset{p}{\wedge} V \to D_S(-p+1) \otimes_{\overline{\mathbb{Q}}} \overset{p-1}{\wedge} V$$

is given by

$$\xi \otimes e_{i_1} \wedge \cdots \wedge e_{i_p} \mapsto \sum_{j=1}^{p} (-1)^j \xi \cdot \partial_{i_j} \otimes e_{i_1} \wedge \cdots \wedge \hat{e}_{i_j} \wedge \cdots \wedge e_{i_p}.$$

It is also the exact sequence in the exact category $\mathrm{MF}_{\mathrm{rh}}(S)$.

Using the exact sequence (3.6) and applying Lemma 3.3, we can see that the right hand side of (3.5) is isomorphic to the cohomology at the middle term of the following complex induced from the algebraic Gauss-Manin connection:

$$F^{r-p+1} H^q_{\mathrm{dR}}(X_S/S) \otimes \Omega^{p-1}_{S/\overline{\mathbb{Q}}} \xrightarrow{\nabla} F^{r-p} H^q_{\mathrm{dR}}(X_S/S) \otimes \Omega^p_{S/\overline{\mathbb{Q}}} \xrightarrow{\nabla} F^{r-p-1} H^q_{\mathrm{dR}}(X_S/S) \otimes \Omega^{p+1}_{S/\overline{\mathbb{Q}}}.$$

Passing to the limit over S, we thus have:

PROPOSITION 3.4. *Let X be a nonsingular projective variety over $\mathbb{C}$. Then there is the following natural map:*

$$\mathrm{Ext}^p_{\underline{\mathrm{M}}(\mathbb{C})}(\mathbb{Q}(0), H^q(X)(r)) \to \Xi_X^{r-p,q-r+p}(p).$$

DEFINITION 3.5. Let p, q, r be integers. The space of *Mumford's infinitesimal invariants* $\Lambda_X^{p,q}(r)$ is defined as the cohomology at the middle term of the following complex induced from the arithmetic Gauss-Manin connection (3.2):

$$H^{q-1}(\Omega^{p+1}_{X/\mathbb{C}}) \otimes \Omega^{r-1}_{\mathbb{C}/\overline{\mathbb{Q}}} \xrightarrow{\overline{\nabla}} H^q(\Omega^p_{X/\mathbb{C}}) \otimes \Omega^r_{\mathbb{C}/\overline{\mathbb{Q}}} \xrightarrow{\overline{\nabla}} H^{q+1}(\Omega^{p-1}_{X/\mathbb{C}}) \otimes \Omega^{r+1}_{\mathbb{C}/\overline{\mathbb{Q}}}.$$

Clearly, there is the natural map,

$$\Xi_X^{p,q}(r) \to \Lambda_X^{p,q}(r). \tag{3.7}$$

By the similar argument to the above, we can see

$$\varinjlim_{X_S} \mathrm{Ext}^p_{\text{graded } \mathrm{gr}^F D_S\text{-mod}}(\mathcal{O}_S, \mathrm{gr}^F H^q_{\mathrm{dR}}(X_S/S)) \simeq \Lambda_X^{r-p,q-r+p}(p),$$

and the map (3.7) is compatible to the map between Yoneda extension groups induced from the exact functor $\mathrm{MF}_{\mathrm{rh}}(S) \to \{\text{graded } \mathrm{gr}^F D_S\text{-mod}\}$, $(M, F) \mapsto \mathrm{gr}^F M$.

Together with Proposition 3.4, we thus have:

PROPOSITION 3.6. *Let X be a nonsingular projective variety over $\mathbb{C}$. Then we have the natural map from the extension groups in $\underline{\mathrm{M}}(\mathbb{C})$ to the space of Mumford's infinitesimal invariants as the composition of the map in Proposition* 3.4 *and* (3.7)*:*

$$\mathrm{Ext}^p_{\underline{\mathrm{M}}(\mathbb{C})}(\mathbb{Q}(0), H^q(X)(r)) \to \Lambda_X^{r-p,q-r+p}(p).$$

4. Algebraic Cycles and Extensions of Arithmetic Hodge Structure

In this section, we study the extension groups in the Abelian category $\underline{\mathrm{M}}(\mathbb{C})$ and algebraic cycles, in particular, the Bloch conjecture.

4.1. Higher Abel-Jacobi Maps. Let X be a nonsingular variety over $\mathbb{C}$. We denote its higher Chow group by $\mathrm{CH}^r(X,m)$. Then we have the following map

$$(4.1)\qquad \begin{array}{ccc} \varinjlim_{X_S} \mathrm{CH}^r(X_S,m) & \longrightarrow & \varinjlim_{X_S} \mathrm{Ext}^{2r-m}_{\mathrm{MHM}(X_S)}(\mathbb{Q}_{X_S}(0), \mathbb{Q}_{X_S}(r)) \\ \| & & \| \\ \mathrm{CH}^r(X,m) & \longrightarrow & \mathrm{Ext}^{2r-m}_{\underline{\mathrm{M}}(X)}(\mathbb{Q}_X(0), \mathbb{Q}_X(r)), \end{array}$$

where the top horizontal arrow is the one in Corollary 2.7. We denote the above map by $c = c^r(X,m)$. On the other hand, there is the Leray spectral sequence

$$(4.2)\qquad E_2^{pq} = \mathrm{Ext}^p_{\underline{\mathrm{M}}(\mathbb{C})}(\mathbb{Q}(0), H^q(X)(r)) \Longrightarrow \mathrm{Ext}^{p+q}_{\underline{\mathrm{M}}(X)}(\mathbb{Q}_X(0), \mathbb{Q}_X(r)),$$

as the inductive limit of the Leray spectral sequence for the model X_S (Corollary 2.6). It degenerates at E_2 terms if X is projective. The spectral sequence (4.2) defines the Leray filtration on the extension group $\mathrm{Ext}^\bullet_{\underline{\mathrm{M}}(X)}(\mathbb{Q}_X(0), \mathbb{Q}_X(r))$, which we denote by $F^\bullet$.

DEFINITION 4.1. Let X be a nonsingular projective variety over $\mathbb{C}$. We define the filtration on the higher Chow group as follows:

$$F^\nu \mathrm{CH}^r(X,m) := c^{-1} F^\nu \mathrm{Ext}^{2r-m}_{\underline{\mathrm{M}}(X)}(\mathbb{Q}_X(0), \mathbb{Q}_X(r)).$$

Then, together with the E_2-degeneration of the spectral sequence (4.2), we obtain:

$$\rho_X^\nu \colon \mathrm{Gr}^\nu_F \mathrm{CH}^r(X,m) \to \mathrm{Ext}^\nu_{\underline{\mathrm{M}}(\mathbb{C})}(\mathbb{Q}(0), H^{2r-m-\nu}(X)(r)).$$

We call the above the *νth higher Abel-Jacobi map.*

By definition, we can easily show Theorem 1.1(1) and (2). For example, Theorem 1.1(2)(d) can be shown by using the argument in [**Mur2**]. Theorem 1.1(2)(e) follows from the hard Lefschetz theorem in the terminology of arithmetic Hodge structure. (See [**A**] for details.)

EXAMPLE 4.2. Let $r = \dim X = n$, $\nu = 2$ and $m = 0$. Then the higher Abel-Jacobi map gives the following map:

$$(4.3)\qquad \rho_X^2 \colon T(X) \to \mathrm{Ext}^2_{\underline{\mathrm{M}}(\mathbb{C})}(\mathbb{Q}(0), H^{2n-2}(X)(n)).$$

We call the above (4.3) the *second Albanese map.*

4.2. Mumford's Infinitesimal Invariants. Composing the higher Abel-Jacobi map and the maps in Propositions 3.4 and 3.6, we have Theorem 1.1(3):

$$\xi_X^\nu\colon \operatorname{Gr}_F^\nu \operatorname{CH}^r(X,m) \to \Xi_X^{r-\nu,r-m}(\nu),$$
$$\delta_X^\nu\colon \operatorname{Gr}_F^\nu \operatorname{CH}^r(X,m) \to \Lambda_X^{r-\nu,r-m}(\nu).$$

When $m=0$, the above maps have already been defined in [**AS**].

We know the map of infinitesimal invariants δ_X is not zero in general. More strongly, the following holds.

THEOREM 4.3 ([**AS**, Theorem (0-6)]). *Let $X \subset \mathbb{P}_{\mathbb{C}}^{m+s}$ be a nonsingular complete intersection variety of degree $(d_1,\dots,d_s)$, with a sufficiently general coefficients. Assume that one of the following conditions holds.*

(1) *$m \geq 2$ and $\min\{d_i\}\cdot s \geq 2m+s$.*
(2) *$m=1$ and $\min\{d_i\}\cdot s \geq 3+s$.*

Then the map of infinitesimal invariant induces an injective map

$$\delta_X^m\colon F^m \operatorname{CH}^m(X)_{\mathrm{gen}} \to \Lambda_X^{0,m}(m). \tag{4.4}$$

Here $F^m \operatorname{CH}^m(X)_{\mathrm{gen}}$ denotes the subgroup of $F^m \operatorname{CH}^m(X)$ generated by 0-cycles in general positions (loc. cit., Chapter II, Definition (5-4-1)). *In particular, we have*

$$F^{m+1} \operatorname{CH}^m(X)_{\mathrm{gen}} = 0.$$

REMARK 4.4. For a nonsingular complete intersection X of dimension m, it is easy to see that

$$F^2 \operatorname{CH}^m(X) = \cdots = F^m \operatorname{CH}^m(X) = T(X)$$

and

$$\delta_X^m(F^{m+1} \operatorname{CH}^m(X)) = 0,$$

where $F^\bullet$ denotes either our filtration or S. Saito's filtration ([**AS**, Chapter II, §2], [**SaS1**], [**SaS2**]). It is conjectured that our filtration coincides with S. Saito's filtration. But I do not have the proof of it.

PROOF (SKETCH). We prove the assertion by the induction of $\dim X$.

We first show the map

$$\delta_C^1\colon F^1 \operatorname{CH}^1(C) \to \Lambda_C^{0,1}(1). \tag{4.5}$$

is injective for a general complete intersection curve $C \subset \mathbb{P}_{\mathbb{C}}^{s+1}$ of degree $(d_1,\dots,d_s)$, which satisfies $\min\{d_i\}\cdot s \geq 3+s$. Let $f\colon C_S \to S$ be a model of C, where we may assume that there is a dominant morphism $S \to \mathcal{M}$ to the moduli space of complete intersection curves of degree $(d_1,\dots,d_s)$, because the coefficients of C is general. Let $\Lambda_{C_S/S}^{0,1}(1) := \Gamma(S, R^1 f_*\mathcal{O}_{C_S} \otimes \Omega^1_{S/\overline{\mathbb{Q}}}/\overline{\nabla}(f_*\Omega^1_{C_S/S}))$. Then the map (4.5) can be obtained by the inductive limit over S of the map $F_S^1 \operatorname{CH}^1(C_S) \to \Lambda_{C_S/S}^{0,1}(1)$. Here $F_S^\bullet \operatorname{CH}^\bullet(C_S)$ denotes the relative filtration on the Chow group (cf. [**AS**, Chapter II, §2]). Let $\Lambda_{C_S^{\mathrm{an}}/S^{\mathrm{an}}}^{0,1}(1) := \Lambda_{C_S/S}^{0,1}(1) \otimes \mathcal{O}_S^{\mathrm{an}} \simeq \Gamma(S^{\mathrm{an}}, R^1 f_*\mathcal{O}_{C_S} \otimes \Omega^1_{S^{\mathrm{an}}/\mathbb{C}}/\overline{\nabla}(f_*\Omega^1_{C_S^{\mathrm{an}}/S^{\mathrm{an}}}))$. In order to prove the injectivity of (4.5), it suffices to show that if a cycle $z \in F_S^1 \operatorname{CH}^1(C_S)$ has a zero-image under the following map

$$\delta_S^1\colon F_S^1 \operatorname{CH}^1(C_S) \to \Lambda_{C_S^{\mathrm{an}}/S^{\mathrm{an}}}^{0,1}(1), \tag{4.6}$$

then $z|_C$ is rationally equivalent to zero. Since

$$f_*\Omega^1_{C_S^{\mathrm{an}}/S} \otimes \Omega^1_{S^{\mathrm{an}}/\mathbb{C}} \to R^1 f_*\mathcal{O}_{C_S} \otimes \Omega^2_{S^{\mathrm{an}}/\mathbb{C}}$$

is injective by the symmetrizer lemma for complete intersections (loc. cit., Chapter I, Corollary (4-5)), we have the injection

$$\Xi^{0,1}_{C_S^{\mathrm{an}}/S^{\mathrm{an}}}(1) := \Gamma(S^{\mathrm{an}}, (R^1 f_*\Omega^\bullet_{C_S^{\mathrm{an}}/S^{\mathrm{an}}}) \otimes \Omega^1_{S^{\mathrm{an}}/\mathbb{C}}/\nabla(f_*\Omega^1_{C_S^{\mathrm{an}}/S^{\mathrm{an}}})) \hookrightarrow \Lambda^{0,1}_{C_S^{\mathrm{an}}/S^{\mathrm{an}}}(1).$$

There is the following exact sequence

$$0 \to \Gamma(S^{\mathrm{an}}, R^1 f_*\mathbb{C}_{C_S^{\mathrm{an}}}/R^1 f_*\mathbb{Q}_{C_S^{\mathrm{an}}}) \to \Gamma(S^{\mathrm{an}}, J(C_S/S)) \to \Xi^{0,1}_{C_S^{\mathrm{an}}/S^{\mathrm{an}}}(1),$$

where we put $J(C_S/S) := R^1 f_*\Omega^\bullet_{C_S^{\mathrm{an}}/S^{\mathrm{an}}}/f_*\Omega^1_{C_S^{\mathrm{an}}/S^{\mathrm{an}}} + R^1 f_*\mathbb{Q}_{C_S^{\mathrm{an}}}$. It is easy to see that $\Gamma(S^{\mathrm{an}}, R^1 f_*\mathbb{C}_{C_S^{\mathrm{an}}}) = 0$ and hence $\Gamma(S^{\mathrm{an}}, R^1 f_*\mathbb{C}_{C_S^{\mathrm{an}}}/R^1 f_*\mathbb{Q}_{C_S^{\mathrm{an}}}) = 0$ because of the dominant morphism $S \to \mathcal{M}$. Thus it follows from the Abel's theorem that if $\delta_S(z) = 0$, then $z|_{f^{-1}(t)} = 0$ for any $t \in S(\mathbb{C})$, in particular, $z|_C = 0$.

Next we prove for any dimensional case. Let X be as in Theorem 4.3 and $m \geq 2$. let

$$X = X_m \supset X_{m-1} \supset \cdots \supset X_2 \supset X_1 = C$$

be a sequence of general hyperplane sections, in which the multi-degree of each X_i satisfies the condition in Theorem 4.3. Then there is the commutative diagram

$$\begin{array}{ccc}
F^m \operatorname{CH}^m(X) & \xrightarrow{\delta^m_X} & \Lambda^{0,m}_X(m) \\
\uparrow & & \uparrow \alpha_{m-1} \\
F^{m-1} \operatorname{CH}^{m-1}(X_{m-1}) & \xrightarrow{\delta^{m-1}_{X_{m-1}}} & \Lambda^{0,m-1}_{X_{m-1}}(m-1) \\
\uparrow & & \uparrow \alpha_{m-2} \\
\vdots & & \vdots \\
\uparrow & & \uparrow \alpha_1 \\
F^1 \operatorname{CH}^1(C) & \xrightarrow{\delta^1_C} & \Lambda^{0,1}_C(1).
\end{array}$$

By the above, δ^1_C is injective. Moreover, each α_i is also injective by the symmetrizer lemma for open complements (loc. cit., Chapter I, Theorem (4-4)). Put

$$F^m \operatorname{CH}^m(X)_{\mathrm{gen}} := \operatorname{Image}\Big(\sum_C F^1 \operatorname{CH}^1(C) \to F^m \operatorname{CH}^m(X)\Big)$$

where C runs over all curves as above. Then we have that the map (4.4)

$$\delta^m_X \colon F^m \operatorname{CH}^m(X)_{\mathrm{gen}} \to \Lambda^{0,m}_X(m)$$

is injective. □

Mumford's infinitesimal invariants have the advantage of explicit calculations. But it is a "coarse" invariant, that is, there are several examples of an algebraic cycle which is not rationally equivalent to zero, but its infinitesimal invariant is zero (cf. [**G2**]). We hope that our higher Abel-Jacobi invariants capture all cycles (cf. Conjecture 4.7 below). In fact, there is an example of a 0-cycle on a surface such that its infinitesimal invariant vanishes, but the higher Abel-Jacobi invariant does not:

THEOREM 4.5. *Let C be a projective nonsingular curve over $\overline{\mathbb{Q}}$ with genus $g \geq 2$, such that* $\operatorname{rank} \mathrm{NS}(C \times C) = 3$.[4] *Let $O \in C(\overline{\mathbb{Q}})$ be a $\overline{\mathbb{Q}}$-valued point such that $K_C - (2g-2) \cdot O$ is not $\mathbb{Q}$-linearly equivalent to 0.* [5] *Let $P \in C(\mathbb{C}) \setminus C(\overline{\mathbb{Q}})$ be any $\mathbb{C}$-valued point which is not $\overline{\mathbb{Q}}$-valued one. Put $X := C_{\mathbb{C}} \times C_{\mathbb{C}}$ and $z := (P,P) - (P,O) - (O,P) + (O,O) \in T(X)$. Then we have*

$$\xi_X^2(z) = 0 \quad in \quad \Xi_X^{0,2}(2),$$

but

$$\rho_X^2(z) \neq 0 \quad in\, \operatorname{Ext}^2_{\underline{\mathrm{M}}(\mathbb{C})}(\mathbb{Q}(0), H^2(X)(2)).$$

The proof of the above theorem will appear in the forthcoming paper [**A**].

4.3. Bloch Conjecture. Recall the famous *Bloch conjecture* [**B1**, Lecture 1]:

CONJECTURE 4.6 (Bloch). *Let X be a projective nonsingular surface over $\mathbb{C}$ with the geometric genus $p_g = \dim H^0(K_{X/\mathbb{C}}) = 0$. Then the (usual) Albanese map*

$$A_0(X) \to \mathrm{Alb}(X)$$

is injective.

This conjecture was observed by S. Bloch with an ingenious insight, and it led to the Bloch-Beilinson formula (1.1). The Bloch conjecture was proved for surfaces which are not of general type [**BKL**]. But, in general, it is still open.

Now we conjecture:

CONJECTURE 4.7. *The second Albanese map* (4.3)

$$\rho_X^2 \colon T(X) \to \operatorname{Ext}^2_{\underline{\mathrm{M}}(\mathbb{C})}(\mathbb{Q}(0), H^2(X)(2))$$

is injective for any projective nonsingular surface X (not necessarily $p_g = 0$).

Or, more generally

CONJECTURE 4.8. *The cycle map $c = c^r(X,m)$ in* (4.1) *is injective for any projective nonsingular variety X over $\mathbb{C}$.*

Conjecture 4.7 implies the Bloch conjecture (use the method in [**B1**,Lecture 1]). Moreover Conjecture 4.8 also has an application to algebraic cycles on varieties over an algebraic number field. Let us explain it.

Let $X_{\overline{\mathbb{Q}}}$ be a nonsingular projective variety over $\overline{\mathbb{Q}}$. By the construction, the higher Abel-Jacobi map factors as follows:

$$\mathrm{Gr}_F^{\nu} \mathrm{CH}^r(X_{\overline{\mathbb{Q}}}) \to \operatorname{Ext}^{\nu}_{\mathrm{MHM}(\overline{\mathbb{Q}})}(\mathbb{Q}(0), H^{2r-\nu}(X)(r)) \to \operatorname{Ext}^{\nu}_{\underline{\mathrm{M}}(\mathbb{C})}(\mathbb{Q}(0), H^{2r-\nu}(X)(r)).$$

Here $F^{\nu} \mathrm{CH}^r(X_{\overline{\mathbb{Q}}})$ denotes the subgroup of $F^{\nu} \mathrm{CH}^r(X_{\mathbb{C}})$ generated by cycles of codimension r defined over $\overline{\mathbb{Q}}$. Since the arithmetic Gauss-Manin connection in $\mathrm{MHM}(\overline{\mathbb{Q}})$ vanishes, we can show any extension groups in $\mathrm{MHM}(\overline{\mathbb{Q}})$ vanishes if $\nu \geq 2$ (cf. Theorem 2.10(2)). Then, due to Conjecture 4.8, we have

$$F^2 \mathrm{CH}^r(X_{\overline{\mathbb{Q}}}) = 0.$$

[4]The proof of the existence of such a curve was taught from S. Mochizuki (the private communication with the author).

[5]There are infinitely many such points O by Raynaud's theorem [**R**].

In particular,

$$T(X_{\overline{\mathbb{Q}}}) = 0.$$

Let X_k be a nonsingular projective variety over an algebraic number field k. Due to the injectivity of the map $\mathrm{CH}_0(X_k) \to \mathrm{CH}_0(X_{\overline{\mathbb{Q}}})$ (note that the Chow groups are always tensored with $\mathbb{Q}$) , we also have $T(X_k) = 0$. Combined with the Mordell-Weil theorem, we obtain the finiteness of the rank of the Chow group:

$$\operatorname{rank} \mathrm{CH}_0(X_k) < \infty.$$

THEOREM 4.9. (1) *Conjecture* 4.7 *implies the Bloch conjecture* 4.6.

(2) *Let X_k be a nonsingular projective variety over an algebraic number field k. Assume that Conjecture* 4.8 *holds. Then we have $F^2\,\mathrm{CH}^r(X_{\overline{\mathbb{Q}}}) = 0$. In particular, it implies that the rank of* $\mathrm{CH}_0(X_k)$ *is finite.*

Further we can show the 'almost' converse of Theorem 4.9(2) for $r = 2$:

THEOREM 4.10 ([**A**]). *Assume the followings hold:*

(1) *The Abel-Jacobi map (cf.* §2.5)

$$\rho\colon\ \mathrm{CH}^2(X_{\overline{\mathbb{Q}}})_{\mathrm{hom}} \to J^2(X_{\mathbb{C}})$$

is injective for any projective nonsingular variety $X_{\overline{\mathbb{Q}}}$ of arbitrary dimension over $\overline{\mathbb{Q}}$. (Here we put $X_{\mathbb{C}} = X_{\overline{\mathbb{Q}}} \otimes \mathbb{C}$.)

(2) *Let $X_{\overline{\mathbb{Q}}}$ and ρ be as above. Let $z \in \mathrm{CH}^2(X_{\mathbb{C}})_{\mathrm{hom}}$ be any algebraic cycle. Then $\rho(z) = 0$ if and only if $\rho(z^\sigma) = 0$ for any $\sigma \in \mathrm{Aut}(\mathbb{C}/\overline{\mathbb{Q}})$. (Note that $\mathrm{Aut}(\mathbb{C}/\overline{\mathbb{Q}})$ acts on $\mathrm{CH}^2(X_{\mathbb{C}})_{\mathrm{hom}}$.)*

Then Conjecture 4.8 *for $r = 2$ and $m = 0$ is true. In particular,* (1) *and*(2) *implies the Bloch conjecture* 4.6.

Theorem 4.10(1) is nothing but the conjecture due to A. Beilinson. Theorem 4.10(2) should be true if $F^2_{\mathcal{M}} = \ker \rho$.

4.4. Finally we propose the following problem about surjectivity of the cycle map.

Let X be a projective nonsingular variety over $\mathbb{C}$, and put

$$J^r(X)_{\mathrm{arith}} = \mathrm{Image}(\mathrm{Ext}^1_{\underline{\mathrm{M}}(\mathbb{C})}(\mathbb{Q}(0), H^{2r-1}(X)(r)) \to J^r(X)_{\mathbb{Q}})$$

for $r \geq 2$ (cf. §2.5).

PROBLEM 4.11. Is the Abel-Jacobi map

$$\mathrm{CH}^r(X)_{\mathrm{hom}} \to J^r(X)_{\mathrm{arith}}$$

surjective?

In general, the Abel-Jacobi map (2.11) is not surjective. For example, when the Hodge number $\sum_{k\geq r+1} h^{k,2r-1-k}(X) \neq 0$, it is not so (or, more strong result, see [**G1**]). It is an interesting problem to give Hodge theoretic characterization of its image.

References

[A] M. Asakura, *Arithmetic Hodge structure and higher Abel-Jacobi maps*, (in preparation).

[AS] M. Asakura and S. Saito, *Filtration on Chow groups and higher Abel-Jacobi maps*, (preprint).

[BBD] A. A. Beilinson, J. Bernstein, and P. Deligne, *Faisceaux pervers*, Analysis and Topology on Singular Spaces. I (Luminy, 1981) Astérisque, vol. 100, Soc. Math. France, Paris, 1983, pp. 5–171.

[B1] S. Bloch, *Lectures on algebraic cycles*, Duke Univ. Math. Ser., vol. 4, Duke Univ. Durham, NC, 1980.

[B2] ———, *Algebraic cycles and higher K-theory*, Adv. in Math. **61** (1986), 267–304.

[BKL] S. Bloch, A. Kas and D. Lieberman, *Zero cycles on surfaces with* $p_g = 0$, Compositio Math. **33** (1976), 135–145.

[Ca] J. Carlson, *Extensions of mixed Hodge structures*, Journées de geométrie algébrique d'Angers 1979, Sijthoff & Noordhoff, Alphen aan den Rijn-Germantown, MD, 1980, pp.107–127.

[D1] P. Deligne, *Théorie de Hodge*. II, Inst. Hautes tudes Sci. Publ. Math. **40** (1971), 5–57.

[D2] ———, *Théorie de Hodge*. III, Inst. Hautes tudes Sci. Publ. Math. **44** (1975), 5–77.

[D3] ———, *Equations différentielles à points singuliers réguliers*, Lecture Notes in Math., vol.163, Springer-Verlag, Berlin-New York, 1970.

[DMOS] P. Deligne, J. S. Milne, A . Ogus, and K.-Y. Shih, *Hodge cycles, motives and Shimura varieties*, Lecture Notes in Math., vol. 900, Springer-Verlag, Berlin-New York, 1982.

[EZ] F. El Zein and S. Zucker, *Extendability of normal functions associated to algebraic cycles*, Topics in tRanscendental Algebraic Geometry (Princeton, NJ, 1981/82) (P. Griffiths, ed.) Ann. of Math. Stud., vol. 106, Princeton Univ. Press, Princeton, NJ, 1984, pp. 269–288.

[G1] M. L. Green, *Griffiths infinitesimal invariant and the Abel-Jacobi map*, J. Differential. Geom. **29** (1989), 545–555.

[G2] ———, *Algebraic cycles and Hodge theory*, (these proceedings).

[H] R. Hartshorne, *On the de Rham cohomology of algebraic varieties*, Inst. Hautes tudes Sci. Publ. Math. **45** (1975), 5–99.

[J1] U. Jannsen, *Mixed motives and algebraic K-theory*, Lecture Notes in Math., vol. 1400, Springer-Verlag, Berlin , 1990.

[J2] ———, *Motivic sheaves and filtration on Chow groups*, Motives. I (Seattle, WA, 1991), Proc. Sympos. Pure Math. vol. 55, Amer.Math. Soc., Providence, RI, 1994, pp. 245–302.

[J3] ———, *Deligne homology, Hodge-$\mathcal{D}$-conjecture, and motives*, Beilinson's Conjectures on Special Values of L-functions, Perspect. Math., vol. 4, Academic Press, Boston, MA, 1988, pp. 305–372.

[K] M. Kashiwara, *A study of variation of mixed Hodge structure*, Publ. Res. Inst. Math. Sci. **22** (1986), 991–1024.

[Me] Z. Mebkhout, *Les formalisme des six opérations de Grothendieck pour les $\mathcal{D}_X$-modules cohérents*, Travaux en Cours, vol. 35, Hermann, Paris 1989.

[M] D. Mumford, *Rational equivalence of 0-cycles on surfaces*, J. Math. Kyoto Univ. **9** (1969), 195-204.

[Mur1] J. P. Murre, *On a conjectural filtration on Chow groups of an algebraic variety*. I. *The general conjectures and some examples*, Indag. Math. (N.S.), **4** (1993), 177–188;II. Verification of the conjectures for threefolds which are the product on a surface and a curve, 189–201.

[Mur2] ———, *On the motives of an algebraic surface*, J. Reine. Angew. Math. **409** (1990), 190–204.

[R] M. Raynaud, *Courbes sur une variété abélienne et points de torsion*, Invent. Math. **71** (1983), 207–233.

[SaM1] M. Saito, *Modules de Hodge polarisables*, Publ. Res. Inst. Math. Sci. **24** (1988), 849–995.

[SaM2] ———, *Mixed Hodge modules*, Publ. Res. Inst. Math. Sci. **26** (1990), 221–333.

[SaM3] ———, *On the formalism of mixed sheaves* (RIMS preprint) .

[SaS1] S. Saito, *Motives and filtration on Chow groups*, Invent. Math. **125** (1996), 149–196.

[SaS2] ———, *Motivic filtration on Chow groups*. II (these proceedings).

[SZ] J. Steenbrink and S. Zucker, *Variation of mixed Hodge structure*, Invent. Math. **80** (1985), 489–542 .

Research Institute for Mathematical Sciences, Kyoto University, Oiwake-cho, Sakyo-ku, Kyoto, 606-8502, Japan

Current address: Graduate School of Mathematics, Kyushu University, Hakozaki Higashi-ku, Fukuoka Japan

E-mail address: asakura@math.kyushu-u.ac.jp

Centre de Recherches Mathématiques
CRM Proceedings and Lecture Notes
Volume **24**, 2000

Hermitian Vector Bundles and Characteristic Classes

José I. Burgos Gil

ABSTRACT. This paper reviews the realization of Beilinson's regulator map using Hermitian differential geometry. This construction is a generalization of Chern-Weil theory of characteristic classes of vector bundles, to higher K-theory.

1. Introduction

Chern classes of complex vector bundles can be constructed using different techniques. For instance, in algebraic topology, one can use the Euler class of an oriented sphere bundle to define the top Chern class and then define the other classes inductively (see [**25**]). Or, in differential geometry, we can use Chern-Weil theory, which produces Chern classes in de Rham cohomology by means of the curvature of a connection (see for example [**5**], [**34**] or [**16**]). Another aproach, due to Grothendieck [**17**], starts with the first Chern class of a line bundle, and then uses an explicit formula for the cohomology of a projective bundle to define higher Chern classes. This approach is very useful in algebraic geometry and can be used to produce Chern classes in the Chow ring, in étale cohomology or in Deligne-Beilinson cohomology.

The Grothendieck method has been generalized by Gillet [**11**] to produce characteristic classes from higher K-theory to any arbitrary cohomology satisfying certain axioms. In the particular case when the characteristic class is the Chern character and the cohomology theory is real Deligne-Beilinson cohomology, the map obtained is Beilinson's regulator map [**2**]. This map is a generalization of Borel's regulator and it is involved in very deep conjectures in Arithmetic Geometry.

Beilinson's regulator is still a very mysterious map, in part because higher K-theory is a rich and complex world. Thus it is useful to have as many approaches to Beilinson's regulator as possible. Gillet and Soulé [**13**] have given a description of Beilinson's regulator for K_1 using Bott-Chern forms. This description can be seen as a generalization of Chern-Weil theory to K_1. In the paper [**8**], S. Wang and the author have extended this description of Beilinson's regulator to higher K-theory.

1991 *Mathematics Subject Classification.* 14C30, 14C25, 14F05, 14C35, 19E15.
Received by the editors September 21, 1998; Accepted October 22, 1998.
Partially supported by Grant PB96-0234.
This is the final form of the paper.

The aim of this paper is to review the construction of characteristic classes for higher K-theory. This paper is meant to be introductory, and so the focus is placed on the basic ideas of the theory. But references are given where the reader can find detailed proofs and general statements.

The main object of study will be algebraic vector bundles over smooth complex algebraic varieties. These objects can be studied from the point of view of complex geometry or from algebraic geometry. Since our objective is to give a complex geometric construction of an object defined in the algebraic geometry setting we will shift from one point of view to the other in different sections.

The plan of the study is as follows. In Section 2, Grothendieck's construction of Chern classes is presented in the particular case of sheaf cohomology with integer coefficients. In Section 3, we recall the Chern-Weil theory of characteristic classes of Hermitian vector bundles. Section 2 is devoted to a simple version of Gillet's construction of characteristic classes for higher K-theory. In Section 5, we review real Deligne cohomology. Bott-Chern forms are the topic of Section 6. In Section 7, we introduce the complex of exact cubes. By a result of R. McCarthy [**24**] the rational homology of this complex is isomorphic to rational K-theory. Finally, Section 8 is devoted to the definition of characteristic classes for higher K-theory using exact cubes of Hermitian vector bundles. For simplicity, we only discuss the case of projective varieties. But note that an important ingredient in the comparison between this construction and Gillet's construction, is the extension of this theory to quasi-projective varieties.

2. Chern Classes of Vector Bundles

There are many different constructions of Chern classes of vector bundles (see for example [**25**], [**34**] or [**18**]). In this section we will review a very general one due to Grothendieck [**17**]. In this construction, the properties of the cohomology theory that are needed to define Chern classes are given as axioms for a cohomology. These axioms are satisfied by many theories, for instance, Chow rings of algebraic varieties. Moreover Grothendieck's construction is the basis of Gillet's construction of Chern classes for higher algebraic K-theory [**11**].

In this section, we will specialize the construction of Chern classes to the case of sheaf cohomology of smooth complex varieties with integer coefficients. We will use the classical topology. To stress this point we will work with holomorphic vector bundles. In Section 4 we will discuss the axiomatic approach in the algebraic geometry context.

Let us introduce the first Chern class of a holomorphic line bundle. This will act as a normalization for the Chern classes. Let X be a complex manifold, and let $\mathcal{O}_X$ be the sheaf of holomorphic functions on X. Let $\mathcal{O}_X^*$ be the sheaf of invertible holomorphic functions. Then there is an isomorphism

$$\left\{\begin{matrix}\text{Isomorphism classes}\\ \text{of holomorphic}\\ \text{line bundles}\end{matrix}\right\} \longleftrightarrow H^1(X, \mathcal{O}_X^*).$$

The exponential sequence

$$0 \to (2\pi i)\mathbb{Z} \to \mathcal{O}_X \xrightarrow{\exp} \mathcal{O}_X^* \to 0$$

gives us a long exact sequence in cohomology

$$H^1(X,\mathcal{O}_X) \to H^1(X,\mathcal{O}_X^*) \xrightarrow{\delta} H^2(X,\mathbb{Z}(1)) \to H^2(X,\mathcal{O}_X),$$

where we have written $\mathbb{Z}(1)$ for $(2\pi i)\mathbb{Z} \subset \mathbb{C}$.

DEFINITION 2.1. The *first Chern class* of a line bundle $\mathcal{L}$, denoted $c_1(\mathcal{L}) \in H^2(X,\mathbb{Z}(1))$, is the image of its class in $H^1(X,\mathcal{O}_X^*)$ by the connection morphism δ.

Once we have defined the first Chern class, the main tool to define higher Chern classes will be the Dold-Thom isomorphism. Let E be a rank n holomorphic vector bundle over X. Let $\mathbb{P}(E)$ be the associated projective bundle and let us denote by $p\colon \mathbb{P}(E) \to X$ the projection. Then the vector bundle p^*E has a tautological subbundle, whose fibre at each point is the line determined by that point. Let us denote by $\mathcal{O}_{\mathbb{P}(E)}(-1)$ this line bundle and let $\mathcal{O}_{\mathbb{P}(E)}(1)$ be the dual line bundle. Let ξ be the first Chern class of $\mathcal{O}_{\mathbb{P}(E)}(1)$. Let us denote by $\mathbb{Z}(p)$ the constant sheaf $(2\pi i)^p\mathbb{Z} \subset \mathbb{C}$.

THEOREM 2.2 (Dold-Thom isomorphism). *For each pair of integers i, m, the map*

$$\sum_{k=0}^{n-1} p^*(\cdot)\cup\xi^k\colon \bigoplus_{k=0}^{n-1} H^{m-2k}(X,\mathbb{Z}(i-k)) \to H^m(\mathbb{P}(E),\mathbb{Z}(i))$$

is an isomorphism.

PROOF. When X is one point, the result is the classical formula for the cohomology of the projective space. The general case follows from the fact that the existence of the global classes ξ^k, implies the triviality of the Leray spectral sequence of the morphism $p\colon \mathbb{P}(E) \to X$. □

This theorem allows us to define Chern classes in the following way.

DEFINITION 2.3. The Chern classes of the vector bundle E are the classes $c_i(E) \in H^{2i}(X,\mathbb{Z}(i))$ determined by the equation

$$p^*(c_n(E)) + p^*(c_{n-1}(E))\cup\xi + \cdots + p^*(c_1(E))\cup\xi^{n-1} + \xi^n = 0. \tag{2.1}$$

The total Chern class is the sum

$$c(E) = 1 + c_1(E) + \cdots + c_n(E).$$

The Chern classes are characterized by the first Chern class and the behaviour under inverse images and exact sequences. This property is very useful, for instance, when comparing different definitions of Chern classes.

THEOREM 2.4. *There exists a unique way to assign, to each holomorphic vector bundle E, a total Chern class $c(E)$ satisfying the following properties:*

1. *Normalization: If $\mathcal{L}$ is a line bundle then $c(\mathcal{L}) = 1 + c_1(\mathcal{L})$, where $c_1(\mathcal{L})$ is defined in Definition 2.1.*
2. *Functoriality: For any morphism of complex varieties $f\colon X \to Y$ we have $c(f^*E) = f^*c(E)$.*
3. *Whitney sum formula: For any exact sequence of vector bundles*

$$0 \to S \to E \to Q \to 0 \tag{2.2}$$

we have $c(E) = c(S)\cup c(Q)$.

SKETCH OF PROOF. Let us start by proving the uniqueness. Assume that there exists a theory of Chern classes satisfying conditions 1, 2 and 3. Let E be a vector bundle over a smooth complex manifold X. Let us write $Q = p^*E/\mathcal{O}_{\mathbb{P}(E)}(-1)$. Then Q is a rank $n-1$ vector bundle over $\mathbb{P}(E)$. The basic idea is to use the exact sequence

$$0 \to \mathcal{O}(-1) \to p^*E \to Q \to 0.$$

First, using induction and properties 1 and 3 one easily proves that $c_i(E) = 0$ for $i > n$. Then, since the first Chern class of $\mathcal{O}_{\mathbb{P}(E)}(-1)$ is $-\xi$, properties 1, 2 and 3 imply that

$$p^*c(E) = (1 - \xi)c(Q).$$

Since $c_n(Q) = 0$ and $(1-\xi)^{-1} = 1 + \xi + \xi^2 + \xi^3 + \dots$, the Chern classes should satisfy the formula (2.1). By the Dold-Thom isomorphism this equation determines the Chern classes.

To prove the existence we need to show that Chern classes, as defined in Definition 2.3 satisfy the conditions 1, 2 and 3. The first condition follows from the definition. The second one is easy. For the third one we need to use the existence and some properties of the Gysin morphism (see [**17**] for more details). □

3. Chern Classes of Hermitian Vector Bundles

In this section we will explain the Chern-Weil construction of Chern classes using complex differential geometry. This theory is explained in many places (see for instance [**5**], [**34**] or [**16**]) and we refer the reader to them for details.

Let B be a subring of $\mathbb{R}$. We will write $B(p) = (2\pi i)^p B$. The symmetric group on n elements, $\mathfrak{S}_n$, acts on $B[T_1, \dots, T_n]$ by permuting the variables. Let $\mathrm{IP}(n)$ be the subset of invariant polynomials. Then $\mathrm{IP}(n) = B[\sigma_1, \dots, \sigma_n]$, where σ_i is the degree i symmetric elementary function in n variables. An analogous result is true if we replace polynomials by formal power series. We will denote by $\widehat{\mathrm{IP}}(n)$ the set of invariant power series in n variables and by $\mathrm{IP}(n)_k$ the space of invariant homogeneous polynomials of degree k.

Let be $\mathfrak{M}_n$ the vector space of $n \times n$ complex matrices. Let $\varphi\colon \mathfrak{M}_n \to \mathbb{C}$ be a map such that $\varphi(A)$ is a homogeneous polynomial of degree k in the entries of A, with coefficients in B. We say that φ is invariant if, for all $A \in \mathfrak{M}_n$ and $g \in \mathrm{GL}_n(\mathbb{C})$ we have

$$\varphi(A) = \varphi(gAg^{-1}).$$

Let us denote by $I_k(\mathfrak{M}_n)$ the space of invariant homogeneous polynomials of degree k. There is an isomorphism $I_k(\mathfrak{M}_n) \to \mathrm{IP}(n)_k$ which sends any φ to its value in the diagonal matrix with entries $T_1, \dots, T_n$. Due to this isomorphism we will identify both spaces.

Let X be a complex manifold. Let $\mathcal{E}^*$ denote the sheaf of complex smooth differential forms, $\mathcal{E}^{p,q}$ the sheaf of (p,q)-forms, $\mathcal{E}^*_{\mathbb{R}}$ the subcomplex of real forms and $\mathcal{E}^*_{\mathbb{R}}(p) = (2\pi i)^p \mathcal{E}^*_{\mathbb{R}}$. Let $\mathcal{E}^*(X)$, $\mathcal{E}^{p,q}(X)$, $\mathcal{E}^*_{\mathbb{R}}(X)$ and $\mathcal{E}^*_{\mathbb{R}}(p)(X)$ denote the corresponding groups of global sections. Let E be a holomorphic vector bundle on X. Then we will denote by $\mathcal{E}^*(E)$ the sheaf of E-valued smooth differential forms and by $\mathcal{E}^*(X, E)$ the corresponding space of global sections. $\mathcal{E}^{p,q}(X, E)$ will denote the space of forms of type (p,q) with values in E.

A connection on E is a $\mathbb{C}$-linear map

$$D\colon \mathcal{E}^0(X,E) \to \mathcal{E}^1(X,E)$$

satisfying the Leibnitz rule

$$D(\phi\xi) = d\phi\xi + \phi D\xi, \tag{3.1}$$

for any $\phi \in \mathcal{E}^0(X)$ and $\xi \in \mathcal{E}^0(X,E)$.

Given any connection D we can extend it to obtain operators

$$D\colon \mathcal{E}^k(X,E) \to \mathcal{E}^{k+1}(X,E)$$

by imposing the Leibnitz rule. But in general $(\mathcal{E}^*(X,E), D)$ is not a complex, because the operator

$$K = D^2\colon \mathcal{E}^0(X,E) \to \mathcal{E}^2(X,E)$$

may be non zero. In fact the operator K is $\mathcal{E}^0(X)$-linear and thus it can be seen as a section $K \in \mathcal{E}^2(X, \mathrm{End}(E))$. This operator is called the curvature of the connection.

Due to the decomposition $\mathcal{E}^1(X,E) = \mathcal{E}^{0,1}(X,E) \oplus \mathcal{E}^{1,0}(X,E)$ we can decompose $D = D^{0,1} + D^{1,0}$, with

$$D^{0,1}\colon \mathcal{E}^0(X,E) \to \mathcal{E}^{0,1}(X,E)$$
$$D^{1,0}\colon \mathcal{E}^0(X,E) \to \mathcal{E}^{1,0}(X,E).$$

Since E is holomorphic, there is a well defined $\bar\partial$-operator, which in a local frame is given by

$$\bar\partial(f_1,\dots,f_n) = (\bar\partial f_1,\dots,\bar\partial f_n).$$

Let us assume that E is provided with a Hermitian metric h. Let us denote by $\langle\cdot,\cdot\rangle$ the corresponding inner product. Then there is a unique connection that is compatible with both the complex structure and the Hermitian metric. That is, there is a unique connection $D = D(h)$ satisfying the following conditions:

1. "Compatibility with the Hermitian metric": For any ξ, $\eta \in \mathcal{E}^0(X,E)$
$$d\langle\xi,\eta\rangle = \langle D\xi,\eta\rangle + \langle\xi, D\eta\rangle.$$
2. "Compatibility with the complex structure": $D^{0,1} = \bar\partial$.

Let us write $K = K(h) = D^2$, the curvature of the metric h. Let f be a local frame, if we denote by $h(f)$ the matrix of the metric h in this frame, then (see [**34**, p. 82]).

$$K = \bar\partial(h^{-1}(f)\partial h(f)). \tag{3.2}$$

From this equation it is clear that $K \in \mathcal{E}^{1,1}(X, \mathrm{End}(E))$. Thus in a local frame it is given by a matrix of $(1,1)$ forms.

Let $\varphi \in \widehat{\mathrm{IP}}(n)$ and let us denote by φ_k its component of degree k. The invariance of φ_k implies that we have a well defined element $\widetilde{\varphi}_k(E,h) = \varphi_k(-K) \in \mathcal{E}^{2k}(X)$. To see this, one first defines $\varphi_k(-K)$ in a local frame, where K is a matrix of $(1,1)$ forms and then one uses the invariance to glue together these local definitions (see for instance [**34**, III.3]). Since $I = \bigoplus_{k\geq 1} \mathcal{E}^k(X)$ is a nilpotent ideal, we have also a well defined form $\widetilde{\varphi}(E,h) = \bigoplus \widetilde{\varphi}_k(E,h)$.

THEOREM 3.1. *Let $E \to X$ be a holomorphic vector bundle and let h be a Hermitian metric. Then*

1. *The form* $\widetilde{\varphi}(E,h)$ *is closed.*
2. *The form* $\widetilde{\varphi}(E,h)$ *satisfies*

$$\widetilde{\varphi}(E,h) \in \bigoplus_{p \geq 0} \mathcal{E}^{2p}_{\mathbb{R}}(p)(X).$$

3. *The class of* $\widetilde{\varphi}(E,h)$ *in de Rham cohomology is independent of the metric* h.

Since any holomorphic vector bundle admits a Hermitian metric the above theorem allows us to use Hermitian metrics to define characteristic classes.

DEFINITION 3.2. Let X be a complex manifold. Let E be a rank n holomorphic vector bundle and let $\varphi \in \mathbb{R}[[T_1,\dots,T_n]]$ be an invariant power series. Let us choose any Hermitian metric h on E. Then the cohomology class of $\widetilde{\varphi}(E,h)$, denoted

$$\varphi^{\mathrm{dR}}(E) \in \bigoplus_{p} H^{2p}_{\mathrm{dR}}(X,\mathbb{R}(p))$$

will be called the de Rham Chern class of E associated to the power series φ. The differential form $\widetilde{\varphi}(E,h)$ will be called the Chern form.

REMARK 3.3. This distinction between Chern classes and de Rham Chern classes is provisional. We will distinguish between them until we see that they agree.

EXAMPLES 3.4.

1. If $\varphi = \sigma_i$, the ith elementary symmetric function then $c^{\mathrm{dR}}_i(E) = \varphi^{\mathrm{dR}}(E)$ is the ith de Rham Chern class of E (see the remark above).
2. If $\varphi = 1 + \sum \sigma_i$ then $c^{\mathrm{dR}}(E) = \varphi^{\mathrm{dR}}(E)$ is called the total de Rham Chern class. Observe that the total Chern form is given by

$$\tilde{c}(E,h) = \det(1-K).$$

3. If B contains the field $\mathbb{Q}$, then the Chern character is defined by the power series

$$\mathrm{ch}(T_1,\dots,T_n) = \sum_{i=1}^{n} \exp(T_i).$$

The natural inclusion $\mathbb{Z}(p) \longrightarrow \mathcal{E}^0_{\mathbb{R}}(p)$ induces a morphism $\psi\colon H^*(X,\mathbb{Z}(p)) \longrightarrow H^*_{\mathrm{dR}}(X,\mathbb{R}(p))$. In order to see that the two definitions of Chern classes are compatible we have to compare $\psi(c(E))$ with $c^{\mathrm{dR}}(E)$. The key to compare both definitions of Chern classes is the Theorem 2.4. Thus we only need to show that the de Rham Chern classes also satisfy the properties given in Theorem 2.4.

THEOREM 3.5.

1. *Let* $f : X \to Y$ *be a morphism of complex manifolds. Let* (E,h) *be a Hermitian vector bundle on* Y. *Then*

$$f^*\tilde{c}(E,h) = \tilde{c}(fE, f^*h).$$

2. *Let* $\mathcal{L}$ *be a holomorphic line bundle. Then*

$$c^{\mathrm{dR}}_1(\mathcal{L}) = \psi\left(c_1(\mathcal{L})\right).$$

3. *Let*

$$0 \to S \to E \to Q \to 0$$

be an exact sequence of holomorphic vector bundles on X. *Then*

$$c(E) = c(S) \wedge c(Q).$$

PROOF. The proof of properties 1 and 3 are standard and can be found in any reference for Chern-Weil theory. For instance, see [**34**]. Since property 2 is more specific of the comparison we are making, we will provide a proof. The idea is to represent $c_1(\mathcal{L})$ as a Čech cocycle and $c_1^{\mathrm{dR}}(\mathcal{L})$ as a differential form. So we will use the comparison between Čech and de Rham cohomologies to compare the two classes.

Let us start representing $c_1(\mathcal{L})$ using Čech cohomology. Let us assume that $\mathfrak{U} = \{U_i\}$ is a good cover of X (see for instance [**6**]). This means that the open sets U_i and all their finite intersections are contractible. This implies that Čech cohomology for the constant sheaf $\mathbb{Z}(1)$ agrees with sheaf cohomology. Given a sheaf of Abelian groups $\mathcal{F}$, we will denote by $C^*(\mathfrak{U}, \mathcal{F})$ the complex of Čech cochains with respect to the open cover $\mathfrak{U}$.

Let $\{(U_i, s_i)\}$ be a trivialization of the line bundle $\mathcal{L}$. Thus s_i is a non vanishing section of $\Gamma(U_i, \mathcal{L})$. For each pair i, j let us write $U_{i,j} = U_i \cap U_j$. Let $g_{ij} = s_j/s_i$ be the transition functions. So g_{ij} belongs to $\Gamma(U_{i,j}, \mathcal{O}^*)$ and $\{g_{ij}\} \in C^1(\mathfrak{U}, \mathcal{O}^*)$ is a 1-cocycle that represents the class of $\mathcal{L}$ in $H^1(X, \mathcal{O}^*)$. We have to apply the connection morphism to this class. Since the open set $U_{i,j}$ is contractible, we can choose a determination of the logarithm $\log g_{ij}$ over $U_{i,j}$. Over $U_{ijk} = U_i \cap U_j \cap U_k$ let us write $t_{ijk} = \log g_{jk} - \log g_{ik} + \log g_{ij}$. Then t_{ijk} is constant and it is an integer multiple of $2\pi i$. The set $\{t_{ijk}\}$ is a Čech 2-cocycle for the sheaf $\mathbb{Z}(1)$ that represents $c_1(\mathcal{L})$.

Next we want to represent $c_1^{\mathrm{dR}}(\mathcal{L})$ as an explicit differential form. Since $\mathcal{L}$ has rank one, the curvature $K \in \mathcal{E}^{1,1}(X, \mathrm{End}(\mathcal{L})) = \mathcal{E}^{1,1}(X)$ is a differential form. By (3.2) this form is given, in each open U_i, by

$$K = \bar{\partial}(h(s_i)^{-1}\partial h(s_i)) = \bar{\partial}\partial \log(h(s_i)).$$

Since this form does not depend on the section s_i it is a global differential form. Thus the first de Rham Chern class is represented by the form $\tilde{c}_1(\mathcal{L}, h) = -K = \partial\bar{\partial}\log(h(s_i))$.

To compare the two classes we will follow the comparison between Čech and de Rham cohomologies given in [**6**]. The main tool is the double complex $C^*(\mathfrak{U}, \mathcal{E}^*)$. It has natural morphisms from $\mathcal{E}^*(X)$ and from $C^*(\mathfrak{U}, \mathbb{C})$ and is quasi-isomorphic to both complexes. Let

$$d' \colon C^p(\mathfrak{U}, \mathcal{E}^q) \to C^{p+1}(\mathfrak{U}, \mathcal{E}^q)$$
$$d'' \colon C^p(\mathfrak{U}, \mathcal{E}^q) \to C^p(\mathfrak{U}, \mathcal{E}^{q+1})$$

denote the differentials, where d' is the differential of Čech cochains and d'' is $(-1)^p$ times the differential of forms.

Then $\{t_{ijk}\} = d'\{\log g_{ij}\}$, and

$$d''\{\log g_{ij}\} = \{-d(g_{ij}/g_{ij})\} = \{-s_i/s_j\, d(s_j/s_i)\}.$$

therefore the Čech cochain $\{s_i/s_j\, d(s_j/s_i)\}$ also represents $c_1(\mathcal{L})$. On the other hand, $-K = -d''\{\partial \log(h(s_i))\}$, and

$$\begin{aligned} d'\{-\partial\log(h(s_i))\} &= \{-\partial \log\left(h(s_j)/h(s_i)\right)\} \\ &= \{-\partial \log\left(s_j\bar{s}_j/s_i\bar{s}_i\right)\} \\ &= \{-s_i/s_j\, d(s_j/s_i)\}. \end{aligned}$$

Therefore $-K$ represents also the class $c_1(\mathcal{L})$. □

As a consequence of this theorem we see that $\psi(c(E)) = c^{\mathrm{dR}}(E)$. In particular this implies that the Chern class c_i corresponds to the elementary symmetric function σ_i. This justifies the following definition.

DEFINITION 3.6. Let X be a complex manifold. Let E be a rank n holomorphic vector bundle and let $\varphi \in B[[T_1, \ldots, T_n]]$ be an invariant power series. Let

$$\varphi(T_1, \ldots, T_n) = \phi(\sigma_i, \ldots, \sigma_n)$$

be the expression of φ in terms of symmetric elementary functions. Then *the Chern class of E associated to the power series φ* is:

$$\varphi(E) = \phi(c_1(E), \ldots, c_n(E)) \in \bigoplus_p H^{2p}(X, B(p)).$$

COROLLARY 3.7. *Let X be a complex manifold and let E be a holomorphic vector bundle of rank n. Let $\varphi \in B[[\sigma_1, \ldots, \sigma_n]]$ be an invariant power series. Then*

$$\varphi^{\mathrm{dR}}(E) = \psi\left(\varphi(E)\right).$$

In view of this result we will drop the superscript ^{dR} from the notation.

The Chern character (see Example 3.4.3) is one of the most interesting power series of characteristic classes. The main advantage of the Chern character class is that it behaves very well under exact sequences and tensor products. For the proof of the next proposition we refer also to [**34**].

PROPOSITION 3.8. 1. *Let $0 \to S \to E \to Q \to 0$ be an exact sequence of vector bundles. Then*

$$\mathrm{ch}(E) = \mathrm{ch}(S) + \mathrm{ch}(Q).$$

2. *Let E and F be vector buncles. Then*

$$\mathrm{ch}(E \otimes F) = \mathrm{ch}(E) \wedge \mathrm{ch}(F).$$

Unlike the Chern character class, the Chern character form does not need to behave additively for exact sequences. Let

$$0 \to (S, h') \to (E, h) \to (Q, h'') \to 0$$

be an exact sequence of Hermitian vector bundles. Let us write $(S, h') \oplus (Q, h'')$ for the orthogonal direct sum. Then

$$\widetilde{\mathrm{ch}}\left((S, h') \oplus (Q, h'')\right) = \widetilde{\mathrm{ch}}(S, h') + \widetilde{\mathrm{ch}}(Q, h'').$$

But in general

$$\widetilde{\mathrm{ch}}(E, h) \neq \widetilde{\mathrm{ch}}\left((S, h') \oplus (Q, h'')\right).$$

At first glance this may seem unfortunate. But, in Section 8, we will see that, as Schechtmann pointed out ([**30**]), the lack of additivity at the level of forms can be used to construct characteristic classes for higher K-theory.

4. Chern Classes for Higher K-Theory

In this section we will review Gillet's construction of Chern classes for higher K-theory [**11**]. We will follow the simplified version given in [**31**]. We will assume that the reader has some familiarity with the language of simplicial objects (see for instance [**23**] or [**14**]).

In this section we will be in the algebraic geometry context. To stress this fact, instead of working with complex numbers, let us fix a ground field k. Let $\mathcal{V}$ be the category of all smooth quasi-projective schemes, equipped with the Zarisky topology. Let us denote by $D(\mathcal{V})$ (resp. $D^+(\mathcal{V})$) the derived category of complexes of Abelian sheaves on $\mathcal{V}$ (resp. which are bounded below). Let us assume that we have a fixed graded complex of sheaves

$$\mathcal{F}^{\cdot}(*) = \bigoplus_{i\in\mathbb{Z}} \mathcal{F}^{\cdot}(i), \quad \text{with } \mathcal{F}^{\cdot}(i) \in D^+(\mathcal{V}).$$

Thus for each smooth quasi-projective scheme $X \in \mathrm{Ob}(\mathcal{V})$ we have the cohomology groups $\mathbb{H}^{\cdot}(X, \mathcal{F}^{\cdot}(*))$. We need to assume certain properties of this cohomology theory in order to mimic the construction of Chern classes of Section 2.

Let us denote by $\mathcal{O}^*$ the sheaf of invertible rational functions. It defines an element of $D^+(\mathcal{V})$ assuming that it is a complex concentrated in degree zero. Then $\mathbb{H}^1(X, \mathcal{O}^*)$ parameterizes isomorphism classes of algebraic line bundles over X. The first property we need for our cohomology theory is:

P1. There is a morphism in $D^+(\mathcal{V})$

$$\mathcal{O}^*[1] \to \mathcal{F}^{\cdot}(1).$$

In particular, for each $X \in \mathrm{Ob}(\mathcal{V})$, we obtain a morphism $c_1 \colon \mathbb{H}^1(X, \mathcal{O}^*) \to \mathbb{H}^2(X, \mathcal{F}^{\cdot}(1))$ which allows us to define the first Chern class of a line bundle.

The next property we need is a multiplicative structure for the cohomology.

P2. For each n, $m \in \mathbb{Z}$, there are homomorphisms in $D^+(\mathcal{V})$

$$\cup \colon \mathcal{F}^{\cdot}(n) \overset{L}{\underset{\mathbb{Z}}{\otimes}} \mathcal{F}^{\cdot}(m) \to \mathcal{F}^{\cdot}(n+m), \quad \text{and } e \colon \mathbb{Z} \to \mathcal{F}^{\cdot}(0)$$

which make $\mathcal{F}^{\cdot}(*)$ an associative and graded commutative (with respect to the first degree) algebra with unit.

The third property we need is a formula for the cohomology of the projective space, in other words, we need the Dold-Thom isomorphism to be satisfied.

P3. For $X \in \mathrm{Ob}(\mathcal{V})$, let $p \colon \mathbb{P}^n_X \to X$ be the n-dimensional projective space over X. Let ξ be the first Chern class of the line bundle $\mathcal{O}(1)$. Then, for each pair of integers i, m, the morphism

$$\sum_{k=0}^{n-1} p^*() \cup \xi^k \colon \bigoplus_{k=0}^{n-1} \mathbb{H}^{m-2k}(X, \mathcal{F}^{\cdot}(p-k)) \to \mathbb{H}^m(\mathbb{P}^n_X, \mathcal{F}^{\cdot}(p))$$

is an isomorphism.

By a Leray spectral sequence argument, from property **P3** it follows that the Dold-Thom isomorphism is satisfied for any projective bundle.

With properties **P1**, **P2** and **P3**, we can repeat the procedure of the Section 2 and define, for each vector bundle E over $X \in \mathrm{Ob}(\mathcal{V})$, Chern classes $c_i(E) \in \mathbb{H}^{2i}(X, \mathcal{F}^{\cdot}(i))$. But in order to have the Whitney sum formula we need to assume a Gysin property (see [**31**] and [**17**]).

P4. Let $i\colon Y \hookrightarrow X$ be a closed immersion in $\mathcal{V}$ of pure codimension 1 and let $[Y] \in H^1(X, \mathcal{O}^*)$ be the class of the divisor Y. Then for any $x \in \mathbb{H}^{2i}(X, \mathcal{F}^{\cdot}(i))$ such that $i^*x = 0$, we have

$$x \cup c_1([Y]) = 0.$$

The next step is to extend the definition of Chern classes to simplicial schemes. Let Δ be the category whose objects are the ordered sets $[n] = \{0, \dots, n\}$, $n \geq 0$ and whose morphisms are ordered maps. Let $\mathcal{SV}$ denote the category of simplicial objects in $\mathcal{V}$, that is, the category of contravariant functors between Δ and $\mathcal{V}$.

DEFINITION 4.1. Let $X. \in \operatorname{Ob} \mathcal{SV}$ be a simplicial scheme. A *rank n vector bundle* over $X.$ is an object of $\mathcal{SV}$, $E.$, together with a morphism of simplicial schemes $\pi.\colon E. \to X.$ satisfying the following conditions

1. For any $k \geq 0$ the morphism $\pi_k\colon E_k \to X_k$ is a rank n vector bundle.
2. For any j, $k \geq 0$ and any $\rho \in \operatorname{Mor}_{\Delta}([j],[k])$ the commutative diagram

$$\begin{array}{ccc} E_k & \xrightarrow{E(\rho)} & E_j \\ \downarrow{\scriptstyle \pi_k} & & \downarrow{\scriptstyle \pi_k} \\ X_k & \xrightarrow{X(\rho)} & X_j \end{array}$$

is a relative morphism of vector bundles.

Alternatively, following [**12**], we can define a vector bundle over $X.$ as a vector bundle E_0 over X_0, together with an isomorphism $\alpha\colon \delta_0^* E_0 \to \delta_1^* E_0$ of vector bundles over E_1, such that $\delta_2^*\alpha \circ \delta_0^*\alpha = \delta_1^*\alpha$.

If $X.$ is a simplicial scheme then again $\mathbb{H}^1(X., \mathcal{O}^*)$ parameterizes isomorphism classes of line bundles (see [**12**, ex 1.1]). Moreover, the property **P3** implies that the Dold-Thom isomorphism is satisfied for arbitrary projective bundles over simplicial schemes (see [**11**, Lemma 2.4]). Therefore Grothendieck's construction of Chern classes can be applied to simplicial schemes. In particular it can be applied in the universal case.

Let $B.\,\mathbf{GL}_n/k$ be the classifying scheme of the group scheme $\mathbf{GL}_n$ over k. The simplicial scheme $B.\,\mathbf{GL}_n/k$ is provided with a universal rank n vector bundle, denoted by E_n. Let us denote by $c_i^{(n)} = c_i(E_n) \in \mathbb{H}^{2i}(B.\,\mathbf{GL}_n/k, \mathcal{F}^{\cdot}(i))$ the ith Chern class of the universal bundle.

The next objective is to explain how classes in the cohomology of the classifying scheme give rise to maps between K-theory and cohomology of schemes.

Let $S.$ be a simplicial set and let X be a scheme. Then we can construct the simplicial scheme $S. \times X$ such that

$$(S. \times X)_n = \coprod_{p \in S_n} \{p\} \times X,$$

and the faces and degeneracies are induced by those of $S.$. Let A be a finitely generated k-algebra such that $U = \operatorname{Spec} A$ is a smooth scheme. Then $B.\,\mathbf{GL}_n(A)$ is a simplicial set. Thus we can construct the simplicial scheme $B.\,\mathbf{GL}_n(A) \times U$. Since an element of $B_j\,\mathbf{GL}_n(A)$ is a morphism between U and $B_j\,\mathbf{GL}_n/k$ we obtain a tautological morphism of simplicial schemes

$$\tau\colon B.\,\mathbf{GL}_n(A) \times U \to B.\,\mathbf{GL}_n/k.$$

Thus, we obtain classes $\tau^*(c_i^{(n)}) \in \mathbb{H}^{2i}(B.\,\mathbf{GL}_n(A) \times U, \mathcal{F}^{\cdot}(i))$.

For a simplicial set $S.$, let us denote by $\mathbb{Z}S.$ the homological complex which, in degree j is the free Abelian group generated by S_j, and whose differential is $d = \sum(-1)^i \delta_i$.

LEMMA 4.2. *Let $S.$ be a simplicial set and let X be a smooth quasi-projective scheme. Then there is a natural short exact sequence*

$$0 \to \prod_{p+q=n-1} \operatorname{Ext}^1_{\mathbb{Z}}(H_p(\mathbb{Z}S.), \mathbb{H}^q(X, \mathcal{F}^{\cdot}(i))) \to \mathbb{H}^n(S. \times X, \mathcal{F}^{\cdot}(i)) \xrightarrow{\alpha} \prod_{p+q=n} \operatorname{Hom}(H_p(\mathbb{Z}S.), \mathbb{H}^q(X, \mathcal{F}^{\cdot})) \to 0.$$

PROOF. This result is the dual of the Künneth formula (see for instance [**29**, Theorem 11.32]). □

Thus, from each class $c_i^{(n)} \in H^{2i}(B.\,\mathbf{GL}_n / k, \mathcal{F}^{\cdot}(i))$ we obtain a family of morphisms

$$\alpha \circ \tau(c_i^{(n)})_j \colon H_j(\mathbb{Z}B.\,\mathbf{GL}_n(A)) \to \mathbb{H}^{2i-j}(U, \mathcal{F}^{\cdot}(i)).$$

Now we want to go to the limit when n goes to infinity. For $m > n$, let us consider the inclusion $i_{n,m}\,\mathbf{GL}_n / k \to \mathbf{GL}_{n+1} / k$ given by

$$A \mapsto \begin{pmatrix} A & 0 \\ 0 & I \end{pmatrix},$$

where I is the identity $(m-n) \times (m-n)$-matrix. Since

$$i_{n,m}^* E_m = E_n \oplus (\text{a trivial vector bundle}),$$

we have that $i_{n,m}^* c_i^{(m)} = c_i^{(n)}$. Therefore, the family of morphisms $\{\psi \circ \varphi(c_i^{(n)})_j\}_{n \geq 1}$ defines a morphism

$$c_{i,j} \colon H_j(\mathrm{GL}(A), \mathbb{Z}) = \varinjlim{}_n H_j(\mathbb{Z}B.\,\mathbf{GL}_n(A)) \to \mathbb{H}^{2i-j}(U, \mathcal{F}^{\cdot}(i)).$$

The K-theory groups of A are the homotopy groups of the $+$-construction of $B.\,\mathrm{GL}(A)$. Since the $+$-construction is acyclic, we have that

$$H_*(B.\,\mathrm{GL}(A)^+, \mathbb{Z}) = H_*(B.\,\mathrm{GL}(A), \mathbb{Z}) = H_*(\mathrm{GL}(A), \mathbb{Z}).$$

DEFINITION 4.3. Let A be a finitely generated k-algebra such that $U = \operatorname{Spec} A$ is a smooth scheme. For each pair of integers i, j, the ith Chern class map in $K_j(A)$ is the composition

$$c_{i,j} \colon K_j(A) = \pi_j(B.\,\mathrm{GL}(A)^+) \xrightarrow{\text{Hurewicz}} H_j(\mathrm{GL}(A), \mathbb{Z}) \xrightarrow{c_{i,j}} \mathbb{H}^{2i-j}(U, \mathcal{F}^{\cdot}(i)).$$

The following result is an easy consequence of the definition.

PROPOSITION 4.4. *Let $f^\sharp \colon A \to B$ be a morphism of finitely generated k algebras such that $U = \operatorname{Spec} A$ and $V = \operatorname{Spec} B$ are smooth schemes. Let $f \colon V \to U$ be the corresponding morphism of k-schemes. Then the following diagram*

$$\begin{array}{ccc} K_j(A) & \xrightarrow{c_{i,j}} & \mathbb{H}^{2i-j}(U, \mathcal{F}.(i)) \\ {\scriptstyle f^*}\downarrow & & \downarrow{\scriptstyle f^*} \\ K_j(B) & \xrightarrow{c_{i,j}} & \mathbb{H}^{2i-j}(V, \mathcal{F}.(i)) \end{array}$$

is commutative.

In order to extend the construction of Chern classes to arbitrary smooth quasi-projective schemes we will require a property of invariance under homotopy. This will allow us to use Jouanolou's trick. But strictly speaking this property is not necessary (see [**11**]).

P5. For any scheme $X \in \mathrm{Ob}(\mathcal{V})$, the natural map $\mathbb{A}^1_X \to X$ induces an isomorphism

$$\mathbb{H}^*(X, \mathcal{F}^{\cdot}(*)) \to \mathbb{H}^*(\mathbb{A}^1_X, \mathcal{F}^{\cdot}(*)).$$

This property implies that, for any fibre bundle $p\colon Y \to X$, with fibre $\mathbb{A}^n$, (not necessarily a vector bundle), the map $p^*\mathbb{H}^*(X, \mathcal{F}^{\cdot}(*)) \to \mathbb{H}^*(Y, \mathcal{F}^{\cdot}(*))$ is an isomorphism.

For any schema $X \in \mathcal{V}$, by Jouanolou's lemma [**20**, Lemma 1.5] there is a fibre bundle $p\colon Y \to X$, with Y affine, which is a torsor over a vector bundle.

DEFINITION 4.5. Let X be a smooth scheme over k. Let us choose an affine torsor $p : Y \to X$. For each pair of integers i, j, the *ith Chern class map for* $K_j(X)$ is the composition

$$c_{i,j}\colon K_j(X) \overset{p^*}{\cong} K_j(Y) \xrightarrow{c_{i,j}} \mathbb{H}^{2i-j}(Y, \mathcal{F}^{\cdot}(i)) \overset{p^*}{\cong} \mathbb{H}^{2i-j}(X, \mathcal{F}^{\cdot}(i)).$$

Thanks to the naturality of the map $c_{i,j}$ in the affine case (Proposition 4.4) and [**20**, Proposition 1.6] , this definition does not depend on the choice of the affine torsor Y.

So far we have constructed Chern classes for higher K-theory. Now we want to construct characteristic classes for arbitrary power series. Let again B be a subring of $\mathbb{R}$ and assume that $\mathcal{F}$ is a sheaf of B-modules. Observe that, replacing T_{n+1} by 0, we have a morphism between the ring of invariant power series

$$B[[T_1, \dots, T_{n+1}]]^{\mathfrak{S}_{n+1}} \to B[[T_1, \dots, T_n]]^{\mathfrak{S}_n}.$$

In terms of elementary symmetric functions the above morphism is the morphism

$$B[[\sigma_1, \dots, \sigma_{n+1}]] \to B[[\sigma_1, \dots, \sigma_n]]$$

that sends σ_{n+1} to 0. We will denote by

$$i^*_{n,m}\colon \widehat{\mathrm{IP}}(m) \to \widehat{\mathrm{IP}}(n)$$

the induced morphisms. These morphisms make $\{\widehat{\mathrm{IP}}(n)\}_n$ an inverse system. Observe that we have a commutative diagram

$$\begin{array}{ccc} \{\widehat{\mathrm{IP}}(m)\}_n & \longrightarrow & \prod_i \mathbb{H}^{2i}(B.\,\mathbf{GL}_m/k, \mathcal{F}^{\cdot}(i)) \\ \downarrow{\scriptstyle i^*_{n,m}} & & \downarrow{\scriptstyle i^*_{n,m}} \\ \{\widehat{\mathrm{IP}}(n)\}_n & \longrightarrow & \prod_i \mathbb{H}^{2i}(B.\,\mathbf{GL}_n/k, \mathcal{F}^{\cdot}(i)) \end{array}$$

where the horizontal arrows send the elementary symmetric function σ_i to the ith Chern class.

DEFINITION 4.6. A *stable invariant power series* is an element

$$\{\varphi^{(n)}\} \in \varprojlim \widehat{\mathrm{IP}}(n).$$

Observe that any stable invariant power series is given by an element

$$\varphi \in B[[\sigma_1, \dots, \sigma_n, \dots]].$$

Proceeding as with the Chern classes, for any invariant stable power series φ, we obtain characteristic classes

$$\varphi : K_j(X) \to \bigoplus \mathbb{H}^{2i-j}(X, \mathcal{F}(i)).$$

EXAMPLE 4.7. Assume that $B \supset \mathbb{Q}$. Let us denote by $\overline{\mathrm{ch}}^{(n)}$ the reduced Chern character of rank n. That is

$$\overline{\mathrm{ch}}^{(n)}(T_1, \dots, T_n) = \mathrm{ch}(T_1, \dots, T_n) - n.$$

Since $\overline{\mathrm{ch}}^{(n+1)}(T_1, \dots, T_n, 0) = \overline{\mathrm{ch}}^{(n)}(T_1, \dots, , T_n)$, then $\{\overline{\mathrm{ch}}^{(n)}\}$ is a stable invariant power series. Thus it defines a reduced Chern character class, denoted $\overline{\mathrm{ch}}$.

DEFINITION 4.8. The Chern character is the map

$$\mathrm{ch} \colon K_i(\cdot) \to \bigoplus_j \mathbb{H}^{2j-i}(\cdot, \mathcal{F}(j)),$$

given by

$$\mathrm{ch} = \begin{cases} \overline{\mathrm{ch}}, & \text{if } i > 0, \\ \mathrm{rank} + \overline{\mathrm{ch}}, & \text{if } i = 0. \end{cases}$$

Recall that there is a product structure [**21**]

$$K_i \otimes K_j \to K_{i+j}$$

that extends the multiplicative structure induced in K_0 by the tensor product. The Chern character is also compatible with this product. For a proof of the following result see [**11**] or [**31**].

THEOREM 4.9. *For $x \in K_i$ and $y \in K_j$ then*

$$\mathrm{ch}(x \cdot y) = \mathrm{ch}(x) \cup \mathrm{ch}(y).$$

5. Real Deligne Cohomology

Let us recall the definition of real Deligne cohomology and some complexes that can be used to compute it. Again let X be a smooth proper complex variety. As in Section 3 let B be a subring of $\mathbb{R}$. Let us write $B(p) = (2\pi i)^p B \subset \mathbb{C}$. We will denote also by $B(p)$ the constant sheaf. Let Ω_X^* be the sheaf of holomorphic differential forms on X.

DEFINITION 5.1. The *(B-)Deligne cohomology* of X, denoted $H_{\mathcal{D}}(X, B(p))$, is the hypercohomology of the complex of sheaves

$$B(p) \to \mathcal{O}_X \to \Omega_X^1 \to \cdots \to \Omega_X^{p-1}.$$

This definition has been extended by Beilinson to the case of smooth complex algebraic varieties, not necessarily proper (see [**2**], [**1**], [**10**] and [**19**]). This extension is known as Deligne-Beilinson cohomology. Beilinson also showed that Deligne-Beilinson cohomology can be written as sheaf cohomology for a sheaf in the Zariski topology, satisfying the properties **P1** to **P5** of last section. Thus we can apply the construction of the last section and obtain characteristic classes from higher K-theory to Deligne-Beilinson cohomology. In particular, if we take $B = \mathbb{R}$ then the

Chern character in real Deligne-Beilinson cohomology is called Beilinson's regulator and it is a generalization of Borel's regulator (see [**4**], [**2**], [**27**]).

Let F denote the Hodge filtration:

$$F^p\Omega_X^* = \bigoplus_{p' \geq p} \Omega_X^{p'}.$$

Then real Deligne cohomology can be defined also as the hypercohomology of the simple complex associated to the morphism of complexes

$$\underline{\mathbb{R}}(p) \oplus F^p\Omega_X^* \to \Omega_X^*.$$

We want to have real Deligne cohomology as the cohomology of an explicit complex, as de Rham cohomology is the cohomology of the complex of differential forms. To this end we will resolve the constant sheaf and the sheaves of holomorphic forms using smooth differential forms. We will use the same notation as in Section 3 for the sheaves and complexes of differential forms.

The Hodge filtration of the complex $\mathcal{E}^*$ is given by

$$F^p\mathcal{E}^n = \bigoplus_{p' \geq p} \mathcal{E}^{p',n-p'}.$$

The sheaves of differential forms are fine, hence acyclic. Since $\mathcal{E}_{\mathbb{R}}^*(p)$ is a resolution of the constant sheaf $\mathbb{R}(p)$, and $\mathcal{E}^*$ is a resolution of Ω_X^*, compatible with the Hodge filtration, we obtain that real Deligne cohomology of X is the cohomology of the simple complex associated to the morphism of complexes

$$u_p \colon \mathcal{E}_{\mathbb{R}}^*(p)(X) \oplus F^p\mathcal{E}^*(X) \to \mathcal{E}^*(X),$$

given by $u_p(r, f) = f - r$. Let us denote the simple complex associated to u_p as

$$s(u_p) = s(\mathcal{E}_{\mathbb{R}}^*(p)(X) \oplus F^p\mathcal{E}^*(X) \to \mathcal{E}^*(X)).$$

An element of $s^n(u_p)$ is given by a triple

$$(r, f, \omega) \in \mathcal{E}_{\mathbb{R}}^n(p)(X) \oplus F^p\mathcal{E}^n(X) \oplus \mathcal{E}^{n-1}(X),$$

and the differential is given by $d(r, f, \omega) = (dr, df, f - r - d\omega)$.

Following Deligne [**9**] we can use a simpler complex to compute real Deligne cohomology.

DEFINITION 5.2. Let $\mathcal{D}^*(X, p)$ denote the complex given by

$$\mathcal{D}^n(X,p) = \begin{cases} \mathcal{E}_{\mathbb{R}}^{n-1}(p-1)(X) \cap \bigoplus\limits_{\substack{p'+q'=n-1 \\ p'<p,\ q'<p}} \mathcal{E}^{p',q'}(X), & \text{for } n \leq 2p-1 \\ \mathcal{E}_{\mathbb{R}}^{n}(p)(X) \cap \bigoplus\limits_{\substack{p'+q'=n \\ p'\geq p,\ q'\geq p}} \mathcal{E}^{p',q'}(X), & \text{for } n \geq 2p. \end{cases}$$

The differential of this complex, denoted by $d_{\mathcal{D}}$, is induced by d in degree greater or equal than $2p$, by $-d$ in degree less or equal than $2p-2$ and is equal to $-2\partial\bar{\partial}$ in degree $2p-1$.

A proof that the cohomology of this complex is Deligne cohomology can be found on [**7**]. This proof is based in the following facts:

1. The morphism $u\colon \mathcal{E}^n_{\mathbb{R}}(p)(X) \oplus F^p\mathcal{E}^n(X) \to \mathcal{E}^n(X)$ is injective for $n \leq 2p-1$ and the cokernel is
$$\mathcal{E}^{n-1}_{\mathbb{R}}(p-1)(X) \cap \bigoplus_{\substack{p'+q'=n-1\\ p'<p,\ q'<q}} \mathcal{E}^{p',q'}(X).$$
2. The above morphism u is surjective for $n \geq 2p-1$ and the kernel is
$$\mathcal{E}^n_{\mathbb{R}}(p)(X) \cap \bigoplus_{\substack{p'+q'=n\\ p'\geq p,\ q'\geq q}} \mathcal{E}^{p',q'}(X).$$
3. In particular, for $n = 2p-1$ the morphism u is an isomorphism. Moreover, if $\omega \in \mathcal{E}^{2p-2}(X)$, then $du^{-1}d\omega = -2\partial\bar{\partial}\omega$.

EXAMPLE 5.3. Observe that $\mathcal{D}^{2p}(X,p) = \mathcal{E}^{2p}_{\mathbb{R}}(p)(X) \cap \mathcal{E}^{p,p}(X)$. Thus if (E,h) is a Hermitian holomorphic vector bundle of rank n and $\varphi \in \mathbb{R}[[T_1,\dots,T_n]]$ is a real invariant power series, then
$$\widetilde{\varphi}(E,h) \in \bigoplus_p \mathcal{D}^{2p}(X,p).$$

In order to write down explicit morphisms between $\mathcal{D}^*(X,p)$ and $s(u_p)$ we need to introduce some notations. Given a differential form a, we will write $a = \sum a^{p,q}$ its decomposition in forms of pure bidegree. Let $F^{p,p}$ be the morphism defined by
$$F^{p,p}a = \sum_{\substack{p'\geq p\\ q'\geq p}} a^{p',q'}.$$
Let π_p be the morphism given by
$$\pi_p a = (a + (-1)^p\bar{a})/2.$$
Observe that π_p is the projection of $\mathcal{E}^*(X)$ onto $\mathcal{E}^*_{\mathbb{R}}(p)(X)$, and that, for $n < 2p$, $\rho_p = \pi_{p-1} \circ F^{n-p+1,n-p+1}$ is the projection of $\mathcal{E}^n(X)$ onto the cokernel of u_p.

Let $\psi\colon s^n(u_p) \to \mathcal{D}^n(X,p)$ and $\varphi\colon \mathcal{D}^n(X,p) \to s^n(u_p)$ be the morphisms given by
$$\psi(r,f,\omega) = \begin{cases} \rho_p(\omega), & \text{for } n \leq 2p-1,\\ F^{p,p}r + 2\pi_p(\partial\omega^{p-1,q+1}), & \text{for } n \geq 2p,\end{cases}$$
$$\varphi(x) = \begin{cases} (\partial x^{p-1,q} - \bar{\partial}x^{q,p-1}, 2\partial x^{p-1,q}, x), & \text{for } n \leq 2p-1,\\ (x,x,0), & \text{for } n \geq 2p,\end{cases}$$
where $q = n-p$. Then ψ and φ are homotopy equivalences inverse to each other (see [**7**]).

Real Deligne cohomology has a product [**2**], [**1**] that can be described, in terms of the complexes $s(u_p)$ in the following way. Let $0 \leq \alpha \leq 1$ be a real number. For $(r,f,\omega) \in s^n(u_p)$ and $(s,g,\eta) \in s^m(u_q)$, let us write
$$(r,f,\omega) \cup_\alpha (s,g,\eta) = (r\wedge s, f\wedge g,\\ \alpha(\omega\wedge s + (-1)^n f\wedge\eta) + (1-\alpha)(\omega\wedge g + (-1)^n r\wedge\eta)).$$
This is a family of products, all of them homotopically equivalent. Moreover, for $\alpha = \frac{1}{2}$ this product is graded commutative, whereas for $\alpha = 0$, 1 this product is associative. Therefore they induce a graded commutative, associative product in

Deligne cohomology. This product induces a product in the complex $\mathcal{D}^*(X,p)$ (see [**7**]). This complex is only associative or graded commutative up to homotopy.

In order to have a complex with a graded commutative and associative product we will use the Thom-Whitney simple, introduced in [**26**]. The Thom-Whitney simple associates, to a strict cosimplicial differential complex, a new differential complex. This new differential complex is homotopically equivalent to the total complex of the original cosimplicial complex. The main interest of the Thom-Whitney simple is that, if we start with a strict cosimplicial graded commutative associative algebra, the complex that we obtain is again a differential graded commutative associative algebra.

In our case, we can consider the morphism u_p a strict cosimplicial complex $u_p^{\cdot}$ writing

$$\begin{aligned} u_p^0 &= \mathcal{E}^*_{\mathbb{R}}(p)(X) \oplus F^p\mathcal{E}^*(X), \\ u_p^1 &= \mathcal{E}^*(X), \\ u_p^i &= 0, \qquad\qquad \text{for } i \geq 2, \end{aligned}$$

with morphisms given by

$$\delta^0(r,f) = f, \quad \delta^1(r,f) = r,$$

and the other morphisms equal to zero.

Let us describe the Thom-Whitney simple in this case. Let L_1^* be the complex of algebraic forms in the affine line $\mathbb{A}^1_{\mathbb{R}}$. That is, $L_1^0 = \mathbb{R}[t]$, and $L_1^1 = \mathbb{R}[t]dt$. Let $\delta_0, \delta_1 \colon L_1^* \to \mathbb{R}$ be the morphisms given by evaluation at 0 and 1 respectively. That is $\delta_0(f(t) + g(t)dt) = f(0)$, and $\delta_1(f(t)+g(t)dt) = f(1)$.

DEFINITION 5.4. The *Thom-Whitney simple* of $u_p^{\cdot}$, denoted $s_{TW}(u_p^{\cdot})$ is the subcomplex of

$$\mathcal{E}^*_{\mathbb{R}}(p)(X) \oplus F^p\mathcal{E}^*(X) \oplus L_1^* \otimes \mathcal{E}^*(X)$$

formed by the elements (r,f,ω) such that

$$\begin{aligned} f &= (\delta_0 \otimes \mathrm{Id})(\omega), \\ r &= (\delta_1 \otimes \mathrm{Id})(\omega). \end{aligned}$$

The differential and the product of the Thom-Whitney simple are given componentwise:

$$d(r,f,\omega) = (dr, df, d\omega)$$

and

$$(r,f,\omega) \wedge (s,g,\eta) = (r \wedge s, f \wedge g, \omega \wedge \eta).$$

With these definitions of differential and product, the direct sum $\bigoplus_p s_{TW}(u_p)$ is a differential graded commutative associative algebra.

We can construct explicit equivalences (see [**26**])

$$s_{TW}(u_p) \underset{E}{\overset{I}{\rightleftarrows}} s(u_p)$$

given by

$$E(r,f,\omega) = (r, f, t \otimes f + (1-t) \otimes r + dt \otimes \omega),$$

and

$$I(r,f,(h(t)+g(t)dt)\otimes\omega)=\left(r,f,\int_0^1 g(t)dt\omega\right).$$

We will write $I'=\psi\circ I: s_{TW}(u_p)\to\mathcal{D}^*(X,p)$ and $E'=E\circ\varphi\colon\mathcal{D}^*(X,p)\to s_{TW}(u_p)$. We will later use that I' and E' is a pair of homotopy equivalences, inverse to each other, between $\mathcal{D}^*(X,*)$ and a complex with an associative and graded commutative product.

6. Bott-Chern Forms

For any stable invariant power series φ and any exact sequence

$$\bar{\xi}\colon 0\to(S,h')\xrightarrow{f}(E,h)\to(Q,h'')\to 0$$

of Hermitian vector bundles, the Whitney sum formula implies that the associated Chern classes satisfy

$$\varphi(E)=\varphi(S\oplus Q). \tag{6.1}$$

But in general this equation is no longer true for the Chern forms. That is, the form

$$\widetilde{\varphi}(E,h)-\widetilde{\varphi}((S,h')\oplus(Q,h''))\in\bigoplus\mathcal{D}^{2p}(X,p)$$

may be non zero. Nevertheless, equation (6.1) and the $\partial\bar{\partial}$-lemma ([**16**]) imply that there exists a differential form $\widetilde{\varphi}_1$ such that

$$\widetilde{\varphi}(E,h)-\widetilde{\varphi}((S,h')\oplus(Q,h''))=-2\partial\bar{\partial}\widetilde{\varphi}_1. \tag{6.2}$$

Since $\mathcal{D}^{2p}(X,p)=\mathcal{E}^{2p}_{\mathbb{R}}(p)(X)\cap\mathcal{E}^{p,p}(X)$ and $-2\partial\bar{\partial}$ is a purely imaginary operator, bihomogeneous of bidegree $(1,1)$, we can choose

$$\widetilde{\varphi}_1\in\bigoplus_p\mathcal{E}^{2p-2}_{\mathbb{R}}(p-1)(X)\cap\mathcal{E}^{p-1,p-1}(X)=\bigoplus_p\mathcal{D}^{2p-1}(X,p).$$

In other words, the form $\widetilde{\varphi}(E,h)-\widetilde{\varphi}((S,h')\oplus(Q,h''))$ is exact in the complex $\bigoplus\mathcal{D}^*(X,p)$ The aim of this section is to solve the equation (6.2) in a functorial way.

We will say that the exact sequence $\bar{\xi}$ is split if (E,h) is the orthogonal direct sum (S,h') and (Q,h'').

THEOREM 6.1 (Gillet and Soulé [**13**]). *Let φ be a stable invariant power series. To each exact sequence of Hermitian vector bundle*

$$\bar{\xi}\colon 0\to(S,h')\xrightarrow{f}(E,h)\to(Q,h'')\to 0$$

we can assign a differential form $\widetilde{\varphi}_1(\bar{\xi})\in\bigoplus_p\mathcal{D}^{2p-1}(X,p)$, called the Bott-Chern form. satisfying the following properties.

1. $-2\partial\bar{\partial}\widetilde{\varphi}_1=\widetilde{\varphi}(E,h)-\widetilde{\varphi}((S,h')\oplus(Q,h''))$.
2. *If $f:X\to Y$ is a morphism of complex manifolds then*

$$\widetilde{\varphi}_1(f^*\bar{\xi})=f^*\widetilde{\varphi}_1(\bar{\xi}).$$

3. *If $\bar{\xi}$ is a split exact sequence of Hermitian bundles, then $\widetilde{\varphi}_1(\bar{\xi})=0$.*

Moreover, these properties characterize $\widetilde{\varphi}_1$ up to $\operatorname{Im}\partial+\operatorname{Im}\bar{\partial}$. That is, up to a boundary in the complex $\bigoplus_p\mathcal{D}^*(X,p)$.

The proof that these properties characterize Bott-Chern forms can be found in [3]. Let us give a construction of Bott-Chern forms. The method we will follow is a modification of the method of Gillet and Soulé that avoids the need to choose a partition of unity. In this way we obtain well defined Bott-Chern forms satisfying the conditions of Theorem 6.1 and not only classes up to $\operatorname{Im}\partial + \operatorname{Im}\bar{\partial}$.

A standard procedure to prove that two differential forms α and β, defined on a differential variety Y, are cohomologous is the following. First, one constructs a geometric homotopy between α and β. That is, a differential form η defined on $Y \times \mathbb{R}$ such that $\eta|_{Y\times\{1\}} = \alpha$ and $\eta|_{Y\times 0} = \beta$. From this homotopy one obtains a primitive for $\alpha - \beta$ by integration:

$$d\int_0^1 \eta = \alpha - \beta. \tag{6.3}$$

Let us denote by Δ^1 the current

$$\Delta^1(\eta) = \int_0^1 \eta.$$

Then the equation (6.3) can be expressed as the equation in currents

$$d\Delta^1 = \delta_1 - \delta_0, \tag{6.4}$$

where δ_0 and δ_1 are the Dirac delta currents centered at 0 and 1 respectively. We will adapt the above procedure to Deligne cohomology.

Let us assume for a while that the Hermitian metric h'' on Q is the Hermitian metric induced by the metric h of E. The first step is to construct a geometric homotopy between the Hermitian vector bundles (E,h) and $(S,h')\oplus(Q,h'')$. This homotopy will be parametrized by the complex projective line instead of by the unit interval. Let $(x:y)$ be homogeneous coordinates of $\mathbb{P}^1 = \mathbb{P}^1_{\mathbb{C}}$. Then x and y are sections of the bundle $\mathcal{O}_{\mathbb{P}^1}(1)$. The standard metric of $\mathbb{C}^2$ induces the Fubini-Study metric on $\mathcal{O}_{\mathbb{P}^1}(1)$. Let us denote by g this metric. Then

$$g(x) = \frac{x\bar{x}}{x\bar{x}+y\bar{y}} \quad \text{and} \quad g(y) = \frac{y\bar{y}}{x\bar{x}+y\bar{y}}.$$

Let $p_1\colon X\times\mathbb{P}^1 \to X$ and $p_2\colon X\times\mathbb{P}^1\to\mathbb{P}^1$ denote the projections. Let us write $E(1) = p_1^*E\otimes p_2^*\mathcal{O}(1)$ and $S(1) = p_1^*S\otimes p_2^*\mathcal{O}(1)$. Let us consider the morphism

$$\begin{aligned}\psi\colon\ S &\longrightarrow S(1)\oplus E(1)\\ s &\longmapsto s\otimes y + f(s)\otimes x.\end{aligned}$$

Observe that the vector bundle $S(1)\oplus E(1)$ has a metric induced by the metrics of S, E and $\mathcal{O}(1)$.

DEFINITION 6.2. The *transgression bundle* associated to the exact sequence $\bar{\xi}$ is the Hermitian vector bundle

$$\operatorname{tr}_1(\bar{\xi}) = \operatorname{coker}(\psi)$$

with the Hermitian metric induced by the metric of $S(1)\oplus E(1)$.

The restrictions of the transgression bundle $\operatorname{tr}_1(\bar{\xi})$ are

$$\operatorname{tr}_1(\bar{\xi})|_{X\times(0:1)} = (E,h), \tag{6.5}$$

$$\operatorname{tr}_1(\bar{\xi})|_{X\times(1:0)} = (S,h')\oplus(Q,h''). \tag{6.6}$$

Thus $\mathrm{tr}_1(\bar{\xi})$ is a geometric homotopy between $(S,h')\oplus(Q,h'')$ and (E,h). Note that to obtain (6.5) we had to assume that h'' is the induced metric.

The second step is to obtain a geometric homotopy of differential forms. This homotopy is given by the form $\widetilde{\varphi}(\mathrm{tr}_1(\bar{\xi}))$. Then the functoriality of the Chern forms and the equations (6.5) and (6.6) imply that

$$\widetilde{\varphi}(\mathrm{tr}_1(\bar{\xi}))|_{X\times(0:1)} = \widetilde{\varphi}((E,h)), \tag{6.7}$$

$$\widetilde{\varphi}(\mathrm{tr}_1(\bar{\xi}))|_{X\times(1:0)} = \widetilde{\varphi}((S,h')\oplus(Q,h'')). \tag{6.8}$$

The third step is to integrate the geometric homotopy to obtain a primitive. Let $t = x/y$ be an absolute coordinate in $\mathbb{P}^1$. Let $[\frac{1}{2}\log t\bar{t}]$ denote the current defined by

$$\left[\frac{1}{2}\log t\bar{t}\right](\eta) = \frac{1}{2\pi i}\int_{\mathbb{P}^1}\frac{1}{2}\log t\bar{t}\eta.$$

The current $[\frac{1}{2}\log t\bar{t}]$ will play the role, in Deligne cohomology, that the current Δ^1 plays in de Rham cohomology. The analogue of equation (6.4) is the Poincaré-Lelong equation (see for instance [**16**])

$$-2\partial\bar{\partial}\left[\frac{1}{2}\log t\bar{t}\right] = \delta_{(0:1)} - \delta_{(1:0)}. \tag{6.9}$$

DEFINITION 6.3. The *Bott-Chern form* associated to the exact sequence $\bar{\xi}$ and the power series φ is the differential form

$$\widetilde{\varphi}_1(\bar{\xi}) = \left[\frac{1}{2}\log t\bar{t}\right](\mathrm{tr}_1(\bar{\xi})). \tag{6.10}$$

Equations (6.7), (6.8) and (6.9) imply the condition 1 of Theorem 6.1. Moreover the functoriality (condition 2) is clear from the construction of Bott-Chern forms.

LEMMA 6.4. *If $\bar{\xi}$ is a split exact sequence then $\widetilde{\varphi}_1(\bar{\xi}) = 0$.*

PROOF. Let us consider the morphism $\iota\colon \mathbb{P}^1 \to \mathbb{P}^1$ given by $\iota(x:y) = (y:x)$. The line bundle $\mathcal{O}(1)$ with the Fubini-Study metric is invariant under ι^*. Since $\bar{\xi}$ is split, the map ψ in the definition of the transgression bundle is

$$\begin{aligned}\psi\colon\ S &\longrightarrow S(1)\oplus S(1)\oplus Q(1)\\ s &\longmapsto s\otimes y + s\otimes x + 0\end{aligned}$$

Therefore, by the invariance of $\mathcal{O}(1)$, $\iota^*\,\mathrm{tr}_1(\bar{\xi})$ is the cokernel of the morphism

$$\begin{aligned}\psi'\colon\ S &\longrightarrow S(1)\oplus S(1)\oplus Q(1)\\ s &\longmapsto s\otimes x + s\otimes y + 0,\end{aligned}$$

which is isometric to $\mathrm{tr}_1(\bar{\xi})$. Therefore $\iota^*\widetilde{\varphi}(\mathrm{tr}_1(\bar{\xi})) = \widetilde{\varphi}(\mathrm{tr}_1(\bar{\xi}))$. Thus it is an even form. On the other hand, the current $[\frac{1}{2}\log t\bar{t}]$ is odd: $\iota_*[\frac{1}{2}\log t\bar{t}] = -[\frac{1}{2}\log t\bar{t}]$. Hence $\widetilde{\varphi}_1(\bar{\xi}) = [\frac{1}{2}\log t\bar{t}](\mathrm{tr}_1(\bar{\xi})) = 0$. □

Let us assume now that the metric h'' of Q is arbitrary. Let h''' be the Hermitian metric on Q induced by the metric h. Then from the exact sequence $\bar{\xi}$ we can define two new exact sequences.

$$\lambda^1\bar{\xi}: 0\to(S,h')\to(E,h)\to(Q,h''')\to 0$$
$$\lambda^2\bar{\xi}: 0\to(Q,h'')\to(Q,h''')\to 0\to 0.$$

In both exact sequences the third metric is induced by the second one.

DEFINITION 6.5. Let

$$\bar{\xi} : 0 \to (S, h') \to (E, h) \to (Q, h'') \to 0$$

be an exact sequence of Hermitian vector bundles. Let φ be a stable invariant power series. Then the *Bott-Chern form* associated to φ anf $\bar{\xi}$ is

$$\widetilde{\varphi}_1(\bar{\xi}) = \widetilde{\varphi}_1(\lambda^1 \bar{\xi}) + \widetilde{\varphi}_1(\lambda^2 \bar{\xi})$$

REMARK 6.6. Since, by Lemma 6.4 the Bott-Chern of the exact sequence

$$0 \to (Q, h) \to (Q, h) \to 0 \to 0$$

is zero, if $\bar{\xi}$ is an exact sequence with the third metric induced by the second one then Definition 6.3 gives the same result as Definition 6.5.

7. Exact Cubes

Let X be a smooth quasi-projective variety over $\mathbb{C}$. Let us fix a small full subcategory $\mathfrak{E} = \mathfrak{E}(X)$ of the category of algebraic vector bundles over X, which is equivalent to it.

Let $\langle -1, 0, 1 \rangle$ be the category associated to the ordered set $\{-1, 0, 1\}$. Let $\langle -1, 0, 1 \rangle^n$ be the nth Cartesian power, and let $\langle -1, 0, 1 \rangle^0$ be the category with one element and one morphism. Following Loday [**22**], we define exact cubes as follows:

DEFINITION 7.1. An exact n-cube of $\mathfrak{E}$ is a functor $\mathcal{F}$ from $\langle -1, 0, 1 \rangle^n$ to $\mathfrak{E}$ such that, for all integers $1 \leq i \leq n$, and all $n-1$-tuples $(\alpha_1, \dots, \alpha_{n-1}) \in \langle -1, 0, 1 \rangle^{n-1}$ the sequence

$$\mathcal{F}_{\alpha_1, \dots, \alpha_{i-1}, -1, \alpha_i, \dots, \alpha_{n-1}} \to \mathcal{F}_{\alpha_1, \dots, \alpha_{i-1}, 0, \alpha_i, \dots, \alpha_{n-1}} \to \mathcal{F}_{\alpha_1, \dots, \alpha_{i-1}, 1, \alpha_i, \dots, \alpha_{n-1}}$$

is a short exact sequence. We have written $\mathcal{F}_{\alpha_1, \dots, \alpha_n}$ for $\mathcal{F}(\alpha_1, \dots, \alpha_n)$.

We will denote by $\underline{C}_n \mathfrak{E}$ the category of exact n-cubes. It is a small exact category. We will write $C_n \mathfrak{E} = \mathrm{Ob}(\underline{C}_n \mathfrak{E})$.

DEFINITION 7.2. Given an exact n-cube $\mathcal{F}$ and integers $i \in \{1, \dots, n\}$, $j \in \{-1, 0, 1\}$, *the face* $\partial_i^j \mathcal{F}$ is the exact $n-1$-cube defined by

$$\left(\partial_i^j \mathcal{F} \right)_{\alpha_1, \dots, \alpha_{n-1}} = \mathcal{F}_{\alpha_1, \dots, \alpha_{i-1}, j, \alpha_i, \dots, \alpha_{n-1}}.$$

EXAMPLES 7.3.
1. An exact 0-cube is an element of $\mathrm{Ob}(\mathfrak{E})$.
2. An exact 1-cube is an exact sequence of objects of $\mathfrak{E}$.
3. For each $i \in \{1, \dots, n\}$, we can see an exact n-cube $\mathcal{F}$ as the exact sequence of exact $n-1$-cubes

$$0 \to \partial_i^{-1} \mathcal{F} \to \partial_i^0 \mathcal{F} \to \partial_i^1 \mathcal{F} \to 0$$

   This exact sequence will be denoted $\partial_i \mathcal{F}$. Note that $\mathcal{F}$ is characterized by any of the exact sequences $\partial_i \mathcal{F}$.

Let $\mathbb{Z}C_n(\mathfrak{E})$ be the free Abelian group generated by $C_n(\mathfrak{E})$. Let us define a differential $d \colon \mathbb{Z}C_n(\mathfrak{E}) \to \mathbb{Z}C_{n-1}(\mathfrak{E})$ by the formula

$$d = \sum_{i=1}^{n} \sum_{j=-1}^{1} (-1)^{i+j+1} \partial_i^j.$$

It is easy to see that $d^2 = 0$; thus we have obtained a homology complex denoted $\mathbb{Z}C_*(\mathfrak{C})$. Since we are using a cubic theory, in order to obtain the right homology, we need to factor out by the degenerate elements.

For each exact $n-1$-cube $\mathcal{F}$, and each integer $i \in \{1, \dots, n\}$ we will denote by $s^i_{-1}\mathcal{F}$ the exact n-cube defined by the exact sequence (see Example 7.3.3)

$$\partial^{\cdot}_i(s^i_{-1}\mathcal{F}) : 0 \to 0 \to \mathcal{F} \overset{\mathrm{Id}}{\to} \mathcal{F} \to 0.$$

Analogously we define s^i_1 by the exact sequence

$$\partial^{\cdot}_i(s^i_1\mathcal{F}) : 0 \to \mathcal{F} \xrightarrow{\mathrm{Id}} \mathcal{F} \to 0 \to 0.$$

The exact cubes in the image of s_{-1} and s_1 are called degenerate n-cubes. Clearly the differential d sends a degenerate cube to a linear combination of degenerate cubes. Therefore the subgroup generated by all degenerated cubes form a subcomplex of $\mathbb{Z}C_*(\mathfrak{C})$ denoted by D_*.

DEFINITION 7.4. The *reduced cubical complex* of the category $\mathfrak{C}$ is

$$\widetilde{\mathbb{Z}}C_*\mathfrak{C} = \mathbb{Z}C_*\mathfrak{C}/D_*.$$

The homology of the complex $\mathbb{Z}C^{\mathrm{red}}_*\mathfrak{C}$ is closely related to the K-theory of X. For instance, let $S_*\mathfrak{C}$ be the Waldhausen space associated with the category $\mathfrak{C}$ [**32**]. Then

$$K_i(X) = \pi_{i+1}|S_*\mathfrak{C}|.$$

Now, as in [**33**], [**8**] or [**24**], one can construct a morphism of complexes

$$\gamma \colon \mathbb{Z}S.\mathfrak{C}[1] \to \widetilde{\mathbb{Z}}C_*\mathfrak{C}.$$

Composing with the Hurewicz morphism one obtains a natural map

$$K_i\mathfrak{C} \to H_i(\widetilde{\mathbb{Z}}C_*\mathfrak{C}).$$

For the purpose of constructing characteristic classes this map is enough (see [**8**]). But R. McCarthy [**24**] has given a precise description of the homology of $\widetilde{\mathbb{Z}}C_*\mathfrak{C}$ that makes this complex much more interesting.

THEOREM 7.5. *The homology of $\widetilde{\mathbb{Z}}C_*\mathfrak{C}$ is the homology of the algebraic K-theory spectrum of the category $\mathfrak{C}$. In particular*

$$K_i(\mathfrak{C}) \otimes \mathbb{Q} \cong H_i(\widetilde{\mathbb{Z}}C_*\mathfrak{C}) \otimes \mathbb{Q}.$$

Moreover, the use of cubes makes $\widetilde{\mathbb{Z}}C_*\mathfrak{C}$ very well behaved to study products.

DEFINITION 7.6. Let $\mathcal{F}$ be an exact n-cube and let $\mathcal{G}$ be an exact m-cube. Then $\mathcal{F} \otimes \mathcal{G}$ is the exact $n+m$-cube given by

$$(\mathcal{F} \otimes \mathcal{G})_{\alpha_1,\dots,\alpha_{n+m}} = \mathcal{F}_{\alpha_1,\dots,\alpha_n} \otimes \mathcal{G}_{\alpha_{n+1},\dots,\alpha_{n+m}}.$$

This product makes $\widetilde{\mathbb{Z}}C_*\mathfrak{C}$ an associative differential algebra which is homotopically commutative. Therefore its homology has the structure of an associative and commutative algebra.

THEOREM 7.7 (R. McCarthy [**24**]). *The morphism*

$$K_*(\mathfrak{C}) \to H_*(\widetilde{\mathbb{Z}}C_*\mathfrak{C})$$

is multiplicative.

We want to introduce Hermitian metrics in the vector bundles.

DEFINITION 7.8. Let $\overline{\mathfrak{E}} = \overline{\mathfrak{E}}(X)$ be the category with

$$\operatorname{Ob} \overline{\mathfrak{E}} = \{(E,h) \mid E \in \operatorname{Ob} \mathfrak{E}, h \text{ Hermitian metric on } E\}$$

and

$$\operatorname{Hom}_{\overline{\mathfrak{E}}}((E,h),(F,g)) = \operatorname{Hom}_{\mathfrak{E}}(E,F).$$

For each vector bundle $E \in \mathfrak{E}$ let us choose a Hermitian metric h_E. This gives us a functor $\mathfrak{F}\colon \mathfrak{E} \to \overline{\mathfrak{E}}$. Let $\mathfrak{G}\colon \overline{\mathfrak{E}} \to \mathfrak{E}$ be the functor *forget the metric*. These functors are equivalences inverse to each other. In particular, this implies the following result.

LEMMA 7.9. *The functors $\mathfrak{F}$ and $\mathfrak{G}$ induce morphisms of complexes*

$$\mathfrak{F}\colon \widetilde{\mathbb{Z}C}_*\mathfrak{E} \to \widetilde{\mathbb{Z}C}_*\overline{\mathfrak{E}},$$
$$\mathfrak{G}\colon \widetilde{\mathbb{Z}C}_*\overline{\mathfrak{E}} \to \widetilde{\mathbb{Z}C}_*\mathfrak{E},$$

which are homotopy equivalences, inverse one of the other.

PROOF. It is clear that $\mathfrak{G}\circ\mathfrak{F} = \operatorname{Id}$. In addition the homotopy $\mathfrak{h}$ between $\mathfrak{F}\circ\mathfrak{G}$ and the identity is given in the following way. Let $\overline{\mathcal{F}}$ be an element of $\widetilde{\mathbb{Z}C}_n\overline{\mathfrak{E}}$. Then $\mathfrak{h}\overline{\mathcal{F}}$ is the exact $n+1$ cube defined by the exact sequence

$$\partial_n^{\cdot}(\mathfrak{h}\overline{\mathcal{F}}) : 0 \to \mathfrak{F}\circ\mathfrak{G}(\overline{\mathcal{F}}) \xrightarrow{\operatorname{Id}} \overline{\mathcal{F}} \to 0 \to 0. \qquad \square$$

To define higher Bott-Chern forms for exact cubes of Hermitian vector bundles, extending the technique of Section 6, we need the third metric in any short exact sequence to be induced by the middle metric. To this end we introduce the following notation.

DEFINITION 7.10. Let $\overline{\mathcal{F}} = \{(E_\alpha, h_\alpha)\}$ be an exact n-cube of Hermitian vector bundles. We say that $\overline{\mathcal{F}}$ is an emi-n-cube, if, for each n-tuple $\alpha = (\alpha_1,\dots,\alpha_n)$, and each i with $\alpha_i = 1$, the metric h_α is induced by the metric $h_{(\alpha_1,\dots,\alpha_{i-1},0,\alpha_{i+1},\dots,\alpha_n)}$.

Let $\mathbb{Z}C^{\mathrm{emi}}\overline{\mathfrak{E}}$ be the subcomplex of $\mathbb{Z}C\overline{\mathfrak{E}}$ generated by the the emi-n-cubes, and let D^{emi} be the subcomplex generated by the degenerate emi-n-cubes. We will write

$$\widetilde{\mathbb{Z}C}^{\mathrm{emi}}\overline{\mathfrak{E}} = \mathbb{Z}C^{\mathrm{emi}}\overline{\mathfrak{E}}/D^{\mathrm{emi}} \subset \widetilde{\mathbb{Z}C}\overline{\mathfrak{E}}.$$

Let us see that the complexes $\widetilde{\mathbb{Z}C}\overline{\mathfrak{E}}$ and $\widetilde{\mathbb{Z}C}^{\mathrm{emi}}\overline{\mathfrak{E}}$ are homotopically equivalent. Let $\overline{\mathcal{F}} = \{(F_\alpha, h_\alpha)\} \in C_n\overline{\mathfrak{E}}$. For $i = 1,\dots,n$ let $\lambda_i^1\overline{\mathcal{F}}$ be defined by

$$\lambda_i^1\overline{\mathcal{F}}_\alpha = \begin{cases} (\mathcal{F}_\alpha, h_\alpha), & \text{if } \alpha_i = -1, 0,\\ (\mathcal{F}_\alpha, h'_\alpha), & \text{if } \alpha_i = 1,\end{cases}$$

where h'_α is the metric induced by $h_{(\alpha_1,\dots,\alpha_{i-1},0,\alpha_{i+1},\dots,\alpha_n)}$. Thus the operator λ_i^1 changes the metrics of the face $\partial_i^1\overline{\mathcal{F}}$ by those induced by the metrics of the face $\partial_i^0\overline{\mathcal{F}}$.

Let $\lambda_i^2\overline{\mathcal{F}}$ be the exact n-cube determined by the exact sequence

$$\partial_i^{\cdot}(\lambda_i^2\overline{\mathcal{F}}) : 0 \to \partial_i^1\overline{\mathcal{F}} \to \partial_i^1\lambda_i^1\overline{\mathcal{F}} \to 0 \to 0.$$

This n-cube measures the difference between $\overline{\mathcal{F}}$ and $\lambda_i^1\overline{\mathcal{F}}$.

Let us write $\lambda_i\overline{\mathcal{F}} = \lambda_i^1\overline{\mathcal{F}} + \lambda_i^2\overline{\mathcal{F}}$, and let us denote by λ the map

$$\lambda\colon\ \mathbb{Z}C_n\overline{\mathfrak{E}} \longrightarrow \mathbb{Z}C_n\overline{\mathfrak{E}}$$

$$\overline{\mathcal{F}} \longmapsto \begin{cases} \lambda_n \dots \lambda_1\overline{\mathcal{F}}, & \text{if } n \geq 1, \\ \overline{\mathcal{F}}, & \text{if } n = 0. \end{cases}$$

The map λ is a morphism of complexes. Moreover, the image of λ lies in the set of emi-n-cubes and λ sends degenerate cubes to degenerate cubes. Therefore the morphism λ induces a morphism of complexes

$$\tilde{\lambda}\colon \widetilde{\mathbb{Z}}C\overline{\mathfrak{E}} \to \widetilde{\mathbb{Z}}C^{\mathrm{emi}}\overline{\mathfrak{E}}.$$

Proposition 7.11. *The morphism $\tilde{\lambda}$ is a homotopy equivalence.*

Proof. Let us denote by

$$\iota\colon \widetilde{\mathbb{Z}}C^{\mathrm{emi}}\overline{\mathfrak{E}} \to \widetilde{\mathbb{Z}}C\overline{\mathfrak{E}}$$

the inclusion. Observe that, if $\overline{\mathcal{F}}$ is an emi-n-cube then

$$\lambda\overline{\mathcal{F}} = \overline{\mathcal{F}} + \text{degenerate elements.}$$

Therefore $\tilde{\lambda} \circ \iota = \mathrm{Id}$. Let $\mathfrak{F}$ and $\mathfrak{G}$ be the morphisms of complexes of Lemma 7.9. Since $\mathfrak{F} \circ \mathfrak{G}$ is homotopically equivalent to the identity, we obtain the equivalence

$$\mathfrak{F} \circ \mathfrak{G} \circ \iota \circ \tilde{\lambda} \sim \iota \circ \tilde{\lambda}.$$

But, since the functor $\mathfrak{G}$ forgets the metric, $\mathfrak{G}(\iota \circ \lambda\overline{\mathcal{F}}) - \mathfrak{G}(\overline{\mathcal{F}})$ consists only in degenerate cubes. Therefore $\mathfrak{F} \circ \mathfrak{G} \circ \iota \circ \tilde{\lambda} = \mathfrak{F} \circ \mathfrak{G}$. In consequence $\iota \circ \tilde{\lambda} \sim \mathrm{Id}$. □

8. Higher Bott-Chern forms

Let X be a smooth complex projective variety. The aim of this section is to give a morphism between the complex of emi-cubes and the complex $\mathcal{D}^*(X,*)$. This morphism will realize the characteristic classes from higher K-theory to real Deligne cohomology.

Observe that, since we want to realize the characteristic classes as a morphism of complexes of Abelian groups, we will obtain a morphism of groups. In particular the induced map

$$K_0(X) \to \bigoplus_p H^{2p}_{\mathcal{D}}(X,p)$$

will be additive. This forces us to choose the Chern character as our characteristic class. Nevertheless, since the Chern classes can be recovered from the components of the Chern character form, the formulae we obtain can be applied to any characteristic class.

The reason we restrict ourselves to projective varieties is to avoid the technical difficulties of the logarithmic singularities at infinity. But note that a main ingredient in the proof that higher Bott-Chern forms give Beilinson's regulator is the extension to quasi-projective varieties (see [**8**])

Let us see that Bott-Chern forms are the degree one step of the morphism of complexes we are looking for. If

$$\bar{\xi}\colon 0 \to (S,h') \xrightarrow{f} (E,h) \to (Q,h'') \to 0$$

is an exact sequence of Hermitian vector bundles, then

$$d\bar{\xi} = (E,h) - (Q,h'') - (S,h').$$

Therefore

$$\begin{aligned} d_{\mathcal{D}}\,\widetilde{\mathrm{ch}}_1(\bar{\xi}) &= -2\partial\bar{\partial}\,\widetilde{\mathrm{ch}}_1(\bar{\xi}) \\ &= \widetilde{\mathrm{ch}}(E,h) - \widetilde{\mathrm{ch}}(Q,h'') - \widetilde{\mathrm{ch}}(S,h') \\ &= \widetilde{\mathrm{ch}}(d\bar{\xi}). \end{aligned}$$

To extend this morphism to higher degrees, we will iterate the definition of Bott-Chern forms.

Let $\overline{\mathcal{F}} = \{\overline{\mathcal{F}}_\alpha\}$ be an emi-n-cube. The first step is to construct a geometric nth order homotopy between the vertexes of $\overline{\mathcal{F}}$. The reason for calling it a homotopy will be clear in proposition 8.3. This homotopy will be a Hermitian vector bundle defined over $X \times (\mathbb{P}^1)^n$ and will be called the nth transgression bundle. Moreover, we want the nth transgression bundle to be an exact functor. Let us define it inductively. If $n = 1$, an emi-1-cube is a short exact sequence $\bar{\xi}$ of Hermitian vector bundles with the third metric induced by the second one. Then, the definition of the first transgression bundle is given in 6.2. It follows from the definition that tr_1 is an exact functor. Assume now that we have defined the functor tr_{n-1} and that it is an exact functor. As in Example 7.3.3, the emi-n-cube $\overline{\mathcal{F}}$ can be seen as an exact sequence of emi-$n-1$-cubes:

$$\partial_n^{\cdot}\overline{\mathcal{F}} \colon \partial_n^{\cdot}(\overline{\mathcal{F}})0 \to \partial_n^{-1}\overline{\mathcal{F}} \to \partial_n^{0}\overline{\mathcal{F}} \to \partial_n^{1}\overline{\mathcal{F}} \to 0.$$

Applying the functor tr_{n-1} to this exact sequence of emi-$n-1$-cubes, we obtain an exact sequence of Hermitian vector bundles on $X \times (\mathbb{P}^1)^{n-1}$, denoted $\mathrm{tr}_{n-1}(\partial_n^{\cdot}\overline{\mathcal{F}})$.

Definition 8.1. Let $\overline{\mathcal{F}}$ be an emi-n-cube. Then the *nth transgression bundle* is

$$\mathrm{tr}_n(\overline{\mathcal{F}}) = \mathrm{tr}_1(\mathrm{tr}_{n-1}(\partial_n^{\cdot}\overline{\mathcal{F}})).$$

Since tr_1 is an exact functor, and by induction hypothesis we may assume that tr_{n-1} is also an exact functor, we obtain that tr_n is also an exact functor.

Remark 8.2. An emi-n-cube can be seen as an exact sequence of emi-$n-1$-cubes in n different ways depending on which faces we take. Thus the above construction may depend, in principle, on the choice of an ordering of the subindexes. Nevertheless, the result is independent of this order. See for instance [**8**, Definition 3.8] for a more symmetric definition or [**28**, Proposition 2.1] for a proof of the invariance under permutations.

The basic property of the transgression bundle is the following.

Proposition 8.3 ([**8**, Proposition 3.9]). *Let $\overline{\mathcal{F}}$ be an emi-n-cube. Let $(x_i : y_i)$ be homogeneous coordinates in the ith factor of $(\mathbb{P}^1)^n$. Then*

$$\mathrm{tr}_n(\overline{\mathcal{F}})|_{\{x_i=0\}} \cong \mathrm{tr}_{n-1}(\partial_i^0\overline{\mathcal{F}}),$$

$$\mathrm{tr}_n(\overline{\mathcal{F}})|_{\{y_i=0\}} \cong \mathrm{tr}_{n-1}(\partial_i^{-1}\overline{\mathcal{F}}) \overset{\perp}{\oplus} \mathrm{tr}_{n-1}(\partial_i^1\overline{\mathcal{F}}).$$

In view of this proposition, the nth transgression bundle of an emi-n-cube is a homotopy between the $n-1$-transgression bundles of its faces.

The second step is to go from a homotopy of vector bundles to a homotopy of differential forms. This step is simple; the required homotopy is $\widetilde{\mathrm{ch}}(\mathrm{tr}_n(\overline{\mathcal{F}}))$ because by the functoriality of the Chern character form and Proposition 8.3 we obtain that

$$\widetilde{\mathrm{ch}}(\mathrm{tr}_n(\overline{\mathcal{F}}))|_{\{x_i=0\}} \cong \widetilde{\mathrm{ch}}(\mathrm{tr}_{n-1}(\partial_i^0\overline{\mathcal{F}})), \tag{8.1}$$

$$\widetilde{\mathrm{ch}}(\mathrm{tr}_n(\overline{\mathcal{F}}))|_{\{y_i=0\}} \cong \widetilde{\mathrm{ch}}(\mathrm{tr}_{n-1}(\partial_i^{-1}\overline{\mathcal{F}})) + \widetilde{\mathrm{ch}}(\mathrm{tr}_{n-1}(\partial_i^1\overline{\mathcal{F}})). \tag{8.2}$$

The third step is to integrate the differential form $\widetilde{\mathrm{ch}}(\mathrm{tr}_n(\overline{\mathcal{F}}))$ defined on $X \times (\mathbb{P}^1)^n$, to obtain a differential form, $\widetilde{\mathrm{ch}}_n(\overline{\mathcal{F}})$, defined on X.

To this end we will introduce some currents on $(\mathbb{P}^1)^n$. Let us introduce the homological analogue of the complex $\mathcal{D}^*$, where these currents will live. For any smooth complex projective variety Y, let $D_*(Y)$ be the complex of currents on Y. That is, $D_n(Y)$ is the topological dual of $\mathcal{E}^n(Y)$. We will denote by $D_*^{\mathbb{R}}(Y)$ the subcomplex of real currents and by $D_{p,q}(Y)$ the currents of type p, q (i.e., the topological dual of $\mathcal{E}^{p,q}(Y)$). We will write $D_*^{\mathbb{R}}(Y)(p) = (2\pi i)^{-p} D_*^{\mathbb{R}}(Y)$.

DEFINITION 8.4. Let $D_*(Y,*)$ be the complex defined by

$$\mathcal{D}_n(Y,p) = \begin{cases} D_n^{\mathbb{R}}(p)(Y) \cap \displaystyle\bigoplus_{\substack{p'+q'=n \\ p'\leq p,\ q'\leq p}} D_{p',q'}(Y), & \text{for } n \leq 2p. \\ D_{n+1}^{\mathbb{R}}(p+1)(X) \cap \displaystyle\bigoplus_{\substack{p'+q'=n+1 \\ p'>p,\ q'>q}} D_{p',q'}(X), & \text{for } n \geq 2p+1. \end{cases}$$

The homology of the above complex is the Deligne homology of Y. If Y is equidimensional of dimension n then, for any form $\omega \in D^j(Y,p)$, we will denote by $[\omega] \in D_{2n-j}(Y, n-p)$ the current defined by

$$[\omega](\eta) = \frac{1}{(2\pi i)^n} \int_Y \omega \wedge \eta. \tag{8.3}$$

This morphism realizes the Poincaré duality. If ω is a locally integrable form, we will use also the notation $[\omega]$ to denote its associated current.

Let us denote by $d_j^i \colon (\mathbb{P}^1)^{n-1} \to (\mathbb{P}^1)^n$, for $i = 1, \dots, n$ and $j = 0, \infty$ the inclusions given by

$$d_0^i(x_1, \dots, x_n) = (x_1, \dots, x_{i-1}, (0:1), x_i, \dots, x_n)$$
$$d_\infty^i(x_1, \dots, x_n) = (x_1, \dots, x_{i-1}, (1:0), x_i, \dots, x_n).$$

The currents we need in order to integrate the form $\widetilde{\mathrm{ch}}(\mathrm{tr}_n(\overline{\mathcal{F}}))$ are provided by the following result.

THEOREM 8.5 ([**33**], see also [**8**] and [**15**]). *There exists a family of currents* $\{[W_n]\}_{n\geq 0}$ *with* $[W_n] \in \mathcal{D}_n((\mathbb{P}^1)^n, 0)$ *such that*

1. $[W_0] = 1$.
2. $d_{\mathcal{D}}[W_n] = \sum_{i=1}^n (-1)^i \left((d_0^i)_*[W_{n-1}] - (d_\infty^i)_*[W_{n-1}]\right)$.

PROOF. By equation (6.9), we can write $[W_1] = [\frac{1}{2}\log t\bar{t}]$. Let $p_i \colon (\mathbb{P}^1)^n \to \mathbb{P}^1$ denote the projection over the ith factor. Let us write $\lambda_i = p_i^*(\frac{1}{2}\log t\bar{t})$. Then λ_i is a locally integrable function over $(\mathbb{P}^1)^n$. Let I' and E' be the homotopy equivalences introduced at the end of Section 5. Then we will write

$$W_n = I'(E'(\lambda_1) \cup \cdots \cup E'(\lambda_n)) \tag{8.4}$$

which is a locally integrable form. The current $[W_n]$ is the associated current. Condition 2 is, formally, consequence of the equation (6.9) and Leibnitz rule (see [**8**, Proposition 6.7] and [**33**]). Explicitly, the forms W_n are given by (compare [**15**, 2.2])

$$W_n = \frac{(-1)^n}{2\,n!}\sum_{i=1}^{n}\sum_{\sigma\in\mathfrak{S}_n}(-1)^{i-1}(-1)^{\sigma}\log(t_{\sigma(1)}\bar{t}_{\sigma(1)}) \\ \frac{dt_{\sigma(2)}}{t_{\sigma(2)}}\wedge\cdots\wedge\frac{dt_{\sigma(i)}}{t_{\sigma(i)}}\wedge\frac{d\bar{t}_{\sigma(i+1)}}{\bar{t}_{\sigma(i+1)}}\wedge\cdots\wedge\frac{d\bar{t}_{\sigma(n)}}{\bar{t}_{\sigma(n)}}. \qquad \square$$

DEFINITION 8.6. Let $\overline{\mathcal{F}}$ be an emi-n-cube. Then the nth Bott-Chern form of $\overline{\mathcal{F}}$ is

$$\widetilde{\mathrm{ch}}_n(\overline{\mathcal{F}}) = [W_n](\widetilde{\mathrm{ch}}(\mathrm{tr}_n(\overline{\mathcal{F}}))) = \frac{1}{(2\pi i)^n}\int_{(\mathbb{P}^1)^n} W_n\wedge\widetilde{\mathrm{ch}}(\mathrm{tr}_n(\overline{\mathcal{F}})).$$

Let $\overline{\mathfrak{E}}$ be a small category of Hermitian vector bundles over X (see Section 7). Let us write $\widetilde{\mathbb{Z}}C^n_{\mathrm{emi}}\mathfrak{E} = \widetilde{\mathbb{Z}}C^{\mathrm{emi}}_{-n}\mathfrak{E}$. Then $\widetilde{\mathbb{Z}}C^*_{\mathrm{emi}}\mathfrak{E}$ is a cohomological complex. The definition of higher Bott-Chern forms induces maps

$$\mathrm{ch}\colon \widetilde{\mathbb{Z}}C^n_{\mathrm{emi}}\mathfrak{E}\to\bigoplus_p\mathcal{D}^n(X,p)[2p].$$

PROPOSITION 8.7. *The induced map*

$$\mathrm{ch}\colon \widetilde{\mathbb{Z}}C^*_{\mathrm{emi}}\mathfrak{E}\to\bigoplus_p\mathcal{D}^*(X,p)[2p].$$

is a morphism of complexes.

PROOF. This proposition is a direct consequence of (8.1) and Theorem 8.5. $\square$

The main result concerning higher Bott-Chern forms is

THEOREM 8.8 ([**8**]). *The composition map*

$$K_i(X)\xrightarrow{\text{Hurewicz}} H^{-i}(\widetilde{\mathbb{Z}}C^*_{\mathrm{emi}}\mathfrak{E})\to\bigoplus_p H^{2p-i}_{\mathcal{D}}(X,\mathbb{R}(p))$$

agrees with Beilinson's regulator map.

REMARK 8.9. The construction of higher Bott-Chern can be made working always with the Thom-Whitney simple. We define the Chern character form in the Thom-Whitney complex as

$$\widetilde{\mathrm{ch}}(\overline{\mathcal{F}})_{TW} = E'(\widetilde{\mathrm{ch}}(\overline{\mathcal{F}})).$$

The analogues of the forms W_n are the forms

$$(W_n)_{TW} = E'(\lambda_1)\cup\cdots\cup E'(\lambda_n).$$

Then, for any emi-n-cube,$\overline{\mathcal{F}}$, the higher Bott-Chern form in the Thom-Whitney complex is

$$\widetilde{\mathrm{ch}}_n(\overline{\mathcal{F}})_{TW} = \frac{1}{(2\pi i)^n}\int_{(\mathbb{P}^1)^n}(W_n)_{TW}\cup\widetilde{\mathrm{ch}}(\mathrm{tr}_n(\overline{\mathcal{F}}))_{TW},$$

where the integral is computed componentwise. In this way we obtain a morphism which is multiplicative at the level of complexes (see [**8**]).

References

1. A. A. Beilinson, *Notes on absolute Hodge cohomology*, Applications of Algebraic K-Theory to Algebraic Geometry and Number Theory, Part I (Boulder, 1983) (S. J. Bloch, ed.), Contemp. Math., vol. 55, Amer. Math. Soc., Providence, RI, 1986, pp. 35–68.
2. ———, *Higher regulators and values of L-functions*, J. Soviet Math. **30** (1985), 2036–2070.
3. J.-M. Bismut, H. Gillet, and C. Soulé, *Analytic torsion and holomorphic determinant bundles. I. Bott-Chern forms and analytic torsion*, Comm. Math. Phys. **115** (1988), 49–78.
4. A. Borel, *Cohomologie de SL_n et valeurs de fonctions zêta aux points entiers*, Ann. Scuola Norm. Sup. Pisa Cl. Sci. (4) **4** (1977), 613–636.
5. R. Bott and S. S. Chern, *Hermitian vector bundles and the equidistribucion of the zeroes of their holomorphic sections*, Acta Math. **114** (1968), 71–112.
6. R. Bott and L. W. Tu, *Differential forms in algebraic topology*, Grad. Text in Math., vol. 82, Springer-Verlag, New York-Berlin, 1982.
7. J. I. Burgos, *Arithmetic Chow rings and Deligne-Beilinson cohomology*, J. Algebraic Geom. **6** (1997), 335–377.
8. J. I.o Burgos and S. Wang, *Higher Bott-Chern forms and Beilinson's regulator*, Invent. Math. **132** (1998), 261–305.
9. P. Deligne, *Le déterminant de la cohomologie*, Current Trends id Arithmetic Algebraic Geometry (Arcata, CA, 1985) (K. A. Ribet, ed.), Contemp. Math., vol. 67, Amer. Math. Soc., Providence, RI, 1987, pp. 93–177.
10. H. Esnault and E. Viehweg, *Deligne-Beilinson cohomology*, Beilinson's Conjectures on Special Values of L-Functions (M. Rappoport, N. Schappacher, and P. Schneider, eds.), Perspect. Math., vol. 4, Academic Press, Boston, MA, 1988, pp. 43–91.
11. H. Gillet, *Riemann-Roch theorems for higher algebraic K-theory*, Adv. in Math. **40** (1981), 203–289.
12. ———, *Universal cycle classes*, Compositio Math. **49** (1983), 3–49.
13. H. Gillet and C. Soulé, *Characteristic classes for algebraic vector bundles with Hermitian metric.* I, Ann. of Math. (2) **131** (1990), 163–203; II, 205–238.
14. P. G. Goerss and J. F. Jardine, *Simplicial homotopy theory*, Preprint 148, K-Theory Preprint Archives, http://www.math.uiuc.edu/K-theory/, 1997.
15. A. B. Goncharov, *Chow polylogarithms and regulators*, Math. Res. Lett. **2** (1995), 95–112.
16. P. Griffiths and J. Harris, *Principles of algebraic geometry*, Wiley Classics Lib., John Wiley & Sons, Inc., New York, 1994.
17. A. Grothendieck, *La théorie des classes de Chern*, Bull. Soc. Math. France **86** (1958), 137–154.
18. F. Hirzebruch, *Topological methods in algebraic geometry*, 3th ed., Grundlehren Math. Wiss., vol. 131, Springer Verlag, New York, 1978.
19. U. Jannsen, *Deligne homology, Hodge-$\mathcal{D}$-conjecture and motives*, Beilinson's Conjectures on Special Values of L-Functions (M. Rappoport, N. Schappacher, and P. Schneider, eds.), Perspect. Math., vol. 4, Academic Press, Boston, MA, 1988, pp. 305–372.
20. J.-P. Jouanolou, *Une suite exacte de Mayer-Vietoris en K-théorie algébrique*, Algebraic K-Theory. I Higher K-Theories (Seattle, WA, 1972) (H. Bass, ed.), Lecture Notes in Math., vol. 341, Springer Berlin, 1973, pp. 293–316.
21. J.-L. Loday, *K-théorie algébrique et représentations de groupes*, Ann. Sci. École Norm. Sup. (4) **9** (1976), 309–377.
22. ———, *Spaces with finitely many non trivial homotopy groups*, J. Pure Appl. Algebra **24** (1982), 179–202.
23. J. P. May, *Simplicial objects in algebraic topology*, Van Nostrand Mathematical Studies, vol. 11, D. Van Nostrand, Princeton-Toronto-London, 1967.
24. R. McCarthy, *A chain complex for the spectrum homology of the algebraic K-theory of an exact category*, Algebraic K-Theory (Toronto, ON, 1996) (V. P Snaith, ed.), Fields Inst. Commun. vol. 16, Amer. Math. Soc., Providence, RI, 1997, pp. 199–220.
25. J. W. Milnor and J. D. Stasheff, *Characteristic classes*, Ann. of Math. Stud., vol. 76, Princeton Univ. Press, Princeton, NJ, 1974.
26. V. Navarro Aznar, *Sur la théorie de Hodge-Deligne*, Invent. Math. **90** (1987), 11–76.
27. M. Rappoport, *Comparison ot the regulators of Beilinson and of Borel*, Beilinson's Conjectures on Special Values of L-Functions (M. Rappoport, N. Schappacher, and P. Schneider, eds.), Perspect. Math., vol. 4, Academic Press, Boston, MA, 1988, pp. 169–192.

28. D. Roessler, *Analytic torsion for cubes of vector bundles and Gillet's Riemann-Roch theorem*, J. Algebraic Geom (to appear).
29. J. J. Rotman, *An introduction to homological algebra*, Pure Appl. Math., vol. 85, Academic Press, NewYork-London, 1979.
30. V. V. Schechtman, *On the delooping of Chern character and Adams operations*, K-theory, Arithmetic and Geometry (Yu. I. Manin, ed.), Lecture Notes in Math., vol. 1289, Springer, Berlin-New York 1987, pp. 265–319.
31. P. Schneider, *Introduction to the Beilinson conjectures*, Beilinson's Conjectures on Special Values of L-Functions (M. Rappoport, N. Schappacher, and P. Schneider, eds.), Perspect. Math., vol. 4, Academic Press, Boston, MA, 1988, pp. 1–35.
32. F. Waldhausen, *Algebraic K-theory of generalized free products*. Part 1, Ann. of Math. **108** (1978), 135–204; Part 2, 205–256.
33. X. Wang, *Higher order characteristic classes in arithmetic geometry*, Ph.D. thesis, Harvard, 1992.
34. R. O. Wells, *Differential analysis on complex manifolds*, Grad. Texts in Math., vol. 65, Springer-Verlag, New York-Berlin, 1980.

DEPARTAMENTO DE ALGEBRA Y GEOMETRÍA, UNIVERSIDAD DE BARCELONA, GRAN VIA 585, 08007 BARCELONA, SPAIN

E-mail address: burgos@cerber.mat.ub.es

Centre de Recherches Mathématiques
CRM Proceedings and Lecture Notes
Volume **24**, 2000

The Mixed Motive of a Projective Variety

Masaki Hanamura

Let $\mathcal{D}(k)$ be the triangulated category of mixed motives over a field k defined in [**Ha**], and $(\mathrm{Proj}\,/k)$ (resp. $(\mathrm{Smooth\,Proj}\,/k)$) the category of projective varieties (resp. smooth projective varieties) over k. There is a functor

$$h\colon (\mathrm{Smooth\,Proj}\,/k)^{\mathrm{opp}} \to \mathcal{D}(k).$$

Assuming any variety admits resolution of singularities, we will prove that the functor extends to a functor (the *cohomological motive functor*)

$$h\colon (\mathrm{Proj}\,/k)^{\mathrm{opp}} \to \mathcal{D}(k).$$

For a projective variety X, $h(X)$ is its cohomological motive, and its dual $h(X)^\vee$ homological motive. Composing h with the cycle complex functor

$$\mathcal{Z}^r\colon \mathcal{D}(k) \to D(\mathbb{Q})$$

where $D(\mathbb{Q})$ is the derived category of $\mathbb{Q}$-vector spaces, one obtains the *cohomological cycle complex* $\mathcal{Z}^r(h(X))$ of X. By definition, the *Chow cohomology* of X is

$$\mathrm{CHC}^r(X,n) := H^{-n}\mathcal{Z}^r(h(X)).$$

The main ingredients of the paper are the theory of higher Chow groups of S. Bloch and the method of cubical hyperresolutions [**GNPP**]. For a given projective variety X, we take its cubical hyperresolution $X_\bullet$ of X, a truncated strict simplicial scheme consisting of smooth projective varieties having the same cohomology as X. We naturally associate to it the object

$$[X_\bullet]^* := [X_0 \xrightarrow{\Gamma_d} X_1 \xrightarrow{\Gamma_d} X_2 \xrightarrow{\Gamma_d} \cdots]$$

in $\mathcal{D}(k)$. Here $\Gamma_d = \sum(-1)^i\Gamma_{d_i}$ is the alternating sum of the graphs of the face maps d_i's. We can show it is independent of the choice of hyperresolution, so we denote it $h(X)$. The association $X \mapsto h(X)$ is shown to be contravariantly functorial.

Throughout this paper we will employ the language of mixed motives defined as diagrams of smooth projective varieties [**Ha**]. Although some results here are formulated in terms of cycle complexes rather than motives, the proofs essentially require the idea of diagrams.

1991 *Mathematics Subject Classification.* 14C25, 14F99, 11G.
Received by the editors October 1, 1998; Accepted January 6, 1999.
This is the final form of the paper.

H. Gillet and Ch. Soulé [**GS**] have independently used hyperresolutions (or more precisely hyperenvelopes) to study motives. See also [**GN**].

We consider quasi-projective varieties over a field k (namely, not necessarily irreducible, reduced quasi-projective schemes over k), and work under the assumption that desingularization exists for any scheme (for example $\operatorname{char} k = 0$).

For a map $f : X \to Y$ of varieties, its graph is the subvariety $\Gamma_f \subset Y \times X$ which is the image of the closed immersion $(f, id)\colon X \to Y \times X$.

For a cycle z on $X \times Y$, its *transpose* ${}^t z$ is the cycle $s_*(z)$ on $Y \times X$ where $s\colon X \times Y \to Y \times X$ is the isomorphism which switches the factors.

$\mathcal{D}(k)$ is the motivic category as defined in [**Ha**]. $\mathbb{Q}(r)$ ($r \in \mathbb{Z}$) are the Tate objects. For an object K of $\mathcal{D}(k)$, its dual is denoted by $K^\vee$, and $K(r) := K \otimes \mathbb{Q}(r)$ (Tate twists). (In fact, all the constructions in this paper take place in the full subcategory $\mathcal{D}_{\text{finite}}(k)$ of motives of finite type. See §2 for the definitions.)

$\mathcal{Z}^0(K, \cdot)$ is the cycle complex of an object K of $\mathcal{D}(k)$. We define

$$\mathcal{Z}^r(K, \cdot) = \mathcal{Z}^0(K(r)[2r], \cdot)$$

and $\mathrm{CH}^r(K, n) = \mathrm{CH}^0(K(r)[2r], n) := H_n \mathcal{Z}^r(K(r)[2r], \cdot)$.

1. Hyperresolutions and Higher Chow Groups

Hyperresolutions [GNPP], see also [Ca]. In the following, we denote by I a *finite ordered category.* By definition, I is a small category satisfying the following conditions:

- (i) The set of objects $\mathrm{Ob}(I)$ is a finite set, and for i, $j \in \mathrm{Ob}(I)$, $\mathrm{Hom}_I(i, j)$ is a finite set.
- (ii) For each object i of I, $\mathrm{Hom}_I(i, i) = \{\mathrm{id}\}$. If $\mathrm{Hom}_I(i, j)$ and $\mathrm{Hom}_I(j, i)$ are both nonempty, then $i = j$.

By an I-scheme we mean a contravariant functor from I into the category of schemes. I-schemes form a category.

For example, if I is a finite ordered set, then its associated category is a finite ordered category; by definition, the associated category has I as the set of objects and $\mathrm{Hom}_I(i, j)$ consists of a single element if $i \le j$, and is empty if $i > j$. Another example of a finite ordered category we will consider is the *truncated strict simplicial category* $(\Delta_{\text{mon}})_p$. Recall the objects of this category are $[n] = \{0, 1, \dots, n\}$, $n \le p$ and the morphisms are strictly increasing maps. A $(\Delta_{\text{mon}})_p$-scheme is nothing but a *truncated strict simplicial scheme.*

We denote by $\square_n^+$ the category associated to the partially ordered set with elements $\alpha = (\alpha_0, \alpha_1, \alpha_2, \dots, \alpha_n)$ with $\alpha_i \in \{0, 1\}$. ($\alpha \le \alpha'$ if and only if $\alpha_i \le \alpha'_i$ for all i.) By $\square_n$ we mean the subset (and subcategory) of $\square_n^+$ consisting of elements (objects) except $(0, 0, \dots, 0)$. A $\square_n^+$- or $\square_n$- scheme is called a cubical scheme.

Let S be a I-scheme, a *2-resolution* of S is a Cartesian square consisting of I-schemes Z_{ab} (a, $b = 0$, 1)

$$\begin{array}{ccc} Z_{11} & \longrightarrow & Z_{01} \\ \downarrow & & \downarrow{\scriptstyle f} \\ Z_{10} & \longrightarrow & Z_{00} \end{array}$$

where

(1) $Z_{00} = S$,

(2) Z_{01} is a smooth I-scheme,
(3) the horizontal arrows are closed immersions of I-schemes,
(4) f is a proper morphism of I-schemes, and
(5) Z_{10} contains the discriminant locus of f.

Let $r \geq 1$ be an integer, $X_\bullet^n$ a $\square_n^+ \times I$-scheme for $1 \leq n \leq r$. Suppose that for all n, $1 \leq n < r$, the $\square_{n-1}^+ \times I$-schemes $X_{00\bullet}^{n+1}$ and $X_{1\bullet}^n$ are the same. Then we define, by induction on r, the $\square_r^+ \times I$-scheme $Z_\bullet = \mathrm{rd}(X_\bullet^1, \dots, X_\bullet^r)$ (called the *reduction*) as follows. If $r = 1$, we define $Z_\bullet = X_\bullet^1$. If $r = 2$, we define $Z_{\bullet\bullet} = \mathrm{rd}(X_\bullet^1, X_{\bullet\bullet}^2)$ by

$$Z_{\alpha\beta} = \begin{cases} X_{0\beta}^1 & \text{if } \alpha = (0,0) \\ X_{\alpha\beta}^2 & \text{if } \alpha \in \square_1 \end{cases}$$

for all $\beta \in \square_0^+$, with the evident morphisms. If $r > 2$, we define

$$Z_\bullet = \mathrm{rd}(\mathrm{rd}(X_\bullet^1, \dots X_\bullet^{r-1}), X_\bullet^r).$$

DEFINITION 1.1. Let S be an I-scheme. A *cubical hyperresolution* of S is a $\square_r^+ \times I$-scheme $Z_\bullet$ such that $Z_\bullet = \mathrm{rd}(X_\bullet^1, \dots, X_\bullet^r)$ where

(1) $X_\bullet^1$ is a 2-resolution of S,
(2) For $1 \leq n < r$, $X_\bullet^{n+1}$ is a 2-resolution of $X_\bullet^n$, and
(3) Z_α is smooth for all $\alpha \in \square_r$.

A cubical hyperresolution $Z_\bullet$ naturally defines a strict simplicial scheme the nth term of which is

$$Z_n = \coprod_{|\alpha|=n+1} Z_\alpha \quad \text{where } |(\alpha_0, \dots \alpha_m)| = \alpha_0 + \cdots + \alpha_m;$$

we denote this by the same $Z_\bullet$, the face maps by d_i and the augmentation map by a.

$$\dots \rightrightarrows Z_1 \underset{d_1}{\overset{d_0}{\rightrightarrows}} Z_0 \xrightarrow{a} S.$$

Under the current assumption that resolutions of singularities exist, a cubical hyperresolution of any variety exists. In the following we refer to cubical hyperresolution as *hyperresolution*.

Cycle Complexes [B1, B2, BL]. Let X be a variety (or a scheme) over a field k. Define the *cycle complex* of X as follows. Let $\square^n = \mathbb{A}_k^n$ with coordinates $(x_1, \dots, x_n)$. Faces of $\square^n$ are intersection of codimension one faces, and the latter is of the form $\square_{i,a}^{n-1} = \{x_i = a\}$ where $a = 0$ or 1. A face of dimension m is canonically isomorphic to $\square^m$. Let Σ_n be the symmetric group of order n acting on $\square^n$ by permutations of coordinates.

Let

$$\mathcal{Z}_s(X,n) = \left\{ \begin{matrix} \mathbb{Q}\text{-cycle } z \text{ of dimen-} \\ \text{sion } s+n \text{ on } X \times \square^n \end{matrix} \;\middle|\; \begin{matrix} \text{each irreducible component of } z \text{ meets each} \\ \text{face of } X \times \square^n \text{ properly and } z \text{ is alternating} \\ \text{with respect to the action of } \Sigma_n. \end{matrix} \right\}$$

The inclusions of a codimension one faces $\delta_{i,a} \colon \square_{i,a}^{n-1} \hookrightarrow \square^n$ induce the map

$$\partial = \sum (-1)^{i+a} \delta_{i,a}^* \colon \mathcal{Z}_s(X,n) \to \mathcal{Z}_s(X, n-1).$$

and $(\mathcal{Z}_s(X,\cdot),\partial)$ is a homology complex. We call this the cycle complex (of codimension r) of X. By definition the (rational) higher Chow groups are the homology groups of this complex:

$$CH_s(X,n) = H_n\mathcal{Z}_s(X,\cdot).$$

For an equi-dimensional variety X we set $\mathcal{Z}^r(X,\cdot) = \mathcal{Z}_{\dim X-r}(X,\cdot)$.

In [**BL**], considering "normalized" $\mathbb{Z}$- cycles in place of $\mathbb{Q}$-cycles, integral higher Chow group is defined, and denoted it by $\mathrm{CH}^r(X,n)$. In this paper we only consider the rational cycle complexes and rational Chow groups.

A proper map of varieties $f\colon X\to Y$ induces a map of cycle complexes $f_*\colon \mathcal{Z}_s(X,\cdot)\to\mathcal{Z}_s(Y,\cdot)$. If Y is smooth, there is a quasi-isomorphic subcomplex $\mathcal{Z}^r(Y,\cdot)'\subset\mathcal{Z}^r(X,\cdot)$ on which $f^*\colon\mathcal{Z}^r(Y,\cdot)\to\mathcal{Z}^r(X,\cdot)$ is defined.

For X and Y smooth and projective, and $f\in\mathcal{Z}^s(X\times Y,\ell)$, consider the partially defined map (of graded vector spaces)

$$f_*\colon \mathcal{Z}^r(X,\cdot) \dashrightarrow \mathcal{Z}^{r+s-\dim X}(Y,\cdot+\ell),\quad f_*(z) = p_{Y*}(f\cdot(z\times Y)).$$

There is a quasi-isomorphic subcomplex of $\mathcal{Z}^r(X,\cdot)$ on which f_* is defined. In fact, more generally the following holds. For a smooth projective variety X and a not necessarily smooth, quasi-projective variety T, there is a collection of *distinguished subcomplexes* of $\mathcal{Z}^r(X\times T,\cdot)$ satisfying:

(i) For a distinguished subcomplex $\mathcal{Z}^r(X\times T,\cdot)'$, the inclusion into $\mathcal{Z}^r(X\times T,\cdot)$ is a quasi-isomorphism. The intersection of a finite collection of distinguished subcomplexes is again distinguished.

(ii) For any cycle $f\in\mathcal{Z}^s(X\times Y,\ell)$ there is a distinguished subcomplex $\mathcal{Z}^r(X\times T,\cdot)'$ on which the map

$$f_*\colon z\mapsto p_{Y\times T*}((f\times T)\cdot(z\times Y))$$

is defined and induces a map $f_*\colon\mathcal{Z}^r(X\times T,\cdot)'\to\mathcal{Z}^{r+s-\dim X}(Y\times T,\cdot+\ell)$.

(iii) Let X, Y, Z be smooth projective, $f\in\mathcal{Z}(X\times Y,\ell)$, $g\in\mathcal{Z}(Y\times Z,m)$ such that $g\circ f$ is defined and $\in\mathcal{Z}(X\times Z,\ell+m)$. Then there are distinguished subcomplexes $\mathcal{Z}(X\times T,\cdot)'$ and $\mathcal{Z}(Y\times T,\cdot)'$ such that the three maps

$$f_*\colon\mathcal{Z}(X\times T,\cdot)'\to\mathcal{Z}(Y\times T,\cdot+\ell)',$$
$$g_*\colon\mathcal{Z}(Y\times T,\cdot)'\to\mathcal{Z}(Z\times T,\cdot+m),$$

and

$$(g\circ f)_*\colon\mathcal{Z}(X\times T,\cdot)'\to\mathcal{Z}(Z\times T,\cdot+\ell+m)$$

are all defined. Similarly for a finite sequence of composable correspondences.

For the proof, see [**B1**] and [**Ha**, II, §1]; there the case T is a point is treated, and the general case is parallel.

THEOREM 1.2 (Localization Theorem [**B2**]). *Let X be a variety and $Z\subset X$ a closed subset. Then the diagram of complexes with the natural maps*

$$\mathcal{Z}_s(Z,\cdot)\to\mathcal{Z}_s(X,\cdot)\to\mathcal{Z}_s(X-Z,\cdot)$$

can be extended to a distinguished triangle in the derived category.

Let U be a quasi-projective variety, and $a\colon U_\bullet \to U$ be a hyperresolution. The double complex

$$\xrightarrow{d_*} \mathcal{Z}_s(U_n,\cdot) \xrightarrow{d_*} \dots \xrightarrow{d_*} \mathcal{Z}_s(U_1,\cdot) \xrightarrow{d_*} \mathcal{Z}_s(U_0,\cdot) \to 0$$

(where $\mathcal{Z}_s(U_0,n)$ is placed in homological degree n and $d_* := \sum(-1)^i d_{i\,*}$) will be denoted by $\mathcal{Z}_s(U_\bullet,\cdot)_*$. The following is a consequence of the localization theorem of cycle complexes.

THEOREM 1.3. *The natural map of complexes* $a_*\colon \mathcal{Z}_s(U_\bullet,\cdot)_* \to \mathcal{Z}_s(U,\cdot)$ *is a quasi-isomorphism.*

PROOF. By induction on the length r of the hyperresolution. Let $\mathrm{rd}(U^1_\bullet,\dots,U^r_\bullet)$ be a hyperresolution. $U^1_\bullet$ is a 2-resolution of U, which is a square:

$$\begin{array}{ccc} W' & \xrightarrow{\alpha} & U' \\ \downarrow{\scriptstyle g} & & \downarrow{\scriptstyle f} \\ W & \xrightarrow{\beta} & U. \end{array}$$

The $\square^+_{r-1} \times \square^+_0$-scheme $\mathrm{rd}(U^2_\bullet,\dots,U^r_\bullet)$ is a hyperresolution of the $\square^+_0$-scheme $g\colon W' \to W$; it consists of $\square_{r-1}$-schemes $W'_\bullet$ and $W_\bullet$, augmented to W' and W, respectively, and a map $W'_\bullet \to W_\bullet$ augmented to $W' \to W$.

$$\begin{array}{ccc} W'_\bullet & \longrightarrow & W' \\ \downarrow & & \downarrow \\ W_\bullet & \longrightarrow & W. \end{array}$$

One has a map of distinguished triangles

$$\begin{array}{ccccccc} \mathcal{Z}_s(W') & \xrightarrow{\alpha_*} & \mathcal{Z}_s(U') & \longrightarrow & \mathcal{Z}_s(U'-W') & \xrightarrow{[1]} & \\ \downarrow{\scriptstyle g_*} & & \downarrow{\scriptstyle g_*} & & \downarrow{\scriptstyle f_*} & & \\ \mathcal{Z}_s(W) & \xrightarrow{\alpha_*} & \mathcal{Z}_s(U) & \longrightarrow & \mathcal{Z}_s(U-W) & \xrightarrow{[1]} & . \end{array}$$

Since $U'-W' \xrightarrow{\sim} U-W$, the third horizontal arrow is a quasi-isomorphism; hence we have the distinguished triangle:

$$\mathcal{Z}_s(W') \xrightarrow{(g_*,\alpha_*)} \mathcal{Z}_s(W) \oplus \mathcal{Z}_s(\tilde{U}) \xrightarrow{\beta_*-f_*} \mathcal{Z}_s(U) \xrightarrow{[1]} .$$

By induction hypothesis, the maps induced by the augmentations

$$\mathcal{Z}_s(W'_\bullet) \to \mathcal{Z}_s(W') \quad \text{and} \quad \mathcal{Z}_s(W_\bullet) \to \mathcal{Z}_s(W)$$

are quasi-isomorphisms. Hence the claim for the given hyperresolution of U. $\square$

We also define the double complex

$$\mathcal{Z}^r(U_\bullet,\cdot)^* := [\mathcal{Z}^r(U_0,\cdot)' \xrightarrow{d^*} \mathcal{Z}^r(U_1,\cdot)' \xrightarrow{d^*} \dots \xrightarrow{d^*} \mathcal{Z}^r(U_n,\cdot)' \xrightarrow{d^*} \cdots]$$

where $\mathcal{Z}^r(U_n,\cdot)'$ are quasi-isomorphic subcomplexes so that the above sequence of maps $d^* := \sum(-i)^i d^*_i$ are defined. We call it the *cohomological cycle complex* of U.

2. The Category $\mathcal{D}_{\text{finite}}(k)$

We collect all that is needed in this paper on the category of mixed motives. We first recall a notion from [**Ha**, II, §3].

C-complexes. A *C-complex* (of Abelian groups) consists of

(i) Complexes $A^m = (A^{m,\bullet}, d_{A^m})$ (each of which is not necessarily bounded) for $m \in \mathbb{Z}$, such that for all but finitely many m's, $A^m = 0$; and

(ii) For $m < n$, maps of graded groups
$$F^{m,n} \colon A^{m,\bullet} \to A^{n,\bullet-(n-m-1)}$$
subject to the condition
$$F^{m,n} \circ (-1)^m d_{A^m} + (-1)^n d_{A^n} \circ F^{m,n} + \sum_{m<\ell<n} F^{\ell,n} \circ F^{m,\ell} = 0$$
as a map $A^{m,\bullet} \to A^{n,\bullet-n+m+2}$.

One can associate with it a complex, the *total complex* $\mathrm{Tot}(A) = (\mathrm{Tot}(A)^\bullet, \mathbf{d})$, defined as:
$$\mathrm{Tot}(A)^p = \bigoplus_{p\in\mathbb{Z}} A^{m,p-m}$$
(which is a finite sum), and the differential $\mathbf{d}$ is the direct sum, for m, of $(-1)^m d_{A^m} + F^{m,n} \colon A^{m,\bullet} \to \bigoplus_{n\geq m} A^{n,\bullet-(n-m-1)}$. The condition in (ii) is equivalent to $\mathbf{d}$ being a differential.

For details on the category $\mathcal{D}(k)$ of mixed motives, we refer the reader to [**Ha**, II]. We will only need the subcategory $\mathcal{D}_{\text{finite}}(k)$ of mixed motives of finite type; the definitions are briefly recalled.

A *finite symbol* is a formal sum
$$\bigoplus_{\alpha\in I} (X_\alpha, r_\alpha)$$
where X_α is a smooth projective variety, I a finite index set and $r_\alpha \in \mathbb{Z}$. We write 0 for the corresponding symbol when I is an empty set.

Define dual, tensor product, and inner Hom of (a) finite symbol(s) as:
$$\big(\oplus(X_\alpha, r_\alpha)\big)^\vee = \oplus(X_\alpha, \dim X_\alpha - r_\alpha);$$
$$(\oplus(X_\alpha, r_\alpha)) \otimes (\oplus(X'_{\alpha'}, r_{\alpha'})) = \oplus(X_\alpha \times X'_{\alpha'}, r_\alpha + r'_{\alpha'});$$
and
$$\underline{\mathrm{Hom}}\left(\bigoplus(X_\alpha, r_\alpha), \bigoplus(X'_{\alpha'}, r_{\alpha'})\right) = \left(\bigoplus(X_\alpha, r_\alpha)\right)^\vee \otimes \left(\bigoplus(X'_{\alpha'}, r_{\alpha'})\right).$$

Define the *cycle complex* of a finite formal symbol by
$$\mathcal{Z}^0\left(\bigoplus(X_\alpha, r_\alpha), \cdot\right) = \bigoplus \mathcal{Z}^{r_\alpha}(X_\alpha, \cdot).$$

Note there is a *partially defined* map
$$\mathcal{Z}^0(\underline{\mathrm{Hom}}((X_1, r_1), (X_2, r_2)), \cdot) \otimes \mathcal{Z}^0(\underline{\mathrm{Hom}}((X_2, (X_3, r_3)), \cdot) \to \mathcal{Z}^0(\underline{\mathrm{Hom}}((X_1, r_1), (X_3, r_3)), \cdot)$$
given by the composition of correspondences
$$u \otimes v \mapsto v \circ u = p_{13*}[(u \times X_3) \cdot (X_1 \times v)].$$

By definition, an object of $\mathcal{D}_{\text{finite}}(k)$ is a set of data $K = (K^m) = (K^m, f^{m,n})$ where

(i) For each integer m, $K^m = \bigoplus_{\alpha \in I(m)} (X_\alpha, r_\alpha)$, a finite symbol.

(ii) For (m, n) with $m < n$, given $f^{m,n} = (f^{m,n}_{\alpha\beta}) \in \mathcal{Z}^0(\underline{\mathrm{Hom}}(K^m, K^n))^{-n+m+1}$, which are subject to the conditions:

For $f^{m_k, m_{k+1}} \in \mathcal{Z}^0(\underline{\mathrm{Hom}}(K^{m_k}, K^{m_{k+1}}))^{-m_{k+1}+m_k+1}$ $(k = 1, 2, \ldots, r)$ one has

$$f^{m_r, m_{r+1}} \circ f^{m_r, m_{r-1}} \circ \cdots \circ f^{m_1, m_2}$$

is defined and is an element of $\mathcal{Z}^0(\underline{\mathrm{Hom}}(K^{m_1}, K^{m_{r+1}}))^{-m_{r+1}+m_1+r}$.

For $m < n$, one has

$$(-1)^n \partial f^{m,n} + \sum_{m<\ell<n} f^{\ell,n} \circ f^{m,\ell} = 0.$$

On the left side the compositions of the correspondences are *required* to be defined.

There is the functor of *cycle complexes* $\mathcal{Z}^0$ from $\mathcal{D}_{\text{finite}}(k)$ to the derived category of $\mathbb{Q}$-vector spaces. To define $\mathcal{Z}^0(K, \cdot)$, for each m and $\alpha \in I(m)$, take a distinguished subcomplex $\mathcal{Z}^0((X_\alpha, r_\alpha), \cdot)'$ so that each $f^{m,n}_{\alpha,\beta}$ induces the map $f^{m,n}_{\alpha,\beta *} \colon \mathcal{Z}^0((X_\alpha, r_\alpha), \cdot)' \to \mathcal{Z}^0((X_\beta, r_\beta), \cdot)'$. We then let

$$\mathcal{Z}^0(K^m, \cdot)' := \bigoplus_\alpha \mathcal{Z}^0((X_\alpha, r_\alpha), \cdot)'$$

and have $f^{m,n}{}_* \colon \mathcal{Z}^0(K^m, \cdot)' \to \mathcal{Z}^0(K^n, \cdot + (n - m - 1))'$ is defined. We define $\mathcal{Z}^0(K, \cdot)$ to be the *total complex* $\mathrm{Tot}(\mathcal{Z}^0(K^m, \cdot), f_*^{m,n})$, namely the complex $(\mathbb{K}, d)$ with

$$\mathbb{K}^i = \bigoplus_{j \geq i} \mathcal{Z}^0(K^j, j - i)',$$

and

$$d^i = \sum_j \Big((-1)^j \partial_j + \sum_{j<\ell} f_*^{j,\ell} \Big).$$

Let (K, f) and (L, g) be objects in $\mathcal{D}_{\text{finite}}(k)$. The *function cycle complex* $\mathcal{Z}^0(\underline{\mathrm{Hom}}(K, L))^\bullet$ is defined as follows. Let $\mathcal{Z}^0(\underline{\mathrm{Hom}}(K^m, L^{m'}), \cdot)'$ be distinguished subcomplexes such that

For $u \in \mathcal{Z}^0(\underline{\mathrm{Hom}}(K^m, L^{m'}), \cdot)$, both $u \circ f^{n,m}$ and $g^{m',n'} \circ u$ are defined.

(This is possible since there are only finitely many non-zero $f^{n,m}$'s and $g^{m',n'}$'s.) The cohomological complex to be defined has the group of N-cochains

$$\mathcal{Z}^0(\underline{\mathrm{Hom}}(K, L))^N = \bigoplus_{-m+m'-p=N} \mathcal{Z}^0(\underline{\mathrm{Hom}}(K^m, L^{m'}), p)'$$

The differential of this complex, which we denote by D, is the sum of the three kinds of maps:

$$(-1)^{p+m'+n+1}(\circ f^{n,m}) \colon \mathcal{Z}^0(\underline{\mathrm{Hom}}(K^m, L^{m'}, p)' \to \mathcal{Z}^0(\underline{\mathrm{Hom}}(K^n, L^{m'}), p+n'-m'-1)',$$

$$(-1)^{m'+n'}(g^{m',n'} \circ) \colon \mathcal{Z}^0(\underline{\mathrm{Hom}}(K^m, L^{m'}), p)' \to \mathcal{Z}^0(\underline{\mathrm{Hom}}(K^m, L^{n'}), p+n'-m'-1)',$$

and

$$(-1)^{m'}\partial\colon \mathcal{Z}^0(\underline{\mathrm{Hom}}(K^m, L^{m'}), \cdot)' \to \mathcal{Z}^0(\underline{\mathrm{Hom}}(K^m, L^{m'}), \cdot - 1)'.$$

Given three objects K, L and M, the *partially defined* composition map

$$\mathcal{Z}^0(\underline{\mathrm{Hom}}(K, L))^\bullet \otimes \mathcal{Z}^0(\underline{\mathrm{Hom}}(L, M))^\bullet \dashrightarrow \mathcal{Z}^0(\underline{\mathrm{Hom}}(K, M))^\bullet$$

$$u \otimes v \longmapsto v \circ u;$$

$$(v \circ u)^{m,n} = \sum_{\ell \in \mathbb{Z}} v^{\ell,n} \circ u^{m,\ell}$$

satisfies the Leibniz formula

$$D(v \circ u) = Dv \circ u + (-1)^{\deg v} v \circ Du,$$

where $\deg v$ is the total degree of v in the cohomological complex. There is a quasi-isomorphic subcomplex of $\mathcal{Z}(\underline{\mathrm{Hom}}(K, L))^\bullet \otimes \mathcal{Z}(\underline{\mathrm{Hom}}(L, M))^\bullet$ on which the composition is defined.

By definition

$$\mathrm{Hom}_{\mathcal{D}_{\mathrm{finite}}(k)}(K, L) = H^0 \mathcal{Z}^0(\underline{\mathrm{Hom}}(K, L))^\bullet.$$

The composition of morphisms is induced from the composition of the function complexes. A morphism $u\colon K \to L$ is represented by $u^{m,n} \in \mathcal{Z}^0(\underline{\mathrm{Hom}}(K^m, L^n), n - m))'$ (non-zero only for $m \leq n$) subject to the condition

$$(-1)^n \partial u^{m,n} - \sum (-1)^{m+\ell} u^{\ell,n} \circ f^{m,\ell} + \sum (-1)^{\ell+n} g^{\ell,n} \circ u^{m,\ell} = 0.$$

It defines the zero morphism if there exist $U^{m,n} \in \mathcal{Z}^0(\underline{\mathrm{Hom}}(K^m, L^n), n - m + 1)'$ (non-zero only for $m \leq n - 1$) such that

$$u^{m,n} = (-1)^n \partial U^{m,n} + \sum (-1)^{m+\ell} U^{\ell,n} \circ f^{m,\ell} + \sum (-1)^{\ell+n} g^{\ell,n} \circ U^{m,\ell}.$$

We have the following (only the functor h, the duality functor and the cycle complex functor will be used in this paper), see [**Ha**, II].

THEOREM 2.1. *The category $\mathcal{D}_{\mathrm{finite}}(k)$ has a structure of triangulated category. Moreover*

(1) *$\mathcal{D}_{\mathrm{finite}}(k)$ has dual, tensor product, inner Hom, the unit object $\mathbb{Q}$, and the Tate objects $\mathbb{Q}(r)$.*
(2) *There is a contravariant functor $h\colon (\mathrm{Smooth\,Proj}/k) \to \mathcal{D}_{\mathrm{finite}}(k)$.*
(3) *If X is smooth and projective, one has*

$$\mathrm{Hom}_{\mathcal{D}_{\mathrm{finite}}(k)}(\mathbb{Q}, h(X)(r)[2r - m]) = K_m(X)^{(r)}_{\mathbb{Q}}.$$

Here the right hand side is an Adams-graded piece of the K-group of X.
(4) *There is the cycle complex functor $\mathcal{Z}^0\colon \mathcal{D}_{\mathrm{finite}}(k) \to D(\mathbb{Q})$.*

3. The Motive of a Projective Variety

We first note that one may speak of correspondences from a projective variety to a smooth projective variety. More generally, we will consider higher correspondences from a projective variety into a diagram K.

Let X be a possibly singular projective variety, and K an object of $\mathcal{D}_{\mathrm{finite}}(k)$. One defines the function cycle complex $\mathcal{Z}^0(\underline{\mathrm{Hom}}(X, K))^\bullet$ (with differential ∂ of degree 1) as follows.

If $K = (Y, r)$ (concentrated in degree 0), then

$$\mathcal{Z}^0(\underline{\mathrm{Hom}}(X,(Y,r)))^\bullet = \mathcal{Z}_{\dim Y - r}(X \times Y, -\bullet)$$

by definition. An element $f \in \mathcal{Z}^0(\underline{\mathrm{Hom}}((Y,r),(Z,s)))^{-\ell} = \mathcal{Z}^{\dim Y + s - r}(Y \times Z, \ell)$ induces a map

$$f_*\colon \mathcal{Z}^0(\underline{\mathrm{Hom}}(X,(Y,r)))'^\bullet \to \mathcal{Z}^0(\underline{\mathrm{Hom}}(X,(Z,s)))^{\bullet-\ell}$$

where $\mathcal{Z}^0(\underline{\mathrm{Hom}}(X,(Y,r)))'^\bullet$ is a distinguished subcomplex.

By linearity, these can be extended to the case K is a formal symbol. One may thus form

$$\mathcal{Z}^0(\underline{\mathrm{Hom}}(X,K))^\bullet := \mathrm{Tot}\left(\mathcal{Z}^0(\underline{\mathrm{Hom}}(X,K^m))^\bullet, f_*^{m,n}\right)$$

Tot meaning the total complex. Here precisely speaking, $\mathcal{Z}^0(\underline{\mathrm{Hom}}(X,K^m))^\bullet$ is to be replaced its distinguished subcomplex. Such a consistent choice is possible by the properties of distinguished subcomplexes. It is well-defined up to canonical quasi-isomorphism. We have the following functorial properties:

If $\psi\colon X' \to X$ is a map of projective varieties there is induced a map of complexes

$$(\psi \times 1_K)_* = (-) \circ \Gamma_\psi : \mathcal{Z}^0(\underline{\mathrm{Hom}}(X',K))^\bullet \to \mathcal{Z}^0(\underline{\mathrm{Hom}}(X,K))^\bullet.$$

It is transitive: $(\psi \times 1)_* \circ (\psi' \times 1)_* = ((\psi \circ \psi') \times 1)_*$. There is a partially defined composition

$$\mathcal{Z}^0(\underline{\mathrm{Hom}}(X,K))^\bullet \otimes \mathcal{Z}^0(\underline{\mathrm{Hom}}(K,K'))^\bullet \dashrightarrow \mathcal{Z}^0(\underline{\mathrm{Hom}}(X,K'))^\bullet,$$

which sends $z \otimes u$ to $u \circ z$. One has Leibniz rule: if $u \in \mathcal{Z}^0(\underline{\mathrm{Hom}}(K,K'))^{-\ell}$,

$$(\partial u) \circ z = \partial(u \circ z) + (-1)^{\ell-1}(u \circ \partial z).$$

The composition is transitive in either argument: $(u' \circ u) \circ z = u' \circ (u \circ z)$ and $(u \circ z) \circ \Gamma_\psi = u \circ (z \circ \Gamma_\psi)$. (The map obtained by passing to 0th cohomology

$$H^0\mathcal{Z}^0(\underline{\mathrm{Hom}}(X,K))^\bullet \otimes H^0\mathcal{Z}^0(\underline{\mathrm{Hom}}(K,K'))^\bullet \dashrightarrow H^0\mathcal{Z}^0(\underline{\mathrm{Hom}}(X,K'))^\bullet,$$

will be also denoted $z \otimes u \mapsto u \circ z$.)

Let X be a projective variety, and $a\colon X_\bullet \to X$ a hyperresolution. There is naturally associated an object of $\mathcal{D}_{\text{finite}}(k)$

$$h(X) = [X_\bullet]^* = [X_0 \xrightarrow{\Gamma_d} X_1 xrightarrow \Gamma_d X_2 \xrightarrow{\Gamma_d} \cdots]$$

(namely, the object $(K^m, f^{m,n})$ with $K^m = X_m$ for $m \geq 0$, $K^m = 0$ for $m < 0$, and $f^{m,m+1} = \Gamma_d$ for $m \geq 0$, $f^{m,n} = 0$ otherwise). Here $\Gamma_d = \sum(-1)^i\Gamma_{d_i}$ is the alternating sum of the graphs of the face maps d_i's. We call this the *cohomological motive* of X. The notation $h(X)$, which does not involve the choice of a hyperresolution, will be justified by showing that this object is well-defined independent of the choice. Note

$$\mathcal{Z}^r([X_\bullet]^*, \cdot) = \mathcal{Z}^r(X_\bullet, \cdot)^*.$$

We also define

$$h(X)^\vee = [X_\bullet]_* = [\dots \xrightarrow{{}^t\Gamma_d} X_2^\vee \xrightarrow{{}^t\Gamma_d} X_1^\vee \xrightarrow{{}^t\Gamma_d} X_0^\vee]$$

to be the *homological motive* of X. (According to the general definition of dual in [**Ha**, II, (4.5)],

$$([X_\bullet]^*)^\vee = [\dots \xrightarrow{-{}^t\Gamma_d} X_2^\vee \xrightarrow{{}^t\Gamma_d} X_1^\vee \xrightarrow{-{}^t\Gamma_d} X_0^\vee],$$

with alternating signs to ${}^t\Gamma_d$; it is isomorphic to the above one.)

Note $\Gamma_a \in \mathcal{Z}^0(\underline{\mathrm{Hom}}(X, [X_\bullet]^*))^0$ is ∂-closed. Let $[\Gamma_a] \in H^0\mathcal{Z}^0(\underline{\mathrm{Hom}}(X, [X_\bullet]^*))$. The pair $([X_\bullet]^*, [\Gamma_a])$ satisfies the condition: for any smooth projective variety Y and an integer r, the map

$$(-)\circ\Gamma_a = (a\times 1)_* \colon \mathcal{Z}^0\left(\underline{\mathrm{Hom}}([X_\bullet]^*, (Y,r))\right) \to \mathcal{Z}^0\left(\underline{\mathrm{Hom}}(X, (Y,r))\right)$$

is a quasi-isomorphism. (Note $\mathcal{Z}^0\left(\underline{\mathrm{Hom}}([X_\bullet]^*, (Y,r))\right) = \mathcal{Z}_{\dim Y - r}(X_\bullet \times Y, \cdot)_*$.)

DEFINITION 3.1. Let X be a projective variety. Suppose an object $S^\bullet$ of $\mathcal{D}_{\mathrm{finite}}(k)$ and $\alpha \in H^0\mathcal{Z}^0(\underline{\mathrm{Hom}}(X, S^\bullet))$ satisfy the following condition: For any object K of $\mathcal{D}_{\mathrm{finite}}(k)$, the map

$$(-)\circ\alpha : H^0\mathcal{Z}^0(\underline{\mathrm{Hom}}(S^\bullet, K)) \to H^0\mathcal{Z}^0(\underline{\mathrm{Hom}}(X, K))^\bullet$$

is an isomorphism. Then we call the pair $(S^\bullet, \alpha)$ a *right resolution* of X. If it exists, a right resolution is unique up to a unique isomorphism: If $(\bar{S}^\bullet, \bar{\alpha})$ is another right resolution, there exists a unique isomorphism $u\colon S^\bullet \to \bar{S}^\bullet$ such that $u\circ\alpha = \bar{\alpha}$.

(The condition may be viewed as a universal property; more specifically, in an appropriate category whose class of objects is the disjoint union of $\mathrm{Ob}(\mathcal{D}_{\mathrm{finite}}(k))$ and $\mathrm{Ob}(\mathrm{Proj}/k)$. We leave the details.)

For a representative $\tilde{\alpha} \in \mathcal{Z}^0(\underline{\mathrm{Hom}}(X, S^\bullet))^0$ of α,

$$(-)\circ\tilde{\alpha}\colon \mathcal{Z}^0(\underline{\mathrm{Hom}}(S^\bullet, K))^\bullet \dashrightarrow \mathcal{Z}^0(\underline{\mathrm{Hom}}(X, K))^\bullet$$

is a partially defined map of complexes. The required condition is equivalent to this map being a quasi-isomorphism for all K.

The pair $([X_\bullet]^*, [\Gamma_a])$ is a right resolution; the condition is satisfied for formal symbols, and hence for diagrams K. The following theorem now follows.

THEOREM 3.2. (1) *The pair $([X_\bullet]^*, [\Gamma_a])$ is a right resolution of X. In particular, if $X_\bullet$ and $X'_\bullet$ are hyperresolutions of the same projective variety X, then the objects $[X_\bullet]^*$ and $[X'_\bullet]^*$ are canonically isomorphic.*

(2) *Let X, X' be projective varieties, and $\psi : X' \to X$ a map. Let $(S^\bullet, \alpha)$ and $(S'^\bullet, \alpha')$ be right resolutions of X and X', respectively. There is a unique morphism $u(\psi)\colon S^\bullet \to S'^\bullet$ such that $u(\psi)\circ\alpha = (\psi\times 1)_*\alpha'$. The association $\psi \mapsto u(\psi)$ is contravariant functorial: $u(\psi\circ\psi') = u(\psi')\circ u(\psi)$.*

COROLLARY. *The complex $\mathcal{Z}^r(X_\bullet, \cdot)^*$ is independent of the choice of a hyperresolotuion. The association $X \mapsto \mathcal{Z}^r(X_\bullet, \cdot)^*$ is contravariantly functorial in X.*

REMARK. In (2) of the theorem, the condition for a representative of $u(\psi)$, $f \in \mathcal{Z}^0(\underline{\mathrm{Hom}}(S^\bullet, S'^\bullet))^0$ is, explicitly: there exists an element $g \in \mathcal{Z}^0(\underline{\mathrm{Hom}}(X, S'^\bullet))^{-1}$ such that

$$f\circ\tilde{\alpha} + \tilde{\alpha'}\circ\Gamma_\psi + \partial g = 0.$$

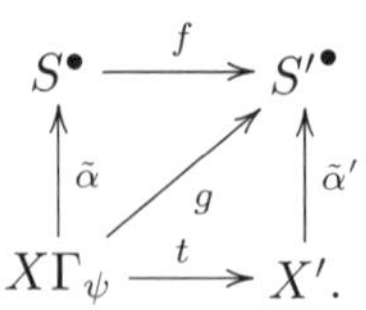

DEFINITION 3.3. For a projective variety X, let

$$\mathrm{CHC}^r(X, n) := H^{-n}\mathcal{Z}^r(h(X)) = H^{-n}\mathcal{Z}^r(X_\bullet, \cdot)^*.$$

This is the *higher Chow cohomology* of X. It is contravariantly functorial in X, and coincides with $CH^r(X, n)$ if X is smooth projective.

References

[B1] S. Bloch, *Algebraic cycles and higher K-theory*, Adv. in Math. **61** (1986), 267–304.

[B2] ———, *The moving lemma for higher Chow groups*, J. Algebraic Geom. **3** (1994), 537–568.

[BL] S. Bloch and M. Levine (book in preparation).

[Ca] J. Carlson, *Polyhedral resolutions of algebraic varieties*, Trans. Amer. Math. Soc. **292** (1985) 595–612.

[De] P. Deligne, *Théorie de Hodge* II., Inst. Hautes Études Sci. Publ. Math. **40** (1971), 5–57; III, **44** (1974), 5–77.

[FV] E. Friedlander and V. Voevodsky, *Bivariant cycle cohomology* (preprint).

[Fu] W. Fulton, *Intersection Theory*, Ergeb. Math. Grenzgeb. (3), vol. 2, Springer-Verlag, Berlin-New York, 1984.

[GS] H. Gillet and C. Soulé, *Descent, motives and K-theory*, J. Reine. Angew. Math. **478** (1996), 127–176.

[GN] F. Guillen and V. Navarro-Aznar, *Un critére d'extension d'un foncteur défini sur les schémas lisses* (preprint).

[GNPP] F. Guillen, V. Navarro-Aznar, P. Pascual-Gainza, and F. Puerta, *Hyperrésolutions cubiques et descente cohomologique*, Lecture Notes in Math., vol. 1335, Springer-Verlag, Berlin-New York, 1988.

[Ha] M. Hanamura, *Mixed motives and algebraic cycles.* I, II, III (preprint).

[Le] M. Levine, *Mixed motives* (book to appear).

[Li] S. Lichtenbaum, *Suslin homology and Deligne* 1*-motives* (preprint).

Graduate School of Mathematics, Kyushu University, Fukuoka 812, Japan
E-mail address: hanamura@math.kyushu-u.ac.jp

Centre de Recherches Mathématiques
CRM Proceedings and Lecture Notes
Volume **24**, 2000

Bloch's Conjecture and the K-Theory of Projective Surfaces

Claudio Pedrini

Abstract. Let X be a smooth irreducible projective variety over a field k. Milnor's pairing in K-theory: $K_i \otimes K_j \to K_{i+j}$ induces maps

$$\tau_{i,j}\colon \mathrm{CH}^i(X) \otimes K_j(k) \to H^i(X, \mathcal{K}_{i+j})$$

The aim of this paper is to give conditions for the maps $\tau_{i,j}$ to be isomorphisms (up to torsion). In particular in the case when $k = \mathbb{C}$, we describe (§1) the kernel and cokernel of the map $\tau_{1,1}$ and relate this with the regulator from K-Theory to Deligne-Beilinson Cohomology. We also show (Theorem 2.9) that, if X is a smooth projective surface over $\mathbb{C}$ such that $\mathrm{CH}^2(X) = \mathbb{Z}$, then the above maps are all isomorphisms, up to torsion. It follows that the K-theory of X is "trivial" (up to torsion) in the sense that there are isomorphisms: $K_n(X)_{\mathbb{Q}} \simeq (K_0(X) \otimes K_n(\mathbb{C}))_{\mathbb{Q}}$, for all n. This result, which was known for rational surfaces and for Enriques surfaces (see [**Coo**]), notably applies to surfaces of general type with $p_g = 0$ for which Bloch's Conjecture holds, i.e., such that $\mathrm{CH}^2(X) = \mathbb{Z}$. We also apply our results to compute $\mathcal{K}$-cohomology of hyperelliptic surfaces (Example 2.10).

Introduction

Let X be a smooth variety over a field k. Milnor's pairing in K-theory:

$$K_i \otimes K_j \to K_{i+j} \tag{1}$$

induces (for every i and j) a sheaf map on X_{Zar}:

$$\Phi_{i,j}\colon \mathcal{K}_i \otimes K_j(k) \to \mathcal{K}_{i+j}$$

where $K_j(k)$ denotes the constant sheaf (associated to the corresponding group).

The above map gives rise to a map in cohomology

$$H^i(X, \mathcal{K}_i) \otimes K_j(k) \to H^i(X, \mathcal{K}_{i+j}).$$

Using the Bloch-Quillen isomorphism $H^i(X, \mathcal{K}_i) \simeq \mathrm{CH}^i(X)$, where $\mathrm{CH}^i(X)$ is the Chow group of codimension i cycles modulo rational equivalence, we get the

---

1991 *Mathematics Subject Classification.* 14C25, 14C35, 19E15.
Received by the editors February 18, 1999; Accepted April 28, 1999.
A member of GNSAGA of CNR, partially supported by MURST and by ECC Science Plan SC1 CT910756.
This is the final form of the paper.

following maps:

$$\tau_{i,j}\colon \mathrm{CH}^i(X)\otimes K_j(k)\to H^i(X,\mathcal{K}_{i+j}) \tag{2}$$

In general the above maps are neither surjective nor injective. The aim of this paper is to give conditions for the maps $\tau_{i,j}$ to be isomorphisms (up to torsion), in particular in the case when X is a complex projective surface, and to apply these results for computing the uniquely divisible part of the $\mathcal{K}$-cohomology of X for some special classes of surfaces.

In a forthcoming paper with C. Weibel [**PW**] we prove that for a smooth projective surface over $\mathbb{C}$ the groups $K_j(X)$ are direct sums of a finite group and a divisible group and we compute their torsion subgroups.

From these results it follows that the torsion part of $K_j(X)$ only depends on the topological invariants of the associated complex manifold $X(\mathbb{C})$ and on the rank of $\mathrm{Pic}\, X$.

Here we show that, on the contrary, the uniquely divisible part of $K_j(X)$ depends on the geometric invariants of the surface X such as p_g and q (see Theorem 2.9 and Example 2.10).

We first recall some known results about the maps $\tau_{i,j}$ and their relation with Bloch's Conjecture.

For an Abelian group G (a sheaf of Abelian groups $\mathcal{F}$), we will denote by $G_{\mathbb{Q}}$ ($\mathcal{F}_{\mathbb{Q}}$) the group $G\otimes\mathbb{Q}$ (the sheaf $\mathcal{F}\otimes\mathbb{Q}$).

A description of the kernel and the cokernel of the map

$$\tau_{1,1}\colon \mathrm{Pic}\, X\otimes k^*\to H^1(X,\mathcal{K}_2)$$

has been given in [**Ped**, Theorem 1.8] when X is a smooth projective variety over an algebraically closed field of characteristic 0. It follows that $\mathrm{Ker}\,\tau_{1,1}$ is uniquely divisible and $\mathrm{Coker}\,\tau_{1,1}$ is an extension of a uniquely divisible group by the torsion subgroup of the Brauer group $\mathrm{Br}\, X$. This result extends a theorem proved by Colliot-Thélène and Raskind [**CTR**] in the case $H^2(X,\mathcal{O}_X)=0$.

Several authors have constructed examples of complex surfaces X with non-zero $H^2(X,\mathcal{O}_X)$ such that the map $\tau_{1,1}\otimes\mathbb{Q}$ is not surjective, i.e., such that the group $H^1(X,\mathcal{K}_2)$ is *indecomposable*: Abelian surfaces (Nori), Jacobian surfaces (Collino), $K3$-surfaces (Voisin), some special quintic surfaces of general type (Müller-Stach). See [**MSt**] for a more complete list of references.

Recently Collino [**Col**] has proved that the map

$$\tau_{m,1}\otimes\mathbb{Q} : (\mathrm{CH}^m(X)\otimes\mathbb{C}^*)_{\mathbb{Q}}\to H^m(X,\mathcal{K}_{m+1})_{\mathbb{Q}}$$

is not surjective in the following cases: (i) $m=1$ and X is a quartic surface in $\mathbb{P}^3$; (ii) $m=2$ and X is a cubic fourfold in $\mathbb{P}^5$.

All the above examples are constructed by means of the so called *regulator map*

$$R\colon H^1(X,\mathcal{K}_2)\to H^3_{\mathcal{D}}(X,\mathbb{Z}(2))$$

where $H^p_{\mathcal{D}}(X,\mathbb{Z}(q))$ are the Deligne cohomology groups (see §1). Our results (Theorem 1.7 and Corollary 1.8) illustrate the relation between the surjectivity of $\tau_{1,1}\otimes\mathbb{Q}$ and the "non-triviality" of R.

Surjectivity of the maps in (2) is related to the vanishing of the Albanese Kernel of X in the following way. Let X be a smooth projective variety over an algebraically closed field k and let $A_0(X)$ be the Chow group of zero cycles of degree 0, i.e., the kernel of the degree map $\mathrm{CH}^2(X)\to\mathbb{Z}$. Let $\alpha : A_0(X)\to \mathrm{Alb}\, X$ be the Abel-Jacobi map where $\mathrm{Alb}\, X$ is the Albanese variety of X. Roitman's Theorem [**Roi**] states that this map induces an isomorphism on torsion subgroups. We will denote by $T(X)$ the *Albanese Kernel* of X, i.e., the kernel of the Abel-Jacobi map.

By a result of Mumford [**Mum**] if X is a smooth projective surface over $\mathbb{C}$ and $H^2(X, \mathcal{O}_X) \neq 0$ then $T(X) \neq 0$, and in fact it is a huge group, that is, it is not finite dimensional in the sense of [**Mum**].

Bloch's Conjecture asserts that, if X is a smooth projective surface over $\mathbb{C}$ such that $p_g = \dim H^2(X, \mathcal{O}_X) = 0$ then the Abel-Jacobi map is an isomorphism, i.e., $T(X) = 0$. Bloch's Conjecture has been proved in [**BKL**] for all surfaces with $p_g = 0$; but for surfaces of general type for which the conjecture is still open. Note that for a surface of general type the condition $p_g = 0$ implies that also $q = 0$ where $q = \dim H^1(X, \mathcal{O}_X)$.

If X is a smooth projective surface over $\mathbb{C}$ such that $T(X) = 0$ (hence $p_g = 0$), then, by results of S. Bloch and V. Srinivas [**BS**] and of K. Coombes and V. Srinivas [**CS**], the map

$$\tau_{1,1} \otimes \mathbb{Q} \colon (\operatorname{Pic} X \otimes \mathbb{C}^*)_{\mathbb{Q}} \to (H^1(X, \mathcal{K}_2)_{\mathbb{Q}}$$

is surjective.

If $p_g = q = 0$, then also the maps:

$$\tau_{1,j} \otimes \mathbb{Q} \colon (\operatorname{Pic} X \otimes K_j(\mathbb{C}))_{\mathbb{Q}} \to H^1(X, \mathcal{K}_{j+1})_{\mathbb{Q}}$$

and

$$\tau_{0,j} \otimes \mathbb{Q} \colon K_i(\mathbb{C}) \otimes \mathbb{Q} \to H^0(X, \mathcal{K}_i)_{\mathbb{Q}}$$

are surjective, for all $j \geq 1$. Since the maps $\tau_{0,j}$ are always injective the maps $\tau_{0,j} \otimes \mathbb{Q}$ are isomorphisms.

Note that, if $q > 0$ then the maps $\tau_{0,j} \otimes \mathbb{Q}$ are in general not surjective for $j > 1$ (see Remark 1.11). Here we show (Example 2.10) that if X is a surface with $p_g = 0$ but $q \neq 0$, then also the map $\tau_{1,2} \otimes \mathbb{Q}$ is in general not surjective.

From the above results it follows that if X is a smooth projective surface over $\mathbb{C}$ such that $\mathrm{CH}^2(X) \simeq \mathbb{Z}$ then the maps $\tau_{0,j} \otimes \mathbb{Q}$ are isomorphisms and $\tau_{1,j} \otimes \mathbb{Q}$ are surjective. Furthermore, a simple argument, based on Gersten's resolution for the sheaves $\mathcal{K}_n$, shows that also $\tau_{2,j} \otimes \mathbb{Q}$ are isomorphisms (see Lemma 2.5).

We also prove (Theorem 1.7(ii) and Proposition 2.3) that if $q = 0$ then the maps $\tau_{1,j}$ are injective.

Summing up all these results plus the fact that the Brown-Gersten spectral sequence degenerates up to torsion at the level of E_2 (Proposition 2.6), we get the following statement (Theorem 2.9).

THEOREM 0.1. *Let X be a smooth irreducible projective surface over $\mathbb{C}$, such that $\mathrm{CH}^2(X) \simeq \mathbb{Z}$. Then all the maps $\tau_{i,j}$ in (2) are isomorphisms up to torsion and the K-theory of X decomposes as follows:*

$$K_n(X) \otimes \mathbb{Q} \simeq K_0(X) \otimes K_n(\mathbb{C})_{\mathbb{Q}}$$

for all n.

It is interesting to relate the above result with the known cases of surfaces with $p_g = q = 0$ for which Bloch's conjecture is known to be true, i.e., rational surfaces and Enriques surfaces.

For rational surfaces the proof that all the maps in (2) are isomorphisms follows immediately from a direct computation of $H^p(\mathbb{P}^2, \mathcal{K}_q)$ and the fact that both sides of (2) change in the same way under blow-ups.

For an Enriques surface the proof given in [**Coo**] is based on the existence of an elliptic fibration $\pi\colon X \to \mathbb{P}^1$ such that the associated Jacobian fibration J is a rational surface. In fact, for any variety X the $\mathcal{K}$-cohomology groups depend only on the "motive" of X [**Sou**, 1.2.3], and for an Enriques surface the motive associated to X is isomorphic, up to torsion, to the motive associated to the rational surface J.

Therefore our result suggests that, if X is a surface of general type with $p_g = 0$, then Bloch's Conjecture should imply that there exists a rational surface Y whose associated motive is isomorphic, up to torsion, to the motive of X.

This paper is organized as follows.

In §1 after introducing some notations and recalling known results on Deligne-Beilinson cohomology, we prove necessary and sufficient conditions for the surjectivity of the map $\tau_{1,1}\colon (\operatorname{Pic} X \otimes \mathbb{C}^*) \to H^1(X, \mathcal{K}_2)$, for a variety X over $\mathbb{C}$ (Theorems 1.2 and 1.5). We also show that if $H^1(X, \mathcal{O}_X) = 0$ then the above map is injective. This is the only result in § 1 that will be used in the proof of Theorem 2.9.

Then we consider the regulator map $R\colon H^1(X, \mathcal{K}_2) \to H^3_{\mathcal{D}}(X, \mathbb{Z}(2))$ and prove a result (Theorem 1.7) which relates the kernel $K(X)$ and the image $R(X)$ of R with the kernel and cokernel of $\tau_{1,1}$, for a smooth projective variety X over $\mathbb{C}$. This in particular illustrates the relation between the surjectivity of $\tau_{1,1}$ and the "non-triviality" of $R(X)$. As a corollary (Corollary 1.10) we get an explicit description of $H^1(X, \mathcal{K}_2)$.

In §2 after showing some results on the surjectivity and injectivity of the maps $\tau_{i,j}$, we prove Theorem 2.9 on the K-theory of complex surfaces X with $\mathrm{CH}^2(X) = \mathbb{Z}$. We also apply some of the results contained in this section to compute the K-cohomology of hyperellyptic surfaces, i.e., a class of surfaces with $p_g = 0$ but $\mathrm{CH}^2(X) \neq \mathbb{Z}$.

We thank C. Weibel and A. Rosenschon for many helpful comments on a preliminary version of this paper.

1. Deligne-Beilinson Cohomology and the Map $\operatorname{Pic} X \otimes \mathbb{C}^* \to H^1(X, \mathcal{K}_2)$

In this section, after introducing the notations and recalling some results on the Deligne-Beilinson cohomology, we will prove our main results on the surjectivity of the map $\tau_{1,1} \otimes \mathbb{Q}$, i.e., on the decomposability of the group $H^1(X, \mathcal{K}_2)_{\mathbb{Q}}$ for a smooth projective variety over $\mathbb{C}$.

We show (Theorem 1.5) that if $H^2(X, \mathcal{O}_X) = 0$ then $\tau_{1,1} \otimes \mathbb{Q}$ is surjective if and only if the $d\log$ map: $d\log\colon \mathcal{K}_2 \to \mathcal{F}^{2,2}_{\mathbb{Z}}$ is surjective, where $\mathcal{F}^{2,2}_{\mathbb{Z}}$ is the integral part of the Deligne-Beilinson cohomology sheaf. Then we consider the regulator map: $R\colon H^1(X, \mathcal{K}_2) \to H^3_{\mathcal{D}}(X, \mathbb{Z}(2))$ and we relate (Theorem 1.7) the image and the kernel of R with the map $\tau_{1,1}$ and with the cohomology of the sheaf $\mathcal{F}^{2,2}_{\mathbb{Z}}$. It follows (Corollary 1.8) that if the $d\log$ map is surjective then the image $R(X)$ of R is trivial up to torsion, i.e., $R(X)_{\mathbb{Q}} = (\mathrm{NS}(X) \otimes \mathbb{C}^*)_{\mathbb{Q}}$ if and only if $\tau_{1,1} \otimes \mathbb{Q}$ is surjective.

Let X be a smooth projective variety over $\mathbb{C}$. The *Deligne complex* $\mathbb{Z}(r)_{\mathcal{D}}$ is the truncated de Rham complex:

$$0 \to \mathbb{Z}(r) \to \mathcal{O}_{X_{\mathrm{an}}} \to \cdots \to \Omega^{r-1}_{X_{\mathrm{an}}} \to 0$$

where X_{an} is the associated complex analytic manifold and $\mathbb{Z}(r) = (2\pi i)^r$.

The *Deligne cohomology groups* $H^n_{\mathcal{D}}(X,\mathbb{Z}(r))$ are the hypercomology groups of the above complex $\mathbb{Z}(r)_{\mathcal{D}}$.

One can also define *Deligne-Beilinson cohomology groups* for a smooth (not necessarily proper) variety X by using a good compactification $\overline{X}$ of X, i.e., such that $D = \overline{X} - X$ is a normal crossing divisor [**EV**, 2.3]. The Deligne-Beilinson cohomology groups of X: $H^n_{\mathcal{D}}(X,\mathbb{Z}(r))$ are then defined as the hypercohomology groups of the *Deligne-Beilinson* complex of $(\overline{X},X)$, which is defined via the de Rham complex $\Omega^{\cdot}_{\overline{X}}(\log D)$ of meromorphic forms on $\overline{X}$, holomorphic on X and with at most logarithmic poles along D. This definition is independent of the good compactification chosen: see [**EV**, 2.6].

By $\mathcal{H}^p_{\mathcal{D}}(\mathbb{Z}(r))$ we will denote the Zariski sheaf associated to Deligne-Beilinson cohomology, i.e., the sheaf on X_{Zar} associated to the presheaf $U \to H^p_{\mathcal{D}}(U,\mathbb{Z}(r))$.

By $\mathcal{H}^p(\mathbb{Z}(r))$ we will denote the Zariski sheaves associated to singular cohomology on X_{an} with $\mathbb{Z}(r)$-coefficients. Similarly for $\mathbb{C}$ and $\mathbb{C}^* \simeq \mathbb{C}/\mathbb{Z}$ coefficients.

There are *Chern class maps*

$$c_i \colon \mathcal{K}_p \to \mathcal{H}^{2i-p}_{\mathcal{D}}(\mathbb{Z}(i))$$

from the Zariski sheaf associated to K-theory to the Deligne-Beilinson cohomology sheaf (see [**EV**]). Moreover, there are natural maps maps from $\mathcal{H}^p_{\mathcal{D}}(\mathbb{Z}(r)) \to \mathcal{H}^p(\mathbb{Z}(r))$: the composition coincides with the Chern class map from K-theory to singular cohomology.

There are short exact sequences:

$$0 \to \mathcal{H}^{n-1}(\mathbb{C}^*(n)) \to \mathcal{H}^n_{\mathcal{D}}(\mathbb{Z}(n)) \to \mathcal{F}^{n,n}_{\mathbb{Z}} \to 0 \tag{3}$$

where $\mathcal{F}^{n,n}_{\mathbb{Z}}$ denotes the image of the sheaf $\mathcal{H}^n_{\mathcal{D}}(\mathbb{Z}(2))$ in $\mathcal{H}^n(\mathbb{Z}(r))$ (see [**Esn1**]).

If X is a smooth projective variety over $\mathbb{C}$ then for every open subset U of X there is an isomorphism (see [**EV**, 2.12]): $\mathcal{O}^*_U \simeq H^1_{\mathcal{D}}(U,\mathbb{Z}(1))$, where $\mathcal{O}^*_U$ denotes the group of algebraic invertible functions on U. Therefore we get an isomorphism

$$\mathcal{H}^1_{\mathcal{D}}(\mathbb{Z}(1)) \simeq \mathcal{O}^*_X \simeq \mathcal{K}_1. \tag{4}$$

For X smooth of dimension d over $\mathbb{C}$ the sheaf $\mathcal{H}^p_{\mathcal{D}}(\mathbb{Z}(r))$ has a Bloch-Ogus resolution which terminates with:

$$\coprod_{x\in X^{d-1}} (i_x)_* \left(H^{p-d+1}_{\mathcal{D}}(\mathbb{C}(x),\mathbb{Z}(1))\right) \to \coprod_{x\in X^d} (i_x)_* \left(H^{p-d}_{\mathcal{D}}(\mathbb{C}(x),\mathbb{Z}(0))\right) \to 0 \tag{5}$$

The above resolution yields an isomorphism

$$H^d(X,\mathcal{H}^d_{\mathcal{D}}(\mathbb{Z}(d))) \simeq \mathrm{CH}^d(X)$$

and

$$H^p(X,\mathcal{H}^q_{\mathcal{D}}(\mathbb{Z}(d))) = 0$$

for $p = d$, $d-1$ and $q \neq d$.

There is a Leray spectral sequence [**Gil2**]:

$$E_2^{p,q} = H^p(X,\mathcal{H}^q_{\mathcal{D}}(\mathbb{Z}(r))) \implies H^{p+q}_{\mathcal{D}}(X,\mathbb{Z}(r))$$

which, in particular, gives rise to the following exact sequence (see [**BV1**, §3]):

$$0 \to H^1(X,\mathcal{H}^2_{\mathcal{D}}(\mathbb{Z}(2))) \xrightarrow{i} H^3_{\mathcal{D}}(X,\mathbb{Z}(2)) \to H^0(X,\mathcal{H}^3_{\mathcal{D}}(\mathbb{Z}(2))) \to \mathrm{CH}^2(X) \tag{6}$$

The Chern class map $c_2\colon \mathcal{K}_2 \to \mathcal{H}^2_{\mathcal{D}}(\mathbb{Z}(2))$ yields a map $h(c_2)\colon H^1(X,\mathcal{K}_2) \to H^1(X,\mathcal{H}^2_{\mathcal{D}}(\mathbb{Z}(2))$ which composed with the injective map i in (6) gives the so-called *regulator map*

$$R\colon H^1(X,\mathcal{K}_2) \to H^3_{\mathcal{D}}(X,\mathbb{Z}(2)).$$

Let X be a smooth projective variety over an algebraically closed field k of characteristic zero and let $\Phi = \Phi_{1,1}\colon \mathcal{K}_1 \otimes k^* \to \mathcal{K}_2$ be the map in (1). Then Φ induces the following cohomology sequence (see [**Ped**, 1.8]):

$$(7)\quad 0 \to D \to H^0(X,\operatorname{Coker}\Phi) \to \operatorname{Pic} X \otimes k^* \xrightarrow{\tau_{1,1}} H^1(X,\mathcal{K}_2) \to H^1(X,\operatorname{Coker}\Phi) \to 0$$

where

$$D \simeq H^0(X,\operatorname{Im}\Phi) \simeq \big(H^0(X,\mathcal{K}_2)/K_2(k)\big)_{\mathbb{Q}}$$

and $\operatorname{Ker}\tau_{1,1}$ is a uniquely divisible group.

The next lemma relates the sheaf $\operatorname{Coker}\Phi$ with the sheaf $\mathcal{F}^{2,2}_{\mathbb{Z}}$ in the case $k=\mathbb{C}$. It will be used in the proof of Theorem 1.2 which gives a necessary and sufficient condition for the map $\tau_{1,1}$ to be surjective.

LEMMA 1.1. *:Let X be a smooth, irreducible, projective variety over $\mathbb{C}$. Let $\Phi : \mathcal{K}_1 \otimes \mathbb{C}^* \to \mathcal{K}_2$ be the sheaf map induced by Milnor's pairing: $\mathcal{K}_1 \otimes \mathcal{K}_1 \to \mathcal{K}_2$.*

Then the Chern class map c_2 induces a map:

$$\gamma_2\colon \operatorname{Coker}\Phi \to \mathcal{F}^{2,2}_{\mathbb{Z}}$$

where $\mathcal{F}^{2,2}_{\mathbb{Z}}$ is the Image of $\mathcal{H}^2_{\mathcal{D}}(\mathbb{Z}(2))$ in $\mathcal{H}^2(\mathbb{Z}(2))$. The kernel and cokernel of γ_2 are constant sheaves and the Image of γ_2 is the subsheaf $\mathcal{F}^{1,1}_{\mathbb{Z}} \wedge \mathcal{F}^{1,1}_{\mathbb{Z}}$ of $\mathcal{F}^{2,2}_{\mathbb{Z}}$.

PROOF. Let θ be the composition map:

$$\theta\colon \mathcal{K}_1 \otimes \mathbb{C}^* \xrightarrow{\Phi} \mathcal{K}_2 \xrightarrow{c_2} \mathcal{H}^2_{\mathcal{D}}(\mathbb{Z}(2)) \xrightarrow{c_2'} \mathcal{H}^2 \to (\mathbb{Z}(2)),$$

where c_2' is induced by the Chern class map from Deligne cohomology to singular cohomology. We claim that θ is the zero map.

Using the exact sequence in (3), for $n=2$, the Chern class map from K-Theory to singular homology $c_1\colon \mathcal{K}_1 \to \mathcal{H}^1(\mathbb{Z}(1))$ and the naturality of Chern classes we immediately see that the following diagram commutes:

$$\begin{array}{ccccc}
\mathcal{K}_1 \otimes \mathbb{C}^* & \xrightarrow{\Phi} & \mathcal{K}_2 & & \\
\downarrow{\scriptstyle c_1} & & \downarrow{\scriptstyle c_2} & & \\
\mathcal{H}^1(\mathbb{C}^*) & \xrightarrow{i} & \mathcal{H}^2_{\mathcal{D}}(\mathbb{Z}(2)) & \xrightarrow{c_2'} & \mathcal{H}^2(\mathbb{Z})
\end{array}$$

We have $ic_2' = 0$ because of (3): this proves the claim.

Hence the Chern class map c_2 induces a sheaf map

$$\gamma_2\colon \operatorname{Coker}\Phi \to \mathcal{F}^{2,2}_{\mathbb{Z}}.$$

Taking $n=1$ in (3) we get a short exact sequence:

$$0 \to \mathcal{H}^0(\mathbb{C}^*(1)) \to \mathcal{H}^1_{\mathcal{D}}(\mathbb{Z}(1)) \to \mathcal{F}^{1,1}_{\mathbb{Z}} \to 0$$

where $\mathcal{F}^{1,1}_{\mathbb{Z}}$ be the image of the map $\mathcal{H}^1_{\mathcal{D}}(\mathbb{Z}(1)) \to \mathcal{H}^1(\mathbb{Z}(1))$. By (4) there are isomorphisms: $\mathcal{H}^1_{\mathcal{D}}(\mathbb{Z}(1)) \simeq \mathcal{O}^*_X \simeq \mathcal{K}_1$. Hence $\mathcal{F}^{1,1}_{\mathbb{Z}} \simeq \mathcal{O}^*_X/\mathbb{C}^*$.

The sheaf $\mathcal{F}_{\mathbb{Z}}^{2,2}$ has the following Gersten-Quillen resolution:

$$\mathcal{F}_{\mathbb{Z}}^{2,2} \to (i_\xi)_* \left(H^2_{\mathcal{D}}(\mathbb{C}(X),\mathbb{Z}(2))/H^1(\mathbb{C}(X),\mathbb{C}^*(2))\right) \\ \to \coprod_{x\in X^1} (i_x)_* \left(H^1_{\mathcal{D}}(\mathbb{C}(x),\mathbb{Z}(1))/H^0(\mathbb{C}(x)^*,\mathbb{C}^*(1))\right) \to \coprod_{x\in X^2} (i_x)_*\mathbb{Z} \to 0$$

where ξ denotes the generic point of X. We have: $H^0(\mathbb{C}(x),\mathbb{C}^*(1)) \simeq \mathbb{C}^*$ and $H^1_{\mathcal{D}}(\mathbb{C}(x),\mathbb{Z}(1)) \simeq \mathbb{C}(x)^*$, because of (4).

On the other hand the sheaf $\operatorname{Coker}\Phi$ has the following Gersten-Quillen resolution (see [**Ped**]):

$$\operatorname{Coker}\Phi \to (i_\xi)_*(K_2(\mathbb{C}(X)/\Phi(\mathbb{C}(X))) \to \coprod_{x\in X^1} (i_x)_*(\mathbb{C}(x)^*/\mathbb{C}^*) \\ \to \coprod_{x\in X^2} (i_x)_*(\mathbb{Z}) \to 0$$

where $\Phi(\mathbb{C}(X))$ is the image of the map $\mathbb{C}(X)^*\otimes\mathbb{C}^* \to K_2(\mathbb{C}(X))$.

Therefore the map γ_2 has kernel and cokernel which are constant sheaves.

The $d\log$ map induces a surjection (see [**Esn1**, §3]):

$$\mathcal{K}_2 \xrightarrow{d\log\wedge d\log} \mathcal{F}_{\mathbb{Z}}^{1,1}\wedge\mathcal{F}_{\mathbb{Z}}^{1,1}.$$

Therefore the image of $\mathcal{K}_1\otimes\mathcal{K}_1$ in $\mathcal{H}^2(\mathbb{Z}(2))$ is the subsheaf $\mathcal{F}_{\mathbb{Z}}^{1,1}\wedge\mathcal{F}_{\mathbb{Z}}^{1,1}$ of $\mathcal{F}_{\mathbb{Z}}^{2,2}$. This implies that the image of γ_2 in $\mathcal{F}_{\mathbb{Z}}^{2,2}$ is the sheaf $\mathcal{F}_{\mathbb{Z}}^{1,1}\wedge\mathcal{F}_{\mathbb{Z}}^{1,1}$. □

THEOREM 1.2. *Let X be as in Lemma 1.1. Then the map $\tau=\tau_{1,1}$ in (2) is surjective up to torsion if and only if $H^1(X,\mathcal{F}_{\mathbb{Z}}^{1,1}\wedge\mathcal{F}_{\mathbb{Z}}^{1,1})\otimes\mathbb{Q}=0$.*

PROOF. By [**Ped**, 1.8] the cokernel of the map $\tau\colon \operatorname{Pic}X\otimes\mathbb{C}^*\to H^1(X,\mathcal{K}_2)$ is isomorphic to $H^1(X,\operatorname{Coker}\Phi)$ where Φ is the map defined in Lemma 1.1. From Lemma 1.1 it follows that there is a short exact sequence of sheaves:

$$0\to \operatorname{Ker}\gamma_2 \to \operatorname{Coker}\Phi \to \operatorname{Im}\gamma_2 \to 0$$

where $H^i(X,\operatorname{Ker}\gamma_2)=0$ for $i>0$.

Therefore we get:

$$H^1(X,\operatorname{Coker}\Phi)\simeq H^1(X,\operatorname{Im}\gamma_2)\simeq H^1(X,\mathcal{F}_{\mathbb{Z}}^{1,1}\wedge\mathcal{F}_{\mathbb{Z}}^{1,1}).$$

Hence $\operatorname{Coker}\tau\otimes\mathbb{Q}\simeq H^1(X,\mathcal{F}_{\mathbb{Z}}^{1,1}\wedge\mathcal{F}_{\mathbb{Z}}^{1,1})\otimes\mathbb{Q}$. □

The following result has been proved independently by A. Rosenschon (see [**Ros**, 5.3]).

COROLLARY 1.3. *Let X be a smooth projective variety over $\mathbb{C}$ and let $\mathcal{F}_{ind}^{2,2}$ be the quotient sheaf $\mathcal{F}_{\mathbb{Z}}^{2,2}/\mathcal{F}_{\mathbb{Z}}^{1,1}\wedge\mathcal{F}_{\mathbb{Z}}^{1,1}$. Then there is an exact sequence:*

$$0\to H^0(X,\mathcal{F}_{ind}^{2,2})\to H^1(X,\operatorname{Coker}\Phi)\to H^1(X,\mathcal{F}_{\mathbb{Z}}^{2,2})\to 0 \tag{8}$$

PROOF. Let us consider the cohomology sequence associated to the short exact sequence of sheaves:

$$0\to\operatorname{Im}\gamma_2\to\mathcal{F}_{\mathbb{Z}}^{2,2}\to\mathcal{F}_{\mathbb{Z}}^{2,2}/\mathcal{F}_{\mathbb{Z}}^{1,1}\wedge\mathcal{F}_{\mathbb{Z}}^{1,1}\to 0.$$

We have: $H^0(X,\mathcal{F}_{\mathbb{Z}}^{2,2})=0$ by [**Esn2**, Theorem 1.3] (see also [**Ros**, 3.11]). Hence $H^0(X,\operatorname{Im}\gamma_2)=0$. From Lemma 1.1 it follows that: $H^1(X,\operatorname{Im}\gamma_2)\simeq H^1(X,\operatorname{Coker}\Phi)$ and

$$H^2(X,\operatorname{Im}\gamma_2)\simeq H^2(X,\operatorname{Coker}\Phi)\simeq H^2(X,\mathcal{F}_{\mathbb{Z}}^{2,2}).$$

Hence we get the claimed exact sequence. □

REMARK 1.4. A. Rosenschon has also proved that the group $H^0(X, \mathcal{F}^{2,2}_{ind})$ is uniquely divisible.

THEOREM 1.5. *Let X be a smooth irreducible projective variety over $\mathbb{C}$. Assume that $H^2(X, \mathcal{O}_X) = 0$. Then the map $\tau_{1,1} \otimes \mathbb{Q}$: $(\operatorname{Pic} X \otimes \mathbb{C}^*)_{\mathbb{Q}} \to H^1(X, \mathcal{K}_2)_{\mathbb{Q}}$ is surjective if and only if*

$$\mathcal{F}^{2,2}_{\mathbb{Z}} = \mathcal{F}^{1,1}_{\mathbb{Z}} \wedge \mathcal{F}^{1,1}_{\mathbb{Z}}$$

i.e., if and only if the $d\log$ map $\mathcal{K}_2 \to \mathcal{F}^{2,2}_{\mathbb{Z}}$ is surjective.

PROOF. The map γ_2 in Lemma 1.1 induces diagram with exact rows as follows:

$$\begin{array}{ccccccccc} 0 & \longrightarrow & H^0(X, \mathcal{H}^2(\mathbb{Q}/\mathbb{Z})) & \longrightarrow & H^1(X, \operatorname{Coker}\Phi) & \longrightarrow & H^1(X, \operatorname{Coker}\Phi) \otimes \mathbb{Q} & \longrightarrow & 0 \\ & & \simeq\downarrow & & h_1(\gamma_2)\downarrow & & \downarrow & & \\ 0 & \longrightarrow & H^0(X, \mathcal{H}^2(\mathbb{Q}/\mathbb{Z})) & \longrightarrow & H^1(X, \mathcal{F}^{2,2}_{\mathbb{Z}}) & \longrightarrow & H^1(X, \mathcal{F}^{2,2}_{\mathbb{Z}}) \otimes \mathbb{Q} & \longrightarrow & 0. \end{array}$$

Exactness of the top row of the above diagram has been proved in [**Ped**, 1.8]. It follows from the exact sequence of sheaves:

$$0 \to \operatorname{Coker}\Phi \to (\operatorname{Coker}\Phi) \otimes \mathbb{Q} \to \mathcal{K}_2 \otimes \mathbb{Q}/\mathbb{Z} \to 0$$

and the isomorphism $\mathcal{K}_2 \otimes \mathbb{Q}/\mathbb{Z} \simeq \mathcal{H}^2(\mathbb{Q}/\mathbb{Z})$, coming from a theorem of Merkuriev-Suslin [**MS**].

The bottom row comes from the following isomorphism: $(H^1(X, \mathcal{F}^{2,2}_{\mathbb{Z}}))_{\text{tors}} \simeq H^0(X, \mathcal{H}^2(\mathbb{Q}/\mathbb{Z}))$ (see [**Ros**, 5.1] and [**BV1**, 3.1]) and from the vanishing $H^1(X, \mathcal{F}^{2,2}_{\mathbb{Z}}) \otimes \mathbb{Q}/\mathbb{Z} = 0$ (see [**BV1**, 3.1]).

Because of the assumption $H^2(X, \mathcal{O}_X) = 0$, we have $H^0(X, \mathcal{H}^2(\mathbb{Q}/\mathbb{Z})) \simeq H^1(X, \mathcal{F}^{2,2}_{\mathbb{Z}}) \simeq H^3(X, \mathbb{Z})_{\text{tors}}$. Therefore $H^1(X, \mathcal{F}^{2,2}_{\mathbb{Z}}) \otimes \mathbb{Q} = 0$. Hence from the above diagram we get that $H^1(X, \operatorname{Coker}\Phi) \otimes \mathbb{Q} = 0$ if and only if the kernel of $H^1(X, \operatorname{Coker}\Phi) \to H^1(X, \mathcal{F}^{2,2}_{\mathbb{Z}})$ vanishes.

By Corollary 1.3, this is equivalent to

$$H^0(X, \mathcal{F}^{2,2}_{\text{ind}}) = 0.$$

The sheaf $\mathcal{F}^{2,2}_{\text{ind}}$ being a constant sheaf we get our statement. □

The following result, which also appears in [**Ros**, 5.15], is an immediate consequence of the above Theorem 1.5 and of [**BS**].

COROLLARY 1.6. *Let X a smooth irreducible projective variety over $\mathbb{C}$. Assume that $H^2(X, \mathcal{O}_X) = 0$ and Bloch's Conjecture holds for X. Then $\mathcal{F}^{2,2}_{\mathbb{Z}} = \mathcal{F}^{1,1}_{\mathbb{Z}} \wedge \mathcal{F}^{1,1}_{\mathbb{Z}}$.*

PROOF. By [**BS**, Theorem 3] the map $\tau_{1,1} \otimes \mathbb{Q}$ is surjective. Then apply Theorem 1.5. □

In order to prove our next result on the regulator map R we need to set up some necessary notations and recall some known facts on the Picard group $\operatorname{Pic} X$ and the Neron-Severi group $\operatorname{NS} X$ of a smooth projective variety X over $\mathbb{C}$.

For a smooth projective variety X over an algebraically closed field k we will denote by $\operatorname{Pic}^0 X$ the subgroup of $\operatorname{Pic} X$ consisting of divisor classes algebraically equivalent to 0, and by $\operatorname{NS} X$ the *Neron-Severi* group of X, i.e., the quotient group $\operatorname{Pic} X / \operatorname{Pic}^0 X$.

The group $\operatorname{Pic}^0 X$ is isomorphic to the group of closed points of the Picard variety of X, hence it is a divisible group. For all $j > 1$ the group $K_j(k)$ is divisible, hence the group $\operatorname{Pic}^0 X \otimes K_j(k)$ is uniquely divisible. Therefore we have a short exact sequence:

$$0 \to \operatorname{Pic}^0 X \otimes K_j(k) \to \operatorname{Pic} X \otimes K_j(k) \xrightarrow{c_1 \otimes \mathrm{id}} \operatorname{NS} X \otimes K_j(k) \to 0 \tag{9}$$

where $c_1\colon \operatorname{Pic} X \to \operatorname{NS} X$.

If $k = \mathbb{C}$ then the exponential sequence yields an isomorphism:

$$\operatorname{Pic}^0 X \simeq H^1(X, \mathcal{O}_X)/H^1(X, \mathbb{Z})$$

where $H^1(X,\mathbb{Z})$ is singular cohomology. The map c_1 is induced by the Chern class map $c_1\colon \mathcal{K}_1 \to \mathcal{H}^1(\mathbb{Z})$ from K-Theory to singular cohomology using the isomorphisms:

$$H^1(X, \mathcal{K}_1) \simeq \operatorname{Pic} X; \quad H^1(X, \mathcal{H}^1(\mathbb{Z})) \simeq \operatorname{NS} X$$

where the second isomorphism comes from the results of Bloch and Ogus [**BO**].

In particular, for $j = 1$ and $k = \mathbb{C}$, the exact sequence in (9) yields the following new exact sequence:

$$0 \to \operatorname{Pic}^0 X \otimes \mathbb{C}^* \to \operatorname{Pic} X \otimes \mathbb{C}^* \xrightarrow{c_1 \otimes \mathrm{id}} \operatorname{NS} X \otimes \mathbb{C}^* \to 0 \tag{10}$$

where $\operatorname{NS} X \otimes \mathbb{C}^* \simeq H^1(X, \mathcal{H}^1(\mathbb{C}^*))$ by [**BO**]

THEOREM 1.7. *Let X be a smooth irreducible projective variety over $\mathbb{C}$. Let $R(X)$ and* $\operatorname{Ker} R$ *be, respectively, the image and the kernel of the regulator map $R\colon H^1(X,\mathcal{K}_2) \to H^3_{\mathcal{D}}(X, \mathbb{Z}(2))$. Then there are exact sequences as follows:*

(i) $0 \to \operatorname{NS}(X) \otimes \mathbb{C}^* \to R(X) \to H^1(X, \mathcal{F}^{2,2}_{\mathbb{Z}}) \to 0$

(ii) $0 \to \operatorname{Ker} \tau \to \operatorname{Pic}^0 X \otimes \mathbb{C}^* \to \operatorname{Ker} R \to H^0(X, \mathcal{F}^{2,2}_{\mathrm{ind}}) \to 0$

where $\tau = \tau_{1,1}\colon \operatorname{Pic} X \otimes \mathbb{C}^* \to H^1(X, \mathcal{K}_2)$.

The group $H^1(X, \mathcal{F}^{2,2}_{\mathbb{Z}})_{\mathbb{Q}} \simeq (R(X)/\operatorname{NS}(X) \otimes \mathbb{C}^*)_{\mathbb{Q}}$ is at most countable.

PROOF. The regulator map R coincides with the composite map

$$H^1(X, \mathcal{K}_2) \xrightarrow{h(c_{2,2})} H^1(X, \mathcal{H}^2_{\mathcal{D}}(\mathbb{Z}(2))) \xrightarrow{i} H^3_{\mathcal{D}}(X, \mathbb{Z}(2))$$

where i is the injective map in the exact sequence (6), and $c_2\colon \mathcal{K}_2 \to \mathcal{H}^2_{\mathcal{D}}(\mathbb{Z}(2))$ is the Chern class map.

The sheaf $\mathcal{K}_2$ has the following Gersten-Quillen resolution

$$0 \to \mathcal{K}_2 \to (i_\xi)_*\big(K_2(\mathbb{C}(X))\big) \to \coprod_{x \in X^1} (i_x)_*(\mathbb{C}(x)^*) \to \coprod_{x \in X^2} (i_x)_*\mathbb{Z} \to 0$$

where ξ is the generic point of X.

The sheaf $\mathcal{H}^2_{\mathcal{D}}(\mathbb{Z}(2))$ has a Bloch-Ogus resolution as follows:

$$\begin{aligned} 0 \to \mathcal{H}^2(\mathbb{Z}(2)_{\mathcal{D}}) \to (i_\xi)_*\big(H^2_{\mathcal{D}}(\mathbb{C}(X), \mathbb{Z}(2))\big) \to \coprod_{x \in X^1} (i_x)_* H^1_{\mathcal{D}}(\mathbb{C}(x), \mathbb{Z}(1)) \\ \to \coprod_{x \in X^2} (i_x)_*\mathbb{Z} \to 0, \end{aligned}$$

where $H^1_{\mathcal{D}}(\mathbb{C}(x), \mathbb{Z}(1)) = \lim H^1_{\mathcal{D}}(U, \mathbb{Z}(1)) \simeq \mathbb{C}(x)^*$, because of the isomorphism in (4).

Therefore the kernel and cokernel of the map $c_2 \colon \mathcal{K}_2 \to \mathcal{H}^2_{\mathcal{D}}(\mathbb{Z}(2))$ are constant sheaves. This implies that the map

$$H^1(X, \mathcal{K}_2) \xrightarrow{h(c_2)} H^1(X, \mathcal{H}^2_{\mathcal{D}}(\mathbb{Z}(2)))$$

is surjective.

The map i being injective we get:

$$R(X) = H^1(X, \mathcal{H}^2_{\mathcal{D}}(\mathbb{Z}(2)))$$

and

$$\operatorname{Ker} R = \ker[h(c_2)].$$

From the exact sequence of sheaves in (3) we get, for $n = 2$, the following cohomology sequence:

$$\cdots \to H^0(X, \mathcal{F}^{2,2}_{\mathbb{Z}}) \to H^1(X, \mathcal{H}^1(\mathbb{C}^*(2))) \to H^1(X, \mathcal{H}^2_{\mathcal{D}}(\mathbb{Z}(2))) \\ \to H^1(X, \mathcal{F}^{2,2}_{\mathbb{Z}}) \to H^2(X, \mathcal{H}^1(\mathbb{C}^*(2)))$$

By the results of Bloch and Ogus [**BO**] we have an isomorphism:

$$H^1(X, \mathcal{H}^1(\mathbb{C}^*)) \simeq \mathrm{NS}(X) \otimes \mathbb{C}^*$$

and $H^2(X, \mathcal{H}^1(\mathbb{C}^*)) = 0$. Moreover $H^0(X, \mathcal{F}^{2,2}_{\mathbb{Z}}) = 0$ by [**Esn2**, Theorem 1.3]. Therefore we get the following short exact sequence:

$$0 \to \mathrm{NS}(X) \otimes \mathbb{C}^* \to H^1(X, \mathcal{H}^2_{\mathcal{D}}(\mathbb{Z}(2))) \to H^1(X, \mathcal{F}^{2,2}_{\mathbb{Z}}) \to 0 \tag{11}$$

which proves (i), because $R(X) = H^1(X, \mathcal{H}^2_{\mathcal{D}}(\mathbb{Z}(2)))$.

In order to prove (ii) we first show that the maps τ and R are compatible in the sense that there is a commutative diagram as follows:

$$\begin{array}{ccc} \operatorname{Pic} X \otimes \mathbb{C}^* & \xrightarrow{\tau} & H^1(X, \mathcal{K}_2) \\ \downarrow{\scriptstyle c_1 \otimes \mathrm{id}} & & \downarrow{\scriptstyle R} \\ \mathrm{NS}\, X \otimes \mathbb{C}^* & \xrightarrow{\sigma} & H^1(X, \mathcal{H}^2_{\mathcal{D}}(\mathbb{Z}(2))) \end{array} \tag{12}$$

where σ is the injective map in (11) and $c_1 \otimes \mathrm{id}$ is the map in (10).

Let $\Phi = \Phi_{1,1}$ be the sheaf map defined in (1). There is a cup product for Deligne-Beilinson cohomology sheaves (see [**EV**, §3]):

$$\mathcal{H}^p_{\mathcal{D}}(\mathbb{Z}(r)) \cup \mathcal{H}^q_{\mathcal{D}}(\mathbb{Z}(s)) \to \mathcal{H}^{p+q}_{\mathcal{D}}(\mathbb{Z}(r+s))$$

and these operations are compatible with the exact sequences in (3) [**EV**, 3.9].

Therefore the following diagram commutes

$$\begin{array}{ccc} \mathcal{K}_1 \otimes \mathbb{C}^* & \xrightarrow{\Phi} & \mathcal{K}_2 \\ \downarrow{\scriptstyle c_1 \otimes \mathrm{id}} & & \downarrow{\scriptstyle c_2} \\ \mathcal{H}^1(\mathbb{C}^*(2) & \xrightarrow{\Psi} & \mathcal{H}^2_{\mathcal{D}}(\mathbb{Z}(2)) \end{array}$$

where the left-hand vertical map comes from the Chern class $c_1 \colon \mathcal{K}_1 \to \mathcal{H}^1(\mathbb{Z})$ and from the isomorphism $\mathcal{H}^1(\mathbb{Z}) \otimes \mathbb{C}^* \simeq \mathcal{H}^1(\mathbb{C}^*)$. It also coincides with the composition map:

$$\mathcal{K}_1 \otimes \mathbb{C}^* \xrightarrow{\simeq} \mathcal{H}^1_{\mathcal{D}}(\mathbb{Z}(1)) \otimes \mathcal{H}^0(\mathbb{C}^*) \to \mathcal{H}^1(\mathbb{C}^*(2)).$$

The map $\Psi\colon \mathcal{H}^1(\mathbb{C}^*(2)) \to \mathcal{H}^2_{\mathcal{D}}(\mathbb{Z}(2))$ is the map in the exact sequence in (1) for $n = 2$.

Then the diagram in (12) arises by applying H^1 to the above diagram and using the isomorphisms $H^1(X, \mathcal{K}_1 \otimes \mathbb{C}^*) \simeq \operatorname{Pic} X \otimes \mathbb{C}^*$ and $H^1(X, \mathcal{H}^1(\mathbb{C}^*)) \simeq \mathbf{NS}X \otimes \mathbb{C}^*$.

From (12) we get following diagram of exact sequences:

$$\begin{array}{ccccccc}
\operatorname{Pic} X \otimes \mathbb{C}^* & \xrightarrow{\tau} & H^1(X, \mathcal{K}_2) & \longrightarrow & \operatorname{Coker} \tau & \longrightarrow & 0 \\
\downarrow {\scriptstyle c_1 \otimes \mathrm{id}} & & \downarrow {\scriptstyle R} & & \downarrow & & \\
\operatorname{NS} X \otimes \mathbb{C}^* & \longrightarrow & H^1(X, \mathcal{H}^2_{\mathcal{D}}(\mathbb{Z}(2))) & \longrightarrow & H^1(X, \mathcal{F}^{2,2}_{\mathbb{Z}}) & \longrightarrow & 0.
\end{array}$$

The bottom row is the exact sequence in (11), the right-hand vertical map is part of the exact sequence in (8), and the left-hand vertical map fits into the exact sequence in (10).

From the above diagram we get an exact sequence:

$$0 \to \operatorname{Ker} \tau \to \operatorname{Pic}^0 X \otimes \mathbb{C}^* \to \operatorname{Ker} R \to \mathbf{H} \to 0,$$

where

$$\mathbf{H} = \operatorname{Ker}[\operatorname{Coker} \tau \to H^1(X, \mathcal{F}^{2,2}_{\mathbb{Z}})].$$

From the exact sequence in (8) it follows that $\mathbf{H} = H^0(X, \mathcal{F}^{2,2}_{\mathrm{ind}})$ and this gives (ii).

The last assertion follows from the following result (see [**MSt**, Proposition 1]): the image of the map

$$\overline{R}\colon H^1(X, \mathcal{K}_2) \to \left(H^3_{\mathcal{D}}(X, \mathbb{Z}(2))/\operatorname{NS}(X) \otimes \mathbb{C}^*\right)_{\mathbb{Q}}$$

obtained from R by composing with the quotient map, is countable.

Using (i) this shows that the group $H^1(X, \mathcal{F}^{2,2}_{\mathbb{Z}}) \otimes \mathbb{Q}$ is countable.

By [**BV1**, 3.1] (see also [**Ros**, 5.1]) we have an isomorphism:

$$(H^1(X, \mathcal{F}^{2,2}_{\mathbb{Z}}))_{\mathrm{tors}} \simeq H^0(X, \mathcal{H}^2(\mathbb{Q}/\mathbb{Z})) \simeq (\mathbb{Q}/\mathbb{Z})^{b_2 - \rho} \oplus (H^3(X, \mathbb{Z}))_{\mathrm{tors}}$$

where b_2 is the second Betti number and ρ the Picard number. Hence $H^1(X, \mathcal{F}^{2,2}_{\mathbb{Z}})$ is countable. □

COROLLARY 1.8. *Let X be as in Theorem 1.7. Then $(\operatorname{Coker} \tau)_{\mathbb{Q}} = 0$ if and only if $H^0(X, \mathcal{F}^{2,2}_{\mathrm{ind}}) = 0$ and $R(X)_{\mathbb{Q}} = (\operatorname{NS}(X) \otimes \mathbb{C}^*)_{\mathbb{Q}}$.*

PROOF. From Corollary 1.3, we get $H^1(X, \operatorname{Coker} \Phi)_{\mathbb{Q}} = (\operatorname{Coker} \tau)_{\mathbb{Q}} = 0$ if and only if $H^0(X, \mathcal{F}^{2,2}_{\mathrm{ind}}) = 0$ and $H^1(X, \mathcal{F}^{2,2}_{\mathbb{Z}})_{\mathbb{Q}} = 0$. The rest follows from Theorem 1.7. □

It might be worthwhile to point out that a similar result does not hold in general for the maps $\tau_{i,j}$ if $i > 1$ in the sense that there exist varieties X such that the image of the regulator map $R_m\colon H^m(X, \mathcal{K}_{m+1}) \to H^{2m+1}_{\mathcal{D}}(X, \mathbb{Q}(m+1))$ (where $R_1 = R$) is trivial, i.e., it lies in the image of $(\mathrm{CH}^m(X) \otimes \mathbb{C}^*)_{\mathbb{Q}}$ via the composite map:

$$(\mathrm{CH}^m(X) \otimes \mathbb{C}^*)_{\mathbb{Q}} \to H^m(X, \mathcal{H}^{m+1}_{\mathcal{D}}(\mathbb{Q}(m+1))) \to H^{2m+1}_{\mathcal{D}}(X, \mathbb{Q}(m+1))$$

but $\operatorname{Coker} \tau_{m,1} \neq 0$. In fact Collino [**Col**, 7.1] has shown that there are varieties X of dimension 3, hypersurfaces of high enough degree in a cubic fourfold, such that R_2 is trivial but $\operatorname{Coker} \tau_{2,1}$ is not finitely generated.

EXAMPLES 1.9. We give some illustrations.

(1) Let $X \subset \mathbb{P}^3$ be a general hypersurface of degree $d \geq 5$. Then $(R(X))_{\mathbb{Q}} \simeq \mathrm{NS}(X) \otimes \mathbb{C}/\mathbb{Q} \simeq \mathbb{C}/\mathbb{Q}$ [**MSt**, 6.1]). Therefore $H^1(X, \mathcal{F}^{2,2}_{\mathbb{Z}})_{\mathbb{Q}} = 0$ and

$$(\operatorname{Coker} \tau)_{\mathbb{Q}} = H^0(X, \mathcal{F}^{2,2}_{\mathrm{ind}}) \simeq \operatorname{Ker} R.$$

Note that in this situation we also have $(\operatorname{Coker} \tau)_{\mathbb{Q}} \simeq H^0(X, \mathcal{G})$ where $\mathcal{G}$ is the constant sheaf associated to the cokernel of the Chern class $c_2 \colon \mathcal{K}_2 \to \mathcal{H}^2_{\mathcal{D}}(\mathbb{Z}(2))$.

(2) Let X be a general quartic K_3-surface. Then $(\operatorname{Coker} \tau)_{\mathbb{Q}} \neq 0$ [**Voi**].

(3) Let X be a general quartic surface. Then $H^1(X, \mathcal{F}^{2,2}_{\mathbb{Z}})_{\mathbb{Q}} \neq 0$ [**Col**].

Corollary 1.10. *Let X be a smooth projective surface over $\mathbb{C}$. Then the group $H^1(X, \mathcal{K}_2)$ decomposes as follows:*

$$H^1(X, \mathcal{K}_2) \simeq B \oplus (\mathbb{Q}/\mathbb{Z})^{b_2} \oplus H^1(X, \mathcal{F}^{2,2}_{\mathbb{Z}})_{\mathbb{Q}} \oplus (\mathrm{NS}\, X \otimes \mathbb{C}^*)_{\mathbb{Q}} \oplus \operatorname{Ker} R$$

where $B = H^2(X, \mathbb{Z})_{\mathrm{tors}}$ is a finite group and $H^1(X, \mathcal{F}^{2,2}_{\mathbb{Z}})_{\mathbb{Q}}$ is at most countable.

Proof. By the results of [**PW**, 6.7.1] it follows that $H^1(X, \mathcal{K}_2) \simeq B \oplus (\mathbb{Q}/\mathbb{Z})^{b_2} \oplus V$ where V is a uniquely divisible group. We also have:

$$(R(X)_{\mathrm{tors}} \simeq (\mathrm{NS}\, X \otimes \mathbb{C}^*)_{\mathrm{tors}} \oplus H^1(X, \mathcal{F}^{2,2})_{\mathrm{tors}} \simeq B \oplus (\mathbb{Q}/\mathbb{Z})^{b_2}.$$

This shows that $\operatorname{Ker} R$ is uniquely divisible. The rest follows from Theorem 1.7. □

Remark 1.11. If $H^1(X, \mathcal{O}_X) \neq 0$ then, in general, $\operatorname{Ker} \tau \neq 0$. If X is an elliptic curve over $\mathbb{C}$, then $\operatorname{Ker} \tau$ contains infinite elements (see [**Roi**]). Also the group $D = (H^0(X, \mathcal{K}_2)/K_2(\mathbb{C}))_{\mathbb{Q}}$ in (7) does not vanish for elliptic curves: Bloch in [**Blo**, Lecture 8] constructs non-trivial elements in D by taking images of elements in $H^0(X, \mathcal{K}_2)$ defined by rational functions f and g having all zeroes and poles on torsion points of X.

The following corollary of Theorem 1.7 has been proved in [**Ras**, 1.1], for curves over an algebraically closed field.

Corollary 1.12. *Let C be a smooth curve over $\mathbb{C}$. Then there is an exact sequence as follows:*

$$0 \to \ker \tau \to \operatorname{Pic}^0 C \otimes \mathbb{C}^* \to K(C) \to 0$$

where $\tau\colon \operatorname{Pic} C \otimes \mathbb{C}^* \to H^1(C, \mathcal{K}_2)$ and

$$K(C) = \operatorname{Ker}[H^1(C, \mathcal{K}_2) \to (\mathbb{C}^*)^r]$$

r being the number of proper irreducible components of C.

Proof. Everything follows from Corollary 1.6 because of the isomorphisms:

$$H^3_{\mathcal{D}}(C, \mathbb{Z}(2)) \simeq \mathrm{NS}(C) \otimes \mathbb{C}^* \simeq (\mathbb{C}^*)^r$$

and the vanishing of $\mathcal{F}^{2,2}_{\mathbb{Z}}$ for a curve. □

2. Surfaces with $\mathrm{CH}^2 = \mathbb{Z}$ and Hyperelliptic Surfaces

In this section we prove some results on the maps $\tau_{i,j}$ in (2), which will be used to compute $\mathcal{K}$-cohomology of surfaces with $\mathrm{CH}^2 = \mathbb{Z}$ (Theorem 2.9) and for hyperelliptic surfaces (Example 2.10).

LEMMA 2.1. *Let X be a smooth irreducible projective surface over an algebraically closed field. Then the cup product:*

$$H^1(X,\mathcal{K}_{j+1}) \otimes H^1(X,\mathcal{K}_1) \xrightarrow{\cup} H^2(X,\mathcal{K}_{j+2})$$

induces a pairing

$$g_j\colon H^1(X,\mathcal{K}_{j+1}) \otimes \operatorname{Pic}(X) \to K_j(k).$$

PROOF. Because of the isomorphisms $H^1(X,\mathcal{K}_1) \simeq \operatorname{Pic} X$ and $H^2(X,\mathcal{K}_2) \simeq \operatorname{CH}^2(X)$ the product:

$$H^1(X,\mathcal{K}_1) \otimes H^1(X,\mathcal{K}_1) \to H^2(X,\mathcal{K}_2)$$

induces, by composition with the degree map $\operatorname{CH}^2(X) \to \mathbb{Z}$ a pairing

$$f\colon \operatorname{Pic} X \otimes \operatorname{Pic} X \to \mathbb{Z}$$

which coincides with the intersection pairing on the surface X. Hence f is trivial on $\operatorname{Pic}^0 X \otimes \operatorname{Pic} X$.

By the projection formula we have a commutative diagram:

$$\begin{array}{ccccc}
H^1(X,\mathcal{K}_1) \otimes H^1(X,\mathcal{K}_1) \otimes K_j(k) & \longrightarrow & H^2(X,\mathcal{K}_2) \otimes K_j(k) & \xrightarrow{\deg} & K_j(k) \\
\downarrow{\scriptstyle \mathrm{id}\otimes\tau_{1,j}} & & \downarrow{\scriptstyle \tau_{2,j}} & & \| \\
H^1(X,\mathcal{K}_1) \otimes H^1(X,\mathcal{K}_{j+1}) & \longrightarrow & H^2(X,\mathcal{K}_{j+2}) & \xrightarrow{s_*} & K_j(k)
\end{array}$$

where $s_*\colon H^2(X,\mathcal{K}_{j+2}) \to H^0(\operatorname{Spec} k,\mathcal{K}_j) \simeq K_j(k)$ comes from the structure map $s\colon X \to \operatorname{Spec} k$.

Hence f gives rise to a map f_j which fits into the following commutative diagram:

$$(13)\qquad \begin{array}{ccc}
\operatorname{Pic} X \otimes \operatorname{Pic} X \otimes K_j(k) & \xrightarrow{f_j} & K_j(k) \\
{\scriptstyle \mathrm{id}\otimes\tau_{1,j}}\downarrow & & \downarrow{\scriptstyle \simeq} \\
\operatorname{Pic} X \otimes H^1(X,\mathcal{K}_{j+1}) & \xrightarrow{g_j} & K_j(k)
\end{array}$$

where g_j is obtained by composing the cup product map with s_*. Therefore g_j induces a pairing:

$$(H^1(X,\mathcal{K}_{j+1}) \otimes \operatorname{Pic}(X) \to K_j(k) \qquad \square$$

REMARK 2.2. In the case $j = 1$, by a result of Gillet [**Gil1**, 1.3.4], we have that $g_1\colon H^1(X,\mathcal{K}_2) \otimes \operatorname{Pic} X \to k^*$ vanishes on $H^1(X,\mathcal{K}_2) \otimes \operatorname{Pic}^0 X$. Hence it factors through a pairing:

$$H^1(X,\mathcal{K}_2) \otimes \operatorname{NS} X \to k^*.$$

The proof in [**Gil1**] is based upon the fact that for a complete curve C over k there is an isomorphism $H^0(C,\mathcal{K}_1) \simeq H^0(C,\mathcal{O}_X^*) \simeq k^*$. However, if $j \geq 2$, in general, $H^0(C,\mathcal{K}_j) \neq K_j(k)$.

The following result extends part of Theorem 1.7.

PROPOSITION 2.3. *Let X be a smooth irreducible projective surface over an algebraically closed field. Then there is an exact sequence:*

$$0 \to \operatorname{Ker} \tau_{1,j} \to \operatorname{Pic}^0 X \otimes K_j(k) \to \operatorname{Ker} \pi_j \to \operatorname{Coker} \tau_{1,j}$$

where the map $\pi_j : H^1(X, \mathcal{K}_{j+1}) \to \operatorname{Hom}(\operatorname{Pic} X, K_j(k))$ is induced by the pairing g_j in Lemma 2.1.

In particular, if $\operatorname{Pic}^0 X = 0$ then $\tau_{1,j}$ is injective.

PROOF. The pairing g_j in Lemma 2.1 induces a map:

$$\pi_j \colon H^1(X, \mathcal{K}_{j+1}) \to \operatorname{Hom}(\operatorname{Pic} X, K_j(k)).$$

The intersection pairing for divisors on the surface X induces a map: $\operatorname{NS} X \otimes \operatorname{NS} X \otimes K_j(k) \to K_j(k)$ which gives rise to

$$\mu_j \colon \operatorname{NS} X \otimes K_j(k) \to \operatorname{Hom}(\operatorname{NS} X, K_j(k)).$$

The group $\operatorname{NS} X$ is finitely generated and $K_j(k)$ is divisible. Hence there are isomorphisms as follows:

$$\operatorname{NS} X \otimes K_j(k) \simeq (K_j(k))^\rho$$

and

$$\operatorname{Hom}(\operatorname{NS} X, K_j(k)) \simeq (K_j(k))^\rho \oplus \mathbf{F}$$

where ρ is the rank of $\operatorname{NS} X$ and $\mathbf{F}$ is a finite group.

Because of the non-degeneracy of the intersection pairing on X, the map

$$\mu_j \colon \operatorname{NS} X \otimes K_j(k) \to \operatorname{Hom}(\operatorname{NS} X, K_j(k))$$

is injective with a finite cokernel.

The exact sequence in (9) yields the following exact sequence, because $K_j(k)$ is divisible:

$$(14) \quad 0 \to \operatorname{Hom}(\operatorname{NS} X, K_j(k)) \to \operatorname{Hom}(\operatorname{Pic} X, K_j(k)) \to \operatorname{Hom}(\operatorname{Pic}^0 X, K_j(k)) \to 0$$

and we therefore get the following commutative diagram:

$$\begin{array}{ccccccccc}
0 & \longrightarrow & \operatorname{Pic}^0 X \otimes K_j(k) & \longrightarrow & \operatorname{Pic} X \otimes K_j(k) & \longrightarrow & \operatorname{NS} X \otimes K_j(k) & \longrightarrow & 0 \\
 & & \downarrow & & \downarrow{\scriptstyle \tau_{i,j}} & & \downarrow{\scriptstyle \mu'_j} & & \\
0 & \longrightarrow & \operatorname{Ker} \pi_j & \longrightarrow & H^1(X, \mathcal{K}_{j+1}) & \xrightarrow{\pi_j} & \operatorname{Hom}(\operatorname{Pic} X, K_j(k)) & &
\end{array}$$

where the top row is the exact sequence in (9) and the map μ'_j is obtained by composing μ_j with the map in (14).

The map μ'_j being injective the map $(\operatorname{Pic}^0 X \otimes K_j(k)) \to (\operatorname{Ker} \pi_j)$ has a kernel isomorphic to $\operatorname{Ker} \tau_{i,j}$. The rest follows from Snake Lemma. □

REMARK 2.4. Under the assumptions of Proposition 2.3, if $j = 1$ and $k = \mathbb{C}$ then the pairing g_1 induces a map $\pi_1 \colon H^1(X, \mathcal{K}_2) \to \operatorname{Hom}(\operatorname{NS} X, \mathbb{C}^*)$ (see Remark 2.2). Hence the map μ'_1 has a finite cokernel. The group $\operatorname{Pic}^0 X \otimes \mathbb{C}^*$ being uniquely divisible the above diagram yields an exact sequence

$$0 \to \mathbf{Ker}\, \tau_{1,1} \to \operatorname{Pic}^0 X \otimes \mathbb{C}^* \to (\mathbf{Ker}\, \pi_1)_{\mathbb{Q}} \to \mathbf{Coker} \tau_{1,1} \otimes \mathbb{Q} \to 0$$

From Theorem 1.7(ii) and from Corollary 1.3 we get:

$$(\operatorname{Ker} \pi_1)_{\mathbb{Q}} \simeq \operatorname{Ker} R \oplus H^1(X, \mathcal{F}_{\mathbb{Z}}^{2,2})_{\mathbb{Q}}.$$

LEMMA 2.5. *Let X be a smooth irreducible projective surface over an algebraically closed field k. Then, for every $j \geq 0$, there is a split exact sequence:*

$$0 \to (A_0(X) \otimes K_j(k))/T_j \to H^2(X, \mathcal{K}_{j+2}) \to K_j(k) \to 0 \tag{15}$$

where $A_0(X)$ is the group of zero cycles of degree zero and

$$T_j = \operatorname{Ker}\left[\operatorname{CH}^2(X) \otimes K_j(k) \overset{\tau_{2,j}}{\to} H^2(X, \mathcal{K}_{j+2})\right].$$

PROOF. Let $\Phi_{2,j} \colon \mathcal{K}_2 \otimes K_j(k) \to \mathcal{K}_{j+2}$ be the map of Zariski sheaves, where $K_j(k)$ denotes the constant sheaf associated to the corresponding group.

The map $\Phi_{2,j}$ induces a map between the Gersten-Quillen resolution of the sheaves $\mathcal{K}_2 \otimes K_j(k)$ and $\mathcal{K}_{j+2}$. Therefore we get a commutative diagram as follows:

$$\begin{array}{ccccc}
\coprod_{x \in X^2} \mathbb{Z} \otimes K_j(k) & \longrightarrow & H^2(X, \mathcal{K}_2) \otimes K_j(k)) & \longrightarrow & 0 \\
\downarrow & & \downarrow & & \\
\coprod_{x \in X^2} K_j(k(x)) & \longrightarrow & H^2(X, \mathcal{K}_{j+2}) & \longrightarrow & 0
\end{array}$$

where the left hand vertical map is an isomorphism because $k(x) \simeq k$.

Moreover: $H^2(X, \mathcal{K}_2) \otimes K_j(k) \simeq \operatorname{CH}^2(X) \otimes K_j(k)$.

Therefore the map:

$$\tau_{2,j} \colon \operatorname{CH}^2(X) \otimes K_j(k) \to H^2(X, \mathcal{K}_{j+2})$$

is surjective.

Let $s :: X \to \operatorname{Spec} k$ be the structure map. Then s being proper, we get a map

$$s_* \colon H^2(X, \mathcal{K}_{j+2}) \to H^0(\operatorname{Spec} k, \mathcal{K}_j) \simeq K_j(k)$$

which fits into the following commutative diagram:

$$\begin{array}{ccccccccc}
0 & \longrightarrow & \mathbf{T}_j & \longrightarrow & \operatorname{CH}^2(X) \otimes K_j(k) & \longrightarrow & H^2(X, \mathcal{K}_{j+2}) & \longrightarrow & 0 \\
 & & \downarrow & & \simeq \downarrow & & s_* \downarrow & & \\
0 & \longrightarrow & A_0(X) \otimes K_j(k) & \longrightarrow & \operatorname{CH}^2(X) \otimes K_j(k) & \xrightarrow{\tau_{2,j}} & K_j(k) & \longrightarrow & 0
\end{array}$$

where the bottom row is obtained by tensoring with $K_j(k)$ the split exact sequence:

$$0 \to A_0(X) \to \operatorname{CH}^2(X) \xrightarrow{\deg} \mathbb{Z} \to 0.$$

The above diagram yields the exact sequence in (15) and a splitting of the map deg gives a splitting of (15). □

The following proposition extends a result proved in [**Coo**, Theorem 4.3], for Enriques surfaces.

PROPOSITION 2.6. *Let X be a smooth, irreducible projective surface over an algebraically closed field k. Assume that the map $\tau_{0,j} \otimes \mathbb{Q} : K_j(k)_{\mathbb{Q}} \to H^0(X, \mathcal{K}_j)_{\mathbb{Q}}$ are all surjective. Then the spectral sequence*

$$E_2^{p,-q} \implies K_{-p-q}(X)$$

degenerates, up to torsion, at the E_2-terms. Therefore the K-theory of X decomposes, up to torsion, as follows:

$$K_j(X)_{\mathbb{Q}} \simeq (\mathcal{K}_j(k)_{\mathbb{Q}} \oplus H^1(X, \mathcal{K}_{j+1})_{\mathbb{Q}} \oplus H^2(X, \mathcal{K}_{j+2})_{\mathbb{Q}}.$$

PROOF. Because X is a surface, the only possible non-zero differentials in the spectral sequence are:

$$d_2 \colon H^0(X,\mathcal{K}_j) \to H^2(X,\mathcal{K}_{j+1}),$$

hence it is enough to show that $d_2 \otimes \mathbb{Q} = 0$. By [**Sus**, 3.6] there are inclusions:

$$K_j(k) \subset H^0(X,\mathcal{K}_j) \subset K_j(k(X)).$$

Hence our hypothesis imply that $K_j(k)_{\mathbb{Q}} = H^0(X,\mathcal{K}_j)_{\mathbb{Q}}$.

The proper maps: $i\colon \operatorname{Spec} k \to X$ and $s\colon X \to \operatorname{Spec} k$, where i is the inclusion of a closed point and s the structure map, yield, by functoriality, an inclusion $K_j(k) \subset K_j(X)$.

The map:

$$K_j(X) \to K_j(k(X))$$

coming from the localization sequence in [**Qui**, 5.4] induces the identity on the subgroup $K_j(k)$. Therefore the maps $K_j(X)_{\mathbb{Q}} \to H^0(X,\mathcal{K}_j)_{\mathbb{Q}}$ are surjective for all j.

From the spectral sequence we get surjective maps

$$K_j(X) \to E_\infty^{0,j} = \operatorname{Ker} d_2 \subset H^0(X,\mathcal{K}_j).$$

Therefore the surjectivity of the maps $K_q(X)_{\mathbb{Q}} \to H^0(X,\mathcal{K}_q)_{\mathbb{Q}}$ for $q = j,\ j+1$ implies that $d_2 \otimes \mathbb{Q} = 0$. □

REMARK 2.7. By [**PW**, 6.3] it follows that the differential $d_2 \colon H^0(X,\mathcal{K}_2) \to H^2(X,\mathcal{K}_3)$ is zero for all smooth surfaces over $\mathbb{C}$.

COROLLARY 2.8. *Assume that all the maps $\tau_{i,j} \otimes \mathbb{Q}$ are isomorphisms. Then*

$$K_j(X)_{\mathbb{Q}} \simeq (K_0(X) \otimes K_j(k))_{\mathbb{Q}}$$

for all $j \geq 0$.

PROOF. From Proposition 2.6, it follows that

$$K_j(X) \otimes \mathbb{Q} \simeq K_j(k)_{\mathbb{Q}} \oplus H^1(X,\mathcal{K}_{j+1})_{\mathbb{Q}} \oplus H^2(X,\mathcal{K}_{j+2})_{\mathbb{Q}}$$

where $\tau_{1,j} \otimes \mathbb{Q}\colon H^1(X,\mathcal{K}_{j+1})_{\mathbb{Q}} \simeq (\operatorname{Pic}(X) \otimes K_j(k))_{\mathbb{Q}}$ and $\tau_{2,j} \otimes \mathbb{Q}\colon H^2(X,\mathcal{K}_{j+2})_{\mathbb{Q}} \simeq (\mathrm{CH}^2(X) \otimes K_j(k))_{\mathbb{Q}}$. The rest immediately follows from the Riemann-Roch formula for K_0. □

THEOREM 2.9. *Let X be a smooth irreducible projective surface over $\mathbb{C}$ such that $\mathrm{CH}^2(X) \simeq \mathbb{Z}$. Then all the maps $\tau_{i,j} \otimes \mathbb{Q}$ are isomorphisms and*

$$K_j(X)_{\mathbb{Q}} \simeq (K_0(X) \otimes K_j(\mathbb{C}))_{\mathbb{Q}}$$

for all $j \geq 0$.

PROOF. From [**BS**, Theorem 3], it follows that the maps $\operatorname{Pic} X \otimes K_j(\mathbb{C})_{\mathbb{Q}}) \to H^1(X,\mathcal{K}_{j+1})_{\mathbb{Q}}$ are surjective and that the maps $K_j(\mathbb{C}) \otimes \mathbb{Q} \to H^0(X,\mathcal{K}_j)_{\mathbb{Q}}$ are isomorphisms. From Proposition 2.3, it follows, since $\operatorname{Pic}^0(X) = 0$, that the maps $\operatorname{Pic} X \otimes K_j(\mathbb{C})_{\mathbb{Q}}) \to H^1(X,\mathcal{K}_{j+1})_{\mathbb{Q}}$ are all isomorphisms. From the exact sequence in Lemma 2.5, it follows, since $A_0(X) = 0$, that the maps

$$\tau_{2,j} \otimes \mathbb{Q} : (\mathrm{CH}^2(X) \otimes K_j(\mathbb{C}))_{\mathbb{Q}} \to H^2(X,\mathcal{K}_{j+2})_{\mathbb{Q}}$$

are isomorphisms.

The rest of the statement follows from Remark 2.7. □

EXAMPLE 2.10. [Hyperelliptic surfaces] In the following example we use some of the results given in §1 and §2 to compute the $\mathcal{K}$-cohomology of hyperelliptic surfaces, i.e., a class of surfaces with $p_g = 0$ and $q = 1$.

In particular, we show that in this case the map $\tau_{1,2} \otimes \mathbb{Q}$ is not surjective.

Let X be an hyperelliptic surface, i.e., a surface of the form $X = E \times F/G$ where E and F are elliptic curves and G is a finite group of translations of E acting on $F \otimes E$ by $g(f,e) = (\phi(g), e+g)$ for a suitable $\phi : G \to \operatorname{Aut} F$ (see [**BM**]).

These surfaces are characterized by having invariants $p_g = 0$, $q = 1$ and $12K = 0$, where K is the canonical divisor. By Noether's formula we get: $b_2 = 2$, $b_1 = 2q = 2$ where b_i are the Betti numbers of X. Hence $\operatorname{NS} X \simeq H^2(X, \mathbb{Z}) \simeq (\mathbb{Z})^2$.

X has an elliptic fibration $\pi\colon X \to E$ with a section σ. We can also assume that π is smooth [**Bea**, VI.7]. The section $\sigma\colon E \to X$ of π has image a locally principal divisor (since $X \to E$ is smooth of relative dimension 1) and hence defines a class $[\sigma] \in \operatorname{Pic} X$ (see [**Gil1**, 3.2]). We also have $F/\phi(G) \simeq \mathbb{P}^1$ hence $\epsilon\colon X \to F/\phi(G)$ is a fibration over $\mathbb{P}^1$ with elliptic fibres.

By results of [**BKL**] we have $\operatorname{Pic}^0 E \simeq \operatorname{Alb} X \simeq A_0(X)$ hence the composite map $\pi^* \cup [\sigma]$:

$$\operatorname{Pic} E \xrightarrow{\pi^*} \operatorname{Pic} X \xrightarrow{\cup[\sigma]} \operatorname{CH}^2(X)$$

is an isomorphism.

Moreover, we have $\operatorname{Pic} X \simeq \operatorname{Pic}^0 X \oplus (\mathbb{Z})^2$ and $\operatorname{Pic}^0 X \simeq \operatorname{Pic}^0 E$. Hence $\operatorname{Pic} X \simeq \operatorname{Pic} E \oplus \mathbb{Z}$.

By tensoring with $K_j(\mathbb{C})$ we get

$$\operatorname{Pic} X \otimes K_j(\mathbb{C}) \simeq (\operatorname{Pic} E \otimes K_j(\mathbb{C})) \oplus K_j(\mathbb{C}) \simeq (\operatorname{CH}^2(X) \otimes K_j(\mathbb{C})) \oplus K_j(\mathbb{C}).$$

The summand $K_j(\mathbb{C})$ corresponds to the fibre $\epsilon^*(P)$ for any point $P \in F/\phi(G)$ because $F/\phi(G)$ is rational.

Therefore we have a commutative diagram:

$$\begin{array}{ccccccccc} 0 & \longrightarrow & \operatorname{Pic} E \otimes K_j(\mathbb{C}) & \longrightarrow & \operatorname{Pic} X \otimes K_j(\mathbb{C}) & \longrightarrow & K_j(\mathbb{C}) & \longrightarrow & 0 \\ & & \downarrow{\scriptstyle \mu_{1,j}} & & \downarrow{\scriptstyle \tau_{1,j}} & & \downarrow & & \\ 0 & \longrightarrow & H^1(E, \mathcal{K}_{j+1}) & \xrightarrow{\pi^*} & H^1(X, \mathcal{K}_{j+1}) & \longrightarrow & G_j & \longrightarrow & 0 \end{array} \tag{16}$$

where the map π^* is injective, because it is split by σ^* and $G_j = \mathit{Coker}\pi^*$. The map $\mu_{1,j}$ is surjective and the map $K_j(\mathbb{C}) \to G_j$ is injective, because it corresponds to the embedding $K_j(\mathbb{C}) \subset \coprod_{x \in X^1} K_n(\mathbb{C}(x))$ relative to the fibre $\epsilon^*(P)$ for a point $P \in F/\phi(G)$.

Then Snake Lemma yields $\operatorname{Ker} \mu_{1,j} \simeq \operatorname{Ker} \tau_{1,j}$ and $G_j = K_j(\mathbb{C}) \oplus F_j$, where $\operatorname{Coker} \tau_{1,j} \simeq F_j$.

Because of the non vanishing of $\operatorname{Ker} \mu_{1,1}$ (see Remark 1.11), $\operatorname{Ker} \tau_{1,1}$ is also non-zero.

If $j \geq 0$ then the group $H^1(E, \mathcal{K}_{j+1})$ is divisible. Hence from the bottom row of (16) we get:

$$H^1(X, \mathcal{K}_{j+1}) \simeq H^1(E, \mathcal{K}_{j+1}) \oplus G_j \simeq H^1(E, \mathcal{K}_{j+1}) \oplus K_j(\mathbb{C}) \oplus F_j \tag{17}$$

By [**PW**, 3.2 and 6.7.1] for a smooth projective surface over $\mathbb{C}$, we have $H^1(X,\mathcal{K}_2)_{\text{tors}} = B \oplus (\mathbb{Q}/\mathbb{Z})^{b_2}$ where $B = H^2(X,\mathbb{Z})_{\text{tors}}$, and

$$H^1(X,\mathcal{K}_{j+1})_{\text{tors}} \cong \begin{cases} (\mathbb{Q}/\mathbb{Z})^{\rho} & \text{if } j \geq 3 \text{ is odd;} \\ (\mathbb{Q}/\mathbb{Z})^{b_3} & \text{if } j \geq 2 \text{ is even.} \end{cases}$$

If C is a curve, then

$$H^1(C,\mathcal{K}_{j+1})_{\text{tors}} \cong \begin{cases} \mathbb{Q}/\mathbb{Z} & \text{if } j \geq 1 \text{ is odd;} \\ 0 & \text{if } j \geq 2 \text{ is even.} \end{cases}$$

In our case since $B = 0$ and $\rho = b_2 = b_1 = b_3 = 2$, from (17) we get:

$$(\operatorname{Coker} \tau_{1,j})_{\text{tors}} = (F_j)_{\text{tors}} \cong \begin{cases} 0 & \text{if } j \geq 1 \text{ is odd;} \\ (\mathbb{Q}/\mathbb{Z})^2 & \text{if } j \geq 2 \text{ is even.} \end{cases}$$

By [**BS**] for $j = 1$ we also have $(\operatorname{Coker} \tau_{1,1})_{\mathbb{Q}} = (F_1)_{\mathbb{Q}} = 0$. Hence

$$H^1(X,\mathcal{K}_2) \simeq H^1(E,\mathcal{K}_2) \oplus \mathbb{C}^* \simeq V(E) \oplus \mathbb{C}^* \oplus \mathbb{C}^*$$

where $V(E)$ is the uniquely divisible group $\operatorname{Ker}\left[H^1(E,\mathcal{K}_2) \to \mathbb{C}^*\right]$. By comparing with the expression in Corollary 1.10 we see that, in this case $V(E) = \operatorname{Ker} R$.

We also have a map $\pi_* \colon H^1(X,\mathcal{K}_{j+1}) \to H^0(E,\mathcal{K}_j)$ whose kernel is isomorhic to $H^1(E,\mathcal{K}_{j+1})$. This follows from the Gersten-Quillen resolutions of the sheaf $\mathcal{K}_{j+1}$ on X and E an the fact that $\pi^*(Q) \in X^1$ for all closed points Q of E. Therefore we get

$$H^0(E,\mathcal{K}_j) \simeq K_j(\mathbb{C}) \oplus F_j$$

In particular, this proves that $(F_2)_{\mathbb{Q}} \neq 0$ because $H^0(E,\mathcal{K}_2)_{\mathbb{Q}} \neq K_2(\mathbb{C})$ (see Remark 1.11). Hence the map:

$$(\operatorname{Pic} X \otimes K_2(\mathbb{C}))_{\mathbb{Q}} \to (H^2(X,\mathcal{K}_3))_{\mathbb{Q}}$$

is not surjective.

A similar argument also proves that $H^2(X,\mathcal{K}_{j+2}) \simeq H^1(E,\mathcal{K}_{j+1})$. In fact we have a surjective map $\pi_* \colon H^2(X,\mathcal{K}_{j+2}) \to H^1(E,\mathcal{K}_{j+1})$ and from the commutative diagram

$$\begin{array}{ccc} \operatorname{Pic} E \otimes K_j(\mathbb{C}) \simeq & \longrightarrow & \mathrm{CH}^2(X) \otimes K_j(\mathbb{C}) \\ \Big\downarrow{\scriptstyle \mu_{1,j}} & & \Big\downarrow{\scriptstyle \tau_{1,j}} \\ H^1(E,\mathcal{K}_{j+1}) \pi^* \cup [\sigma] & \longrightarrow & H^2(X,\mathcal{K}_{j+2}) \end{array}$$

where $\operatorname{Ker} \mu_{1,j} \simeq \ker \tau_{1,j}$, it follows that the map π_* is an isomorphism.

References

[BV1] L. Barbieri Viale, *On the Deligne-Beilinson cohomology sheaves*, alg.geom/9412006, 1994.

[Bea] A. Beauville, *Complex algebraic surfaces*, London Math. Soc. Lect. Note Ser., vol. 68, Cambridge Univ. Press, 1983.

[BM] C. Bennet and R. Miranda, *The automorphism group of the hyperelliptic surfaces*, Rocky Mountain J. Math. **20** (1990), 31–37.

[Blo] S. Bloch, *Lectures on algebraic cycles*, Duke Univ. Math. Ser., vol. 4, Duke Univ. Press, Durham, 1980.

[BKL] S. Bloch, A. Kas, and D. Lieberman, *Zero cycles on surfaces with $p_g = 0$*, Compositio Math. **33** (1976), 135–145.

[BO] S. Bloch and A. Ogus, *Gersten's conjecture and the homology of schemes*, Ann. Sci. École Norm. Sup. (4) **7** (1974), 181–201.

[BS] S. Bloch and V. Srinivas, *Remarks on correspondences and algebraic cycles*, Amer. J. Math. **105** (1983), 1235–1253.

[Col] A. Collino, *Infinite generation for the group of indecomposable cohomology classes on quartic surfaces in $\mathbb{P}^3$ and on cubic hypersurfaces in $\mathbb{P}^5$*, preprint, 1997.

[Coo] K. Coombes, *The K-cohomology of Enriques surfaces*, Contemp. Math. **126** (1992), 47–57.

[CS] K. Coombes and V. Srinivas, *A remark on K_1 of a surface*, Math. Ann. **265** (1983), 335–342.

[CTR] J.-L. Colliot-Thélène and W Raskind, *$\mathcal{K}_2$-cohomology and the second Chow group*, Math. Ann. **270** (1985), 165–199.

[Esn1] H. Esnault, *A note on the cycle map*, J. Reine Angew. Math. **411** (1990), 51–65.

[Esn2] ———, *Un remarque sur la cohomologie du faisceau de Zariski de la K-théorie de Milnor sur un varieté lisse complexe*, Math. Z. **205** (1990), 373–378.

[EV] H. Esnault and E. Viehweg, *Deligne-Beilinson cohomology*, Beilinson's Conjectures on Special Values of L-Functions, Perspect. Math., vol. 4, Academic Press, Boston, MA, 1988, pp. 43–92.

[Gil1] H. Gillet, *Deligne homology and the Abel-Jacobi map*, Bull. Amer. Math. Soc. (N.S.) **10** (1984), 285–288.

[Gil2] H. Gillet, *On the K-theory of surfaces with multiple curves and a conjecture of Bloch*, Duke Math. J. **51** (1984), 195–233.

[MS] A. Merkurjev and A. Suslin, *K-comology of Severi-Brauer varieties and the norm residue homomorphism*, Math. USSR-Izv. **21** (1983), 307–340.

[MSt] S. Müller-Stach, *Constructing indecomposable motivic cohomology classes on algebraic surfaces*, J. Algebraic Geom. **6** (1997), 513–543

[Mum] D. Mumford, *Rational equivalence of zero cycles on surfaces*, J. Math. Kyoto Univ. **9** (1968), 195–204.

[Ped] C. Pedrini, *Finiteness results for Chow groups of varieties over number field*, Arithmetic Geometry (Cortona, 1994), Sympos. Math., vol. 37, Cambridge Univ. Press, Cambridge, 1997, pp. 235–259.

[PW] C. Pedrini and C. Weibel, *The higher K-theory of a complex surface*, preprint, 1998.

[Qui] D. Quillen, *Higher algebraic K-theory*. I, Algebraic K-theory. I. Higher K-theories (Seattle, WA., 1972), Lecture Notes in Math., vol. 341, Springer, Berlin, 1973, pp. 85–147.

[Ras] W. Raskind, *On K_1 of curves over global fields*, Math. Ann. (1990), 179–193.

[Rob] L. Roberts, *Symbols on elliptic curves*, Comm. Algebra **4** (1976), 787–791.

[Roi] A. A. Roitman, *The torsion of the group of zero cycles modulo rational equivalence*, Ann. of Math. (2) **111** (1980), 553–569

[Ros] A. Rosenschon, *Indecomposable elements in K_1 of a smooth projective variety*, K-Theory **16** (1999), 185–199.

[Sou] C. Soulé, *Opérations en K-théorie algébrique*, Canad. J. Math. **37** (1985), 488–550.

[Sus] A. Suslin, *Torsion on K_2 of fields*, K-Theory **1** (1987), 5–29

[Voi] C. Voisin, *Remarks on zero-cycles of self products of varieties*, Moduli of Vector Bundles (Sanda, 1994; Kyoto, 1994), Lecture Notes in Pure and Appl. Math., vol. 179, Dekker, New York, 1996, pp. 265–285.

Dipartimento di Matematica, Universitá di Genova, Via Dodecaneso 35, 16146 Genova, Italy

E-mail address: pedrini@dima.unige.it

Centre de Recherches Mathématiques
CRM Proceedings and Lecture Notes
Volume **24**, 2000

From Jacobians to One-Motives: Exposition of a Conjecture of Deligne

Niranjan Ramachandran

Abstract. Deligne has conjectured that certain mixed Hodge theoretic invariants of complex algebraic invariants are motivic. This conjecture specializes to an algebraic construction of the Jacobian for smooth projective curves, which was done by A. Weil. The conjecture (and one-motives) are motivated by means of Jacobians, generalized Jacobians of Rosenlicht, and Serre's generalized Albanese varieties. We discuss the connections with the Hodge and the generalized Hodge conjecture. We end with some applications to number theory by providing partial answers to questions of Serre, Katz and Jannsen.

Parmi toutes les choses mathématiques que j'avais eu le privilège de découvrir et d'amener au jour, cette realité des motifs m'apparaît encore comme la plus fascinante, la plus chargée de mystère — au cœur même de l'identité profonde entre la « géométrie » et l'« arithmétique ». Et le « yoga des motifs » auquel m'a conduit cette réalité longtemps ignorée est peut-être le plus puissant instrument de découverte que j'ai dégagé dans cette première période de ma vie de mathématicien.

Alexandre Grothendieck
Récoltes et semailles

One of the most profound contributions of Grothendieck to mathematics is the concept of *motives*. Even though Grothendieck himself wrote very little on this subject, the philosophy of motives played a very important role in his research. While the influence and importance of motives is clear, it is hard to fathom the true extent of their potential impact. Motives provide fascinating bridges among the mathematical trinity: algebra, geometry, and analysis ([**49**] is a beautiful introduction). The question *What is a motive*? eludes an answer even today (cf. *Problems of present day mathematics* in [**10**, pp. 39–42]).

1991 *Mathematics Subject Classification.* 11Gxx, 14C25, 14C30, 14F99.
Key words and phrases. mixed motives, Hodge theory, étale cohomology, finite fields, independence of ℓ, weight filtration, Picard variety.
Received by the editors November 23, 1998; Accepted; April 27, 1999.
This research was funded, in part, by grants from Hewlett-Packard and the National Science Foundation.
This is the final form of the paper.

The prototype of motives are Abelian varieties. One aspect of the vision of Grothendieck, namely that motives (= pure motives) are attached to smooth projective varieties, is itself inspired by the theory of the Picard and Albanese varieties of smooth projective varieties. The latter theory can be viewed as a *purely algebraic* definition of the first cohomology and homology groups of smooth projective varieties. One thinks of the Picard and Albanese varieties as the motivic h^1 and h_1 of smooth projective varieties. The amazingly potent nature of a motive is already evident from the rich theory of smooth projective curves and their Jacobians. We begin by reviewing briefly this classical episode, surely one of the most beautiful edifices of nineteenth century mathematics. We consider generalizations of Albanese varieties to smooth varieties which naturally leads us to semi-Abelian varieties. From there, it is a short step to one-motives. Deligne's conjecture is motivated as a counterpart for arbitrary complex varieties of the generalized Hodge conjecture for smooth projective complex varieties. We end with a few arithmetical applications of the proof of Deligne's conjecture.

Even though one-motives were defined by Deligne in 1972, they do not seem to be that well known. Abelian varieties provide a testing ground for many conjectures in the theory of pure motives (just as toric varieties do for higher dimensional algebraic geometry). In a similar vein, one could argue (and successfully so!) that one-motives provide a fertile testing ground for the theory of mixed motives (a theory that remains mysterious even today!). Deligne's view (cf. §6) is that one-motives are precisely the *mixed motives of level one*.[1] In particular, these should enable one to describe the H^1 of arbitrary varieties thereby generalizing the description by Abelian varieties of the H^1 of smooth projective varieties. Many of the hypothesized properties of mixed motives can be verified for one-motives [**26**]. The role of one-motives in the development of mixed Hodge structures underscores their importance. Deligne admits that it was exactly the close relation between one-motives over $\mathbb{C}$ and mixed Hodge structures that convinced him of the validity of the philosophy of mixed Hodge structures [**17**, 2.1]:

> *Pour aller plus loin, il fallait se convaincre que tout motif a une filtration par le poids W, croissante, avec $\mathrm{Gr}_i^W(M)$ pur de poids i (= facteur direct de $H^i_{\mathrm{mot}}(X)$ pour X projectif non singulier). C'est sur le H^1 des courbes, i.e., sur les 1-motifs que je m'en suis convaincu, et le premier test qu'a dû passer la définition des structures de Hodge mixtes est qu'elles redonnent comme cas particulier les 1-motifs sur $\mathbb{C}$.*

The appearance of one-motives in the cohomology of algebraic varieties is not limited to H^1. In fact, Deligne's conjecture (the focus of this paper) predicts their appearance in the higher cohomology of higher dimensional complex varieties.

There has been a slow but surely growing interest in the theory of one-motives. We list a few instances of their appearance: Mumford-Tate groups and special values of G-functions (Y. André [**1, 2**]), transcendence issues (D. Bertrand [**7**], D. Masser [**35**], K. Ribet [**42**]), automorphic forms and Shimura varieties (J.-L. Brylinski [**11**], C. Brinkmann [**9**]), crystalline cohomology (J. M. Fontaine-K. Joshi (*in preparation*)), geometric monodromy (M. Raynaud [**41**]), Tate curves and rigid analytic uniformization (Raynaud [**17**, 2.1]), moduli spaces of Abelian varieties and cubical structures (C. L. Chai and G. Faltings [**20**], L. Breen [**8**]), Fourier

---

[1] This is the reason for the terminology of one-motives.

transform and geometric Langlands correspondence (G. Laumon [**32**], A. Beilinson and V. Drinfeld), function field analogue of Stark's conjecture (J. Tate [**51**]).

There are no proofs in this paper; its intention is to be purely expository and motivational. For details and proofs, we refer to the bibliography at the end of the paper.

1. Jacobians

Consider a smooth projective complex curve C, i.e., a compact Riemann surface C^{an}. Associated with it is an Abelian variety (its Jacobian variety J), usually constructed via the period mapping (a transcendental construction):

$$J = \frac{H^0(C^{\mathrm{an}};\Omega^1_{an})^*}{\operatorname{Im} H_1(C^{\mathrm{an}};\mathbb{Z})}.$$

where $H^0(C^{\mathrm{an}};\Omega^1_{an})$ denotes the complex vector space of global differential forms of degree one and $*$ denotes the dual complex vector space.

There is a morphism from the curve to its Jacobian; this is unique up to translations in the Jacobian. It induces an isomorphism on $H_1(C;\mathbb{Z}) \xrightarrow{\sim} H_1(J;\mathbb{Z})$. This is actually an isomorphism of polarized Hodge structures. One can even show that the polarized Hodge structure on $H_1(C;\mathbb{Z})$ uniquely determines the curve up to isomorphism (Torelli's theorem).

The first amazing fact (proved by Weil) is that the Jacobian variety admits a *purely algebraic construction*. What does one mean by the phrase "purely algebraic construction"? We mean that this invariant can be constructed without ever leaving the universe of algebraic geometry. One is allowed any (and every) tool—geometric, cohomological, sheaf-theoretic, cycle theoretic methods—strictly in the realm of algebraic geometry. One is not allowed to use, for example, the classical topology of the complex numbers. Such a construction for *a priori* transcendental invariants lends their definition a wider scope, that is, a setting far more general than originally envisioned or intended. For instance, the "purely algebraic construction" of the Jacobian allows it to be defined in positive characteristic, the starting point of Weil's proof of the Riemann hypothesis for curves over finite fields; as is well known, it was the Riemann hypothesis that later developed into the celebrated Weil conjectures (aka Deligne's theorem).

Other consequences of the algebraic construction of the Jacobian include

(1) a definition of the Jacobian variety (an Abelian variety over k) valid for a smooth projective curve over any field k.
(2) generalization of Torelli's theorem to any perfect field: The canonically polarized Jacobian of a smooth projective curve determines the curve up to isomorphism.
(3) Galois representations on the ℓ-adic Tate module of the Jacobian, a powerful tool in the study of rational points of curves over number fields, e.g., Mordell's conjecture, modular curves.
(4) *geometric class field theory* for curves (M. Rosenlicht, S. Lang [**31**], and J.-P. Serre [**48**]).

The classical theory of differentials of the first, second and third kind on a complex curve C are subsumed in the theory of the Jacobian [**52, 36**] (cf. Section 3.1).

From the Jacobian J of a smooth projective curve C over a field k, one can obtain all the cohomological invariants associated with the curve by applying "realization" functors. Let us be more explicit as to what this means.

Hodge. For every embedding $\iota\colon k \hookrightarrow \mathbb{C}$, we get a Hodge structure $H_\iota := H^1(C_\iota;\mathbb{Z}(1))^2$ by considering the compact Riemann surface C_ι^{an} associated to the complex curve C_ι. This Hodge structure is of type $\{(-1,0),(0,-1)\}$; it is polarizable, by Poincaré duality. A classical theorem of Riemann asserts that every such structure H corresponds to an essentially unique complex Abelian variety $A(H)$. More precisely, one has :

THEOREM 1. *The functor $A \mapsto H_1(A;\mathbb{Z})$ provides an equivalence of categories between the category of complex Abelian varieties and torsion-free polarizable Hodge structures of type $\{(-1,0),(0,-1)\}$.*

The functor $A \mapsto H^1(A;\mathbb{Z})$ provides an equivalence of categories between the category of complex Abelian varieties and torsion-free polarizable Hodge structures of type $\{(1,0),(0,1)\}$.

The complex Abelian variety J_ι, obtained by base change via ι, is canonically isomorphic to the $A(H_\iota)$ corresponding to H_ι; this isomorphism is induced by the exponential sequence on C_ι. These are valid for every $\iota\colon k \hookrightarrow \mathbb{C}$.

Even though the process of obtaining the Hodge structure H_ι is a transcendental one (it uses the classical topology on the complex curve C_ι), one can separate the algebro-geometric step—that of associating the Jacobian J with C which makes no reference to ι—in this process from the transcendental step (which depends on ι). One has a factorization

$$C \to J \xrightarrow{\iota} J_\iota \leftrightarrows H^1(C_\iota;\mathbb{Z}(1))$$

where the first association is completely algebraic (and independent of ι) and the last association is given by Theorem 1.

Étale. Let $G = \mathrm{Gal}(\bar{k}/k)$ be the Galois group of an algebraic closure $\bar{k}$ of k. Let ℓ be a prime different from the characteristic p of k. One has the étale cohomology group $H^1_{\mathrm{et}}(C \times_k \bar{k};\mathbb{Z}_\ell(1))$ for each ℓ; this is a representation of G.

From every Abelian variety A over k, one can fabricate an ℓ-adic representation of G by considering the ℓ-adic Tate module $T_\ell(A)$, the projective limit of the ℓ-power torsion points of $A(\bar{k})$. This construction, applied to the Jacobian J, yields the Galois representation $T_\ell J$, canonically isomorphic to the Galois representation $H^1_{\mathrm{et}}(C \times_k \bar{k};\mathbb{Z}_\ell(1))$, via the Kummer sequence.

Here too, the main point is that the association $C \mapsto J$ is the common step in obtaining any of the étale cohomology groups $H^1_{\mathrm{et}}(C \times_k \bar{k};\mathbb{Z}_\ell(1))$.

de Rham. Assume that the characteristic of k is zero. Consider the first de Rham cohomology group $H^1_{\mathrm{dR}}(C)$ of C; it is a vector space over k. One can obtain this as well from the Jacobian J: We may assume that C has a rational point and we use it to get a morphism α from C to J. The induced map $\alpha^*\colon H^1_{\mathrm{dR}}(J) \to H^1_{\mathrm{dR}}(C)$ on the de Rham cohomology groups, independent of the rational point, is an isomorphism. This isomorphism is usually stated as follows: *every global*

[2]Here $\mathbb{Z}(1) = 2\pi i\mathbb{Z}$ is the Tate Hodge structure; it has rank one and of type $(-1,-1)$. We have shifted from homology to cohomology, for ease of exposition.

differential one-form on C is the pullback, via α^*, of a unique translation-invariant differential one-form on J.

There are various relations between the above cohomology theories; these are formalized as "compatibility isomorphisms." Consider a smooth projective curve C over $\mathbb{Q}$. The de Rham cohomology group $H^1_{\mathrm{dR}}(C)$ is a vector space over $\mathbb{Q}$. The complex curve C_ι obtained from C for the unique embedding $\iota\colon \mathbb{Q} \hookrightarrow \mathbb{C}$ ("the infinite prime") provides $H^1(C_\iota;\mathbb{Q})$, a vector space over $\mathbb{Q}$, using the usual singular homology theory. One has an isomorphism $H^1_{\mathrm{dR}}(C)\otimes_{\mathbb{Q}}\mathbb{C} \xrightarrow{\sim} H^1(C_\iota;\mathbb{Q})\otimes_{\mathbb{Q}}\mathbb{C}$ of $\mathbb{C}$-vector spaces. This isomorphism usually does *not* preserve the underlying $\mathbb{Q}$-vector spaces. The determinant of the isomorphism (= period) is usually a transcendental number and arises in deep arithmetic questions about C [**19**].

2. Intermediate Jacobians and the Hodge Conjecture

One may ask whether the Jacobian of a smooth projective curve discussed in the previous section admits a generalization to higher dimensions. It turns out that it admits two such generalizations.

Associated with any smooth projective complex variety X are two Abelian varieties: the Albanese variety $\mathrm{Alb}(X)$ and the Picard variety $\mathrm{Pic}(X)$. Let X^{an} be the compact complex analytic manifold associated with X. Let Ω^i_{an} be the locally free sheaf of i-forms on X^{an} and $\Omega^0_{\mathrm{an}} = \mathcal{O}_{X^{\mathrm{an}}}$, the structure sheaf of regular functions. The cohomology groups of X^{an} admit a canonical decomposition as complex vector spaces:

$$H^n(X^{\mathrm{an}};\mathbb{C}) = \bigoplus_{i=0}^{n} H^{n-i}(X^{\mathrm{an}};\Omega^i_{\mathrm{an}})$$

This decomposition is known as the Hodge decomposition of the cohomology of X^{an}. One says that $H^n(X^{\mathrm{an}};\mathbb{Z})$ is a Hodge structure of weight n and of type $\{(0,n),\dots,(n,0)\}$.

The pairing of (homological) cycles and differential forms on X^{an} gives us, by Stokes' theorem, a natural map from $H_1(X^{\mathrm{an}};\mathbb{Z})$ to $H^0(X^{\mathrm{an}};\Omega^1_{\mathrm{an}})^*$; it is an injection (mod torsion). We denote its image by $\mathrm{Im}\,H_1(X^{\mathrm{an}};\mathbb{Z})$. Let $\mu\colon \mathbb{Z}(1) \to \mathcal{O}_{X^{\mathrm{an}}}$ be the natural map of sheaves on X^{an}. We denote the image of the induced map $\mu_* :: H^1(X^{\mathrm{an}};\mathbb{Z}(1)) \to H^1(X^{\mathrm{an}};\mathcal{O}_{X^{\mathrm{an}}})$ by $\mathrm{Im}\,H^1(X^{\mathrm{an}};\mathbb{Z}(1))$.

One possible description of the Albanese variety and the Picard variety of X is:

$$\mathrm{Alb}(X) = \frac{H^0(X^{\mathrm{an}};\Omega^1_{\mathrm{an}})^*}{\mathrm{Im}\,H_1(X^{\mathrm{an}};\mathbb{Z})}$$

$$\mathrm{Pic}^0(X) = \frac{H^1(X^{\mathrm{an}};\mathcal{O}_{X^{\mathrm{an}}})}{\mathrm{Im}\,H^1(X^{\mathrm{an}};\mathbb{Z}(1))}$$

It follows that one has isomorphisms

$$H_1(X;\mathbb{Z})/\mathrm{torsion} \xrightarrow{\sim} H_1(\mathrm{Alb}(X);\mathbb{Z}) \quad H_1(\mathrm{Pic}(X);\mathbb{Z}) \xrightarrow{\sim} H^1(X;\mathbb{Z}(1))$$

of polarizable Hodge structures. Therefore, it is possible to recover the polarizable Hodge structure on $H_1(X;\mathbb{Z})/\mathit{torsion}$ and $H^1(X;\mathbb{Z})$ from the Albanese and the Picard variety. The key point here is Theorem 1.

The second amazing fact is that the Albanese variety and the Picard variety also admit a *purely algebraic construction.* Therefore it is possible to develop a theory of the Albanese and the Picard variety for a smooth projective variety over any field. This has been used by S. Lang to extend *geometric class field theory* to higher dimensions. The theory of Albanese and Picard varieties should be viewed as an algebraic version of the classical theory of the first homology and cohomology groups for smooth complex projective varieties. For example, a classical observation (due to J. Igusa and Weil) is that the Albanese and Picard varieties of any smooth projective variety over a field are *dual* Abelian varieties. This can be viewed as a special case of a universal coefficient formula. The Albanese and the Picard varieties of a smooth projective curve are both isomorphic to the Jacobian. One obtains the

THEOREM (Abel-Jacobi). *The Jacobian of a smooth projective curve is a self-dual Abelian variety.*

The properties of the Jacobian about the realizations in the various cohomologies of the curve carry over *mutatis mutandis* to the Picard variety of a smooth projective variety.

These previous remarks indicate a strong tie between the theory of Abelian varieties and the cohomology of smooth projective varieties over a field. The question that poses itself is whether the entire cohomology $H^*(X)$ (with its Hodge structure) of a smooth complex projective variety X admits a *purely algebraic* construction. It turns out that, in this generality, the answer is no. However, one can still hope that parts of the cohomology admit an algebraic construction. The Hodge conjecture and its generalization formulate the expectations.

2.1. Hodge Conjecture. Let X be a smooth projective complex variety. Let d be the dimension of X. For the even-dimensional cohomology groups, one has the

CONJECTURE (W. V. D. Hodge **[25]**). *The $\mathbb{Q}$-vector space*

$$H^{i,i}_{\mathbb{Q}}(X) := (H^{2i}(X;\mathbb{Q}) \cap H^{i,i}) \subset H^{2i}(X;\mathbb{C})$$

is generated by the image of codimension i algebraic cycles on X under the cycle class map.

COROLLARY (of the conjecture). *Let V be a smooth projective variety over a field k. Fix a natural number i. The dimension of the $\mathbb{Q}$-vector space $H^{i,i}_{\mathbb{Q}}(V \times_\iota \mathbb{C})$ is independent of the complex imbedding $\iota\colon k \hookrightarrow \mathbb{C}$.*

The only cases in which the above conjecture is known for any complex smooth projective variety are $i = 1, d-1$; this is due to S. Lefschetz, K. Kodaira and D. C. Spencer. For $i = 1$, one obtains that the $\mathbb{Q}$-vector space $NS(X)\otimes_{\mathbb{Z}}\mathbb{Q}$ obtained from the Néron-Severi group of X is isomorphic to $H^{1,1}_{\mathbb{Q}}(X)$, the famous $(1,1)$-theorem of Lefschetz. This yields the

THEOREM 2. *For any smooth projective variety V over any field k, the dimension of the $\mathbb{Q}$-vector space $H^{1,1}_{\mathbb{Q}}(V \times_\iota \mathbb{C})$ is independent of the complex imbedding $\iota\colon k \hookrightarrow \mathbb{C}$.*

For the odd-dimensional cohomology groups, one has the theory of intermediate Jacobians due to P. Griffiths **[23]**; these compact complex tori however are *not* always Abelian varieties. A famous problem asks for a direct and purely algebraic construction of the maximal Abelian varieties contained in them. This may be reformulated as the purely algebraic construction of the Abelian variety $A^i(X)$ corresponding to the maximal $\mathbb{Z}$-Hodge structure of level 1 (i.e., of type $\{(i -$

$1, i), (i, i-1)\})$ contained in $H^{2i-1}(X;\mathbb{C})$. This problem is completely answered in general only for $i = 1$, d: one has $A^1(X) = \mathrm{Pic}(X)$ and $A^d(X) = \mathrm{Alb}(X)$. For other i, it would follow from the *generalized Hodge conjecture* (as corrected by Grothendieck [**24**]). Conjecturally, $A^i(X)$ is the image under the Abel-Jacobi map of codimension i cycles on X which are algebraically equivalent to zero. Results in this direction can be found in [**44**], [**37**].

A counterpart of these famous problems for *arbitrary* complex varieties is furnished by the conjecture of Deligne. This asks for a purely algebraic construction of algebraic objects (one-motives) that correspond to certain *mixed Hodge structures of level one* contained in the cohomology of complex algebraic varieties.

3. One-Motives

In envisaging motives, Grothendieck naturally did not restrict himself to smooth and projective varieties. His vision was that mixed motives are attached to arbitrary varieties. Since Poincaré duality does not hold in this generality, one must differentiate between cohomology mixed motives and homology mixed motives. Grothendieck did provide a precise definition of pure motives but not of mixed motives. His definition allows one to interpret *Abelian varieties* as examples of pure motives. As regards mixed motives, one still does not have a satisfactory definition (let alone a satisfactory theory!) for them.

Despite the sad state of affairs in mixed motives, we do have a precise definition (by Deligne [**15**, §10] in 1972) of one-motives. It is hoped that one-motives are prototypes of mixed motives. One believes that the properties of one-motives are true for mixed motives in general. Namely, it is supposed that

- there exists an increasing weight filtration W on mixed motives
- W is strict for morphisms of mixed motives, i.e., if $M \xrightarrow{f} N$ is a morphism of mixed motives, then $f(W_i(M)) = f(M) \cap W_i(N)$
- the graded pieces of the weight filtration are (pure) motives.

As we shall see, one-motives over $\mathbb{C}$ give rise to mixed Hodge structures.

Before I turn to the actual definition of one-motives, let me dwell a little bit longer on the theme of Albanese and Picard varieties. If one wants a purely algebraic description of the mixed Hodge structure $H_1(X;\mathbb{Z})$/torsion of a smooth complex variety X, then one sees that complex Abelian varieties do not suffice: the possible weights on mixed Hodge structure on $H_1(X;\mathbb{Z})/\textit{torsion}$ are -1 and -2. It turns out that the weight -2 part is a direct sum of the Tate Hodge structures $\mathbb{Z}(1)$. The mixed Hodge structure $H := H_1(X;\mathbb{Z})$/torsion defines an extension

$$(**) \qquad 0 \to H' \to H \to H'' \to 0$$

where H' is of type $(-1,-1)$ (= a direct sum of Tate Hodge structures) and H'' is a polarizable Hodge structure (pure of weight -1) of type $\{(-1,0),(0,-1)\}$. Polarizability of the graded pieces of mixed Hodge structures on the cohomology of complex algebraic varieties is a consequence of Chow's lemma and Hironaka's resolution of singularities. The identity $H_1(\mathbb{C}^*;\mathbb{Z}) = \mathbb{Z}(1)$ suggests bringing algebraic tori into the picture; one is lead by the extension $(**)$ to the consideration of semi-Abelian varieties (= extensions of Abelian varieties by algebraic tori). In general, these extensions do not split (they may not split even after isogeny). The correspondence of Riemann (cf. Theorem 1) extends to

THEOREM 3. *The functor* $G \mapsto H_1(G;\mathbb{Z})$ *from the category of complex semi-Abelian varieties to the category of torsion-free mixed Hodge structures* $\mathcal{H}$ *of type*

$$\{(-1,-1),(-1,0),(0,-1)\}$$

(with $\mathrm{Gr}^W_{-1}\mathcal{H}$ *polarizable) is an equivalence of categories.*

Any mixed Hodge structure $\mathcal{H}$ as in this theorem comes from an essentially unique complex semi-Abelian variety $A(\mathcal{H})$. In addition, the pure Hodge structure $\mathrm{Gr}^W_{-1}\mathcal{H}$ corresponds to the maximal Abelian variety quotient A of the semi-Abelian variety $A(\mathcal{H})$. The complex dimension of the maximal torus of $A(\mathcal{H})$ is the rank of the group $\mathrm{Gr}^W_{-2}\mathcal{H}$.

The main point of mixed Hodge structures is that there are *nontrivial extensions.* One can ask for the effect of the correspondence in Theorem 3 on the extension groups. Let the pure Hodge structure H' of type $(-1,-1)$ correspond to an algebraic torus T. Let H'' of type $\{(-1,0),(0,-1)\}$ correspond to a complex Abelian variety A. We assume that H'' and H' are torsion-free. Denote by $\mathrm{Ext}^1_{\mathrm{MHS}}(H'',H')$ the set of isomorphism classes of extensions $(**)$ in the category of mixed Hodge structures. Let us denote by $\mathrm{Ext}^1(A,T)$ the set of isomorphism classes of extensions G of A by T in the category of commutative complex algebraic groups (G is a semi-Abelian variety). These two sets of isomorphism classes form an Abelian group under Yoneda addition. The correspondence of Theorem refRi2 actually induces an isomorphism

$$\varepsilon\colon \mathrm{Ext}^1(A,T) \xrightarrow{\sim} Ext^1_{\mathrm{MHS}}(H'',H') \quad G \mapsto H_1(G;\mathbb{Z}).$$

THEOREM (Weil-Barsotti). *Let* k *be an algebraically closed field. Let* $\mathcal{A}$ *be an Abelian variety over* k *and let* $\mathcal{A}^*$ *be the dual Abelian variety. Let* $\mathcal{T}$ *be a torus (a variety) over* k *and let* $\mathcal{D}$ *be the character group of* $\mathcal{T}$*, i.e.,* $\mathcal{D} = \mathrm{Hom}(\mathcal{T},\mathbb{G}_m)$.

The group $\mathrm{Ext}^1(\mathcal{A},\mathcal{T})$ *of isomorphism classes of extensions in the category of commutative group varietes is naturally isomorphic to the group* $\mathrm{Hom}(\mathcal{D},\mathcal{A}^*)$ *of homomorphisms from* $\mathcal{D}$ *to* $\mathcal{A}^*$.

In particular, when $k=\mathbb{C}$, A is a complex Abelian variety, and $\mathcal{T}=\mathbb{G}_m$, we obtain an isomorphism $\mathrm{Ext}^1(A,\mathbb{G}_m) \xrightarrow{\sim} \mathrm{Hom}(\mathbb{Z},A^*) \xrightarrow{\sim} A^*(\mathbb{C})$ of Abelian groups. So, for the mixed Hodge structures in Theorem 3, we obtain not only that the objects and their morphisms are algebraic in nature but that the extension groups also are algebraic.

PROBLEM 1. Provide a purely algebraic construction of the semi-Abelian variety $A(H)$ associated with the mixed Hodge structure $H := H_1(X;\mathbb{Z})/\mathrm{torsion}$ of a complex smooth variety X.

For X assumed to be also projective (i.e., $H_1(X)$ is pure of weight -1), this semi-Abelian variety is the classical Albanese variety $\mathrm{Alb}(X)$ of X. Recall that the Albanese variety $\mathrm{Alb}(X)$ enjoys a universal property (which uniquely characterizes it): given a base point x of X, there is a unique morphism $f_x\colon X \to \mathrm{Alb}(X)$ sending x to the identity point e of $\mathrm{Alb}(X)$; this morphism f_x is universal for morphisms from X to Abelian varieties A which send x to the identity point e_A of A.

3.1. Generalized Jacobians. We consider Problem 1 for smooth curves. Let U be a smooth complex curve (assumed to be connected) and let C be the unique smooth compactification of U. (A good example to have in mind is that of $C=E$, an elliptic curve.) Let $\{P_i\}$ be the set $C(\mathbb{C}) - U(\mathbb{C})$ of points ("at infinity"). Put

$H := H_1(U;\mathbb{Z})$. We wish to describe the semi-Abelian variety $A(H)$. The Hodge structure $\mathrm{Gr}^W_{-1} H$, isomorphic to $H_1(C;\mathbb{Z})$, corresponds to the Jacobian J. So the semi-Abelian variety $A(H)$ is an extension of J by an algebraic torus T.

Since J is characterized as being universal for morphisms from C into Abelian varieties (modulo the choice of a base-point), one is tempted to pose a variant for U using semi-Abelian varieties. Rosenlicht [**43**] has defined (purely algebraically) a semi-Abelian variety for any given modulus (= a divisor) on a smooth projective curve. These are known as the generalized Jacobians of Rosenlicht. Consider the generalized Jacobian J_m corresponding to the modulus $m = \sum_i P_i$ on C. Any base point x in U determines a canonical map $g_x\colon U \to J_m$. One can show [**48**] that g_x is universal for morphisms from U into semi-Abelian varieties G which send x to the identity e_G of G. Furthermore, it can be shown that $A(H) = J_m$! In fact, one has the

COROLLARY. *The map $g_x : U \to J_m$ induces an isomorphism*

$$g_x\colon H_1(U;\mathbb{Z}) \xrightarrow{\sim} H_1(J_m;\mathbb{Z}).$$

Rosenlicht's work thus answers Problem 1 for curves.

Let d denote the cardinality of the set $D := C(\mathbb{C}) - U(\mathbb{C})$. We have an exact sequence

$$0 \to \mathrm{Gr}^W_{-2} H \to H = H_1(U;\mathbb{Z}) \to H_1(C;\mathbb{Z}) \to 0.$$

One sees easily that the rank of $\mathrm{Gr}^W_{-2} H$ is $d-1$. Heuristically speaking, deleting one point from C does not change the rank of the first homology and every subsequent point that is removed increases the rank by one.

Consider the complex vector space $H_D := H^0(C;\Omega^1(D))$ of differential forms regular on U and allowed to have only logarithmic poles along D. These are the classical *differential forms of the third kind.*[3] The dimension of H_D is $g + d - 1$ where g is the genus of the curve C. Given any regular one-form ω on J_m, one can certainly pull it back via g_x to obtain a regular one-form $g_x^*(\omega)$ on U. If ω is invariant under translations of J_m, it turns out that the poles of $g_x^*(\omega)$ on C are logarithmic which yields that

$g_x^*(\omega) \in H_D$. The de Rham analog of the previous corollary is encapsulated in [**48**]: *Every element of H_D is the pullback via g_x of a unique translation-invariant differential one-form on J_m.*

Other natural questions in this context are: How does one describe T? How does one determine the extension J_m? When is J_m isomorphic to $J \times T$?

Rosenlicht's work provides the answer, as summarized in the next proposition. We have a natural map from $\mathbb{Z}(D)$ (the free Abelian group on the elements of D) to $H^2(C;\mathbb{Z}(1))$ by sending elements of D (viewed as divisors on C) to their degree. Let B_D denote the kernel of this map; it is a free Abelian group of rank $d-1$. There is a natural map $\phi : B_D \to J$ by $a \mapsto \mathcal{O}(a)$.

PROPOSITION. *With notations as above, one has the following*:

(i) $T = \mathrm{Hom}(B_D, \mathbb{G}_m)$.

(ii) *The map ϕ corresponds to the extension J_m under the isomorphism in the Weil-Barsotti theorem.*

(iii) *The semi-Abelian variety J_m is isomorphic to the direct product $J \times T$ if and only if ϕ is the zero map.*

[3] Weil [**52**] was the first to notice the connection between these and extensions of J by $\mathbb{G}_m$.

From (iii), we see that J_m is generally a nontrivial extension.[4] We also get an answer to when J_m is isogenous to a direct product $J \times T$: this happens exactly when the image of ϕ is torsion.

If U' is obtained by deleting a finite number of points from U, then one has a surjection $H' := H_1(U'; \mathbb{Z}) \twoheadrightarrow H_1(U; \mathbb{Z})$; hence the rank of H' is larger than that of $H_1(U; \mathbb{Z})$. Translated to Abelian varieties, we obtain a surjection $A(H') \twoheadrightarrow A(H)$.

3.2. Generalized Albaneses. The existence of universal morphisms into semi-Abelian varieties for smooth varieties (over algebraically closed fields) has been proved by Serre [**46**]; these are the "generalized Albanese varieties". For any smooth complex variety X, one can show

PROPOSITION ([**39**]). *The generalized Albanese* $\mathrm{Alb}(X)$ *corresponds to the mixed Hodge structure* $H_1(X; \mathbb{Z})/\mathrm{torsion}$.

Therefore, the semi-Abelian variety $A(H)$ associated with the mixed Hodge structure $H := H_1(X; \mathbb{Z})/\mathrm{torsion}$ of any *smooth* complex algebraic variety X admits a purely algebraic construction!

A few words about Serre's construction of the generalized Albanese are in order. Let V be a smooth projective complex variety. Let D_i be a finite number of effective integral divisors on V. Let U be the open subvariety of V corresponding to the complement of the support of the divisor $D := \Sigma_i D_i$. Consider the group B_D of divisors which are

(a) algebraically equivalent to zero and
(b) supported on D.

Let T_D be the torus $\mathrm{Hom}(B_D, \mathbb{G}_m)$ obtained from B_D (a free Abelian group of finite rank). The natural homomorphism

$$\phi_D \colon B_D \to Pic(V) \quad a \mapsto \mathcal{O}(a)$$

determines, by the Weil-Barsotti theorem, a complex semi-Abelian variety G_D. The group variety G_D, an extension of $\mathrm{Alb}(C)$ by the torus T_D, is the generalized Albanese of U. Any base point x of U determines a canonical map $g_x \colon U \to G_D$. it induces an isomorphism

$$g_x \colon H_1(U; \mathbb{Z})/\mathrm{torsion} \xrightarrow{\sim} H_1(G_D; \mathbb{Z}).$$

Consider $H_D := H^0(V; \Omega^1(D))$ the complex vector space of differential one-forms which are regular on U and which are allowed to have logarithmic poles along D. Every element of H_D is the pullback of a unique translation-invariant differential one-form on G_D [**46**].

So far, we have concentrated on extending the Albanese variety to smooth varieties. Generalizing the Picard variety to smooth varieties cannot be handled without bringing in one-motives. In fact, they are hidden in the map ϕ_D.[5] To motivate this, let us look at the mixed Hodge structure $H := H^1(U; \mathbb{Z}(1))$, notations as above. The possible weights on it are 0 and -1. It sits in an exact sequence:

$$0 \to H^1(V; \mathbb{Z}(1)) \to H \to B_D \to 0.$$

[4] A consequence of the fact that the universal morphism for C, i.e., $g_x : C \to J$ is an imbedding.

[5] Amusingly enough, ϕ_D is the starting point of Serre's construction of the generalized Albanese; in other words, the Picard 1-motive of U precedes the generalized Albanese of U.

Therefore, H is an extension of B_D by $H^1(V;\mathbb{Z}(1))$. Since $H^1(V;\mathbb{Z}(1))$ corresponds to Pic(V) under Theorem 1, we see that the algebraic object that we need to describe H must be an extension of B_D by $Pic(V)$. If we look closely at ϕ_D and think of

$$B_D \xrightarrow{\phi_D} \mathrm{Pic}(V)$$

as a (two-term) complex M of group schemes, then we see that M provides such an extension. This turns out to be the desired one. The complex M is an example of a one-motive. Indeed, it is the Picard one-motive of the smooth variety U **[39]**.

In general, the mixed Hodge structures $H_1(X;\mathbb{Z})$/torsion and $H^1(X;\mathbb{Z}(1))$ of a complex algebraic variety X are

(a) of level one with weights -2, -1, 0.
(b) of type $\{(0,0),(-1,0),(0,-1),(-1,-1)\}$ and
(c) their graded weight -1 piece Gr^W_{-1} is polarizable.

PROBLEM 2. Do the mixed Hodge structures $H_1(X;\mathbb{Z})$/torsion and $H^1(X;\mathbb{Z}(1))$ admit a purely algebraic construction?

As we shall see next, one-motives enable us to answer this question.

3.3. One-Motives [15, §10]. Let k be a perfect field and let $\mathbb{G}$ be the Galois group of an algebraic closure $\bar{k}$ of k. A 1-motive M over k consists of a semi-Abelian variety G over k, a finitely generated torsion-free Abelian group B with a structure of a $\mathbb{G}$-module, and a homomorphism $u\colon B \to G(\bar{k})$ of $\mathbb{G}$-modules. In particular, if k is algebraically closed, then u is a homomorphism of Abelian groups. We write the 1-motive M as $[B \xrightarrow{u} G]$. A morphism ϕ between 1-motives M and M' consists of a pair of morphisms $\phi_1\colon B \to B'$ (of $\mathbb{G}$-modules) and $\phi_2\colon G \to G'$ (of group schemes) satisfying $\phi_2 u = u'\phi_1$. It is convenient to regard B as corresponding to a group scheme, locally constant in the étale topology on Spec k. The category of 1-motives over k is additive (but not Abelian).

A morphism ϕ (defined over k) consisting of $\phi_1\colon B \to B'$ and $\phi_2\colon G \to G'$ is termed an *isogeny* if ϕ_1 is injective with finite cokernel and ϕ_2 is surjective with finite kernel. One can consider the $\mathbb{Q}$-linear category (an Abelian category) of isogeny classes of 1-motives over k obtained by formally inverting all isogenies defined over k. For any 1-motive M, let us denote by $M \otimes \mathbb{Q}$ its isogeny class; we call this an isogeny 1-motive. The objects of the category of isogeny 1-motives are the same as those of 1-motives but the morphisms have changed: $\mathrm{Hom}(M\otimes\mathbb{Q}, N\otimes\mathbb{Q}) = \mathrm{Hom}(M,N)\otimes_{\mathbb{Z}}\mathbb{Q}$. In effect, isogenies of 1-motives have been transformed into isomorphisms of isogeny 1-motives.

One-motives have weight filtrations: Let $M := [B \xrightarrow{u} G]$ be a 1-motive. The semi-Abelian variety G is an extension of an Abelian variety A by a torus T. The weight filtration is defined as $W_{-3}M = 0$, $W_{-2}M = [0 \to T]$, $W_{-1}M = [0 \to G]$, $W_0 M = M$. The graded quotients are T, A and B. Of course, the triple (T, A, B) does not determine M because of the main point in mixed motives: there are nontrivial extensions in the game!

Deligne has defined various realization functors from the category of 1-motives over k including the Hodge realization $\mathfrak{T}_{\mathbb{Z}}$ (for each imbedding of k into $\mathbb{C}$), étale realization $\mathfrak{T}_\ell$ (for each prime ℓ different from the characteristic of k), and the de Rham realization $\mathfrak{T}_{\mathrm{dR}}$ (when the characteristic of k is zero). One deduces realization functors for isogeny 1-motives: Hodge realization $\mathfrak{T}_{\mathbb{Z}}$ in the category of $\mathbb{Q}$-mixed

Hodge structures, étale realization $\mathfrak{T}_\ell$ in the category of $\mathbb{Q}_\ell$-vector spaces (together with a $\mathbb{G}$-action), de Rham realization $\mathfrak{T}_{\mathrm{dR}}$ in k-vector spaces.

All these realization functors generalize the familiar constructions with Abelian varieties such as forming the Tate module. Cartier duality of tori (sending tori to their character group and vice versa) and duality of Abelian varieties can be simultaneously generalized to a theory of duality of one-motives. This is compatible with the realization functors.

4. The Conjecture

The Hodge realization is the one most relevant to our discussion. In [**15**, §10.1], Deligne has shown that the Hodge realization embeds the category of 1-motives over $\mathbb{C}$ as a full subcategory of the category of mixed Hodge structures and he provided a description of the image. Namely, he has shown that every torsion-free mixed Hodge structure H of the form

$$(*) \qquad \{(0,0),(-1,0),(0,-1),(-1,-1)\},$$

with $\mathrm{Gr}^W_{-1} H$ polarizable, arises from an essentially unique 1-motive $I(H)$ over $\mathbb{C}$. (Note the usual convention that a mixed Hodge structure always refers to a $\mathbb{Z}$-mixed Hodge structure but here the weight filtration W is on $H_{\mathbb{Q}} := H \otimes_{\mathbb{Z}} \mathbb{Q}$.) The association $I(H)$ with H is the functor inverse to the Hodge realization $\mathfrak{T}_{\mathbb{Z}}$. The Hodge realization preserves the natural weight filtrations on both categories.

Problem 2 can be rephrased as

PROBLEM 3. Do the one-motives associated with the mixed Hodge structures $H^1(X;\mathbb{Z}(1))$ and $H_1(X;\mathbb{Z})/\text{torsion}$ of an arbitrary complex algebraic variety X admit a purely algebraic description?

In fact, mixed Hodge structures of the type $(*)$ can appear in higher cohomology groups as well and one can pose the same problem for the associated one-motives. To do so, let us begin with the

THEOREM (Deligne [**16**, §7]). *Let V be a complex algebraic variety. Denote by d the dimension of V.*

(i) *The cohomology groups $H^n(V;\mathbb{Z}(1))$ are endowed with a mixed Hodge structure (functorial for morphisms).*
(ii) *The possible weights on $H^n(V;\mathbb{Z}(1))$ are $-2, -1, 0, \ldots, 2n-2$ and the possible Hodge numbers are $(p,q) \in [-1, n-1] \times [-1, n-1]$.*
(iii) *If $n \geq d$, then the possible Hodge numbers on $H^n(V;\mathbb{Z}(1))$ are $(p,q) \in [n-d-1, d-1] \times [n-d-1, d-1]$.*
(iv) *If V is proper, then $p+q \leq n-2$.*
(v) *If V is smooth, then $p+q \geq n-2$.*

By virtue of the previous theorem, one can define the maximal mixed Hodge substructure $t^n(V;\mathbb{Z}(1))$ of type $(*)$ of $H^n(V;\mathbb{Z}(1))/\text{torsion}$. From the previous theorem we deduce that $t^n(V;\mathbb{Z}(1)) = 0$ in these cases:

- V is smooth and $n > 2$.
- $n > 1 + d$.

Let us consider $t^n(X;\mathbb{Z}(1))$ in the case of a smooth projective complex variety X. It is enough to look at the cases $n = 0, 1, 2$. For $n = 0, 1$, we obtain $t^n(X;\mathbb{Z}(1)) = H^n(X;\mathbb{Z}(1))$. For the case $n = 0$, we get $H^0(X;\mathbb{Z}(1)) = \mathbb{Z}(1)$.

This Hodge structure corresponds to the torus $C^* = H^0(X;\mathbb{G}_m)$, where the last cohomology group can be taken to be the étale cohomology group (or the Zariski cohomology). For the case $n = 1$, the Hodge structure $H^1(X;\mathbb{Z}(1))$ is pure of weight -1 and is of type $\{(-1,0),(0,-1)\}$. We saw earlier that this structure corresponds under Theorem 1 to $\mathrm{Pic}(X)$, the Picard variety of X. The structure $t^2(X;\mathbb{Z}(1))$ is the $(0,0)$-part of $H^2(X)$. In other words, we have $t^2(X;\mathbb{Z}(1)) = \mathrm{Hom}_{\mathrm{MHS}}(\mathbb{Z}, H^2(X;\mathbb{Z}(1)))$. The theorem of Lefschetz identifies it as the (maximal torsion-free quotient of the) Néron-Severi group $\mathrm{NS}(X)/\mathrm{torsion}$ of X.

Therefore, in the case of smooth projective varieties X, the mixed Hodge structures $t^n(X;\mathbb{Z}(1))$ correspond to algebraic objects and these algebraic objects admit a purely algebraic construction. One is led to the following

CONJECTURE (Deligne, ibid. 10.4.1, 1973). *For any complex algebraic variety V, the 1-motives $I^n(V)$ associated with $t^n(V;\mathbb{Z}(1))$ admit a purely algebraic construction.*

REMARK. For integers $n > 1 + \dim V$, the conjecture is *vacuously true* by weight considerations as discussed earlier (cf. (iii) of previous theorem, [**16**]).

REMARK. For smooth projective complex varieties X, we have the Hodge (resp. the generalized Hodge conjecture) which provides a purely algebraic description of the maximal pure Hodge substructure of level zero (resp. one) contained in $H^{2n}(X)$ (resp. $H^{2n-1}(X)$). We see that the weight of the Hodge substructure in question is $2n$ (resp. $2n-1$). Deligne's conjecture asks for a purely algebraic description of the maximal mixed Hodge structure of level one contained in the low weight part, contained in $W_0H^n(V;\mathbb{Z}(1))$ to be precise, of the mixed Hodge structure $H^n(V;\mathbb{Z}(1))$ for an *arbitrary* complex algebraic variety. This is why we view Deligne's conjecture as a counterpart to the Hodge conjecture. These two conjectures, considered for smooth projective complex varieties, overlap for small values of n.

4.1. Examples. For a smooth proper variety X, the conjecture is equivalent to an algebraic description of the Picard scheme Pic_X. The 1-motive $I^1(X)$ is $[0 \to \mathrm{Pic}^0_X]$; the neutral component Pic^0_X of Pic_X is classically known as the Picard variety $\mathrm{Pic}(X)$ of X. The 1-motive $I^2(X)$ is $[\mathrm{NS}(X)/\mathrm{torsion} \to 0]$.

For a variety Y assumed to be proper, $I^1(Y)$ is of the form $[0 \to G]$. This is because the mixed Hodge structure $H^1(Y;\mathbb{Z}(1))$ has only negative weights. The semi-Abelian variety G turns out to be the maximal semi-Abelian quotient of the neutral component Pic^0_Y of the Picard scheme of Y.

We resume notations as in Section 3.2. The 1-motive $I^1(U)$ is $[B_D \xrightarrow{\phi_D} \mathrm{Pic}(V)]$; its dual is the 1-motive $[0 \to G_D]$ where G_D is the "generalized Albanese" of U (Section 3.2). If $\dim U = 1$ i.e. U is a smooth curve, then one obtains the generalized Jacobian J_D $(= G_D)$ of Rosenlicht (cf. Section 3.1).

For these cases, Deligne's conjecture is clearly true. In the case of curves, one has the following

THEOREM (Deligne [**15**, §10.3]). *The 1-motive $I^1(X)$ of an arbitrary curve X over $\mathbb{C}$ admits a purely algebraic construction.*

This theorem was, in fact, provided by Deligne as evidence for the possible veracity of the conjecture. Other special cases of the conjecture which were proved earlier:

- J. Carlson has proved the conjecture for H^2 of surfaces **[12]** (assumed to be either projective or smooth). He has also announced results for projective normal crossing schemes **[13]** which remain unpublished.
- The proof of the conjecture for H^1 of arbitrary schemes can be found in **[40]**, **[39]** where we also formulated a homological version of Deligne's conjecture and proved it for H_1 (**[5]** contains the same results and more).

These new motivic invariants (i.e. the one-motives I^n) are generally nontrivial for singular schemes; normal crossing schemes are typical examples. For related results, I refer to the work of S. Lichtenbaum **[33, 34]** on the connections of one-motives of curves with Suslin homology, of H. Önsiper **[38]** on the generalized Albanese and zero-cycles, of L. Barbieri-Viale, C. Pedrini and C. Weibel **[3]** on the one-motive $I^3(S)$ of a projective surface, and of L. Barbieri-Viale and V. Srinivas **[4]** on the Néron-Severi group of a projective surface.

CONSTRUCTION (**[40]**). For any algebraic scheme U over a perfect field k, one defines (in a purely algebraic manner) isogeny one-motives $L^n(U) \otimes \mathbb{Q}$. These are contravariant functorial for arbitrary morphisms. If d is the dimension of U, then $L^n(U) \otimes \mathbb{Q} = 0$ for $n > d+1$.

The main result of **[40]** is the following:

THEOREM (Deligne's conjecture up to isogeny). *For any complex algebraic variety V, the (Hodge-theoretic) isogeny one-motive $I^*(V) \otimes \mathbb{Q}$ is canonically isomorphic to the (purely algebraic) $L^*(V) \otimes \mathbb{Q}$. The isomorphism is furnished by the exponential sequence.*

REMARK. It seems likely that the methods of H.Gillet-C.Soulé **[22]** should suffice to prove Deligne's conjecture integrally and thereby refine this theorem. However, details have not yet been worked out.

I would like to mention a few points regarding the proof of the previous theorem. The main idea is inspired by **[12]** and can be explained easily for a complex *projective* variety V. Consider the construction by Deligne of the mixed Hodge structure $H^n(V; \mathbb{Z}(1))$. One takes a smooth proper hypercovering of V, in other words, a map $\alpha \colon X_\bullet \to V$ from a smooth projective simplicial scheme $X_\bullet$ which induces an isomorphism

$$\alpha^* \colon H^*(V; \mathbb{Z}(1)) \xrightarrow{\sim} H^*(X_\bullet; \mathbb{Z}(1))$$

of mixed Hodge structures. The existence of smooth hypercoverings depend on the resolution of singularities. There is a spectral sequence which calculates the cohomology of V in terms of the cohomology of the various constituents X_i. One finds that the Hodge structure $\mathrm{Gr}_k^W H^n(V; \mathbb{Z}(1))$ is a subquotient of $H^{k+2}(X_{n-k-2}; \mathbb{Z}(1))$. Since the conjecture deals with only $k = -2, -1, 0$, we see immediately that we need subquotients of $H^0(X_n; \mathbb{Z}(1))$, $H^1(X_{n-1}; \mathbb{Z}(1))$ and $H^2(X_{n-2}; \mathbb{Z}(1))$. Furthermore, since we are interested in $t^n(V; \mathbb{Z}(1))$, instead of $H^2(X_{n-2}; \mathbb{Z}(1))$, we need to consider a subquotient of $t^2(X_{n-2}; \mathbb{Z}(1))$. Since each X_j is smooth projective, our previous discussion tells us that the mixed Hodge structures $H^0(X_n; \mathbb{Z}(1))$, $H^1(X_{n-1}; \mathbb{Z}(1))$ and $t^2(X_{n-2}; \mathbb{Z}(1))$ admit a purely algebraic description. Namely, they are described by the torus $H^0(X_n; \mathbb{G}_m)$, the Abelian variety $\mathrm{Pic}(X_{n-1})$ and the group $\mathrm{NS}(X_{n-2})$. It can be checked that the relevant differentials in the spectral sequence mentioned earlier are actually algebraic in origin; they are either induced by actual geometric maps or Gysin maps. Therefore, we obtain that the required subquotients of $H^0(X_n;\mathbb{Z}(1))$, $H^1(X_{n-1};\mathbb{Z}(1))$ and $t^2(X_{n-2};\mathbb{Z}(1))$ can be obtained by taking subquotients of the torus $H^0(X_n;\mathbb{G}_m)$, the Abelian variety $\mathrm{Pic}(X_{n-1})$ and the group $\mathrm{NS}(X_{n-2})$.

What we have done so far is to describe the graded pieces of the 1-motive $L^n(X_\bullet)$. One still needs to figure out the extensions to complete the description. This can be done elegantly using the Picard scheme of truncated simplicial schemes. A difficult problem is to show that the $L^n(X_\bullet)$ is isomorphic to $I^n(X_\bullet) \xrightarrow{\sim} I^n(V)$. The degeneration (with $\mathbb{Z}$-coefficients) of the spectral sequence mentioned earlier is enough to deduce this isomorphism. However, the degeneration is known at present only with rational coefficients **[15]**; this weaker degeneration is enough to assure that $L^n(X_\bullet)$ and $I^n(X_\bullet)$ are isogenous whereby the proof of the previous theorem for projective varieties. There is another problem, namely that of showing that the one-motives $L^n(X_\bullet)$ do not depend upon the choice of the hypercovering $X_\bullet$ of V. Here also one obtains that, given another hypercovering $Y_\bullet$ of V, the one-motive $L^n(X_\bullet)$ is isogenous to the one-motive $L^n(Y_\bullet)$. Therefore, the isogeny one-motives $L^n(X_\bullet)\otimes\mathbb{Q}$ depend only on V and deserve to be denoted by $L^n(V)\otimes\mathbb{Q}$. This finishes the brief description of the construction for a projective variety V. The arguments are just a little bit harder for an arbitrary variety.[6]

5. Applications

The construction in the previous section has several consequences. The last theorem of the previous section is about the Hodge realization of these isogeny one-motives. Considerations of their ℓ-adic realizations lead to "independence of ℓ" results for varieties over number fields or finite fields.

5.1. A Lefschetz (1,1) Theorem for Complex Varieties[7]. For any complex variety X, the previous theorem provides a *purely algebraic* characterization of the $(1,1)$ part of $H^n(X;\mathbb{C})$. For a complex projective variety V, the $(1,1)$-part of $H^2(V;\mathbb{C})$ is provided in terms of divisors on any smooth resolution $wide\tilde{V}$ of V. This is a higher-dimensional generalization of the results in **[12]**, **[4]** on the Néron-Severi group of complex projective surfaces.

As a byproduct of the purely algebraic description, one obtains the following **[40]** (compare with Theorem 2)

THEOREM. *For any variety V over a field k and any integer n, the rank of $H^{1,1}_{\mathbb{Q}}(V_\iota)^n := \mathrm{Hom}_{\mathrm{MHS}}(\mathbb{Q}(0), \mathrm{Gr}_0^W H^n(V_\iota;\mathbb{Q}(1)))$ is independent of the complex imbedding $\iota\colon k\hookrightarrow\mathbb{C}$.*

The usual Néron-Severi group is obtained for $n=2$ and V projective.

---

[6]One can try the same over a perfect field k of positive characteristic using de Jong's weak resolution of singularities. At some point in the proof, one has to isolate the Néron-Severi in the ℓ-adic H^2. Since the Tate conjecture for divisors is not known, one needs a trick (suggested by M. Marcolli): take inverse limits of the isogeny 1-motives of the various hypercoverings of a given variety, use the finiteness properties of the ℓ-adic cohomology groups to show that this inverse limit is represented by an actual isogeny 1-motive. Thus, the positive characteristic case does not quite parallel the characteristic zero case (cf. **[40]**).

[6]The terminology is inspired by a talk of V. Srinivas at the ICM'98 Satellite conference on Arithmetic Geometry in Essen on recent results about the $(1,1)$-part of H^2 of normal projective complex varieties **[6]**. The methods, compared to mine, have the great advantage of being completely intrinsic. Further, the results do not neglect torsion.

5.2. Independence of ℓ for Parts of Étale Cohomology. Let k be a field. Let ℓ be a prime different from p, the characteristic of k. Let $G = \mathrm{Gal}(\bar{k}/k)$ be the Galois group. Let V be an *arbitrary* variety over k with $\bar{V} := V \times_k \bar{k}$. The ℓ-adic étale cohomology groups $H^i_{\mathrm{et}}(\overline{V}; \mathbb{Q}_\ell)$, for varying ℓ, furnish a system of ℓ-adic representations of G. There are many similarities amongst these ℓ-adic cohomology theories (which would be explained by the theory of motives). Examples of such similarities: (i) the "rationality" of Galois representations **[47]** over number fields. (ii) the "independence of ℓ" of the characteristic polynomial of Frobenius, over finite fields.

If k is a number field or a finite field, one has a G-equivariant weight filtration W_j on the étale cohomology $H^i_{\mathrm{et}}(\bar{V}; \mathbb{Q}_\ell)$ for any prime $\ell \neq p$; in other words, the $\mathbb{Q}_\ell$-subspace $W_j H^i_{\mathrm{et}}(\bar{V}; \mathbb{Q}_\ell)$ is a subrepresentation of G **[16]**. They provide new systems of ℓ-adic representations of G.

5.2.1. *k Is a Number Field.*

QUESCTION (Serre**[47, 50]**). For fixed V and i, is the system of Galois representations $H^i_{\mathrm{et}}(\bar{V}; \mathbb{Q}_\ell)$ "rational"?

Deligne's proof of the Weil conjectures **[16]** provides an affirmative answer for smooth projective V (see **[29]** for related results). We prove **[40]**

THEOREM. *The system of Galois representations $W_j H^i_{\mathrm{et}}(\overline{V}; \mathbb{Q}_\ell)$ is "rational" for $j = 0$, 1.*

5.2.2. *k Is a Finite Field of Characteristic p.* The Galois group G has a canonical generator Φ, the Frobenius element. Let Φ^*_ℓ denote the endomorphism of $H^i_{\mathrm{et}}(\bar{V}; \mathbb{Q}_\ell)$ induced by Φ. For smooth and proper varieties V, it is known by Deligne's proof of the Weil conjectures **[16]** that the system $H^i_{\mathrm{et}}(\bar{V}; \mathbb{Q}_\ell)$ of ℓ-adic representations of G satisfy:

(1) the dimension of the $\mathbb{Q}_\ell$-vector space $H^i_{\mathrm{et}}(\bar{V}; \mathbb{Q}_\ell)$ is independent of ℓ.
(2) the characteristic polynomial of Φ^*_ℓ has $\mathbb{Q}$-coefficients independent of ℓ.

QUESCTION (N. Katz **[30]**). Are (1) and (2) true for the ℓ-adic cohomology of arbitrary varieties? Are they true for the weight graded pieces of the cohomology of arbitrary varieties?

The second question clearly refines the first. I remark that neither is answered by a formal application of Deligne's proof of the Weil conjectures and de Jong's "weak resolution" **[28]**. Neither is answered even under the assumption of "strong resolution of singularities". Genuinely new ingredients are needed for an answer.

However, utilizing the one-motive theoretic interpretation of parts of ℓ-adic cohomology, one proves **[40]** the

THEOREM. (a) *For any variety V and any integer i, the following systems of ℓ-adic representations of G satisfy* (1) *and* (2) *above*: (i) $W_0 H^i_{\mathrm{et}}(\bar{V}; \mathbb{Q}_\ell)$; (ii) $\mathrm{Gr}^W_1 H^i_{\mathrm{et}}(\bar{V}; \mathbb{Q}_\ell)$.
(b) *the dimension of the $\mathbb{Q}_\ell$-vector space $\mathrm{Gr}^W_1 H^i_{\mathrm{et}}(\bar{V}; \mathbb{Q}_\ell)$ is even.* (*cf.* **[16**, 8.1**]**)

The Hodge analog of (b) over $\mathbb{C}$ is trivial: Hodge structures of odd weight are of even dimension. In this context, the dimension in (b) turns out to be the rank of the Tate module of an Abelian variety and this rank is always twice the dimension of the Abelian variety.

Cases of folklore conjectures about mixed motives **[26]** can also be proved **[40]**.

6. Afterwords

We have seen that one-motives provide a description of several *mixed motives of level one* of varieties over perfect fields. For instance, they provide an algebraic description of the motivic H^1 of arbitrary varieties over perfect fields. Doubts have been expressed by Grothendieck about extending these to the case of imperfect fields. In a letter to L. Illusie (reproduced as an appendix to [**27**]), he writes:

> *En caractéristique $p > 0$, j'ai des doutès tres sérieux pour l'existence d'une théorie des motifs pas iso du tout à cause des phénomènes de p-torsion (surtout pour les schémas qui ne sont projectifs et lisses). Ainsi, si on admet la description de Deligne des « motifs mixtes » de niveau 1 comme le genre de chose permettant de définir un H^1 motivique d'un schema pas projectif ou pas lisse, on voit déjà pour une courbe algébrique sur un corps imparfait k, la construction ne peut fournir en général qu'un objet du type voulu sur la clôture parfaite de k.*

The method of hypercoverings (which we use to construct the isogeny one-motives $L^n \otimes \mathbb{Q}$) may not appeal to everyone (certainly it is not my preferred method). However, to this day, one has not found a completely intrinsic description of the mixed Hodge structure on the cohomology of a complex algebraic variety V. Until this is achieved, one cannot even dream of providing a purely algebraic construction of the one-motives $I^n(V)$. Needless to say, an intrinsic construction of the mixed Hodge structure on the cohomology of V is one of the outstanding problems in Hodge theory. The folklore analogy [**14**] between Galois modules and Hodge structures exhibits the weak points of the latter [**21**]:

> *In the mid-sixties, analogies between (p,q)-decompositions and representations of the Galois group in étale cohomology was understood. The formal side of this similarity is quite simple: (p,q)-decomposition is the action of the torus $\mathbb{C}^*$ while class field theory identifies the multiplicative group of a non-Archimedean local field with the maximal Abelian quotient of its Galois group, so that the (p,q)-decomposition is the Archimedean analog of its Galois representation. On the other hand, the non-formal part of this symmetry is mysterious: Galois representations result from the Galois symmetries in the étale topology, while the roots of the Hodge structures are lost in darkness: hidden "Hodge symmetries" generating these structures are still unknown.*
>
> *... a natural construction (of the mixed Hodge structure on the cohomology of complex algebraic varieties) is still unknown; in particular, it is not known what kind of analysis lies behind the notion of a mixed Hodge structure.*[8]

Acknowledgments

I would like to acknowledge the encouragement and enthusiasm of the participants at the meeting in Banff where these results were first announced. The extremely kind comments of the referee on the exposition and the mathematics were immensely useful; I heartily thank the referee. I would also like to express

[8]Note the pioneering work of M. Saito [**45**] in this direction.

my gratitude to S. Lichtenbaum for his constant support; the key idea for the main result described here arose during discussions with him.

References

1. Y. André, *Mumford-Tate groups of mixed Hodge structures and the theorem of the fixed part*, Compositio Math. **82** (1992), 1–24.
2. ______, *Théorie des motifs et interprétation géométrique des valeurs p-adiques de G-functions (une introduction)*, Number Theory (Paris, 1992–1993), London Math. Soc. Lecture Note Ser., vol. 215, Cambridge Univ. Press, Cambridge, 1995, pp. 37–60.
3. L. Barbieri Viale, C. Pedrini, and C. Weibel, *Roitman's theorem for singular complex projective surfaces*, Duke Math. J. **84** no. 1 (1996), 155–190.
4. L. Barbieri Viale and V. Srinivas, *The Néron-Severi group and the mixed Hodge structure on* H^2, J. Reine Angew. Math. **450** (1994), 37–42.
5. ______, *Albanese and Picard 1-motives*, C. R. Acad. Sci. Paris Sér. I Math. **326** (1998), 1397–1401.
6. J. Biswas and V. Srinivas, *A Lefschetz* $(1,1)$ *theorem for normal projective complex varieties*, math.AG/9904086
7. D. Bertrand, *Relative splitting of one-motives*, Number Theory (Tiruchirapalli, 1996), Contemp. Math., vol. 210, Amer. Math. Soc., Providence, RI, 1998, pp. 3–17..
8. L. Breen, *The cube structure on the determinant bundle*, Theta Functions—Bowdoin 1987, Part 1 (Brunswick, ME, 1987), Proc. Sympos. Pure Math., vol. 49, Amer. Math. Soc., Providence, RI, 1989, pp. 663–673..
9. C. Brinkmann, *Die Andersonextension und 1-Motive.* Bonner Math. Schriften, vol. 223, Universität Bonn, Mathematisches Institut, Bonn, 1991.
10. F. E. Browder (ed.), *Mathematical developments arising from Hilbert problems*, Proc. Sympos. Pure Math., vol. 28, Part 1, Amer. Math. Soc., Providence, RI, 1976.
11. J.-L. Brylinski, "*1-motifs*" *et formes automorphes (théorie arithmétique des domaines de Siegel)*, Conference on Automorphic Theory (Dijon, 1981), Publ. Math. Univ. Paris VII, vol. 15, Univ. Paris VII, Paris, 1983, pp. 43–106.
12. J. A. Carlson, *The one-motif of an algebraic surface*, Compositio Math. **56** (1985), 271–314.
13. ______, *The geometry of the extension class of a mixed Hodge structure*, Proc. Sympos. Pure Math., vol. 46, Part 2, Amer. Math. Soc., Providence, RI, 1987, pp. 199–222.
14. P. Deligne, *Théorie de Hodge.* I, Actes du Congrès International des Mathématiciens (Nice, 1970), vol. 1, Gauthiers-Villars, Paris, 1971, pp. 425-430..
15. ______, *Théorie de Hodge.* II, Inst. Hautes Études Sci. Publ. Math. **40** (1971), 5–57; III, **44** (1974), 5–77.
16. ______, *Poids dans la cohomologie des variétés algébriques*, Proceedings of the International Congress of Mathematicians (Vancouver, BC, 1974). Canad. Math. Congress, Montréal, QC, 1975, pp. 79-85.
17. ______. *A quoi servent les motifs?* Motives (Seattle, WA, 1991), Proc. Symp. Pure Math. **55**, Part 1, Amer. Math. Soc., Providence, RI., 1994, pp. 143-161.
18. ______, letter to the author, April 15, 1998.
19. P. Deligne, J. Milne, A. Ogus and K.-Y. Shih, *Hodge cycles, motives and Shimura varieties*, Lecture Notes in Mathematics, vol. 900, Springer-Verlag, Berlin-New York, 1982.
20. G. Faltings and C.-L. Chai, *Degeneration of Abelian varieties.* Ergeb. Math. Grenzgeb. (3), vol. 22, Springer-Verlag, Berlin, 1990.
21. S. I. Gelfand and Y. I. Manin, *Algebra. V. Homological algebra*, Encyclopaedia Math. Sci., vol. 38, Springer, Berlin, 1994.
22. H. Gillet and C. Soulé, *Descent, motives and K-theory*, J. Reine Angew. Math. **478** (1996), 127–176.
23. P. Griffiths, *Some results on algebraic cycles on algebraic manifolds*, Algebraic Geometry (Bombay, 1968), Oxford Univ. Press, London, 1969, pp. 93–191.
24. A. Grothendieck, *Hodge's general conjecture is false for trivial reasons*, Topology **8** (1969), 299–303.
25. W. V. D. Hodge, *Topological invariants of algebraic varieties*, Proceedings of the International Congress of Mathematicians (Cambridge, MA, 1950), vol. 1, Amer. Math. Soc., Providence, RI, 1952, pp. 182–192.

26. U. Jannsen, *Mixed motives, motivic cohomology,* Ext-*groups*, Proceedings of the International Congress of Mathematicians, Vol. 1, 2 (Zürich, 1994), Birkhaüser, Basel, 1995, pp. 667–679.
27. ______, *Motivic sheaves and filtrations on Chow groups*, Motives (Seattle, WA, 1991), Proc. Symp. Pure Math. vol. 55, Part 1, Amer. Math. Soc., Providence, RI., 1996, pp. 245–302.
28. A. J. de Jong, *Smoothness, semi-stability and alterations*, Inst. Hautes Études Sci. Publ. Math. **83** (1996), 51–93.
29. N. M. Katz, *Wild ramification and some problems of "independence of l"*, Amer. J. Math. **105** (1983), 201–227.
30. ______, *Review of l-adic cohomology*, Motives (Seattle, WA, 1991), Proc. Symp. Pure Math., vol. 55, Part 1, Amer. Math. Soc., Providence RI., 1996, pp. 21–30.
31. S. Lang, *Unramified class field theory over function fields in several variables*, Ann. of Math. (2) **64** (1956), 285–325.
32. G. Laumon, *Transformation de Fourier généralisée*, math.AG/9603004
33. S. Lichtenbaum, *Suslin homology and Deligne* 1-*motives*, Algebraic K-Theory and Algebraic Topology (Lake Louise, AB, 1991), NATO Adv. Sci. Inst. Ser. C Math. Phys. Sci., vol. 407, Kluwer Acad. Publ., Dordrecht, 1993, pp. 189–196.
34. ______, *Quasi-motives of curves*, Algebraic K-Theory (Toronto, ON, 1996), Fields Inst. Commun., vol. 16, Amer. Math. Soc., Providence, RI, 1997 pp. 185–197.
35. D. Masser, *Specializations of finitely generated subgroups of Abelian varieties*, Trans. Amer. Math. Soc. **311** (1989) , 413–424.
36. W. Messing, *Differentials of the first, second and third kinds*, Algebraic Geometry (Arcata, CA., 1974), Proc. Sympos. Pure Math., vol. 29, Amer. Math. Soc., Providence, RI, 1975, pp. 547–561.
37. J. P. Murre, *Applications of algebraic K-theory to the theory of algebraic cycles*, Algebraic geometry, (Sitges, 1983), Lecture Notes in Math., vol. 1124, Springer, Berlin-New York, 1985, pp. 216–261..
38. H. Önsiper, *On glueing curves on surfaces and zero cycles*, Atti Accad. Naz. Lincei Cl. Sci. Fis. Mat. Natur. Rend. Lincei (9) Mat. Appl. **5** (1994), 5–9.
39. N. Ramachandran, *Albanese and Picard motives of schemes*, math.AG/9804042.
40. ______, *A conjecture of Deligne on one-motives*, math.AG/9806117.
41. M. Raynaud, *1-motifs et monodromie géométrique*, Périodes p-adiques (Bures-sur-Yvette, 1988). Astérisque, vol. 223, Soc. Math. France, Paris, 1994, pp. 295–319.
42. K. Ribet, *Cohomological realization of a family of* 1-*motives*, J. Number Theory **25** (1987), 152–161.
43. M. Rosenlicht, *Generalized Jacobian varieties*, Ann. of Math. (2) **59** (1954) 505–530.
44. H. Saito, *Abelian varieties attached to cycles of intermediate dimension*, Nagoya Math. J. **75** (1979), 95–119.
45. M. Saito, *On the theory of mixed Hodge modules*, Selected Papers on Number Theory, Algebraic Geometry, and Differential Geometry, Amer. Math. Soc. Transl. Ser. 2, vol. 160, Amer. Math. Soc., Providence, RI, 1994, pp. 47–61.
46. J.-P. Serre, *Morphismes universels et différentiels de troisième espèce*, Exp. 11, Séminaire C. Chevalley. 3e année. 1958/59. Variétés de Picard, École Normale Supérieure, Paris, 1960.
47. ______, *Abelian l-adic representations and elliptic curves*, W. A. Benjamin, New York-Amsterdam., 1968.
48. ______, *Algebraic groups and class fields*, Grad. Text in Math., vol. 117, Springer-Verlag, New York-Berlin, 1988.
49. ______, *Motifs*, Journées Arithmétiques, 1989 (Luminy, 1989), Astérisque, vol. 198-200, Soc. Math. France, Paris, 1991, pp. 333–349.
50. ______, *Propriétés conjecturales des groupes de Galois motiviques et représentations l-adique*, Motives (Seattle, WA, 1991), Proc. Symp. Pure Math. , vol. 55, Part 1, Amer. Math. Soc., Providence, RI, 1996, pp. 377-400.
51. J. Tate, *Les conjectures de Stark sur les fonctions L d'Artin en $s = 0$*, Progr. Math., vol. 47, Birkhäuser, Boston, MA, 1984.
52. A. Weil, *Variétés abéliennes*, Algèbre et théorie des nombres, Colloques Internationaux du Centre National de la Recherche Scientifique, no. 24, Centre National de la Recherche Scientifique, Paris, 1950, pp. 125–127.

Department of Mathematics, University of Michigan, Ann Arbor MI 48109, USA.
E-mail address: atma@math.lsa.umich.edu, atma@math.umd.edu

Centre de Recherches Mathématiques
CRM Proceedings and Lecture Notes
Volume **24**, 2000

Motives, Algebraic Cycles and Hodge Theory

Shuji Saito

CONTENTS

Introduction	235
1. Filtrations of Bloch-Beilinson Type	237
2. Refinement of Mumford's Theorem	241
3. Higher Abel-Jacobi Maps and the Main Results	244
4. Zero-Cycles on Complete Intersections	247
5. Higher Griffiths Groups	249
References	252

Introduction

The purpose of the following article is to survey the works [**26**], [**27**], [**28**] and [**1**]. The main themes are to realize the filtration on Chow groups conjectured by Bloch and Beilinson based on the Beilinson's philosophy of mixed motives and to find a Hodge theoretical invariant which identifies algebraic cycles lying in deeper levels of the filtration. The story goes back to 19th century when the Abel-Jacobi theorem, the monumental achievement of function theory in one variable, was established.

THEOREM 0.1. *Let X be a Riemann surface and let $A_0(X)$ be the group of divisors of degree zero on X modulo the subgroup of divisors of meromorphic functions on X. Then the Abel-Jacobi map*

$$\rho_X\colon A_0(X) \to J(X)$$

is an isomorphism, where $J(X) = \Gamma(X, \Omega_X^1)^*/H_1(X, \mathbb{Z})$ is the Jacobian variety of X.

We recall the definition of the Abel-Jacobi map. Express an element α of $A_0(X)$ in the form

$$(*) \qquad \alpha = \sum_{i=1}^{n} n_i([x_i] - [o])$$

Received by the editors October 7, 1998; Accepted November 11, 1998.
This is the final form of the paper.

where $o \in X$ is a fixed base point and $n_i \in \mathbb{Z}$ and $x_i \in X$. Then

$$\rho_X(\alpha)(\omega) = \sum_{i=1}^{n} n_i \int_{\gamma_i} \omega \quad \text{for } \omega \in \Gamma(X, \Omega_X^1),$$

where γ_i is a path on X from o to x_i. Thus Theorem 0.1 implies the following.

THEOREM 0.2. *For a given α as $(*)$ the following conditions are equivalent.*

1. *There exists a meromorphic function on X whose divisor is equal to α.*
2. *There exists $\gamma \in H_1(X, \mathbb{Z})$ such that*

$$\sum_{i=1}^{n} n_i \int_{\gamma_i} \omega = \int_{\gamma} \omega \quad \textit{for } \forall \omega \in \Gamma(X, \Omega_X^1).$$

The above results have been generalized to the case of divisors (namely algebraic cycles of codimension one) on a smooth projective variety X. The natural question is how one can extend the result to algebraic cycles of codimension> 1 on X. To be more precise let

$$\alpha = \sum_{i=1}^{n} n_i([V_i])$$

be an algebraic cycle on X with V_i a closed subvariety of codimension $r > 1$ in X. The question is if one can have an analogous result to Theorem 0.2 which gives a criterion for α to be rationally equivalent to zero. The first important step has been made by Weil and Griffiths who constructed the so-called Griffiths Abel-Jacobi map (cf. [**11**])

$$\rho_X^r \colon \mathrm{CH}^r(X)_{\mathrm{hom}} \to J^r(X).$$

for a smooth projective variety X over $\mathbb{C}$. Here $\mathrm{CH}^r(X)$ is the Chow group of algebraic cycles of codimension r on X modulo rational equivalence and $\mathrm{CH}^r(X)_{\mathrm{hom}}$ is the subgroup of cycles homologically equivalent to zero. $J^r(X)$ is a complex torus which is determined by the Hodge structure on the cohomology of X:

$$J^r(X) = H^{2r-1}(X, \mathbb{C})/F^r H^{2r-1}(X, \mathbb{C}) + H^{2r-1}(X, \mathbb{Z}).$$

The map ρ_X^r is defined in the similar manner to the case of curves by using periods of integrals.

In higher codimension, as the Mumford's theorem of non-representability [**19**] reveals, the story is not so simple as the case of codimension one (cf. Theorem 2.1). According to the result, ρ_X^r is not injective in general. Indeed its kernel may be too gigantic to be represented by such a usual algebro-geometric object as a complex torus. S. Bloch [**3**] is the first to find an important clue to bring order into the chaos of the world of algebraic cycles of codimension> 1. In case X is a surface and $r = 2$ he conjectured that even though it may be gigantic, $\mathrm{Ker}(\rho_X^2)$ should be controlled in some sense by the Hodge structure on $H^2(X, \mathbb{Q})$, philosophically speaking $\mathrm{Ker}(\rho_X^2)$ provides a geometric interpretation of the category of polarized Hodge structure of weight two in the same way that Abelian varieties do for weight one (cf. [**3**, p. 0.5]). The pioneering idea of Bloch was fortified by the Beilinson's conjectural theory of mixed motives which implies the existence of filtration on Chow groups

$$\mathrm{CH}^r(X) \supset F^1\,\mathrm{CH}^r(X) \supset F^2\,\mathrm{CH}^r(X) \supset \cdots \supset F^\nu\,\mathrm{CH}^r(X) \supset \cdots,$$

with $F^1\,\mathrm{CH}^r(X) = \mathrm{CH}^r(X)_{\mathrm{hom}}$ and $F^2\,\mathrm{CH}^r(X) = \mathrm{Ker}(\rho_X^r)$. Beilinson predicts the celebrated formula

$$\mathrm{Ext}^{\nu}_{\mathcal{MM}}(\mathbf{1}, h^{2r-\nu}(X)(r)) \simeq \mathrm{Gr}^{\nu}_F\,\mathrm{CH}^r(X) \otimes \mathbb{Q}$$

holds where $\mathrm{Gr}^{\nu}_F\,\mathrm{CH}^r(X) = F^{\nu}\,\mathrm{CH}^r(X)/F^{\nu+1}\,\mathrm{CH}^r(X)$. It expresses each graded quotient of the filtration as an extension group in a certain conjectural category of mixed motives $\mathcal{MM}$ (See the remark after Definition 1.1 for more detail).

Even though it is still conjectural, the formula provides a comprehensive explanation of various phenomena appearing in the world of algebraic cycles. This is one of the main themes of [**27**] where are given two candidates of the conjectural filtration and studied the relation between them. This is recalled in §1 of this paper. There are related works by Jannsen [**15**] and Murre [**20**].

In §2 we recall the result of [**26**] on the generalization and refinement of Mumford's theorem of non-representability of Chow groups in the context of the filtration of Chow groups. The results tell how each graded quotient of the filtration on Chow groups of X is influenced by the Hodge structure on the cohomology groups of X and provides a framework to realize the Bloch's dream to construct a geometric interpretation of Hodge structures of higher weight by using algebraic cycles.

Now that we have a candidate of the conjectural filtration with a list of basic properties, we come to the problem of extending the Griffiths Abel-Jacobi map $\rho_X^r\colon \mathrm{CH}^r(X)_{\mathrm{hom}} \to J^r(X)$ to the *higher Abel-Jacobi map*

$$\rho_X^{r,\nu}\colon F^{\nu}\,\mathrm{CH}^r(X) \to J^{r,\nu}(X).$$

This is the main theme of the works [**1**] and [**28**] where the construction of the higher Abel-Jacobi map is actually done in the relative situation. Moreover we show that our higher Abel-Jacobi map is able to identify algebraic cycles on generic complete intersections in $\mathbb{P}^n$ and thus provides new results on algebraic cycles. The results of [**1**] concern mainly on zero-cycles on complete intersections and those of [**28**] on k-planes on hypersurfaces. These results are recalled in §4 and §5 after reviewing the definition of the higher Abel-Jacobi map in §3. We remark that what actually plays an important role in the proof of the main results in [**1**] and [**28**] is the *infinitesimal invariant of the higher Abel-Jacobi map.* Its prototype has already appeared in the seminal work by Voisin [**30**]. The quite similar construction has been made independently by M. Green and P. Griffiths (cf. [**10**]).

This article is based on the author's lectures at NATO ASI/1998 CRM Summer School at Banff, Canada. He would like to thank J. Lewis, N. Yui, B. Gordon, and S. Müller-Stach for giving me the great opportunity. He would also like to thank the audience at the conference for their interest and patience.

1. Filtrations of Bloch-Beilinson Type

In this section we fix a ground field k and let $\mathcal{C}$ be the category of smooth projective varieties over k. For $X \in \mathcal{C}$ and an integer $r \geq 0$, $\mathrm{CH}^r(X)$ denotes the group of algebraic cycles of codimension r in X modulo rational equivalence. We fix a Weil cohomology theory (cf. [**16**, §1.2]). Namely, fixing the field Q of characteristic zero, we are given a functor

$$H^*\colon \mathcal{C} \to (Q\text{-vector spaces}); \quad X \rightsquigarrow H^*(X) = \bigoplus_{i=0}^{2\dim(X)} H^i(X)$$

from $\mathcal{C}$ to the category of finite dimensional, graded, anti-commutative Q-algebras, which satisfies a certain list of axioms. Here we list a part of the axioms which we will use.

(W1) For a morphism $f\colon Y \to X$ we have the functorial homomorphisms

$$f^*\colon H^i(X) \to H^i(Y),$$
$$f_*\colon H^{i-2t}(Y) \to H^i(X),$$

where $t = \dim(X) - \dim(Y)$.

(W2) We are given the cycle maps

$$\gamma_X^r\colon \mathrm{CH}^r(X) \to H^{2r}(X)$$

and the intersection products on cycles and the products on cohomology groups are compatible.

(W3) The hard Lefschetz theorem holds. Namely, for $X \in \mathcal{C}$ of dimension d and for an integer $\nu \geq 0$, we have the isomorphism

$$\lambda_\ell^\nu\colon H^{d-\nu}(X) \simeq H^{d+\nu}(X),$$

where $\lambda_\ell\colon H^*(X) \to H^{*+2}(X)$ is given by the product with the cohomology class $\ell \in H^2(X)$ of an ample line bundle.

Let V, $X \in \mathcal{C}$ and for $\Gamma \in \mathrm{CH}^q(W \times X)$ and let

$$\Gamma_*\colon \mathrm{CH}^s(W) \to \mathrm{CH}^r(X),$$
$$\varphi_\Gamma^i\colon H^{i-2(r-s)}(W) \to H^i(X)$$

denote the homomorphisms induced by Γ as an algebraic correspondence, where $s = r - q + \dim(W)$. These are given by the formulae

$$\Gamma_*(\alpha) = (p_X)_*((p_W)^*(\alpha) \cdot \Gamma) \qquad \text{for } \alpha \in \mathrm{CH}^s(W),$$
$$\varphi_\Gamma^i(\beta) = (p_X)_*((p_W)^*(\beta) \cdot \gamma_{W\times X}^q(\Gamma)) \qquad \text{for } \beta \in H^{i-2(r-s)}(W),$$

where $p_X : W \times X \to X$ and $p_W : W \times X \to W$ are the projections.

DEFINITION 1.1. A filtration $F^\cdot\,\mathrm{CH}^*$ of Bloch-Beilinson type (B-B type for short) is being given a descending filtration on $\mathrm{CH}^r(X)$,

$$F^\cdot\,\mathrm{CH}^r(X) \quad \text{for each } X \in \mathcal{C} \text{ and } r \geq 0,$$

satisfying the following conditions:

(BB0) $F^0\,\mathrm{CH}^r(X) = \mathrm{CH}^r(X)$.

(BB1) $F^\cdot\,\mathrm{CH}^*$ is respected by algebraic correspondences: For W, $X \in \mathcal{C}$ and $\Gamma \in \mathrm{CH}^q(W \times X)$,

$$\Gamma_*(F^\nu\,\mathrm{CH}^s(W)) \subset F^\nu\,\mathrm{CH}^r(X) \quad \text{for every } \nu \geq 0,$$

where $r - s + \dim(W)$. Let

$$\mathrm{Gr}_F^\nu\,\Gamma_*\colon \mathrm{Gr}_F^\nu\,\mathrm{CH}^s(W) \to \mathrm{Gr}_F^\nu\,\mathrm{CH}^r(X)$$

be the induced homomorphism. Here

$$\mathrm{Gr}_F^\nu\,\mathrm{CH}^r(X) = F^\nu\,\mathrm{CH}^r(X)/F^{\nu+1}\,\mathrm{CH}^r(X),$$

(BB2) Under the above assumption we have the implication

$$\varphi_\Gamma^{2r-\nu} = 0 \implies \mathrm{Im}(\mathrm{Gr}_F^\nu\,\Gamma_*) \text{ is torsion},$$

where $\varphi_\Gamma^{2r-\nu}\colon H^{2s-\nu}(W) \to H^{2r-\nu}(X)$ is defined as above.

The motivation to consider the above filtration is the Beilinson's conjectural theory [**2**] of mixed motives, where Chow groups are expressed as *Hom* groups in the derived category of $\mathcal{MM}_k$, the (conjectural) category of mixed motives. The theory implies the existence of a filtration $F_{\mathcal{M}}$ of B-B type on Chow groups which satisfies the following conjectural formula

$$\mathrm{Ext}^{\nu}_{\mathcal{MM}_k}(\mathbf{1}, h^{2r-\nu}(X)(r)) \simeq \mathrm{Gr}^{\nu}_{F_{\mathcal{M}}} \mathrm{CH}^r(X) \otimes_{\mathbb{Z}} \mathbb{Q},$$

where $h^i(X)$ denotes a Grothendieck motive associated to X and $h^i(X)(r)$ is its Tate twist and $\mathbf{1}$ is the motive of $\mathrm{Spec}(k)$. In this framework the condition (BB2) is a consequence of the commutative diagram

$$\begin{array}{ccc} \mathrm{Gr}^{\nu}_{F_{\mathcal{M}}} \mathrm{CH}^r(W) \otimes_{\mathbb{Z}} \mathbb{Q} & \cong & \mathrm{Ext}^{\nu}_{\mathcal{MM}_k}(\mathbf{1}, h^{2s-\nu}(W)(s)) \ , \\ \downarrow \Gamma_* & & \downarrow \Gamma_* \\ \mathrm{Gr}^{\nu}_{F_{\mathcal{M}}} \mathrm{CH}^r(X) \otimes_{\mathbb{Z}} \mathbb{Q} & \cong & \mathrm{Ext}^{\nu}_{\mathcal{MM}_k}(\mathbf{1}, h^{2r-\nu}(X)(r)) \end{array}$$

where the right vertical arrow is induced by $\Gamma_* \colon h^{2s-\nu}(W)(s) \to h^{2r-\nu}(X)(r)$, the map induced by Γ on the motives associated to W and X (cf. [**14**]).

Now we define two particular filtrations of B-B type which play the most important role among other filtrations of B-B type.

DEFINITION 1.2. We define the filtration $F^{\cdot}_{\mathrm{B}} \mathrm{CH}^*$ (resp. $F^{\cdot}_{\mathrm{BM}} \mathrm{CH}^*$) of B-B type in the following inductive way.

(i) $F^0_{\mathrm{B}} \mathrm{CH}^s(W) = \mathrm{CH}^s(W)$ for every $W \in \mathcal{C}$ and integer $s \geq 0$.
(ii) Let $\nu \geq 0$ be an integer. Assume that we have defined $F^{\nu}_{\mathrm{B}} \mathrm{CH}^s(W)$ for every $W \in \mathcal{C}$ and for every integer $s \geq 0$. Then we define

$$F^{\nu+1}_{\#} \mathrm{CH}^r(X) = \sum_{W,q,\Gamma} \mathrm{Im}(\Gamma_* \colon F^{\nu}_{\#} \mathrm{CH}^{r+d_W-q}(W) \to \mathrm{CH}^r(X))(\# = \mathrm{B} \text{ or } \mathrm{BM})$$

where W, q and Γ range over the following data.

(a) $W \in \mathcal{C}$ of dimension d_W.
(b) an integer q such that $r \leq q \leq r + d_W$.
(c) $\Gamma \in \mathrm{CH}^q(W \times X)$, satisfying the condition $s = r - q + d_W$

$$\varphi^{2r-\nu}_{\Gamma} = 0 \quad (\text{resp. } \varphi^{2r-\nu}_{\Gamma}(H^{2s-\nu}(W)) \subset N^{r-\nu+1} H^{2r-\nu}(X))$$

For integers p, $i \geq 0$, $N^p H^i(X)$ is the p-th coniveau filtration:

$$N^p H^i(X) = \sum_{\substack{Y \hookrightarrow X \\ \mathrm{codim}_X Y \geq p}} \{\mathrm{Ker}(H^i(X) \to H^i(X - Y))\},$$

where $Y \hookrightarrow X$ ranges over all closed immersion with $\mathrm{codim}(Y) \geq p$.

It is clear from the definition that $F^{\cdot}_{\mathrm{B}} \mathrm{CH}^*$ and $F^{\cdot}_{\mathrm{BM}} \mathrm{CH}^*$ are filtrations of B-B type and that for any filtration $F^{\cdot} \mathrm{CH}^*$ of B-B type we have

$$F^{\nu}_{\mathrm{B}} \mathrm{CH}^r(X) \subset F^{\nu} \mathrm{CH}^r(X) \quad \text{for every } X \in \mathcal{C} \text{ and } r, \nu \geq 0.$$

The semi-purity for the Weil cohomology theory implies $N^{r-\nu+1} H^{2r-\nu}(X) = 0$ for $\nu \leq 1$ and this implies $F^{\nu}_{B} \mathrm{CH}^r(X) = F^{\nu}_{BM} \mathrm{CH}^r(X)$ for $\nu \leq 2$ (cf. Proposition 1.2 below). It is easy to see (cf. [**26**, Lemma 2.5]) that $F^1_{\mathrm{B}} \mathrm{CH}^r(X) = \mathrm{CH}^r(X)_{\mathrm{hom}}$, the subgroup of cycles classes homologically equivalent to zero. When the ground field

$k = \mathbb{C}$, $F^2_B \mathrm{CH}^r(X)$ is related to the kernel of the Abel-Jacobi map (cf. Proposition 2.1 below). We also have the following results Proposition 1.1, Proposition 1.2 and Proposition 1.3 which are taken from [**27**].

PROPOSITION 1.1. *Assume that the following condition (cf. [**16**, §2]) is satisfied for any $X \in \mathcal{C}$:*

C(X) *the Künneth components of the cohomology class of the diagonal $\Delta_X \hookrightarrow X \times X$ are all $\mathbb{Q}$-algebraic.*

Then the intersection of cycles induces

$$F^\nu_{\mathrm{B}} \mathrm{CH}^r(X) \times F^\mu_{\mathrm{B}} \mathrm{CH}^s(X) \to F^{\nu+\mu}_{\mathrm{B}} \mathrm{CH}^{r+s}(X).$$

PROPOSITION 1.2 ([**27**, Theorem 1-6 and Lemma 4.8]). *Let $F^\cdot \mathrm{CH}^*$ be any filtration of B-B type. Assume that* B(W) *(cf. see Definition 1.3 below) holds for any $W \in \mathcal{C}$ and that the numerical and homological equivalences coincide. For W, $X \in \mathcal{C}$ and $\Gamma \in \mathrm{CH}^q(W \times X)$, let*

$$\mathrm{Gr}^\nu_F \Gamma_* \colon \mathrm{Gr}^\nu_F \mathrm{CH}^s(W) \to \mathrm{Gr}^\nu_F \mathrm{CH}^r(X),$$
$$\varphi^{2r-\nu}_\Gamma \colon H^{2s-\nu}(W) \to H^{2r-\nu}(X),$$

be the homomorphisms induced by Γ, where $s = r - q + \dim(W)$. If we have

$$\mathrm{Im}(\varphi^{2r-\nu}_\Gamma) \subset N^{r-\nu+1} H^{2r-\nu}(X),$$

then $\mathrm{Im}(\mathrm{Gr}^\nu_F \Gamma_*)$ *is torsion. In particular we have*

$$F^\nu_{\mathrm{B}} \mathrm{CH}^r(X) \otimes_{\mathbb{Z}} \mathbb{Q} = F^\nu_{\mathrm{BM}} \mathrm{CH}^r(X) \otimes_{\mathbb{Z}} \mathbb{Q} \quad \textit{for every } X \in \mathcal{C} \textit{ and } r, \nu \geq 0.$$

DEFINITION 1.3. Let $W \in \mathcal{C}$. For an integer $\nu \leq d = \dim(W)$ let

$$\lambda^\nu_\ell \colon H^{d-\nu}(W) \simeq H^{d+\nu}(W)$$

be the hard Lefschetz isomorphism (cf. (W3)). We define the condition

B(W) For each $\nu \leq d$ the inverse of λ^ν_ℓ is $\mathbb{Q}$-algebraic: There exists $\Gamma \in \mathrm{CH}^\nu(W \times W) \otimes_{\mathbb{Z}} \mathbb{Q}$ which induces the inverse of λ^ν_ℓ.

Recall that B(W) is the so-called hard Lefschetz conjecture and is a part of the standard conjectures. B(W) holds if W is obtained under successive operations of product and hyperplane section from projective smooth varieties of the following types; a curve, a surface, an Abelian variety, a smooth complete intersection in $\mathbb{P}^N$ or a generalized flag variety G/P (cf. [**16**]).

DEFINITION 1.4. A filtration $F^\cdot \mathrm{CH}^*$ of B-B type is B-discrete if

$$D^r_{F_{\mathrm{B}}}(X) = D^r_F(X) \quad \text{for every } X \text{ and } r \geq 0.$$

Here, for a filtration $F^\cdot \mathrm{CH}^*$ of B-B type, we write

$$D^r_F(X) = \bigcap_{\nu > 0} F^\nu \mathrm{CH}^r(X).$$

It is easy to see that $D^r_{F_{\mathrm{B}}}(X) = F^{2r+1} \mathrm{CH}^r(X)$ and $D^r_{F_{\mathrm{BM}}}(X) = F^{r+1} \mathrm{CH}^r(X)$.

PROPOSITION 1.3 ([**27**, Theorem 1-10]). *Assume* C(X). *For any B-discrete filtration $F^\cdot \mathrm{CH}^*$ of B-B type, we have*

$$F^\nu_{\mathrm{B}} \mathrm{CH}^r(X) \otimes_{\mathbb{Z}} \mathbb{Q} = F^\nu \mathrm{CH}^r(X) \otimes_{\mathbb{Z}} \mathbb{Q} \quad \textit{for every } r, \nu \geq 0.$$

Now we introduce the important conjecture due to Bloch and Beilinson.

CONJECTURE 1.1. There exists a filtration $F^{\cdot}\,\mathrm{CH}^*$ of B-B type which satisfies the following additional condition.

(BB3) For each $X \in \mathcal{C}$ and $r \geq 0$ there exists an integer $N > 0$ such that $F^N\,\mathrm{CH}^r(X) \otimes_{\mathbb{Z}} \mathbb{Q} = 0$.

Noting that B(X) implies C(X) (cf. [**16**, Theorem2.9]), Proposition 1.2 and Proposition 1.3 mply the following.

PROPOSITION 1.4. *Assume that* B(X) *holds for any* $X \in \mathcal{C}$ *and that the numerical and homological equivalence coincide. Then Conjecture* 1.1 *is equivalent to the following Conjecture* 1.2. *Moreover, if they hold,* $F^{\cdot}_{\mathrm{BM}}\,\mathrm{CH}^*$ *is the only filtration which possibly satisfies the conditions* (BB0) *through* (BB3).

CONJECTURE 1.2. For any $X \in \mathcal{C}$ and $r \geq 0$ we have

$$F^{r+1}_{\mathrm{BM}}\,\mathrm{CH}^r(X) \otimes_{\mathbb{Z}} \mathbb{Q} = 0.$$

We note that Conjecture 1.2 is a consequence of the Beilinson's conjectural theory mixed motives and that it implies several pending conjectures on algebraic cycles as is observed in [**27**].

2. Refinement of Mumford's Theorem

The purpose of this section is to recall the results of [**26**] which generalize the following seminal result of Mumford on *non-representability* of Chow groups.

THEOREM 2.1 ([**19**]). *Let X be a projective smooth surface over $\mathbb{C}$ with*

$$H^0(X, \Omega_X^2) \neq 0.$$

Let $A_0(X)$ be the group of zero-cycles of degree zero modulo rational equivalence. Then $A_0(X)$ is not finite-dimensional, which means that there exists no morphism $C \to X$ with C, a (possibly reducible) projective smooth curve such that the induced map $A_0(C) \to A_0(X)$ is surjective. In particular $A_0(X)$ can not be given the structure of an Abelian variety.

Let the notation be as in §1. In this section we take the ground field k to be $\mathbb{C}$, the field of complex numbers. We adopt the rational Betti cohomology theory

$$\mathcal{C} \ni X \rightsquigarrow H^*(X, \mathbb{Q}) = H^*_B(X(\mathbb{C}), \mathbb{Q})$$

as our Weil cohomology theory. First we fix a filtration $F^{\cdot}\,\mathrm{CH}^*$ of B-B type which will be used for the rest of this paper.

DEFINITION 2.1. The filtration $F^{\cdot}\,\mathrm{CH}^*$ on Chow groups is defined exactly in the same way as $F^{\cdot}_{\mathrm{BM}}\,\mathrm{CH}^*$ except that we replace the condition (c) of Definition 1.2 by the following.

(c)′ $\Gamma \in \mathrm{CH}^q(W \times X)_{\mathbb{Q}}$, satisfying the condition

$$\varphi_\Gamma^{2r-\nu}(H^{2s-\nu}(W, \mathbb{Q})) \subset N_H^{r-\nu+1} H^{2r-\nu}(X, \mathbb{Q}) \quad (s = r - q + d_W).$$

For integers p, $i \geq 0$, $N_H^p H^i(X, \mathbb{Q})$ is the maximal sub-Hodge structure of $H^i(X, \mathbb{Q})$ contained in the Hodge filtration $F^p H^i(X, \mathbb{C})$. It is known that

$$N^p H^i(X, \mathbb{Q}) \subset N_H^p H^i(X, \mathbb{Q})$$

and the Grothendieck's generalized Hodge conjecture predicts that they coincide (cf. [**12**]). Thus $F^{\cdot}\,\mathrm{CH}^*$ is considered as the Hodge theoretic analogue of $F^{\cdot}_{\mathrm{BM}}\,\mathrm{CH}^*$

and the conjecture implies that they coincide. The main reason why we adopt $F^{\cdot}\,\mathrm{CH}^*$ rather than $F^{\cdot}_{\mathrm{BM}}\,\mathrm{CH}^*$ is to avoid the use of the hard Lefschetz conjecture in the proof of Theorem 2.2 below.

The following are some basic properties of $F^{\cdot}\,\mathrm{CH}^*$.

PROPOSITION 2.1. (1) *$F^{\cdot}\,\mathrm{CH}^*$ is a filtration of B-B type satisfying*

$$F^{\nu}\,\mathrm{CH}^r(X) = F^{\nu+1}\,\mathrm{CH}^r(X) \quad \text{for every } \nu > r \text{ and } X \in \mathcal{C}.$$

(2) $F^1\,\mathrm{CH}^r(X) = \mathrm{CH}^r(X)_{\mathrm{hom}}$, *the subgroup of cycles homologically equivalent to zero.*

(3) *Let $\rho^r_X\colon \mathrm{CH}^r(X)_{\mathrm{hom}} \to J^r(X)$ be the Abel-Jacobi map, where $J^r(X)$ is the intermediate Jacobian of X. We have $F^2\,\mathrm{CH}^r(X) \subset \mathrm{Ker}(\rho^r_X)$ and*

$$(\mathrm{Ker}(\rho^r_X) \cap \mathrm{CH}^r(X)_{\mathrm{alg}})/(F^2\,\mathrm{CH}^r(X) \cap \mathrm{CH}^r(X)_{\mathrm{alg}})$$

is torsion.

Here $\mathrm{CH}^r(X)_{\mathrm{alg}} \subset \mathrm{CH}^r(X)$ is the subgroup of cycles algebraically equivalent to zero.

In order to state our result on non-representability of Chow groups, we need introduce the notion of the *rank of a family of algebraic cycles*, which generalizes the notion of finite-dimensionality used in Theorem 2.1. For an Abelian group M we denote $M_{\mathbb{Q}} = M \otimes_{\mathbb{Z}} \mathbb{Q}$.

DEFINITION 2.2. Let $X \in \mathcal{C}$ and fix an integer $r \geq 0$ and a subgroup $\Xi \subset \mathrm{CH}^r(X)$.

(1) Let $S \subset \mathrm{CH}^r(X)_{\mathrm{alg}}$ be a subgroup.
  (a) S is of rank 0 mod Ξ if we have $\dim_{\mathbb{Q}}(S + \Xi/\Xi)_{\mathbb{Q}} < \infty$.
  (b) For an integer $\mu \geq 1$, S is of rank $\leq \mu \bmod \Xi$ if there exists Y, (possibly reducible) smooth projective of dimension $\leq \dim(X) - r + \mu$ and a morphism $f\colon Y \to X$ such that

$$S_{\mathbb{Q}} \subset \Xi_{\mathbb{Q}} + \mathrm{Im}\left(f_*\colon \mathrm{CH}^s(Y)_{\mathrm{alg}} \to \mathrm{CH}^r(X)_{\mathrm{alg}}\right)_{\mathbb{Q}},$$

  where $s = \dim(Y) - \dim(X) + r$. Let $r(S \bmod \Xi)$ be the minimal integer μ for which S is of rank $\leq \mu$. In case $S \supset \Xi$ we also denote $r(S/\Xi) = r(S \bmod \Xi)$. Note that by definition we have $0 \leq r(S \bmod \Xi) \leq r$.

(2) $\Gamma \in \mathrm{CH}^r(W \times X)$ is of rank$\leq \mu \bmod \Xi$ if so is $\mathrm{Im}(\Gamma_*)$, where

$$\Gamma_*\colon \mathrm{CH}_0(W)_{\mathrm{alg}} \to \mathrm{CH}^r(X)_{\mathrm{alg}}$$

is induced by Γ as an algebraic correspondence. We denote $r(\Gamma \bmod \Xi) = r(\mathrm{Im}(\Gamma_*) \bmod \Xi)$.

By definition we have the equivalence of conditions (recall $A_0(X) = \mathrm{CH}_0(X)_{\mathrm{alg}}$)

$$r(A_0(X)) \leq 1 \Leftrightarrow A_0(X) \text{ is finite dimensional.}$$

We also note the following implication (cf. [**26**, §5])

$$\begin{aligned}\mathrm{Ker}(\rho^r_X) \cap \mathrm{CH}^r(X)_{\mathrm{alg}} \subset \Xi &\implies \mathrm{CH}^r(X)_{\mathrm{alg}} \text{ is weakly representable modulo } \Xi \\ &\implies r(\mathrm{CH}^r(X)_{\mathrm{alg}} \bmod \Xi) \leq 1.\end{aligned}$$

DEFINITION 2.3. Let W, $X \in \mathcal{C}$ and let $r \geq \nu \geq 0$ be integers. For $\Gamma \in \mathrm{CH}^r(W \times X)$ we put

$$Z_\Gamma \operatorname{Gr}_F^\nu \mathrm{CH}^r(X) = \operatorname{Im}\big(\operatorname{Gr}_F^\nu \Gamma_*\colon \operatorname{Gr}_F^\nu \mathrm{CH}_0(W) \to \operatorname{Gr}_F^\nu \mathrm{CH}^r(X)\big)$$

(cf. Definition 1.1 (BB1)). We define

$$Z_0 \operatorname{Gr}_F^\nu \mathrm{CH}^r(X) = \sum_{W,\Gamma} Z_\Gamma \operatorname{Gr}_F^\nu \mathrm{CH}^r(X) \subset \operatorname{Gr}_F^\nu \mathrm{CH}^r(X),$$

where the sum is taken over W, Γ as above.

Let W and Γ be as in Definition 2.3. Γ induces the homomorphisms also on cohomology groups

$$\psi_\Gamma^{2r-\nu}\colon H^{2d-\nu}(W,\mathbb{Q}) \to H^{2r-\nu}(X,\mathbb{Q}),$$
$$\varphi_\Gamma^\nu\colon H^{2m+\nu}(X,\mathbb{Q}) \to H^\nu(W,\mathbb{Q})$$

where $d = \dim(W)$ and $m = \dim(X) - r$. Note that $\psi_\Gamma^{2r-\nu}$ is the dual of φ_Γ^ν with respect to Poincaré duality of cohomology groups. We note (cf. [**26**, §1])

FACT 2.1. (1) $\operatorname{Im}(\psi_\Gamma^{2r-\nu}) \subset N^{r-\nu}H^{2r-\nu}(X,\mathbb{Q})$.
(2) Assuming $\mathrm{B}(W)$ for any $W \in \mathcal{C}$ we have

$$N^{r-\nu}H^{2r-\nu}(X,\mathbb{Q}) = \sum_{W,Q} \operatorname{Im}(\psi_\Gamma^{2r-\nu}).$$

The following result is a generalization of several works on non-representability of Chow groups ([**19**], [**3**, §1 Appendix], [**24**], [**13**], [**17**], [**25**], [**29**]). One will see in the remark below Corollary 2.1 that Theorem 2.2 refines Theorem 2.1.

THEOREM 2.2. *Let the notation be as above. Let $\nu \geq 1$ be an integer. Consider the following conditions*
(i) $r(\Gamma \bmod F^{\nu+1}\mathrm{CH}^r(X)) \leq \nu - 1$.
(ii) $r(Z_\Gamma \operatorname{Gr}_F^\nu \mathrm{CH}^r(X)) \leq \nu - 1$.
(iii) $Z_\Gamma \operatorname{Gr}_F^\nu \mathrm{CH}^r(X) = 0$.
(iv) $\operatorname{Im}(\psi_\Gamma^{2r-\nu}) \subset N_H^{r-\nu+1}H^{2r-\nu}(X,\mathbb{Q})$.
(v) $\operatorname{Im}(\varphi_\Gamma^\nu) \subset N_H^1 H^\nu(W,\mathbb{Q})$.
Then we have the implications

$$\text{(i)} \implies \text{(v)} \iff \text{(iv)} \implies \text{(iii)} \implies \text{(ii)}.$$

All the above conditions are equivalent assuming either $\nu \leq 2$ or that there exist a dominant rational map $\widetilde{W} \dashrightarrow W$ such that $\mathrm{B}(\widetilde{W})$ holds (cf. Definition 1.3).

Taking $W = X$ and $\Gamma \subset X \times X$ to be the diagonal, Theorem 2.2 implies immediately the following result on zero-cycles.

COROLLARY 2.1. *Let $X \in \mathcal{C}$ with $n = \dim(X)$. Assume that there there exist a dominant rational map $\widetilde{X} \dashrightarrow X$ such that $\mathrm{B}(\widetilde{X})$ holds. For an integer $\nu \geq 1$ the following conditions are equivalent.*
(i) $r(A_0(X)/F^{\nu+1}\mathrm{CH}_0(X)) \leq \nu - 1$.
(ii) $r(\operatorname{Gr}_F^\nu \mathrm{CH}_0(X)) \leq \nu - 1$.
(iii) $\operatorname{Gr}_F^\nu \mathrm{CH}_0(X) = 0$.
(iv) $H^0(X,\Omega_X^\nu) = 0$.

We remark that the Mumford's theorem Theorem 2.1 is equivalent to the implication

$$r(A_0(X)) \leq 1 \implies H^0(X, \Omega_X^2) = 0$$

for a projective smooth surface X over $\mathbb{C}$ (for which $\mathrm{B}(X)$ is known to hold). On the other hand our theorem implies the equivalence of the conditions for such X

$$r(A_0(X)/F^3\,\mathrm{CH}_0(X)) \leq 1 \iff H^0(X, \Omega_X^2) = 0.$$

Implications of Theorem 2.2 in non-representability of Chow group are given in [**26**, §5]. Another noteworthy implication is, as is first speculated by Bloch [**3**], that one might hope that algebraic cycles play the role of the *geometric interpretation* of Hodge structure of higher weight as polarized Abelian varieties does for polarized Hodge structure of weight one as found by Riemann. See [**26**, §10], for more discussion on this point. The following corollary also indicates the existence of an inflexible hierarchy of the world of algebraic cycles which is controlled by Hodge structures of cohomology groups of varieties.

COROLLARY 2.2. *Let X be a smooth projective variety over $\mathbb{C}$ and assume that the generalized Hodge conjecture holds for any smooth projective variety W. Let* $\dim(X) \geq r \geq \nu \geq 1$ *be integers and put* $m = \dim(X) - r$. *Then the following conditions are all equivalent.*

(i) $r(Z_0\,\mathrm{Gr}_F^\nu\,\mathrm{CH}^r(X)) \leq \nu - 1$.
(ii) $Z_0\,\mathrm{Gr}_F^\nu\,\mathrm{CH}^r(X) = 0$.
(iii) $N_H^{r-\nu} H^{2r-\nu}(X, \mathbb{Q}) = N_H^{r-\nu+1} H^{2r-\nu}(X, \mathbb{Q})$.
(iv) $N_H^m H^{2m+\nu}(X, \mathbb{Q}) = N_H^{m+1} H^{2m+\nu}(X, \mathbb{Q})$.

3. Higher Abel-Jacobi Maps and the Main Results

In this section we take the ground field k to be $\mathbb{C}$, the field of complex numbers and use the same notation as §1. So far we have introduced candidates for the filtration on Chow groups conjectured by Bloch-Beilinson. They are defined in the similar way and all are conjectured to be equal. In §2 we have chosen a particular filtration $F^{\cdot}\,\mathrm{CH}^*$ (cf. Definition 2.1) which is the most pragmatic to work with and we have refined the theorem of Mumford on non-representability of Chow groups by using the filtration.

Now we come to the following

QUESTION 3.1. Is there a *Hodge theoretical construction of the higher Abel-Jacobi maps*, namely for each $\nu \geq 1$ the map

$$\rho_X^{r,\nu} : F^\nu\,\mathrm{CH}^r(X) \to J^{r,\nu}(X)$$

which satisfies the following.

(i) For $\nu = 1$, $J^{r,1}(X) = J^r(X)$, the intermediate Jacobian of X and $\rho_X^{r,1}$ is the Abel-Jacobi map defined by Griffiths.
(ii) $\mathrm{Ker}(\rho_X^{r,\nu})$ coincides with $F^{\nu+1}\,\mathrm{CH}^r(X)$ modulo torsion.

The question should be hard since if such maps existed, $J^{r,\nu}(X)$ for $\nu \geq 2$ could not carry such usual algebro-geometric structure as a complex torus in view of non-representability result Theorem 2.1. M. Green has made remarkable progress on this problem (cf. [**9**]).

Our trick to avoid the difficulty is to consider the relative situation. We start with reviewing the theory of normal functions associated to families of algebraic cycles developed by Griffiths.

Let $f\colon X \to S$ be a smooth projective morphisms of non-singular algebraic varieties over $\mathbb{C}$. Let $F_S^1 \operatorname{CH}^r(X) \subset \operatorname{CH}^r(X)$ be the subgroup of cycle classes whose restrictions to each fiber X_s of f over $s \in S$ are homologically equivalent to zero. Let

$$J^r_{X/S} = \bigcup_{s\in S} J^r(X_s)$$

be the analytic complex torus fibration of intermediate Jacobian $J^r(X_s)$ of fibers X_s. Let $\Gamma(S, J^r_{X/S})$ be the space of holomorphic sections of $\pi\colon J^r_{X/S} \to S$. One defines the map

$$\rho^r_{X/S}\colon F_S^1 \operatorname{CH}^r(X) \to \Gamma(S, J^r_{X/S})\colon \alpha \to (s \to \rho^r_{X_s}(\alpha_{|X_s})),$$

where $\rho^r_{X_s}\colon \operatorname{CH}^r(X_s)_{\mathrm{hom}} \to J^r_{X_s}$ is the Abel-Jacobi map for X_s. An important fact that Griffiths discovered is that for $\alpha \in F_S^1 \operatorname{CH}^r(X)$, $\rho^r_{X/S}(\alpha)$ satisfies a certain differential equation coming from Gauss-Manin connection. To explain more we introduce some more notations.

Let $H^q_{\mathbb{Z}}(X/S)(r) = R^q f_* \mathbb{Z}(r)$ and (resp. $H^q_{\mathbb{C}}(X/S) = R^q f_* \mathbb{C}$) be the local system on S with the fiber $H^q(X_s, \mathbb{Z}(r))$ (resp. $H^q(X_s, \mathbb{C})$) over $s \in S$, where $X_s = f^{-1}(s)$. Put

$$H^q_{\mathcal{O}}(X/S) = H^q_{\mathbb{C}}(X/S) \otimes_{\mathbb{C}} \mathcal{O}_S = R^q f_*(f^{-1}\mathcal{O}_S)$$

and let $F^r H^q_{\mathcal{O}}(X/S) \subset H^q_{\mathcal{O}}(X/S)$ be the Hodge subbundle. As a sheaf on S_{an}

$$J^r_{X/S} = \operatorname{Coker}\big(H^{2r-1}_{\mathbb{Z}}(X/S)(r) \to H^{2r-1}_{\mathcal{O}}(X/S)/F^r H^{2r-1}_{\mathcal{O}}(X/S)\big).$$

We define

$$DJ^r_{X/S} = \operatorname{Ker}(J^r_{X/S} \xrightarrow{\nabla} \Omega^1_S \otimes_{\mathcal{O}_S} H^{2r-1}_{\mathcal{O}}(X/S)/F^{r-1} H^{2r-1}_{\mathcal{O}}(X/S))$$

where $\nabla\colon H^q_{\mathcal{O}}(X/S) \to \Omega^1_S \otimes_{\mathcal{O}_S} H^q_{\mathcal{O}}(X/S)$ is the Gauss-Manin connection. We call an element of $\Gamma(S, DJ^r_{X/S})$ a normal function. Griffiths showed that $\rho^r_{X/S}$ induces

$$\rho^r_{X/S}\colon F_S^1 \operatorname{CH}^r(X) \to \Gamma(S, DJ^r_{X/S}).$$

To generalize the above story we first define $F_S^\nu \operatorname{CH}^r(X)$ with $\nu \geq 0$, a relative version of $F^\cdot \operatorname{CH}^*$, which is a straightforward generalization of Definition 2.1. We fix a smooth variety S and let $\mathcal{C}_S$ be the category of smooth projective schemes over S with connected fibers.

DEFINITION 3.1. We define the filtration $F_S^\cdot \operatorname{CH}^*$ on Chow groups of objects of $\mathcal{C}_S$ in the following inductive way.

(i) $F_S^0 \operatorname{CH}^s(W) = \operatorname{CH}^s(W)$ for every $W \in \mathcal{C}_S$ and integer $s \geq 0$.

(ii) Let $\nu \geq 0$ be an integer. Assume that we have defined $F_S^\nu \operatorname{CH}^s(W)$ for every $W \in \mathcal{C}_S$ and for every integer $s \geq 0$. Then we define

$$F_S^{\nu+1} \operatorname{CH}^r(X) = \sum_{W,q,\Gamma} \operatorname{Im}\big(\Gamma_*\colon F_S^\nu \operatorname{CH}^{r+d_W-q}(W) \to \operatorname{CH}^r(X)\big),$$

where W, q and Γ range over the following data.

(a) $W \in \mathcal{C}_S$ of relative dimension d_W.

(b) an integer q such that $r \leq q \leq r + d_W$.

(c) $\Gamma \in \mathrm{CH}^q(W \times_S X)$, satisfying the condition

$$\varphi_{\Gamma_s}^{2r-\nu}(H^{2t-\nu}(W_s, \mathbb{Q})) \subset N_H^{r-\nu+1} H^{2r-\nu}(X_s, \mathbb{Q}) \quad \text{for } \forall s \in S$$

Here $t = r - q + d_W$ and for $s \in S$, W_s, X_s and Γ_s denote the fibers over s.

REMARK 3.1. It is easy to see that by the above definition $F_S^1 \mathrm{CH}^r(X)$ coincides with the subgroup of those cycles whose restriction to each fiber X_s lies in $F^1 \mathrm{CH}^r(X_s)$, namely is homologically equivalent to zero on X_s. For $\nu \geq 2$, $\alpha \in F_S^\nu \mathrm{CH}^r(X)$ implies that $\alpha_s \in F^\nu \mathrm{CH}^r(X_s)$ for each $s \in S$. To the contrary, if $\alpha_s \in F^\nu \mathrm{CH}^r(X_s)$ for each $s \in S$ then there exists an etale morphisms $\pi \colon T \to S$ such that the pull-back $\pi^*(\alpha) \in \mathrm{CH}^r(X_T)$ lies in $F_T^\nu \mathrm{CH}^r(X_T)$, where $X_T = X \times_S T$. This follows from [**1**, Chapter II, Lemma 2-9].

Now we have come to the first main result (cf. [**1**] and [**28**]).

THEOREM 3.1. *There exists a natural map*

$$\rho_{X/S}^{r,\nu} \colon F_S^\nu \mathrm{CH}^r(X) \to \Gamma(S, DJ_{X/S}^{r,\nu}),$$

where $DJ_{X/S}^{r,\nu}$ is the sheaf (of Abelian groups) defined below and the right hand side denotes the space of the sections of the sheaf. It has the following properties.

(i) *$\rho_{X/S}^{r,\nu}$ is functorial with respect to any base change $T \to S$ with T a smooth variety.*
(ii) $\rho_{X/S}^{r,\nu}(F_S^{\nu+1} \mathrm{CH}^r(X)) = 0$.
(iii) *For $\nu = 1$ $DJ_{X/S}^{r,1} = DJ_{X/S}^r$ and $\rho_{X/S}^{r,1}$ coincides with $\rho_{X/S}^r$, (the relative version of) the Abel-Jacobi map of Griffiths.*

The sheaf $DJ_{X/S}^{r,\nu}$ for $\nu \geq 2$ is defined to be the homology of the following complex

$$\Omega_S^{\nu-2} \otimes_{\mathcal{O}_S} H_{\mathcal{O}}^{2r-\nu}(X/S)/F^{r-\nu+2} \xrightarrow{\nabla} \Omega_S^{\nu-1} \otimes_{\mathcal{O}_S} H_{\mathcal{O}}^{2r-\nu}(X/S)/F^{r-\nu+1}$$
$$\xrightarrow{\nabla} \Omega_S^{\nu} \otimes_{\mathcal{O}_S} H_{\mathcal{O}}^{2r-\nu}(X/S)/F^{r-\nu},$$

where the maps are induced by the Gauss-Manin connection. In case $\nu = 1$ we replace the sheaf on the left hand side by $H_{\mathbb{Z}}^{2r-1}(X/S)(r)$ and define the map $H_{\mathbb{Z}}^{2r-1}(X/S)(r) \to H_{\mathcal{O}}^{2r-1}(X/S)/F^r$ to be induced by the natural inclusion

$$H_{\mathbb{Z}}^q(X/S)(r) \to H_{\mathcal{O}}^q(X/S)$$

and then get $DJ_{X/S}^{r,1} = DJ_{X/S}^r$ by the same definition. We call a section of $DJ_{X/S}^{r,\nu}$ a *higher normal function of type (r, ν) for X/S* and $\rho_{X/S}^{r,\nu}$ the *higher Abel-Jacobi map*.

We expect that the higher Abel-Jacobi map plays an important role in the attempt to answer the question (3-1) in case f is a "meaningful" fibration (note that in case S is a point $\rho_{X/S}^{r,\nu}$ is the zero map for $\nu \geq 2$). In the papers [**1**] and [**28**] the higher Abel-Jacobi map is investigated for the universal family of smooth complete intersections in a projective space.

In the end of this section we introduce the infinitesimal invariants of higher normal functions.

DEFINITION 3.2. (1) For each integer $\nu \geq 1$ we define $\nabla J^{r,\nu}_{X/S}$ to be the homology of the complex

$$\Omega^{\nu-1}_S \otimes_{\mathcal{O}_S} F^{r-\nu+1} H^{2r-\nu}_{\mathcal{O}}(X/S) \xrightarrow{\nabla} \Omega^{\nu}_S \otimes_{\mathcal{O}_S} F^{r-\nu} H^{2r-\nu}_{\mathcal{O}}(X/S) \xrightarrow{\nabla} \Omega^{\nu+1}_S \otimes_{\mathcal{O}_S} F^{r-\nu-1} H^{2r-\nu}_{\mathcal{O}}(X/S)$$

(2) Let $\delta_X : DJ^{r,\nu}_{X/S} \to \nabla J^{r,\nu}_{X/S}$ be the map defined as follows. For a local section $\alpha \in DJ^{r,\nu}_{X/S}$, let $\tilde{\alpha} \in \Omega^{\nu-1}_S \otimes_{\mathcal{O}_S} H^{2r-\nu}_{\mathcal{O}}(X/S)$ be its local lifting. By the transversality of the Gauss-Manin connection we see

$$\nabla(\tilde{\alpha}) \in \operatorname{Ker}\big(\Omega^{\nu}_S \otimes_{\mathcal{O}_S} F^{r-\nu} H^{2r-\nu}_{\mathcal{O}}(X/S) \to \Omega^{\nu+1}_S \otimes_{\mathcal{O}_S} H^{2r-\nu}_{\mathcal{O}}(X/S)\big).$$

We define $\delta_X(\alpha)$ to be the class of $\nabla(\tilde{\alpha})$ in $\nabla J^{r,\nu}_{X/S}$ which we can see that does not depend on the choice of the lifting $\tilde{\alpha}$.

(3) Let

$$\psi^{r,\nu}_{X/S} \colon F^{\nu}_S \operatorname{CH}^r(X) \to \Gamma(S, \nabla J^{r,\nu}_{X/S}).$$

be the composite of $\rho^{r,\nu}_{X/S}$ and δ_X. We call $\psi^{r,\nu}_{X/S}$ the infinitesimal invariant of $\rho^{r,\nu}_{X/S}$.

The definition of $\rho^{r,\nu}_{X/S}$ is based on the cycle map in Deligne cohomology (cf. [**7**]) which is a transcendental cohomology theory. On the other hand $\psi^{r,\nu}_{X/S}$ can be defined in a purely algebraic manner which is independent of $\rho^{r,\nu}_{X/S}$ (cf. [**28**, §2]). It is related to the cycle map defined by El Zein [**6**].

4. Zero-Cycles on Complete Intersections

Fix $\mathbf{d} = (d_1, \dots d_s)$ with d_i positive integers. Let

$$M(\mathbf{d}) \subset [\operatorname{Proj}(H^0(\mathbb{P}^n, \mathcal{O}(d_1))) \times \cdots \times \operatorname{Proj}(H^0(\mathbb{P}^n, \mathcal{O}(d_s)))]/\operatorname{PGL}_{n+1}(\mathbb{C}) \ltimes Q(\mathbf{d})$$

be the moduli space of smooth complete intersection of multi-degree $\mathbf{d}$ in $\mathbb{P}^n$, where

$$Q(\mathbf{d}) = \operatorname{Aut}(\bigoplus_{i=1}^{s} \mathcal{O}(d_i)).$$

Let $f \colon \mathcal{X}(\mathbf{d}) \to M(\mathbf{d})$ be the universal family. Let S be a smooth variety over $\mathbb{C}$ with a morphism $\pi \colon S \to M(\mathbf{d})$ and let $\mathcal{X}_S/S$ be the base change of $\mathcal{X}(\mathbf{d})/M(\mathbf{d})$ via π. Due to the Lefschetz theory one easily sees the following (cf. [**1**, Chapter II §5]).

LEMMA 4.1. *If* $2r - \nu \neq m := n - s$, *then* $\operatorname{Gr}^{\nu}_{F_S} \operatorname{CH}^r(\mathcal{X}_S) = 0$ *and* $DJ^{r,\nu}_{\mathcal{X}_S/S} = 0$.

Thus the only interesting case is that $2r - \nu = m$ for which we have the following.

THEOREM 4.1 (cf. [**1**, Chapter II, Theorem 5-3 and Remark 5-3-1]). *Assume that*

$$\pi \colon S \to M(\mathbf{d})$$

is dominant and smooth. Let r, $\nu \geq 1$ *be integers and assume either* $2r - \nu \neq m := n - s$ *or*

$$(*) \qquad d_{\min}(n - r - 1) \geq 2r + s + 1 \quad \text{with } d_{\min} = \min\{d_i | 1 \leq i \leq s\}$$

If $(r,\nu) \neq (m,m)$ *then* $\Gamma(S, DJ^{r,\nu}_{\mathcal{X}_S/S}) = 0$*, namely there exists no non-zero higher normal functions of type* (r,ν) *for* $\mathcal{X}_S/S$.

The case $\nu = 1$ of Theorem 4.1 is due to Green and Voisin [**8**] and Nagel [**21**] (cf. Theorem 5.2 in §5). The above result suggests us to expect that when $r \neq m$, $\mathrm{CH}^r(X) \otimes_{\mathbb{Z}} \mathbb{Q} \cong \mathbb{Q}$ for a general smooth complete intersection X of dimension m and multidegree $\mathbf{d}$ satisfying $(*)$. Nori [**22**] has given a quite similar conjecture based on his theorem of cohomological connectivity. Some more conjectures in this direction are given in [**28**, §6]. On the other hand we have seen (cf. Theorem 2.1 and Corollary 2.1) that $\mathrm{CH}^m(X)$ is fairly large for (not only general but also) any X as above. One naturally asks if the higher Abel-Jacobi map is able to capture zero-cycles on a general complete intersection. The following result partially answer this question. We note that Lemma 4.1 implies $F^m_S\,\mathrm{CH}^m(\mathcal{X}_S) = F^1_S\,\mathrm{CH}^m(\mathcal{X}_S)$ for $\mathcal{X}_S/S$ as above.

THEOREM 4.2 (cf. [**1**, Chapter I,I Theorem 0-6]). *Assume that* $\pi\colon S \to M(\mathbf{d})$ *is dominant and smooth. Let* $\alpha \in F^1_S\,\mathrm{CH}^m(\mathcal{X}_S)$ *with* $m = n - s$*. Assume* $m \geq 2$ *and that either of the following conditions holds.*

(i) α *is in general position* (*cf.* [**1**, Chapter II, Definition 5-4-1]) *and* $d_{\min}s \geq 2m + s$.

1. *Let* $\mathcal{X}(S)$ *be the set of sections of* $\mathcal{X}_S \to S$*. We can write*

$$\alpha = \sum_{i=1}^{\ell} n_i[p_i(S)] \in F^1_S\,\mathrm{CH}^m(\mathcal{X}_S), \quad p_i \in \mathcal{X}(S) \text{ for } 1 \leq i \leq \ell,$$

where $n_i \in \mathbb{Z}$ *with* $\sum_{i=1}^{\ell} n_i = 0$ *and we have*

$$(**) \qquad d_{\min}(s + \mu - 1) \geq 2m + s - \mu + 1 + \ell\mu \quad \text{for } 1 \leq \forall \mu \leq m - 1.$$

Then the assumption $\psi^{m,m}_{\mathcal{X}_S/S}(\alpha) = 0$ *implies that there exists an etale morphism* $\pi\colon U \to S$ *such that* $\pi^*\alpha = 0 \in \mathrm{CH}^m(\mathcal{X}_S \times_S U)$.

Theorem 4.2 implies the following corollary which is a modest evidence for Conjecture 1.2.

THEOREM 4.3. *Let* X *be a general complete intersection of dimension* $m \geq 2$ *and of multi-degree* $\mathbf{d}$*. Let* $\alpha \in \mathrm{CH}^m(X)$*. Assume either of the following conditions.*

(i) α *is in general position and* $d_{\min}s \geq 2m + s$.

(ii) α *is supported in a set of* ℓ *points of* X *and* $(**)$ *holds.*

If $\alpha \in F^{m+1}\,\mathrm{CH}^m(X)$ *then* $\alpha = 0$.

See [**1**, Chapter II, Definition 5-7-1] for the definition of an zero-cycle on X being in general position. For example, if X_o is a model of X over a field $k \subset \mathbb{C}$ which is finitely generated over $\mathbb{Q}$ and if $p_1, \ldots, p_t \in X(\mathbb{C})$ are a set of points whose images on X_o are independent generic points of X_o over k, then the class of a zero-cycle of the form

$$\sum_{i=1}^{t} n_i[p_i] \quad \text{with } n_i \in \mathbb{Z}$$

is in general position (cf. [**1**, Chapter II, Lemma 5-7-2]).

5. Higher Griffiths Groups

In this section we presents more examples of the higher Abel-Jacobi maps identifying algebraic cycles on hypersurfaces in $\mathbb{P}^n$. We first recall the following famous result which is a consequence of the pioneering work of Griffiths on normal functions. Recall that the Griffiths group for a projective smooth variety X is defined by

$$\mathrm{Grif}^r(X) = \mathrm{CH}^r(X)_{\mathrm{hom}}/\mathrm{CH}^r(X)_{\mathrm{alg}},$$

where $\mathrm{CH}^r(X)_{\mathrm{hom}}$ (resp. $\mathrm{CH}^r(X)_{\mathrm{alg}}$) denotes the subgroup of cycles homologically (resp. algebraically) equivalent to zero.

THEOREM 5.1 ([**11**]). *Let X be a general quintic 3-fold in $\mathbb{P}^4$. Then $\mathrm{Grif}^2(X)$ is not torsion.*

In [**28**] a generalization of Griffiths group is introduced and called the higher Griffiths group. We recall the definition. We adopt the filtration on Chow groups in Definition2.1. Let X be a projective smooth variety over $\mathbb{C}$.

DEFINITION 5.1. For integers $\nu \geq 0$ one defines

$$Z_0F^\nu \mathrm{CH}^r(X) = \sum_{W,\Gamma} \mathrm{Im}\big(\Gamma_*\colon F^\nu \mathrm{CH}_0(W) \to \mathrm{CH}^r(X)\big),$$

where W ranges over all smooth projective varieties and Γ over $\mathrm{CH}^r(W \times X)$. We have $Z_0F^\nu \mathrm{CH}^r(X) \subset F^\nu \mathrm{CH}^r(X)$. For $r = 1$ and we have

$$Z_0F^1 \mathrm{CH}^r(X) = \mathrm{CH}^r(X)_{\mathrm{alg}}.$$

We define

$$\mathrm{Grif}^{r,\nu}(X) := F^\nu \mathrm{CH}^r(X)/Z_0F^\nu \mathrm{CH}^r(X) + F^{\nu+1} \mathrm{CH}^r(X)$$

which we call a higher Griffiths group. When $\nu = 1$ $\mathrm{Grif}^{r,1}(X)$ is the quotient of the Griffiths group $\mathrm{Grif}^r(X)$ by the image of $F^2 \mathrm{CH}^r(X)$.

By Definition 2.3 we may write

$$\mathrm{Grif}^{r,\nu}(X) = \mathrm{Gr}_F^\nu \mathrm{CH}^r(X)/Z_0 \mathrm{Gr}_F^\nu \mathrm{CH}^r(X).$$

As we have seen in Corollary 2.2 the vanishing of $Z_0 \mathrm{Gr}_F^\nu \mathrm{CH}^r(X) \otimes_{\mathbb{Z}} \mathbb{Q}$ is (conjecturally) equivalent to that of

$$N_H^{r-\nu} H^{2r-\nu}(X,\mathbb{Q})/N_H^{r-\nu+1} H^{2r-\nu}(X,\mathbb{Q})$$

and the structure of $Z_0 \mathrm{Gr}_F^\nu \mathrm{CH}^r(X) \otimes_{\mathbb{Z}} \mathbb{Q}$ is expected to be determined in some way by the Hodge structure on the cohomology group (cf. [**3**, p. 0.5] and [**26**, §10]). On the other hand the relation between $\mathrm{Grif}^{r,\nu}(X)$ and Hodge structure seems rather mysterious. It is easy to see that $\mathrm{Grif}^{r,\nu}(X) \otimes_{\mathbb{Z}} \mathbb{Q}$ vanishes if

$$H^{2r-\nu}(X,\mathbb{Q})/N_H^{r-\nu} H^{2r-\nu}(X,\mathbb{Q})$$

does while the converse is not true due to Theorem 5.1 and the following theorem of Green and Voisin.

THEOREM 5.2 (cf. [**8**]). *Let $X \subset \mathbb{P}^4$ be a general hypersurface of degree > 5. Then the image of the Abel-Jacobi map $\mathrm{CH}^2(X)_{\mathrm{hom}} \to J^2(X)$ is torsion.*

REMARK 5.1. For a general hypersurface $X \subset \mathbb{P}^4$ of degree ≥ 5,

$$H^3(X,\mathbb{Q})/N_H^1 H^3(X,\mathbb{Q}) \neq 0.$$

Now we state some results from [**28**]. Fix integers n, $d > 0$. For a hypersurface $X \subset \mathbb{P}^n$ of degree d and a line ℓ contained in X we write

$$\alpha_\ell = d \cdot [\ell] - [X \cap \mathbb{P}^2] \in \mathrm{CH}_1(X) = \mathrm{CH}^{n-2}(X).$$

By Lemma 4.1 we see $\alpha_\ell \in F^{n-3}\,\mathrm{CH}^{n-2}(X)$.

THEOREM 5.3. (1) *Assume $n \geq 4$ and $2n-3 \geq d \geq n+1$. For a general hypersurface $X \subset \mathbb{P}^n$ of degree d, there exist a line $\ell \subset X$ such that α_ℓ is a non-torsion element of* $\mathrm{Grif}^{n-2,n-3}(X)$.

(2) *Let X be a general hypersurface of degree $d = 2n-3$ in $\mathbb{P}^n$. For any line $\ell \subset X$ and for any integer $N > 0$, $N \cdot \alpha_\ell$ is not contained in the subgroup*

$$\mathrm{CH}^{n-2}(X)_{\mathrm{alg}} + F^{n-2}\,\mathrm{CH}^{n-2}(X) \subset \mathrm{CH}^{n-2}(X).$$

(3) *Let $X \subset \mathbb{P}^6$ be a general cubic 5-fold. Let $Z \subset X$ be the section by a general hypersurface of degree $d \geq 4$. The restriction map induces the map*

$$\mathrm{CH}^3(X)_{\mathrm{alg}} \to \mathrm{Grif}^{3,2}(Z)$$

which is injective modulo torsion.

Note that Theorem 5.3(2) in case $n = 4$ is what Griffiths showed. Theorem 5.3(1) and (2) are compared to the results [**23**, Proposition3.3] and [**29**, Theorem 0.7]. We note that for X as in Theorem 5.3(3) the Abel-Jacobi map induces the isomorphism (cf. [**5**])

$$\mathrm{CH}^3(X)_{\mathrm{alg}} \xrightarrow{\cong} J^3(X)$$

where $J^3(X)$ is the intermediate Jacobian of X and is an Abelian variety of dimension 21 in this case (cf. [**18**, Lecture 9 (9.10)]).

The key to the proof of Theorem 5.3 is the following Theorem 5.4 concerning the non-vanishing of the higher Abel-Jacobi map. We fix an integer $k > 0$ with $1 \leq k \leq n-2$. Let

$$H_n \subset \mathrm{Proj}(H^0(\mathcal{O}_{\mathbb{P}^n}(d)))/\,\mathrm{PGL}_{n+1}$$

be the moduli space of smooth hypersurfaces $X \subset \mathbb{P}^n$ of degree d modulo projective equivalence and let

$$LH_n \subset [\mathrm{Proj}(H^0(\mathcal{O}_{\mathbb{P}^n}(d))) \times \mathrm{Gr}(k, \mathbb{P}^n)]/\,\mathrm{PGL}_{n+1}$$

be the moduli space of pairs (X, L) modulo projective equivalence where $L \subset \mathbb{P}^n$ is a k-plane and $X \subset \mathbb{P}^n$ is a smooth hypersurface of degree d containing L. Let

$$\begin{array}{ccc} (\mathcal{X} \hookleftarrow \mathcal{L}) & \longrightarrow & \mathcal{X} \\ \downarrow & & \downarrow \\ LH_n & \longrightarrow & H_n \end{array}$$

be the universal families over the moduli spaces, where $LH_n \to H_n; (X, L) \to X$. Put

$$\alpha_n = d \cdot [\mathcal{L}] - [\mathcal{C}_n] \in \mathrm{CH}^{n-1-k}(\mathcal{X}) \quad \text{with } \mathcal{C}_n = \iota^{-1}(\mathbb{P}^{k+1} \times LH_n),$$

where $\iota \colon \mathcal{X} \hookrightarrow \mathbb{P}^n \times LH_n$ is a fixed LH_n-embedding.

Let $\pi \colon S \to LH_n$ be a flat morphism and write $\mathcal{X}_S = \mathcal{X} \times_{LH_n} S$. Let $\pi^*(\alpha_n) \in \mathrm{CH}^{n-1-k}(\mathcal{X}_S)$ be the pull back of α_n. By Lemma 4.1 we have

$$\pi^*(\alpha_n) \in F_S^1\,\mathrm{CH}^{n-1-k}(\mathcal{X}_S) = F_S^{n-1-2k}\,\mathrm{CH}^{n-1-k}(\mathcal{X}_S).$$

THEOREM 5.4. *For any integer* m *write* $\delta_m = \binom{d+k}{k} - (k+1)(m-k)$. *Assume the following conditions.*

1. $n \geq 2k+2$.
2. $(k+1)d \geq \max\{0, \delta_{m-1}\} + 2m - 1 - 2k$ *for* $n \geq \forall m \geq 2k+2$.

For any dominant and flat morphism $\pi\colon S \to LH_n$, *we have* $\psi^{r,\nu}_{\mathcal{X}_S/S}(\pi^*(\alpha_n)) \neq 0$. *Here (cf. Definition 3.2)*

$$\psi^{r,\nu}_{\mathcal{X}_S/S}\colon F^\nu_S \operatorname{CH}^r(\mathcal{X}_S) \to \Gamma(S, \nabla J^{r,\nu}_{\mathcal{X}_S/S})$$

with $r = n-1-k$, $\nu = n-1-2k$.

REMARK 5.2. Assume $k = 1$. The conditions of Theorem 5.4 are equivalent to $n \geq 4$ and $d \geq n-1$.

We introduce additional inputs which are necessary for the proof of Theorem 5.3.

DEFINITION 5.2. Let S be a non-singular algebraic variety over $\mathbb{C}$ and let $\mathcal{X} \to S$ be a dominant smooth morphisms with geometrically connected fibers.

(1) We define

$$A_S \operatorname{CH}^r()X = \sum_{C,\Gamma} \operatorname{Im}\left(\Gamma_*\colon F^1_S \operatorname{CH}^1(C) \to \operatorname{CH}^r(\mathcal{X})\right),$$

where C ranges over proper smooth curves over S and Γ over $\operatorname{CH}^r(C \times_S \mathcal{X})$.

(2) We define

$$Z_0 F^\nu_S \operatorname{CH}^r()X = \sum_{W,\Gamma} \operatorname{Im}\left(\Gamma_*\colon F^\nu_S \operatorname{CH}^{d_W}(W) \to \operatorname{CH}^r(\mathcal{X})\right),$$

where W ranges over proper smooth schemes of relative dimension d_W over S and Γ over $\operatorname{CH}^r(W \times_S \mathcal{X})$.

The proof of the following fact is standard.

LEMMA 5.1. *Let* H *and* $\mathcal{X} \to H$ *be as in the beginning of this section. Let* X *be a general hypersurface of degree* d *in* $\mathbb{P}^n$ *and let* $\alpha \in \operatorname{CH}^r(X)$. *Assume* $\alpha \in \operatorname{CH}^r(X)_{\mathrm{alg}}$ *(resp.* $\alpha \in Z_0 F^\nu \operatorname{CH}^r(X)$*). Then there exists a dominant smooth morphism* $S \to H$, *a cycle* $\alpha_S \in Z_0 F^\nu_S \operatorname{CH}^r(\mathcal{X}_S)$ *(resp.* $\alpha_S \in A_S \operatorname{CH}^r(\mathcal{X}_S)$*) with* $\mathcal{X}_S = \mathcal{X} \times_H S$ *and a morphism* $\iota : \operatorname{Spec}(\mathbb{C}) \to S$ *such that* (X, α) *is isomorphic to the base change of* $(\mathcal{X}_S, \alpha_S)$ *via* ι.

Now another key ingredient of the proof of Theorem 5.3 is the following. Let H and $\mathcal{X}/H$ be as before. Let $\pi\colon S \to H_n$ be a morphism and $\mathcal{X}_S/S$ be the base change via π of $\mathcal{X}/H$. Fixing an integer $1 \leq k \leq (n-1)/2$ we consider the higher Abel-Jacobi map

$$\rho^{n-1-k,n-1-2k}_{\mathcal{X}/S}\colon F^{n-1-2k}_S \operatorname{CH}^{n-1-k}(\mathcal{X}_S) \to \Gamma(S, DJ^{n-1-k,n-1-2k}_{\mathcal{X}_S/S}).$$

Note that by Lemma 4.1 we have

$$F^\nu_S \operatorname{CH}^{n-1-k}(\mathcal{X}_S) = F^{\nu+1}_S \operatorname{CH}^{n-1-k}(\mathcal{X}_S) \quad \text{if } \nu \neq n-1-2k$$

so that $\rho^{n-1-k,\nu}_{\mathcal{X}_S/S}$ is the zero map unless $\nu = n-1-2k$.

PROPOSITION 5.1 ([**28**, §4 and §5]). *Assume that* π *is dominant and smooth.*

(1) *If* $dk \geq 2(n-1-k)+1$ *then*

$$\rho_{\mathcal{X}/S}^{n-1-k,n-1-2k}(A_S \operatorname{CH}^{n-1-k}(\mathcal{X}_S)) = 0.$$

(2) *If* $2k < n-1$ *and* $dk \geq n+1$ *then*

$$\rho_{\mathcal{X}/S}^{n-1-k,n-1-2k}(Z_0 F_S^{n-1-2k} \operatorname{CH}^{n-1-k}()X_S) = 0.$$

Theorem 5.3 now follows from Theorem 5.4 and Proposition 5.1 together with the result of [**4**].

References

1. M. Asakura and S. Saito, *Filtration on Chow groups and higher normal functions* (preprint).
2. A. A. Beilinson, *Height pairings between algebraic cycles*, *K*-theory, Arithmetic and Geometry (Moscow, 1984–1986), Lecture Notes in Math., vol. 1289 Springer, Berlin-New York, 1987, pp. 1–25.
3. S. Bloch, *Lectures on algebraic cycles*, Duke Univ. Math. Series, vol. 4, Duke University, Mathematics Department, Durham, 1980.
4. C. Borcea, *Deforming varieties of k-planes of projective complete intersections*, Pacific J. Math. **143** (1990), 25–36.
5. A. Collino, *The Abel-Jacobi isomorphism for the cubic fivefold*, Pacific J. Math. **122** (1986), 43–55.
6. F. El Zein, *Complexe dualisant et applications à la classe foundamentale d'un cycle*, Bull. Soc. Math. France, Mém., vol. 58, 1978.
7. H. Esnault and E. Viehveg, *Deligne-Beilinson cohomology*, Beilinson's Conjectures on Special Values of *L*-Functions, (M. Rapoport, M. Schappacher and P. Schneider, eds.), Perspect. Math., vol. 4, Academic Press, Boston, MA, 1988, pp. 43–91.
8. M. Green, *Griffiths infinitesimal invariant and the Abel-Jacobi map*, J. Differential Geom. **29** (1989), 545–555.
9. ______, *What comes after the Abel-Jacobi map?* (preprint).
10. ______, *Algebraic cycles and Hodge theory* (lectures at Banff), 1998.
11. P. Griffiths, *Periods of certain rational integrals.* I, Ann. of Math. (2) **90** (1969), 460–495; II, 496–541.
12. A. Grothendieck, *Hodge's general conjecture is false for trivial reasons*, Topology **8** (1969), 299–303.
13. U. Jannsen, *Mixed motives and algebraic K-theory*, Lecture Notes in Math., vol. 1400, Springer-Verlag, Berlin, 1990.
14. ______, *Motives, numerical equivalence, and semi-simplicity*, Invent. Math. **107** (1992), 447–452.
15. ______ *Mixed sheaves and filtrations on Chow groups*, Motives (U. Jannsen, S. Kleiman, and J.-P. Serre, eds.), Proc. Sympos. Pure Math., vol. 55, Part1, Amer. Math. Soc., Providence, RI, 1994 pp. 245–302..
16. S. Kleiman, *Algebraic cycles and the Weil conjectures*, Dix exposés sur la cohomologie des schémas, Adv. Stud. Pure Math., vol. 3 , North-Holland, Amsterdam, 1968, pp. 359–386.
17. J. D. Lewis, *Towards a generalization of Mumford theorem*, J. Math. Kyoto Univ. **29** (1989), 267–272.
18. ______, *A survey of the Hodge conjecture*, Université de Montral, Centre de recherches mathématiques, Montréal, QC, 1991.
19. D. Mumford, *Rational equivalence of 0-cycles on surfaces*, J. Math. Kyoto Univ. **9** (1969), 195–204.
20. J. P. Murre, *On a conjectural filtration on the Chow groups of algebraic variety*, 1992 (preprint).
21. J. Nagel, *The Abel-Jacobi map for complete intersections*, Indag. Math. (N.S.) **8** (1997), 95–113.
22. M. V. Nori, *Algebraic cycles and Hodge-theoretic connectivity*, Invent. Math. 349-373.
23. K. Paranjape, *Cohomological and cycle-theoretic connectivity*, Ann. of Math. (2) **139** (1994), 641–660.
24. A. A. Roitman, *Rational equivalences of zero-cycles*, Math. USSR-Sb. **18** (1974), 571–588.

25. H. Saito, *Generalization of Abel's theorem and some finiteness Property of zero-cycles on surfaces*, Compositio Math. **84** (1992), 289–332.
26. S. Saito, *Motives and filtration on Chow groups*, Invent. Math. **125** (1996), 149–196.
27. ______, *Motives and filtration on Chow groups*. II (preprint).
28. ______, *Filtration on Chow groups and Griffiths groups* (preprint).
29. C. Schoen, *On Hodge structures and nonrepresentability of Chow groups*, Compositio Math. **88** (1993), 285–316.
30. C. Voisin, *Variations de structure de Hodge et zéro-cycles sur les surfaces générales*, Math. Ann. **299** (1994), 77–103.

Department of Mathematics, Faculty of Sciences, Tokyo Institute of Technology, Oh-okayama, Meguro-ku, Tokyo, 152, JAPAN
E-mail address: sshuji@math.titech.ac.jp

Arithmetic Methods

Centre de Recherches Mathématiques
CRM Proceedings and Lecture Notes
Volume **24**, 2000

Picard-Fuchs Uniformization: Modularity of the Mirror Map and Mirror-Moonshine

Charles F. Doran

ABSTRACT. Motivated by a conjecture of Lian and Yau concerning the mirror map in string theory [**LY2, LY1**], we determine when the mirror map q-series of certain elliptic curve and K3 surface families are Hauptmoduln (genus zero modular functions). Our geometric criteria for modularity characterize orbifold uniformization properties of the Picard-Fuchs equations, effectively demystifying the *Mirror-Moonshine phenomenon*. The ingredients of the proof are largely classical in nature.

A longer, more comprehensive treatment of these results can be found in [**Dor2**]. For a detailed look at several explicit examples of this phenomenon, see the article by Verrill and Yui in this volume [**VY**].

CONTENTS

1.	Introduction	257
2.	Calabi-Yau Manifolds and the Mirror Map	260
3.	Orbifold Uniformization	263
4.	The Elliptic Curve Case	268
5.	M_n-polarized K3 surface case	272
6.	Generalizations	279
References		279

1. Introduction

Although the *mirror map* was introduced into string theoretic physics only in the last decade, whence it has lead to mathematical predictions of curve counts on certain Calabi-Yau threefolds, it is in fact reminiscent of standard mathematical constructions of the 19th century dating back at least to Legendre's work on elliptic functions. In a similarly classical vein such maps have long been of mathematical interest through their connection with modular functions, e.g., the Hauptmoduln of

1991 *Mathematics Subject Classification.* Primary 14D05, 11F03, 14J32; Secondary 30F35, 14J28, 83E30.

Received by the editors November 23, 1998; Accepted May 1, 1998.

This is the final form of the paper.

Fricke-Klein. Stimulated by the Mirror-Moonshine Conjecture of Lian-Yau ([**LY2, LY1**], Conjecture 5.15 below), in this paper we characterize families of elliptic curves and K3 surfaces for which the mirror map is *modular*, i.e., is a Hauptmodul.

The Lian-Yau conjecture was originally stated just for a small class of K3 surface pencils constructed by an "orbifold construction," but it was rapidly realized that their *Mirror-Moonshine* observations apply to a larger class of "torically defined" K3 surface families [**LY3**]. In fact, as we show here, the modularity phenomenon they observed can occur quite generally in families of elliptic curves with section and families of lattice polarized K3 surfaces. In particular, the phenomenon is not "toric" in origin, and can be characterized by other means. The strategy we follow to do so makes use of the existence of a one-dimensional (coarse) moduli space and a global Torelli theorem for each type of family. In the elliptic curve case the moduli space is just the J-line and the Torelli theorem is classical. For families of K3 surfaces with a lattice polarization such moduli and Torelli results were recently formalized by Dolgachev for his own investigations of mirror-symmetric families of K3 surfaces motivated by Arnold's "strange duality" [**Dol**].

We begin in §2 by introducing Calabi-Yau manifolds, Picard-Fuchs equations, and the mirror map. The periods of the holomorphic top forms for a family of Calabi-Yau manifolds define multivalued solutions to the associated *Picard-Fuchs differential equation*, a homogeneous Fuchsian ordinary differential equation on the base of the family. The *mirror map* is then a "local inverse" to the projectivized period mapping at a regular singular point with maximal unipotent monodromy.

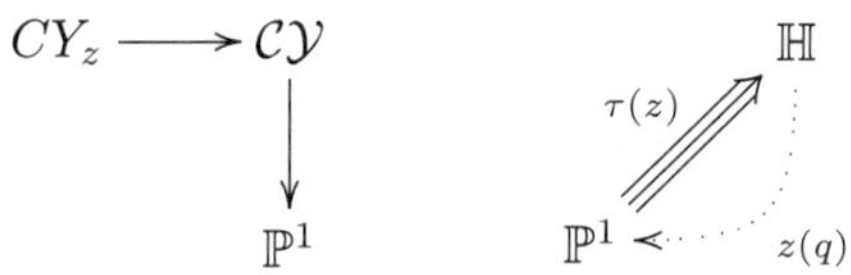

$q := e^{2\pi\imath\tau}$

In fact, since the mirror map depends only on the *projective* periods, it is insensitive to simultaneous rescalings of these periods. Thus it is really the *projective normal form* of the Picard-Fuchs differential equation which determines the mirror map.

Second order homogeneous Fuchsian ordinary differential equations in projective normal form have a particular connection with automorphic functions, dating back at least to Poincaré and Fricke-Klein. An automorphic function f, for a genus zero Fuchsian group of the first kind Γ, can always be realized as the single-valued inverse of the ratio of two multi-valued functions $u_1(x)$ and $u_2(x)$ on the curve $\Gamma\backslash\mathbb{H}^* \simeq \mathbb{P}^1$.

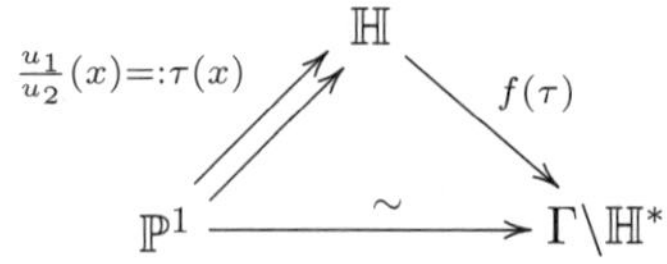

The functions u_i satisfy a second order Fuchsian ordinary differential equation in projective normal form that is naturally associated to f ([**Leh**] and §3.2 below). This equation is sometimes called the "Schwarzian equation" or (*orbifold*) *uniformizing differential equation* attached to the automorphic function. Note this description of f mimics the definition of the mirror map given above. The key observation for families of Calabi-Yau manifolds with order two Picard-Fuchs equations thus follows:

The mirror map is an automorphic function if and only if the projective normal form of the Picard-Fuchs differential equation is an orbifold uniformizing differential equation.

In fact, families of elliptic curves with section do have second order Picard-Fuchs equations (§4.2). For the families of M_n-polarized K3 surfaces, Picard-Fuchs equations are of order three, but they take a particular form as the *symmetric square* of an associated second order equation, a process which preserves the period ratio and hence the mirror map (§5.2). Thus, to solve the Mirror-Moonshine modularity problem we need only characterize, in terms of a "geometric" invariant of the family of elliptic curves with section (respectively M_n-polarized K3 surfaces), when the projective normal form of the Picard-Fuchs equation (respectively its *square root*) is a uniformizing differential equation attached to an automorphic function. We call this *Picard-Fuchs uniformization.*

In the case of elliptic curves, the natural geometric invariant to consider is Kodaira's *functional invariant* $\mathcal{J}(z)$, defined as the composition of the multi-valued period morphism $\tau(z) = \omega_1(z)/\omega_0(z)$ to the upper half plane and the elliptic modular function $J(\tau)$ to the J-line. In other words, the functional invariant associates to each point of the base of the family the J-invariant of its elliptic curve fiber.

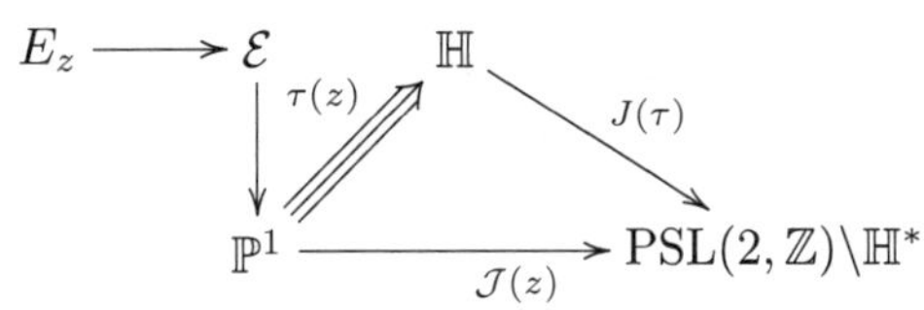

Kodaira proves that $\mathcal{J}(z)$ is a rational function. We consider the "pullback by $\mathcal{J}$" of the uniformizing differential equation Λ for $\mathrm{PSL}(2,\mathbb{Z})$. By the rationality of $\mathcal{J}(z)$, the resulting differential equation $\mathcal{J}^*(\Lambda)$ agrees with the Picard-Fuchs equation up to projective equivalence (Proposition 4.8). Thus we need only determine the conditions on the rational function $\mathcal{J}(z)$ for the projective normal form of $\mathcal{J}^*(\Lambda)$ to be a uniformizing differential equation. We accomplish this by applying our general criterion, Theorem 3.26 and Corollary 3.28, which characterizes rational functions that pullback one projective equivalence class of uniformizing differential equations to another. The main result is

THEOREM 4.10. *The projective normal form of the Picard-Fuchs equation of a Weierstrass elliptic surface will orbifold uniformize if and only if there are no apparent singularities and the functional invariant $\mathcal{J}(t)$ has zeros to orders $= 1$ or $\equiv 0 \bmod 3$ and ones to orders $= 1$ or $\equiv 0 \bmod 2$.*

Next, in §5, we extend our criterion to the case of M_n-polarized families of K3 surfaces. Dolgachev has shown that the (coarse) moduli space for M_n-polarized K3 surfaces is uniformized by the group $\Gamma_0(n)^*$ obtained by adding the Fricke involution to the congruence subgroup $\Gamma_0(n)$ [**Dol**]. By analogy with the elliptic curve case, we introduce the *generalized functional invariant* $\mathcal{H}_n(z)$ as the composition of the multi-valued truncated period morphism $\tau(z) = \Omega_1(z)/\Omega_0(z)$ to the upper half plane and the normalized Hauptmodul $H_n(\tau)$ for $\Gamma_0(n)^*$, and establish its rationality (§5.4).

$$\begin{array}{ccccc} K3_z & \longrightarrow & \mathcal{K}3 & & \mathbb{H} & & \\ & & \Big\downarrow & \overset{\tau(z)}{\nearrow\!\!\!\nearrow\!\!\!\nearrow} & & \overset{H(\tau)}{\searrow} & \\ & & \mathbb{P}^1 & & \xrightarrow[\mathcal{H}(z)]{\qquad\qquad} & & \Gamma_H \backslash \mathbb{H}^* \end{array}$$

The criterion from §3.3 then allows us to determine the conditions on $\mathcal{H}_n(z)$ to pullback the uniformizing differential equation for $\Gamma_0(n)^*$ to another uniformizing differential equation up to projective equivalence. As a consequence we obtain the desired characterization of the Lian-Yau Mirror-Moonshine phenomenon

THEOREM 5.20. *The square root of the projective normal form of the Picard-Fuchs equation of a one parameter family of M_n-polarized K3 surfaces (with genus zero $\Gamma_0(n)^*$) will orbifold uniformize if and only if it has no apparent singularities, and for each elliptic point $p \in X_0(n)^*$, $\mathcal{H}_n(z) - p$ vanishes to permissible orders as listed in Table* 1.

Finally, in §6, a number of natural generalizations of this work are discussed. Our Picard-Fuchs uniformization results for M_n-polarized K3 surface families extend *mutatis mutandis* to the general rank 19 lattice polarized case. We note that the genus-zero restriction on both the uniformizing groups and the base of the families of elliptic curves and K3 surfaces involved can be removed. Moreover, there is an extension of our modularity criteria to certain multi-parameter families of Abelian varieties and K3 surfaces of various Picard ranks. In fact, there are one-parameter families of Picard rank 18 K3 surfaces whose mirror maps are "bimodular." These and other generalizations are treated in [**Dor2**].

REMARK 1.1. This subject is inherently multi-disciplinary, and our proof requires that we recall a number of results from several classical theories. Throughout this paper we assume some basic familiarity with the theory of Fuchsian ordinary differential equations: regular singular points and their classification into types (i.e., generic, logarithmic, apparent), characteristic exponents, Frobenius' method, etc. A favorite reference is Yoshida's book [**Yos**] which also includes the theory of (orbifold) uniformization. Other nice presentations of the classical Fuchsian theory, with quite different applications, can be found in the survey paper by Varadarajan [**Var**] and the book by Anosov and Bolibruch [**AB**].

REMARK 1.2. This article is intended to complement the explicit constructions described by Helena Verrill and Noriko Yui in this proceedings [**VY**]. By providing a simple, geometric criterion for the Lian-Yau mirror-moonshine phenomenon which applies equally well to both elliptic curve and K3 surface families, we hope at once to "demystify" such examples of this phenomenon, and to highlight the importance of explicit constructions in answering questions which remain. Taking advantage of the opportunity such a pairing affords, we refer the reader to their paper for several explicit examples and extra discussion of related subject matter.

2. Calabi-Yau Manifolds and the Mirror Map

In this section we quickly review Calabi-Yau manifolds, Picard-Fuchs equations, and the mirror map. The definitions and concepts are stated without proof, and a single motivational example is provided: that of the universal family of elliptic curves over the J-line. Perhaps the chief novelty here is the emphasis on the role played by the "projective normal form" Fuchsian ordinary differential equation. Although our discussion applies quite generally to families of Calabi-Yau varieties, we will only apply it in this paper to the elliptic curve and K3 surface cases.

2.1. Calabi-Yau Manifolds.

DEFINITION 2.1. A *Calabi-Yau manifold* M is a compact complex manifold with trivial canonical bundle.

EXAMPLE 2.2. A one dimensional Calabi-Yau manifold is an elliptic curve. A simply connected Calabi-Yau manifold of dimension two is a K3 surface.

Up to complex scaling, a Calabi-Yau n-fold has a unique holomorphic n-form

$$\omega_{n,0} \in H^n(M, \mathbb{C}).$$

DEFINITION 2.3. The *periods* of this holomorphic n-form are the numbers

$$p_\gamma(M) := \int_\gamma \omega_{n,0}$$

as γ runs through the n-cycles of M. Fix an ordered basis of n-cycles $\{\gamma_1, \ldots, \gamma_k\}$, for k the nth Betti number of M. Then the periods of M define the vector

$$[p_{\gamma_i}(M)]_{i=1}^k \in \mathbb{P}^{k-1}$$

the (projective) *period point* of M.

EXAMPLE 2.4. Consider the elliptic curve $y^2 = 4x^3 - g_2 x - g_3$ in Weierstrass form. Here $n = 1$ and $k = 2$. The holomorphic one form is the usual differential of the first kind dx/y.

2.2. Period Mapping and Picard-Fuchs Equations.

DEFINITION 2.5. Given a family $\pi\colon X \to S$ of Calabi-Yau manifolds X_s, by continuous extension of the basis of n-cycles the periods of the Calabi-Yau fibers define a vector of k multi-valued functions on the base manifold S

$$[p_{\gamma_i(s)}(X_s)]_{i=1}^k \colon S \to \mathbb{P}^{k-1}$$

the *period mapping* associated with the family.

EXAMPLE 2.6. Consider the family $\mathcal{E}$ of elliptic curves over $\mathbb{P}^1$ defined by the equation

$$\mathcal{E} : y^2 = 4x^3 - \frac{27s}{s-1}x - \frac{27s}{s-1}.$$

The periods of the form dx/y may be given in terms of the hypergeometric function ${}_2F_1$ (see [**Sti**, pp. 232–233] for explicit expressions).

Suppose now that $S = \mathbb{P}^1$. The functions $p_{\gamma(s)}(X_s)$ satisfy an ordinary differential equation with regular singular points (i.e., a *Fuchsian* ODE) [**BP**]

$$\frac{d^k f}{ds^k} + P_1(s)\frac{d^{k-1} f}{ds^{k-1}} + \cdots + P_k(s) f = 0, \quad P_i(s) \in \mathbb{C}(s). \tag{2.1}$$

DEFINITION 2.7. The differential equation (2.1) is called the *Picard-Fuchs differential equation* of the family $\pi\colon X \to \mathbb{P}^1$.

EXAMPLE 2.8. For the family $\mathcal{E}$ in Example 2.6, the Picard-Fuchs equation is

$$\frac{d^2 f}{ds^2} + \frac{1}{s}\frac{df}{ds} + \frac{(31/144)s - 1/36}{s^2(s-1)^2} f = 0.$$

There is a basis of solutions with local monodromies G_0, G_1, G_∞ about the regular singular points $\{0, 1, \infty\}$ respectively, where

$$G_0 = \begin{pmatrix} 1 & 1 \\ -1 & 0 \end{pmatrix}, \quad G_1 = \begin{pmatrix} 0 & -1 \\ 1 & 0 \end{pmatrix}, \quad G_\infty = \begin{pmatrix} 1 & 1 \\ 0 & 1 \end{pmatrix}.$$

2.3. Points of Maximal Unipotent Monodromy and the Mirror Map.

DEFINITION 2.9. A regular singular point of a Fuchsian ordinary differential equation is called a *point of maximal unipotent monodromy* if the local monodromy matrix G is such that $G - I_k$ is nilpotent with exact order k.

In a neighborhood of a point of maximal unipotent monodromy, Frobenius' method tells us that there is a basis of solutions such that the first is holomorphic at the point, the second has logarithmic behavior, the next behaves like $\log^2$, ... , up to $\log^{k-1}$.

EXAMPLE 2.10. The point of maximal unipotent monodromy in Example 2.8 is $\infty \in \mathbb{P}^1$.

REMARK 2.11. Not every family of Calabi-Yau manifolds will have a Picard-Fuchs differential equation with a point of maximal unipotent monodromy. For a class of *torically constructed* hypersurfaces in toric Fano varieties, the existence of points of maximal unipotent monodromy was established by Hosono, Lian, and Yau [**HLY**].

Consider a family of Calabi-Yau manifolds $\pi\colon X \to \mathbb{P}^1$, whose Picard-Fuchs equation has a point of maximal unipotent monodromy. In a neighborhood of such a point consider the truncated period vector consisting only of the holomorphic solution and the logarithmic solution

$$[p_{\text{hol}}(s) : p_{\log}(s)]\colon \mathbb{P}^1 \to \mathbb{P}^1.$$

If the image lies in the upper half plane $\mathbb{H} \subset \mathbb{P}^1$, then, possibly after composition with projective linear transformations so that the singular point lies at $0 \in \mathbb{P}^1$ and maps to $\imath\infty \in \mathbb{H}^* \subset \mathbb{P}^1$, we can consider the q-series for the local inverse mapping

$$z(q(\tau))\colon \mathbb{H} \to \mathbb{P}^1, \quad q(\tau) = e^{2\pi\imath\tau}.$$

DEFINITION 2.12. This q-series $z(q)$ is called the *mirror map* of the family $\pi\colon X \to \mathbb{P}^1$ about the point of maximal unipotent monodromy.

EXAMPLE 2.13. For the family $\mathcal{E}$ of Example 2.6 the mirror map is quite familiar. Since the maximal unipotent monodromy point is at ∞, we change variables first to $z = 1/s$. The single-valued local inverse to the period mapping is then the reciprocal of the q-series for the elliptic modular function $J(q)$

$$J(q) = \frac{1}{q} + 744 + 196884q + 21493760q^2 + O(q^3),$$

$$z(q) = \frac{1}{J(q)} = q - 744q^2 + 356652q^3 - 140361152q^4 + O(q^5).$$

REMARK 2.14. The period mapping is defined as a map to projective space. If one is interested in the mirror map it is often preferable to consider the Picard-Fuchs differential equation only up to "projective equivalence."

DEFINITION 2.15. The *projective normal form* of a Fuchsian ordinary differential equation (e.g., that in equation (2.1) above) is the unique Fuchsian ordinary differential equation without a $(k-1)$st order derivative

$$\frac{d^k g}{ds^k} + R_2(s)\frac{d^{k-2}g}{ds^{k-2}} + \cdots + R_k(s)g = 0, \quad R_i(s) \in \mathbb{C}(s)$$

whose fundamental solutions define the same projective period map as that of equation (2.1).

It is always possible to pass to the projective normal form differential equation by rescaling each fundamental solution of the original equation by the kth root of the Wronskian.

EXAMPLE 2.16. Suppose now that $k = 2$, i.e., the initial differential equation is

$$\frac{d^2 f}{ds^2} + P_1(s)\frac{df}{ds} + P_2(s)f = 0,$$

then the projective normal form of this differential equation takes the particularly simple form

$$\frac{d^2 g}{ds^2} + \left(P_2(s) - \frac{1}{2}P_1{}'(s) - \frac{1}{4}P_1(s)^2\right) g = 0.$$

As the process of taking the projective normal form does not alter the position or type of a maximal unipotent monodromy point, and as the projective solution determines the mirror map there, we find

LEMMA 2.17. *The mirror map of a family of Calabi-Yau manifolds about a point of maximal unipotent monodromy of the Picard-Fuchs equation is determined by the projective normal form of this differential equation.*

3. Orbifold Uniformization

In this section we recall the classical theory of *orbifold uniformization* of Riemann surfaces by Fuchsian second order ordinary differential equations, including some basic facts about the Schwarzian derivative, orbifold uniformizing differential equations, and automorphic functions. In particular, there is a classical characterization of orbifold uniformizing differential equations in terms of characteristic exponent differences (Lemma 3.22). The characteristic exponents of a "pullback" Fuchsian second order differential equation $\mathcal{R}^*(L_2)$ are determined by those of the original differential equation L_2 and the ramification of the rational function $\mathcal{R}(z)$ (Theorem 3.26). Hence one can characterize rational functions which pullback one projective equivalence class of uniformizing differential equations to another.

3.1. Uniformization of Orbifolds.

DEFINITION 3.1. Let X be a complex manifold, $Y \subset X$ a hypersurface, $Y = \bigcup_j Y_j$ its decomposition into irreducible components. Let the numbers b_j, either ∞ or an integer ≥ 2, be the *weights* attached to the corresponding Y_j. The triple $(X, Y, \vec{b})$ is called an *orbifold* if for every point in $X \setminus \cup_j\{Y_j \mid b_j = \infty\}$ there is an open neighborhood U and a covering manifold which ramifies along $U \cap Y$ with the given indices $\vec{b}$.

DEFINITION 3.2. The orbifold $(X, Y, \vec{b})$ is called *uniformizable* if there is a simply connected global covering manifold, or *uniformization*, of X with the given ramification data $(Y, \vec{b})$.

REMARK 3.3. Often when one speaks of uniformization in classical settings it is understood that all the $b_j = \infty$ (i.e., only cusps occur). We distinguish this from the definition given above by calling this more restrictive notion *classical uniformization*. For emphasis we will often refer to uniformization where b_j is allowed to be finite as well as ∞ as *orbifold uniformization*.

DEFINITION 3.4. Let $(X, Y, \vec{b})$ be an orbifold and M its uniformization. The multivalued inverse map $X \to M$ of the projection $M \to X$ is called the *developing map*, uniquely determined up to the group $\mathrm{Aut}(M)$ of holomorphic automorphisms of M.

For our application, we take $X = \mathbb{P}^1(\mathbb{C})$, $Y =$ a finite set of points, $M =$ the upper half plane $\mathbb{H}$, and $\mathrm{Aut}(M) = \mathrm{PSL}(2, \mathbb{R}) \simeq \mathrm{PU}(1,1)$.

3.2. Schwarzian Derivatives, Uniformizing Differential Equations, and Automorphic Functions.

DEFINITION 3.5. The *Schwarzian derivative* $\{w; x\}$ of a nonconstant smooth function $w(x)$, with respect to x, is defined to be

$$\{w; x\} = \frac{2w'w''' - 3(w'')^2}{2(w')^2}.$$

PROPOSITION 3.6 ([**Yos**, Proposition 4.1.1]). *The Schwarzian derivative satisfies "*$\mathrm{PGL}(2, \mathbb{C})$*-invariance":*

$$\left\{\frac{aw+b}{cw+d}; x\right\} = \{w; x\}, \quad \textit{for all } \begin{pmatrix} a & b \\ c & d \end{pmatrix} \in \mathrm{PGL}(2, \mathbb{C})$$

DEFINITION 3.7. A function $w(x)$ is called $\mathrm{PGL}(2, \mathbb{C})$*-multivalued* if any two branches of $w(x)$ are related by a projective transformaion.

An immediate corollary of Proposition 3.6 is

COROLLARY 3.8. *If a function $w = w(x)$ is $\mathrm{PGL}(2, \mathbb{C})$-multivalued, then the Schwarzian derivative $\{w; x\}$ is single valued.*

The relationship of the Schwarzian derivative to orbifolds is given by the following result of Poincaré:

PROPOSITION 3.9 ([**For**, §4.4, Theorem 15]). *Let x and w be the coordinates of X and M respectively. There are two linearly independent solutions u_1 and u_2 of the equation*

$$\frac{d^2u}{dx^2} + \frac{\{w(x); x\}}{2w'(x)^2} u = 0 \tag{3.1}$$

with single valued coefficients such that $w(x) = u_1(x)/u_2(x)$.

DEFINITION 3.10. The differential equation (3.1) is called the *(orbifold) uniformizing differential equation* of the orbifold $(X, Y, \vec{b})$.

An immediate corollary of Proposition 3.9 is

COROLLARY 3.11. *The projective solution of the uniformizing differential equation of the orbifold* $(X, Y, \vec{b})$ *is the developing map* $w(x)$.

Let $\Gamma \subset \mathrm{PSL}(2, \mathbb{R})$ be a discrete subgroup such that $\Gamma \backslash \mathbb{H}^*$ is compact (i.e., a Fuchsian group of the first kind). In our applications we will be primarily interested in a particular class of orbifold uniformizing differential equations associated to such groups Γ with $\Gamma \backslash \mathbb{H}^*$ of genus zero.

DEFINITION 3.12. A function $f(\tau)$ on the upper half plane $\mathbb{H}$ is an *automorphic function* for Γ if

$$f(\gamma\tau) = f(\tau), \quad \text{for all } \gamma \in \Gamma \text{ and for all } \tau \in \mathbb{H},$$

(i.e., invariance) and there are well defined limits of $f(\tau)$ as τ approaches parabolic vertices (i.e., $f(\tau)$ is meromorphic at each cusp of Γ).

DEFINITION 3.13. A univalent automorphic function f for a genus zero Fuchsian group is called a *Hauptmodul*.

EXAMPLE 3.14. The covering map for the uniformization $\mathbb{H} \to \Gamma \backslash \mathbb{H}^*$ is a Hauptmodul if Γ is genus zero. In particular, for the genus zero group $\mathrm{PSL}(2, \mathbb{Z})$ the modular function J is a Hauptmodul. The orbifold uniformizing differential equation is $\Lambda u = 0$, where

$$\Lambda := \frac{d^2}{dx^2} + \frac{36x^2 - 41x + 32}{144x^2(x-1)^2}.$$

REMARK 3.15. The ambiguity in choice of a Hauptmodul is sometimes fixed by choosing a *normalized* Hauptmodul. For example, if the transformation $\tau \mapsto \tau + 1$ is in Γ, then the unique Hauptmodul with q-series about ∞ of the form

$$\frac{1}{q} + \sum_{n=1}^{\infty} a_n q^n$$

is commonly chosen. In particular, for the genus zero group $\mathrm{PSL}(2, \mathbb{Z})$ the normalized Hauptmodul is $J(q) - 744$.

THEOREM 3.16 ([**Leh**, Theorem 6A]). *Let* Γ *be a Fuchsian group of the first kind. If* f, g *are automorphic functions for* Γ, *then they satisfy an algebraic equation*

$$\mathcal{P}(f, g) = 0$$

with complex coefficients.

DEFINITION 3.17. Such an algebraic equation satisfied by a pair of automorphic functions f and g for Γ is known classically as a *modular equation*. Warning: Often in the literature this name is reserved for the particular case of modular equations relating the modular function $J(\tau)$ and $J(n \cdot \tau)$, $n \in \mathbb{N}$ (see [**Coh, CY**] for explicit examples).

An easy corollary of Theorem 3.16 is

COROLLARY 3.18 ([**Leh**, Theorem 6B]). *If there is a univalent automorphic function* f *for* Γ *(i.e., one to one on the fundamental domain), then every automorphic function for* Γ *is a rational function of* f.

Specializing to the case of Hauptmoduln we find

COROLLARY 3.19. *Let $\Phi \subseteq \Gamma$ (finite index) be a pair of genus zero Fuchsian groups of the first kind, with respective Hauptmoduln f, g. Then the rational function $\mathcal{R}(z)$ from Corollary 3.18 expresses g in terms of f:*

$$g = \mathcal{R}(f).$$

DEFINITION 3.20. The rational function $\mathcal{R}(z)$ is called the *modular relation* between the Hauptmoduln f and g.

For several examples of modular relations between $J(\tau)$ and Hauptmoduln of certain modular groups with level structure, see [**CY**].

REMARK 3.21. The orbifold uniformizing differential equation is, up to projective equivalence, the "pullback by $\mathcal{R}$" of that for $\Gamma\backslash\mathbb{H}^*$. In the next subsection we describe necessary and sufficient conditions on a rational function for it to pullback one uniformizing differential equation to (the projective equivalence class of) another.

3.3. Regular Singular Points and Orbifold Uniformization. There is a particularly simple characterization of projective normal form equations which orbifold uniformize, given in terms of their characteristic exponents:

LEMMA 3.22 ([**Yos**, p. 50]). *A second order Fuchsian ordinary differential equation in projective normal form is an orbifold uniformizing differential equation if and only if at each of its regular singular points the difference in characteristic exponents is either 0 or $1/b$ for some $b \geq 2 \in \mathbb{Z}$.*

DEFINITION 3.23. A regular singular point satisfying the conditions of Lemma 3.22 is said to be of *orbifold type.*

As we have already indicated, it is important that we understand the behavior of uniformizing differential equations under pullback. Suppose then that we pullback a second order Fuchsian equation $L_2 f = 0$, with

$$L_2 = \frac{d^2}{dx^2} + P_1(x)\frac{d}{dx} + P_2(x),$$

by a rational function $x = \mathcal{R}(z)$.

PROPOSITION 3.24. *The regular singular points of the pulled back equation $\mathcal{R}^*(L_2)$ lie among the inverse image points of the regular singular points of the original equation L_2 and the extra ramification points of the rational function $\mathcal{R}(z)$.*

PROOF. This is a straightforward computation:

$$\frac{df}{dx} = \frac{df}{dz} \Big/ \frac{d\mathcal{R}}{dz}$$

and

$$\frac{d^2 f}{dx^2} = \left(\frac{d\mathcal{R}}{dz}\right)^{-2} \left(\frac{d^2 f}{dz^2} - \frac{df}{dz}\left(\frac{d^2\mathcal{R}}{dz^2} \Big/ \frac{d\mathcal{R}}{dz}\right)\right)$$

so that the pullback $\mathcal{R}^*(L_2) = 0$ of the equation $L_2 f = 0$ takes the form

$$\frac{d^2 f}{dz^2} + \frac{df}{dz}\left(P_1(\mathcal{R}(z))\frac{d\mathcal{R}}{dz} - \left(\frac{d^2\mathcal{R}}{dz^2} \Big/ \frac{d\mathcal{R}}{dz}\right)\right) + f\left(P_2(\mathcal{R}(z))\left(\frac{d\mathcal{R}}{dz}\right)^2\right) = 0.$$

Thus the only poles in the coefficients can occur at points where $P_1(\mathcal{R}(z))$ has one, or where $P_2(\mathcal{R}(z))$ has one, or at other places where $d\mathcal{R}/dz = 0$ (i.e., the *extra* ramification points of the $\mathcal{R}$ map). □

LEMMA 3.25. *The characteristic exponents of a point in the inverse image as above are exactly those of the point being pulled back times the order of ramification. The characteristic exponents at an extra ramification point differ by an integer.*

PROOF. For simplicity assume that the singular point of L_2 under consideration is at $x = 0$. In general (in the absence of ramification) there are $\deg \mathcal{R}(z)$ points in the inverse image of 0, each with the same characteristic exponents as has L_2 at 0. If ramification of $\mathcal{R}(z)$ occurs at an inverse image point z_0, then $\mathcal{R}(z)$ has a multiple root there with multiplicity equal to the order of ramification. Thus the characteristic exponents at z_0 are those of L_2 at 0 times this order of ramification. At an *extra* ramification point z_1, i.e., a point in the inverse image of an ordinary point of L_2 (characteristic exponents $0, 1$), the characteristic exponents are thus 0 and the ramification index of $\mathcal{R}(z)$ at z_1. □

In light of Lemmas 3.22 and 3.25, we have

THEOREM 3.26. *Characterization of types of regular singular points of a pull-back of an orbifold uniformizing differential equation by a rational function:*

- *The extra ramification points of the rational function are all apparent singularities.*
- *The points in the inverse image of logarithmic singularities of orbifold type (i.e., those with equal characteristic exponents) are logarithmic singularities of orbifold type.*
- *For a regular singular point in the inverse image of a finite order orbifold point with characteristic exponent difference $1/b$: if the characteristic exponent difference is integral (i.e., if b divides the ramification index), then the regular singular point is apparent; if the characteristic exponent difference is not integral (i.e., if b does not divide the ramification index), then the singular point is generic.*

REMARK 3.27. Some comments on the proof of this theorem: The regular singular points which are extra ramification points have characteristic exponent difference an integer ≥ 2 (if the difference were equal to 1 this would be an ordinary point [**Yos**, p. 23]). The characteristic exponent difference is rescaled by the degree of the rational function; if this difference is zero it remains so. In the last case, we can say more about the characteristic exponent difference: by Lemma 3.25 we know that this is the ramification degree r divided by b. The reduced form of this fraction has numerator $r/\gcd(r, b)$.

COROLLARY 3.28. *Let L_2 be an orbifold uniformizing differential equation, $\mathcal{R}(z)$ a rational function, and z_0 a generic regular singular point of $\mathcal{R}^*(L_2)$. The ramification degree r of $\mathcal{R}$ at z_0 divides the order of the orbifold point $\mathcal{R}(z_0)$ for L_2 if and only if z_0 is a point of orbifold type of $\mathcal{R}^*(L_2)$ with order b/r.*

Theorem 3.26 can be interpreted as providing a characterization of the modular relations between Hauptmoduln of a given genus zero Fuchsian group of the first kind and its finite index subgroups.

4. The Elliptic Curve Case

We are now ready to consider the problem of orbifold uniformization by Picard-Fuchs equations for families of elliptic curves—the first case of *Picard-Fuchs uniformization.* We will obtain a simple characterization of elliptic surfaces with modular mirror maps in terms of Kodaira's *functional invariant.* This rational function assigns to each point in the base of an elliptic fibration the J-invariant of the corresponding elliptic curve fiber. Moreover, the projective normal form of the pullback by the functional invariant of the uniformizing differential equation for the elliptic modular function $J(\tau)$ is the projective normal form of the Picard-Fuchs equation of the elliptic fibration. Thus the mirror map is determined by the functional invariant, and using Theorem 3.26 we obtain a criterion for modularity of the mirror map in terms of properties of the functional invariant alone.

4.1. Weierstrass Elliptic Surfaces and Kodaira Theory.

DEFINITION 4.1. Let $\pi\colon X \to C$ be a flat proper map from a reduced irreducible $\mathbb{C}$-scheme X to a complete, smooth curve C, such that every geometric fiber is an irreducible curve of arithmetic genus one, i.e., each fiber is one of

1. an elliptic curve, or
2. a rational curve with a node, or
3. a rational curve with a cusp.

The total space X is normal, and the generic fiber of π is smooth. Assume further that a section $s : C \to X$ is given, not passing through the nodes or cusps of the fibers. Call the collection of such data $(\pi\colon X \to C, s)$ a *Weierstrass fibration* over C. We may resolve the singularities of X to obtain an elliptic surface with section $\bar{\pi}\colon \overline{X} \to C$, called the *induced elliptic surface.*

In fact there is a *canonical form* for such a Weierstrass fibration, exhibiting X as a divisor in a $\mathbb{P}^2$-bundle over the base curve C.

THEOREM 4.2 ([**Mir1**, Theorem 2.1]). *Let Σ denote the given section of π, i.e., $\Sigma = s(C)$, a divisor on X which is taken isomorphically onto C by π. Let $\mathcal{L} = \pi_*[\mathcal{O}_X(\Sigma)/\mathcal{O}_X]$. Suppose that the general fiber of π is smooth. Then $\mathcal{L}$ is invertible and X is isomorphic to the closed subscheme of $\mathbb{P} = \mathbb{P}(\mathcal{L}^{\otimes 2} \oplus \mathcal{L}^{\otimes 3} \oplus \mathcal{O}_Y)$ defined by*

$$y^2 z = 4x^3 - g_2 x z^2 - g_3 z^3,$$

where

$$g_2 \in \Gamma(C, \mathcal{L}^{\otimes -4}), \quad g_3 \in \Gamma(C, \mathcal{L}^{\otimes -6}),$$

and $[x, y, z]$ is the global coordinate system of $\mathbb{P}$ relative to $(\mathcal{L}^{\otimes 2}, \mathcal{L}^{\otimes 3}, \mathcal{O}_C)$. Moreover the pair (g_2, g_3) is unique up to isomorphism, and the discriminant

$$g_2^3 - 27g_3^2 \in \Gamma(C, \mathcal{L}^{\otimes -12})$$

vanishes at a point $q \in C$ precisely when the fiber X_q is singular.

DEFINITION 4.3. Let $\mathcal{J}$ denote the composition of the period morphism $\omega_1/\omega_0\colon C \to \mathbb{H}$ and the morphism $J\colon \mathbb{H} \to \mathbb{P}^1$ extending the classical modular function, i.e., $\mathcal{J} = J \circ \omega_1/\omega_0$. Kodaira calls this the *functional invariant* of $\pi\colon X \to C$.

Kodaira has classified the singular fiber types which can arise in Weierstrass fibered elliptic surfaces.

THEOREM 4.4 ([**Kod**]). *The singular fibers which appear in a smooth minimal elliptic surface fall into "types":* I_n $(n \geq 0)$, II, III, IV, I_n^* $(n \geq 0)$, IV^*, III^*, *and* II^*. *Let* I_0 *denote a smooth elliptic fiber. The fiber of type* I_1 *is a rational curve with a single node. More generally, fibers of type* I_n *consist of a* n*-cycle of intersecting rational curves for* $n \geq 1$. *A fiber of type* II *is just a rational curve with a single cusp. Type* III *fibers consist of two rational curves with a single point of tangency. Fibers of type* IV *consist of three rational components intersecting at a single point. There are also fibers of types* $\mathrm{I}_D n^*$, $n \geq 0$, IV^*, III^*, *and* II^*, *whose dual intersection graphs, minus in each case a multiplicity one component, correspond to those graphs of Dynkin types* D_{n+4}, E_6, E_7, *and* E_8 *respectively.*

We now recall how the Kodaira fiber types correlate with the ramification behavior of the $\mathcal{J}$-map.

LEMMA 4.5 ([**Mir2**, Lemma IV.4.1]). *Let* $F = X_q$ *be the fiber of* π *over* $q \in C$, *and let* $m(\mathcal{J})$ *be the multiplicity of the functional invariant at* q.

1. *If* F *has type* II, IV, IV^*, *or* II^*, *then* $\mathcal{J}(q) = 0$. *Conversely, suppose that* $\mathcal{J}(q) = 0$. *Then*
   - F *has type* I_0 *or* I_0^* *if and only if* $m(\mathcal{J}) \equiv 0 \bmod 3$
   - F *has type* II *or* IV^* *if and only if* $m(\mathcal{J}) \equiv 1 \bmod 3$
   - F *has type* IV *or* II^* *if and only if* $m(\mathcal{J}) \equiv 2 \bmod 3$.
2. *If* F *has type* III *or* III^*, *then* $\mathcal{J}(q) = 1$. *Conversely, suppose that* $\mathcal{J}(q) = 1$. *Then*
   - F *has type* I_0 *or* I_0^* *if and only if* $m(\mathcal{J}) \equiv 0 \bmod 2$
   - F *has type* III *or* III^* *if and only if* $m(\mathcal{J}) \equiv 1 \bmod 2$.
3. F *has type* I_n *or* I_n^* *with* $n \geq 1$ *if and only if* $\mathcal{J}$ *has a pole at* q *of order* n.

4.2. Picard-Fuchs Equations of Weierstrass Elliptic Surfaces. In this subsection we will describe the Picard-Fuchs differential equations for a Weierstrass elliptic curve fibration. We first recall Griffiths' approach, which starts with a first order regular system from which the second order equation can be determined. This method is computationally convenient if all one wants is the Picard-Fuchs equation, but it will not be so useful for our later purposes. Ultimately, we will only care about properties of the projective normal form of the Picard-Fuchs differential equation. For that reason we also present Stiller's approach by "pulling back by $\mathcal{J}$" which suffices to determine this.

THEOREM 4.6 ([**Sas**, p. 304]). *Griffiths' approach to computing the Picard-Fuchs equation of a Weierstrass elliptic surface yields*

$$\frac{d}{dz}\begin{pmatrix}\eta_1\\ \eta_2\end{pmatrix} = \begin{bmatrix} -\frac{d\Delta/dz}{12\Delta} & \frac{3\delta}{2\Delta} \\ \frac{-g_2\delta}{8\Delta} & \frac{d\Delta/dz}{12\Delta}\end{bmatrix}\begin{pmatrix}\eta_1\\ \eta_2\end{pmatrix}$$

where

$$\Delta(z) = g_2(z)^3 - 27g_3(z)^2$$

$$\delta(z) = 3g_3(z)\frac{dg_2(z)}{dz} - 2g_2(z)\frac{dg_3(z)}{dz}$$

and

$$\eta_1 = \int_\gamma \frac{dx}{y}, \quad \eta_2 = \int_\gamma \frac{x\,dx}{y}.$$

Now we turn to Stiller's "pulling back by the $\mathcal{J}$-map" approach.

Recall from Example 3.14 that the uniformizing differential equation for the group $\mathrm{PSL}(2,\mathbb{Z})$ is

$$\Lambda f = \frac{d^2 f}{dx^2} + \frac{36x^2 - 41x + 32}{144x^2(x-1)^2} f = 0,$$

the projective normal form of the Picard-Fuchs equation of the family $\mathcal{E}$ of Example 2.6.

DEFINITION 4.7. Denote by $\Lambda_{\mathcal{J}}$ the projective normal form of the pullback equation $\mathcal{J}^*(\Lambda)$.

By an explicit computation using Theorem 4.6 one can show that

PROPOSITION 4.8. *Given a Weierstrass fibration with functional invariant $\mathcal{J}$, the projective normal form of the Picard-Fuchs differential equation and the equation $\Lambda_{\mathcal{J}}$ are identical.*

Actually, this can also be seen indirectly through Stiller's classification of K-*equations* [**Sti**].

We now can recast the results of Lemma 4.5 above as describing the singular fiber types associated to various regular singular points of the projective normal form of Picard-Fuchs $\Lambda_{\mathcal{J}}$, using Theorem 3.26.

PROPOSITION 4.9. *A regular singular point of $\Lambda_{\mathcal{J}}$ corresponds to a fiber of type*

- I_n *or* I_n^*, $n \geq 1$ $\iff$ *the characteristic exponents are equal*
- I_0 *or* I_0^* $\iff$ *the difference in characteristic exponents is a nonnegative integer*
- II *or* IV* $\iff$ *the difference in characteristic exponents is* $\frac{1}{3}$ *modulo integers*
- III *or* III* $\iff$ *the difference in characteristic exponents is* $\frac{1}{2}$ *modulo integers*
- IV *or* II* $\iff$ *the difference in characteristic exponents is* $\frac{2}{3}$ *modulo integers.*

4.3. Characterization of Elliptic Surfaces with Modular Mirror Maps. Our main result for elliptic curve families is now an immediate consequence of the criterion for orbifold uniformization (Theorem 3.26 and Corollary 3.28) in the elliptic surface setting. We let $L_2 = \Lambda$, $\mathcal{R}(z) = \mathcal{J}(z)$, and apply Proposition 4.8 above.

THEOREM 4.10. *The projective normal form of the Picard-Fuchs equation of a Weierstrass elliptic surface will orbifold uniformize if and only if there are no apparent singularities and the functional invariant $\mathcal{J}(t)$ has zeros to orders $= 1$ or $\equiv 0 \bmod 3$ and ones to orders $= 1$ or $\equiv 0 \bmod 2$.*

This already generalizes results of Stiller [**Sti**] on classical uniformization by Picard-Fuchs differential equations, and Shioda [**Shi**] on elliptic modular surfaces. In addition, this also answers a question of Harnad and McKay regarding the geometric nature of modular solutions to generalized equations of Halphen type [**HM**].

Furthermore, we have this purely geometric characterization of modularity:

COROLLARY 4.11. *The mirror map $z(q)$ of a Weierstrass elliptic surface over $\mathbb{P}^1$ is modular if and only if the surface has functional invariant with* no *excess ramification, and it contains* no *singular fibers of types* IV *or* II*.

Persson and Miranda [**Per, Mir3**] have cataloged the singular fiber types of all rational elliptic surfaces over $\mathbb{C}$. Applying Corollary 4.11 to the list of rational elliptic surfaces over $\mathbb{P}^1$ we obtain:

THEOREM 4.12. *Here is the complete list of rational elliptic modular surfaces, i.e., elliptic surfaces over* $\mathbb{P}^1$ *with Euler characteristic* 12, *non-constant functional invariant, and modular mirror maps, listed according to their singular fiber types*:

I_1, II, III*	I_2, I_4, III, III	I_2, I_3, I_4, III
I_2, II, IV*	I_3, I_3, III, III	I_1, I_1, I_8, II
I_1, I_2, III*	I_1, I_6, II, III	I_1, I_2, I_7, II
I_3, III, III, III	I_2, I_5, II, III	I_1, I_4, I_5, II
I_1, I_3, IV*	I_3, I_4, II, III	I_2, I_3, I_5, II
I_4, II, III, III	I_1, I_7, II, II	I_1, I_1, I_1, I_9
I_5, II, II, III	I_2, I_6, II, II	I_1, I_1, I_2, I_8
I_1, I_1, I_4^*	I_4, I_4, II, II	I_1, I_2, I_3, I_6
I_2, I_2, I_2^*	I_1, I_1, I_7, III	I_1, I_1, I_5, I_5
I_6, II, II, II	I_1, I_2, I_6, III	I_2, I_2, I_4, I_4
I_1, I_5, III, III	I_1, I_3, I_5, III	I_3, I_3, I_3, I_3

PROOF. Miranda's list is particularly convenient to use for this application because it indicates as well the amount of excess ramification of the functional invariant associated to each combination of singular fibers. Between the constraint on the singular fibers and lack of excess ramification of $\mathcal{J}$, we determine the above list of 33 types. □

Explicit expressions for the mirror map in each case can be derived from the Weierstrass canonical form of each elliptic surface. Since the number of singular fibers present is 3 or 4, explicit Weierstrass presentations are available in the literature already: see Schmickler-Hirzebruch for the 3 singular fibers cases [**SH**], and Herfurtner for the surfaces with 4 singular fibers [**Her**].

REMARK 4.13. There is an even more "geometric" interpretation for the excess ramification of the functional invariant. The excess ramification points parametrize deformations of Weierstrass elliptic surfaces which preserve the Kodaira singular fiber types. See [**Dor1**] for more on this.

It seems appropriate to make some additional comments on the construction of explicit algebraic elliptic modular surfaces over $\mathbb{P}^1$ as an application of these results.

Associated to any elliptic surface meeting the equivalent criteria of Theorem 4.10 and Corollary 4.11, there is a mirror map which is a Hauptmodul (unnormalized!) of a genus zero (finite index) subgroup of $PSL(2, \mathbb{Z})$. If we denote this Hauptmodul by H, then we know that the functional invariant $\mathcal{J}$ expresses the modular function J in terms of H, $J(q) = \mathcal{J}(H(q))$, and the functional invariant defines a modular relation between the two genus zero modular groups.

We know that the uniformizing differential equation Λ for $PSL(2, \mathbb{Z})$ is also the projective normal form of the Picard-Fuchs equation of the family of elliptic curves $\mathcal{E}$ in Example 2.6. We can identify the base parameter s of that family with the modular function J. The base parameter of the "pullback surface" obtained by replacing J everywhere with $\mathcal{J}(H)$ in the equation for $\mathcal{E}$ is likewise identified with H. Since we are provided with an explicit algebraic elliptic modular surface with mirror map a Hauptmodul for $PSL(2, \mathbb{Z})$, it is a simple matter to determine an explicit algebraic elliptic surface with mirror map a Hauptmodul H for any specified modular subgroup. We will return to this sort of "inverse construction" at the end of the next section in the context of K3 surfaces and a question of Dolgachev.

5. M_n-polarized K3 surface case

The next case of Picard-Fuchs uniformization is by one parameter families of certain *lattice polarized* K3 surfaces. The direct motivation for considering this case is of course the Mirror-Moonshine Conjecture of Lian-Yau. In the first subsection we recall the definition of M_n-polarized K3 surfaces and Dolgachev's results on their moduli. Next we show how classical Fuchsian ordinary differential equation facts imply that the Picard-Fuchs equation of a family of such surfaces is a "symmetric square." After a brief overview of the Mirror-Moonshine Conjecture, we carefully formulate the question we want to answer in §5.3. A crucial tool, the generalized functional invariant, is introduced in the next subsection. Finally, in §5.5 we prove our main result, Theorem 5.20, on modularity of the mirror map in the M_n-polarized K3 surface case.

5.1. M_n-polarized K3 surfaces. We briefly summarize here the theory of moduli of M_n-polarized K3 surfaces. See the paper in this volume by Verrill and Yui [**VY**, §§3 and 4] for a general discussion of lattice polarized K3 surfaces after Dolgachev [**Dol**].

Let L be the *K3 lattice*

$$L = U \perp U \perp U \perp -\mathrm{E}_8 \perp -\mathrm{E}_8,$$

i.e., the lattice of integral (1,1)-forms on a K3 surface. Let $M = \langle 2n \rangle$. Since the lattice U represents any even integer, we can assume $M \subset U$.

DEFINITION 5.1. The orthogonal complement of M in the K3 lattice is

$$M_n := U \perp U \perp -\mathrm{E}_8 \perp -\mathrm{E}_8 \perp \langle -2n \rangle \simeq M^{\perp}.$$

This lattice is denoted by $\mathcal{M}$ in [**VY**].

The global monodromy group of the "universal family" of K3 surfaces over the (coarse) moduli space of K3 surfaces polarized by a rank 19 lattice M_n has been determined by Dolgachev. Define to begin with

$$\Gamma_0(n) := \left\{ \begin{pmatrix} a & b \\ c & d \end{pmatrix} \in \mathrm{PSL}(2,\mathbb{Z}) \;\middle|\; c \equiv 0 \bmod n \right\} \subset \mathrm{PSL}(2,\mathbb{Z}).$$

DEFINITION 5.2. The element of order two

$$F_n = \begin{pmatrix} 0 & -1/\sqrt{n} \\ \sqrt{n} & 0 \end{pmatrix} \in \mathrm{PSL}(2,\mathbb{R})$$

in the normalizer of $\Gamma_0(n)$ in $\mathrm{PSL}(2,\mathbb{R})$ is the *Fricke involution.*

DEFINITION 5.3. The *Fricke modular group* of level n is the subgroup of $\mathrm{PSL}(2,\mathbb{R})$ generated by $\Gamma_0(n)$ and F_n, and denoted by $\Gamma_0(n)^*$.

THEOREM 5.4 ([**Dol**, Theorem 7.1]). *The coarse moduli space of K3 surfaces polarized by M_n is isomorphic to $X_0(n)^* := \Gamma_0(n)^* \backslash \mathbb{H}$.*

For later convenience we recall now Dolgachev's description of the normalized Hauptmodul H_n for $\Gamma_0(n)^*$.

PROPOSITION 5.5 ([**Dol**, Theorem 7.5]). *If $X_0(n)^*$ is rational. Then there exists a unique holomorphic function*

$$H_n : \mathbb{H} \to \mathbb{C}$$

such that

1. *H_n is invariant with respect to $\Gamma_0(n)^*$*
2. *H_n has a Fourier expansion*

$$H_n(\tau) = \frac{1}{q} + \sum_{k=1}^{\infty} c_k q^k, \; q = e^{2\pi\imath\tau}$$

3. *the coefficients of the Fourier expansion are all integers*
4. *as a meromorphic function on $X_0(n)^*$, the function H_n has a simple point at the cusp at ∞ and generates the field of meromorphic functions on $X_0(n)^*$.*

5.2. Picard-Fuchs Equations of M_n-Polarized K3 Surface Families. Next we establish the symmetric square structure of the Picard-Fuchs equations of M_n-polarized K3 surface families. This is crucial for reducing the study of a mirror map involving a third order Picard-Fuchs equation. We begin by recalling the classical theory of symmetric products of homogeneous Fuchsian ordinary differential equations. This yields in particular a simple criterion, Corollary 5.9, for when a third order equation is a symmetric square.

THEOREM 5.6 ([**Sin**, Lemma 3.1(b)]). *Let $L_1(y)$ and $L_2(y)$ be homogeneous linear differential polynomials with coefficients in $\mathbb{C}(t)$. Then there exists a homogeneous linear differential equation $L_3(y) = 0$ with coefficients in $\mathbb{C}(t)$ and solution space the $\mathbb{C}$-span of*

$$\{\nu_1\nu_2 \mid L_1(\nu_1) = 0 \text{ and } L_2(\nu_2) = 0\}.$$

DEFINITION 5.7. We call the operator $L_3(y)$ constructed above the *symmetric product* of L_1 and L_2, and denote it by $L_1 \circledS L_2$. In fact, the operation is associative, and we may further define $L^{\circledS n}$ for $n \geq 1$ by $L^{\circledS 1} = L$ and $L^{\circledS n} = L^{\circledS n-1} \circledS L$. We call $\mathrm{Sym}^n(L) = L^{\circledS n}$ the *nth symmetric power* of L; conversely, L is the *nth root* of $L^{\circledS n}$.

LEMMA 5.8 ([**Sin**, Lemma 4, p. 129]). *Let $L(y)$ be a homogeneous linear differential polynomial with coefficients in $\mathbb{C}(t)$. Then $L(y) = L_2^{\circledS n}(y)$ for some second order homogeneous linear differential polynomial $L_2(y)$ with coefficients in $\mathbb{C}(t)$ if and only if there exists a fundamental set of solutions $\{y_1, \dots, y_{n+1}\}$ of $L(y) = 0$ such that*

$$y_i y_{i+2} - y_{i+1}^2 = 0, \quad i = 1, \dots, n-1.$$

COROLLARY 5.9. *Let $L(y) = 0$ be a third order homogeneous linear equation with coefficients in $\mathbb{C}(t)$. If there exists a nondegenerate homogeneous polynomial P of degree 2 with constant coefficients and a fundamental set of solutions $\{y_1, y_2, y_3\}$ of $L(y) = 0$ such that $P(y_1, y_2, y_3) = 0$, then $L(y)$ is the second symmetric power of a second order homogeneous linear differential equation with coefficicnts in $\mathbb{C}(t)$.*

PROOF. This follows easily from Lemma 5.8. By assumption, the fundamental set of solutions satisfies a nondegenerate quadratic relation. Since all such quadrics in $\mathbb{P}^2(\mathbb{C})$ are projectively equivalent to

$$y_1 y_3 - y_2^2 = 0$$

the criterion of the lemma applies and $L(y)$ is a symmetric square. □

Assume that we have a second order Fuchsian ordinary differential equation $L_2 f = 0$ where

$$L_2 = \frac{d^2}{dt^2} + P_1(t)\frac{d}{dt} + P_2(t).$$

The second order equation $L_2 f = 0$ is equivalent to the system of first order differential equations

$$\begin{cases} f' = g \\ g' = -P_2 f - P_1 g \end{cases}$$

with $\{f, g\}$ as fundamental solutions. Observe that

$$\{f^n, f^{n-1}g, \ldots, fg^{n-1}, g^n\}$$

forms a set of fundamental solutions for the nth symmetric power $L = L_2^{\circledS n}$. The following result describes a system of first order differential equations for L with these fundamental solutions.

THEOREM 5.10 ([**Lee**, Theorem 2]). *If $\{f, g\}$ satisfy a first order 2×2 differential system*

$$\frac{d}{dt}\begin{pmatrix} f \\ g \end{pmatrix} = \begin{pmatrix} 0 & 1 \\ -P_2 & -P_1 \end{pmatrix}\begin{pmatrix} f \\ g \end{pmatrix},$$

then $\{f^n, f^{n-1}g, \ldots, fg^{n-1}, g^n\}$ satisfy the $(n+1) \times (n+1)$ system

$$\frac{d}{dt}\begin{pmatrix} f^n \\ f^{n-1}g \\ \vdots \\ fg^{n-1} \\ g^n \end{pmatrix} = A \begin{pmatrix} f^n \\ f^{n-1}g \\ \vdots \\ fg^{n-1} \\ g^n \end{pmatrix},$$

where $A = (a_{ij})$ is an $(n+1) \times (n+1)$ matrix such that

$$\begin{aligned} a_{k,k} &= (1-k)P_1, & 1 \le k \le n+1, \\ a_{k,k+1} &= n+1-k, & 1 \le k \le n, \\ a_{k+1,k} &= -kP_2, & 1 \le k \le n, \\ a_{i,j} &= 0, & i > j+1 \text{ or } j > i+1. \end{aligned}$$

EXAMPLE 5.11. In particular, when $n = 2$, the case for a symmetric square, one may rewrite the system as a single third order equation

$$\mathrm{Sym}^2(L_2)f = f''' + 3P_1 f'' + (2{P_1}^2 + 4P_2 + {P_1}')f' + (4P_1P_2 + 2{P_2}')f = 0.$$

In this form, using the expression for the projective normal form of a second order Fuchsian differential equation given in Example 2.16, it is easy to check that:

PROPOSITION 5.12. *Let L_2 be as above a second order Fuchsian ordinary differential operator, and let $L = \mathrm{Sym}^2(L_2)$ be its symmetric square. Then the projective normal form of L is the symmetric square of the projective normal form of L_2.*

REMARK 5.13. In fact, it is possible to provide an explicit description of the relationship between the monodromy matrices of the second order "square root" equation and those of the third order symmetric square equation. This is provided by the faithful representation of $SL(2,\mathbb{C})$ in $SL(3,\mathbb{C})$ via the symmetric square representation [**Sin**].

Finally, we see the relevance of all of this for Picard-Fuchs equations of our M_n-polarized K3 surface families:

THEOREM 5.14. *The Picard-Fuchs equation of a family of M_n-polarized K3 surfaces is the symmetric square of a second order homogeneous linear Fuchsian ordinary differential equation.*

PROOF. To begin with, the order of the Picard-Fuchs equation is equal to the rank of the transcendental lattice, i.e., $22 - 19 = 3$. By Dolgachev's Torelli theorem for lattice polarized K3 surfaces (specifically, the proof of Theorem 5.4 above) the period domain lies on a nondegenerate quadric in $\mathbb{P}^2$. Thus, Corollary 5.9 implies that the third order Picard-Fuchs differential equation is in fact a symmetric square. □

There is another approach to proving this result which takes advantage of the special geometric properties of M_n-polarized K3 surfaces, namely their presentation as *Shioda-Inose surfaces* coming from a product of two elliptic curves linked by an n-isogeny. See [**Pet**] and [**VY**, §4.8] for more details.

5.3. The Mirror-Moonshine Phenomenon. In their first systematic investigations of mirror symmetry for one parameter families of Calabi-Yau manifolds constructed via the "orbifold construction" [**LY2**], Lian and Yau discovered that the reciprocal of the mirror maps for the K3 surfaces they were studying agreed, up to an additive constant, with some of the *McKay-Thompson normalized q-series* in the lists of Conway-Norton [**CN**]. The evidence was sufficiently strong that they formulated

CONJECTURE 5.15 (Lian-Yau [**LY2, LY1**]). *If $z(q)$ is the mirror map for a one parameter family of algebraic K3 surfaces from an orbifold construction which has a third order Picard-Fuchs equation, then, for some $c \in \mathbb{Z}$, the q-series*

$$\frac{1}{z(q)} + c$$

is a McKay-Thompson series $T_g(q)$ for some element g in the Monster.

In [**LY3**], Lian and Yau compute many more toric examples (including over a dozen complete intersection examples), and note that the correspondence to monstrous groups persists. This suggested that the hypothesis regarding the "orbifold construction" should perhaps be weakened to the hypothesis "torically constructed."

As noted in the proof of Theorem 5.14, for a family of lattice polarized K3 surfaces the condition of having a third order Picard-Fuchs equation is equivalent to the generic member possessing a polarization by a lattice of rank 19.

Furthermore, a McKay-Thompson series is in particular a Hauptmodul for some "monstrous" genus zero arithmetic group Γ, and the various equivalent Hauptmoduln are well-defined as generators of the function field of the rational curve $\Gamma\backslash\mathbb{H}^*$ only up to action of Γ. We see that in Conjecture 5.15 an equivalent conclusion is that the mirror map is itself a Hauptmodul (unnormalized!) for some monstrous Γ.

Before Conjecture 5.15 was even formulated, Beukers, Peters, and Stienstra had computed the Picard-Fuchs equation of a particular family of M_n-polarized

K3 surfaces [**BP, PS**]. The mirror map was determined by Verrill and Yui [**VY**, §7.3]. Although it is a Hauptmodul, this q-series does not satisfy the conclusion of the Mirror-Moonshine Conjecture. Thus it provides a counterexample to a "monstrous" generalization of the Mirror-Moonshine Conjecture for torically constructed families.

This suggests the following two questions:

Commensurability: *When is the mirror map commensurable to a McKay-Thompson series?*

Modularity: *Can we characterize the families of M_n-polarized K3 surfaces whose mirror maps are Hauptmoduln for genus zero subgroups of $\Gamma_0(n)^*$?*

We will provide answers to each of these questions in §5.5.

5.4. The Generalized Functional Invariant.

DEFINITION 5.16. By analogy with the functional invariant of an elliptic surface, to any family of M_n-polarized K3 surfaces over a curve C we associate the composition of the truncated period morphism

$$\begin{aligned} \Omega\colon C &\to \mathbb{H} \\ z &\mapsto \Omega_1(z)/\Omega_0(z) \end{aligned}$$

and the canonical projection onto the coarse moduli space

$$H_n\colon \mathbb{H} \to \Gamma_0(n)^*\backslash\mathbb{H},$$

and call it the *generalized functional invariant* $\mathcal{H}_n = H_n \circ \Omega$.

LEMMA 5.17. *The generalized functional invariant $\mathcal{H}_n$ is meromorphic.*

PROOF. This follows directly from the regularity of the singular points of Fuchsian differential equations and the meromorphicity of the Hauptmodul. The question is local; we will always assume we work with charts centered about 0. Consider the second order (projective normal form of square root of) Picard-Fuchs differential equation on $\mathbb{P}^1$ with coordinate z. Let a_ρ denote the regular singular points. Let $\Omega(z)$ be the multivalued holomorphic function from the complement of the a_ρ to the upper half plane, H_n be the Hauptmodul viewed as a function from $\mathbb{H}$ to $\mathbb{P}^1$, and $\mathcal{H}_n$ be the generalized functional invariant. There are two cases to consider: the points a_ρ of finite or of infinite order monodromy.

If a point has finite order m monodromy, then as a function of $z = \sigma^m$ the period $\Omega(\sigma^m)$ is a single-valued holomorphic function of σ in a neighborhood $0 < \sigma < \epsilon$, where $\Im(\Omega(\sigma^m)) > 0$. In particular by Picard's theorem the singularity is not essential.

If the monodromy about an a_ρ is of infinite order, we can argue that the singularity is not essential by bounding the rate of growth of $\mathcal{H}_n(z) = H_n(\Omega(z))$. For appropriately chosen coordinate w centered about the singularity, by Frobenius' method we know that $w = O(z^{-m}\log z)$, so $|w| < O(z^{-(m+1)})$ as $z \to 0$. Because $H_n(w)$ is a meromorphic function on $\Gamma_0(n)^*\backslash\mathbb{H}^*$ (and so on $\mathbb{H}^*$), near a singular point of H_n the rate of growth of $H_n(w)$ equals $O(w^{-r})$ for some finite $r > 0$. Thus $|H_n(\Omega(z))| < O(z^{-r(m+1)})$, and $\mathcal{H}_n(z)$ does not have an essential singularity.

Thus the singularity is either removable or a pole, and $\mathcal{H}_n(z)$ is meromorphic. □

5.5. Commensurability and Modularity Results. We can now address the two main questions regarding the Mirror-Moonshine conjecture for any family of M_n-polarized K3 surfaces over $\mathbb{P}^1$: The first result explains the relationship of the mirror map to McKay-Thompson series. The second characterizes the modularity properties of the mirror map.

5.5.1. *Commensurability.* By checking the tables of Conway-Norton [**CN**], we find: If $\Gamma_0(n)^*$ is a genus zero group with $n \neq 49, 50$, then it is monstrous, i.e., its Hauptmodul is a McKay-Thompson series.

Suppose we are given an algebraic family of M_n-polarized K3 surfaces over $\mathbb{P}^1$. In particular the genus of $\Gamma_0(n)^*$ must be zero. The generalized functional invariant is a rational function, $\mathcal{H}_n(z)$, such that

$$H_n(q) = \mathcal{H}_n(z(q)),$$

i.e., it expresses the Hauptmodul for $\Gamma_0(n)^*$ as a rational function of the mirror map $z(q)$. If $n \neq 49, 50$, this means that there is a rational expression for a McKay-Thompson series $H_n(q)$ in terms of the mirror map $z(q)$. In particular this shows that:

LEMMA 5.18. *The mirror map of a family of M_n-polarized K3 surfaces over $\mathbb{P}^1$ is commensurable with a McKay-Thompson series when $n \neq 49, 50$.*

The groups $\Gamma_0(49)^*$ and $\Gamma_0(50)^*$ are two of Conway-Norton's "ghosts," i.e., non-monstrous genus zero subgroups of $\Gamma_0(n)^*$ containing $\Gamma_0(n)$. Their Hauptmoduln are however still "replicable functions" in the sense of Conway-Norton, and lie among the list of "monster-like" groups of Ferenbaugh [**Fer**]. So we can say:

LEMMA 5.19. *The mirror map of any family of M_n-polarized K3 surfaces over $\mathbb{P}^1$ is commensurable with a replicable function (i.e., a Hauptmodul of a monster-like group).*

5.5.2. *Modularity.* We can now address our main question regarding the Mirror-Moonshine Conjecture for any family of M_n-polarized K3 surfaces over $\mathbb{P}^1$.

By analogy with the elliptic curve case, apply the general criterion for pulling back orbifold uniformizing differential equations (Theorem 3.26 and Corollary 3.28) to M_n-polarized K3 surface families. The role previously played by the functional invariant is filled here by the rational function $\mathcal{H}_n(z)$, our generalized functional invariant. The differential equation $\Lambda_{\mathcal{J}}$ is replaced by the projective normal form of the square root of the Picard-Fuchs differential equation of our one parameter family of M_n-polarized K3 surfaces.

The uniformizing group for the "target" is now $\Gamma_0(n)^*$. What order elliptic points must be considered? In [**Fer**, p. 55] it is indicatedthat for the broader class of groups $G = \Gamma_0(n|h) + e_1, e_2, \ldots$, an elliptic point can only have order 2, 3, 4, or 6. We need to check that the "order of value" p of $\mathcal{H}_n(z)$ (i.e., the vanishing order of $\mathcal{H}_n(z) - p$) for each elliptic point satisfies the conditions of Corollary 3.28.

THEOREM 5.20. *The square root of the projective normal form of the Picard-Fuchs equation of a one parameter family of M_n-polarized K3 surfaces (with genus zero $\Gamma_0(n)^*$) will orbifold uniformize if and only if it has no apparent singularities, and for each elliptic point $p \in X_0(n)^*$, $\mathcal{H}_n(z) - p$ vanishes to permissible orders as listed in Table* 1.

PROOF. The argument here is virtually identical to that for elliptic curves. To begin with we work with the square root of the projective normal form of Picard-Fuchs, a second order Fuchsian ordinary differential equation. Replace $J(\tau)$ with the normalized hauptmodul $H_n(\tau)$ for the appropriate $\Gamma_0(n)^*$. Replace the functional invariant $\mathcal{J}$ with the generalized functional invariant $\mathcal{H}_n$. The inverse image of a cusp under the generalized functional invariant is still a cusp and these are all the cusps (finite index condition). A regular singular point in the inverse image of an elliptic point is either a generic regular singular point or an apparent singularity. By Corollary 3.28 the generic singularities are permitted if and only if they are of the types listed in Table 1. As before, the "extra" points at which $d\mathcal{H}_n/dz = 0$ are just the apparent singularities outside of the inverse image locus, or equivalently those for which the characteristic exponents are distinct. Positivity and surjectivity of the period morphism, and the fact that the universal cover of the complement of the (non-apparent) singularities is $\mathbb{H}$ all follow from the same properties of the $X_0(n)^*$ orbifold. □

We have thus characterized the Mirror-Moonshine phenomenon completely. This characterization has the advantage that it is expressed purely in terms of properties of the generalized functional invariant, and applies equally well to any example family, toric or not.

REMARK 5.21. As an application we can construct families of M_n-polarized K3 surfaces which orbifold uniformize with specified monodromy groups. The only constraint is that we must already have at hand an algebraic construction for a family of M_n-polarized K3 surfaces with mirror map a Hauptmodul for the full group $\Gamma_0(n)^*$. This thus provides a partial answer to a question of Dolgachev regarding the "inverse problem" (see [**VY**, §8.4]), reducing it to the case $G = \Gamma_0(n)^*$. In other words, the problem can be solved completely just by finding explicit algebraic constructions of families of M_n-polarized K3 surfaces with mirror maps equal to $H_n(q)$.

As in the case of elliptic curve families, the starting point is that we know how to pullback from the "universal family" with a given lattice polarization, to another family with the same lattice polarization which orbifold uniformizes with a proper subgroup as monodromy group. Assume for now that the subgroup is also of genus zero.

As long as we have an algebraic presentation (e.g., as hypersurfaces, complete intersections) for the universal family in the first place, then we can construct the pullback family. The necessary pullback function is the rational function which

TABLE 1. Vanishing orders permitting modular mirror maps

Order of elliptic point $p \in X_0(n)^*$	Vanishing order $\mathcal{H}_n(z) - p$
2	$\equiv 0 \bmod 2$ or $= 1 \bmod 2$
3	$\equiv 0 \bmod 3$ or $= 1 \bmod 3$
4	$\equiv 0 \bmod 4$ or $= 1$ or $2 \bmod 4$
6	$\equiv 0 \bmod 6$ or $= 1$, 2, or $3 \bmod 6$

defines the modular relation for the Hauptmodul of the universal family as a rational function in the Hauptmodul of the second family (i.e., for the finite index subgroup). Once again the general fact (Corollary 3.18 and subsequent remarks) relating automorphic functions for genus zero groups provides this rational function.

REMARK 5.22. As was the case with elliptic curves, a necessary condition on the projective normal form Picard-Fuchs equation for orbifold uniformization is that *no apparent singularities* be present. There is a great deal of evidence to support the conjecture that no apparent singularities arise in any torically constructed family of Calabi-Yau manifolds. Aside from the lack of counterexamples, there are other reasons to believe such a conjecture (see the author's work on *geometric isomonodromic deformations*, [**Dor1**]).

6. Generalizations

We comment now on generalizations of these Picard-Fuchs uniformization results in various directions, involving changes in the natures of both the fiber and the base of the families considered.

To begin with, we note that our results for M_n-polarized K3 surface families immediately extend to the general rank 19 lattice polarized case. In particular, the arguments from Fuchsian theory for the symmetric square structure of the Picard-Fuchs equation are independent of the particular rank 19 lattice chosen. What is lacking in the general setting is a complete classification of the rank 19 lattices and a specific description of monodromy groups.

It is possible to lift the convenient *genus zero* restriction on both the uniformizing groups and the bases of the families under consideration. Most of the arguments given in determining the Picard-Fuchs uniformization criteria apply immediately to the case of families of elliptic curves and M_n-polarized K3 surfaces over higher genus curves. These criteria still imply that the mirror map is an automorphic function for a (no longer genus zero) subgroup of the genus zero groups $\mathrm{PSL}(2,\mathbb{Z})$ or $\Gamma_0(n)^*$ respectively, even though it is no longer a Hauptmodul. Modification of the arguments to allow for $\Gamma_0(n)^*$ of genus higher than zero can be made as well.

Furthermore, there is an extension of our modularity criteria to certain multiparameter families of Abelian varieties and K3 surfaces of various Picard ranks. In particular, this can be applied to two-parameter families of K3 surfaces with Picard rank 18 as studied in [**Nis, VY**]. This case is particularly nice, as it is possible to apply another old result of Fano to conclude that generic one-parameter "slices" of such families will possess Picard-Fuchs equations of order four, built from two distinct second order equations by *tensor product*. When the two-parameter family is "modular," with respect to the appropriate Hilbert modular group, the Picard-Fuchs equations on the one-parameter slices reflect this by being "bimodular," i.e., each second order factor equation is independently modular. The example of Hosono, [**VY**, §10.2-3], provides a nice illustration of this phenomenon.

For more on all of this, together with a discussion of the considerably more subtle case of Calabi-Yau threefold families, see [**Dor2**].

References

[AB] D. V Anosov and A. A. Bolibruch, *The Riemann-Hilbert problem*, Aspects Math, vol. E22, Friedr. Vieweg & Sohn, Braunschweig, 1994.

[BP] J. Bertin and C. Peters, *Variations de structures de Hodge, variétés de Calabi-Yau et symétrie miroir*, Introduction à la théorie de Hodge, Panor. Synthèses, vol. 3, Soc. Math. France, Paris, 1996, pp. 169–256.

[CY] I. Chen and N. Yui, *Singular values of Thompson series*, Groups, Difference Sets, and the Monster (Columbus, 1993), Ohio State Univ. Math. Res. Inst. Publ., vol. 4, de Gruyter, Berlin, 1996, pp. 255–326

[Coh] H. Cohn, *Singular Moduli and Modular Equations for Fricke's Cases*, Theta functions—Bowdoin 1987 (Brunswick, ME, 1987), Proc. Sympos. Pure Math., vol. 49 , Part 2 mer. Math. Soc., Providence, RI, 1989, pp. 233–249.

[CN] J. Conway and S. Norton, *Monstrous moonshine*, Bull. London Math. Soc. **11** (1979), 308–339.

[Dol] I. V. Dolgachev, *Mirror symmetry for lattice polarized $K3$ surfaces*, Algebraic Geometry, vol. 4. J. Math. Sci. 81 (1996), 2599–2630.

[Dor1] C. F. Doran, *Picard-Fuchs uniformization and geometric isomonodromic deformations: Modularity and variation of the mirror map*, Ph.D. thesis, Harvard University, 1999.

[Dor2] ———, *Picard-Fuchs uniformization and modularity of the mirror map* (in preparation).

[Fer] C. R. Ferenbaugh, *The genus-zero problem for $n|h$-type groups*, Duke Math. J. **72** (1993), 31–63.

[For] L. R. Ford, *Automorphic functions*, 2nd edition, Chelsea Publishing Co., New York, NY (1951).

[HM] J. Harnad and J. McKay, *Modular solutions to equations of generalized Halphen type*, solv-int/9804006.

[Her] S. Herfurtner, *Elliptic surfaces with four singular fibres*, Math. Ann. **291** (1991), 319–342.

[HLY] S. Hosono, B. H. Lian and S.-T. Yau, *Maximal degeneracy points of GKZ systems*, J. Amer. Math. Soc. **10** (1997), 427–443.

[Kod] K. Kodaira, *On compact analytic surfaces.* II, Ann. of Math. (2) **77** (1963), 563–626.

[Lee] M.-H. Lee, *Picard-Fuchs equations for elliptic modular varieties*, Appl. Math. Letters **4** (1991), 91–95.

[Leh] J. Lehner, *Discontinuous groups and automorphic functions*, Mathematical Surveys, Number VIII, Amer. Math. Soc., 1964.

[LY1] B. H. Lian and S.-T. Yau, *Mirror symmetry, rational curves on algebraic manifolds and hypergeometric series*, XIth International Congress of Mathematical Physics (Paris, 1994), Internat. Press, Cambridge, MA, 1995, pp. 163–184.

[LY2] ———, *Arithmetic properties of mirror map and quantum coupling*, Comm. Math. Phys. **176** (1996), 163–191.

[LY3] ———, *Mirror maps, modular relations and hypergeometric series.* II, S-Duality and Mirror Symmetry (Trieste, 1995), Nuclear Phys. B. Proc. Suppl. **46** (1996),. 248–262.

[Mir1] R. Miranda, *The moduli of Weierstrass fibrations over $\mathbb{P}^1$*, Math. Ann. **255** (1981) 379–394.

[Mir2] ———, *The basic theory of elliptic surfaces*, Dottorato Ric. Mat., ETS Editrice, Pisa, 1989.

[Mir3] ———, *Persson's list of singular fibers for a rational elliptic surface*, Math. Z. **205** (1990), 191–211.

[Nis] S. Nishiyama, *Appell's hypergeometric function F_2 and periods of certain elliptic $K3$ surfaces*, Tokyo J. Math. 10 (1987), 33–67.

[Per] U. Persson, *Configurations of Kodaira fibers on rational elliptic surfaces*, Math. Z. **205** (1990), 1–47.

[Pet] C. Peters, *Monodromy and Picard-Fuchs equations for families of $K3$-surfaces and elliptic curves*, Ann. Sci. École Norm. Sup. (4) **19** (1986), 583–607.

[PS] C. Peters and J. Stienstra, *A pencil of $K3$-surfaces related to Apéry's recurrence for $\zeta(3)$ and Fermi surfaces for potential zero*, Arithmetic of Complex Manifolds (Erlangen, 1988), Lecture Notes in Math., vol. 1399, Springer, Berlin, 1989, pp. 110–127.

[Sas] T. Sasai, *Monodromy representations of homology of certain elliptic surfaces*, J. Math. Soc. Japan **26** (1974), 296–305.

[SH] Schmickler-Hirzebruch, *Elliptische Flächen über $\mathbb{P}_1(\mathbb{C})$ mit drei Ausnahmefasern und die hypergeometrische Differentialgleichung*, Schriftenreihe Math. Inst. Univ. Munster 2. Ser., vol. 33, Universität Münster, Mathematisches Institut, Münster, 1985.

[Shi] T. Shioda, *On elliptic modular surfaces*, J. Math. Soc. Japan **24**, (1972), 20–59.

[Sin] M. Singer, *Algebraic relations among solutions of linear differential equations: Fano's theorem*, Amer. J. Math. **110** (1988), 115–143.

[Sti] P. F. Stiller, *On the uniformization of certain curves*, Pacific J. Math. **107** (1983), 229–244.

[Var] V. S. Varadarajan, *Meromorphic differential equations*, Exposition. Math. **9** (1991), 97–188.

[VY] H. Verrill and N. Yui, *Thompson series, and the mirror maps of pencils of $K3$ surfaces* (in these proceedings).

[Yos] M. Yoshida, *Fuchsian differential equations. With special emphasis on the Gauss-Schwarz theory*, Aspects Math, vol. E11. Friedr. Vieweg & Sohn, Braunschweig 1987.

Department of Mathematics, Harvard University, Cambridge, Massachusetts 02138
E-mail address: doran@math.harvard.edu

Centre de Recherches Mathématiques
CRM Proceedings and Lecture Notes
Volume **24**, 2000

Hilbert Modular Varieties in Positive Characteristic

Eyal Z. Goren

ABSTRACT. This survey is an expanded version of my lecture at the summer school. It surveys some recent developements concerning Hilbert modular varieties in positive characteristic and suggests further questions.

CONTENTS

1. Introduction 283
2. Stratifications of Hilbert Modular Varieties 286
3. The Structure of the p-Torsion 292
4. Hasse Invariants 297
5. Stratifications of a Class of Shimura Varieties 299
References 301

1. Introduction

This paper is an expanded version of my lectute at the summer school. The lecture surveyed some recent developements in the theory of Hilbert modular varieties in positive characteristic in which I was involved. The paper suffers from the same imbalance as the original lecture, in that that it focuses on some particular results and neglects to fully survey our knowledge on these varieties. I can only regert that the scope of the paper does not permit proper discussion of the important and foundational work done by others.

1.1. Definition of HBAS. Let L be a totally real field of degree g over $\mathbb{Q}$, with discriminant d_L, ring of integers $\mathcal{O}_L$ and different ideal $\mathcal{D}_L = \mathcal{D}_{L/\mathbb{Q}}$. Let $\mathrm{Cl}(L)$ (resp. $\mathrm{Cl}(L)^+$) denote the class group (resp. strict class group) of L, and let h_L (resp. h_L^+) denote its order. Recall that the group $\mathrm{Cl}(L)^+$ consists of classes of (fractional) ideals of L, where two ideals $\mathfrak{a}$, $\mathfrak{b}$ are equivalent if and only if there exists $\lambda \in L$ such that $\lambda \gg 0$ and $\lambda\mathfrak{a} = \mathfrak{b}$. Every ideal $\mathfrak{a}$ has a natural positive cone $\mathfrak{a}^+$ consisting of the

1991 *Mathematics Subject Classification.* 11F41, 14G35, 14K10.
Received by the editors November 9, 1998; Accepted January 30, 1999.
I thank H. Darmon for a careful reading of the manuscript and useful comments on it.
This is the final form of the paper.

totally positive elements of $\mathfrak{a}$. We may thus regard $\mathrm{Cl}(L)^+$ as equivalence classes of fractional ideals together with their positive cones; equivalently, of projective, rank one $\mathcal{O}_L$-modules $\mathcal{M}$, endowed with a notion of positivity, i.e., an isomorphism of $\mathbb{R}$-vector spaces, up to multiplication by a positive scalar, is given between $\mathcal{M} \otimes_{\mathcal{O}_L} \mathbb{R}$ and $\mathbb{R}$, for every embedding $L \hookrightarrow \mathbb{R}$. Let $\mathcal{M}^+$ be the positive cone of such a module. Let us fix representatives $\{(\mathfrak{a}, \mathfrak{a}^+)\}$ for the ideal classes in $\mathrm{Cl}(L)^+$.

Throughout this paper L is fixed and one should understand all the definitions below (e.g., HBAS) with respect to this fixed choice.

Let S be a scheme. A *HBAS over S* (read: Hilbert-Blumenthal Abelian scheme) is a triple $(A, \iota, m)/S$ consisting of an Abelian scheme $A \to S$ of relative dimension g and a ring injection, $\iota\colon \mathcal{O}_L \hookrightarrow \mathrm{End}(A/S)$, such that the following holds: Let $(\mathcal{M}_A, \mathcal{M}_A^+)$ denote the $\mathcal{O}_L$-module, $\mathcal{M}_A$, of $\mathcal{O}_L$-linear, symmetric homomorphisms from A to A^t (the dual Abelian scheme), with its natural positive cone, $\mathcal{M}_A^+$, consisting of the $\mathcal{O}_L$-polarizations. We assume $A \otimes_{\mathcal{O}_L} \mathcal{M}_A \cong A^t$. We remark that if d_L is invertible on S, this definition is equivalent to Rapoport's original definition [**29**, Definition 1.1] where one requires the relative tangent space $\mathfrak{t}_{A/S}$ to be a locally free (on the base) $\mathcal{O}_L \otimes \underline{\mathcal{O}}_S$ – module of rank 1.

Furthermore, $m : (\mathcal{M}_A, \mathcal{M}_A^+) \to (\mathfrak{a}, \mathfrak{a}^+)$ is a given isomorphism, where $(\mathfrak{a}, \mathfrak{a}^+)$ is one of the fixed representatives. We refer to [**3**, Sections 1, 2] for more details.

If $S = \mathrm{Spec}(R)$, for some ring R, one says that $(A, \iota, m)/S$ is a HBAS over R, and if $R = k$ is a field, one says that $(A, \iota, m)/S$ is a HBAV over k (read: Hilbert-Blumenthal Abelian variety).

Denote by Fr the Frobenius morphism and by Ver the Verschiebung morphism, when they are defined. The domain and target will be clear from the context. Denote a field of p^r elements by $\mathbb{F}_{p^r}$.

We refer to [**29**] for many of the facts on HBAV's and their moduli spaces, which we cite without specific reference, and to [**5**] as a general reference for the same topics.

1.2. Levels. A *level n structure* on $(A, \iota, m)/S$ is an isomorphism of constant group scheme over S with $\mathcal{O}_L$-action: $\alpha\colon (\mathcal{O}_L/n\mathcal{O}_L)^2_{/S} \to A[n]$.

Let $n \geq 3$. There exists a fine moduli scheme, $\mathcal{M}_L^n \to \mathrm{Spec}(\mathbb{Z}[n^{-1}])$, parameterizing isomorphism classes of HBAS's with level n structure. It is smooth, of relative dimension g, over $\mathrm{Spec}(\mathbb{Z}[(nd_L)^{-1}])$. (See [**3**] and [**1**] for singularities at primes dividing d_L). Let $\mathcal{M}_L^{n\prime}$ be the scheme over $\mathrm{Spec}(\mathbb{Z}[n^{-1}])$, obtained from $\mathcal{M}_L^n$ as a quotient by "forgetting the level structure". The schemes $\mathcal{M}_L^{n\prime}$ and $\mathcal{M}_L^{m\prime}$ agree over $\mathrm{Spec}(\mathbb{Z}[(mn)^{-1}])$. We get, therefore, a scheme $\mathcal{M}_L$ over $\mathrm{Spec}(\mathbb{Z})$. It is a coarse moduli space for HBAS.

The scheme $\mathcal{M}_L$ is a disjoint union of components, $\mathcal{M}_L(\mathfrak{a})$, where $\mathfrak{a}$ runs over $\mathrm{Cl}(L)^+$. The scheme $\mathcal{M}_L^n$ is a disjoint union of components, $\mathcal{M}_L^n(\mathfrak{a}, *)$, where $\mathfrak{a}$ runs over $\mathrm{Cl}(L)^+$ and $*$ is an additional data controlling the level n structure. We have $\mathcal{M}_L(\mathfrak{a})(\mathbb{C}) \cong \mathrm{PSL}(\mathcal{O}_L \oplus \mathfrak{a})\backslash\mathfrak{H}^g$, where $\mathfrak{H}$ denotes the complex upper half plane. (If one does not *fix* the isomorphism m but merely requires its existence, then one gets the space $\mathrm{PGL}(\mathcal{O}_L \oplus \mathfrak{a})^+\backslash\mathfrak{H}^g$). Here

$$\mathrm{GL}(\mathcal{O}_L \oplus \mathfrak{a}) = \{\gamma \in \mathrm{GL}_2(L) : \gamma^t(\mathcal{O}_L \oplus \mathfrak{a}) = {}^t(\mathcal{O}_L \oplus \mathfrak{a})\},$$

and the subgroup $\mathrm{GL}(\mathcal{O}_L \oplus \mathfrak{a})^+$ (resp. $\mathrm{SL}(\mathcal{O}_L \oplus \mathfrak{a})$) consists of elements with totally positive determinant (resp. deteminant 1). Writing (temporarily) $\mathrm{Emb}(L, \mathbb{C}) =$

$\{\phi_1, \dots, \phi_g\}$ the action of the element γ of $\mathrm{GL}(\mathcal{O}_L \oplus \mathfrak{a})$ is given by

$$\gamma(z_1, \dots, z_g) = (\phi_1(\gamma)z_1, \dots, \phi_g(\gamma)z_g),$$

where the action on each coordinate is as Möbius transformations.

More precisely, consider the lattice $\mathcal{O}_L \oplus \mathfrak{a}^{-1}$ in $\mathbb{R}^{2g}$. Every $\mathcal{O}_L$-linear symplectic pairing on it is of the form

$$E_\lambda\left((x_1, y_1), (x_2, y_2)\right) = \mathrm{Tr}_{L/\mathbb{Q}}\left(\lambda(x_1 y_2 - x_2 y_1)\right), \tag{1.1}$$

for some $\lambda \in L$. It is integral iff $\lambda \in \mathfrak{a}\mathcal{D}_L^{-1}$. Under the analytic uniformization given by associating to z in $\mathfrak{H}^g$ the HBAV $\mathbb{C}^g/(\mathcal{O}_L \cdot z + \mathfrak{a}^{-1})$, E_λ induces a polarization iff $\lambda \in \mathfrak{a}\mathcal{D}_L^{-1}$ and $\lambda \gg 0$. Note that under our conventions, the polarization module of $\mathcal{M}_L(\mathfrak{a})$ is $\left(\mathfrak{a}\mathcal{D}_L^{-1}, (\mathfrak{a}\mathcal{D}_L^{-1})^+\right)$.

Fix a strict ideal class $(\mathfrak{a}, \mathfrak{a}^+)$. The moduli problem of $\Gamma_0(p)$-level structure on $\mathcal{M}_L(\mathfrak{a})$ is of parameterizing a pair of Abelian schemes, $(A, \iota, m)/S$ and $(A', \iota', m')/S$, parameterized by $\mathcal{M}_L(\mathfrak{a})$, and an S-isogeny $f\colon (A, \iota, m) \to (A', \iota', m')$, such that the following diagram commutes:

$$\begin{array}{ccc} (\mathcal{M}_{A'}, \mathcal{M}_{A'}^+) & \xrightarrow{m'} & (\mathfrak{a}, \mathfrak{a}^+) \\ {\scriptstyle f^*}\downarrow & & \downarrow{\scriptstyle \times p} \\ \mathcal{M}_A, \mathcal{M}_A^+) & \xrightarrow{m} & (\mathfrak{a}, \mathfrak{a}^+). \end{array} \tag{1.2}$$

Clearly, $\mathrm{Ker}(f)$ determines (A', ι', m') uniquely and $\mathrm{Ker}(f)$ is an $\mathcal{O}_L$-invariant, isotropic subgroup of $A[p]$ of order p^g (with respect to any $\mathcal{O}_L$-linear polarization). To see that it is totally isotropic, note that we may find an $\mathcal{O}_L$-linear polarization of degree prime to p on A. By assumption $p\lambda$ descends to $A/\mathrm{Ker}(f)$. Hence $\mathrm{Ker}(f)$ is an isotropic subgroup with respect to the Mumford pairing induced by $p\lambda$ on $\mathrm{Ker}(p\lambda)$. Since λ is of degree prime to p, it induces a perfect pairing on $A[p]$. It follows that $\mathrm{Ker}(f)$ is isotropic with respect to every $\mathcal{O}_L$-linear polarization.

Conversely, let H be an $\mathcal{O}_L$-invariant isotropic subgroup of order p^g of $A[p]$ with respect to every $\mathcal{O}_L$-linear polarization. There exists a unique $\mathcal{O}_L$-structure ι' on $A' = A/H$ such that $f\colon (A, \iota) \to (A', \iota')$ (the canonical map) is $\mathcal{O}_L$-equivariant. Suppose that H has the property that $f^*(\mathcal{M}_{A'}) = p\mathcal{M}_A$. Then there exists a unique $m'\colon (\mathcal{M}_{A'}, \mathcal{M}_{A'}^+) \to (\mathfrak{a}, \mathfrak{a}^+)$ such that f becomes a $\Gamma_0(p)$-level structure. We elaborate more on this definition in Section 3.2 below.

Let $n \geq 3$ be prime to pd_L. Consider level n structure on $\Gamma_0(p)$-level structure. That is, the moduli space of $(f\colon (A, \iota, m) \to (A', \iota', m'), \alpha)/S$, where S is a scheme over $\mathbb{Z}[1/n]$, and α is a level n structure. This moduli space was studied in [**27**] and [**30**]. We shall denote this moduli space by $\mathcal{M}_0^n(p)$. It is a fine moduli space, with singularities supported over primes q such that $q|pd_L$.

The singularities of $\mathcal{M}_L^n$ were studied in [**3**] and of $\mathcal{M}_0^n(p)$ in [**27**] and [**30**]. The picture was complemented for $g = 2$ in [**1**] (In fact [**27**] deals with more general level structures). The results are as follows:

THEOREM 1.1 ([**3**, Théorème 2.2]). *The morphism $\mathcal{M}_L^n \to \mathrm{Spec}(\mathbb{Z}[n^{-1}])$ is flat, relative local complete intersection, of relative dimension g, and is smooth over $\mathrm{Spec}(\mathbb{Z}[(nd_L)^{-1}])$. For $p|d_L$ the singular locus of $\mathcal{M}_L^n \otimes \mathbb{F}_p$ is of codimension 2.*

THEOREM 1.2 ([**27**, Theorem 2.2.2], [**30**, Theorem 1]). *The morphism $\mathcal{M}_0^n(p)$ $\to \mathrm{Spec}(\mathbb{Z}[1/n])$ is flat, relative local complete intersection, of relative dimension g*

and is smooth over $\mathrm{Spec}(\mathbb{Z}[(pnd_L)^{-1}])$. *The singularities of the fibre over* p *are of codimension* 1.

THEOREM 1.3 ([**1**, Theorem 6.2]). *Let* L *be a real quadratic field and* $p|d_L$. *The singularities of the surface* $\mathcal{M}_L^n \otimes \mathbb{F}_p$ *are equal to the singular points of the supersingular locus. Each is ordinary with* $p+1$ *branches. Each component of the supersingular locus is a rational curve, and there are* $p+1$ *singular points on each component.*

We remark that in the last theorem each singular point is superspecial, but not vice-versa.

2. Stratifications of Hilbert Modular Varieties

2.1. Stratifications. There are various ways in which one may attempt to stratify moduli spaces of Abelian varieties in characteristic p. Some of the main problems in the theory of moduli spaces of Abelian varieties are to understand the local nature of these stratifications and how they relate to each other.

2.1.1. *The Ekedahl-Oort Stratification.* This stratification was defined and studied in [**4**] (see also [**26**], [**16**], [**6**]). It is defined by considering the isomorphism type of the p-torsion subgroup scheme. In the references cited, it is defined for the moduli space of principally polarized Abelian varieties, but certainly one can define a similar stratification for other PEL moduli problems.

The main idea is the following: Let $(A,\lambda)/k$ be a principally polarized Abelian variety over a perfect field k of characteristic p. The polarization λ induces a perfect alternating pairing of group schemes

$$\langle\ ,\ \rangle_\lambda : A[p]\times A[p] \rightarrow \mathbb{G}_m.$$

Letting $(A,\langle\ ,\ \rangle_\lambda)$ vary, one finds that there are finitely many isomorphism types of pairs $(A,\langle\ ,\ \rangle_\lambda)$, over an algebraically closed field. One may define a locally closed scheme carrying all the moduli points x corresponding to pairs (A,λ) with $(A[p],\langle\ ,\ \rangle_\lambda)$ of a fixed isomorphism type over the algebraic closure. Collecting these locally closed schemes to a stratification is more delicate and involves the following combinatorics:

The group $A[p]$ is endowed with a canonical filtration defined by taking a minimal flag of subgroup schemes stable under Ver and under taking perpendiculars with respect to the pairing induced by λ. One can (non-canonically) complete it to a maximal flag of group schemes

$$0 = G_0 \subseteq G_1 \subseteq G_2 \subseteq \cdots \subseteq G_{2g} = A[p],$$

where G_i has order p^i and the flag is stable under Ver and taking perpendiculars. One then has a function,

$$\varphi\colon \{0,\dots,2g\} \rightarrow \{0,\dots,2g\},$$

given by

$$\mathrm{Ver}(G_i) = G_{\varphi(i)}.$$

The function φ is independent of the way one completes the canonical flag to a maximal flag.

THEOREM 2.1 ([**4, 26**]). *The function* φ *determines the isomorphism type of the pair* $(A[p],\langle\ ,\ \rangle_\lambda)$ *over an algebraically closed field.*

We remark that φ is determined by its restriction to $\{1, \dots, g\}$ and it is a monotone function satisfying $\varphi(i+1) \leq \varphi(i)+1$ for $1 \leq i \leq g$. Conversely, every such φ is obtained from a suitable pair $(G, \langle\ ,\ \rangle)$, where $G/\bar{k}$ is a finite commutative group scheme of order p^{2g} killed by p, and $\langle\ ,\ \rangle\colon G \times G \to \mathbb{G}_m$ is a perfect alternating pairing.

Some data about G may be conveniently read from φ. Let $a(G)$ denote $\dim \operatorname{Hom}(\alpha_p, G)$, and let $f(G)$ denote the p-rank of the maximal étale quotient $G^{\text{ét}}$ of G, that is $G^{\text{ét}}$ has order $p^{f(G)}$. One has $a(G) = g - \varphi(g)$ and $f(G)$ is the i such that $\varphi(i) = i = \varphi(i+1)$.

The stratification defined by Ekadahl-Oort is the following. Let S_φ be the locally closed subscheme carrying the moduli points of (A, λ) such that $(A[p], \langle\ ,\ \rangle_\lambda)$ is of the isomorphism class corresponding to φ. Let $|\phi| = \sum_{i=1}^{g} \varphi(i)$. Let

$$\mathbb{S}_d = \bigcup_{|\varphi|=d} S_\varphi.$$

THEOREM 2.2 ([**4, 26**]). *The sets $\mathbb{S}_d$ for $0 \leq d \leq g(g+1)/2$ are smooth upon taking level $n \geq 3$. They define a stratification of the moduli space of principally polarized g-dimensional Abelian varieties (with level n structure), $\mathcal{A}_g$, over $\mathbb{F}_p$. The dimension of $\mathbb{S}_d$ is d, $\mathcal{A}_g = \bigcup_{d=0}^{g(g+1)/2} \mathbb{S}_d$, and $\overline{\mathbb{S}_d} = \bigcup_{i=0}^{d} \mathbb{S}_i$.*

2.1.2. *The Newton Polygon Stratification.* The main idea here is to examine the variation of the isogeny class of a family of p-divisible groups. Given a g-dimensional Abelian variety A over a perfect field k, its p-divisible group $A(p)$ falls into one of finitely many isogeny types which may be described by polygons. These polygons—the Newton polygons—are convex polygons starting at $(0,0)$ and ending at $(2g, g)$, having breaking points with integer coordinates and slopes that satisfy the symmetry condition: a slope α appears if and only if a slope $1-\alpha$ appears, and with the same multiplicity (Manin's symmetry condition). See [**18**, Chapter IV].

The set of Newton polygons for a fixed g is a partially ordered set under the following definition: $\alpha \leq \beta$ if no point of the polygon α lies over the polygon β. We remark that different conventions are sometimes used, e.g., in [**25**].

The key fact is that *under specialization Newton polygons go up.* See [**14**, Theorem 2.3.1], [**15**], [**9**]. Fixing a Newton polygon α, we may consider the closed subscheme W_α "where the Newton polygon is greater or equal to α". Precise conjectures on the behaviour of this stratification appear in [**24**]. Important progress on them was recently made in [**10**].

2.2. The Type. Returning to the setting of HBAV's over $\mathbb{F}_p$, where $p \nmid d_L$, we define another stratification following [**8**]. This stratification seems to be more amenable to arithmetic applications. For that reason we suggest a more general setting in Section 5.

Let $\mathbb{F}$ be the minimal finite field of characteristic p such that

$$\mathcal{O}_L \otimes_{\mathbb{Z}} W(\mathbb{F}) = \bigoplus_{\sigma \in \operatorname{Emb}(\mathcal{O}_L, W(\mathbb{F}))} W(\mathbb{F}), \tag{2.1}$$

where $W(\mathbb{F})$ denotes the ring infinite Witt vectors with coordinates in $\mathbb{F}$. Thus, if

$$p = \mathfrak{p}_1 \dots \mathfrak{p}_r, \quad f_i = \deg(\mathfrak{p}_i), \tag{2.2}$$

then $[\mathbb{F} : \mathbb{F}_p] = \operatorname{lcm}(f_1, \ldots, f_r)$. *From now on we consider $\mathcal{M}_L^n$ as a scheme over* Spec$(\mathbb{F})$. Let

$$\{e_{(j,i)} : 1 \leq j \leq r, 1 \leq i \leq f_j\}$$

be the idempotents corresponding to equation (2.1).. Thus, in equation (2.1), an element $a \in \mathcal{O}_L$ is mapped to $(e_{(j,i)}a)$. We may also interprete the $e_{(j,i)}$'s as $W(\mathbb{F})$-valued characters on $\mathcal{O}_L$, and then we denote them by $\chi_{(j,i)}$.

We write accordingly,

$$\{\sigma_{(j,i)} : 1 \leq j \leq r, 1 \leq i \leq f_j\},$$

for the embeddings of $\mathcal{O}_L$ into $W(\mathbb{F})$, and we assume the indexing is chosen so that

$$\sigma \circ \sigma_{(j,i)} = \sigma_{(j,i+1)}, \quad \forall j, i,$$

where σ is the Frobenius homomorphism on $W(\mathbb{F})$ and, by definition, $(j, f_j + 1) = (j, 1)$. We shall abuse notation and write $e_{(j,i)}, \sigma_{(j,i)}$ and $\chi_{(j,i)}$ for the corresponding objects mod p.

Let $k \supseteq \mathbb{F}$ be a perfect field. Let (A, ι) be a HBAV over k. Let

$$\alpha(A) = \ker(\mathrm{Fr}\colon A \to A^{(p)}) \cap \mathrm{Ker}(\mathrm{Ver}\colon A \to A^{(1/p)}).$$

It is a characteristic subgroup scheme of A of order p^a, where $a = \dim_k \operatorname{Hom}(\alpha_p, A)$. Let $\mathbb{D}_\alpha$ denote the contravariant Dieudonné module of $\alpha(A)$. Since $\alpha(A)$ is $\mathcal{O}_L$-invariant, the module $\mathbb{D}_\alpha$ is invariant under the induced action of $\mathcal{O}_L \otimes k$. Let us denote the indices of the characters appearing in the action of $\mathcal{O}_L$ on the a-dimensional k-vector space $\mathbb{D}_\alpha$ by

$$\tau(A) = (\tau_1(A), \ldots, \tau_r(A)), \quad \tau_j(A) \subseteq \{(j,i) : 1 \leq i \leq f_j\}. \tag{2.3}$$

That is, $\chi_{(j,i)}$ appears in the action of $\mathcal{O}_L$ on $\mathbb{D}_\alpha$ iff $(j,i) \in \tau_j(A)$. We remark that by [**3**, Corollaire 2.9], if $\chi_{(j,i)}$ appears, then it appears with multiplicity 1. We call $\tau(A)$ the *type* of A.

Given a vector $\tau = (\tau_1, \ldots, \tau_r)$, where $\tau_j \subseteq \{(j,i) : 1 \leq i \leq f_j\}$, we let W_τ be the minimal closed subscheme (with the reduced induced structure) of $\mathcal{M}_L^n$ (or $\mathcal{M}_L$) parameterizing pairs (A, ι) over a perfect field k, such that $\tau(A)$ contains τ (i.e., $\tau_i(A)$ contains τ_i for all i). To ease notation, write $W_{(j,i)}$ for W_τ, when τ is the type $(\tau_1, \ldots, \tau_r)$ with $\tau_s = \varnothing$ if $s \neq j$ and $\tau_j = \{(j,i)\}$.

2.3. Reduction to the Inert Case. In [**8**], a stratification of the moduli space of principally polarized Abelian schemes in characteristic p with real multiplication was defined under the assumption that p is inert in L. We indicate how to generalize that to any p not dividing d_L.

Given a pro-representable functor, G, on local Artinian k-algebras, with $G(k)$ a singleton, we denote by $\mathcal{F}G(k)$ the corresponding formal scheme representing it. Recall that a principal quasi-polarization on a p-divisible group, H, is a symmetric isomorphism $H \to H^t$. Note the following three points:

- Fix (A, ι) over k with some level n structure, which we shall suppress from the notation. Let $\lambda \in \mathcal{M}_A^+$ with $(\lambda, p) = 1$. Consider the functor of local deformations of (A, ι) (resp. $(A, \iota, (\mathcal{M}_A, \mathcal{M}_A^+))$; resp. (A, ι, λ)) into HBAS (resp. HBAS with an ordered polarization module; resp. polarized HBAS). Then the natural maps

$$\mathcal{F}(A, \iota, \lambda) \to \mathcal{F}(A, \iota, (\mathcal{M}_A, \mathcal{M}_A^+) \to \mathcal{F}(A, \iota), \tag{2.4}$$

are isomorphisms.

• Let $A(p)$ be the p-divisible group of A. Let $i(p)\colon \mathcal{O}_L \to \mathrm{End}(A(p))$ and $\lambda(p)$ be the induced endomorphism structure and principal quasi-polarization on $A(p)$, respectively. Let $i(p)\otimes 1 : \mathcal{O}_L \otimes \mathbb{Z}_p \to \mathrm{End}(A(p))$ be the canonical map. Consider the functor of local deformation of (A,ι,λ) (resp. $(A(p),\iota(p),\lambda(p))$; resp. $(A(p),\iota(p)\otimes 1,\lambda(p)))$ into polarized HBAS (resp. principally quasi-polarized p-divisible groups with real multiplication by $\mathcal{O}_L$; resp. p-divisible groups with $\mathcal{O}_L\otimes\mathbb{Z}_p$ action and $\mathcal{O}_L\otimes\mathbb{Z}_p$ linear principal quasi-polarization). Then the following maps are isomorphisms:

$$\mathcal{F}(A,\iota,\lambda) \to \mathcal{F}(A(p),\iota(p),\lambda(p)) \to \mathcal{F}(A(p),\iota(p)\otimes 1,\lambda(p)). \tag{2.5}$$

• The decomposition (2.2) induces a decomposition

$$\mathcal{O}_L \otimes \mathbb{Z}_p = \bigoplus_{j=1}^{r} W(\mathbb{F}_{p^{f_j}}),$$

which, in turn, induces p-divisible groups A_j, homomorphisms

$$\iota_j\colon W(\mathbb{F}_{p^{f_j}}) \to \mathrm{End}(A_j),$$

and principal quasi-polarizations λ_j on A_j, such that

$$A(p) = A_1 \times \cdots \times A_r, \tag{2.6}$$

and under this isomorphism

$$\iota(p)\otimes 1 = \iota_1 \times \cdots \times \iota_r; \quad ;\lambda(p) = \lambda_1 \times \dots \lambda_r.$$

We have

$$\mathcal{F}(A(p),\iota(p),\lambda(p)) = \mathcal{F}(A_1,\iota_1,\lambda_1) \underset{\mathrm{Spec}(k)}{\times} \cdots \underset{\mathrm{Spec}(k)}{\times} \mathcal{F}(A_r,\iota_r,\lambda_r).$$

We remark that the first point can be proved using covariant Dieudonné theory by writing down explicitly the deformation matrices (see, e.g., [**8**] for such computations), or by crystalline techniques (see [**29**, Proposition 1.9]). The second point follows from the Serre-Tate Theorem, while the third is purely formal.

We further remark that the decomposition of the p-divisible group in (2.6) need not correspond to any decomposition of the Abelian variety itself. This is clear if one bears in mind that a HBAV is isotypical while in general the p-divisible groups A_i in (2.6) are not isogenous.

2.4. The Main Results. The deformation theory of (A_j,ι_j,λ_j) was studied in detail in [**8**]. It is easy now to draw the following conclusions for $\mathcal{M}_L^n$ (and with small modifications for $\mathcal{M}_L$):

- The scheme W_τ is regular.
- The generic point of every component of W_τ has type τ.
- The dimension of W_τ is $g-|\tau|$, where $|\tau| := \sum_{j=1}^r |\tau_j|$.
- We have $W_\tau \cap W_\rho = W_{\tau\cup\rho}$, where $\tau\cup\rho := (\tau_1\cup\rho_1,\dots,\tau_r\cup\rho_r)$.
- The subscheme W_τ^0, defined as

$$W_\tau \setminus \bigcup_{\substack{\rho\supset\tau,\\ \rho\neq\tau}} W_\rho,$$

  is quasi-affine.
- The components of all the W_τ's have a natural structure of a simplicial complex.

- Let $\mathbb{E} = \det \pi_* \mathbf{t}_{\mathfrak{X}_L^n,0}$ be the Hodge bundle derived from the universal object $\pi\colon \mathfrak{X}_L^n \to \mathcal{M}_L^n$. Let
$$W_1 = \bigcup_{\substack{1\le j\le r,\\ 1\le i\le f_j}} W_{(j,i)}.$$
Then, in the Chow ring of $\mathcal{M}_L^n$,
$$W_1 = (p-1)\cdot c_1(\mathbb{E}).$$
- Let (A_1,ι_1), (A_2,ι_2) be two HBAV's over an algebraically closed field k of characteristic p. We have $A_1[p]\cong A_2[p]$ as groups schemes with $\mathcal{O}_L$-action if and only if $\tau(A_1)=\tau(A_2)$.
- The p-divisible groups, up to isogeny, are precisely those of the form
$$\bigoplus_{j=1}^{r} A_j,$$
where A_j is of the form $G_{\ell/f_j}+G_{(f_j-\ell)/f_j}$ or $f_j\cdot G_{1/2}$. The Newton polygons will be denoted accordingly: $\beta_{f_j,0}<\beta_{f_j,1}<\cdots$. Thus, given a HBAV over k, we can associate to it an r-tuple of Newton polygons $(\gamma_1,\dots,\gamma_r)$; each γ_j is equal to $\beta_{f_j,t}$ for an appropriate t depending on j.
- Let $\lambda(\tau_j)=\max\{|\rho| : \rho\subseteq\tau_j, \rho \text{ spaced}\}$ (ρ "spaced" means $i\in\rho \implies i+1\notin\rho$), unless $|\tau_j|=f_j$ is odd and then we let $\lambda(\tau_j)=(f_j+1)/2$.

  The Newton polygon of the generic point of every component of W_τ is
$$(\beta_{f_1,\lambda(\tau_1)},\dots,\beta_{f_r,\lambda(\tau_r)}).$$

2.5. Examples.

2.5.1. *The Case $g=1$.* In this case $L=\mathbb{Q}$. This is the case of the moduli space of elliptic curves. The types are subsets of $\{(1,1)\}$. The stratification amounts to $W_\varnothing$ being the whole space, and $W_{(1,1)}$ being the supersingular locus. Some of the claims made above are: (i) An elliptic curve is supersingular if and only if one may embedd α_p in it; (ii) An elliptic curve is supersingular if and only if its p-divisible group is isogenous to $G_{1,1}$; (iii) The supersingular locus is reduced; (iv) The possible p-divisible groups for elliptic curves are isogenous to $G_{1,0}\oplus G_{0,1}$ and $G_{1,1}$; (v) One can relate the genus to the number of supersingular points.

2.5.2. *The Case $g=2$* [**1, 27, 30**].

(A) The Prime p Is Inert in L. In this case the types are subsets of the set $\{(1,1),(1,2)\}$. Evidently $W_\varnothing$ is the whole space. One may verify that $W_{(1,1)}\cup W_{(1,2)}$ is precisely the supersingular locus. In particular the p-divisible group is isogenous to $2\cdot G_{1,1}$. The intersection points $W_{(1,1)}\cap W_{(1,2)}$ are the reduced zero dimensional cycle $W_{\{(1,1),(1,2)\}}$, which is in fact exactly the superspecial points.

THEOREM 2.3 ([**1**, Theorem 5.3], [**30**, Theorem 3]). *Take level $n\ge 3$ and consider the case of principal polarization. That is the component $\mathcal{M}_L^n(\mathcal{D}_L)\otimes\mathbb{F}_{p^2}$. Each $W_{(1,i)}$ is a disjoint union of smooth rational curves. The number of components of $W_{(1,i)}$ is*
$$\frac{1}{2}\cdot[\mathcal{M}_L^n(\mathcal{D}_L):\mathcal{M}_L(\mathcal{D}_L)]\cdot\zeta_L(-1).$$

(B) The Prime p Is Split in L. In this case the types are subsets of $\{(1,1),(2,1)\}$. The scheme $W_\varnothing$ is the whole space. We have that $W_{(i,1)}$ is a disjoint union of smooth curves if the level $n \geq 3$ and that the intersection $W_{(1,1)} \cap W_{(2,1)} = W_{\{(1,1),(2,1)\}}$ is reduced and equal to the superspecial locus (p split implies that every supersingular point is superspecial). The generic point of $W_{(i,1)}$ has a p-divisible group isogenous to $G_{1,0} \oplus G_{0,1} \oplus G_{1,1}$.

QUESTION 1. In case (A), can one cover the supersingular locus by reduction of Shimura curves "coming from" the generic fibre? (Those will necessarily be associated to quaternion algebras over $\mathbb{Q}$ in which p ramifies).

QUESTION 2. In case (B), it is easy to see that every Shimura curve will intersect $W_{(1,1)} \cup W_{(2,1)}$ in a finite number of points. Can one find an explicit description of at least one component of $W_{(1,1)} \cup W_{(2,1)}$?

QUESTION 3. What can one say about the intersection graphs of $W_{(1,1)} \cup W_{(1,2)}$ in case (A) and $W_{(1,1)} \cup W_{(2,1)}$ in case (B)?

2.5.3. *The Case* $g = 3$. We assume that p is inert (the other cases being simpler), and that the level $n \geq 3$. The types are subsets of $\{(1,1),(1,2),(1,3)\}$, which we abbreviate by $\{1,2,3\}$.

The scheme $W_\varnothing$ is the whole space. The schemes W_i are regular surfaces, having generically p-divisible group isogenous to $G_{1,2} \oplus G_{2,1}$. The schemes W_{ij} for $\{i,j\}$ equal $\{1,2\},\{1,3\}$ or $\{2,3\}$, are a disjoint union of smooth rational curves. They are precisely the locus of those Abelian surfaces with formal group isomorphic to $G_{1,2} \oplus G_{2,1}$ or $3G_{1,1}$. The supersingular locus can be shown to consist of rational smooth curves. These curves intersect transversely at the points of $W_{\{1,2,3\}}$ and conversely, every such superspecial point is the intersection of three components of the supersingular locus (one of each type $\{1\},\{2\},\{3\}$). In short, the supersingular locus is a disjoint union of bouquets of 3 rational curves of different types. Their total number is given by a class number of a quaternion algebra.

QUESTION 4. Can one explicitly parameterize and determine the local structure of the supersingular locus in $\mathcal{M}_L^n$? It is likely that the techniques of [**17**] will be of importance.

2.6. Some Indications About the Proofs. The main technique is to decompose every canonical linear or semi-linear object as an $\mathcal{O}_L \otimes W(\mathbb{F})$ module. In particular, given (A,ι) over a perfect field k of characteristic p, this can be carried out for the covariant Dieudonné module of its p-divisible group $A(p)$, which we denote by $\mathbb{D}$, and for the covariant Dieudonné module of $A[p]$, which we denote by $\mathbb{D}_p$. It is the reduction of $\mathbb{D}$ modulo p.

We assume for the rest of this section that p is inert in L. The general case is similar enough to leave to the reader. Recall from Section 2.2 that $\mathbb{F} = \mathbb{F}_{p^g}$ in this case and write $\{\sigma_1,\dots,\sigma_g\}$ (cyclically ordered) for $\mathrm{Emb}(L, W(\mathbb{F}))$ and we let e_i and χ_i denote the corresponding idempotents and characters, respectively.

To discuss the local structure we found it convenient to choose an $\mathcal{O}_L$-linear polarization of degree prime to p on (A,ι). Since the corresponding map between the moduli spaces is formally étale this is justified.

Using equation (2.1) we get a decomposition

$$\mathbb{D} = \bigoplus_{i\in\mathbb{Z}/g\mathbb{Z}} \mathbb{D}_i, \tag{2.7}$$

with the following properties:

(i) $F(\mathbb{D}_i) \subseteq \mathbb{D}_{i+1}$, $\dim_k(\mathbb{D}_{i+1}/F(\mathbb{D}_i)) = 1$.
(ii) $V(\mathbb{D}_i) \subseteq \mathbb{D}_{i-1}$, $\dim_k(\mathbb{D}_{i-1}/V(\mathbb{D}_i)) = 1$.
(iii) $\mathcal{O}_L$ acts on $\mathbb{D}_i$ via σ_i.
(iv) The pairing $\langle \cdot, \cdot \rangle$ induces on each $\mathbb{D}_i$ a perfect alternating pairing $\langle \cdot, \cdot \rangle_i$ and $\mathbb{D}_i \perp \mathbb{D}_j$ for $i \neq j$.

DEFINITION 2.4. An admissible basis for $\mathbb{D}$, given the polarization and the $\mathcal{O}_L$-structure, is a $W(k)$-basis $\{X_1, \ldots, X_g, Y_1, \ldots, Y_g\}$ for $\mathbb{D}$, such that:
(i) $\{X_i, Y_i\}$ is a symplectic basis for $\mathbb{D}_i$;
(ii) $Y_i \in V(\mathbb{D}_{i+1})$.

Let

$$\begin{pmatrix} A & B \\ C & D \end{pmatrix} \tag{2.8}$$

be the corresponding display for such a basis. Note that for suitable a_i, b_i, c_i, d_i, we have:

$$\begin{aligned} A &= \mathfrak{d}_2(a_2, \ldots, a_g, a_1), \quad & B &= \mathfrak{d}_2(b_2, \ldots, b_g, b_1), \\ C &= \mathfrak{d}_2(c_2, \ldots, c_g, c_1), \quad & D &= \mathfrak{d}_2(d_2, \ldots, d_g, d_1), \end{aligned} \tag{2.9}$$

where the notation $\mathfrak{d}_2(a_2, \ldots, a_g, a_1)$ indicates a matrix $A = (a_{ij})$ with zero entries except for $a_{21} = a_2$, $\ldots$, $a_{g,g-1} = a_g$ and $a_{1g} = a_1$.

We refer the reader for the theory of displays to [**20**], [**21**] and [**32**]. We refer for the specific claims made and a survey of the theory needed to [**8**, Section 2]. We comment that:
(i) The matrix $\left(\begin{smallmatrix} A & pB \\ C & pD \end{smallmatrix}\right)$ is the matrix of Frobenius with respect to the admissible basis.
(ii) The matrix A is the Hasse-Witt matrix.
(iii) The universal local deformation of (A, ι, λ) is given by $\left(\begin{smallmatrix} A+TC & B+TD \\ C & D \end{smallmatrix}\right)$, where $T = \mathrm{diag}(T_1, \ldots, T_g)$ (the T_i being Teichmuller lifts of the free variables t_i). It "lives" over $\mathrm{Spec}(k[[t_1, \ldots, t_g]])$. By the theory of displays, such a matrix defines a Dieudonné module in "displayed form", which we claim is the Dieudonné module of the universal local deformation of (A, ι, λ).

Granted these facts, one proves that i belongs to $\tau(A)$ if and only if $a_i = 0 \pmod p$, and following the same reasoning, that the closed subscheme of the formal deformation space cut by the condition "i belongs to the type" is given by the equation $t_i = 0$.

To see this, note that under the covariant Dieudonné functor, $\mathrm{Ker}(\mathrm{Ver}\,|_{A[p]})$ is mapped to $\mathrm{Ker}(\mathrm{Fr}\,|_{\mathbb{D}_p})$, which is spanned by the images of $Y_1, \ldots, Y_g$. Evidently

$$\mathbb{D}(\alpha(A))_i \neq 0 \implies \mathrm{Ker}(\mathrm{Fr}\,|_{\mathbb{D}_p})_i = \mathrm{Ker}(\mathrm{Ver}\,|_{\mathbb{D}_p})_i.$$

This happens if and only if $\mathrm{Fr}(\mathbb{D}_{p,i-1}) = \mathrm{Span}_k\{Y_i\}$. That is, if and only if $a_i = 0$.

3. The Structure of the p-Torsion

3.1. The Type Determines $A[p]$. We keep the simplifying assumption that p is inert in L and remark again that the general case is similar. Return to the decomposition (2.7). It gives upon reduction (mod p):

$$\mathbb{D}_p = \bigoplus_{i \in \mathbb{Z}/g\mathbb{Z}} \mathbb{D}_{p,i}. \tag{3.1}$$

Each $\mathbb{D}_{p,i}$ is a two dimensional vector space over k equiped with a perfect alternating pairing. Note that for *every* $\mathcal{O}_L$-invariant subgroup $H \subseteq A[p]$, we have $\mathbb{D}(H) = \oplus_i \mathbb{D}(H)_i$. Moreover, H is isotropic iff $\dim \mathbb{D}(H)_i \leq 1$ for all i. Both groupschemes $\mathrm{Ker}(\mathrm{Fr})$ and $\mathrm{Ker}(\mathrm{Ver})$ are examples of such isotropic subgroups. Note that since we are using the covariant theory, $\mathbb{D}(\mathrm{Ker}(\mathrm{Fr})) = \mathrm{Ker}(\mathrm{Ver})$, where the right hand side denotes that submodule of $\mathbb{D}_p$ killed by Ver.

Recall the theorem: *The isomorphism class of* $(A[p], \iota, \langle\ ,\ \rangle_\lambda)$ *depends only on* $\tau(A)$. We illustrate how this works in the case when $|\tau(A)|$ is even. Assume without loss of generality that $1 \in \tau(A)$ and write $\tau(A) = \{1, i_1, \ldots, i_a\}$. Consider the following choice of basis for $\mathbb{D}(A)$:

Start by taking some non-zero $y_1 \in \mathrm{Ker}(F)$ and x_1 such that $\langle x_1, y_1 \rangle = 1$. Hence $x_1 \notin \mathrm{Ker}(F)$. Then choose:

definition of x_i	**definition of y_i**	i
x_1	$y_1 \in \mathrm{Ker}(F)$	1
$x_2 = \mathrm{Fr}(x_1)$	$y_2 \in \mathrm{Ker}(F)\colon\ \mathrm{Ver}(y_2) = y_1$	2
$\vdots$	$\vdots$	$\vdots$
$x_{i_2-1} = \mathrm{Fr}(x_{i_2-2})$	$y_{i_2-1} \in \mathrm{Ker}(F)\colon\ \mathrm{Ver}(y_{i_2-1}) = y_{i_2-2}$	$i_2 - 1$
$x_{i_2} = \mathrm{Fr}(x_{i_2-1})$	$y_{i_2}\colon\ \mathrm{Ver}(y_{i_2}) = y_{i_2-1}$	i_2
$x_{i_2+1} \in \mathrm{Ker}(\mathrm{Fr})\colon\ \mathrm{Ver}(x_{i_2+1}) = x_{i_2}$	$y_{i_2+1} = \mathrm{Fr}(y_{i_2})$	$i_2 + 1$
$x_{i_2+2} \in \mathrm{Ker}(\mathrm{Fr})\colon\ \mathrm{Ver}(x_{i_2+2}) = x_{i_2+1}$	$y_{i_2+2} = \mathrm{Fr}(y_{i_2+1})$	$i_2 + 2$
$\vdots$	$\vdots$	$\vdots$
		$i_3 - 1$
$x_{i_3}\colon\ \mathrm{Ver}(x_{i_3}) = x_{i_3-1}$	$y_{i_3} = \mathrm{Fr}(y_{i_3-1})$	i_3
$\vdots$	$\vdots$	$\vdots$
$x_{i_a} = \mathrm{Fr}(x_{i_a-1})$	$y_{i_a}\colon\ \mathrm{Ver}(y_{i_a}) = y_{i_a-1}$	i_a
$x_{i_a+1} \in \mathrm{Ker}(\mathrm{Fr})\colon\ \mathrm{Ver}(x_{i_a+1}) = x_{i_a}$	$y_{i_a+1} = \mathrm{Fr}(y_{i_a})$	i_{a+1}
$\vdots$	$\vdots$	$\vdots$
$x_g \in \mathrm{Ker}(\mathrm{Fr})\colon\ \mathrm{Ver}(x_g) = x_{g-1}$	$y_g = \mathrm{Fr}(y_{g-1})$	g
$x_1'\colon\ \mathrm{Ver}(x_1') = x_g$	$y_1' = \mathrm{Fr}(y_g)$	

Note that for every r we have, by induction, $\langle x_r, y_r \rangle = 1$. Hence, x_r and y_r form a symplectic basis for $\mathbb{D}(A[p])_r$. Note that

$$y_1' \in \mathrm{Im}\left(\mathrm{Fr}\,|_{\mathbb{D}(A[p])_g}\right) = \mathrm{Ker}\left(\mathrm{Ver}\,|_{\mathbb{D}(A[p])_1}\right) = \mathrm{Ker}\left(\mathrm{Fr}\,|_{\mathbb{D}(A[p])_1}\right)$$

Therefore, using that k is algebraically closed, and taking a suitable multiple of y_1 we may assume that $y_1' = y_1$. Since $\langle x_1', y_1 \rangle = \langle x_1, y_1 \rangle = 1$ we have $x_1 - x_1' \in \mathrm{Ker}(\mathrm{Fr})$. Thus we may also assume that $x_1' = x_1$.

This procedure of choosing bases gives immediately that the type determines the isomorphism as polarized $\mathcal{O}_L$-groups. It also shows that the isomorphism type of $(A[p], \iota, \langle\ ,\ \rangle)$, for an $\mathcal{O}_L$-linear perfect pairing, depends only on $(A[p], \iota)$ and hence the theorem agrees with the previous formulation given in Section 2.4. The converse is much easier.

QUESTION 5. Assume that p is inert and we are given (A_1,ι_1,λ_1) and (A_2,ι_2,λ_2) over an algebraically closed field of charcateristic p such that $(A_1[p], \langle\ ,\ \rangle_{\lambda_1})$ is isomorphic to $(A_2[p], \langle\ ,\ \rangle_{\lambda_2})$ (the isomorphism needs not preserve the $\mathcal{O}_L$-action!). Is it true that $\tau(A_1)$ and $\tau(A_2)$ are equal up to a cyclic shift? If so, one could succintly say that "the Ekedahl-Oort stratification is obtained from the type stratification by recalling the $\mathbb{F}_p$ structure".

3.2. $\Gamma_0(p)$-Level Structure. To simplify notation, we retain the assumption that p is inert in L. We tacitly assume that we work with level $n \geq 3$ structure, hence our objects have no automorphisms. The level n structure will be suppressed from the notation.

The notion of $\Gamma_0(p)$ level structure was defined in Section 1.2 above. It was studied in [**27**] and [**30**]. We supplement them by explaining how the structure of the fibres of the morphism

$$\mathcal{M}_0^n(p) \otimes \overline{\mathbb{F}}_p \to \mathcal{M}_L^n \otimes \overline{\mathbb{F}}_p$$

may be easily read from the type. As the reader will see, the structure of the fibres is quite complicated. Usually the components of a fibre have different dimensions, and the fibres themselves depend very strongly on the type. Thus

QUESTION 6. Is there another notion of level p yielding "well behaved" fibres mod p?

We shall only treat the case where our $\Gamma_0(p)$-level structures are on triples (A, ι, m), where m is an isomorphism $(\mathcal{M}_A, \mathcal{M}_A^+) \to (\mathcal{D}_L^{-1}, \mathcal{D}_L^{-1+})$.

A given (A, ι) possesses such an m if and only if it has an $\mathcal{O}_L$-linear *principal* polarization λ_0. On the other hand, we saw in Section 1.2 that giving such a $\Gamma_0(p)$-level, $f : (A, \iota, m) \to (A', \iota', m')$, is equivalent to giving a totally isotropic $\mathcal{O}_L$-invariant subgroup H of $A[p]$, of order p^g, having the property that the pull back of the $\mathcal{O}_L$-linear symmetric homomorphisms of A/H are precisely $p \cdot \mathcal{M}_A$. In the case at hand, this is equivalent to $p\lambda_0$ descending to A/H, which always holds because H is isotropic.

To sum up: The $\Gamma_0(p)$-level structures on $(A, \iota, m\colon (\mathcal{M}_A, \mathcal{M}_A^+) \to (\mathcal{D}_L^{-1}, \mathcal{D}_L^{-1+}))$ are given by $\mathcal{O}_L$-invariant subgroups of $A[p]$, of order p^g and isotropic with respect to every $\mathcal{O}_L$-linear polarization.

Let $f\colon (A, \iota, m) \to (A', \iota', m')$ be a $\Gamma_0(p)$-level structure of HBAV's over a perfect field k of characteristic p. In [**30**] we find the following definition: Let f' be the unique isogeny $f'\colon A' \to A$ such that $f' \circ f = p$. An index i is called *critical* if $\mathrm{Lie}(f)_i = \mathrm{Lie}(f')_i = 0$ as maps.

We may identify $\mathrm{Lie}(A)$ with the dual of $\mathcal{D}(\mathrm{Ker}(\mathrm{Fr}\,|_{A[p]}))$, i.e., with $\mathrm{Ker}(\mathrm{Ver}\,|_{\mathbb{D}_p})$. Thus, $\mathrm{Lie}(f)_i = 0$ iff $\mathbb{D}(\mathrm{Ker}(f))_i = \mathrm{Ker}(\mathrm{Ver}\,|_{\mathbb{D}_p})_i$, that is if $\mathrm{Ver}(\mathbb{D}(\mathrm{Ker}(f))_i) = 0$. Similarly, $\mathrm{Lie}(f')_i = 0$ iff $\mathrm{Ver}(\mathbb{D}(\mathrm{Ker}(f'))_i) = 0$. However, $\mathbb{D}(\mathrm{Ker}(f')) = \mathbb{D}_p/\mathbb{D}(\mathrm{Ker}(f))$. Thus, $\mathrm{Lie}(f')_i = 0$ if and only if $\mathrm{Ver}(\mathbb{D}_{p,i}) = \mathbb{D}(\mathrm{Ker}(f))_{i-1}$, i.e., iff $\mathbb{D}(\mathrm{Ker}(f))_{i-1} = \mathrm{Ker}(\mathrm{Fr}\,|_{\mathbb{D}_p})_{i-1}$. To recapitulate: i is a critical index iff $\mathbb{D}(\mathrm{Ker}(f))_{i-1} = \mathrm{Ker}(\mathrm{Fr}\,|_{\mathbb{D}_p})_{i-1}$ and $\mathbb{D}(\mathrm{Ker}(f))_i = \mathrm{Ker}(\mathrm{Ver}\,|_{\mathbb{D}_p})_i$.

THEOREM 3.1 ([**30**, Theorem 3.5]). *Let S be the set of critical indices of f and $J = \mathbb{Z}/g\mathbb{Z} \setminus S$. The completion of the local ring of the k-rational point of $\mathcal{M}_0^n(p)$ defined by f is isomorphic to*

$$W(k)[[Z_j]]_{j \in J}[[X_s, Y_s]]_{s \in S}/(X_s Y_s - p : s \in S).$$

To gain some insight about the structure of the *fibres* of

$$\mathcal{M}_0^n(p) \otimes \overline{\mathbb{F}}_p \to \mathcal{M}_L^n(\mathcal{D}_L) \otimes \overline{\mathbb{F}}_p,$$

we propose the following:

Construct a directed graph as follows:

- For every $i \in \mathbb{Z}/g\mathbb{Z}$ the following vertices:
  - $(i, \circ)$, (i, V), (i, F) if $i \notin \tau(A)$.
  - $(i, \circ)$, (i, VF) if $i \in \tau(A)$.
- The arrows are the following:
  - From $(i, \circ)$ to $(i+1, V)$ or $(i+1, VF)$.
  - From (i, V) to $(i+1, V)$ or $(i+1, VF)$.
  - From $(i, \circ)$ to $(i-1, F)$ or $(i-1, VF)$.
  - From (i, F) to $(i-1, F)$ or $(i-1, VF)$.

(The target depends only on the type. For example, in the first case: If $i+1 \notin \tau(A)$ then the target of the arrow from $(i, \circ)$ is necessarily $(i+1, V)$, and if $i+1 \in \tau(A)$ the target of that arrow is necessarily $(i+1, VF)$).

An *admissible path* (AP) is a collection of vertices and arrows such that:

- If an arrow is in the collection so is its origin and target.
- If a vertex a is in the collection so is every arrow *emanating* from a.
- For every i there is exactly one vertex $(i, *)$ in the collection.

The set of AP's forms a partially ordered set under the following relation: An AP α with vertices $\{(i, \alpha(i))\}$ is greater or equal to an AP β with vertices $\{(i, \beta(i))\}$, if for every i either $\alpha(i) = \beta(i)$ or $\alpha(i) = \circ$. We write $\alpha \geq \beta$. We have the notion of a *maximal admissible path* (MAP). Put:

$$\dim(\alpha) = |\{i : \alpha(i) = \circ\}|.$$

PROPOSITION 3.2. (i) *MAP's correspond bijectively to components of the fibre of $\mathcal{M}_0^n(p) \to \mathcal{M}^n$ over the point correponding to (A, ι). Let $C(\alpha)$ be the component associated to α. We have* $\dim(C(\alpha)) = \dim(\alpha)$. *More precisely there exists an isomorphism*

$$(\mathbb{P}^1)^{\dim(\alpha)} \cong C(\alpha).$$

(ii) *Similarly, one associates to every AP α a closed subset of the fibre $C(\alpha)$. Then,*

$$C(\alpha) \cap C(\beta) = C(\inf\{\alpha, \beta\}).$$

COROLLARY 3.3. (i) *The fibre over (A, ι) depends only on the type $\tau(A)$.*
(ii) *If $\tau(A)$ is not empty (i.e. A is not ordinary), the fibre is connected.*

PROOF. (i) is immediate. To prove (ii), we first remark that the minimal elements of the strata have the following description:

Let $K_1, \ldots, K_r$ be the *blocks* of $\tau(A)$. I.e., the maximal sets of consecutive elements in $\tau(A)$. If $r = 1$ then K_1 is $\{1, \ldots, g\}$ and the AP $\{(i, VF) : 1 \leq i \leq g\}$ is the unique minimal element of the strata. Hence, the strata is connected.

If $r > 1$, let $B_1, \ldots, B_r$ be the *anti-blocks* of $\tau(A)$. That is, the blocks of the complement of $\tau(A)$. Write

$$K_i = \{b_i, \ldots, e_i\}.$$

Then

$$B_i = \{e_i + 1, \ldots, b_{i+1} - 1\}.$$

Given a vector $c = (c_1, \dots, c_r)$ with $c_i \in B_i \cup \{e_i\}$, define an AP $\{(i, \alpha_c(i)) : 1 \leq i \leq g\}$ by

$$\alpha_c(i) = \begin{cases} VF & i \in K_j \\ F & i \in B_j, i \leq c_j \\ V & i \in B_j, i > c_j \end{cases}.$$

To prove that the fibre is connected, it is enough to exhibit a connected set containing all the AP's α_c. It is enough to show that given $c = (c_1, \dots, c_r)$ and j such that $c_j < b_{j+1} - 1$, there exists a connected set containing α_c and $\alpha_{c'}$, where $c' = (c'_1, \dots, c'_r)$ with $c'_i = c_i$ except for $c'_j = c_j + 1$.

Consider the AP β given by

$$\beta(c'_j) = \circ; \quad \beta(i) = \alpha_c(i) \text{ for } i \neq c'_j.$$

Clearly $\beta \geq \alpha_c$, $\alpha_{c'}$. □

EXAMPLE 3.4. Consider the case $g = 4$ and the two types $\tau = \{1, 2\}$ and $\sigma = \{1, 3\}$. The MAP's for τ are given by:

	1	2	3	4
I	VF	VF	$\circ$	V
II	VF	VF	F	$\circ$
III	VF	$\circ$	V	V
IV	$\circ$	VF	F	F

That is: I is the MAP $\{(1, VF), (2, VF), (3, \circ), (4, V)\}$ etc.

The fibre over a point with type τ is thus a connected curve with 4 components, such that each component is a smooth rational curve, components intersect transversely and the jacobian is trivial. On the other hand, the MAP's for σ are

	1	2	3	4
I	VF	$\circ$	VF	$\circ$
II	VF	F	$\circ$	V
III	$\circ$	V	VF	F

The fibre over a point with type σ consists of a surface isomorphic to $\mathbb{P}^1 \times \mathbb{P}^1$ and two copies of $\mathbb{P}^1$, which do not intersect each other and intersect the surface transversely at one point.

QUESTION 7. Work out the complete picture in terms of the type.

3.3. $\Gamma_0(p)$-Level and Moret-Bailly Families. The notion of Moret-Bailly families was introduced in [**19**] and used then extensively, e.g., in [**22**], [**11**], [**12**], [**1**] and [**17**]. In its most simplest version the idea is this:

For a given Abelian variety A/k over a field k of characteristic p, the set

$$\mathrm{Emb}_k(\alpha_p/k, A),$$

has a natural structure of a projective space. It is isomorphic to $\mathbb{P}_k^{a-1}$ and a is called the a-number of A. Dividing A by a "variable" α_p gives a non-isotrivial family $\mathfrak{X} \to \mathbb{P}_k^{a-1}$. Each fibre, $\mathfrak{X}_t$, of this Abelian scheme is canonically the quotient of A by the embedding of α_p corresponding to t.

It is clear how to generalize this construction. One typically considers the family of subgroup schemes of A satisfying a given set of conditions. For example, if A is a HBAV with RM by $\mathcal{O}_L$ and of polarization module $(\mathcal{D}_L^{-1}, \mathcal{D}_L^{-1+})$, then we may consider the family of all subgroups schemes of A, which are $\mathcal{O}_L$-invariant, of order p^g, killed by p and are isotropic with respect to any $\mathcal{O}_L$-linear pairing. That is, we consider the fibre $\mathcal{F}_x$ of the morphism $M_0^n(p) \to \mathcal{M}_L^n$ over x—the moduli point of A!

One gets an Abelian scheme, principally polarized and with RM by $\mathcal{O}_L$:

$$\mathfrak{X} \to \mathcal{F}_t.$$

In this way, understanding the fibres of

$$M_0^n(p) \times \overline{\mathbb{F}}_p \to \mathcal{M}_L^n \times \overline{\mathbb{F}}_p,$$

allows us to construct complete subvarieties of $\mathcal{M}_L^n \times \overline{\mathbb{F}}_p$.

LEMMA 3.5. *Given (A, ι) over a field k of positive characteristic, the map from the fibre of $\mathcal{M}_0^n(p) \times \overline{\mathbb{F}}_p$ over (A, ι) to $\mathcal{M}_L^n \times \overline{\mathbb{F}}_p$ given by*

$$(A, \iota, H) \mapsto (A/H, j),$$

where j is the canonical induced $\mathcal{O}_L$-action, is a finite map.

PROOF. We need just prove quasi-finite. It is enough to prove that for any integer s and for any pair of HBAV (A, ι) and (B, j) over an algebraically closed field k, there are finitely many isogenies

$$f\colon (A, \iota) \to (B, j),$$

such that $\deg(f) \leq s$. One reduces to the case $(A, \iota) = (B, j)$ by composing with an isogeny g from (B, j) to (A, i) (if an f exists, a g exists).

Let B be the centralizer of L in $\mathrm{End}^0(A)$ and $\mathcal{O} = B \cap \mathrm{End}(A)$. The assertion amounts to proving that for every integer s there are finitely many elements of $\mathcal{O}$ such that $\deg(f) \leq s$. It follows from [**2**, Lemma 6] that B is either L, a CM field K with $K^+ = L$, or a quaternion algebra over L which is ramified at all infinite places of L and all places ν of L over p of odd relative degree and is unramified elsewhere. It follows that B has a unique positive involution that we denote by $x \mapsto x^*$. Note that $\deg(x) = \deg(x^*)$, because the involution is induced from any Roasti involution coming from an $\mathcal{O}_L$-linear polarization. Also $\deg(xx^*) = \deg(x)\deg(x^*)$.

Using these facts, one verifies that in fact $\deg(x) = \mathrm{Norm}_{L/\mathbb{Q}}(xx^*)$. It follows that for any integer s, any latice in B contains finitely many elements f such that $\deg(f) \leq s$. □

It is often the case that this procedure can be reversed, in particular for families of supersingular Abelian varieties. The reason is that all supersingular Abelian varieties are isogenous (and tend to stay so with any other structure one tries to impose, e.g. with RM and principal polarization (see [**8**])).

4. Hasse Invariants

Let $p > 2$ be a prime unramified in L. We define "partial Hasse invariants" and a "total Hasse invariant". This generalizes the classical Hasse invariant for elliptic curves. The possibility of lifting the total Hasse invariant is related to an arithmetic condition, which we dub as:

***C*(*L*, *p*)** $\qquad$ *$\zeta_L(2-p)$ is not p-integral.*

We remark that by a theorem of Siegel, $\zeta_L(2-p)$ is a rational number. Further comments on this condition are given below.

We recall the definition of "a modular form of weight χ". For more details see [**13**, Section 1.2].

Fix a ground ring R_0, and let χ be an algebraic character of the torus $\coprod_{\mathcal{O}_L/\mathbb{Z}}\mathbb{G}_m$. A Hilbert modular form of weight χ is an h_L^+-tuple of modular forms $(f_{\mathfrak{a}})_{\mathfrak{a}}$, where $\mathfrak{a}$ runs over ideals forming a set of representatives for $Cl(L)^+$. For every $\mathfrak{a}$, $f_{\mathfrak{a}}$ is a rule associating to a triple (A,ι,ω)—defined over an R_0-algebra R and consisting of a HBAS (A,ι) over R and an $\mathcal{O}_L\otimes R$-basis ω for $\omega_{A/R}$, such that $(\mathcal{M}_A,\mathcal{M}_A^+)$ is in the class of $\mathfrak{a}\mathcal{D}_L^{-1}$—an element

$$f_{\mathfrak{a}}(A,\iota,\omega)\in R.$$

The element $f_{\mathfrak{a}}(A,\iota,\omega)$ depends only on the R-isomorphism class of (A,ι,ω) and the rule $f_{\mathfrak{a}}$ commutes with base-change. Furthermore, for every $b\in(\mathcal{O}_L\otimes R)^\times$,

$$f_{\mathfrak{a}}(A,\iota,b\omega)=\chi(b)^{-1}f_{\mathfrak{a}}(A,\iota,\omega).$$

THEOREM 4.1. *Let p be a rational prime which is unramified at L.*

(1) *For every (j,i), where $1\le j\le r$ and $1\le i\le f_j$, there exists a Hilbert modular form $H_{(j,i)}$ over $\mathbb{F}$ of weight $\chi_{(j,i)}^p\chi_{(j,i+1)}^{-1}$ and level 1, whose divisor is $W_{(j,i)}$. The q-expansion of $H_{(j,i)}$ at every cusp is 1.*

(2) *Let*

$$H_{L,p}=\prod_{j=1}^{r}\prod_{i=1}^{f_j}H_{(j,i)}.$$

The divisor of $H_{L,p}$ is the complement W_1 of the ordinary locus. If $p>2$, and $p>3$ if $g=1$, and if $\boldsymbol{C(L,p)}$ holds, then $H_{L,p}$ is the reduction mod p of an Eisenstein series, $E_{L,p}$, of equi-weight $p-1$.

We explain the definition of $H_{(j,i)}$, refering to [**7**, Section 2] for the proof.

Let (A,ι) be a HBAS over S, where S is an $\mathbb{F}$-scheme. Denote by $f\colon A\to S$ the structure map. By [**3**, Corollaire 2.9] $f_*\Omega^1_{A/S}$ is a locally free $\mathcal{O}_L\otimes\mathcal{O}_S$-module of rank 1. This implies that the usual definition of modular forms (i.e., as sections of $\det(\mathbb{E})$ to some power) agrees with the definition above. It is therefore enough to define $H_{(j,i)}$ on quadruples (A,ι,α,ω), over an $\mathbb{F}$-algebra R, where α is a level n structure, and ω is a generator for the free $\mathcal{O}_L\otimes R$ module $f_*\Omega^1_{A/R}$, where $f\colon A\to\mathrm{Spec}(R)$ is the structural morphism. It will be clear that the definition does not depend on α.

Choose an $\mathcal{O}_L$-linear polarization λ on (A,ι) of degree prime to p. The pairing induced by λ,

$$H^1_{dR}(A/R)\times H^1_{dR}(A/R)\to R,$$

is a perfect alternating pairing. The elements,

$$\omega_{(j,i)}=e_{(j,i)}\omega,\quad 1\le j\le r, 1\le i\le f_j,$$

form a basis for $f_*\Omega^1_{A/R}$. Let $\{\eta_{(j,i)}\}$ be the basis for $R^1f_*\mathcal{O}_A$ dual to the basis $\{\omega_{(j,i)}\}$. Let

$$H_{(j,i)}(A,\iota,\alpha,\omega)=\frac{\mathrm{Fr}(\eta_{(j,i)})}{\eta_{(j,i+1)}}.$$

I.e., $H_{(j,i)}(A,\iota,\alpha,\omega)$ is the unique element of R, such that

$$\mathrm{Fr}(\eta_{(j,i)})=H_{(j,i)}(A,\iota,\alpha,\omega)\cdot\eta_{(j,i+1)}.$$

One may verify that

$$H_{(j,i)}(A,\iota,\alpha,\omega)=0 \Leftrightarrow (j,i)\in\tau_j(A)$$

(see Section 2.6).

Let $s\in(\mathcal{O}_L\otimes_{\mathbb{Z}}R)^\times$. Using the decomposition,

$$\mathcal{O}_L\otimes_{\mathbb{Z}}R=(\mathcal{O}_L\otimes_{\mathbb{Z}}\mathbb{F})\otimes_{\mathbb{F}}R=\bigoplus(j,i)R,$$

and denoting the image of s by $(\chi_{(j,i)}(s))$, we get that the basis for $f_*\Omega^1_{A/R}$ obtained from $s\omega$ is:

$$e_{(j,i)}s\omega=\chi_{(j,i)}(s)\omega_{(j,i)},\quad 1\le j\le r, 1\le i\le f_j.$$

The dual basis is therefore,

$$\chi_{(j,i)}(s)^{-1}\eta_{(j,i)},\quad 1\le j\le r, 1\le i\le f_j.$$

Thus, we find,

$$\frac{\operatorname{Fr}(\chi_{(j,i)}(s)^{-1}\eta_{(j,i)})}{\chi_{(j,i+1)}(s)^{-1}\eta_{(j,i+1)}}=\left(\chi^p_{(j,i)}\chi^{-1}_{(j,i+1)}\right)(s^{-1})\times\frac{\operatorname{Fr}(\eta_{(j,i)})}{\eta_{(j,i+1)}}.$$

That is, $H_{(j,i)}$ is a modular form of weight $\chi^p_{(j,i)}\chi^{-1}_{(j,i+1)}$.

Note that $H_{(j,i)}$ is independent of the polarization λ we have chosen. Indeed, any other choice is of the form λ/m for some $m\in L$ prime to p. The basis dual to $e_{(j,i)}\omega$ is then $m\eta_{(j,i)}$ and one checks that $\operatorname{Fr}(m\eta_{(j,i)})=m\operatorname{Fr}(\eta_{(j,i)})$.

REMARK 4.2. (1) One can show that for an Abelian totally real field L, condition $\boldsymbol{C(L,p)}$ is equivalent to $L_p(1,\chi)$ being a p-adic unit for every non-trivial Dirichlet character χ associated with L [**7**]. See [**31**, Chapter 5] for the definition of $L_p(\cdot,\chi)$ employed here.

(2) One can prove that if L has class number one, then $\boldsymbol{C(L,p)}$ is a necessary and sufficient condition for lifting $H^r_{L,p}$ for some positive integer r prime to p to an Eisenstein series [**7**].

QUESTION 8. Can one formulate a geometric condition providing a necessary and sufficient condition for lifting the Hasse invariant $H_{L,p}$ to characteristic zero?

QUESTION 9. Consider the subring $\mathcal{T}$ of the Chow ring generated by the tautological classes $W_{(j,i)}$. It is naturally a Hecke module. What is the structure of $\mathcal{T}$ as a Hecke module?

QUESTION 10. One can show that the stratification we define by the type has a natural structure of a simplicial complex (see [**8**, Section 4]). What is the homology of this complex?

5. Stratifications of a Class of Shimura Varieties

Fix a prime p, an integer $g\ge 1$ and positive integers $d_1,\dots,d_g$ that are prime to p and satisfy $d_i\mid d_{i+1}$ for $1\le i\le g-1$.

Let K be either a totally real field or a CM field. Fix an algebra B, central over K, which is either K itself, or a quaternion algebra over K. Fix an order R of B which is p-maximal. That is, for every $\mathfrak{p}\mid p$ in K, we have that $R\otimes_{\mathcal{O}_K}\mathcal{O}_{K,\mathfrak{p}}$ is maximal in $B\otimes_K K_{\mathfrak{p}}$. Fix a positive anti-involution $x\mapsto x^*$ of B.

Let $\mathbb{F}$ be the minimal finite field of characteristic p over which every irreducible representation of R/pR can be defined. Thus the following are fixed:

$$\{(p, g, d_1, \ldots, d_g), (B, R, *), \mathbb{F}\}.$$

We consider the functor M on $\mathbb{F}$-schemes given by:

$$S \mapsto \text{ isomorphism classes of } (A, \iota, \lambda)/S,$$

where:

- A is a g-dimensional Abelian scheme over S.
- λ is a polarization of A with elementary divisors $d_1, \ldots, d_g$. Denote by $x \mapsto x^\lambda$ the Rosati involution defined by λ.
- ι is a ring embedding $\iota : R \hookrightarrow \mathrm{End}(A)$ such that $\iota(r)^\lambda = \iota(r^*)$ for every $r \in R$.

Let $\mathcal{M}$ be the corresponding stack.

EXAMPLE 5.1. (A) $B = \mathbb{Q}$, $R = \mathbb{Z}$, $*$ is the identity and $d_1 = \cdots = d_g = 1$. Then $\mathcal{M} = \mathcal{A}_g$ is the stack of principally polarized Abelian varieties of dimension g over $\mathbb{F}_p$-schemes. There is no restriction on p.

(B) $B = K$ a totally real field of degree g, $R = \mathcal{O}_K$ and $*$ is the identity. We take $d_1 = \cdots = d_g = 1$. This is essentially the case of Hilbert modular varieties discussed above, except that we allow p to be ramified in K.

(C) $B = B_{q,\infty}$ a definite quaternion algebra over $\mathbb{Q}$ ramified at a prime q. We let R be a maximal order and $*$ the unique positive involution on B. Let $g = 1$ and $d_1 = 1$.

If $p \neq q$ then $\mathcal{M}$ is empty. If $p = q$ then $\mathcal{M}$ is the reduced zero dimensional scheme, parameterizing supersingular elliptic curves with endomorphism ring isomorphic to R (thus $\mathcal{M}$ consists of one point if the j-invariant is in $\mathbb{F}_p$ and of two points otherwise).

One would like to follow the same steps taken in the case of Hilbert modular varieties and p unramified. We note that "every" canonical linear or semi-linear object decomposes under the action of $R \otimes W(\mathbb{F})$ and often under $R/pR \otimes \mathbb{F}$. In this case we call the factors of the Jordan-Hölder series the *canonical factors* of the corresponding object. One may pose the following questions:

(1) What are the canonical factors of $\mathrm{Lie}(A)$ and $H^1_{dR}(A)$? Are they constant along the components of $\mathcal{M}$?
(2) Given (A, ι, λ) over a field, we associate to it its *type*, which is defined to be the canonical factors of $\mathbb{D}(\alpha(A))$ (taking multiplicity into account!). What is the local "behaviour" of the type? Can one use this invariant to get a stratification of $\mathcal{M}$?
(3) Determine the Newton polygon of the generic point of a component of such strata.
(4) Are there modular forms $\pmod p$ related to this stratification?
(5) What is the structure of the algebra generated by the "tautological classes" over an appropriate Hecke ring?
(6) Study the Ekedahl-Oort stratification, by taking into consideration also the R/pR-structure and study its relation to the type.

EXAMPLE 5.2. We refer to the numbering of the examples above.

(A) This example is essentially the content of [**6**]. In this case the type amounts to the a-number. The stratification is obtained by setting T_r to be the closed subscheme of $\mathcal{A}_g$ "where the a-number is greater or equal to r". G. van der Geer proved in loc. cit. that T_r is irreducible for $r < g$. By a conjecture of Oort [**23**, 15.8] proved in [**28**], given an isogeny type of a p-divisible group

$$\bigoplus_{i=1}^{r}(G_{m_i,n_i} \oplus G_{m_i,n_i}) \oplus s \cdot G_{1,1} \oplus f \cdot (G_{1,0} \oplus G_{0,1}), \tag{5.1}$$

for every a satisfying $1 \leq a \leq s + 2\sum_{i=1}^{r} \min\{m_i, n_i\}$ we may find an Abelian variety A with $a(A) = a$ and such that the p-divisible group of A is as in (5.1). In particular, $\dim(A) = s + f + \sum_{i=1}^{r}(m_i + n_i)$.

We explain how we may get the maximal possible a-number, retaining the principal polarization:

It follows from [**8**], and its generalization to the case where p is an unramified prime, that given an isogeny type as in (5.1), we may find a totally real field L in which p is unramified, such that a p-divisible group isogenous to the one in (5.1)is obtained from a principally polarized HBAV with RM by $\mathcal{O}_L$.

Again, to simplify, we assume that p is inert, but the argument generalizes easily. In this case the possible Newton polygons are $\beta_0 < \beta_1 < \cdots < \beta_s$ ($s = [(g+1)/2]$), where β_0 is the ordinary polygon, β_s is the supersingular polygon and each polygon has just two slopes. The corresponding groups are $g \cdot (G_{1,0} \oplus G_{0,1})$, $G_{1,g-1} \oplus G_{g-1,1}$, etc.

Now consider the type $\tau(\ell) := \{1, 2, \ldots, 2\ell\}$ for $\ell < s$. The a-number is 2ℓ and the Newton polygon of $W_{\tau(\ell)}$ is generically β_ℓ (see Section 2.4). For the supersingular polygon take the type $\tau = \{1, \ldots, g\}$.

Coming back to our example, it follows that the isogeny type of the p-divisible group of the generic point of T_a is the infimum of all the polygons of

$$\bigoplus(G_{m_i,n_i} \oplus G_{m_i,n_i}) \oplus s \cdot G_{1,1} \oplus f \cdot (G_{1,0} \oplus G_{0,1}),$$

such that $s + 2\sum \min\{m_i, n_i\} = a$. This is just

$$(g-a) \cdot (G_{1,0} \oplus G_{0,1}) \oplus a \cdot G_{1,1}.$$

(B) If p is unramified, this is the content of the previous sections. If p is ramified, the situation is much subtler due to the fact that the tangent space of $(A, \iota, \lambda)/k$ need not be a free $\mathcal{O}_L \otimes k$ module. However, the situation is not the worst:

PROPOSITION 5.3 ([**3**, 2.7, 2.8]). *In the Grothendieck group of $\mathcal{O}_L \otimes k$-modules of finite type,* Lie(A) *has the same class as a free $\mathcal{O}_L \otimes k$ of rank* 1. *The module $H^1_{dR}(A)$ is free of rank* 2 *over $\mathcal{O}_L \otimes k$.*

Let $g = 2$. In the light of this result, we see that the stratification we propose amounts in this case to the stratification by the a-number. However, for higher dimensions the concepts diverge.

References

1. E. Bachmat and E. Z. Goren, *On the non ordinary locus in Hilbert-Blumenthal surfaces*, Math. Ann. (to appear).
2. C.-L. Chai, *Every ordinary symplectic isogeny class in positive characteristic is dense in the moduli*, Invent. Math. **121** (1995), 439–479.

3. P. Deligne and G. Pappas, *Singularités des espaces de modules de Hilbert, en les caractéristiques divisant le discriminant*, Compositio Math. **90** (1994), 59–79.
4. T. Ekedahl, T. and F. Oort, *Connected subsets of a moduli space of Abelian varieties* (in preparation).
5. G. van der Geer, *Hilbert modular surfaces*, Ergeb. Math. Grenzgeb. (3), vol. 16, Springer-Verlag Berlin-New York, 1988.
6. ———, *Cycles on the moduli space of Abelian varieties*, alg-geom/9605011.
7. E. Z. Goren, *Hasse invariants for Hilbert modular varieties*, CICMA report 1998-10, Concordia University, Montréal, 1998.
8. E. Z. Goren and F. Oort, *Stratifications of Hilbert modular varieties*, J. Algebraic Geom. (to appear).
9. A. Grothendieck, *Groupes de Barsotti-Tate et cristaux de Dieudonné*, Séminaire de Mathématiques Supérieures, vol. 45 (1970), Les Presses de l'Université de Montréal, Montréal, 1974.
10. J. A. de Jong and F. Oort, *Purity of the stratification by Newton polygon*, 1998 (preprint).
11. T. Katsura and F. Oort, *Families of supersingular Abelian surfaces*, Compositio Math. **62** (1987), 107–167.
12. ———, *Supersingular Abelian varieties of dimension two or three and class numbers*, Algebraic Geometry, (Sendai, 1985) Adv. Stud. Pure Math., vol. 10, North-Holland, Amsterdam-New York, 1987, pp. 253–281.
13. N. M. Katz, *p-adic L-functions for CM fields*, Invent. Math. **49** (1978), 199-297.
14. ———, *Slope filtration of F-crystals*, Journées de géométrie algébrique (Rennes, 1978), Astérisque, vol. 63, Soc. Math. France, Paris, 1979, pp. 113–163.
15. N. Koblitz, *p-adic variation of the zeta-function over families of varieties defined over finite fields*, Compositio Math. **31** (1975), 119–218.
16. H. P. Kraft and F. Oort, *Group schemes annihilated by p* (in preparation).
17. K.-Z. Li and F. Oort, *Moduli of supersingular Abelian varieties*, Lecture Notes in Math., vol. 1680, Springer Verlag, Berlin, 1998.
18. Y. Manin, *The theory of commutative formal groups over a field of finite characteristic*, Russian Math. Surveys **18** (1963), 1–80.
19. L. Moret-Bailly, *Familles de courbes et de variétés abéliennes sur* $\mathbb{P}^1$, Astérisque, vol. 86, Soc. Math. France, Paris, 1981, pp. 109–140.
20. P. Norman, *An algorithm for computing local moduli of Abelian varieties*, Ann. of Math. (2) **101** (1975), 499–509.
21. P. Norman and F. Oort, *Moduli of Abelian varieties*, Ann. of Math. (2) **112** (1980), 413–439.
22. T. Oda and F. Oort, *Supersingular Abelian varieties*,Proceedings of the International Symposium on Algebraic Geometry (Kyoto, 1977), Kinokuniya Book Store, Tokyo, 1978, pp. 595–621.
23. F. Oort *Commutative group schemes*, Lecture Notes in Math., vol. 15, Springer-Verlag, Berlin-New York, 1966.
24. ———, *Moduli of Abelian varieties and Newton polygons*, C. R. Acad. Sci. Paris Sér. I Math. **312** (1991), 385–389.
25. bysame, *Moduli of Abelian varieties in positive characteristic*, Barsotti Symposium in Algebraic Geometry (Abano Terme, 1991), Perspect. Math., vol. 15, Academic Press, San Diego, CA, 1994, pp. 253–276.
26. ———, *A stratification of a moduli space of polarized Abelian varieties in positive characteristic*, Preprint 997, Dept. of Mathematics, Utrecht University, 1997.
27. G. Pappas, *Arithmetic models for Hilbert modular varieties*, Compositio Math. **98** (1995), 43–76.
28. M. Poletti, *Differentiali esatti di prima specie su varietà abeliane*,Ann. Scuola Norm. Sup. Pisa (3) **21** (1967), 107–110.
29. M. Rapoport, *Compactifications de l'espace de modules de Hilbert-Blumenthal*, Compositio Math. **36** (1978), 255–335.
30. H. Stamm, *On the reduction of the Hilbert-Blumenthal-moduli scheme with* $\Gamma_0(p)$*-level structure*, Forum Math. **9** (1997), 405–455.
31. L. C. Washington, *Introduction to cyclotomic fields*, Grad. Texts in Math., vol. 83, Springer-Verlag, New York Berlin, 1982.
32. T. Zink, *The display of a formal p-divisible group*, Preprint 98-017, Universität Bielefeld, 1998.

Department of Mathematices and Statistics, McGill University, 805 Sherbrooke St. W., Montreal, Qc H3A 2K6, Canada
E-mail address: goren@math.mcgill.ca

Centre de Recherches Mathématiques
CRM Proceedings and Lecture Notes
Volume **24**, 2000

On the Néron-Severi Groups of Some $K3$ Surfaces

Yasuhiro Goto

ABSTRACT. We discuss algebraic cycles on the $K3$ surfaces defined as the minimal resolutions of some deformations of weighted Delsarte surfaces. Such $K3$ surfaces have a lot of exceptional divisors which appear in the process of desingularizations. Making use of these trivial algebraic cycles, we obtain several families of $K3$ surfaces whose Néron-Severi groups over $\mathbb{Z}$ can be investigated explicitly.

Introduction

The Néron-Severi group of a smooth projective surface over an algebraically closed field is known to be a finitely generated Abelian group; its $\mathbb{Z}$-rank is called the *Picard number* of the surface and it is bounded by the second Betti number of the surface. One has by now a large number of surfaces whose Picard numbers can be expressed in a closed formula. Such surfaces include rational surfaces, Abelian surfaces, elliptic surfaces, Fermat surfaces and Delsarte surfaces (cf. [**9**], [**26**], [**28**], [**29**], [**30**]). Compared with the computations of the Picard numbers, the determination of $\mathbb{Z}$-bases for the Néron-Severi groups or calculations of the discriminant of their intersection forms are rather difficult problems.

The purpose of this paper is to give several families of $K3$ surfaces for which we can calculate the discriminant of the Néron-Severi groups over $\mathbb{Z}$ numerically. In some cases, we also determine $\mathbb{Z}$-bases for the Néron-Severi groups. While computations of such quantities have their own merit, our results can be applied to investigating the arithmetic of $K3$ surfaces over finite fields (and possibly over number fields); see [**9**], [**10**], [**12**] and [**31**] for details and further discussions.

In this paper, we consider $K3$ surfaces given as the minimal resolutions of either weighted Delsarte surfaces

$$\sum_{i=0}^{3} x_0^{m_{i0}} \cdots x_3^{m_{i3}} = 0 \subset \mathbb{P}^3(q_0, q_1, q_2, q_3)$$

1991 *Mathematics Subject Classification.* Primary 14J28; Secondary 14C25.
Received by the editors August 27, 1998; Accepted October 6, 1998.
This is the final form of the paper.

or their one-parameter deformations

$$\sum_{i=0}^{3} x_0^{m_{i0}} \cdots x_3^{m_{i3}} - \lambda x_0 x_1 x_2 x_3 = 0 \subset \mathbb{P}^3(q_0, q_1, q_2, q_3)$$

in a weighted projective 3-space, where λ is a constant (see Section 1 for precise definitions). Other types of deformations can be dealt with similarly. Since we work over an algebraically closed field, we fix the coefficients of the monomials in the defining equations as above; but, if we were going to study the surfaces, say, over number fields, we would consider also non-trivial coefficients since they have various effects on the arithmetic of the surfaces (e.g. description of the field of definition for each singularity and calculations of the L-series for the surfaces).

Let X be the minimal resolution of a weighted Delsarte $K3$ surface or of its deformation. (There are essentially 95 weights such that X become $K3$ surfaces.) Write $\mathrm{NS}(X)$ for the Néron-Severi group of X and $\operatorname{disc}\mathrm{NS}(X)$ for the discriminant of the intersection form on $\mathrm{NS}(X)$. The $\mathbb{Z}$-rank of $\mathrm{NS}(X)$, i.e. the Picard number of X, is denoted by $\rho(X)$. There are two types of $K3$ surfaces: supersingular $K3$ surfaces and non-supersingular $K3$ surfaces. Supersingular $K3$ surfaces exist only over a field of positive characteristic and the discriminant $\operatorname{disc}\mathrm{NS}(X)$ of such surfaces can be computed from the values of their Artin invariant (cf. [**1**]).

One of the reasons for introducing weighted Delsarte surfaces and their deformations is to obtain a large number of families of surfaces convenient for explicit calculations of various algebraic and geometric quantities, especially those related to algebraic cycles of surfaces. This is possible mainly due to the fact that a weighted Delsarte surface is birational to a finite quotient of a Fermat surface. By examining the associated quotient map, we can extract many properties of weighted Delsarte surfaces from those of Fermat surfaces. For instance, as in the case of Fermat surfaces, we are able to determine the supersingularity of a weighted Delsarte $K3$ surface by pieces of information on the exponents $\{m_{ij}\}$; further, there is a simple method of computing $\operatorname{disc}\mathrm{NS}(X)$ if X is a supersingular weighted Delsarte $K3$ surface (see [**10**] and Section 4). For non-supersingular weighted Delsarte $K3$ surfaces, on the other hand, one can compute the rank of $\mathrm{NS}(X)$ using the algorithm of [**26**], but the description of their $\mathbb{Z}$-bases is a subtler problem which requires further analysis.

Another reason for using the weighted notion is that weighted projective surfaces have a lot of singularities and they give trivial algebraic cycles on the resolutions of the surfaces; the resolution graphs can be described explicitly by using several quantities associated with Q. At the same time, these trivial cycles form part of $\mathbb{Z}$-bases for the Néron-Severi groups of the minimal resolutions and thereby help us to investigate their geometric properties. Even hyperplane sections on a weighted projective surface sometimes become more useful than those on a projective surface in the usual projective 3-space $\mathbb{P}^3$. For example, every hyperplane section on a quartic surface

$$x_0^4 + x_1^4 + x_2^4 + x_3^4 = 0$$

in $\mathbb{P}^3$ has self-intersection number 4, whereas hyperplane sections on a quartic surface in $\mathbb{P}^3(Q)$ with $Q \neq (1,1,1,1)$ have different self-intersection numbers because of the effect of the weight.

The main results of this paper are summarized as follows:

(i) We give non-trivial upper bounds for the Artin invariant of the minimal resolutions of *one-parameter deformations of weighted diagonal $K3$ surfaces*

$$x_0^{m_0} + x_1^{m_1} + x_2^{m_2} + x_3^{m_3} - \lambda x_0 x_1 x_2 x_3 = 0 \subset \mathbb{P}^3(q_0, q_1, q_2, q_3)$$

(when they are supersingular). Such deformations of weighted diagonal $K3$ surfaces provide typical examples of mirror $K3$ surfaces constructed by Roan [**23**]. Our result may thus have some relation with arithmetic aspects of mirror $K3$ surfaces.

(ii) We compute the discriminant of the Néron-Severi group $\mathrm{NS}(X)$ when X is the minimal resolution of a *one-parameter deformation of a weighted quasi-diagonal $K3$ surface*:

$$x_0^{n_0} + x_0 x_1^{n_1} + x_2^{n_2} + x_3^{n_3} - \lambda x_0 x_1 x_2 x_3 = 0 \subset \mathbb{P}^3(q_0, q_1, q_2, q_3).$$

In particular, for 20 distinct choices of $(q_0, \ldots, q_3; n_0, \ldots, n_3)$, we show that $\operatorname{disc} \mathrm{NS}(X)$ is given by

$$\operatorname{disc} \mathrm{NS}(X) = \frac{\det(H_i, E_1, \ldots, E_r)}{q_i^2} \quad (i = 1, 2, 3),$$

where H_i is the strict transform of the hyperplane section $X \cap \{x_i = 0\}$; E_1, ..., E_r are the exceptional divisors on X (the number r of E_j's depends on the weight) and $\det(*)$ denotes the determinant of the intersection matrix. In some cases, we also describe $\mathbb{Z}$-bases for $\mathrm{NS}(X)$.

In Section 1, we recall some fundamental properties of weighted hypersurfaces in $\mathbb{P}^3(Q)$ which we need in the subsequent discussions. In Section 2, we define weighted Delsarte surfaces and their one-parameter deformations. We show that weighted Delsarte surfaces are birational to finite quotients of Fermat surfaces. Using this property, we describe the cohomology of the minimal resolutions of weighted Delsarte surfaces and establish the validity of the Tate conjecture for them. In Section 3, we consider supersingular $K3$ surfaces and their finite quotients. After giving preliminary discussions (including the definition of the Artin invariant of supersingular $K3$ surfaces), we present an upper bound for the value of the Artin invariant in terms of the number of exceptional divisors. In Section 4, we review the main result of [**10**] for computing the Artin invariant of supersingular weighted Delsarte $K3$ surfaces. In Section 5, we apply the result of Section 3 to the one-parameter deformations of weighted diagonal $K3$ surfaces and give non-trivial upper bounds for the values of their Artin invariant. In Section 6, we define weighted quasi-diagonal $K3$ surfaces and their one-parameter deformations. We describe their singular loci and compute the Picard numbers of their minimal resolutions. In Section 7, we choose a certain type of weighted quasi-diagonal $K3$ surfaces and calculate the discriminant of their Néron-Severi groups over $\mathbb{Z}$.

1. Quasi-Smooth Hypersurfaces in $\mathbb{P}_k^3(Q)$

In this section, we discuss some geometry of quasi-smooth weighted hypersurfaces in weighted projective 3-spaces. We describe their singular loci and give a condition for their minimal resolutions to be $K3$. Here a $K3$ surface means a smooth projective surface (over an algebraically closed field) with a trivial canonical divisor and having irregularity zero.

Let k be an algebraically closed field of characteristic $p \geq 0$. Let q_i $(0 \leq i \leq 3)$ be positive integers and put $Q = (q_0, q_1, q_2, q_3)$. Write $k[x_0, x_1, x_2, x_3]$ for the polynomial algebra over k graded by the condition $\deg(x_i) = q_i$ for $0 \le i \le 3$. The algebraic threefold $\mathbb{P}^3_k(Q) := \operatorname{Proj} k[x_0, x_1, x_2, x_3]$ is called the *weighted projective 3-space over k of type Q* (cf. [**3**], [**7**]). We assume that

$$p \nmid q_i \text{ for } 0 \le i \le 3 \quad (\text{if } p > 0) \tag{1.1}$$

$$\gcd(q_\alpha, q_\beta, q_\gamma) = 1 \text{ for every } \{\alpha, \beta, \gamma\} \subset \{0, 1, 2, 3\}. \tag{1.2}$$

Condition (1.2) is not a restriction (see [**3**, Proposition 1.3]); it is useful when we describe various properties of weighted projective surfaces.

Let X be a hypersurface in $\mathbb{P}^3_k(Q)$. X is called *quasi-smooth* (cf. [**7**]) if its affine quasi-cone is smooth outside the origin; more explicitly, if we denote by $\iota\colon \mathbb{A}^4_k \setminus \{(0,0,0,0)\} \to \mathbb{P}^3_k(Q)$ the usual projection map, then the Zariski closure of $\iota^{-1}(X_k)$ in $\mathbb{A}^4_k$ is smooth outside the origin. For instance, weighted diagonal surfaces in $\mathbb{P}^3_k(Q)$ are quasi-smooth (cf. Lemma 5.2). Quasi-smooth surfaces may be singular, but they have at most cyclic quotient singularities of type A which can be resolved by successive blowings-up (cf. [**13**]). Apart from the existence of singularities, quasi-smooth surfaces are very similar to smooth surfaces: with a little modification, we can generalize many results on smooth surfaces to those on quasi-smooth surfaces (see [**7**] for the details). Thus, quasi-smooth surfaces are useful objects and we often confine ourselves to such surfaces in what follows.

Another notion we need to mention about weighted hypersurfaces is well-formedness (cf. [**5**], [**8**]); it is used especially when we describe the singular loci of weighted projective varieties. A hypersurface X in $\mathbb{P}^3_k(Q)$ is said to be *well-formed* if $\operatorname{codim}_X(X \cap \operatorname{Sing}(\mathbb{P}^3_k(Q))) \ge 2$, where $\operatorname{Sing}(\mathbb{P}^3_k(Q))$ denotes the singular locus of $\mathbb{P}^3_k(Q)$.

LEMMA 1.1. (a) *For a point $P = (x_0 : x_1 : x_2 : x_3)$ in $\mathbb{P}^3_k(Q)$, let I_P be a subset of indices corresponding to non-zero x_i's: $I_P = \{i \mid 0 \le i \le 3, x_i \ne 0\}$. Then*

$$\operatorname{Sing}(\mathbb{P}^3_k(Q)) = \{P \in \mathbb{P}^3_k(Q) \mid \gcd(q_i; i \in I_P) \ge 2\}.$$

In particular, every $P \in \operatorname{Sing}(\mathbb{P}^3_k(Q))$ has at least two coordinates being equal to zero.

(b) *Let X be a well-formed quasi-smooth hypersurface in $\mathbb{P}^3_k(Q)$. Then*

$$\operatorname{Sing}(X) = X \cap \operatorname{Sing}(\mathbb{P}^3_k(Q)).$$

Further, all the singularities of X are cyclic quotient singularities.

PROOF. See [**6**] and [**5**] for (a) and (b), respectively. In particular, it follows from condition (1.2) that at least two coordinates of $P \in \operatorname{Sing}(\mathbb{P}^3_k(Q))$ are zero. □

LEMMA 1.2. *Let X be a well-formed quasi-smooth projective hypersurface of degree m in $\mathbb{P}^3_k(q_0, q_1, q_2, q_3)$. Then the dualizing sheaf of X is isomorphic to*

$$\omega_X \cong \mathcal{O}_X(m - q_0 - q_1 - q_2 - q_3).$$

In particular, the minimal resolution of X is a $K3$ surface if $m = q_0 - q_1 - q_2 - q_3$.

PROOF. See [**7**]. □

It has been proved in [**22**] (see also [**33**]) that there are exactly 95 weights Q satisfying conditions (1.1), (1.2) and $m = q_0 - q_1 - q_2 - q_3$ such that generic hypersurfaces of degree m in $\mathbb{P}^3_k(Q)$ become quasi-smooth and well-formed. Such a weight Q may be called a $K3$ *weight*. A complete list of $K3$ weights can be found in [**8**] and [**33**].

2. Weighted Delsarte Surfaces and Their Deformations

In this section, we define weighted Delsarte surfaces and their one-parameter deformations. We show that weighted Delsarte surfaces are birational to finite quotients of Fermat surfaces. Using this information, we describe the cohomology of the minimal resolutions of weighted Delsarte surfaces and establish the validity of the Tate conjecture for them.

Let k be an algebraically closed field of characteristic $p \geq 0$. Fix a weight $Q = (q_0, q_1, q_2, q_3)$ and let $\mathbb{P}^3_k(Q)$ be the weighted projective 3-space over k of type Q. Choose a positive integer m and a 4×4 matrix $M = (m_{ij})$ of integer entries satisfying the conditions:

$$(2.1) \qquad \begin{cases} m_{ij} \geq 0 & \text{for every } (i,j) \\ p \nmid \det M & \text{if } p > 0 \\ \sum_{j=0}^{3} q_j m_{ij} = m & \text{for } 0 \leq i \leq 3 \\ \text{given } j, m_{ij} = 0 & \text{for some } i. \end{cases}$$

Let X_M be the surface in $\mathbb{P}^3_k(Q)$ defined by the equation:

$$X_M : \sum_{i=0}^{3} x_0^{m_{i0}} \cdots x_3^{m_{i3}} = 0 \subset \mathbb{P}^3_k(Q).$$

We call X_M the *weighted Delsarte surface in $\mathbb{P}^3_k(Q)$ of degree m with matrix M*.

This type of surfaces were considered, first in the case $q_0 = q_1 = q_2 = q_3 = 1$ (i.e., in the usual projective 3-space $\mathbb{P}^3_k$), by Delsarte [**4**] when he generalized a result of Weil [**32**] about the zeta-functions of diagonal varieties over a finite field. Then Shioda [**26**] investigated Delsarte surfaces in $\mathbb{P}^3_k$ to demonstrate the effectiveness of his algorithm for computing the Picard numbers of surfaces over k; furthermore, he used Delsarte surfaces to construct examples of supersingular $K3$ surfaces with Artin invariant between 1 and 9 with only the exception of 7 (cf. [**27**]). Finally the weighted version of Delsarte surfaces was introduced in [**10**] to give examples of supersingular $K3$ surfaces with Artin invariant 10.

Let M be a 4×4 matrix defined above. Choose $m = q_0 + q_1 + q_2 + q_3$. For $\lambda \in k$, let $X_{M,\lambda}$ be a one-parameter deformation of a weighted Delsarte surface X_M defined by the equation:

$$X_{M,\lambda} : \sum_{i=0}^{3} x_0^{m_{i0}} \cdots x_3^{m_{i3}} - \lambda x_0 x_1 x_2 x_3 = 0 \subset \mathbb{P}^3(q_0, q_1, q_2, q_3).$$

We call $X_{M,\lambda}$ simply a *deformation of X_M*. Deformations of weighted Delsarte surfaces will be considered in Sections 5–7. In this paper, wherever we discuss weighted Delsarte surfaces and their deformations, we assume them to be quasi-smooth and well-formed.

REMARK 2.1. Without the assumption that weighted Delsarte surfaces are quasi-smooth or well-formed, their geometric properties become more complicated. But, even in this case, one knows that the surfaces have only toric singularities (see [**16**]) and every toric singularity can be resolved by blowings-up.

We now show that weighted Delsarte surfaces are quotients of Fermat surfaces. Let X_M be a weighted Delsarte surface with matrix M. Put $d = \det M$ and denote by $\mathcal{F}_d$ the Fermat surface of degree d in $\mathbb{P}_k^3$:

$$\mathcal{F}_d : y_0^d + y_1^d + y_2^d + y_3^d = 0.$$

Define $\Gamma = (\mu_d \times \mu_d \times \mu_d \times \mu_d)/(\text{diagonal elements})$ and

$$\Gamma_M = \left\{ \gamma = \left(\prod_{j=0}^{3} \lambda_j^{m_{0j}}, \prod_{j=0}^{3} \lambda_j^{m_{1j}}, \prod_{j=0}^{3} \lambda_j^{m_{2j}}, \prod_{j=0}^{3} \lambda_j^{m_{3j}} \right) \in \Gamma \;\middle|\; (\lambda_0, \lambda_1, \lambda_2, \lambda_3) \in \Gamma \right\}.$$

Then Γ_M is a subgroup of Γ and it acts on $\mathcal{F}_d$ by

$$\begin{aligned} &\gamma \cdot (y_0 : y_1 : y_2 : y_3) \\ &\quad = \left(\left(\prod_{j=0}^{3} \lambda_j^{m_{0j}} \right) y_0 : \left(\prod_{j=0}^{3} \lambda_j^{m_{1j}} \right) y_1 : \left(\prod_{j=0}^{3} \lambda_j^{m_{2j}} \right) y_2 : \left(\prod_{j=0}^{3} \lambda_j^{m_{3j}} \right) y_3 \right) \end{aligned}$$

for $\gamma \in \Gamma_M$ and $(y_0 : y_1 : y_2 : y_3) \in \mathcal{F}_d$. Let C be the cofactor matrix of M and b_{ij} $(0 \le i \le 3, 0 \le j \le 3)$ the entries of the transpose of C; i.e., $C^T = (b_{ij})$. Then we have the following result.

LEMMA 2.2. *Let X_M be a weighted Delsarte surface in $\mathbb{P}_k^3(Q)$ with matrix M. Then X_M is birational to the quotient $\mathcal{F}_d/\Gamma_M$ and there exists a dominant rational map $f \colon \mathcal{F}_d \cdots \to X_M$ defined by*

$$f((y_0 : y_1 : y_2 : y_3)) = \left(\prod_{j=0}^{3} y_j^{b_{0j}} : \prod_{j=0}^{3} y_j^{b_{1j}} : \prod_{j=0}^{3} y_j^{b_{2j}} : \prod_{j=0}^{3} y_j^{b_{3j}} \right).$$

Furthermore the minimal resolution of X_M is also birational to $\mathcal{F}_d/\Gamma_M$.

PROOF. See [**10**]. □

Given a smooth projective variety X over k, let $H^n(X, *)$ denote the ℓ-adic étale cohomology $H^n(X_{\text{et}}, \mathbb{Z}_\ell)$ or the crystalline cohomology $H^n(X/W)$, where ℓ is a prime number different from $p = \operatorname{char} k$ and W is the ring of Witt vectors over k. Put

$$\mathfrak{A}(\mathcal{F}_d) = \left\{ \boldsymbol{\alpha} = (\alpha_0, \alpha_1, \alpha_2, \alpha_3) \;\middle|\; \alpha_i \in \mathbb{Z}/d\mathbb{Z}, \alpha_i \neq 0 \ (0 \le i \le 3), \sum_{i=0}^{3} \alpha_i = 0 \right\}$$

$$V(\boldsymbol{\alpha}) = \{ v \in H^2_{\text{prim}}(\mathcal{F}_d, *) \mid \overline{\gamma}(v) = \gamma_0^{\alpha_0} \gamma_1^{\alpha_1} \gamma_2^{\alpha_2} \gamma_3^{\alpha_3} \cdot v, \forall \gamma = (\gamma_0, \gamma_1, \gamma_2, \gamma_3) \in \Gamma \}.$$

$$\mathfrak{A}(X_M) = \left\{ \boldsymbol{\alpha} = (\alpha_0, \alpha_1, \alpha_2, \alpha_3) \in \mathfrak{A}(\mathcal{F}_d) \;\middle|\; \sum_{i=0}^{3} m_{ij} \alpha_i = 0 \ (0 \le j \le 3) \right\}$$

where $H^2_{\text{prim}}(\mathcal{F}_d, *)$ denotes the primitive part of the cohomology and $\overline{\gamma}$ is the endomorphism of $H^2_{\text{prim}}(\mathcal{F}_d, *)$ induced from γ.

PROPOSITION 2.3. *Let X_M be a weighted Delsarte surface in $\mathbb{P}_k^3(Q)$ with matrix M. Write $\widetilde{X}_M$ for the minimal resolution of X_M. Let $\mathcal{F}_d$ be the Fermat surface of degree $d = \det M$ in $\mathbb{P}_k^3$. Then*

$$H^2(\widetilde{X}_M, *) \approx H^2(\mathcal{F}_d, *)^{\Gamma_M} \cong V(0) \oplus \bigoplus_{\boldsymbol{\alpha} \in \mathfrak{A}(X_M)} V(\boldsymbol{\alpha})$$

where "$\approx$" means that the difference between the two cohomology groups are algebraic cycles associated to blowings-up and blowings-down.

PROOF. See [**10**]. □

For a later use, we discuss weighted Delsarte surfaces over a finite field $\mathbb{F}_q$ of q elements. To the end of this section, we assume that all the surfaces are defined over $\mathbb{F}_q$.

Let $\widetilde{X}_M$ be the minimal resolution of X_M. For $0 \leq i \leq 4$, it is known (see [**2**]) that $P_i(\widetilde{X}_M, t) := \det(1 - \Phi t \mid H^2_{\text{et}}(\widetilde{X}_M \times \overline{\mathbb{F}_q}, \mathbb{Q}_\ell))$ is a polynomial with integer coefficients and its reciprocal roots have absolute value $q^{i/2}$, where Φ is the endomorphism induced from the Frobenius automorphism of $\widetilde{X}_M$ and ℓ is a prime such that $\ell \nmid q$. We may write

$$P_2(\widetilde{X}_M, t) = (1 - qt)^{\rho'(\widetilde{X}_M)} \prod_{\epsilon_j \neq 1} (1 - \epsilon_j q t)$$

for some integer $\rho'(\widetilde{X}_M)$ with $1 \leq \rho'(\widetilde{X}_M) \leq b_2$, where ϵ_j is a unit in $\mathbb{C}^\times$ different from 1. If $\rho(\widetilde{X}_M)$ denotes the Picard number of $\widetilde{X}_M$ over $\mathbb{F}_q$ (i.e. the $\mathbb{Z}$-rank of the Néron-Severi group of $\widetilde{X}_M$ over $\mathbb{F}_q$), then the Tate conjecture (cf. [**30**]) asserts that

$$\rho(\widetilde{X}_M) = \rho'(\widetilde{X}_M).$$

THEOREM 2.4. *Let X_M be a weighted Delsarte surface over $\mathbb{F}_q$ with matrix M. Then the Tate conjecture holds for the minimal resolution $\widetilde{X}_M$.*

PROOF. This follows from the fact that the Tate conjecture holds for Fermat surfaces over finite fields (see [**28**]) and $\widetilde{X}_M$ is birational to a quotient of a Fermat surface. Details may be found in [**9**]. □

3. Supersingular $K3$ Surfaces

In this section, we define supersingular surfaces and the Artin invariant. Our purpose is to discuss supersingular $K3$ surfaces and their finite quotients.

Let k be an algebraically closed field of characteristic $p > 0$. Write $\rho(X)$ for the Picard number of X, i.e. the $\mathbb{Z}$-rank of the Néron-Severi group $\mathrm{NS}(X)$ of X over k. One knows that $\rho(X)$ is bounded from above by the second Betti number $b_2(X)$ of X. The surface X may be called *supersingular* if $\rho(X) = b_2(X)$ (cf. [**25**]). If X is supersingular, then $\mathrm{NS}(X) \otimes \mathbb{Z}_\ell \cong H^2_{\text{et}}(X, \mathbb{Z}_\ell)$ for primes $\ell \neq p$ and hence the discriminant of the Néron-Severi group of X is given by

$$\operatorname{disc} \mathrm{NS}(X) = (-1)^{b_2 - 1} p^{2\sigma_0(X)}$$

for some non-negative integer $\sigma_0(X)$ (cf. [**1**]).

The integer $\sigma_0(X)$ is called the *Artin invariant* of X when X is a $K3$ surface. In this paper, however, we use the same terminology for arbitrary supersingular surfaces.

There are several ways of characterizing $\sigma_0(X)$. Let X be a supersingular surface over k (not necessarily $K3$). Let W be the ring of Witt vectors over k and K an algebraic closure of the fraction field of W, fixed once and for all. Write $H^2_{cris}(X/W)$ for the crystalline cohomology of X, which is a W-module of rank $b_2(X)$. Denote by $W\Omega^\bullet_X := \varprojlim_n W_n\Omega^\bullet_X$ the de Rham-Witt complex of X that is defined in [**14**]; it is given operations V (Verschiebung), F (Frobenius) and d (differential) satisfying the relation $d = FdV$ among others properties.

First the slope spectral sequence on $W\Omega_X^\bullet$ induces a surjective homomorphism:

$$d = d_1^{0,2}\colon H^2(X, W\mathcal{O}) \to H^2(X, W\Omega^1).$$

Put $E_2^{0,2}(X) = \ker d$. Then the W-module $E_2^{0,2}$ fits into the exact sequence:

$$0 \to \mathrm{NS}(X) \otimes W \to H^2_{cris}(X/W) \to E_2^{0,2}(X) \to 0. \tag{3.1}$$

Since X is supersingular, we have $\mathrm{rank}_W \mathrm{NS}(X) \otimes W = \mathrm{rank}_W H^2_{cris}(X/W) = b_2(X)$. Hence by identifying W/pW with k, we may regard $E_2^{0,2}$ as a finite dimensional k-vector space. As the intersection pairing on $H^2_{cris}(X/W)$ is perfect, we see that $\sigma_0(X) = \dim_k E_2^{0,2}(X)$ (cf. [**20**]).

On the other hand, let $V : H^2(X, W\mathcal{O}) \to H^2(X, W\mathcal{O})$ be the homomorphism induced from the Verschiebung morphism. By the relation $d = FdV$, we obtain a decreasing chain of W-modules:

$$E_2^{0,2} = \ker d \supset \ker dV \supset \cdots \supset \ker dV^{i-1} \supset \ker dV^i \supset \cdots \tag{3.2}$$

which is, in fact, a chain of k-vector spaces. Suppose now that X has irregularity 0, i.e. $\dim_k H^1(X, \mathcal{O}) = 0$. Then there are exact sequences:

$$\begin{array}{ccccccccc} 0 & \longrightarrow & H^2(X, W\mathcal{O}) & \xrightarrow{V} & H^2(X, W\mathcal{O}) & \longrightarrow & H^2(X, \mathcal{O}) & \longrightarrow & 0 \\ & & \uparrow & & \uparrow & & \uparrow & & \\ 0 & \longrightarrow & \ker dV^i & \xrightarrow{V} & \ker dV^{i-1} & \longrightarrow & A_i & \longrightarrow & 0 \end{array} \tag{3.3}$$

where $A_i := \ker dV^{i-1}/\ker dV^i$ for $i \geq 1$ and the vertical arrows are inclusion maps. Here in addition, if X is $K3$, then $H^2(X, \mathcal{O}_X) \cong k$. Hence noting that $\ker d$ is a σ_0-dimensional k-vector space, we have

$$A_i \cong \begin{cases} k & \text{if } 1 \leq i \leq \sigma_0 \\ 0 & \text{if } i > \sigma_0 \end{cases}$$

(see [**20**] for details). In summary, if X is a supersingular $K3$ surface, then the chain (3.2) becomes

$$E_2^{0,2} = \ker d \supset \ker dV \supset \cdots \supset \ker dV^{\sigma_0(X)-1} \supset \ker dV^{\sigma_0(X)} = \{0\} \tag{3.4}$$

with $\ker dV^{i-1}/\ker dV^i \cong k$ for $1 \leq i \leq \sigma_0(X)$.

Another characterization of σ_0 is given in terms of the Frobenius endomorphism. Let $c_1\colon \mathrm{NS}(\widetilde{X}_M) \otimes W \to H^2_{cris}(\widetilde{X}_M/W)$ be the cycle class map of $\widetilde{X}_M$. Write F for the endomorphism of $H^2_{cris}(\widetilde{X}_M/W)$ induced from the Frobenius automorphism of $\widetilde{X}_M$. Then the image of c_1 is known to be the largest sub-F-crystal, L, such that $F(L) \subset pL$ (cf. [**14**]). Hence by the Poincaré duality, $\sigma_0(\widetilde{X}_M)$ is equal to the W-length of the cokernel of c_1.

LEMMA 3.1. *Let X and X' be supersingular surfaces over k. Assume that X is birational to X'. Then $\sigma_0(X) = \sigma_0(X')$.*

PROOF. It suffices to show that σ_0 is invariant under a blow-up. This is so because a blow-up of X increases both $\mathrm{NS}(X) \otimes W$ and $H^2_{cris}(X/W)$ by W (i.e., by rank 1). Hence by (3.1), the k-dimension of $E_2^{0,2}(X)$ does not change. □

Assume now that X is a supersingular $K3$ surface over k (i.e., X is a smooth projective surface over k with a trivial canonical divisor, irregularity 0 and $\rho(X) = b_2 = 22$). In [**1**], E. Artin proved that $\operatorname{disc} \mathrm{NS}(X) = -p^{2\sigma_0(X)}$ for some integer $\sigma_0(X)$ with $1 \leq \sigma_0(X) \leq 10$ and $\sigma_0(X)$ does take all the integer values between 1 and 10 (cf. [**1**], [**19**]).

REMARK 3.2. For a $K3$ surface X over k, the following statements are conjectured to be equivalent (Artin [**1**], Shioda [**25**]):

(i) X is unirational.
(ii) X is supersingular.
(iii) The height, $\operatorname{ht} \hat{\mathrm{Br}}_X$, of the formal Brauer group of X is infinite; equivalently, $H^2(X, W\mathcal{O}_X)$ is torsion (cf. [**20**]).

In [**1**], Artin uses (iii) as a definition of supersingular $K3$ surfaces. The equivalence between the three statements has been verified for several types of surfaces; see for instance [**1**], [**14**], [**24**], [**28**]. Further, it is known that (i) implies (ii) (cf. [**25**]) and (ii) implies (iii) (cf. [**1**]).

LEMMA 3.3. *Let X be a $K3$ surface over k. Assume that X has Picard number greater than or equal to* 5. *Then conditions* (ii) *and* (iii) *of Remark* 3.2 *for X are equivalent.*

PROOF. It is proved in [**1**] that if X has a pencil of elliptic curves, then (ii) and (iii) are equivalent for X. Further, X has a pencil of elliptic curves if and only if there exists a non-zero divisor D on X satisfying $D^2 = 0$ (see [**21**, §3, Corollary 3 to Theorem 1]). Then by an elementary argument on quadratic forms, such a divisor D exists if the Picard number of X is greater than or equal to 5. $\square$

From here to the end of this section, we fix the situation as follows: Z is a smooth proper surface over k whose minimal model, $\widetilde{Z}$, is a $K3$ surface. G is a finite subgroup of the automorphism group of Z with $\gcd(p, |G|) = 1$. Put $X = Z/G$ and let X' be the minimal resolution of X. Define Z' to be the G-equivariant resolution of Z. We assume that the minimal model $\widetilde{X}$ of X' is also a $K3$ surface. The situation is illustrated as follows:

(3.5)
$$\begin{array}{ccccc} Z & \longleftarrow & Z' & \approx & \widetilde{Z}\ (K3) \\ \downarrow & & \downarrow & & \\ Z/G = X & \longleftarrow & X' & \approx & \widetilde{X}\ (K3) \end{array}$$

where X' is isomorphic to Z'/G and "$\approx$" means that the two surfaces are birationally equivalent.

LEMMA 3.4. *With the notation and hypotheses as in* (3.5), *assume that $\widetilde{Z}$ is supersingular. Then $\widetilde{X}$ is supersingular.*

PROOF. The supersingularity is preserved under birational morphisms. Hence Z' is supersingular and $\mathrm{NS}(Z') \otimes K \cong H^2_{cris}(Z'/W) \otimes K = H^2_{cris}(Z'/K)$. This implies

$$\mathrm{NS}(X') \otimes K \cong (\mathrm{NS}(Z') \otimes K)^G \cong H^2_{cris}(Z'/K)^G \cong H^2_{cris}(X'/K).$$

Therefore $\widetilde{X}$ is also supersingular. $\square$

REMARK 3.5. With Artin's definition for supersingular $K3$ surfaces (i.e. definition (iii) in Remark (3.2)), it holds that $H^2(\widetilde{Z}, W\mathcal{O})$ is torsion if and only if $H^2(\widetilde{X}, W\mathcal{O})$ is torsion; that is, supersingularity is equivalent between $\widetilde{Z}$ and $\widetilde{X}$. The proof can be carried out by using a trace map on the sheaves of Witt vectors associated with the non-singular part of the quotient $Z \to X$, i.e. over a non-singular open subset of X. (I thank Professor L. Illusie for the idea of proof.)

PROPOSITION 3.6. *The notation and hypotheses of* (3.5) *remain in force. Assume that $\widetilde{Z}$ is supersingular and that the action of G is commutative with the differential map d on $W\mathcal{O}_Z$. Then $\widetilde{X}$ is also supersingular and both $\widetilde{X}$ and $\widetilde{Z}$ have the same Artin invariant.*

PROOF. From Lemma 3.4, $\widetilde{X}$ is supersingular. By Lemma 3.1, the equality $\sigma_0(\widetilde{X}) = \sigma_0(\widetilde{Z})$ is equivalent to showing that $\sigma_0(X') = \sigma_0(Z')$.

It follows from the definition of $W\Omega^\bullet_Z$ (cf. [**14**]) that the action of G commutes with the Verschiebung morphism V on $W\Omega^\bullet_Z$, thus in particular, on $W\mathcal{O}_Z$. Write $\psi : Z' \to Z$ for the G-equivariant resolution. Then $\psi^{-1}W\Omega^\bullet_Z$ is isomorphic to $W\Omega^\bullet_{Z'}$ and ψ is commutative with d, V and G. Hence the action of G is commutative with d and V on $W\mathcal{O}_{Z'}$.

Consider the chain (3.4) for Z':

$$E_2^{0,2}(Z') = \ker d \supset \ker dV \supset \cdots \supset \ker dV^{\sigma_0 - 1} \supset \ker dV^{\sigma_0} = \{0\} \tag{3.6}$$

where $d\colon H^2(Z', W\mathcal{O}) \to H^2(Z', W\Omega^1)$ and $\sigma_0 = \sigma_0(Z')$. Since G commutes with d and V on $W\mathcal{O}_{Z'}$, G acts on $\ker dV^i$ for $0 \le i \le \sigma_0$. It follows from the definition of $E_2^{0,2}$ and the assumption $\gcd(p, |G|) = 1$ that $E_2^{0,2}(Z')^G \cong E_2^{0,2}(X')$. Hence by taking G-invariant pieces of (3.6), we obtain

$$E_2^{0,2}(X') = (\ker d)^G \supset (\ker dV)^G \supset \cdots \supset (\ker dV^{\sigma_0 - 1})^G \supset (\ker dV^{\sigma_0})^G = \{0\}.$$

On the other hand, (3.3) implies that $\ker dV^i$ is a k-vector space for each i. Noting also that $\gcd(p, |G|) = 1$, we find the isomorphisms

$$(\ker dV^i)^G / (\ker dV^{i+1})^G \cong (\ker dV^i / \ker dV^{i+1})^G \cong H^2(Z', \mathcal{O})^G \cong H^2(X', \mathcal{O}) \cong k$$

for $0 \le i \le \sigma_0(Z') - 1$. Therefore $\dim_k E_2^{0,2}(X') = \sigma_0$; i.e. $\sigma_0(Z') = \sigma_0(X')$. □

We give a relation between the Artin invariant σ_0 and the number of exceptional divisors.

THEOREM 3.7. *Let X be a normal projective surface over k (smooth or singular). Assume that*

(i) *a resolution $X' \to X$ is obtained by a finite sequence of blow-ups,*
(ii) *p is relatively prime to the determinant of the intersection matrix of the exceptional divisors on X' and*
(iii) *X' is supersingular (not necessarily $K3$).*

Let e be the number of irreducible components of the exceptional divisors on X'. Then $2\sigma_0(X') \le b_2(X') - e$.

PROOF. Let $\mathbb{E}$ denote the free $\mathbb{Z}$-submodule of $\mathrm{NS}(X')$ generated by the exceptional divisors, say E_i's, on X'. By considering the intersection pairing on $\mathbb{E}$, we have an embedding $\mathbb{E} \mapsto \mathbb{E}^* := \mathrm{Hom}(\mathbb{E}, \mathbb{Z})$ and its cokernel is killed by the determinant $\det((E_i.E_j))$ (cf. [**18**, §14]). From the assumption (ii), $\det((E_i.E_j))$ is not divisible by p. Hence if $\mathbb{E}_W$ denotes $\mathbb{E} \otimes_{\mathbb{Z}} W$, then $\mathbb{E}_W \cong \mathbb{E}^*_W$. Together with the assumption (i), this isomorphism implies that $\mathrm{NS}(X')_W := \mathrm{NS}(X') \otimes W$ has an orthogonal decomposition

$$\mathrm{NS}(X')_W \cong N \perp \mathbb{E}_W$$

for some W-submodule N, where N and $\mathbb{E}_W$ are orthogonal with respect to the intersection pairing on $\mathrm{NS}(X')$.

On the other hand, since X' is supersingular, we have

$$\mathrm{NS}(X')^* \supset \mathrm{NS}(X') \supset p\,\mathrm{NS}(X')^*$$

with $(\mathrm{NS}(X')^* : \mathrm{NS}(X')) = p^{2\sigma_0}$ and $(\mathrm{NS}(X') : p\,\mathrm{NS}(X')^*) = p^{b_2-2\sigma_0}$ (here $\sigma_0 = \sigma_0(X')$ and $b_2 = b_2(X')$); see [**1**, Corollary 4.4]. Tensoring these $\mathbb{Z}$-modules with W and noting that $\mathbb{E}_W \cong \mathbb{E}_W^*$, $\mathrm{NS}(X')_W^* \cong N^* \perp \mathbb{E}_W$ and $p\,\mathrm{NS}(X')_W^* \cong pN^* \perp p\mathbb{E}_W$, we obtain

$$N^* \oplus \mathbb{E}_W \supset N \oplus \mathbb{E}_W \supset pN^* \oplus p\mathbb{E}_W.$$

As W/pW is a k-vector space and $\dim_k W/pW = 1$,

$$b_2 - 2\sigma_0 = \dim_k((N \oplus \mathbb{E}_W)/(pN^* \oplus p\mathbb{E}_W)) \geq \dim_k \mathbb{E}_W/p\mathbb{E}_W = e.$$

Therefore $b_2 - 2\sigma_0 \geq e$. This proves the assertion. □

4. Supersingular Weighted Delsarte $K3$ Surfaces

In this section, we give an exposition of the result of [**10**] for computing the discriminant of the Néron-Severi groups of supersingular weighted Delsarte $K3$ surfaces. Refer also to [**27**] for some underlying ideas.

Let k be an algebraically closed field of characteristic $p > 0$. Let X_M be a weighted Delsarte surface in $\mathbb{P}_k^3(Q)$ with matrix M. Write $\mathcal{F}_d$ for the Fermat surface of degree $d = \det M$ in $\mathbb{P}_k^3$. The notation of Section 2 remains in force. Recall Proposition 2.3 that

$$H^2(\widetilde{X}_M, *) \approx H^2(\mathcal{F}_d/\Gamma_M, *) \cong V(0) \oplus \bigoplus_{\boldsymbol{\alpha} \in \mathfrak{A}(X_M)} V(\boldsymbol{\alpha}).$$

Choose Q to be one of the 95 $K3$ weights defined at the end of Section 1. Assume that X_M is well-formed, quasi-smooth and of degree $m = q_0 + q_1 + q_2 + q_3$. Let $\widetilde{X}_M$ be the minimal resolution of X_M. Then $\widetilde{X}_M$ is a $K3$ surface and there exists a unique element $(\alpha_0, \alpha_1, \alpha_2, \alpha_3) =: \boldsymbol{\alpha}_{\mathrm{ss}} \in \mathfrak{A}(X_M)$ such that $V(\boldsymbol{\alpha}_{\mathrm{ss}})$ is of type $(2,0)$ (in the Hodge decomposition of $H^2_{cris}(\widetilde{X}_M/W) \otimes \mathbb{C}$); if we assume $1 \leq \alpha_i < d$ for $0 \leq i \leq 3$, then $V(\boldsymbol{\alpha}_{\mathrm{ss}})$ being of type $(2,0)$ is equivalent to $\alpha_0 + \alpha_1 + \alpha_2 + \alpha_3 = d$. Given $\boldsymbol{\alpha}_{\mathrm{ss}} = (\alpha_0, \alpha_1, \alpha_2, \alpha_3)$ in $\mathfrak{A}(X_M)$, we define

$$f_M = d/\gcd(\alpha_0, \alpha_1, \alpha_2, \alpha_3, d). \tag{4.1}$$

Using the integer f_M, we can determine the supersingularity of $\widetilde{X}_M$ and compute the discriminant of its Néron-Severi group (but we do not know yet how to describe a $\mathbb{Z}$-basis for $\mathrm{NS}(\widetilde{X}_M)$).

PROPOSITION 4.1. *Let X_M be a weighted Delsarte surface with matrix M. Assume that $\widetilde{X}_M$ is $K3$. Then $\widetilde{X}_M$ is supersingular if and only if $p^\mu \equiv -1 \pmod{f_M}$ for some integer $\mu \geq 1$.*

PROOF. See [**10**]. □

THEOREM 4.2. *Let $Q = (q_0, q_1, q_2, q_3)$ be a $K3$ weight. Let X_M be a well-formed quasi-smooth weighted Delsarte surface in $\mathbb{P}^3_k(Q)$ of degree $m = q_0 + q_1 + q_2 + q_3$ with matrix M. Write $\widetilde{X}_M$ for the minimal resolution of X_M. Assume that $p^\mu \equiv -1 \pmod{f_M}$ for some positive integer μ, where f_M is the integer defined in (4.1); let μ_0 be the smallest such μ. Then $\widetilde{X}_M$ is a supersingular $K3$ surface and the Artin invariant $\sigma_0(\widetilde{X}_M)$ of $\widetilde{X}_M$ is equal to μ_0.*

PROOF (Details can be found in [**10**].) Recall that the Artin invariant $\sigma_0(\widetilde{X}_M)$ is equal to the W-length of the cokernel of the cycle class map c_1 (see Section 3). Put $\boldsymbol{\alpha}_0 := -\boldsymbol{\alpha}_{\rm ss}$, where $\boldsymbol{\alpha}_{\rm ss}$ is an element of $\mathfrak{A}(X_M)$ defined above. Then $\boldsymbol{\alpha}_0 \in \mathfrak{A}(X_M)$ and it is the (unique) element in $\mathfrak{A}(X_M)$ of Hodge type $(0,2)$. Further, μ_0 is the smallest positive integer such that $p^{\mu_0}\boldsymbol{\alpha}_0 = \boldsymbol{\alpha}_{\rm ss}$. If we define

$$I_0 = \{p\boldsymbol{\alpha}_0, p^2\boldsymbol{\alpha}_0, \ldots, p^{\mu_0}\boldsymbol{\alpha}_0\ (= \boldsymbol{\alpha}_{\rm ss})\}.$$

then we obtain

$$c_1(\mathrm{NS}(\widetilde{X}_M) \otimes W) \cong V(0) \oplus \bigoplus_{\boldsymbol{\alpha} \in I_0} pV(\boldsymbol{\alpha}) \oplus \bigoplus_{\boldsymbol{\alpha} \in \mathfrak{A}(X_M) \setminus I_0} V(\boldsymbol{\alpha}) \oplus \mathbb{E}$$

where $\mathbb{E}$ denotes the classes of exceptional cycles arising in the desingularization of X_M. Hence the W-length of the image of c_1 is equal to μ_0. Therefore $\sigma_0(\widetilde{X}_M) = \mu_0$. □

5. Supersingular Deformations of Weighted Delsarte $K3$ Surfaces

In this section, we consider the minimal resolutions of one-parameter deformations of weighted Delsarte $K3$ surfaces. Assuming that they are supersingular $K3$ surfaces, we give non-trivial upper bounds for their Artin invariant. The calculation of the exact values of their Artin invariant is left as a problem for further consideration.

Let k be an algebraically closed field of positive characteristic $p > 0$. Choose $Q = (q_0, q_1, q_2, q_3)$ to be one of the 95 $K3$ weights (cf. Section 1). Let

$$X_{M,\lambda}\colon \sum_{i=0}^{3} x_0^{m_{i0}} \cdots x_3^{m_{i3}} - \lambda x_0x_1x_2x_3 = 0 \subset \mathbb{P}^3_k(Q)$$

be a deformation of a weighted Delsarte $K3$ surface of degree $m = q_0 + q_1 + q_2 + q_3$ with matrix M. Assume $X_{M,\lambda}$ to be quasi-smooth and well-formed. Write $\widetilde{X}_{M,\lambda}$ for the minimal resolution of $X_{M,\lambda}$. Then $\widetilde{X}_{M,\lambda}$ is a $K3$ surface.

Let G be a finite subgroup of the automorphism group of $X_{M,\lambda}$. Consider a diagram of surfaces

$$\begin{array}{ccccccc} X_{M,\lambda} & \longleftarrow & X'_{M,\lambda} & \longleftarrow & Z & \longleftarrow & Z' \approx \widetilde{X}_{M,\lambda} \\ \downarrow & & \downarrow & & \downarrow & & \downarrow \\ X_{M,\lambda}/G =: Y & \longleftarrow & \widetilde{Y} & & Z/G =: \mathcal{Y} & \longleftarrow & \widetilde{\mathcal{Y}} \approx \widetilde{Y} \end{array} \tag{5.1}$$

where the surfaces involved are defined as follows: $\widetilde{Y}$ is the minimal resolution of Y, $X'_{M,\lambda}$ is the G-equivariant resolution of $X_{M,\lambda}$ (which may be singular), Z is the minimal resolution of $X'_{M,\lambda}$, $\widetilde{\mathcal{Y}}$ is the minimal resolution of $\mathcal{Y} := Z/G$ and Z' is the

G-equivariant resolution of Z. Then Z' is smooth and birational to $\widetilde{X}_{M,\lambda}$ and $\widetilde{\mathcal{Y}}$ is smooth and birational to $\widetilde{Y}$.

PROPOSITION 5.1. *Assume in diagram* (5.1) *that* $\widetilde{Y}$ *is a* $K3$ *surface and* $\widetilde{X}_{M,\lambda}$ *is supersingular. Further, assume that the action of* G *is commutative with the differential map* d *on* $W\mathcal{O}_{X_{M,\lambda}}$. *Then the following assertions hold*:

(a) $\widetilde{Y}$ *is supersingular.*
(b) $\widetilde{X}_{M,\lambda}$ *and* $\widetilde{Y}$ *have the same Artin invariant*: $\sigma_0(\widetilde{X}_{M,\lambda}) = \sigma_0(\widetilde{Y})$.
(c) *Let* e_X *and* e_Y *denote the numbers of exceptional divisors arising in the resolutions* $\widetilde{X}_{M,\lambda} \to X_{M,\lambda}$ *and* $\widetilde{Y} \to Y$, *respectively. Then*

$$\sigma_0(\widetilde{X}_{M,\lambda}) \le \min\Big\{11 - \frac{e_X}{2},\quad 11 - \frac{e_Y}{2}\Big\}.$$

PROOF. (a) Apply Lemma 3.4to the right-hand side of diagram (5.1).
(b) This follows from Proposition (3.6).
(c) The inequality is an immediate consequence of Theorem 3.7 and the assertion (b). □

We give some examples of $X_{M,\lambda}$ and G such that the assumptions of Proposition 5.1 are satisfied.

Let M be a diagonal matrix

$$M_0 = \begin{bmatrix} m_0 & 0 & 0 & 0 \\ 0 & m_1 & 0 & 0 \\ 0 & 0 & m_2 & 0 \\ 0 & 0 & 0 & m_3 \end{bmatrix}.$$

such that the minimal resolution of X_{M_0} of degree m $(= q_0 + q_1 + q_2 + q_3)$ in $\mathbb{P}^3_k(Q)$ becomes a $K3$ surface. There are 14 pairs of m and Q giving such $K3$ surfaces (see Table 1): indeed, we just need to choose $K3$ weights Q satisfying $q_i \mid m$ for $0 \le i \le 3$ (hence $m_i = m/q_i$). The surface X_{M_0} may be called a *weighted diagonal* $K3$ *surface* (cf. [**9**]) and such a weight Q will be called a *diagonal* $K3$ *weight.*

In what follows, we consider a (1-*parameter*) *deformation* $X_{M_0,\lambda}$ *of a weighted diagonal* $K3$ *surface* defined by the equation

$$X_{M_0,\lambda}\colon x_0^{m_0} + x_1^{m_1} + x_2^{m_2} + x_3^{m_3} - \lambda x_0x_1x_2x_3 = 0 \subset \mathbb{P}^3_k(Q)$$

with $\lambda \in k$. Deformations of weighted diagonal $K3$ surfaces and their finite quotients provide typical examples of mirror $K3$ surfaces constructed by Roan. Mirror symmetry for $K3$ surfaces is a correspondence that relates a family of certain $K3$ surfaces with another by interchanging their refined Hodge structures (details on this subject may be found in [**17**], [**23**]).

LEMMA 5.2. $X_{M_0,\lambda}$ *is quasi-smooth if and only if* $\lambda^m \ne m_0^{q_0} \dots m_3^{q_3}$; *further, in this case,* $X_{M_0,\lambda}$ *is well-formed.*

PROOF. Taking the partial derivatives of the defining equation of $X_{M_0,\lambda}$, we see that $X_{M_0,\lambda}$ is quasi-smooth if and only if $(0,0,0,0)$ is the only solution to the equations:

$$\begin{cases} m_0x_0^{m_0-1} - \lambda x_1x_2x_3 = 0 \\ m_1x_1^{m_1-1} - \lambda x_0x_2x_3 = 0 \\ m_2x_2^{m_2-1} - \lambda x_0x_1x_3 = 0 \\ m_3x_3^{m_3-1} - \lambda x_0x_1x_2 = 0. \end{cases}$$

TABLE 1. Upper bound for the Artin invariant of $\widetilde{X}_{M_0,\lambda}$.

d	Q	e_X	e_Y	$[11 - e_Y/2]$	diagonal case
4	$(1,1,1,1)$	0	18	2	1
6	$(1,1,1,3)$	0	18	2	1
6	$(1,1,2,2)$	3	17	2	1
8	$(1,1,2,4)$	2	17	2	1
10	$(1,2,2,5)$	5	17	2	2
12	$(1,1,4,6)$	1	17	2	1
12	$(1,2,3,6)$	6	15	3	1
12	$(1,3,4,4)$	9	15	3	1
12	$(2,3,3,4)$	11	15	3	1
18	$(1,2,6,9)$	5	15	3	3
20	$(1,4,5,10)$	9	13	4	2
24	$(1,3,8,12)$	7	13	4	2
30	$(2,3,10,15)$	11	11	5	2
42	$(1,6,14,21)$	9	9	6	2

[ ] is the Gaussian symbol, 'diagonal case' means the maximum value of σ_0 for a weighted diagonal $K3$ surface (i.e., $\lambda = 0$) over all supersingular characteristic p and e_X denotes the number of exceptional divisors on $\widetilde{X}_{M_0,\lambda}$.

By straightforward calculations, we find that this is equivalent to $\lambda^m \neq m_0^{q_0} \dots m_3^{q_3}$. Further, it follows immediately from Lemma 1.1(a) that $X_{M_0,\lambda}$ is well-formed. □

To the end of this section, we assume $\lambda^m \neq m_0^{q_0} \cdots m_3^{q_3}$ so that $X_{M_0,\lambda}$ is quasi-smooth and well-formed.

Let SD_Q be the group:

$$SD_Q = \left\{ (\gamma_0, \gamma_1, \gamma_2, \gamma_3) \in (k^\times)^4 \;\middle|\; \gamma_i^{m_i} = 1 \ (0 \le i \le 3) \text{ and } \prod_{i=0}^{3} \gamma_i = 1 \right\}.$$

By conditions (1.1) and (2.1) on Q and M_0, we see that $p \nmid m$. Hence $k^\times$ contains a primitive mth root of unity and SD_Q contains a cyclic subgroup

$$C = \{(\zeta^{q_0}, \zeta^{q_1}, \zeta^{q_2}, \zeta^{q_3}) \mid \zeta \text{ ranges over all } m\text{-th roots of unity in } k^\times\}.$$

The group SD_Q acts on $\mathbb{P}_k^3(Q)$ diagonally (i.e. $z_i \mapsto \gamma_i z_i$) and the action of C on any projective subvariety of $\mathbb{P}_k^3(Q)$ is trivial. As $p \nmid m$, the order of SD_Q is relatively prime to p. Because of the condition $\prod_{i=0}^{3} \gamma_i = 1$, SD_Q acts on $X_{M_0,\lambda}$ diagonally for every λ.

PROPOSITION 5.3. *Let $Q = (q_0, q_1, q_2, q_3)$ be a diagonal $K3$ weight with $m = q_0 + q_1 + q_2 + q_3$. Let $X_{M_0,\lambda}$ be a deformation of a weighted diagonal $K3$ surface of degree m. Let G be a subgroup of SD_Q containing C. Then the following assertions hold*:

(a) *The resolution $\widetilde{X}_{M_0,\lambda}$ can be obtained by a finite sequence of blow-ups of $X_{M_0,\lambda}$.*

(b) *p is relatively prime to the determinant of the intersection matrix of the exceptional divisors on $\widetilde{X}_{M_0,\lambda}$.*

(c) *The morphism on sheaves* $W\mathcal{O}^i_{X_{M,\lambda}}$ $(i \geq 0)$ *induced by the diagonal action of* G *is commutative with the differential map* d.

PROOF. (a) Since $X_{M_0,\lambda}$ has only cyclic quotient singularities of type A, the resolution can be obtained by finite blow-ups.

(b) It follows from the conditions $q_i \mid m$ $(0 \leq i \leq 3)$ and $m = q_0 + q_1 + q_2 + q_3$ that every singularity on X is of type $A_{n,n-1}$ for some integer $n \geq 2$ and n is a divisor of some q_i (cf. [**9**]). By condition (1.1), we have $p \nmid n$. On the other hand, the intersection determinant of the exceptional divisors arising from a singularity of type $A_{n,n-1}$ is equal to n. Hence the determinant of the intersection matrix of all the exceptional divisors on $\widetilde{X}$ is relatively prime to p.

(c) Fix $i \geq 0$. It is straight-forward to see that G is commutative with d on $W_1\Omega^i_Z = \Omega^i_Z$. Working directly with the definition of $W\Omega^\bullet_Z = \varprojlim_n W_n\Omega^\bullet_Z$ and by induction on n, we find that G commutes with d on $W_n\Omega^i_Z$ for $n \geq 1$. Then taking limit on n, we obtain the commutativity in the assertion. □

THEOREM 5.4. *Let* $Q = (q_0, q_1, q_2, q_3)$ *be a diagonal* $K3$ *weight with* $m = q_0 + q_1 + q_2 + q_3$. *Let* $X_{M_0,\lambda}$ *be a deformation of a weighted diagonal* $K3$ *surface of degree* m. *Denote by* e_Y *the numbers of exceptional divisors on* $\widetilde{Y}$ *in diagram* (5.1). *Assume that the minimal resolution* $\widetilde{X}_{M_0,\lambda}$ *is supersingular. Then*

$$\sigma_0(\widetilde{X}_{M_0,\lambda}) \leq 11 - \frac{e_Y}{2}$$

(*The actual value of the right-hand side is shown in Table* 1.)

PROOF. Choose G to be SD_Q (because SD_Q is maximal) and consider diagram (5.1). Since actual calculations can be carried out easily, we only sketch the proof.

For every pair (m, Q), the singular locus of $X_{M_0,\lambda}$ can be determined in the same way as in the case of weighted diagonal surfaces; see [**9**, Table 7] for a list of singularities of $X_{M_0,\lambda}$ together with their types. As $\widetilde{X}_{M_0,\lambda}$ is supersingular, $\widetilde{Y}$ in diagram (5.1) is also supersingular and $\sigma_0(\widetilde{X}_{M_0,\lambda}) = \sigma_0(\widetilde{Y})$ by Proposition 5.1. By looking at the action of SD_Q on $X_{M_0,\lambda}$ (or use the mirror correspondence given in [**23**]), we can calculate the number of exceptional divisors on $\widetilde{Y}$. Then the inequality of Theorem 3.7 gives the required upper bound for the value of the Artin invariant of $\widetilde{Y}$. □

REMARK 5.5. In view of Lemma 3.4, we know that if $\widetilde{X}_{M_0,\lambda}$ is supersingular, so is $\widetilde{Y}$. But the converse is also true if $e_X \geq 4$ (there are 9 cases in Table 1). In fact, if $\widetilde{Y}$ is supersingular, then $H^2(\widetilde{Y}, W\mathcal{O}_{\widetilde{Y}})$ is torsion and so is $H^2(\widetilde{X}_{M_0,\lambda}, W\mathcal{O}_{\widetilde{X}})$; see Remarks 3.2 and 3.5. On the other hand, the inequality $e_X \geq 4$ implies that $\widetilde{X}_{M_0,\lambda}$ has Picard number greater than or equal to 5 since there is always a hyperplane section on it. Hence by Lemma 3.3, $\widetilde{X}_{M_0,\lambda}$ is supersingular.

REMARK 5.6. In addition to the one-parameter deformations of weighted diagonal $K3$ surfaces, there are more families of $K3$ surfaces for which SD_Q acts diagonally. For instance, if $Q = (1, 6, 14, 21)$, then $\mathrm{SD}_Q = C$ and every deformation of a weighted Delsarte surface in $\mathbb{P}^3_k(1, 6, 14, 21)$ is endowed with a diagonal SD_Q-action.

Another example is a two-parameter family of surfaces

$$z_0^6 + z_1^6 + z_2^3 + z_3^3 + \mu z_0^3 z_1^3 + \lambda z_0 z_1 z_2 z_3 = 0$$

in $\mathbb{P}^3_k(1,1,2,2)$ with $\mu, \lambda \in k$ satisfying $(27\mu + \lambda^2)^2 \neq 2^2 \cdot 3^6$. These surfaces are quasi-smooth and they have a diagonal SD_Q-action. Discussions in this section are valid for their minimal resolutions. Consequently, if they are supersingular $K3$ surfaces, then their Artin invariant is bounded by 2.

6. Weighted Quasi-Diagonal $K3$ Surfaces

From this section on, we consider another special family of weighted Delsarte $K3$ surfaces and also their deformations. In this section, we describe the singular loci of these surfaces and compute their Picard numbers. This is an exposition of [**11**].

Let k be an algebraically closed field of arbitrary characteristic $p \geq 0$. Let $Q = (q_0, q_1, q_2, q_3)$ be one of the 95 $K3$ weights such that there exist positive integers n_i $(0 \leq i \leq 3)$ satisfying

$$q_0 n_0 = q_0 + q_1 n_1 = q_2 n_2 = q_3 n_3 = q_0 + q_1 + q_2 + q_3 =: m. \tag{6.1}$$

Put

$$M_1 = \begin{bmatrix} n_0 & 0 & 0 & 0 \\ 1 & n_1 & 0 & 0 \\ 0 & 0 & n_2 & 0 \\ 0 & 0 & 0 & n_3 \end{bmatrix}.$$

Then $X_{M_1,\lambda}$ is a projective surface in $\mathbb{P}^3_k(q_0, q_1, q_2, q_3)$ defined by the equation

$$X_{M_1,\lambda} \colon x_0^{n_0} + x_0 x_1^{n_1} + x_2^{n_2} + x_3^{n_3} - \lambda x_0 x_1 x_2 x_3 = 0$$

with $\lambda \in k$ (we assume condition (2.1)). Since the case $\lambda = 0$ gives an almost diagonal equation, we call $X_{M_1,0}$ a *quasi-diagonal $K3$ surface of degree* $m(= q_0 + q_1 + q_2 + q_3)$. Likewise, $X_{M_1,\lambda}$ will be called a *(one-parameter) deformation of a weighted quasi-diagonal $K3$ surface.* Note that $n_i \geq 2$ for $0 \leq i \leq 3$ and hyperplane sections $X_{M_1,\lambda} \cap \{x_i = 0\}$ are irreducible for $1 \leq i \leq 3$ (for the case $i = 0$, see Lemma 6.7 below).

LEMMA 6.1. *There exist* 51 *weights Q satisfying condition* (6.1). *From these weights, we have* 85 *choices for* (n_0, n_1, n_2, n_3).

PROOF. Both assertions can be verified by looking through Fletcher's table of $K3$ weights (see [**8**]). An actual list of 85 quadruples (n_0, n_1, n_2, n_3) can be found in [**11**]. □

Let $X_{M_1,\lambda}$ be a deformation of a weighted quasi-diagonal $K3$ surface of degree m. Let $\mathfrak{I} := \{(0,1), (0,2), (0,3), (2,3)\}$. For each $(i,j) \in \mathfrak{I}$, put

$$\mathcal{P}_{ij} := \{P = (x_0 : x_1 : x_2 : x_3) \in X_{M_1,\lambda} \mid x_i x_j \neq 0,\ x_h = 0 \text{ for } h \neq i,j\}$$

and

$$d_{ij} := \gcd(q_i, q_j).$$

(Note $\mathcal{P}_{ij} = \varnothing$ if $(i,j) = (1,2)$ or $(1,3)$.) Write $\mathfrak{I}_1 := \{(i,j) \in \mathfrak{I} \mid d_{ij} \geq 2\}$.

PROPOSITION 6.2. *Let $X_{M_1,\lambda}$ be a deformation of a weighted quasi-diagonal $K3$ surface of degree m in $\mathbb{P}^3_k(q_0, q_1, q_2, q_3)$. Assume the inequality* $\lambda^{m(m-q_0)} \neq n_0^{q_0(m-q_0-q_1)} n_1^{(m-q_0)(q_0+q_1)} n_2^{q_2(m-q_0)} n_3^{q_3(m-q_0)} / (n_1 - 1)^{q_0(m-q_0-q_1)}$. *Then the following assertions hold*:

(a) *$X_{M_1,\lambda}$ is a well-formed quasi-smooth weighted projective surface.*

(b) *Put* $e_{ij} = \mathrm{lcm}(q_i, q_j)$. *Then for* $(i,j) \in \mathfrak{I}$,

$$\#\mathcal{P}_{ij} = \begin{cases} (m - q_0)/e_{01} & \text{if } (i,j) = (0,1) \\ m/e_{ij} & \text{otherwise.} \end{cases}$$

(c) *Write* $\mathrm{Sing}(X_{M_1,\lambda})$ *for the singular locus of* $X_{M_1,\lambda}$. *Then*

$$\mathrm{Sing}(X_{M_1,\lambda}) = \begin{cases} \bigcup_{(i,j)\in\mathfrak{I}_1} \mathcal{P}_{ij} & \text{if } q_1 = 1 \\ \{(0:1:0:0)\} \cup \bigcup_{(i,j)\in\mathfrak{I}_1} \mathcal{P}_{ij} & \text{if } q_1 \geq 2. \end{cases}$$

(d) *If* $(i,j) \in \mathfrak{I}_1$, *then every point in* $\mathcal{P}_{ij}$ *is a cyclic quotient singularity of type* $A_{d_{ij},d_{ij}-1}$.

(e) *Assume* $q_1 \geq 2$. *Then* $(0:1:0:0) \in X_{M_1,\lambda}$ *is a cyclic quotient singularity of type* A_{q_1,q_1-1}.

PROOF. (a) can be seen in the same way as in Lemma 5.2. Assertions (b)–(e) can be proved similarly as in the case of weighted diagonal surfaces (cf. [**9**]); details may be found in [**11**]. □

Let $X_{M_1} (= X_{M_1,0})$ be a weighted quasi-diagonal $K3$ surface. We will describe the cohomology of $\widetilde{X}_{M_1}$ and calculate its Picard number. First we compute the zeta-function of X_{M_1} over a finite field.

PROPOSITION 6.3. *Let* X_{M_1} *be a weighted quasi-diagonal* $K3$ *surface of degree* m *in* $\mathbb{P}^3(q_0, q_1, q_2, q_3)$ *defined over a finite field* $\mathbb{F}_q$ *of* q *elements. Let*

$$N = \mathrm{lcm}(n_1, n_2, n_3)$$
$$N_i = N/n_i$$

for $1 \leq i \leq 3$. *Assume* $q \equiv 1 \pmod{N}$ *and fix a character,* χ, *of* $\mathbb{F}_q^\times$ *of exact order* N. *Then the zeta-function of* X_{M_1} *is described as follows:*

$$Z(X_{M_1}, t) = \frac{1}{(1-t)P_2(X_{M_1}, t)(1-q^2t)}$$

where

$$P_2(X_{M_1}, t) = (1 - qt) \prod_{a \in \mathcal{V}(X_{M_1})} (1 - q\chi^{aN_3}t) \prod_{\boldsymbol{\alpha} \in \mathcal{W}(X_{M_1})} (1 - j(\boldsymbol{\alpha})t)$$

$$\mathcal{V}(X_{M_1}) = \{a \in \mathbb{Z} \mid 0 < a < n_3,\ q_3 a \equiv 0 \pmod{q_2}\}$$

$$\mathcal{W}(X_{M_1} = \left\{ \boldsymbol{\alpha} = (\alpha_0, \alpha_1, \alpha_2, \alpha_3) \;\middle|\; \begin{array}{l} \alpha_0 \in \mathbb{Z}/N\mathbb{Z}, \alpha_i \in N_i\mathbb{Z}/N\mathbb{Z}\ (1 \leq i \leq 3) \\ \alpha_i \neq 0\ (0 \leq i \leq 3), \sum_{i=0}^{3} \alpha_i = 0 \\ n_0(\alpha_1 + \alpha_2 + \alpha_3) = \alpha_1 \end{array} \right\}$$

$j(\boldsymbol{\alpha})$ *is the Jacobi sum associated with* $\boldsymbol{\alpha}$ *relative to* χ.

PROOF. The formulae and relations can be obtained by using results of [**32**] and [**15**]; see [**11**] for the details. The results can be deduced also from Proposition 2.3. □

REMARK 6.4. In the definition of $\mathcal{W}(X_{M_1})$, $\alpha_i \neq 0$ $(1 \leq i \leq 3)$ implies $\alpha_0 \neq 0$.

Choosing a prime $\ell \neq \mathrm{char}\, k$, let $H^n(X, \mathbb{Q}_\ell)$ denote the ℓ-adic étale cohomology (tensored with $\mathbb{Q}_\ell$) of a smooth projective variety X over k. Recall the definition of the set $\mathfrak{A}(X_{M_1})$ in Section 2.

COROLLARY 6.5. *Let X_{M_1} be a weighted quasi-diagonal K3 surface of degree m in $\mathbb{P}^3_k(Q)$. Write e_X for the number of exceptional divisors on the minimal resolution $\widetilde{X}_{M_1}$. Then $\mathcal{W}(X_{M_1})$ is a subset of $\mathfrak{A}(X_{M_1})$ and $H^2(\widetilde{X}_{M_1}, \mathbb{Q}_\ell)$ has the following decomposition:*

$$H^2(\widetilde{X}_M, \mathbb{Q}_\ell) \cong V(0) \oplus \overbrace{\mathbb{Q}_\ell \oplus \cdots \oplus \mathbb{Q}_\ell}^{\#\mathcal{V}(X_{M_1})} \oplus \bigoplus_{\boldsymbol{\alpha} \in \mathcal{W}(X_{M_1})} V(\boldsymbol{\alpha}) \oplus \overbrace{\mathbb{Q}_\ell \oplus \cdots \oplus \mathbb{Q}_\ell}^{e_X}$$

where $V(\boldsymbol{\alpha}) \cong \mathbb{Q}_\ell$ as a $\mathbb{Q}_\ell$-vector space.

PROOF. When $k = \mathbb{F}_q$, the decomposition follows from the description of $P_2(X_{M_1}, t)$ in Proposition 6.3. Then by a specialization argument, we obtain the same decomposition of $H^2(\widetilde{X}_M, \mathbb{Q}_\ell)$ for an arbitrary k with $\operatorname{char} k = p > 0$ (note that every monomial in the defining equation of X_{M_1} has coefficient 1). When $\operatorname{char} k = 0$, the assertion can be proved first for $k = \mathbb{C}$ by using a comparison theorem. The case for a general k then follows from it by a specialization argument. □

COROLLARY 6.6. *Let X_{M_1} be a weighted quasi-diagonal K3 surface of degree m in $\mathbb{P}^3_k(Q)$. Then the element $\boldsymbol{\alpha}_{\mathrm{ss}} \in \mathcal{W}(X_{M_1})$ defined in Section 4 is given by*

$$\boldsymbol{\alpha}_{\mathrm{ss}} = (N - N_1 - N_2 - N_3, N_1, N_2, N_3)$$

and $\gcd(N - N_1 - N_2 - N_3, N_1, N_2, N_3) = 1$.

PROOF. Using (6.1) and $m = q_0 + q_1 + q_2 + q_3$, we find

$$\begin{aligned} N_1 + N_2 + N_3 = N\left(\frac{1}{n_1} + \frac{1}{n_2} + \frac{1}{n_3}\right) &= N\left(\frac{q_1}{m - q_0} + \frac{q_2}{m} + \frac{q_3}{m}\right) \\ &= N\left\{1 - \frac{q_2 + q_3}{n_0(m - q_0)}\right\}. \end{aligned}$$

This implies

$$n_0(N_1 + N_2 + N_3) - N_1 = n_0 N\left\{1 - \frac{q_2 + q_3}{n_0(m - q_0)}\right\} - \frac{N q_1}{m - q_0} = N(n_0 - 1).$$

Hence $n_0(N_1 + N_2 + N_3) = N_1$ in $\mathbb{Z}/N\mathbb{Z}$ and $(N - N_1 - N_2 - N_3, N_1, N_2, N_3)$ is an element of $\mathfrak{A}(X_{M_1})$. As $q_2 + q_3 < q_1 + q_2 + q_3 = m - q_0$, we have $0 < N_1 + N_2 + N_3 < N$. Hence $\boldsymbol{\alpha}_{\mathrm{ss}} = (N - N_1 - N_2 - N_3, N_1, N_2, N_3)$. The last part of the assertion follows from $\gcd(N_1, N_2, N_3) = 1$. □

LEMMA 6.7. *Put $l_{23} := \operatorname{lcm}(q_2, q_3)$. Then the cardinality of $\mathcal{V}(X_{M_1})$ is equal to*

$$\#\mathcal{V}(X_{M_1}) = \frac{m}{l_{23}} - 1$$

and m/l_{23} is equal to the number of irreducible components of the hyperplane section $X_{M_1,\lambda} \cap \{x_0 = 0\}$.

PROOF. Noting $m = n_3 q_3$, we can calculate $\#\mathcal{V}(X_{M_1})$ by easy arithmetic. There are m/l_{23} points on $x_2^{n_2} + x_3^{n_3} = 0$ in $\mathbb{P}^1_k(q_2, q_3)$ and the hyperplane section $\{x_0 = 0\}$ has the same number of irreducible components. □

THEOREM 6.8. *Let k be an algebraically closed field of characteristic $p \geq 0$. Let X_{M_1} be a weighted quasi-diagonal K3 surface of degree m in $\mathbb{P}^3_k(Q)$. Write e_X for the number of exceptional divisors arising in the resolution $\widetilde{X}_{M_1} \to X_{M_1}$. Put*

$$\mathcal{W}_{al}(X_{M_1}) := \{\boldsymbol{\alpha} \in \mathcal{W}(X_{M_1}) \mid j(\boldsymbol{\alpha})/q \text{ is a root of unity}\}$$

where $j(\boldsymbol{\alpha})$ is the Jacobi sum defined above. Then the Picard number of $\widetilde{X}_{M_1}$ (over k) is equal to

$$\begin{aligned}\rho(\widetilde{X}_{M_1}) &= e_X + \frac{m}{l_{23}} + \#\mathcal{W}_{al}(X_{M_1}) \\ &= \begin{cases} 22 & \text{if } p^\mu \equiv -1 \pmod{N} \text{ for some } \mu \geq 1 \\ 22 - \phi(N) & \text{otherwise} \end{cases}\end{aligned}$$

where $\phi(N)$ is the Euler number of N.

PROOF. We begin by showing the first equality. Suppose $p > 0$. Then we may argue $\rho(\widetilde{X}_{M_1})$ by choosing $k = \mathbb{F}_q$ (then use specialization for general cases). In this case, the Tate conjecture holds for $\widetilde{X}_{M_1}$ (cf. Theorem 2.4) and the corresponding statement is nothing but the asserted formula. Suppose now that $p = 0$. Then we may choose $k = \mathbb{C}$, where one knows that the Tate conjecture holds for $\widetilde{X}_{M_1}$ by the validity of the Hodge conjecture. Hence the Picard number is given also by the same formula.

To prove the second equality, note first that $\widetilde{X}_{M_1}$ is supersingular if and only if $p^\mu \equiv -1 \pmod{N}$ for some integer $\mu \geq 1$ (this can be seen by showing $f_{M_1} = N$ or directly applying a result of [**28**] on $\mathcal{W}(X_{M_1})$). Then by definition, $\rho(\widetilde{X}_{M_1}) = b_2 = 22$ if $\widetilde{X}_{M_1}$ is supersingular.

It remains to look at the non-supersingular case. Let $U(\mathbb{Z}/N\mathbb{Z})$ be the group of units in $\mathbb{Z}/N\mathbb{Z}$. We can define an action of $U(\mathbb{Z}/N\mathbb{Z})$ on $\mathcal{W}(X_{M_1})$ by assigning $g \cdot \boldsymbol{\alpha} = (g\alpha_0, g\alpha_1, g\alpha_2, g\alpha_3)$ for $g \in U(\mathbb{Z}/N\mathbb{Z})$ and $\boldsymbol{\alpha} = (\alpha_0, \alpha_1, \alpha_2, \alpha_3)$. Suppose that $\widetilde{X}_{M_1}$ is not a supersingular $K3$ surface. Then using a result of [**28**] (cf. also [**29**], [**9**]), we see that the set $\mathcal{W} \setminus \mathcal{W}_{al}$ is equal to the $U(\mathbb{Z}/N\mathbb{Z})$-orbit of $\boldsymbol{\alpha}_{\text{ss}}$, where $\boldsymbol{\alpha}_{\text{ss}}$ is the element of $\mathcal{W}(X_{M_1})$ described in Corollary 6.6. Hence it follows from Corollary 6.6 that

$$\#\mathcal{W}_{al}(X_{M_1}) = \#\mathcal{W}(X_{M_1}) - \phi(N).$$

On the other hand, we have

$$22 = 1 + \#\mathcal{V} + \#\mathcal{W} + e_X = \frac{m}{l_{23}} + \#\mathcal{W} + e_X.$$

Applying these relations to the first equality, we obtain $\rho(\widetilde{X}_{M_1}) = 22 - \phi(N)$. This proves the theorem. □

REMARK 6.9. If λ is chosen to be generic (especially in the case $k = \mathbb{C}$), then $\rho(\widetilde{X}_{M_1,\lambda})$ is equal to the sum of the number of exceptional divisors on $\widetilde{X}_{M_1,\lambda}$ and the number of irreducible components of hyperplane sections on $X_{M_1,\lambda}$. Thus, Lemma 6.7 implies

$$\rho(\widetilde{X}_{M_1,\lambda}) = e_X + \frac{m}{l_{23}}$$

for a generic choice of λ, where e_X denotes the number of exceptional divisors on $\widetilde{X}_{M_1,\lambda}$. This corresponds to the case where $\#\mathcal{W}_{al}(X_{M_1}) = 0$ in Theorem 6.8.

7. Néron-Severi Groups of Some Non-Supersingular $K3$ Surfaces

In this section, we consider some special families of weighted quasi-diagonal $K3$ surfaces and their deformations. Our goal is to compute the discriminant of their Néron-Severi groups explicitly and to describe $\mathbb{Z}$-bases for the Néron-Severi groups.

Let k be an algebraically closed field of characteristic $p \geq 0$. Let $X_{M_1,\lambda}$ be a deformation of weighted quasi-diagonal $K3$ surface of degree m in $\mathbb{P}^3_k(q_0, q_1, q_2, q_3)$. Throughout the section, we assume the inequality given in Proposition (6.2) so that $X_{M_1,\lambda}$ is quasi-smooth and well-formed. Write e_X for the number of exceptional divisors on $\widetilde{X}_{M_1,\lambda}$.

In this section, we confine ourselves to the case:

$$\rho(\widetilde{X}_{M_1,\lambda}) = e_X + 1. \tag{7.1}$$

If $\lambda = 0$, then the assumption means that $m = l_{23}$ and $\#\mathcal{W}_{al} = 0$ in the notation of Theorem 6.8. We will compute the discriminant of the Néron-Severi group of $\widetilde{X}_{M_1,\lambda}$ with property (7.1). Note that such a $K3$ surface is not supersingular and the exceptional divisors form a part of a $\mathbb{Z}$-basis for $\mathrm{NS}(\widetilde{X}_{M_1,\lambda})$ (cf. [**9**], [**18**]).

LEMMA 7.1. *There exist* 18 *weights* Q *such that weighted quasi-diagonal* $K3$ *surfaces in* $\mathbb{P}^3_k(Q)$ *satisfy property* (7.1). *There are a total of* 20 *such surfaces* (*see Table* 2).

PROOF. One knows from Lemma 6.1 that there are 51 weights Q and 85 choices for (n_0, n_1, n_2, n_3) that define weighted quasi-diagonal $K3$ surfaces. From these surfaces, we choose the surfaces with $m = l_{23}$ and $\#\mathcal{W}_{al}(X_{M_1}) = 0$. The first condition can be checked easily. The second condition can be dealt with by describing the set $\mathcal{W}(X_{M_1})$ and using a result of [**28**] for determining $\mathcal{W}_{al}(X_{M_1})$. Then by case-by-case analyses, we obtain the result.

TABLE 2. Discriminant of the Néron-Severi group of $\widetilde{X}_{M_1,\lambda}$.

m	Q	(n_0, n_1, n_2, n_3)	N	e_X	$\mathrm{disc}\,\mathrm{NS}(\widetilde{X}_{M_1,\lambda})$
10	$(2,1,2,5)$	$(5,8,5,2)$	40	5	-2^4
12	$(1,1,4,6)$	$(12,11,3,2)$	66	1	-1
14	$(2,3,2,7)$	$(7,4,7,2)$	28	9	-2^6
15	$(5,2,3,5)$	$(3,5,5,3)$	15	13	-5^2
15	$(3,4,3,5)$	$(5,3,5,3)$	15	13	-3^4
18	$(2,1,6,9)$	$(9,16,3,2)$	48	5	-2^2
20	$(2,9,4,5)$	$(10,2,5,4)$	20	13	-2^4
21	$(1,10,3,7)$	$(21,2,7,3)$	42	9	-1
24	$(2,11,3,8)$	$(12,2,8,3)$	24	13	-2^2
28	$(1,9,4,14)$	$(28,3,7,2)$	42	9	-1
28	$(7,3,4,14)$	$(4,7,7,2)$	14	15	-7
30	$(2,7,6,15)$	$(15,4,5,2)$	20	13	-2^4
36	$(1,5,12,18)$	$(36,7,3,2)$	42	9	-1
36	$(3,11,4,18)$	$(12,3,9,2)$	18	15	-3
40	$(5,7,8,20)$	$(8,5,5,2)$	10	17	-5
42	$(6,1,14,21)$	$(7,36,3,2)$	36	9	-1
42	$(14,1,6,21)$	$(3,28,7,2)$	28	9	-1
42	$(21,1,6,14)$	$(2,21,7,3)$	21	9	-1
42	$(2,5,14,21)$	$(21,8,3,2)$	24	13	-2^2
48	$(3,5,16,24)$	$(16,9,3,2)$	18	15	-3

Alternatively, we may evaluate $\#\mathcal{W}_{al}(X_{M_1}) = 0$ as follows: assume $m = l_{23}$. Then Theorem 6.8 implies

$$e_X + \phi(N) + \#\mathcal{W}_{al}(X_{M_1}) = 21.$$

Hence $\#\mathcal{W}_{al}(X_{M_1}) = 0$ if and only if $e_X + \phi(N) = 21$. Computations of e_X and $\phi(N)$ are straight-forward and therefore we can readily determine whether or not $\#\mathcal{W}_{al}(X_{M_1}) = 0$. □

DEFINITION 7.2. Let X be a smooth surface over k. If $\mathbb{E}$ denotes a subgroup of $\mathrm{NS}(X)$ generated by a finite number of divisors $E_1, \ldots, E_n$ on X, then we write $\det(D, \mathbb{E})$ for the determinant of the following $(n+1) \times (n+1)$ intersection matrix

$$\begin{pmatrix} (D \,.\, D) & (D \,.\, E_1) & \ldots & (D \,.\, E_n) \\ (E_1 \,.\, D) & (E_1 \,.\, E_1) & \ldots & (E_1 \,.\, E_n) \\ \vdots & \vdots & \ddots & \vdots \\ (E_n \,.\, D) & (E_n \,.\, E_1) & \ldots & (E_n \,.\, E_n) \end{pmatrix}$$

where $(E_i \,.\, E_j)$ denotes the intersection number between E_i and E_j on X; $(D \,.\, D)$ and $(D \,.\, E_i) = (E_i \,.\, D)$ are defined to be the intersection numbers if $D \in \mathrm{Div}(X)$ and are considered to be variables with values in $\mathbb{Z}$ if $D \notin \mathrm{Div}(X)$.

LEMMA 7.3. *Let X be a normal surface over k. Let $\widetilde{X} \to X$ be the minimal resolution of X and $E_1, \ldots, E_r$ the exceptional divisors on $\widetilde{X}$ with $r \geq 0$. Assume $\rho(\widetilde{X}) = 1 + r$ and that there exist a finite number of divisors, $D_1, \ldots, D_n$, on $\widetilde{X}$ such that* $\gcd(\det(D_1, \mathbb{E}), \ldots, \det(D_n, \mathbb{E})) \mid \det(D, \mathbb{E})$ *for every D. Then*

$$\operatorname{disc} \mathrm{NS}(\widetilde{X}) = \gcd(\det(D_1, \mathbb{E}), \ldots, \det(D_n, \mathbb{E}))$$

and there exists some divisor D_0 in the linear combinations of $D_1, \ldots, D_n$ such that $\{D_0, E_1, \ldots, E_n\}$ forms a $\mathbb{Z}$-basis for $\mathrm{NS}(\widetilde{X})$.

PROOF. Since $\mathbb{E}$ forms a part of a $\mathbb{Z}$-basis for $\mathrm{NS}(\widetilde{X})$, we have

$$\gcd(\det(D, \mathbb{E}); D \text{ arbitrary}) \mid \operatorname{disc} \mathrm{NS}(\widetilde{X})$$
$$\operatorname{disc} \mathrm{NS}(X) \mid \gcd(\det(D_1, \mathbb{E}), \ldots, \det(D_n, \mathbb{E})).$$

By the assumption,

$$\gcd(\det(D_1, \mathbb{E}), \ldots, \det(D_n, \mathbb{E})) \mid \gcd(\det(D, \mathbb{E}); D \text{ arbitrary}).$$

Hence $\operatorname{disc} \mathrm{NS}(\widetilde{X})$ is equal to $\gcd(\det(D_1, \mathbb{E}), \ldots, \det(D_n, \mathbb{E}))$ and $E_1, \ldots, E_n$ together with some linear combination of $D_1, \ldots, D_n$ give a $\mathbb{Z}$-basis for $\mathrm{NS}(\widetilde{X})$. □

LEMMA 7.4. *Let $X_{M_1,\lambda}$ be a deformation of a weighted quasi-diagonal $K3$ surface in $\mathbb{P}^3_k(Q)$ of degree m. For $0 \leq i \leq 3$, denote by H_i the hyperplane section $\{x_i = 0\} \cap X_{M_1,\lambda}$ on $X_{M_1,\lambda}$ and by $\widetilde{H}_i$ the proper transform of H_i on $\widetilde{X}_{M_1,\lambda}$. Then the following assertions hold*:

(a) *H_0 is a union of m/l_{23} singular rational curves with a cusp singularity.*

(b) *H_i and $\widetilde{H}_i$ are smooth projective curves for $i = 1, 2, 3$.*

PROOF. (a) H_0 is isomorphic to a scheme in $\mathbb{P}^2_k(q_1, q_2, q_3) := \operatorname{Proj} k[x_1, x_2, x_3]$ defined by the equation $x_2^{n_2} + x_3^{n_3} = 0$. Since x_1 is free, this is a union of m/l_{23} projective curves of genus 0 that intersect each other at $(x_1 : x_2 : x_3) = (1 : 0 : 0)$. Localizing at $x_1 \neq 0$, we find that these curves have a unique cusp singularity at $(1 : 0 : 0)$.

(b) H_1 is isomorphic to a projective curve in $\mathbb{P}^2_k(q_0, q_2, q_3) := \operatorname{Proj} k[x_0, x_2, x_3]$ defined by the equation $x_0^{n_0} + x_2^{n_2} + x_3^{n_3} = 0$. The expression of H_1 may be simplified if the weight (q_0, q_2, q_3) can be reduced; but in any event, we see that H_1 is quasi-smooth. Now one knows that quasi-smooth weighted projective varieties are normal (cf. [**7**]). Hence, being a curve, H_1 is smooth. This implies also the smoothness of $\widetilde{H}_1$.

H_2 is isomorphic to a projective curve in $\mathbb{P}^2_k(q_0, q_1, q_3) := \operatorname{Proj} k[x_0, x_1, x_3]$ defined by the equation $x_0^{n_0} + x_0 x_1^{n_1} + x_3^{n_3} = 0$. Possibly simplifying the defining equation by reducing the weight, we find that H_2 is quasi-smooth. Hence for the same reason as above, H_2 and $\widetilde{H}_2$ are smooth projective curves. Similarly, we conclude that H_3 and $\widetilde{H}_3$ are smooth projective curves over k. □

Recall that e_X is the number of exceptional divisors on $\widetilde{X}_{M_1,\lambda}$.

THEOREM 7.5. *Let X_{M_1} be a weighted quasi-diagonal $K3$ surface in $\mathbb{P}^3_k(Q)$ of degree m satisfying $\rho(\widetilde{X}_{M_1}) = e_X + 1$. Let $X_{M_1,\lambda}$ be a deformation of X_{M_1}. Assume $\rho(\widetilde{X}_{M_1,\lambda}) = e_X + 1$. Then the discriminant of the Néron-Severi group of $\widetilde{X}_{M_1,\lambda}$ is equal to*

$$\operatorname{disc} \mathrm{NS}(\widetilde{X}_{M_1,\lambda}) = \frac{\det(\widetilde{H}_i, \mathbb{E})}{q_i^2}$$

where $i = 1, 2, 3$ and $\mathbb{E}$ is a subgroup of $\mathrm{NS}(\widetilde{X}_{M_1,\lambda})$ generated by the exceptional divisors on $\widetilde{X}_{M_1,\lambda}$. A $\mathbb{Z}$-basis for $\mathrm{NS}(\widetilde{X}_{M_1,\lambda})$ is given by the exceptional divisors on $\widetilde{X}_{M_1,\lambda}$ and a linear combination of $\widetilde{H}_i$'s ($i = 1, 2, 3$). The actual value of the discriminant is shown in Table 2.

PROOF. This involves a large amount of numerical calculations, so we indicate only the outline of the proof. First note that $\rho(\widetilde{X}_{M_1,\lambda}) = e_X + 1$ means that $\mathrm{NS}(\widetilde{X}_{M_1,\lambda})$ is generated over $\mathbb{Z}$ by the set of all exceptional divisors plus a single divisor. We compute $\det(\widetilde{H}_i, \mathbb{E})$ for $1 \le i \le 3$ and describe the form of $\det(D, \mathbb{E})$ for a general D (cf. Definition 7.2). Then we find that $\det(\widetilde{H}_i, \mathbb{E}) = q_i^2 R$ for some integer R and that $\det(D, \mathbb{E})$ is divisible by R for arbitrary D's. Therefore by Lemma 7.3 and the relation $\gcd(q_1, q_2, q_3) = 1$, we obtain the asserted formula. □

REMARK 7.6. If $\rho(\widetilde{X}_{M_1}) = e_X + 1$, then $\rho(\widetilde{X}_{M_1,\lambda}) = e_X + 1$ for a generic choice of λ.

It is not too difficult but rather a tedious task to find a linear combination of $\widetilde{H}_i$'s that gives a part of a $\mathbb{Z}$-basis for $\mathrm{NS}(\widetilde{X}_{M_1,\lambda})$. Here we just mention a special case which is an immediate consequence of the theorem.

COROLLARY 7.7. *With the assumptions and notation of Theorem 7.5, let $q_1 = 1$. Then $\widetilde{H}_1$ and the exceptional divisors on $\widetilde{X}_{M_1,\lambda}$ give a $\mathbb{Z}$-basis for $\mathrm{NS}(\widetilde{X}_{M_1,\lambda})$.*

Acknowledgments

The results in Section 3 were obtained while I was a post-doctoral fellow at the School of Mathematics, Tata Institute of Fundamental Research in Bombay; I thank the institute for its hospitality. I am grateful to Kirti Joshi and Noriko Yui for useful comments and suggestions on earlier versions of the paper. Many thanks are due to the referee of the paper for numerous comments and suggestions that improved the exposition of the paper.

References

1. M. Artin, *Supersingular $K3$ surfaces*, Ann. Sci. École Norm. Sup. (4) **7** (1974), 543–567.
2. P. Deligne, *La conjecture de Weil. I*, Inst. Hautes Études Sci. Publ. Math. **43** (1974), 237–307.
3. C. Deloıme, *Espaces projectifs anisotropes*, Bull. Soc. Math. France **103** (1975), 203–223.
4. J. Delsarte, *Nombre de solutions des équations polynomiales sur un corps fini*, Exp. 39, Séminaire Bourbaki: Volume 1950/1951–1951/1952, Exposés 33–67, W. A. Benjamin, Inc., New York-Amsterdam 1966.
5. A. Dimca, *Singularities and coverings of weighted complete intersections*, J. Reine Angew. Math. **366** (1986), 184–193.
6. A. Dimca and S. Dimiev, *On analytic coverings of weighted projective spaces*, Bull. London Math. Soc. **17** (1985), 234–238.
7. I. Dolgachev, *Weighted projective varieties*, Group Actions and Vector Fields (Vancouver, BC, 1981), Lecture Notes in Math., vol. 956, Springer, Berlin-New York, 1982, pp. 34–71.
8. A. R. Fletcher, *Working with weighted complete intersections*, Preprint no. MPI89-35, Max-Planck-Institut für Mathematik, Bonn, 1989.
9. Y. Goto, *Arithmetic of weighted diagonal surfaces over finite fields*, J. Number Theory **59** (1996), 37–81.
10. ______, *The Artin invariant of supersingular weighted Delsarte K3 surfaces*, J. Math. Kyoto Univ. **36** (1996), 359–363.
11. ______, *On some arithmetic quantities of weighted quasi-diagonal surfaces*, Preprint no. MPI 95-111, Max-Planck-Institut für Mathematik, Bonn, 1995.
12. F. Gouvêa and N. Yui, *Arithmetic of diagonal hypersurfaces over finite fields*, London Math. Soc. Lecture Note Ser., vol. 209,, Cambridge Univ. Press, Cambridge, 1995.
13. F. Hirzebruch, *Über vierdimensionale Riemannsche Flächen mehrdeutiger analytischer Funktionen von zwei Veränderlichen*, Math. Ann. **126** (1953), 1–22.
14. L. Illusie, *Complexe de de Rham-Witt et cohomologie cristalline*, Ann. Sci. École Norm. Sup. (4) **12** (1979), 501–661.
15. K. F. Ireland and M. I. Rosen, *A classical introduction to modern number theory*, Grad. Texts in Math., vol. 84, Springer-Verlag, New York-Berlin, 1982.
16. M. Ishida, *The algebraic surface defined by a sum of four monomials*, Comment. Math. Uni. St. Paul. **40** (1991), 39–51.
17. M. Kobayashi, *Duality of weights, mirror symmetry and Arnold's strange duality*, 1996 (preprint).
18. J. Lipman, *Rational singularities with applications to algebraic surfaces and unique factorization*, Inst. Hautes tudes Sci. Publ. Math. **36** (1969), 195–279.
19. J. S. Milne, *Duality in the flat cohomology of a surface*, Ann. Sci. École Norm. Sup. (4) **9** (1976), 171–201.
20. N. Nygaard, *A p-adic proof of the nonexistence of vector fields on $K3$ surfaces*, Ann. of Math. (2) **110** (1979), 515–528.
21. I. I. Pjateckiĭ-Šapiro and I. R. Šafarevič, *Torelli theorem for algebraic surfaces of type $K3$*, Izv. Akad. Nauk SSSR Ser. Mat. **35** (1971), 530–572 (Russian).
22. M. Reid, *Canonical 3-folds*, Journées de Géométrie Algébrique (Angers, 1979) Sijthoff & Noordhoff, lphen aan den Rijn-Germantown, MD, 1980, pp. 273–310.
23. S.-S. Roan, *Mirror symmetry and Arnold's duality*, MPI 92-86, Max-Planck-Institut für Mathematik, Bonn, 1992.
24. A. N. Rudakov and I. R. Šafarevič, *Supersingular $K3$ surfaces over fields of characteristic* 2, Izv. Akad. Nauk SSSR, Ser. Mat. **42** (1978), 848–869 (Russian); English transl., Math. USSR-Izv. **42** (1978), 147–165.
25. T. Shioda, *On Unirationality of Supersingular Surfaces*, Math. Ann. **225** (1977), 155–159.
26. ______, *An explicit algorithm for computing the Picad number of certain algebraic surfaces*, Amer. J. Math. **108** (1986), 415–432.
27. ______, *Supersingular $K3$ surfaces with big Artin invariant*, J. Reine Angew. Math. **381** (1987), 205–210.
28. T. Shioda and T. Katsura, *On Fermat varieties*, Tôhoku Math. J. (2) **31** (1979), 97–115.
29. N. Suwa and N. Yui, *Arithmetic of certain algebraic surfaces over finite fields*, Number Theory (New York, 1985/1988), Lecture Notes in Math., vol. 1383, Springer, Berlin-New York, 1989, pp. 186–256.

30. J. Tate, *Algebraic cycles and poles of zeta functions*, Arithmetic Algebraic Geometry (Purdue Univ. 1963), Harper & Row, New York, 1965, pp. 93–110.
31. ———, *On the conjecture of Birch and Swinnerton-Dyer and a geometric analogue*, Dix exposés sur la cohomologie des schémas, Adv. Stud. Pure Math., vol. 3, North-Holland, Amsterdam, 1968, pp. 189–214.
32. A. Weil, *Number of solutions of equations over finite fields*, Bull. Amer. Math. Soc. **55** (1949), 497–508.
33. T. Yonemura, *Hypersurface simple $k3$ singularities*, Tôhoku Math. J. (2) **42** (1990), 351–380.

RESEARCH INSTITUTE FOR MATHEMATICAL SCIENCES, KYOTO UNIVERSITY, SAKYO-KU, KYOTO 606-8502 JAPAN

Current address: Department of Mathematics Hokkaido Univ. of Education 1-2 Hachiman-cho, Hakodate, Hokkaido, 040-8567 Japan

E-mail address: ygoto@cc.hokkyodai.ac.jp.ac.jp

Centre de Recherches Mathématiques
CRM Proceedings and Lecture Notes
Volume **24**, 2000

Torsion Zero-Cycles and the Abel-Jacobi Map Over the Real Numbers

Joost van Hamel

ABSTRACT. This is a study of the torsion in the Chow group of zero-cycles on a variety over the real numbers. The first section recalls important results from the literature. The rest of the paper is devoted to the study of the Abel-Jacobi map $\alpha\colon A_0(X) \to \mathrm{Alb}(X)(\mathbb{R})$ restricted to torsion subgroups. Using Roitman's theorem over the complex numbers and a version of Bloch's cohomological Abel-Jacobi map over the real numbers, it is shown that this map can be described completely in terms of étale cohomology. For some examples (products of curves, Abelian varieties, certain fibre bundles) the torsion in the kernel and cokernel of the Abel-Jacobi map α is computed explicitly.

Introduction

Consider the Abel-Jacobi map

$$\alpha\colon A_0(X) \to \mathrm{Alb}(X)(k)$$

from the group of zero-cycles of degree 0 modulo rational equivalence into the k-points of the Albanese variety of a nonsingular projective geometrically irreducible variety X over a field k. If k is algebraically closed, then α is well-known to be surjective, but it need not be injective. However, Roitman's theorem on torsion zero-cycles asserts that the restriction

$$\alpha_{\mathrm{tors}}\colon A_0(X)_{\mathrm{tors}} \to \mathrm{Alb}(X)(k)_{\mathrm{tors}}$$

to the torsion subgroups is an isomorphism when k is algebraically closed (with the p-part in characteristic p due to Milne). In particular, we get for k algebraically closed of characteristic zero that $A_0(X)_{\mathrm{tors}} \simeq (\mathbb{Q}/\mathbb{Z})^{2q}$, where $q = \dim_k H^1(X, \mathcal{O}_X)$.

For $k = \mathbb{R}$, and $X(\mathbb{R}) \neq \varnothing$, it was shown by Colliot-Thélène and Scheiderer that

$$A_0(X)_{\mathrm{tors}} \simeq (\mathbb{Q}/\mathbb{Z})^q \times (\mathbb{Z}/2)^{s-1},$$

with q as above and s the number of connected components of the set of real points $X(\mathbb{R})$ (for the Euclidean topology; let me state here that in this paper the set of real points $X(\mathbb{R})$ and the set of complex points $X(\mathbb{C})$ will always be equipped with

---

1991 *Mathematics Subject Classification.* 14C25, 14C30, 14P05.

Received by the editors July 31, 1998; Accepted January 14, 1999.
This is the final form of the paper.

the Euclidean topology). However, the methods of [**CTS96**] do not give precise information on the Abel-Jacobi map.

In general the map α_{tors} is not an isomorphism over $\mathbb{R}$: it is easy to construct varieties having $q = 0$, but a non-connected set of real points (for example, suitable hypersurfaces of dimension ≥ 2 in projective space). In that case $\mathrm{Alb}(X)$ is trivial, so α_{tors} is not injective. Another easy example is an Abelian variety X the set of real points of which has more than two connected components (see Example 5.2). Surjectivity of α_{tors} can fail as well; examples are curves of odd genus without real points (see Section 1.2), but also certain types of 'twisted' fibre bundles that do have real points (see Example 5.1).

A standard trace argument, combined with Roitman's theorem on α_{tors} over $\mathbb{C}$, implies that the kernel and cokernel of α_{tors} over $\mathbb{R}$ are purely 2-torsion. In particular, the restriction

$$\alpha_{\text{tors,div}} \colon A_0(X)_{\text{tors,div}} \to \mathrm{Alb}(X)_{\text{tors,div}}$$

to the maximal divisible torsion subgroups is surjective. Since both groups are isomorphic to $(\mathbb{Q}/\mathbb{Z})^q$ one might be led to hope that at least $\alpha_{\text{tors,div}}$ is an isomorphism, but even this analogue of Roitman's theorem fails: see again Example 5.1.

In this paper we will see that over $\mathbb{R}$ the torsion Abel-Jacobi map α_{tors} is determined by the étale cohomology of X with coefficients in $\mathbb{Q}/\mathbb{Z}(j) = \varinjlim_n \mu_n^{\otimes j}$ for $j \in \mathbb{Z}$. More precisely, we have the following result, which is a combination of Theorem 3.3.i and Proposition 4.1.

THEOREM. *Let X be a nonsingular projective geometrically irreducible variety over $\mathbb{R}$ of dimension d.*

(i) *The image of α_{tors} is canonically isomorphic to the image of the natural mapping*

$$H^{2d-1}_{\text{ét}}(X, \mathbb{Q}/\mathbb{Z}(d)) \to H^{2d-1}_{\text{ét}}(X_{\mathbb{C}}, \mathbb{Q}/\mathbb{Z}(d))^{\mathrm{Gal}(\mathbb{C}/\mathbb{R})}.$$

(ii) *The kernel of $\alpha_{\text{tors,div}}$ is isomorphic to the cokernel of the natural mapping*

$$H^{2d-1}_{\text{ét}}(X, \mathbb{Q}/\mathbb{Z}(d+1)) \to H^{2d-1}_{\text{ét}}(X_{\mathbb{C}}, \mathbb{Q}/\mathbb{Z}(d+1))^{\mathrm{Gal}(\mathbb{C}/\mathbb{R})}.$$

Given the image of α_{tors}, the kernel of $\alpha_{\text{tors,div}}$, and the number of connected components of $X(\mathbb{R})$, it is of course a simple calculation to find the $\mathbb{Z}/2$-dimension of the kernel of α_{tors}. The formula is given in Theorem 3.3.iii.

My method of proving the above results is strongly inspired by Bloch's proof of Roitman's theorem as presented in [**CT93**]. There the key step is the construction of a cohomological Abel-Jacobi map

$$\lambda_i \colon \mathrm{CH}^i(X)_{\text{tors}} \to H^{2i-1}(X, \mathbb{Q}/\mathbb{Z}(d))$$

for the torsion of the Chow group of cycles of codimension i. This map is then shown to be an isomorphism in the case of zero-cycles (i.e., $i = d$) by inspection of the Bloch-Ogus spectral sequence.

When the ground field is not algebraically closed, the construction of λ_i usually runs into trouble. We will see in Section 3.1, however, that over $\mathbb{R}$ there are no problems, thanks to the finiteness of the absolute Galois group. Contrary to the case of an algebraically closed field, we do not get that λ_d is an isomorphism, but an analysis of the upper part of the Bloch-Ogus spectral sequence carried out in

Section 2 does give us a short exact sequence

$$0 \to \mathrm{CH}_0(X)_{\mathrm{tors}} \xrightarrow{\lambda_d} H^{2d-1}_{\text{ét}}(X, \mathbb{Q}/\mathbb{Z}(d)) \to \bigoplus_{i>0} H^{d-2i}(X(\mathbb{R}), \mathbb{Z}/2) \to 0$$

(see Theorem 3.2), and also the fact that the image of λ_d maps surjectively onto the image of the base change map from $H^{2d-1}_{\text{ét}}(X,\mathbb{Q}/\mathbb{Z}(d))$ to $H^{2d-1}_{\text{ét}}(X_{\mathbb{C}},\mathbb{Q}/\mathbb{Z}(d))^{\mathrm{Gal}(\mathbb{C}/\mathbb{R})}$. From these two facts it is not so hard to derive the main theorem.

Structure of the Paper. In the first section I will present a few existing results on real zero-cycles and I will discuss some of the special features of étale cohomology and the Bloch-Ogus spectral sequence for varieties over $\mathbb{R}$.

Section 2 contains a new result on the Bloch-Ogus spectral sequence for varieties over $\mathbb{R}$. Only Corollary 2.2 will be used in the rest of the paper.

The cohomological Abel-Jacobi map will be constructed and applied to the study of the real Abel-Jacobi map in Section 3 with Theorem 3.2 and Theorem 3.3 as main results. Also the filtration on $\mathrm{CH}_0(X)_{\mathrm{tors}}$ induced by the cohomological Abel-Jacobi map and the Hochschild-Serre spectral sequence is introduced.

In Section 4 some of the criteria of Theorem 3.3 are translated into criteria that are easier to check in practice. This section is by nature quite technical; from a conceptual point of view, it does not add anything to the insight provided by Section 3. The reader might prefer to skip it at first reading (or to read the introduction only) and go directly to the examples of Section 5 where the criteria of Section 4 are actually put to work.

Remark. Undoubtedly, all results in this paper will be valid over arbitrary real closed fields when rephrased properly, and the proofs should be relatively straightforward to adapt. However, I lacked the courage and the expertise to carry this out myself.

1. Known Results

This section contains an overview of some known results on zero-cycles, Albanese varieties and étale cohomology over the real numbers. Throughout the paper a *variety* will be a reduced, separated, but not necessarily irreducible scheme of finite type over a field.

Sources and Further Reading. The theory of real Abelian varieties as used in Section 1.1 is essentially due to Comessatti and put into a modern framework by Silhol; the reader is referred to [**Si89**, Chapter IV] for more information. The results on divisors on nonsingular projective curves over $\mathbb{R}$ in Section 1.2 go back to Harnack, Weichold, Klein and Witt; see [**PW91**] for a historical overview with precise references, and also for results concerning non-projective and/or singular curves. In [**CTS96**] the reader can find more about the Chow group of zero-cycles of (not necessarily complete or nonsingular) varieties of higher dimensions.

The relation between étale cohomology and the equivariant cohomology of the set of complex points of a variety over $\mathbb{R}$ discussed in Section 1.5 seems to be established first in [**Cox79**]. The relation between equivariant cohomology and the cohomology of the set of fixed points for an action of a group of prime order lies at the very origin of the equivariant cohomology theory, as developed by A. Borel (see [**Bo60**], [**Hs75**], [**AP93**]) after pioneering work of P. A. Smith. For us it will be convenient to follow Grothendieck's algebraic approach as described in [**Gr57**, Chapter V], rather than using Borel's geometric definition. In [**Sch94**] the equivariant cohomology theory is set up in the context of topos theory, allowing in particular to obtain analogues of the results in Sections 1.5 and 1.6 for varieties over arbitrary real closed fields. For the readers who wish to compare the topological and the topos-theoretic approach I have included references to analogous results in [**Sch94**] whenever I was able to locate them. A basic paper on the application of equivariant cohomology of $X(\mathbb{C})$ to real algebraic geometry is [**Kr83**]; see also [**Si89**, Chapter I]. The treatment of the Bloch-Ogus spectral sequence over the real numbers in Section 1.7 is based on [**Sch94**, Chapter 19].

1.1. Albanese Varieties Over $\mathbb{R}$. The Albanese variety $\mathrm{Alb}(X)$ of a nonsingular projective geometrically irreducible variety X over $\mathbb{R}$ is an Abelian variety over $\mathbb{R}$ admitting a universal regular Galois-equivariant homomorphism from the group of zero-cycles of degree 0 on $X_{\mathbb{C}}$ into the set of complex points of $\mathrm{Alb}(X)$ (see for example [**Mu94**] for the definition of a regular homomorphism, and [**Ra98**] for a modern definition in terms of flat sheaves over the ground field). The universal regular homomorphism induces the Abel-Jacobi map

$$\alpha\colon A_0(X) \to \mathrm{Alb}(X)(\mathbb{R}),$$

that we considered in the introduction.

We have a canonical isomorphism

$$\mathrm{Alb}(X)(\mathbb{C}) \simeq \mathrm{Hom}(H^0(X(\mathbb{C}),\Omega^1),\mathbb{C})/(H_1(X(\mathbb{C}),\mathbb{Z})/\mathrm{tors})$$

of complex tori with an anti-holomorphic involution. Here the involution on $\mathrm{Hom}(H^0(X(\mathbb{C}),\Omega^1),\mathbb{C})$ and on $H_1(X(\mathbb{C}),\mathbb{Z})$ is induced by the anti-holomorphic involution on $X(\mathbb{C})$ and complex conjugation on $\mathbb{C}$. Taking $q = \dim_{\mathbb{C}} H^0(X(\mathbb{C}),\Omega^1)$ and $\Lambda = H_1(X(\mathbb{C}),\mathbb{Z})/\mathrm{tors}$, we get that

$$\mathrm{Alb}(X)(\mathbb{R}) \simeq (\mathbb{C}^q/\Lambda)^G.$$

From the theory of lattices with a G-action we see that Λ is of the form

$$\Lambda \simeq \mathbb{Z}^a \times \mathbb{Z}(1)^a \times \mathbb{Z}[G]^{q-a},$$

for some $0 \le a \le q$. Here $\mathbb{Z}(1)$ is $\mathbb{Z}$ with the involution $z \mapsto -z$, and $\mathbb{Z}[G]$ is the free Abelian group generated by the elements of G with G acting in the obvious way. Since $(\mathbb{R}[G]/\mathbb{Z}[G])^G \simeq \mathbb{R}/\mathbb{Z}$ and $(\mathbb{R}(1)/\mathbb{Z}(1))^G \simeq \mathbb{Z}/2$, we get an isomorphism

$$\mathrm{Alb}(X)(\mathbb{R}) \simeq (\mathbb{R}/\mathbb{Z})^q \times (\mathbb{Z}/2)^a.$$

It is clear from the above construction that the torsion of $\mathrm{Alb}(X)(\mathbb{R})$ is isomorphic to $H_1(X(\mathbb{C}),\mathbb{Q}/\mathbb{Z})^G$; Poincaré duality gives an isomorphism

$$\mathrm{Alb}(X)(\mathbb{R})_{\mathrm{tors}} \simeq H^{2d-1}(X(\mathbb{C}),\mathbb{Q}/\mathbb{Z}(d))^G,$$

where $d = \dim(X)$ and $\mathbb{Q}/\mathbb{Z}(d)$ is $\mathbb{Q}/\mathbb{Z}$ with involution $x \mapsto (-1)^d x$. The d-fold twist of the coefficients comes from the fact that the fundamental class of $X(\mathbb{C})$ lies in the G-invariant part of $H_{2d}(X(\mathbb{C}),\mathbb{Z}(d))$, since complex conjugation on $X(\mathbb{C})$ preserves the orientation if and only if d is even.

1.2. Divisors on curves. Zero-cycles on nonsingular curves are divisors, so the Albanese variety of a nonsingular projective geometrically irreducible curve X coincides with the Picard variety; it is better known as the Jacobian variety $\mathrm{Jac}(X)$ of X. From general theory (valid over arbitrary fields) it follows that the mapping

$$\alpha\colon A_0(X) \to \mathrm{Jac}(X)(\mathbb{R})$$

is injective for X as above and surjective if $X(\mathbb{R}) \neq \varnothing$.

If X has genus g and $X(\mathbb{R})$ has $s > 0$ connected components, then

$$A_0(X) \simeq \operatorname{Jac}(X)(\mathbb{R}) \simeq (\mathbb{R}/\mathbb{Z})^g \times (\mathbb{Z}/2)^{s-1}.$$

More precisely, we have that the class of a divisor Z of degree zero lies in the maximal divisible subgroup of $A_0(X)$ if and only if Z is an integral linear combination of

(i) real points with multiplicity 2,
(ii) closed points with complex residue field,
(iii) divisors of the form $[P_1] - [P_2]$ with P_1 and P_2 real points in the same connected component of $X(\mathbb{R})$.

This also holds when $X(\mathbb{R}) = \varnothing$, so then

$$A_0(X) \simeq (\mathbb{R}/\mathbb{Z})^g.$$

However, in this case we actually have

$$\operatorname{Jac}(X)(\mathbb{R}) \simeq \begin{cases} (\mathbb{R}/\mathbb{Z})^g & \text{if } g \text{ is even,} \\ (\mathbb{R}/\mathbb{Z})^g \times \mathbb{Z}/2 & \text{if } g \text{ is odd,} \end{cases}$$

so when g is odd and X has no real points, then the mapping α has cokernel $\mathbb{Z}/2$. Otherwise α is surjective.

1.3. Zero-Cycles on Higher Dimensional Varieties. For a nonsingular projective geometrically irreducible variety X over $\mathbb{R}$ of arbitrary dimension it was shown in [**CTI81**], using the calculation for curves given above, that $A_0(X)$ modulo the maximal divisible subgroup $A_0(X)_{\text{div}}$ is a $\mathbb{Z}/2$-vector space of dimension $s-1$ if $X(\mathbb{R})$ has $s > 0$ connected components, and that $A_0(X)$ is divisible if X has no real points. In fact, the criteria for a zero-cycle to be divisible modulo rational equivalence are precisely the same as the criteria for a divisor on a curve given above.

In [**CTS96**] it was shown using Roitman's theorem over $\mathbb{C}$ and a trace argument, that the torsion of $A_0(X)_{\text{div}}$ is isomorphic to $(\mathbb{Q}/\mathbb{Z})^q$ with $q = \dim_{\mathbb{R}} H^1(X, \mathcal{O}_X)$. Together with the above expression for $A_0(X)/A_0(X)_{\text{div}}$, this gives the formula of the introduction:

$$A_0(X)_{\text{tors}} \simeq (\mathbb{Q}/\mathbb{Z})^q \times (\mathbb{Z}/2)^{s-1}$$

when $X(\mathbb{R}) \neq \varnothing$ (see [**CTS96**, Theorem 1.6.b]). The group $A_0(X)/A_0(X)_{\text{tors}}$ is uniquely divisible, which follows from the analogous statement for varieties over $\mathbb{C}$ and a trace argument (see [**CTS96**, Theorem 1.3]).

1.4. Étale and Galois Cohomology. The group $G = \operatorname{Gal}(\mathbb{C}/\mathbb{R}) = \mathbb{Z}/2$ does not have finite cohomological dimension for 2-torsion coefficients, so $\operatorname{Spec}\mathbb{R}$ does not have finite étale cohomological dimension, nor does any variety X over $\mathbb{R}$ with $X(\mathbb{R}) \neq \varnothing$.

As a graded ring, the cohomology ring $H^*(G, \mathbb{Z})$ is isomorphic to the (commutative!) ring $\mathbb{Z}[\eta^2]/(2\eta^2)$, where η^2 has degree 2. Also, $H^*(G, \mathbb{Z}/2) \simeq \mathbb{Z}/2[\eta]$, with η of degree 1. The notation already indicates what the natural map $H^*(G, \mathbb{Z}) \to H^*(G, \mathbb{Z}/2)$ looks like. The cohomology of G is *periodic*: cup product with η^2 induces for any $q \geq 0$ and any G-module M a surjection

$$H^q(G, M) \to H^{q+2}(G, M),$$

which is an isomorphism if $q > 0$. If M is a $\mathbb{Z}/2$-module, cup product with η induces for any $q \geq 0$ a surjection $H^q(G, M) \to H^{q+1}(G, M)$ which is an isomorphism if $q > 0$.

For the étale cohomology of a variety X over $\mathbb{R}$ this implies that when $q > 2\dim(X)$ we have for any torsion sheaf $\mathcal{F}$ on X that $H^q_{\text{ét}}(X, \mathcal{F}) \simeq H^{q+2}_{\text{ét}}(X, \mathcal{F})$, and even $H^q_{\text{ét}}(X, \mathcal{F}) \simeq H^{q+1}_{\text{ét}}(X, \mathcal{F})$ if $\mathcal{F}$ is a 2-torsion sheaf, as can be seen from the *Hochschild-Serre spectral sequence*

$$E_2^{p,q} = H^p(G, H^q_{\text{ét}}(X_{\mathbb{C}}, \mathcal{F})) \implies H^{p+q}_{\text{ét}}(X, \mathcal{F})$$

(see Section 1.6 for a similar result with more details of the proof). In other words, the étale cohomology of varieties over $\mathbb{R}$ is stable in degree $> 2\dim(X)$. This stable part has a very natural interpretation in terms of the cohomology of $X(\mathbb{R})$. For example, in the case $\mathcal{F} = \mathbb{Z}/2$ we have for every $q > 2\dim(X)$ an isomorphism

$$H^q_{\text{ét}}(X, \mathbb{Z}/2) \simeq \bigoplus_{i=0}^{\dim X} H^i(X(\mathbb{R}), \mathbb{Z}/2),$$

which was first constructed by D. Cox in [**Cox79**]. In the next sections we will see how this isomorphism can be obtained from a comparison theorem between étale cohomology over $\mathbb{R}$ and equivariant cohomology, combined with a localization theorem in equivariant cohomology.

1.5. Comparison Between Étale and Equivariant Cohomology. To any Abelian torsion group M with an action of $G = \mathbb{Z}/2$ we can associate on the one hand a locally constant sheaf on the étale site of a variety X over $\mathbb{R}$, and on the other hand a locally constant G-sheaf on the space $X(\mathbb{C})$, equipped with the Euclidean topology and the canonical $G = \text{Gal}(\mathbb{C}/\mathbb{R})$-action. It was first proved in [**Cox79**] using étale homotopy theory that for every $q \geq 0$ there is an isomorphism

$$H^q_{\text{ét}}(X, M) \simeq H^q(X(\mathbb{C}); G, M). \tag{1}$$

Here the right hand side denotes equivariant cohomology in the sense of either Grothendieck (see [**Gr57**, Chapter V]) or Borel (see [**Bo60**])—both theories are equivalent since G is finite. In fact, if we take Grothendieck's definition of $H^q(X(\mathbb{C}); G, -)$ as the qth right derived functor of taking G-invariant global sections, then the isomorphism (1) follows directly from the usual comparison between the étale cohomology of $X_{\mathbb{C}}$ and the cohomology of $X(\mathbb{C})$ as given in [**AGV73**, Exp. XVI] and the fact that étale cohomology of $\operatorname{Spec}\mathbb{R}$ corresponds to Galois cohomology (see also [**Sch94**, (15.3)], [**Ni94**], and the comments in [**CTS96**, §2.3]).

Observe that the G-sheaf associated to the G-module $\mathbb{Q}/\mathbb{Z}(j)$ defined in Section 1.1 corresponds to the étale sheaf $\mathbb{Q}/\mathbb{Z}(j) = \varinjlim_n \mu_n^{\otimes j}$.

1.6. Equivariant Cohomology. We will consider equivariant sheaf cohomology, as defined in [**Gr57**, Chapter V], for $G = \text{Gal}(\mathbb{C}/\mathbb{R})$ acting on $V = X(\mathbb{C})$ or $X(\mathbb{R})$, where X is a variety over $\mathbb{R}$. Coefficients will always be taken in (the locally constant G-sheaf on V associated to) a torsion G-module M, and we will mainly concentrate on the cases $M = \mathbb{Z}/2$ or $\mathbb{Q}/\mathbb{Z}(j)$. We denote by $i\colon V^G \hookrightarrow V$ the inclusion of the set of fixed points (note that here $V^G = X(\mathbb{R})$). By $\pi\colon V \to V/G$ we denote the quotient map.

By [**Gr57**, Theorem 5.2.1] we have two spectral sequences converging to equivariant cohomology: the *Hochschild-Serre spectral sequence*

$$E_2^{p,q} = H^p(G, H^q(V, M)) \implies H^{p+q}(V; G, M), \tag{2}$$

(in this context also known as the *Borel-Serre spectral sequence*) and the spectral sequence

$$E_2^{p,q} = H^p(V/G, \mathcal{H}^q(G, M)) \implies H^{p+q}(V; G, M), \tag{3}$$

where $\mathcal{H}^q(G, M)$ is the sheaf on V/G associated to the presheaf

$$U \mapsto H^q(G, \pi_* M(U)),$$

by [**Gr57**, Proposition 5.2.2]. It is not hard to check (see [**Gr57**, Theorem 5.3.1]) that for $q > 0$ the support of $\mathcal{H}^q(G, M)$ is contained in $\pi(V^G)$, so that $H^p(V/G, \mathcal{H}^q(G, M)) \simeq H^p(V^G, \mathcal{H}^q(G, M))$ for $q > 0$. Hence the morphism of spectral sequences (3) that corresponds to the restriction homomorphism

$$i^* \colon H^n(X(\mathbb{C}); G, M) \to H^n(X(\mathbb{R}); G, M) \tag{4}$$

is an isomorphism on the $E_2^{p,q}$-level for $q > 0$.

1.6.1. *Localization.* Since $X(\mathbb{C})/G$ has finite cohomological dimension (see [**Qu71**, Proposition A.11], compare [**Sch94**, Corollary 7.18]), we have that the $E_2^{p,0}$-term $H^p(V/G, \mathcal{H}^0(G, M))$ of spectral sequence (3) is zero for $V = X(\mathbb{C})$ or $X(\mathbb{R})$, and all sufficiently large p. Hence the restriction map (4) is an isomorphism for all sufficiently large n.

In the modern theory of transformation groups this is usually rephrased in terms of *localization* (compare [**Qu71**, §4], [**Hs75**, §III.2]). First we consider the case $M = \mathbb{Z}/2$. Let $H^*(V; G, \mathbb{Z}/2)[\eta^{-1}]$ be the localization of the $H^*(G, \mathbb{Z}/2)$-module $H^*(V; G, \mathbb{Z}/2)$ with respect to the multiplicative subset of $H^*(G, \mathbb{Z}/2) = \mathbb{Z}/2[\eta]$ generated by η. Then the above implies that the restriction map induces an isomorphism of graded $\mathbb{Z}/2[\eta, \eta^{-1}]$-algebras

$$H^*(X(\mathbb{C}); G, \mathbb{Z}/2)[\eta^{-1}] \xrightarrow{\sim} H^*(X(\mathbb{R}); G, \mathbb{Z}/2)[\eta^{-1}] \tag{5}$$

(compare [**Sch94**, Corollary 7.19]).

In the case $M = \mathbb{Q}/\mathbb{Z}(j)$ for some $j \in \mathbb{Z}$ we localize with respect to the multiplicative subset of $H^*(G, \mathbb{Z}) = \mathbb{Z}[\eta^2]/(2\eta^2)$ generated by η^2. Note that $H^*(G, \mathbb{Z})[\eta^{-2}] \simeq \mathbb{Z}/2[\eta^2, \eta^{-2}]$. Hence the isomorphism

$$H^*(X(\mathbb{C}); G, \mathbb{Q}/\mathbb{Z}(j))[\eta^{-2}] \xrightarrow{\sim} H^*(X(\mathbb{R}); G, \mathbb{Q}/\mathbb{Z}(j))[\eta^{-2}] \tag{6}$$

induced by the restriction map is an isomorphism of graded $\mathbb{Z}/2[\eta^2, \eta^{-2}]$-modules.

Later we will also need to know that the natural homomorphisms

$$H^n(X(\mathbb{C}); G, \mathbb{Z}/2) \underset{\mathbb{Z}/2}{\otimes} H^*(G, \mathbb{Z}/2)[\eta^{-1}] \to H^*(X(\mathbb{C}); G, \mathbb{Z}/2)[\eta^{-1}] \tag{7}$$

and

$$H^n(X(\mathbb{C}); G, \mathbb{Q}/\mathbb{Z}(j)) \underset{\mathbb{Z}}{\otimes} H^*(G, \mathbb{Z})[\eta^{-2}] \to H^{n \,(\mathrm{mod}\, 2)}(X(\mathbb{C}); G, \mathbb{Q}/\mathbb{Z}(j))[\eta^{-2}] \tag{8}$$

are isomorphisms for n sufficiently large. We obtain good lower bounds for n by considering the Hochschild-Serre spectral sequence (2). For $M = \mathbb{Z}/2$ we have that on the $E_2^{p,q}$-level the cup product with η coincides with the homomorphism $E_2^{p,q} \to E_2^{p+1,q}$ given by cup product with η in Galois cohomology. Hence the periodicity of the cohomology of G implies that if for some $D \geq 0$ we have that $H^q(X(\mathbb{C}), \mathbb{Z}/2) = 0$ for all $q > D$, then cup product with η induces a surjection

$$H^n(X(\mathbb{C}); G, \mathbb{Z}/2) \to H^{n+1}(X(\mathbb{C}); G, \mathbb{Z}/2)$$

for $n \geq D$ which is an isomorphism if $n > D$. We can always take $D = 2\dim(X)$, and even $D = \dim(X)$ if X is affine. We obtain that with such D the homomorphism (7) is a surjection for $n \geq D$ and an isomorphism if $n > D$ (compare [**Sch94**, Corollaries 7.19 and 7.20]). By similar reasoning for $M = \mathbb{Q}/\mathbb{Z}(j)$ we obtain the same result for the homomorphism (8).

1.6.2. *Cohomology of the Set of Real Points.* We will now analyse the equivariant cohomology of $X(\mathbb{R})$. For $M = \mathbb{Z}/2$ the $E_2^{p,q}$-terms of the spectral sequence (3) for $V = X(\mathbb{R})$ have the form

$$E_2^{p,q} = H^p(X(\mathbb{R}), \mathbb{Z}/2)$$

for all p, $q \geq 0$. The edge morphisms $E_2^{p,0} = H^p(X(\mathbb{R}), \mathbb{Z}/2) \to H^p(X(\mathbb{R}); G, \mathbb{Z}/2)$ for $p \geq 0$, and the cup product induce an isomorphism

$$H^*(X(\mathbb{R}), \mathbb{Z}/2) \underset{\mathbb{Z}/2}{\otimes} H^*(G, \mathbb{Z}/2) \xrightarrow{\sim} H^*(X(\mathbb{R}); G, \mathbb{Z}/2) \tag{9}$$

of graded $H^*(G, \mathbb{Z}/2)$-algebras (see [**Gr57**, Theorem 4.4.1, Corollary 5.4.1], compare [**Sch94**, Corollary 6.3.2]). In particular, the spectral sequence (3) is trivial for $V = X(\mathbb{R})$ and $M = \mathbb{Z}/2$ (see [**Gr57**, Theorem 4.4.1]).

Let $\mathfrak{m} \subset H^*(G, \mathbb{Z}/2)$ be the maximal ideal generated by $(1 - \eta)$. Combining the isomorphisms (9) and (5) we get an isomorphism of rings (not of graded rings)

$$H^*(X(\mathbb{C}); G, \mathbb{Z}/2)/\mathfrak{m}H^*(X(\mathbb{C}); G, \mathbb{Z}/2) \xrightarrow{\sim} H^*(X(\mathbb{R}), \mathbb{Z}/2)$$

(compare [**Sch94**, Corollary 7.19]). The composition of this isomorphism with the quotient map will be denoted by

$$\beta = \beta_{\mathbb{Z}}/2 \colon H^*(X(\mathbb{C}); G, \mathbb{Z}/2) \to H^*(X(\mathbb{R}), \mathbb{Z}/2).$$

Note that our results on the map (7) imply that the map

$$H^n(X(\mathbb{C}); G, \mathbb{Z}/2) \to H^*(X(\mathbb{R}), \mathbb{Z}/2)$$

obtained by restricting β to the cohomology in degree n is surjective for $n \geq 2\dim(X)$ (resp. $n \geq \dim(X)$ if X is affine) and bijective for $n > 2\dim(X)$ (resp. $n > \dim(X)$ if X is affine), compare [**Sch94**, Corollary 7.20]. Under the comparison isomorphism (1) this map corresponds for $n > 2\dim(X)$ to the isomorphism given in Section 1.4.

For $M = \mathbb{Q}/\mathbb{Z}(j)$ with $j \in \mathbb{Z}$ we have that the $E_2^{p,q}$-terms of the spectral sequence (3) for $V = X(\mathbb{R})$ have the following form:

$$E_2^{p,q} = \begin{cases} H^p(X(\mathbb{R}), \mathbb{Q}/\mathbb{Z}) & \text{if } q = 0 \text{ and } j \text{ is even,} \\ H^p(X(\mathbb{R}), \mathbb{Z}/2) & \text{if } q \geq 0 \text{ and } j \not\equiv q \pmod 2, \\ 0 & \text{otherwise.} \end{cases}$$

The coefficient map $\mathbb{Z}/2 \to \mathbb{Q}/\mathbb{Z}(j)$ induces a homomorphism of $E_2^{p,q}$-terms of the spectral sequence (3) which is an isomorphism for all $p \geq 0$, $q \geq 0$ and $j \not\equiv q \pmod 2$. Therefore the spectral sequence (3) is trivial for $M = \mathbb{Q}/\mathbb{Z}(j)$ and $V =$

$X(\mathbb{R})$, and the isomorphism (9) combined with the coefficient map $\mathbb{Z}/2 \to \mathbb{Q}/\mathbb{Z}(j)$ induces for $n > \dim(X)$ an isomorphism

$$\bigoplus_{i \not\equiv n+j \pmod 2} H^i(X(\mathbb{R}), \mathbb{Z}/2) \xrightarrow{\sim} H^n(X(\mathbb{R}); G, \mathbb{Q}/\mathbb{Z}(j)).$$

Hence, taking $\mathfrak{m}_{\mathbb{Z}} \subset H^*(G, \mathbb{Z})$ to be the maximal ideal generated by $(1 - \eta^2)$, we get an isomorphism of groups

$$H^*(X(\mathbb{C}); G, \mathbb{Q}/\mathbb{Z}(j))/\mathfrak{m}_{\mathbb{Z}} H^*(X(\mathbb{C}); G, \mathbb{Q}/\mathbb{Z}(j)) \xrightarrow{\sim} H^*(X(\mathbb{R}), \mathbb{Z}/2).$$

which in this case preserves (resp. reverses) the natural $\mathbb{Z}/2$-gradings when j is odd (resp. even). Note that here the grading modulo 2 on the equivariant cohomology of $X(\mathbb{C})$ descends to the quotient since $\mathfrak{m}_{\mathbb{Z}}$ is generated by the element $1 - \eta^2$, which is purely of even degree. The composition of the above isomorphism with the quotient map will be denoted by

$$\beta = \beta_{\mathbb{Q}/\mathbb{Z}(j)} \colon H^*(X(\mathbb{C}); G, \mathbb{Q}/\mathbb{Z}(j)) \to H^*(X(\mathbb{R}), \mathbb{Z}/2).$$

The restriction of β to the cohomology in degree n gives a surjection

$$H^n(X(\mathbb{C}); G, \mathbb{Q}/\mathbb{Z}(j)) \to \bigoplus_{i \not\equiv n+j \pmod 2} H^i(X(\mathbb{R}), \mathbb{Z}/2)$$

for $n \geq 2\dim(X)$ (resp. $n \geq \dim(X)$ if X is affine) which is an isomorphism if $n > 2\dim(X)$ (resp. $n > \dim(X)$ if X is affine); compare [**Sch94**, Theorems 20.2.11 and 20.2.13]).

The notations β, $\beta_{\mathbb{Z}}/2$, and $\beta_{\mathbb{Q}/\mathbb{Z}(j)}$ will also be used for the composition of the comparison isomorphism (1) with the mappings defined above.

REMARK 1.1. In order to give some idea of the nature of the mappings β, let me mention here that for X nonsingular the étale cycle map $\mathrm{cl}\colon \mathcal{Z}^q(X) \to H^{2q}_{\text{ét}}(X, \mathbb{Z}/2)$ composed with $\beta_{\mathbb{Z}}/2$ and the ith projection gives for every pair i, $q \geq 0$ a mapping

$$\mathcal{Z}^q(X) \to H^i(X(\mathbb{R}), \mathbb{Z}/2)$$

which is zero outside the range $q \leq i \leq 2q$, and which coincides for $i = q$ with the cohomological real algebraic cycle map

$$\mathrm{cl}_{\mathbb{R}} \colon \mathcal{Z}^q(X) \to H^q(X(\mathbb{R}), \mathbb{Z}/2)$$

that sends a subvariety $V \subset X$ of codimension q to the Poincaré dual of the class in the 'Borel-Moore' homology group $H_{\dim(X)-q}(X(\mathbb{R}), \mathbb{Z}/2)$ represented by the set of real points $V(\mathbb{R})$, as defined by Borel and Haefliger in [**BH61**]. See [**Kr94**, Theorem 0.6], [**Sch95**, Remark 3.5], or see [**vH96**, §§5.1, 5.2] for a purely topological proof.

1.7. The Bloch-Ogus Spectral Sequence. The mapping from the étale site to the Zariski site induces for any variety X over $\mathbb{R}$ and any étale sheaf $\mathcal{F}$ on X the *local-to-global spectral sequence*

$$E_2^{p,q} = H^p_{\text{Zar}}(X, \mathcal{H}^q(\mathcal{F})) \implies H^{p+q}_{\text{ét}}(X, \mathcal{F}), \tag{10}$$

where $\mathcal{H}^q(\mathcal{F})$ is the Zariski sheaf associated to the presheaf

$$U \mapsto H^q_{\text{ét}}(U, \mathcal{F}).$$

For X nonsingular and $\mathcal{F}$ a locally constant torsion sheaf this spectral sequence is often called the *Bloch-Ogus spectral sequence*, since (by [**BO74**, Remark 6.4]) it coincides with the coniveau spectral sequence studied by Bloch and Ogus in their fundamental paper [**BO74**]. They showed in particular that with the above hypotheses on X and $\mathcal{F}$ the groups $H^p_{\mathrm{Zar}}(X, \mathcal{H}^q(\mathcal{F}))$ vanish for all $p > q$. In the rest of this section we will not need to assume that X is nonsingular.

For X over an algebraically closed field and $\mathcal{F}$ a torsion sheaf we have that $\mathcal{H}^q(\mathcal{F}) = 0$ for $q > \dim(X)$. In view of Section 1.4, this does not hold for X over $\mathbb{R}$ with $X(\mathbb{R})$ nonempty and $\mathcal{F}$ not 2-divisible. However, Colliot-Thélène and Parimala showed in [**CTP90**] for X nonsingular that the sheaves $\mathcal{H}^q(\mathbb{Z}/2)$ do admit a simple description for $q > \dim(X)$. Scheiderer then showed how to derive their result (and the analogue when X is singular) from the cohomological facts presented in Sections 1.5 and 1.6. Moreover, his approach, which we will follow here, immediately gave an easy description of the cohomology groups $H^p(X, \mathcal{H}^q(\mathbb{Z}/2))$ for $q > \dim(X)$, showing that these 'extra' $E_2^{p,q}$-terms of the local-to-global spectral sequence over $\mathbb{R}$ are not so hard to understand.

The homomorphism $\beta_{\mathbb{Z}/2}$ defined in Section 1.6.2 induces for $q > d$ an isomorphism $\tilde{\beta}_{\mathbb{Z}/2}$ from $\mathcal{H}^q(\mathbb{Z}/2)$ to the Zariski-sheaf associated to the presheaf

$$U \mapsto \bigoplus_i H^i(U(\mathbb{R}), \mathbb{Z}/2)$$

and $\beta_{\mathbb{Q}/\mathbb{Z}(j)}$ induces for $q > d$ an isomorphism $\tilde{\beta}_{\mathbb{Q}/\mathbb{Z}(j)}$ from $\mathcal{H}^q(\mathbb{Q}/\mathbb{Z}(j))$ to the Zariski-sheaf associated to the presheaf

$$U \mapsto \bigoplus_{i \not\equiv q+j \pmod 2} H^i(U(\mathbb{R}), \mathbb{Z}/2).$$

It is a remarkable fact that the graded pieces of the above presheaves of degree $i > 0$ vanish after sheafifying for the Zariski topology (Lemma 1.2.i). This observation is due to Scheiderer and it was the key result that allowed him to give a different proof of the result of Colliot-Thélène and Parimala, as well as the easy description referred to above of the higher $E_2^{p,q}$-terms in the local-to-global spectral sequence (see Corollary 1.3).

LEMMA 1.2. *Let X be a variety over $\mathbb{R}$. Let $\varphi\colon X(\mathbb{R}) \to X_{\mathrm{Zar}}$ be the obvious mapping of topological spaces. Let $\mathcal{F}$ be a locally constant sheaf on $X(\mathbb{R})$.*

(i) *For $i > 0$ we have*

$$R^i\varphi_*\mathcal{F} = 0.$$

(ii) *For any $p \geq 0$ we have*

$$H^p_{\mathrm{Zar}}(X, \varphi_*\mathcal{F}) = H^p(X(\mathbb{R}), \mathcal{F}).$$

PROOF. (i) This is the analogue of [**Sch94**, Theorem 19.2] for the Euclidean topology, so it follows from comparison between the cohomology of $X(\mathbb{R})$ and the cohomology of the real spectrum of X. Scheiderer's earlier approach (in [**Sch90**]) gives a direct proof: it is sufficient to find for every Zariski-neighbourhood U of $P \in X(\mathbb{R})$ a smaller Zariski-open neighbourhood $V \subset U$ of P, such that the restriction

$$H^i(U(\mathbb{R}), \mathcal{F}) \to H^i(V(\mathbb{R}), \mathcal{F})$$

is zero. Such a V is easily constructed from the data that give a semi-algebraic triangulation of $U(\mathbb{R})$ (see [**BCR87**, Theorem 9.2.1]).

(ii) This follows from the first statement by general homological algebra. □

COROLLARY 1.3. *Let X be a variety over $\mathbb{R}$ of dimension d. Let $q > d$.*

(i) *The homomorphism $\beta_{\mathbb{Z}/2}$ induces an isomorphism*

$$\tilde{\beta}_{\mathbb{Z}/2}\colon \mathcal{H}^q(\mathbb{Z}/2) \xrightarrow{\sim} \varphi_*\mathbb{Z}/2,$$

hence for any $p \geq 0$ an isomorphism

$$H^p_{\mathrm{Zar}}(X, \mathcal{H}^q(\mathbb{Z}/2)) \xrightarrow{\sim} H^p(X(\mathbb{R}), \mathbb{Z}/2).$$

(ii) *For $j \in \mathbb{Z}$ the homomorphism $\beta_{\mathbb{Q}/\mathbb{Z}(j)}$ induces an isomorphism*

$$\tilde{\beta}_{\mathbb{Q}/\mathbb{Z}(j)}\colon \mathcal{H}^q(\mathbb{Q}/\mathbb{Z}(j)) \xrightarrow{\sim} \begin{cases} 0 & \text{if } q \equiv j \pmod 2, \\ \varphi_*\mathbb{Z}/2 & \text{if } q \not\equiv j \pmod 2, \end{cases}$$

hence for every $p \geq 0$ an isomorphism

$$H^p_{\mathrm{Zar}}(X, \mathcal{H}^q(\mathbb{Q}/\mathbb{Z}(j))) \xrightarrow{\sim} \begin{cases} 0 & \text{if } q \equiv j \pmod 2, \\ H^p(X(\mathbb{R}), \mathbb{Z}/2) & \text{if } q \not\equiv j \pmod 2. \end{cases}$$

(iii) *For $q \not\equiv j \pmod 2$ the isomorphisms $\tilde{\beta}_{\mathbb{Z}}/2$ and $\tilde{\beta}_{\mathbb{Q}/\mathbb{Z}(j)}$ are compatible with the homomorphism $\mathcal{H}^q(\mathbb{Z}/2) \to \mathcal{H}^q(\mathbb{Q}/\mathbb{Z}(j))$ induced by the coefficient map $\mathbb{Z}/2 \hookrightarrow \mathbb{Q}/\mathbb{Z}(j)$.*

PROOF. Immediate from Lemma 1.2, the above descriptions of $\mathcal{H}^q(\mathbb{Z}/2)$ and $\mathcal{H}^q(\mathbb{Q}/\mathbb{Z}(j))$, and the definition of $\beta_{\mathbb{Z}}/2$ and $\beta_{\mathbb{Q}/\mathbb{Z}(j)}$. □

The fact that $\mathcal{H}^q(\mathbb{Z}/2)$ is isomorphic to $\varphi_*\mathbb{Z}/2$ for $q > d$ is (for X nonsingular) the main theorem of Colliot-Thélène and Parimala (see [**CTP90**, Theorem 2.3.1]). The full result is due to Scheiderer (see [**Sch94**, Proposition 19.4, Corollary 19.5.1]).

Of course, the relation between the higher $E_2^{p,q}$-terms of the local-to-global spectral sequence and the cohomology of the set of real points is not only induced by β on the local level, but globally as well. Heuristically, this means that the mapping β maps the $E_2^{p,q}$-terms for $q > \dim(X)$ isomorphically to $H^p(X(\mathbb{R}), \mathbb{Z}/2)$ or 0. In order to make this statement well-defined, we follow Scheiderer's approach.

PROPOSITION 1.4. *Let X be a variety defined over $\mathbb{R}$ of pure dimension d. Let M be a torsion G-module. The comparison isomorphism* (1) *and the inclusion $i\colon X(\mathbb{R}) \hookrightarrow X(\mathbb{C})$ induce a morphism from the local-to-global spectral sequence*

$$E_2^{p,q} = H^p_{\mathrm{Zar}}(X, \mathcal{H}^q(M)) \implies H^{p+q}_{\text{ét}}(X, M),$$

to the spectral sequence (3) *for $V = X(\mathbb{R})$:*

$$E_2^{p,q} = H(X(\mathbb{R}), \mathcal{H}^q(G, M)) \implies H^{p+q}(X(\mathbb{R}); G, M).$$

On the limit terms this morphism is the comparison isomorphism (1) *followed by the restriction map* (4). *When $M = \mathbb{Z}/2$ or $\mathbb{Q}/\mathbb{Z}(j)$ for some $j \in \mathbb{Z}$, then on the $E_2^{p,q}$-terms this morphism is for $p \geq 0$, and $q > d$ the isomorphism of Corollary* 1.3.

PROOF. This is the analogue of [**Sch94**, Proposition 19.7]. Let ψ_*^G be the functor sending a G-sheaf $\mathcal{F}$ of Abelian groups on $X(\mathbb{C})$ to the sheaf on X_{Zar} associated to the presheaf

$$U \mapsto \mathcal{F}(U)^G.$$

The comparison between étale cohomology and equivariant cohomology gives that the spectral sequence (10) is canonically isomorphic to the spectral sequence associated to the decomposition of derived functors

$$R\Gamma^G_{X(\mathbb{C})} = R\Gamma_{X_{\mathrm{Zar}}} \circ R\psi^G_*, \tag{11}$$

with notations as in [**Gr57**, §5.2]. The inclusion $i\colon X(\mathbb{R}) \hookrightarrow X(\mathbb{C})$ induces an isomorphism of functors

$$\psi^G_* \circ i_* = \varphi_* \circ \mathcal{H}^0(G, -),$$

with φ as in Lemma 1.2 and $\mathcal{H}^0(G,-)$ as in Section 1.6. Hence the canonical morphism $M \to i_*M$ of G-sheaves on $X(\mathbb{C})$ and the above isomorphisms induce a map from the local-to-global spectral sequence with coefficients in M into the spectral sequence associated to the decomposition of derived functors

$$R\Gamma^G_{X(\mathbb{R})} = R\Gamma_{X_{\mathrm{Zar}}} \circ R(\varphi_* \circ \mathcal{H}^0(G, -)) \tag{12}$$

applied to the G-sheaf M on $X(\mathbb{R})$. Since $\Gamma_{X_{\mathrm{Zar}}} \circ \varphi_* = \Gamma_{X(\mathbb{R})}$ and φ_* is exact by Lemma 1.2, the spectral sequence associated to (12) coincides with the spectral sequence (3) for $V = X(\mathbb{R})$, which is the spectral sequence associated to the decomposition

$$R\Gamma^G_{X(\mathbb{R})} = R\Gamma_{X(\mathbb{R})} \circ \mathcal{H}^0(G, -).$$

This proves the first two statements. The last statement follows immediately from the definition of β. □

2. Vanishing of Differentials in the Bloch-Ogus Spectral Sequence

A crucial point in the study of the Bloch-Ogus spectral sequence over $\mathbb{R}$ is the vanishing of the differentials

$$d^{p,q}_r\colon E^{p,q}_r \to E^{p+r,q-r+1}_r$$

for large q. See also the historical remarks below. We will prove in this section that for X nonsingular all differentials with source $E^{p,q}_r$ are zero when $q > \dim(X)$ (part (i) of Theorem 2.1). In other words, the 'extra' $q > \dim(X)$ part of the Bloch-Ogus spectral sequence is completely degenerate. For many purposes, and certainly for our purposes here, this means that the Bloch-Ogus spectral sequence for varieties over $\mathbb{R}$ is no more complicated than the Bloch-Ogus spectral sequence over $\mathbb{C}$. Also, we will show that the $q > \dim(X)$ part maps to zero, under the pull-back mapping in étale cohomology associated to base change from $\mathbb{R}$ to $\mathbb{C}$. This is part (ii) of Theorem 2.1. In the rest of the paper we will not use the theorem in its full strength; we will only need Corollary 2.2.

The theorem is proved by means of an auxiliary degenerate spectral sequence that maps into the Bloch-Ogus spectral sequence. This will be the local-to-global spectral sequence for equivariant cohomology with supports in $X(\mathbb{R})$. It is defined and studied in Section 2.1. Then in Section 2.2 we study the map into the Bloch-Ogus spectral sequence, and prove the theorem.

It should be said that, using Proposition 1.4, it is actually not very hard to derive Theorem 2.1 from a result of Krasnov in [**Kr94**] concerning the image of the mapping β (Theorem 2.8 in the present paper). This, and further connections with existing results, will be treated in Section 2.3.

THEOREM 2.1. *Let X be a nonsingular variety over $\mathbb{R}$ of dimension d. Let $M = \mathbb{Z}/2$ or $\mathbb{Q}/\mathbb{Z}(j)$ for some $j \in \mathbb{Z}$ and let $n \geq 0$. Consider the Bloch-Ogus spectral sequence with coefficients in M, and let $F^\bullet$ be the associated descending filtration on $H^n_{\text{ét}}(X, M)$, with pth graded piece equal to $E^{p,n-p}_\infty$.*

(i) *For any $p \in \mathbb{Z}$, $q > d$ and $r \geq 2$ the differential $d^{p,q}_r \colon E^{p,q}_r \to E^{p+r,q-r+1}_r$ is zero.*

(ii) *The kernel of the canonical mapping*

$$\pi^* \colon H^n_{\text{ét}}(X, M) \to H^n_{\text{ét}}(X_{\mathbb{C}}, M)$$

maps surjectively onto

$$H^n_{\text{ét}}(X, M)/F^{n-d}H^n_{\text{ét}}(X, M).$$

PROOF. See Section 2.2. □

COROLLARY 2.2. *Let X be a nonsingular variety over $\mathbb{R}$ of dimension d.*

(i) *We have a short exact sequence*

$$0 \to H^{d-1}_{\text{Zar}}(X, \mathcal{H}^d(\mathbb{Q}/\mathbb{Z}(d))) \to H^{2d-1}_{\text{ét}}(X, \mathbb{Q}/\mathbb{Z}(d)) \xrightarrow{\beta'} \bigoplus_{i>0} H^{d-2i}(X(\mathbb{R}), \mathbb{Z}/2) \to 0,$$

where β' is the mapping β followed by the projection.

(ii) *The image of the base change map*

$$\pi^* \colon H^{2d-1}_{\text{ét}}(X, \mathbb{Q}/\mathbb{Z}(d)) \to H^{2d-1}_{\text{ét}}(X_{\mathbb{C}}, \mathbb{Q}/\mathbb{Z}(d))$$

is generated by the image of $H^{d-1}_{\text{Zar}}(X, \mathcal{H}^d(\mathbb{Q}/\mathbb{Z}(d)))$.

PROOF. (i) The exactness follows from part (i) of the theorem and Proposition 1.4.

(ii) Part (ii) of the theorem says in the case of $n = 2d - 1$ that the cokernel of the inclusion $H^{d-1}_{\text{Zar}}(X, \mathcal{H}^d(\mathbb{Q}/\mathbb{Z}(d))) \hookrightarrow H^{2d-1}_{\text{ét}}(X, \mathbb{Q}/\mathbb{Z}(d))$ maps to zero under π^*. □

Historical Remarks. The first result on vanishing of differentials of high degree in the Bloch-Ogus spectral sequence over $\mathbb{R}$ is due to Colliot-Thélène and Parimala, who proved the vanishing of all differentials in the Bloch-Ogus spectral sequence for a nonsingular geometrically irreducible surface over $\mathbb{R}$ having a compact set of real points (see [**CTP90**, Proposition 3.11]). Inspired by this result, Scheiderer proved for a (possibly singular) variety X of dimension d over an arbitrary real closed field the vanishing of all differentials having target $E^{p,q}_r$ with $r \geq 2$, $p + q \geq 2d$ and $(p, q) \neq (d, d)$ (see [**Sch94**, Proposition 19.8]). Together with Colliot-Thélène he improved the bounds on (p, q) to $p \in \mathbb{Z}$ and $q > d$ (see [**CTS96**, Theorem 3.1.b, Remark 3.1.1]). For X nonsingular they proved in the same paper that the differentials having target $E^{d,d}_r$ vanish as well (see [**CTS96**, Theorem 3.2.b]). Actually, all these results are stated for coefficients in $\mathbb{Z}/2$ only, but the case of coefficients in $\mathbb{Q}/\mathbb{Z}(j)$ is an easy consequence.

2.1. Equivariant Cohomology with Supports. The equivariant cohomology groups of $X(\mathbb{C})$ with supports in $X(\mathbb{R})$ and coefficients in $\mathbb{Z}/2$ needed in the proof of Theorem 2.1 can be defined by

$$H^q_{X(\mathbb{R})}(X(\mathbb{C}); G, \mathbb{Z}/2) := R^q \operatorname{Hom}_{G,\mathbb{Z}/2}(i_*\mathbb{Z}/2, \mathbb{Z}/2).$$

Here $i\colon X(\mathbb{R}) \hookrightarrow X(\mathbb{C})$ denotes the inclusion and the subscript appended to $R^q \operatorname{Hom}$ indicates that we are working in the derived category of G-sheaves of $\mathbb{Z}/2$-modules on $X(\mathbb{C})$. This is the obvious equivariant version of ordinary cohomology with supports:

$$H^q_{X(\mathbb{R})}(X(\mathbb{C}), \mathbb{Z}/2) = R^q \operatorname{Hom}_{\mathbb{Z}/2}(i_*\mathbb{Z}/2, \mathbb{Z}/2)$$

(see [**Iv86**, §II.9]). If X is nonsingular of dimension d, then $X(\mathbb{R})$ is a submanifold of $X(\mathbb{C})$ of pure codimension d, so we have for every $n \in \mathbb{Z}$ a canonical isomorphism

$$H^n(X(\mathbb{R}), \mathbb{Z}/2) \xrightarrow{\sim} H^{n+d}_{X(\mathbb{R})}(X(\mathbb{C}), \mathbb{Z}/2),$$

known as the *Thom isomorphism* (see [**Iv86**, Theorem VIII.2.3]). The situation is analogous in the equivariant situation:

PROPOSITION 2.3 (Thom isomorphism). *Let X be a nonsingular variety over $\mathbb{R}$ of pure dimension d. We have for any $n \in \mathbb{Z}$ a canonical isomorphism*

$$\tau\colon H^n(X(\mathbb{R}); G, \mathbb{Z}/2) \xrightarrow{\sim} H^{n+d}_{X(\mathbb{R})}(X(\mathbb{C}); G, \mathbb{Z}/2).$$

PROOF. The map τ is the isomorphism

$$\begin{aligned} H^n(X(\mathbb{R}), \mathbb{Z}/2) &= R^n \operatorname{Hom}_{G,\mathbb{Z}/2}(\mathbb{Z}/2, i_*\mathbb{Z}/2) = R^{n+d} \operatorname{Hom}_{G,\mathbb{Z}/2}(\mathbb{Z}/2, i_*\mathbb{Z}/2[-d]) \\ &\xrightarrow[\text{[Iv86, VIII.1]}]{\sim} R^{n+d} \operatorname{Hom}_{G,\mathbb{Z}/2}(\mathbb{Z}/2, R\mathcal{H}om_{\mathbb{Z}/2}(i_*\mathbb{Z}/2, \mathbb{Z}/2)) \\ &= R^{n+d} \operatorname{Hom}_{G,\mathbb{Z}/2}(i_*\mathbb{Z}/2, \mathbb{Z}/2) = H^{n+d}_{X(\mathbb{R})}(X(\mathbb{C}); G, \mathbb{Z}/2). \quad \square \end{aligned}$$

The decomposition of derived functors (11) applied to the complex of G-sheaves $R\mathcal{H}om_{\mathbb{Z}/2}(i_*\mathbb{Z}/2, \mathbb{Z}/2)$ gives the local-to-global spectral sequence

$$E_2^{p,q} = H^p_{\mathrm{Zar}}(X, \mathcal{H}^q_{X(\mathbb{R})}) \implies H^{p+q}_{X(\mathbb{R})}(X(\mathbb{C}); G, \mathbb{Z}/2), \tag{13}$$

where $\mathcal{H}^q_{X(\mathbb{R})}$ is the Zariski-sheaf associated to the presheaf

$$U \mapsto H^q_{U(\mathbb{R})}(U(\mathbb{C}); G, \mathbb{Z}/2).$$

This is the auxiliary spectral sequence mentioned in the introduction of this section.

LEMMA 2.4. *Let X be a nonsingular variety over $\mathbb{R}$ of pure dimension d. Consider the spectral sequence* (3) *for $V = X(\mathbb{R})$ and $M = \mathbb{Z}/2$ shifted vertically by d positions:*

$$E_2^{p,q} = H^p(X(\mathbb{R}), \mathcal{H}^{q-d}(G, \mathbb{Z}/2)) \implies H^{p+q-d}(X(\mathbb{R}); G, \mathbb{Z}/2).$$

There is an isomorphism from this spectral sequence into the spectral sequence (13), *corresponding to the Thom isomorphism τ on limit terms.*

PROOF. We saw in the proof of Proposition 1.4 that the spectral sequence (3) for $V = X(\mathbb{R})$ and $M = \mathbb{Z}/2$ corresponds to the decomposition of derived functors

$$R\Gamma_{X(\mathbb{R})} = R\Gamma_{X_{\mathrm{Zar}}} \circ (R\psi_*^G \circ i_*).$$

Therefore the required isomorphism is induced by the quasi-isomorphism

$$i_*\mathbb{Z}/2[d] \xrightarrow{\sim} R\mathcal{H}om_{\mathbb{Z}/2}(i_*\mathbb{Z}/2, \mathbb{Z}/2)$$

which was used in the construction of the Thom isomorphism. $\square$

COROLLARY 2.5. *Let X be a nonsingular variety over $\mathbb{R}$ of pure dimension d. Consider the spectral sequence* (13) .

(i) *The differentials* $d_r^{p,q}$ *are zero for every* p, $q \in \mathbb{Z}$ *and* $r \geq 2$.
(ii) *We have for every* $p \geq 0$ *and every* $r \geq 2$:

$$E_r^{p,q} \simeq \begin{cases} 0 & \text{for } q < d, \\ H^p(X(\mathbb{R}), \mathbb{Z}/2) & \text{for } q \geq d \end{cases}$$

(iii) *Cup product with the nontrivial element* $\eta \in H^1(G, \mathbb{Z}/2)$ *induces for every* $q \geq d$, *every* $p \in \mathbb{Z}$ *and every* $r \geq 2$ *an isomorphism* $E_r^{p,q} \xrightarrow{\sim} E_r^{p,q+1}$.

PROOF. This follows immediately from Lemma 2.4 and the corresponding properties of the spectral sequence (3) for $V = X(\mathbb{R})$ and $M = \mathbb{Z}/2$. □

2.2. Proof of Theorem 2.1. First we will study the canonical mapping

$$\psi\colon H^n_{X(\mathbb{R})}(X(\mathbb{C}); G, \mathbb{Z}/2) \to H^n(X(\mathbb{C}); G, \mathbb{Z}/2)$$

obtained by forgetting the supports.

LEMMA 2.6. *Let* X *be a variety defined over* $\mathbb{R}$ *of dimension* d.
(i) *The mapping* ψ *is an isomorphism in degree* $n > 2d + 1$.
(ii) *If* X *is nonsingular, then* ψ *is injective in every degree* $n \geq 0$.

Note that in the nonequivariant setting part (i) is trivially true, since both groups are zero in degree $n > 2d+1$. However, the nonequivariant analogue of the last statement does not hold (not even for $X = \mathbb{A}^N$ with $N > 0$).

PROOF. (i) This follows from the long exact sequence

$$\cdots \to H^n_{X(\mathbb{R})}(X(\mathbb{C}); G, \mathbb{Z}/2) \to H^n(X(\mathbb{C}); G, \mathbb{Z}/2) \\ \to H^n(X(\mathbb{C}) - X(\mathbb{R}); G, \mathbb{Z}/2) \to \cdots$$

which is obtained by applying $R\operatorname{Hom}_{G,\mathbb{Z}/2}(-, \mathbb{Z}/2)$ to the short exact sequence

$$0 \to i_*\mathbb{Z}/2 \to \mathbb{Z}/2 \to \mathbb{Z}/2_{X(\mathbb{C})-X(\mathbb{R})} \to 0$$

of G-sheaves on $X(\mathbb{C})$ (compare [**Iv86**, §II.9]). Since G acts freely on $X(\mathbb{C})-X(\mathbb{R})$, we have that $H^n(X(\mathbb{C}) - X(\mathbb{R}); G, \mathbb{Z}/2)$ is canonically isomorphic to the ordinary nth cohomology group of the quotient space, which vanishes for $n > 2d$.

(ii) Consider the following commutative diagram:

$$\begin{array}{ccc} H^n_{X(\mathbb{R})}(X(\mathbb{C}); G, \mathbb{Z}/2) & \xhookrightarrow{\cup \eta^N} & H^{n+N}_{X(\mathbb{R})}(X(\mathbb{C}); G, \mathbb{Z}/2) \\ \downarrow{\psi} & & \downarrow{\psi} \\ H^n(X(\mathbb{C}); G, \mathbb{Z}/2) & \xrightarrow{\cup \eta^N} & H^{n+N}(X(\mathbb{C}); G, \mathbb{Z}/2) \end{array}$$

For any $N \geq 0$ the top horizontal arrow is injective by Corollary 2.5.iii. Taking N large enough we get that the right hand vertical arrow is an isomorphism by statement (i); the injectivity of the left hand vertical arrow now is obvious. □

COROLLARY 2.7. *Let* X *be a nonsingular variety defined over* $\mathbb{R}$ *of dimension* d. *The mapping* ψ *corresponds to a morphism form the local-to-global spectral sequence* (13) *into the Bloch-Ogus spectral sequence* (10) *with coefficients in* $\mathbb{Z}/2$, *that induces on the* $E_2^{p,q}$*-level for any* $p \in \mathbb{Z}$ *and* $q > d$ *an isomorphism*

$$H^p_{\text{Zar}}(X, \mathcal{H}^q_{X(\mathbb{R})}) \xrightarrow{\sim} H^p_{\text{Zar}}(X, \mathcal{H}^q(\mathbb{Z}/2)).$$

PROOF. The existence of the morphism of spectral sequences is clear from the definitions. For the last statement, take $q > d$. We will show that ψ induces an isomorphism

$$\widetilde{\psi}\colon \mathcal{H}^q_{X(\mathbb{R})} \xrightarrow{\sim} \mathcal{H}^q(\mathbb{Z}/2).$$

We see from part (ii) of the above lemma that $\widetilde{\psi}$ is injective. Since both $\mathcal{H}^q_{X(\mathbb{R})}$ and $\mathcal{H}^q(\mathbb{Z}/2)$ are isomorphic to $\varphi_*\mathbb{Z}/2$ by Proposition 2.3 and Corollary 1.3.i, this implies that $\widetilde{\psi}$ is an isomorphism. □

PROOF OF THEOREM 2.1. Consider the mapping of spectral sequences associated to the injection $\mathbb{Z}/2 \hookrightarrow \mathbb{Q}/\mathbb{Z}(j)$. By Corollary 1.3, this induces a surjection $H^p_{\mathrm{Zar}}(X, \mathcal{H}^q(\mathbb{Z}/2)) \to H^p_{\mathrm{Zar}}(X, \mathcal{H}^q(\mathbb{Q}/\mathbb{Z}(j)))$ for any $q > d$. Therefore it is sufficient to prove the theorem for $M = \mathbb{Z}/2$.

(i) The first statement follows immediately from Corollary 2.5.i and Corollary 2.7.

(ii) From the Hochschild-Serre spectral sequence for $H^i_{\text{ét}}(X, \mathbb{Z}/2)$ we see that cup product with the nontrivial element $\eta \in H^1(G, \mathbb{Z}/2)$ induces a surjection from $H^{i-1}_{\text{ét}}(X, \mathbb{Z}/2)$ to the kernel of the the pull-back map $H^i_{\text{ét}}(X, \mathbb{Z}/2) \to H^i_{\text{ét}}(X_{\mathbb{C}}, \mathbb{Z}/2)$. Therefore it is sufficient to prove that cup product with η induces for every $q > d$ a surjection on the $E^{p,q}_\infty$-level of the Bloch-Ogus spectral sequence

$$E^{p,q-1}_\infty \xrightarrow{\cup\eta} E^{p,q}_\infty = E^{p,q}_2 = H^p_{\mathrm{Zar}}(X, \mathcal{H}^q(\mathbb{Z}/2)).$$

This follows immediately from Corollary 2.5.iii and Corollary 2.7. More precisely, it follows from the following commutative diagram, where $q > d$ and the vertical arrows are induced by the morphism of spectral sequences of Corollary 2.7:

$$\begin{array}{ccc} H^p_{\mathrm{Zar}}(X, \mathcal{H}^{q-1}_{X(\mathbb{R})}) & \xrightarrow{\cup\eta} & H^p_{\mathrm{Zar}}(X, \mathcal{H}^q_{X(\mathbb{R})}) \\ \downarrow & & \downarrow \\ E^{p,q-1}_\infty & \xrightarrow{\cup\eta} & H^p_{\mathrm{Zar}}(X, \mathcal{H}^q(\mathbb{Z}/2)) \end{array}$$

Here the upper horizontal arrow is an isomorphism by Corollary 2.5.iii and the right hand vertical arrow is an isomorphism by Corollary 2.7. □

2.3. Connections With Known Results. Consider the following result:

THEOREM 2.8. *Let X be a nonsingular variety over $\mathbb{R}$ of dimension d. For every $k \geq 0$, the homomorphism β followed by the projection to the k lowest degree factors gives a surjection*

$$H^{d+k}_{\text{ét}}(X, \mathbb{Z}/2) \to \bigoplus_{i=0}^{k} H^i(X(\mathbb{R}), \mathbb{Z}/2).$$

For X projective this is due to Krasnov, as a corollary to a more technical, but much stronger result (see [**Kr94**, Theorem 3.1, Corollary 3.2]). It also follows (for X complete) from the general result [**AP93**, Proposition 5.3.7] on the relation between β and the Gysin morphism between the cohomology of Poincaré duality spaces, in this case applied to the inclusion $i\colon X(\mathbb{R}) \hookrightarrow X(\mathbb{C})$. Both proofs are purely topological: the important fact is that $X(\mathbb{R})$ is a submanifold of $X(\mathbb{C})$ of codimension $\geq d$ (with equality if X has pure dimension d). In fact, the only reason for the projectiveness condition in Krasnov's result is that in the proof it is used that $X(\mathbb{R})$ is a finite cell complex. As far as I can follow his argument, it seems that this condition can be relaxed in order to allow for arbitrary nonsingular X of pure dimension d. On the other hand, in the approach of Allday and Puppe the compactness of $X(\mathbb{R})$ and $X(\mathbb{C})$ is essential, since Poincaré duality (in the sense of nondegeneracy of the cup product pairing) forms the heart of their methods.

PROPOSITION 2.9. *For any variety X over $\mathbb{R}$ of dimension d the statement of Theorem 2.1 is equivalent to the statement of Theorem 2.8.*

PROOF. As remarked in the proof of Theorem 2.1, we can restrict ourselves to coefficients in $\mathbb{Z}/2$. Then, in view of Proposition 1.4, we have that part (i) of Theorem 2.1 is equivalent to the statement that for $k \geq 0$ the map

$$\beta' \colon H^{d+k+1}_{\text{ét}}(X, \mathbb{Z}/2) \to \bigoplus_{i=0}^{k} H^i(X(\mathbb{R}), \mathbb{Z}/2)$$

given by β followed by the projection is surjective. This is a weaker version of Theorem 2.8. Once we have part (i) of Theorem 2.1, we use the fact that cup product with η induces a surjection from $H^{i-1}_{\text{ét}}(X, \mathbb{Z}/2)$ to the kernel of the map $H^i_{\text{ét}}(X, \mathbb{Z}/2) \to H^i_{\text{ét}}(X_{\mathbb{C}}, \mathbb{Z}/2)$ (as we saw in the proof). Since β respects cup product and sends η to the unit element of $H^*(X(\mathbb{R}), \mathbb{Z}/2)$, we get the following commutative diagram:

$$\begin{array}{ccc} H^{d+k}_{\text{ét}}(X, \mathbb{Z}/2) & \xrightarrow{\ \cup\eta\ } & H^{d+k+1}_{\text{ét}}(X, \mathbb{Z}/2) \\ & \searrow \qquad \swarrow{\scriptstyle \beta'} & \\ & \bigoplus_{i=0}^{k} H^i(X(\mathbb{R}), \mathbb{Z}/2) & \end{array}$$

By Proposition 1.4 we have that the subgroup $F^{k+1}H^{d+k+1}_{\text{ét}}(X, \mathbb{Z}/2)$ maps to zero under the mapping β' defined above, so (since we have part (i) of Theorem 2.1), the diagram gives an immediate equivalence between part (ii) of Theorem 2.1 and the statement of Theorem 2.8. □

It is also possible to give a direct proof of Theorem 2.8 with the methods used here.

SKETCH OF PROOF OF THEOREM T2.8. Without loss of generality we may assume X to be of pure dimension d. Consider the composite mapping

$$\theta \colon H^k(X(\mathbb{R}), \mathbb{Z}/2) \xrightarrow{\tau} H^{d+k}_{X(\mathbb{R})}(X(\mathbb{C}); G, \mathbb{Z}/2)$$
$$\xrightarrow{\psi} H^{d+k}(X(\mathbb{C}); G, \mathbb{Z}/2) \xrightarrow{i^*} H^{d+k}(X(\mathbb{R}); G, \mathbb{Z}/2).$$

This map corresponds to a morphism θ^{**}_* from the shifted spectral sequence of Lemma 2.4 to the spectral sequence (3) for $V = X(\mathbb{R})$, $M = \mathbb{Z}/2$. In order to prove the theorem, it is sufficient to prove that θ^{**}_* is an isomorphism on the $E^{p,q}_\infty$-level for $q \geq d$.

Since both spectral sequences are trivial, it follows from Lemma 2.4, Corollary 2.7 and Proposition 1.4 that $\theta^{p,q}_r$ is an isomorphism for every $r \geq 2$, $p \in \mathbb{Z}$ and $q > d$; the case $q = d$ follows from Corollary 2.5.iii and the analogue for the spectral sequence (3). □

REMARK 2.10. The use of Corollary 2.7 and Proposition 1.4, hence of the Zariski topology, in the above proof is a purely matter of convenience. With a little more work (using arguments similar to the ones used in the proof of Lemma 2.6 and Corollary 2.7) we could have given a direct proof of the fact that $\theta_r^{p,q}$ is an isomorphism for every $r \geq 2$, $p \in \mathbb{Z}$ and $q \geq d$.

3. The Abel-Jacobi Maps for Torsion Cycle Classes

In this section we will see that Bloch's construction of a cohomological Abel-Jacobi map for torsion cycles actually works over the real numbers. Combined with the results of Section 2, this gives us in Section 3.2 precise information on the Abel-Jacobi mapping α_{tors} in terms of étale cohomology. The cohomological Abel-Jacobi map also induces a filtration on $\mathrm{CH}_0(X)_{\mathrm{tors}}$, which will be treated in Section 3.3.

3.1. Bloch's Cohomological Abel-Jacobi Map. We will analyse Bloch's construction of a cohomological Abel-Jacobi map, as presented in [**CT93**, §§3, 4]. Let X be a nonsingular variety over a field k of characteristic zero. Let $\mathcal{K}_i$ be the Zariski-sheaf associated to the presheaf $U \mapsto K_i(\mathcal{O}(U))$ of algebraic K-groups. The Gersten conjecture and the Merkurjev-Suslin theorem on the K_2 of fields give for every $i \geq 0$ a diagram

$$\begin{array}{ccccccc} 0 \longrightarrow H^{i-1}_{\mathrm{Zar}}(X,\mathcal{K}_i)\otimes\mathbb{Q}/\mathbb{Z} & \xrightarrow{\varphi_i} & H^{i-1}_{\mathrm{Zar}}(X,\mathcal{H}^i(\mathbb{Q}/\mathbb{Z}(i))) & \xrightarrow{\psi_i} & \mathrm{CH}^i(X)_{\mathrm{tors}} \longrightarrow 0 \\ & & \downarrow{\scriptstyle\gamma_i} & & \\ & & H^{2i-1}_{\text{ét}}(X,\mathbb{Q}/\mathbb{Z}(i)) & & \end{array} \tag{14}$$

with exact top row (see [**CT93**, §3.2]).

When k is algebraically closed and X is nonsingular and projective, it can be shown that $\gamma_i \circ \varphi_i = 0$ (see [**CT93**, proof of Theorem 4.3]) and ψ_i and γ_i then induce Bloch's cohomological Abel-Jacobi map

$$\lambda_i\colon \mathrm{CH}^i(X)_{\mathrm{tors}} \to H^{2i-1}_{\text{ét}}(X,\mathbb{Q}/\mathbb{Z}(i)).$$

If k is not algebraically closed, then $\gamma_i \circ \varphi_i$ need not be zero; for example, in the case $i = 1$ the diagram simplifies to the short exact sequence

$$0 \to k^* \otimes \mathbb{Q}/\mathbb{Z} \xrightarrow{\gamma_i\circ\varphi_i} H^1_{\text{ét}}(X,\mathbb{Q}/\mathbb{Z}(1)) \to \mathrm{CH}^1(X)_{\mathrm{tors}} \to 0,$$

but $k^* \otimes \mathbb{Q}/\mathbb{Z} \neq 0$ when k is a number field or a p-adic field. When $k = \mathbb{R}$, we do have $k^* \otimes \mathbb{Q}/\mathbb{Z} = 0$, hence $\gamma_1 \circ \varphi_1 = 0$. By the following theorem this generalizes to higher codimensions.

THEOREM 3.1. *Let X be a nonsingular projective geometrically irreducible variety over $\mathbb{R}$ of dimension d. For any $i \geq 0$ the mapping*

$$\gamma_i \circ \varphi_i\colon H^{i-1}_{\mathrm{Zar}}(X,\mathcal{K}_i)\otimes\mathbb{Q}/\mathbb{Z} \to H^{2i-1}_{\text{ét}}(X,\mathbb{Q}/\mathbb{Z}(i))$$

is zero. Hence the cohomological Abel-Jacobi mapping

$$\lambda_i\colon \mathrm{CH}^i(X)_{\mathrm{tors}} \to H^{2i-1}_{\text{ét}}(X,\mathbb{Q}/\mathbb{Z}(i))$$

is defined. The image of λ_i coincides with the image of the map

$$\gamma_i\colon H^{i-1}_{\mathrm{Zar}}(X,\mathcal{H}^i(\mathbb{Q}/\mathbb{Z}(i))) \to H^{2i-1}_{\text{ét}}(X,\mathbb{Q}/\mathbb{Z}(i)).$$

PROOF. In view of diagram (14) it is sufficient to prove that $\gamma_i \circ \varphi_i = 0$. Consider the commutative diagram associated to the base change $\pi\colon X_{\mathbb{C}} \to X$:

$$\begin{array}{ccc} H^{i-1}_{\mathrm{Zar}}(X,\mathcal{K}_i)\otimes\mathbb{Q}/\mathbb{Z} & \xrightarrow{\gamma_i\circ\varphi_i} & H^{2i-1}_{\text{ét}}(X,\mathbb{Q}/\mathbb{Z}(i)) \\ \downarrow & & \downarrow \pi^* \\ H^{i-1}_{\mathrm{Zar}}(X_{\mathbb{C}},\mathcal{K}_i)\otimes\mathbb{Q}/\mathbb{Z} & \xrightarrow{\gamma_i^{\mathbb{C}}\circ\varphi_i^{\mathbb{C}}} & H^{2i-1}_{\text{ét}}(X_{\mathbb{C}},\mathbb{Q}/\mathbb{Z}(i)) \end{array}$$

Since $\gamma_i^{\mathbb{C}} \circ \varphi_i^{\mathbb{C}} = 0$ by [**CT93**, proof Theorem 4.3] and $H^{i-1}_{\mathrm{Zar}}(X,\mathcal{K}_i)\otimes\mathbb{Q}/\mathbb{Z}$ is divisible, it is sufficient to show that the kernel of π^* does not contain any nonzero divisible subgroup. Since π is étale of degree 2, we have that $\pi_*\circ\pi^*$ is multiplication by 2, hence the kernel of π^* is purely 2-torsion. □

Note that if γ_i is injective, then λ_i maps $\mathrm{CH}^i(X)_{\mathrm{tors}}$ isomorphically onto the image of γ_i. For X over any field k, the mappings γ_1 and γ_2 are injective for trivial reasons: there is no differential in the Bloch-Ogus spectral sequence that could kill $H^{i-1}_{\mathrm{Zar}}(X,\mathcal{H}^i(\mathbb{Q}/\mathbb{Z}(i)))$ when $i = 1, 2$. For $i = d = \dim(X)$ the situation is different in general, since we have for every j smaller or equal to the cohomological dimension of k and satisfying $0 < j \le d-2$ a potentially nontrivial differential

$$d_{j+1}^{d-2-j,d+j}\colon E_{j+1}^{d-2-j,d+j} \to E_{j+1}^{d-1,d}.$$

However, we proved in Section 2 that these differentials vanish when $k = \mathbb{R}$. We also determined the cokernel of γ_d, so we obtain the following result, which could be considered as a cohomological alternative that works over $\mathbb{R}$ for Roitman's theorem on torsion zero-cycles over algebraically closed fields.

THEOREM 3.2. *Let X be a nonsingular, projective, geometrically irreducible variety over $\mathbb{R}$ of dimension d. We have a short exact sequence*

$$0 \to \mathrm{CH}_0(X)_{\mathrm{tors}} \xrightarrow{\lambda_d} H^{2d-1}_{\text{ét}}(X,\mathbb{Q}/\mathbb{Z}(d)) \xrightarrow{\beta'} \bigoplus_{i>0} H^{d-2i}(X(\mathbb{R}),\mathbb{Z}/2) \to 0,$$

where β' is the mapping β followed by the projection.

PROOF. By Theorem 3.1 this is an immediate consequence of Corollary 2.2.i. □

3.2. Kernel and Image of the Abel-Jacobi Map. Consider the Abel-Jacobi map α restricted to the torsion subgroups:

$$\alpha_{\mathrm{tors}}\colon A_0(X)_{\mathrm{tors}} \to \mathrm{Alb}(X)(\mathbb{R})_{\mathrm{tors}}$$

and the restriction $\alpha_{\mathrm{tors,div}}$ to the maximal divisible torsion subgroups. We define:

$T(X)_{\mathrm{tors}} := \mathrm{Ker}(\alpha_{\mathrm{tors}})$,
$T(X)^0 := \mathrm{Ker}(\alpha_{\mathrm{tors,div}})$,
$T(X)^{\mathrm{top}} := T(X)_{\mathrm{tors}}/T(X)^0$,
$A_0(X)^{\mathrm{top}} := A_0(X)_{\mathrm{tors}}/A_0(X)_{\mathrm{tors,div}}$ $(= A_0(X)/A_0(X)_{\mathrm{div}})$,
$\mathrm{Im}(\alpha)^{\mathrm{top}} := \mathrm{Im}(\alpha_{\mathrm{tors}})/\mathrm{Im}(\alpha_{\mathrm{tors,div}})$ $(= \mathrm{Im}(\alpha)/\mathrm{Im}(\alpha)_{\mathrm{div}})$.

The equalities between parentheses follow from the fact that $A_0(X)/A_0(X)_{\mathrm{tors}}$ is uniquely divisible (see Section 1.3). By definition we have short exact sequences

$$0 \to T(X)^{\mathrm{top}} \to A_0(X)^{\mathrm{top}} \to \mathrm{Im}(\alpha)^{\mathrm{top}} \to 0 \tag{15}$$

and

$$0 \to T(X)^0 \to T(X)_{\mathrm{tors}} \to T(X)^{\mathrm{top}} \to 0. \tag{16}$$

These are short exact sequences of finite dimensional $\mathbb{Z}/2$-vector spaces, since $T(X)_{\mathrm{tors}}$ is a finite dimensional $\mathbb{Z}/2$-vector space (by [**CTS96**, Theorem 1.6.a] and a trace argument) and

$$A_0(X)^{\mathrm{top}} \simeq \begin{cases} 0 & \text{if } s = 0, \\ (\mathbb{Z}/2)^{s-1} & \text{otherwise,} \end{cases}$$

where s is the number of connected components of X (see Section 1.3). We will now give cohomological descriptions of the above groups.

Over $\mathbb{C}$ we have a commutative diagram

$$\begin{array}{ccc} & A_0(X_{\mathbb{C}})_{\mathrm{tors}} & \\ {\scriptstyle \alpha^{\mathbb{C}}_{\mathrm{tors}}} \swarrow & & \searrow {\scriptstyle \lambda^{\mathbb{C}}_d} \\ \mathrm{Alb}(X)(\mathbb{C})_{\mathrm{tors}} & \xrightarrow{\sim} & H^{2d-1}_{\text{ét}}(X_{\mathbb{C}}, \mathbb{Q}/\mathbb{Z}(d)) \end{array}$$

in which all arrows are isomorphisms by Roitman's theorem (cf. [**CT93**, Theorem 4.2]). Over $\mathbb{R}$ we have the diagram

$$\begin{array}{ccc} A_0(X)_{\mathrm{tors}} & \overset{\lambda_d}{\hookrightarrow} & H^{2d-1}_{\text{ét}}(X, \mathbb{Q}/\mathbb{Z}(d)) \\ \Big\downarrow {\scriptstyle \alpha_{\mathrm{tors}}} & & \Big\downarrow {\scriptstyle \pi^*} \\ \mathrm{Alb}(X)(\mathbb{R})_{\mathrm{tors}} & \xrightarrow{\sim} & H^{2d-1}_{\text{ét}}(X_{\mathbb{C}}, \mathbb{Q}/\mathbb{Z}(d))^G \end{array} \tag{17}$$

in which the lower horizontal arrow is an isomorphism.

THEOREM 3.3. *Let X be a nonsingular projective geometrically irreducible variety over $\mathbb{R}$ of dimension d.*

(i) *The isomorphism*

$$\mathrm{Alb}(X)(\mathbb{R})_{\mathrm{tors}} \simeq H^{2d-1}_{\text{ét}}(X_{\mathbb{C}}, \mathbb{Q}/\mathbb{Z}(d))^G$$

sends the image of α_{tors} isomorphically onto the image of

$$\pi^* \colon H^{2d-1}_{\text{ét}}(X, \mathbb{Q}/\mathbb{Z}(d)) \to H^{2d-1}_{\text{ét}}(X_{\mathbb{C}}, \mathbb{Q}/\mathbb{Z}(d))^G.$$

(ii) *The injection*

$$\lambda_d \colon A_0(X)_{\mathrm{tors}} \hookrightarrow H^{2d-1}_{\text{ét}}(X, \mathbb{Q}/\mathbb{Z}(d)$$

sends the group $T(X)^0$ isomorphically onto the kernel of π^* restricted to the maximal divisible subgroup of $H^{2d-1}_{\text{ét}}(X, \mathbb{Q}/\mathbb{Z}(d))$.

(iii) *If $X(\mathbb{R}) = \varnothing$, then $T(X)_{\mathrm{tors}} \simeq T(X)^0$. Otherwise, we have an isomorphism*

$$T(X)_{\mathrm{tors}} \simeq (\mathbb{Z}/2)^{c+s-1-t},$$

where s is the number of connected components of $X(\mathbb{R})$, t is the $\mathbb{Z}/2$-dimension of $\mathrm{Im}(\alpha)^{\mathrm{top}}$, and c is the $\mathbb{Z}/2$-dimension of $T(X)^0$.

PROOF. (i) Immediate from diagram (17), Theorem 3.1 and Corollary 2.2.ii.

(ii) Immediate from diagram (17) and the fact that λ_d induces an isomorphism

$$A_0(X)_{\mathrm{tors,div}} \simeq H^{2d-1}_{\text{ét}}(X, \mathbb{Q}/\mathbb{Z}(d))_{\mathrm{div}}$$

by Theorem 3.2.

(iii) Immediate from (i) and (ii) and the short exact sequences (15) and (16). □

COROLLARY 3.4. *Let X_1, X_2 be nonsingular projective geometrically irreducible varieties over $\mathbb{R}$, such that $X_1(\mathbb{C})$ and $X_2(\mathbb{C})$ are equivariantly homeomorphic. The groups $A_0(X_i)_{\text{tors}}$, $\text{Im}(\alpha_{\text{tors}})$, $T(X_i)_{\text{tors}}$, $T(X_i)^0$, $T(X_i)^{\text{top}}$, $A_0(X)^{\text{top}}$, and $\text{Im}(\alpha_{\text{tors}})^{\text{top}}$ are isomorphic for $i = 1, 2$.*

PROOF. Immediate from Theorem 3.3 and the isomorphism (1). □

REMARK 3.5. The two easy examples of varieties with $T(X)_{\text{tors}} \neq 0$ given in the introduction have in fact $T(X)_{\text{tors}} = T(X)^{\text{top}}$. Indeed, the property $T(X)^{\text{top}} \neq 0$ is so common, that it can hardly be considered to be a 'bad' case. On the other hand, we will see in Example 5.2 that the property $T(X)^0 \neq 0$ does not occur for varieties like products of curves or Abelian varieties (nor does it occur, trivially, for complete intersections). It corresponds to certain nontrivial differentials in the Hochschild-Serre spectral sequence, which could be considered to be 'bad'.

3.3. A Cohomological Filtration on Torsion Cycle Classes. The filtration on étale cohomology associated to the Hochschild-Serre spectral sequence

$$E_2^{p,q} = H^p(G, H^q_{\text{ét}}(X_{\mathbb{C}}, \mathbb{Q}/\mathbb{Z}(i))) \implies H^{p+q}_{\text{ét}}(X, \mathbb{Q}/\mathbb{Z}(i))$$

induces for every i a descending filtration

$$0 = F^{2i} \subset \cdots \subset F^1 \subset F^0 = H^{2i-1}_{\text{ét}}(X, \mathbb{Q}/\mathbb{Z}(i))$$

with the pth graded piece given by $E_\infty^{p,2i-1-p}$. Pulling back to $\text{CH}^i(X)_{\text{tors}}$ via the cohomological Abel-Jacobi map

$$\lambda_i \colon \text{CH}^i(X)_{\text{tors}} \to H^{2i-1}_{\text{ét}}(X, \mathbb{Q}/\mathbb{Z}(i))$$

gives a descending filtration on $\text{CH}^i(X)_{\text{tors}}$, which is nondegenerate (i.e., $F^N = 0$ for $N \gg 0$) if and only if λ_i is injective. Note that for $i = \dim(X)$ we get

$$F^1 \, \text{CH}_0(X)_{\text{tors}} = T(X)_{\text{tors}}.$$

We will see in Section 4.2 that the induced filtration on the maximal divisible subgroup $\text{CH}_0(X)_{\text{tors,div}}$ can be determined from the Hochschild-Serre spectral sequence. In Example 5.2.1 we will see that for Abelian varieties this filtration coincides with the filtration coming from the Pontryagin product.

REMARK 3.6. A priori the filtration on $\text{CH}_0(X)_{\text{tors}}$ can go as deep as F^{2d-1}, but I have no examples of an X of dimension d with $F^{d+1} \, \text{CH}_0(X)_{\text{tors}} \neq 0$. For the induced filtration on $A_0(X)^{\text{top}} = A_0(X)_{\text{tors}}/A_0(X)_{\text{tors,div}}$ it can be shown, using Theorem 2.8 and Poincaré duality (compare the discussion in [**MvH98**, §2]), that $F^{d+1} A_0(X)^{\text{top}} = 0$ for any nonsingular projective geometrically irreducible X of dimension d.

4. Methods of Calculation

In Section 3.2 we proved that the image of

$$\alpha_{\text{tors}} \colon A_0(X)_{\text{tors}} \to \text{Alb}(X)(\mathbb{R})_{\text{tors}}$$

coincides with the image of the pull-back

$$H^{2d-1}(X, \mathbb{Q}/\mathbb{Z}(d)) \to H^{2d-1}(X_{\mathbb{C}}, \mathbb{Q}/\mathbb{Z}(d))^G,$$

which can be computed explicitly from the Hochschild-Serre spectral sequence. In this section we will see in Proposition 4.1 that the kernel $T(X)^0$ of the restriction of

α_{tors} to the maximal divisible subgroup of $\mathrm{CH}_0(X)_{\text{tors}}$ admits a similar description. As a corollary we get that if the Hochschild-Serre spectral sequence with $\mathbb{Q}/\mathbb{Z}$-coefficients is trivial, then α_{tors} is surjective and $T(X)^0$ is trivial.

By Poincaré duality (Lemma 4.3) the same conclusion holds if the Hochschild-Serre spectral sequence with coefficients in $\mathbb{Z}$ is trivial (in other words, when X is a *$\mathbb{Z}$-GM-variety*). This will be used in Example 5.2. The fact that α_{tors} (or, in fact, α) is surjective for a $\mathbb{Z}$-GM-variety was originally proved by V. A. Krasnov in [**Kr84**] without using a cohomological Abel-Jacobi map. For the calculations in Example 5.1 it will be useful that Poincaré duality allows us to transform calculations involving high degree cohomology with coefficients in $\mathbb{Q}/\mathbb{Z}$ into calculations involving low degree cohomology with coefficients in $\mathbb{Z}$ (see Corollary 4.6).

In Section 4.2 we use the methods of Proposition 4.1 for calculating the cohomological filtration on $T(X)^0$ in terms of the Hochschild-Serre spectral sequence.

4.1. Calculating the Kernel of the Abel-Jacobi Mapping.

PROPOSITION 4.1. *Let X be a nonsingular projective geometrically irreducible variety over $\mathbb{R}$ of dimension d. Let $T(X)^0$ be the kernel of the Abel-Jacobi mapping α restricted to the maximal divisible subgroup of $A_0(X)_{\text{tors}}$. Then $T(X)^0$ is isomorphic to the cokernel of the base change map*

$$\pi^*\colon H^{2d-1}_{\text{ét}}(X,\mathbb{Q}/\mathbb{Z}(d+1)) \to H^{2d-1}_{\text{ét}}(X_{\mathbb{C}},\mathbb{Q}/\mathbb{Z}(d+1))^G.$$

PROOF. Let $F^\bullet$ be the descending filtration on étale cohomology associated to the Hochschild-Serre spectral sequence. Since $F^1H^{2d-1}_{\text{ét}}(X,\mathbb{Q}/\mathbb{Z}(d))$ is the kernel of the base change map π^* for coefficients in $\mathbb{Q}/\mathbb{Z}(d)$, the intersection $F^1H^{2d-1}_{\text{div}}$ with the maximal divisible subgroup of $H^{2d-1}_{\text{ét}}(X,\mathbb{Q}/\mathbb{Z}(d))$ is isomorphic to $T(X)^0$ by Theorem 3.3.

I claim that $F^1H^{2d-1}_{\text{div}}$ is precisely the kernel of the map

$$F^1H^{2d-1}_{\text{ét}}(X,\mathbb{Q}/\mathbb{Z}(d)) \xrightarrow{\cup\eta} F^2H^{2d}_{\text{ét}}(X,\mathbb{Q}/\mathbb{Z}(d+1)) \tag{18}$$

induced by cup-product with the nontrivial element $\eta \in H^1(G,\mathbb{Z}(1))$. Since on the E_2-level of the Hochschild-Serre spectral sequence this gives an isomorphism

$$H^p(G,H^q_{\text{ét}}(X_{\mathbb{C}},\mathbb{Q}/\mathbb{Z}(d))) \simeq H^{p+1}(G,H^q_{\text{ét}}(X_{\mathbb{C}},\mathbb{Q}/\mathbb{Z}(d+1)))$$

for any $q \geq 0$ and any $p > 0$, we see that the kernel of (18) can be identified with the elements that are killed by differentials from $H^0(G,H^{2d}_{\text{ét}}(X_{\mathbb{C}},\mathbb{Q}/\mathbb{Z}(d+1)))$. More precisely, we have an exact sequence

$$\begin{aligned}H^{2d-1}_{\text{ét}}(X,\mathbb{Q}/\mathbb{Z}(d+1)) &\xrightarrow{\pi^*} H^{2d-1}(X_{\mathbb{C}},\mathbb{Q}/\mathbb{Z}(d+1))^G\\ &\to F^1H^{2d-1}_{\text{ét}}(X,\mathbb{Q}/\mathbb{Z}(d)) \xrightarrow{\cup\eta} F^2H^{2d}_{\text{ét}}(X,\mathbb{Q}/\mathbb{Z}(d+1)),\end{aligned}$$

which gives an isomorphism between the kernel of (18) and the cokernel of π^*.

In order to prove that the kernel of (18) coincides with $F^1H^{2d-1}_{\text{div}}$, hence with $T(X)^0$, it is sufficient to prove that the kernel of the homomorphism

$$H^{2d-1}_{\text{ét}}(X,\mathbb{Q}/\mathbb{Z}(d)) \xrightarrow{\cup\eta} H^{2d}_{\text{ét}}(X,\mathbb{Q}/\mathbb{Z}(d+1)) \tag{19}$$

is the maximal divisible subgroup. Since the image of (19) is contained in $F^1H^{2d}_{\text{ét}}(X,\mathbb{Q}/\mathbb{Z}(d+1))$, which is purely 2-torsion (compare the end of the proof of Theorem 3.1), it is sufficient to prove that the kernel of the homomorphism (19) is divisible.

This we will deduce from the long exact sequence

$$(20)\quad \cdots \to H^{2d-1}(X_{\mathbb{C}}, \mathbb{Q}/\mathbb{Z}) \to H^{2d-1}_{\text{ét}}(X, \mathbb{Q}/\mathbb{Z}(d)) \xrightarrow{\delta} H^{2d}_{\text{ét}}(X, \mathbb{Q}/\mathbb{Z}(d+1)) \to \cdots$$

obtained from the short exact sequence $0 \to \mathbb{Q}/\mathbb{Z}(d+1) \to \pi_*\mathbb{Q}/\mathbb{Z}_{X_{\mathbb{C}}} \to \mathbb{Q}/\mathbb{Z}(d) \to 0$ of étale sheaves on X (compare [**Sch94**, (7.8.1)]). I claim that the boundary map δ coincides with the homomorphism (19) (compare [**Sch94**, Lemma 7.10.1]). Assuming this, the proposition follows, since the group $H^{2d-1}(X_{\mathbb{C}}, \mathbb{Q}/\mathbb{Z})$ is divisible.

In order to prove the last claim, observe that the above short exact sequence is the pull-back of a short exact sequence of G-modules. Moreover, the boundary map δ can be considered as the Yoneda product with an element of $[\delta] \in \operatorname{Ext}^1_G(\mathbb{Q}/\mathbb{Z}(d), \mathbb{Q}/\mathbb{Z}(d+1)) \simeq \mathbb{Z}/2$ (this is Ext in the category of G-modules). Since tensor product with $\mathbb{Q}/\mathbb{Z}(d)$ induces a surjection $H^1(G, \mathbb{Z}(1)) = \operatorname{Ext}^1_G(\mathbb{Z}, \mathbb{Z}(1)) \twoheadrightarrow \operatorname{Ext}^1_G(\mathbb{Q}/\mathbb{Z}(d), \mathbb{Q}/\mathbb{Z}(d+1))$, we get that δ is either cup product with η or trivial. We rule out this last possibility by considering the long exact sequence (20) for $\operatorname{Spec}\mathbb{R}$ instead of X (without changing d). Since $H^i(\operatorname{Spec}\mathbb{C}, \mathbb{Q}/\mathbb{Z}) = 0$ for all $i > 0$ and $H^{d+1}(\operatorname{Spec}\mathbb{R}, \mathbb{Q}/\mathbb{Z}(d)) = \mathbb{Z}/2 = H^{d+2}(\operatorname{Spec}\mathbb{R}, \mathbb{Q}/\mathbb{Z}(d+1))$, we see that δ is nontrivial. □

COROLLARY 4.2. *Let X be a nonsingular projective geometrically irreducible variety over $\mathbb{R}$. If the Hochschild-Serre spectral sequence with coefficients in $\mathbb{Q}/\mathbb{Z}$ is trivial, then the mapping*

$$\alpha_{\text{tors}} \colon A_0(X)_{\text{tors}} \to \operatorname{Alb}(X)(\mathbb{R})_{\text{tors}}$$

is surjective and the mapping

$$\alpha_{\text{tors,div}} \colon A_0(X)_{\text{tors,div}} \to \operatorname{Alb}(X)(\mathbb{R})_{\text{tors,div}}$$

is an isomorphism.

PROOF. If the Hochschild-Serre spectral sequence with coefficients in $\mathbb{Q}/\mathbb{Z}$ is trivial, then the Hochschild-Serre spectral sequence with coefficients in $\mathbb{Q}/\mathbb{Z}(1)$ is trivial as well, as we see using cup product with the nontrivial element $\eta \in H^1(G, \mathbb{Z}(1))$ and the periodicity of the cohomology of G. The result now follows immediately from Proposition 4.1, Theorem 3.3.i and the surjectivity of $\alpha_{\text{tors,div}}$. □

LEMMA 4.3. *Let M be a compact oriented manifold of dimension m with an action of $G = \mathbb{Z}/2$. Let $r \geq 2$, let $p \geq 0$, let $j \in \mathbb{Z}$, let $p' \geq r$ with $p' \not\equiv p \pmod 2$ and let $j' \in \mathbb{Z}$ with $j' \equiv j \pmod 2$ if the G-action preserves the orientation and $j' \not\equiv j \pmod 2$ if the G-action reverses the orientation. The image of the differential*

$$d(\mathbb{Q}/\mathbb{Z}(j)) \colon E_r^{p,q}(\mathbb{Q}/\mathbb{Z}(j)) \to E_r^{p+r,q-r+1}(\mathbb{Q}/\mathbb{Z}(j))$$

in the Hochschild-Serre spectral sequence of M with coefficients in $\mathbb{Q}/\mathbb{Z}(j)$ is isomorphic to the image of the differential

$$d(\mathbb{Z}(j')) \colon E_r^{p'-r,m-q+r-1}(\mathbb{Z}(j')) \to E_r^{p',m-q}(\mathbb{Z}(j'))$$

in the Hochschild-Serre spectral sequence with coefficients in $\mathbb{Z}(j')$.

PROOF. This is a formal consequence of Poincaré duality for M and G: the cup product pairing induces for any $k \in \mathbb{Z}$ a natural isomorphism of G-modules

$$H^k(X(\mathbb{C}), \mathbb{Q}/\mathbb{Z}(j)) \simeq \operatorname{Hom}(H^{m-k}(X(\mathbb{C}), \mathbb{Z}(j')), \mathbb{Q}/\mathbb{Z}),$$

and for any G-module A, any $i > 0$, and any odd $N > i$, the cup product pairing

$$H^i(G, \mathrm{Hom}(A, \mathbb{Q}/\mathbb{Z})) \otimes H^{N-i}(G, A) \to H^N(G, \mathbb{Q}/\mathbb{Z}) \simeq \mathbb{Z}/2 \hookrightarrow \mathbb{Q}/\mathbb{Z}$$

induces an isomorphism

$$H^i(G, \mathrm{Hom}(A, \mathbb{Q}/\mathbb{Z})) \simeq \mathrm{Hom}(H^{N-i}(G, \mathrm{Hom}(A, \mathbb{Q}/\mathbb{Z})), \mathbb{Q}/\mathbb{Z}).$$

Combining these isomorphism, we get for i, k, N as above an isomorphism

$$H^i(G, H^k(M, \mathbb{Q}/\mathbb{Z}(j))) \simeq \mathrm{Hom}(H^{N-i}(G, H^{m-k}(M, \mathbb{Z}(j'))), \mathbb{Q}/\mathbb{Z}). \tag{21}$$

The first two Poincaré duality morphisms given above can be obtained from natural mappings of complexes. This enables us to construct a map Φ from the Hochschild-Serre spectral sequence with coefficients in $\mathbb{Q}/\mathbb{Z}(j)$ into the (shifted) Pontryagin dual

$$\begin{aligned} \widehat{E}_2^{i,k}[N] = \mathrm{Hom}(H^{N-i}(G, H^{m-k}(M, \mathbb{Z}(j'))), \mathbb{Q}/\mathbb{Z}) \\ \Longrightarrow \mathrm{Hom}(H^{N+m-i-k}(M; G, \mathbb{Z}(j')), \mathbb{Q}/\mathbb{Z}) \end{aligned} \tag{22}$$

of the Hochschild-Serre spectral sequence with coefficients in $\mathbb{Z}(j')$, with the property that Φ induces the isomorphism (21) on the $E_2^{i,k}$-level (still for $i > 0$ and $N > i$ odd). Unfortunately, I do not know any references to the literature for this type of result. Let me just say here that on the one hand this mapping of spectral sequences can be constructed explicitly by writing down a map between well-chosen double complexes. On the other hand, it can be obtained from the following transformation of derived functors (with Γ^G as in [**Gr57**, Ch. V]):

$$F_1 := R\Gamma^G(R\,\mathrm{Hom}(-, \mathbb{Q}/\mathbb{Z})) \to R\,\mathrm{Hom}(R\Gamma^G(-), \mathbb{Q}/\mathbb{Z}[-N]) =: F_2,$$

since the composite derived functor $F_1 \circ \Gamma_M(-)$ applied to the sheaf $\mathbb{Z}(j')[m]$ induces the Hochschild-Serre spectral sequence with coefficients in $\mathbb{Q}/\mathbb{Z}(j)$, and $F_2 \circ \Gamma_M(-)$ induces the spectral sequence (22), when applied to $\mathbb{Z}(j')[m]$.

Taking $i = p$, $N = p' + p$ and $k = q$, we get for every $r \geq 2$ a commutative diagram

$$\begin{array}{ccc} E_r^{p,q}(\mathbb{Q}/\mathbb{Z}(j)) & \xrightarrow{d(\mathbb{Q}/\mathbb{Z}(j))} & E_r^{p+r,q-r+1}(\mathbb{Q}/\mathbb{Z}(j)) \\ \downarrow \Phi & & \downarrow \Phi \\ \widehat{E}_r^{p,q}[N] & \xrightarrow{\hat{d}} & \widehat{E}_r^{p+r,q-r+1}[N] \\ \| & & \| \\ \mathrm{Hom}(E_r^{p',m-q}(\mathbb{Z}(j')), \mathbb{Q}/\mathbb{Z}) & \xrightarrow{d(\mathbb{Z}(j'))^\vee} & \mathrm{Hom}(E_r^{p'-r,m-q+r+1}(\mathbb{Z}(j')), \mathbb{Q}/\mathbb{Z}) \end{array}$$

where $d(\mathbb{Z}(j'))^\vee$ denotes the Pontryagin dual of $d(\mathbb{Z}(j'))$. For $p > 0$ and $p' > r$ the vertical arrows are isomorphisms, and all groups in the diagram are finite, so we have proved the statement. The remaining cases $p = 0$ and/or $p' = r$ follow from the periodicity of the cohomology of G. □

For M as above, we deduce that all differentials in the Hochschild-Serre spectral sequence with coefficients in $\mathbb{Q}/\mathbb{Z}$ are trivial if and only if all differentials in the Hochschild-Serre spectral sequence with coefficients in $\mathbb{Z}$ are trivial.

DEFINITION 4.4 ([**Kr83**]). Let X be a nonsingular projective variety over $\mathbb{R}$. We say that X is a *$\mathbb{Z}$-GM-variety* if all differentials are zero in the Hochschild-Serre spectral sequence converging to equivariant cohomology of $X(\mathbb{C})$ with coefficients in $\mathbb{Z}$.

In fact, Krasnov uses GM$\mathbb{Z}$ instead of $\mathbb{Z}$-GM; the notation $\mathbb{Z}$-GM comes from Silhol, and should be read as '$\mathbb{Z}$-Galois-Maximal'.

COROLLARY 4.5. *Let X be a nonsingular projective geometrically irreducible $\mathbb{Z}$-GM-variety over $\mathbb{R}$. Then the mapping α_{tors} is surjective and the mapping $\alpha_{\text{tors,div}}$ is an isomorphism.*

PROOF. Immediate from Corollary 4.2, Lemma 4.3. □

The part of the above corollary concerning the surjectivity of α_{tors} is equivalent to Theorem 3.2 in [**Kr84**].

COROLLARY 4.6. *Let X be a nonsingular projective geometrically irreducible variety over $\mathbb{R}$.*

(i) *If $X(\mathbb{R}) \neq \varnothing$, then the group $T(X)^0$ is isomorphic to the kernel of the composite mapping*

$$H^1(G, H^1(X(\mathbb{C}), \mathbb{Z}(1))) \to H^2(X(\mathbb{C}); G, \mathbb{Z}(1)) \to H^1(X(\mathbb{R}), \mathbb{Z}/2),$$

where the first map is induced by the Hochschild-Serre spectral sequence, and the second map is induced by β.

(ii) *The group $T(X)^0$ is isomorphic to the kernel of the mapping*

$$\operatorname{Pic}^0(X)/\operatorname{Pic}^0(X)_{\text{div}} \to H^1(X(\mathbb{R}), \mathbb{Z}/2)$$

induced by the cycle map. Here $\operatorname{Pic}^0(X)$ is the kernel of the composite map

$$\operatorname{Pic}(X) \to \operatorname{Pic}(X_{\mathbb{C}}) \to H^2(X(\mathbb{C}), \mathbb{Z}(1)).$$

PROOF. (i) Since $X(\mathbb{R}) \neq \varnothing$, the differential $E_2^{2d+1,1}(\mathbb{Z}(1)) \to E_2^{2d+3,0}(\mathbb{Z}(1)) = H^{2d+3}(G, H^0(X(\mathbb{C}), \mathbb{Z}(1)))$ in the Hochschild-Serre spectral sequence is zero by [**Kr83**, Lemma 2.2]. It then follows from Lemma 4.3 (with $p = 0$, $q = 2d-1$, $j = d+1$, $p' = 2d+1$, $j' = 1$, and $2 \leq r \leq 2d$) that the cokernel of the mapping π^* of Proposition 4.1 is isomorphic to the kernel of the quotient map $E_2^{2d+1,1}(\mathbb{Z}(1)) \to E_\infty^{2d+1,1}(\mathbb{Z}(1))$. Since β maps $E_\infty^{2d+1,1}(\mathbb{Z}(1))$ injectively into $H^1(X(\mathbb{R}), \mathbb{Z}/2)$, the statement follows from the periodicity isomorphism $H^1(G, H^1(X(\mathbb{C}), \mathbb{Z}(1))) = E_2^{1,1}(\mathbb{Z}(1)) \simeq E_2^{2d+1,1}(\mathbb{Z}(1))$.

(ii) This follows from (i) if $X(\mathbb{R}) \neq \varnothing$, since then the equivariant cycle map $\operatorname{Pic}(X) \to H^2(X(\mathbb{C}); G, \mathbb{Z}(1))$ (see [**Kr94**, §4]) induces an isomorphism

$$\operatorname{Pic}^0(X)/\operatorname{Pic}^0(X)_{\text{div}} \simeq H^1(G, H^1(X(\mathbb{C}), \mathbb{Z}(1)))$$

which is compatible with the mappings of statements (i) and (ii). For $X(\mathbb{R}) = \varnothing$, we have that the cycle map induces an isomorphism $\operatorname{Pic}^0(X)/\operatorname{Pic}^0(X)_{\text{div}} \simeq E_3^{1,1}(\mathbb{Z}(1))$ (see [**vH98**, Proposition 3.3]) and we adapt the argument of (i). □

REMARKS 4.7. (i) There is an analogue of Corollary 4.6 for the cokernel of α_{tors}, but it is more cumbersome to state, and we will not need it in the examples, since there the image of α is quite easy to determine.

(ii) I have no method to determine the 'topological quotient' $T(X)^{\text{top}}$ of the kernel of α_{tors} from the Hochschild-Serre spectral sequence without using a priori information about the topology of $X(\mathbb{R})$; it seems unlikely that such a method should exist in general.

4.2. Calculating the Cohomological Filtration on the Divisible Subgroup. Let $F^\bullet$ be the filtration defined on $\mathrm{CH}_0(X)_{\text{tors}}$ in Section 3.3. After restricting the filtration to the maximal divisible subgroup $\mathrm{CH}_0(X)_{\text{tors,div}}$ we can determine the graded pieces from the Hochschild-Serre spectral sequence.

PROPOSITION 4.8. *Let X be a nonsingular projective geometrically irreducible variety over $\mathbb{R}$ of dimension d. We have for any $p \geq 0$ that*

$$\mathrm{Gr}^p F^\bullet \mathrm{CH}_0(X)_{\text{tors,div}} \simeq \mathrm{Ker}\{E_\infty^{p,2d-1-p}(\mathbb{Q}/\mathbb{Z}(d)) \xrightarrow{\cup\eta} E_\infty^{p+1,2d-1-p}(\mathbb{Q}/\mathbb{Z}(d+1))\},$$

where $E_\infty^{p,q}(\mathbb{Q}/\mathbb{Z}(j))$ denotes the $E_\infty^{p,q}$-term of the Hochschild-Serre spectral sequence with coefficients in $\mathbb{Q}/\mathbb{Z}(j)$ and $\eta \in H^1(G, \mathbb{Z}(1))$ is the nontrivial element.

PROOF. This follows immediately from the fact that the cohomological Abel-Jacobi mapping induces an isomorphism between $\mathrm{CH}_0(X)_{\text{div}}$ and the kernel of the mapping $H_{\text{ét}}^{2d-1}(X, \mathbb{Q}/\mathbb{Z}(d)) \xrightarrow{\cup\eta} H_{\text{ét}}^{2d}(X, \mathbb{Q}/\mathbb{Z}(d+1))$, as we saw in the proof of Proposition 4.1. □

REMARK 4.9. I do not know whether there are varieties X with nonzero $F^p\,\mathrm{CH}_0(X)_{\text{tors,div}}$ for $p > \dim(X)$. In other words, with nonzero differentials

$$E_r^{0,2d-1}(\mathbb{Q}/\mathbb{Z}(d+1)) \to E_r^{r,2d-r}(\mathbb{Q}/\mathbb{Z}(d+1))$$

for $r > \dim(X) + 1$ in the Hochschild-Serre spectral sequence (compare [**vH98**, Remark 3.5.ii]).

5. Examples

Stock examples in the study of zero-cycles are products of curves, Abelian varieties and conic bundles. Here we will consider them over the real numbers. We will compute the groups defined in Section 3.2 and in some cases the filtration defined in 3.3 as well. Since the Abel-Jacobi map for torsion zero-cycles is completely determined by the equivariant topology of the complex points, we can in fact replace the geometric conditions by topological conditions. Thus Example 5.1 is an immediate generalization of the example of a conic bundle over a curve of which the real part does not map surjectively onto the real part of the curve (it was already shown in [**Si89**, §V.4] that a smooth conic bundle with that property is not a $\mathbb{Z}$-GM-variety). Similarly, the results for products of elliptic curves in Example 5.2 immediately generalize to Abelian varieties in Example 5.2.1.

5.1. Fibrations Over Curves. Let C be a nonsingular projective geometrically irreducible curve over $\mathbb{R}$ of genus g such that $C(\mathbb{R})$ has $c > 0$ connected components. Recall that $c \leq g+1$ (cf. Section 1.2). Let X be a nonsingular projective geometrically irreducible variety over $\mathbb{R}$ such that $X(\mathbb{R})$ has $s > 0$ connected components and let $\varphi\colon X \to C$ be a dominant morphism satisfying the following two conditions:

(a) φ induces an isomorphism $\varphi^*\colon \mathrm{Pic}^0(C) \xrightarrow{\sim} \mathrm{Pic}^0(X)$.
(b) The map $X(\mathbb{R}) \to C(\mathbb{R})$ induced by φ is not surjective.

Let $a > 0$ be the number of connected components V of $C(\mathbb{R})$ such that $\varphi^{-1}(x)$ has real points for some $x \in V$. We denote by h the $\mathbb{Z}/2$-dimension of the kernel of the pull-back map $\varphi^*\colon H^1(C(\mathbb{R}), \mathbb{Z}/2) \to H^1(X(\mathbb{R}), \mathbb{Z}/2)$. Hypothesis (2) implies that $h \geq 1$. Note that it is very easy to construct examples with $a < s$, $a < c$, and/or $h > 1$.

It follows from hypothesis (1) that φ induces an isomorphism $\mathrm{Alb}(X) \xrightarrow{\sim} \mathrm{Jac}(C)$. This isomorphism is compatible with the push-forward of cycles and the Abel-Jacobi map. We see from Sections 1.2 and 1.3 that the image of the composite map

$$A_0(X) \to A_0(C) \to \mathrm{Jac}(C)(\mathbb{R})/\,\mathrm{Jac}(C)(\mathbb{R})_{\mathrm{div}} \simeq (\mathbb{Z}/2)^{c-1}$$

is generated by zero-cycles of the form $[\varphi(P)] - [\varphi(Q)]$ with $P, Q \in X(\mathbb{R})$, and that for the Albanese map $\alpha\colon A_0(X) \to \mathrm{Alb}(X)$ we have:

(i) $\mathrm{Coker}\,\alpha \simeq \mathbb{Z}/2^{c-a}$,

and

(ii) $\mathrm{Im}(\alpha)^{\mathrm{top}} \simeq (\mathbb{Z}/2)^{a-1}$.

The exact sequence (15) then implies that

(iii) $T(X)^{\mathrm{top}} \simeq (\mathbb{Z}/2)^{s-a}$.

We will determine the group $T(X)^0$ using Corollary 4.6.ii. For this, consider the following commutative diagram:

$$\begin{array}{ccc} \mathrm{Pic}^0(X) & \xrightarrow{\mathrm{cl}_{X(\mathbb{R})}} & H^1(X(\mathbb{R}), \mathbb{Z}/2) \\ \wr\big\uparrow \varphi^* & & \big\uparrow \varphi^* \\ \mathrm{Pic}^0(C) & \xrightarrow{\mathrm{cl}_{C(\mathbb{R})}} & H^1(C(\mathbb{R}), \mathbb{Z}/2) \end{array}$$

Here the horizontal arrows are the cycle maps (cf. Remark 1.1) and the left hand vertical arrow is an isomorphism by hypothesis (1). It follows from the discussion in Section 1.2 that the kernel of $\mathrm{cl}_{C(\mathbb{R})}$ is divisible, and that the kernel of $\varphi^* \circ \mathrm{cl}_{C(\mathbb{R})}$ modulo its maximal divisible subgroup is isomorphic to $(\mathbb{Z}/2)^{h-1}$ (since $h > 0$). Hence the commutativity of the diagram and Corollary 4.6.ii imply

(iv) $T(X)^0 \simeq (\mathbb{Z}/2)^{h-1}$.

Adding up we obtain

(v) $T(X)_{\mathrm{tors}} \simeq (\mathbb{Z}/2)^{s-a+h-1}$.

The cohomological filtration on $T(X)_{\mathrm{tors}}$ not only depends on the invariants defined so far, but also on the topology of the fibres of φ.

5.1.1. *Smooth Conic Bundles.* Let C be as above and let $\varphi\colon X \to C$ be a conic bundle, then hypothesis (1) is satisfied. If, moreover, we assume φ to be smooth, then for every connected component $V \subset C(\mathbb{R})$ we have that $\varphi^{-1}(V)(\mathbb{R}) \subset X(\mathbb{R})$ is either empty or an S^1-bundle over V. So $a = s$ and $h = c - s$, which implies that if hypothesis (2) is satisfied, then

$$T(X)_{\mathrm{tors}} \simeq T(X)^0 \simeq (\mathbb{Z}/2)^{c-s-1}.$$

Moreover, Proposition 4.8, Lemma 4.3 and a closer examination of the Hochschild-Serre spectral sequence with coefficients in $\mathbb{Z}$ (which will not be carried out here) yield that

$$\operatorname{Gr}^j F^\bullet \operatorname{CH}_0(X)_{\text{tors}} \simeq \begin{cases} (\mathbb{Q}/\mathbb{Z})^g \times (\mathbb{Z}/2)^{s-1} & \text{if } j=0, \\ 0 & \text{if } j=1, \\ (\mathbb{Z}/2)^{c-s-1} & \text{if } j=2, \\ 0 & \text{if } j>2. \end{cases}$$

Note that for every choice of g, c, s with $0<s<c\leq g+1$ it is easy to construct a smooth conic bundle as above (compare [**Si89**, §V.4]).

5.2. Products of Curves and Abelian Varieties. For $i=1,\dots,d$, let C_i be a nonsingular projective geometrically irreducible curve over $\mathbb{R}$ of genus g_i such that $C_i(\mathbb{R})$ has $s_i>0$ connected components. Take $X=C_1\times\cdots\times C_d$. We have $\operatorname{Alb}(X)\simeq\operatorname{Alb}(C_1)\times\cdots\times\operatorname{Alb}(C_d)$, so

$$\operatorname{Alb}(X)(\mathbb{R}) \simeq (\mathbb{R}/\mathbb{Z})^{\sum_i g_i} \times (\mathbb{Z}/2)^{\sum_i (s_i-1)}.$$

On the other hand, $X(\mathbb{R})$ has $s=\prod_i s_i$ connected components, so

$$A_0(X)_{\text{tors}} \simeq (\mathbb{Q}/\mathbb{Z})^{\sum_i g_i} \times (\mathbb{Z}/2)^{s-1},$$

which implies that the kernel $T(X)_{\text{tors}}$ of α_{tors} is nontrivial if $s-1=(\prod_i s_i)-1>\sum_i(s_i-1)$; in other words, if $s_i>1$ for more than one index i. We will see that the converse holds as well.

By [**Kr83**, Propositions 3.6 and 5.6] we have that X is a $\mathbb{Z}$-GM-variety (see Definition 4.4), so it follows from Corollary 4.5 that

(i) α_{tors} *is surjective.*
(ii) $T(X)^0=0$.

From the exact sequences (15) and (16) we then see:

(iii) $T(X)_{\text{tors}}=T(X)^{\text{top}}\simeq(\mathbb{Z}/2)^{\prod_i s_i-\sum_i(s_i-1)-1}$.
(iv) $T(X)_{\text{tors}}$ *is nontrivial if and only if* $C_i(\mathbb{R})$ *is not connected for more than one index* i.

It will be some more work to determine the cohomological filtration $F^\bullet$ on $\operatorname{CH}_0(X)_{\text{tors}}$. Since the total cohomology $H^*(X(\mathbb{C}),\mathbb{Z})$ is torsion free, the boundary map

$$H^{2d-1}(X(\mathbb{C});G,\mathbb{Q}/\mathbb{Z}(d)) \to H^{2d}(X(\mathbb{C});G,\mathbb{Z}(d))$$

obtained from the short exact sequence $0\to\mathbb{Z}(d)\to\mathbb{Q}(d)\to\mathbb{Q}/\mathbb{Z}(d)\to 0$ induces for $p>0$ an isomorphism

$$H^p(G,H^{2d-1-p}(X(\mathbb{C}),\mathbb{Q}/\mathbb{Z}(d))) \simeq H^{p+1}(G,H^{2d-1-p}(X(\mathbb{C}),\mathbb{Z}(d)))$$

on the E_2-level of the Hochschild-Serre spectral sequences. Since X is a $\mathbb{Z}$-GM-variety, this is still an isomorphism on the E_∞-level. It follows that up to a shift by 1 in the indices the cohomological filtration $F^\bullet$ on $T(X)_{\text{tors}}$ is the same filtration as the one induced by the descending filtration $F^\bullet_{\mathbb{Z}}$ of $H^{2d}(X(\mathbb{C});G,\mathbb{Z}(d))$ associated to the Hochschild-Serre spectral sequence. In other words,

$$F^jT(X)_{\text{tors}} \simeq F^{j+1}_{\mathbb{Z}}H^d(X(\mathbb{R}),\mathbb{Z}/2)$$

for any $j>0$, where $F^j_{\mathbb{Z}}H^d(X(\mathbb{R}),\mathbb{Z}/2)$ is the image of $F^j_{\mathbb{Z}}H^{2d}(X(\mathbb{C});G,\mathbb{Z}(d))$ under β followed by the projection onto $H^d(X(\mathbb{R}),\mathbb{Z}/2)$.

We have for any $j \geq 0$ a commutative diagram

$$\begin{array}{ccc} \displaystyle\bigoplus_{\substack{0\leq j_1,\dots,j_d\leq 2\\ \sum_i j_i=j}} \bigotimes_i F_{\mathbb{Z}}^{j_i} H^2(C_i(\mathbb{C}); G, \mathbb{Z}(1)) & \xrightarrow{\beta} & \displaystyle\bigotimes_i H^*(C_i(\mathbb{R}), \mathbb{Z}/2) \\ \downarrow & & \downarrow \\ F_{\mathbb{Z}}^j H^{2d}(X(\mathbb{C}); G, \mathbb{Z}(d)) & \xrightarrow{\beta} & H^*(X(\mathbb{R}), \mathbb{Z}/2) \end{array} \tag{23}$$

in which, by the Künneth formula, the vertical arrow on the right is an isomorphism and the vertical arrow on the left is a surjection (compare [**Kr83**, §5.6]).

Since $F_{\mathbb{Z}}^2 H^1(C_i(\mathbb{R}), \mathbb{Z}/2) = 0$ for every i, this implies:

(v) $F^j T(X)_{\text{tors}} \simeq \displaystyle\bigoplus_{\substack{0\leq j_1,\dots,j_d\leq 1\\ \sum_i j_i=j+1}} \bigotimes_i F_{\mathbb{Z}}^{j_i} H^1(C_i(\mathbb{R}), \mathbb{Z}/2)$ *for* $j > 0$.

Observe that

$$F_{\mathbb{Z}}^{j_i} H^1(C_i(\mathbb{R}), \mathbb{Z}/2) = \begin{cases} H^1(C_i(\mathbb{R}), \mathbb{Z}/2) & \text{if } j_i = 0, \\ H^1(C_i(\mathbb{R}), \mathbb{Z}/2)^0 & \text{if } j_i = 1, \end{cases}$$

where $H^1(C_i(\mathbb{R}), \mathbb{Z}/2)^0$ is the kernel of the trace map $H^1(C_i(\mathbb{R}), \mathbb{Z}/2) \twoheadrightarrow \mathbb{Z}/2$. Hence $F_{\mathbb{Z}}^1 H^1(C_i(\mathbb{R}), \mathbb{Z}/2) = 0$ if and only if $C_i(\mathbb{R})$ is connected. We deduce for $j > 0$:

(vi) $F^j T(X)_{\text{tors}} = 0$ *if and only if* $j > \#\{i\colon s_i > 1\} - 1$.

If $s_i \in \{1, 2\}$ for all i (e.g., if X is a product of elliptic curves), then $X(\mathbb{R})$ has 2^a connected components, where a is the number of indices for which $C_i(\mathbb{R})$ is not connected, and we get the following result.

(vii) *With* X *and* a *as above, we have that* $\dim_{\mathbb{Z}}/2T(X)_{\text{tors}} = 2^a - a - 1$ *and* $\dim_{\mathbb{Z}}/2(\operatorname{Gr}^j F^\bullet \operatorname{CH}_0(X)_{\text{tors}}) = \binom{a}{j+1}$ *for* $j > 0$. *In particular, we have for* $j > 0$ *that* $F^j T(X)_{\text{tors}} = 0$ *if and only if* $j > a - 1$.

5.2.1. *Abelian Varieties.* When X is an Abelian variety of dimension d defined over $\mathbb{R}$, then $X(\mathbb{C})$ is equivariantly homeomorphic to the set of complex points of a product of d elliptic curves over $\mathbb{R}$. If $X(\mathbb{R})$ has 2^a connected components, we can take a product of a copies of an elliptic curve of which the real part has 2 connected components and $d - a$ copies of an elliptic curve of which the real part has 1 connected component. Therefore by Corollary 3.4 the statements (1), (2), and (7) above hold for X, so we have a complete picture of $T(X)_{\text{tors}}$ in this case.

We can also rephrase the results on the filtration in terms of the *Pontryagin product*

$$*\colon \operatorname{CH}_0(X) \times \operatorname{CH}_0(X) \to \operatorname{CH}_0(X)$$

defined by $\gamma * \tau = \mu_*(\gamma \times \tau)$, where $\mu\colon X \times X \to X$ is the multiplication (cf. [**Bl76**]). It can be checked that if $\gamma \in F^j \operatorname{CH}_0(X)_{\text{tors}}$ and $\tau \in F^{j'} \operatorname{CH}_0(X)_{\text{tors}}$, then $\gamma * \tau \in F^{j+j'+1} \operatorname{CH}_0(X)_{\text{tors}}$. In other words, the j-fold Pontryagin power $\operatorname{CH}_0(X)_{\text{tors}}^{*j} \subset \operatorname{CH}_0(X)_{\text{tors}}$, generated by elements of the form $\gamma_1 * \cdots * \gamma_j$, is contained in $F^{j-1} \operatorname{CH}_0(X)_{\text{tors}}$. From the representation (5) of $F^\bullet \operatorname{CH}_0(X)_{\text{tors}}$ it is a matter of combinatorics to deduce that

(viii) $F^j \operatorname{CH}_0(X)_{\text{tors}} = \operatorname{CH}_0(X)_{\text{tors}}^{*(j+1)}$ *for any* $j \geq 0$.

Acknowledgements

I would like to thank J.-L. Colliot-Thélène and C. Scheiderer for the stimulating discussions. I also would like to thank the latter for his help with the historical remarks and the references to the literature in the first two sections. Finally, I would like to thank the referees for the careful reading and detailed comments that helped me to improve the exposition, and for pointing out an error in the original version of Example 5.1.

References

[AP93] C. Allday and V. Puppe, *Cohomological methods in transformation groups*, Cambridge Stud. Adv. Math., vol. 32, Cambridge Univ. Press, Cambridge, 1993.

[AGV73] M. Artin, A. Grothendieck, and J.-L. Verdier (eds.), *Théorie des topos et cohomologie étale des schémas*, vol. 3, Lecture Notes in Math., vol. 305, Springer-Verlag, Berlin-New York, 1973.

[Bl76] S. Bloch, *Some elementary theorems about algebraic cycles on Abelian varieties*, Invent. Math. **37** (1976), 215–228.

[BO74] S. Bloch and A. Ogus, *Gersten's conjecture and the homology of schemes*, Ann. Sci. École Norm. Sup. (4) **7** (1974), 181–201.

[BCR87] J. Bochnak, M. Coste, and M.-F. Roy, *Géométrie algébrique réelle*, Ergeb. Math. Grenzgeb. (3), vol. 12, Springer-Verlag, Berlin-New York, 1987; *Real algebraic geometry*, Ergeb. Math. Grenzgeb. (3), vol. 36, Springer-Verlag, Berlin, 1998.

[Bo60] A. Borel, *Seminar on transformation groups*, Ann. of Math. Stud., vol. 46, Princeton Univ. Press, Princeton, NJ, 1960.

[BH61] A. Borel and A. Haefliger, *La classe d'homologie fondamentale d'un espace analytique*, Bull. Soc. Math. France **89** (1961), 461–513.

[CT93] J.-L. Colliot-Thélène, *Cycles algébriques de torsion et K-théorie algébrique*, Arithmetic algebraic geometry (E. Ballico, ed.), Lecture Notes in Math., vol. 1553, Springer-Verlag, Berlin, 1993, pp. 1–49.

[CTI81] J.-L. Colliot-Thélène and F. Ischebeck, *L'équivalence rationnelle sur les cycles de dimension zéro des variétés algébriques réelles*, C. R. Acad. Sci. Paris Sér. I Math. **292** (1981), 723–725.

[CTP90] J.-L. Colliot-Thélène and R. Parimala, *Real components of algebraic varieties and étale cohomology*, Invent. Math. **101** (1990), 81–99.

[CTS96] J.-L. Colliot-Thélène and C. Scheiderer, *Zero-cycles and cohomology on real algebraic varieties*, Topology **35** (1996), 533–559.

[Cox79] D. Cox, *The étale homotopy type of varieties over $\mathbb{R}$*, Proc. Amer. Math. Soc. **76** (1979), 17–22.

[Gr57] A. Grothendieck, *Sur quelques points d'algèbre homologique*, Tôhoku Math. J. (2) **9** (1957), 119–221.

[vH96] J. van Hamel, *Equivariant Borel-Moore homology and Poincaré duality for discrete transformation groups*, Tech. Report WS-463, Vrije Universiteit Amsterdam, 1996.

[vH98] J. van Hamel, *Divisors on real algebraic varieties without real points*, Manuscripta Math (to appear).

[Hs75] W.-Y. Hsiang, *Cohomology theory of topological transformation groups*, Ergeb. Math. Grenzgeb., vol. 85, Springer-Verlag, New York-Heidelberg, 1975.

[Iv86] B. Iversen, *Cohomology of sheaves*, Universitext, Springer-Verlag, Berlin-New York, 1986.

[Kr83] V.A. Krasnov, *Harnack-Thom inequalities for mappings of real algebraic varieties*, Izv. Akad. Nauk SSSR Ser. Mat. **47** (1983), 268–297 (Russian); English transl., Math. USSR Izv. **22** (1983) 247–275.

[Kr84] ———, *Albanese map for $GM\mathbb{Z}$-varieties*, Mat. Zametki **35** (1984), 739–747 (Russian); English transl., Math. Notes **35** (1984), 391–396.

[Kr94] ———, *On equivariant Grothendieck cohomology of a real algebraic variety, and its applications*, Izv. Ross. Akad. Nauk Ser. Mat. **58** (1994), 36–52 (Russian); English transl., Russian Acad. Sci. Izv. Math. **44** (1995), 461–477.

[MvH98] F. Mangolte and J. van Hamel, *Algebraic cycles on real Enriques surfaces*, Compositio Math. **110** (1998), 215–237.

[Mu94] J. P. Murre, *Algebraic cycles and algebraic aspects of cohomology and K-theory*, Algebraic Cycles and Hodge Theory (Torino, 1993) (A. Albano and F. Bardelli, eds.), Lecture Notes in Math., vol. 1594, Springer, Berlin , 1994, pp. 93–152.

[Ni94] V. V. Nikulin, *On the Brauer group of real algebraic surfaces*, Algebraic Geometry and Its Applications (Yaroslavl, 1992) (A. Tikhomirov and A. Tyurin, eds.), Aspects Math., vol. E25, Vieweg, Braunschweig, 1994, pp. 113–136.

[PW91] C. Pedrini and C. Weibel, *Invariants of real curves*, Rend. Sem. Mat. Univ. Politec. Torino **49** (1991), 139–173.

[Qu71] D. Quillen, *The spectrum of an equivariant cohomology ring.* I, Ann. of Math. (2) **94** (1971), 549–572.

[Ra98] N. Ramachandran, *Albanese and Picard one-motives of schemes*, math.AG/9804042, 1998.

[Sch90] C. Scheiderer, *A remark on the paper* Real components of real algebraic varieties and étale cohomology *by J.-L. Colliot-Thélène and R. Parimala*, 1990 (unpublished)

[Sch94] ———, *Real and étale cohomology*, Lecture Notes Math., vol. 1588, Springer-Verlag, Berlin, 1994.

[Sch95] ———, *Purity theorems for real spectra and applications*, Real Analytic and algebraic geometry (Trento, 1992) (F. Broglia, M. Galbiati, and A. Tognoli, eds.), de Gruyter, Berlin, 1995, pp. 229–250.

[Si89] R. Silhol, *Real algebraic surfaces*, Lecture Notes in Math., vol. 1392, Springer-Verlag, Berlin, 1989.

Mathematisch Instituut, Universiteit Utrecht, Budapestlaan 6, 3584 CD Utrecht, The Netherlands

E-mail address: `vanhamel@math.uu.nl`

Centre de Recherches Mathématiques
CRM Proceedings and Lecture Notes
Volume **24**, 2000

A Remark on the Griffiths Groups of Certain Product Varieties

Kenichiro Kimura

Let k be a field and let W be a projective smooth variety over k. Let $Z^2(W)$ be the free Abelian group generated by irreducible codimension 2 subvarieties of W. We write $Z^2(W)_{\text{hom}}$ (resp. $Z^2(W)_{\text{alg}}$) for the subgroup of $Z^2(W)$ which is homologically (resp. algebraically) equivalent to 0. The Griffiths group of codimension 2 of W is then defined as follows:

$$\operatorname{Gr}^2(W) := Z^2(W)_{\text{hom}}/Z^2(W)_{\text{alg}}.$$

In this note we specify W to be of the following form:

$$W = S \times_{\mathbb{Q}} C$$

where S is a projective smooth surface over $\mathbb{Q}$ and C is a curve over $\mathbb{Q}$ described as follows: Take any integer $m \in \mathbb{Z}_{\geq 1}$. The curve C is the normalization of the projective completion of the affine curve

$$y^m = f(u), \quad f(u) \in \mathbb{Q}[u]$$

which we assume to be geometrically irreducible.

Let μ_m be the group of mth roots of unity and let $K := \mathbb{Q}(\mu_m)$. Fix a primitive mth root of unity $\zeta \in \mu_m$, and let $\sigma \in \operatorname{Aut}(C)$ be defined by

$$\sigma(u, y) = (u, \zeta y),$$

that is, the group μ_m acts on W on the second factor. When the field of definition for W ought to be specified, we use the notation $W_{\mathbb{Q}}, W_K$, etc.

First we prove:

LEMMA 1. *Let R be the ring*

$$R := \mathbb{Q}[x] \Big/ \left(\sum_{i=0}^{m-1} x^i \right).$$

Then $\operatorname{Gr}^2(W_K)_{\mathbb{Q}} := \operatorname{Gr}^2(W_K) \otimes \mathbb{Q}$ is an R module.

1991 *Mathematics Subject Classification.* 14C25, 14D10, 14F20.

Received by the editors August 7, 1998; Accepted December 12, 1998.

The author was supported in part by grant-in -aid for scientific research, ministry of education, science and culture, Japan.

This is the final form of the paper.

PROOF. We let R act on $\mathrm{Gr}^2(W_K)_{\mathbb{Q}}$ in such a way that for any element $\alpha \in \mathrm{Gr}^2(W_K)_{\mathbb{Q}}$ and $f(x) \in R$,

$$f(x)\alpha = f(\sigma)\alpha.$$

For this to be well defined we need to show that

$$\sum_{i=0}^{m-1} \sigma^i \alpha = 0.$$

Let

$$\pi \colon C \to \mathbb{P}^1 \quad (u, y) \mapsto u$$

be the projection. Then we see that

$$\sum_{i=0}^{m-1} \sigma^k \alpha = \pi^*(\pi_* \alpha) \in \pi^*(\mathrm{Gr}^2(S \times \mathbb{P}^1) \otimes \mathbb{Q}).$$

The Chow group decomposes as:

$$CH^2(S \times \mathbb{P}^1) \simeq pr_S{}^* CH^1(S) \oplus pr_S{}^* CH^2(S)$$

from which we obtain the decomposition:

$$CH^2(S \times \mathbb{P}^1)_{\mathrm{hom}} \simeq pr_S{}^* CH^1(S)_{\mathrm{hom}} \oplus pr_S{}^* CH^2(S)_{\mathrm{hom}}.$$

Since the homological and algebraic equivalence coincide modulo torsion for zero-cycles and divisors, it follows that

$$\mathrm{Gr}^2(S \times \mathbb{P}^1) \otimes \mathbb{Q} = 0.$$

So the lemma follows. □

There is the decomposition

$$R = \oplus_{n|m} K_n$$

where K_n runs through the subfields $\mathbb{Q}(\mu_n)$ for $n|m$. Note that K_m coincides with the field K previously defined. This gives rise to the decomposition of the Griffiths group:

$$\mathrm{Gr}^2(W_K)_{\mathbb{Q}} = \oplus_{n|m} G_n$$

where $G_n := K_n \mathrm{Gr}^2(W_K)_{\mathbb{Q}}$. The Galois group $\mathrm{Gal}(K/\mathbb{Q})$ acts on $\mathrm{Gr}^2(W_K)_{\mathbb{Q}}$ and each direct factor G_n is stable under this action.

We need the following fact.

LEMMA 2. *Let*

$$\phi \colon W_K \to W_{\mathbb{Q}}$$

be the natural projection. Then we have

$$\mathrm{Gr}^2(W_K)_{\mathbb{Q}}^{\mathrm{Gal}(K/\mathbb{Q})} = \phi^* \mathrm{Gr}^2(W_{\mathbb{Q}})_{\mathbb{Q}}.$$

PROOF. The projection ϕ induces the projection on the Griffiths groups:

$$\phi_* \colon \mathrm{Gr}^2(W_K)_{\mathbb{Q}} \to \mathrm{Gr}^2(W_{\mathbb{Q}})_{\mathbb{Q}}$$

with the property that

$$\phi_* \phi^* = [K : \mathbb{Q}]$$

and

$$\phi^*\phi_*(\alpha) = \sum_{\delta \in \mathrm{Gal}(K/\mathbb{Q})} \delta(\alpha).$$

Hence the lemma follows. □

Since the pullback $\phi^*\colon \mathrm{Gr}^2(W_{\mathbb{Q}})_{\mathbb{Q}} \to \mathrm{Gr}^2(W_K)_{\mathbb{Q}}$ is injective, we may identify $\mathrm{Gr}^2(W_{\mathbb{Q}})_{\mathbb{Q}}$ with $\phi^* \mathrm{Gr}^2(W_{\mathbb{Q}})_{\mathbb{Q}}$.

Now we can state the main result of this paper.

THEOREM 3. *Let G_m be the direct factor of $\mathrm{Gr}^2(W_K)_{\mathbb{Q}}$ for $K_m = K$. Then*

$$G_m = (G_m \cap \mathrm{Gr}^2(W_{\mathbb{Q}})_{\mathbb{Q}}) \otimes_{\mathbb{Q}} K.$$

PROOF. We can write G_m as the inductive limit

$$G_m = \varinjlim V$$

where Vs are finite dimensional sub-K-vector spaces of G_m which are stable under $\mathrm{Gal}(K/\mathbb{Q})$ action. Thus it suffices to show that

$$V = V^{\mathrm{Gal}(K/\mathbb{Q})} \otimes_{\mathbb{Q}} K$$

for each such V. The action of $\mathrm{Gal}(K/\mathbb{Q})$ is semilinear with respect to the K-vector space structure, so this is the result of a Galois decent which is for example described in [**De**]. □

REMARK 4. The proof of the theorem only uses the fact that G_m is a K-vector space for a Galois extension K with a semilinear action of $\mathrm{Gal}(K/\mathbb{Q})$. For the other direct factors G_n, $n < m$ we cannot argue immediately in the same way because they are not necessarily $\mathrm{Gal}(K_n/\mathbb{Q})$ modules.

We will explain the significance of this fact in connection with the conjecture of Beilinson and Bloch on the special values of L-functions of W at $s = 2$.

Given projective smooth curve X over $\mathbb{Q}$ and an algebraic correspondence $\Gamma \in Z^2(X \times_{\mathbb{Q}} W)$ there is a map

$$\Gamma_*\colon, H^1(X)(-1) \to H^3(W).$$

Let I be the submotive of $H^3(W)$ defined by

$$I := \sum_{X,\Gamma} \Gamma_* H^1(X)(-1)$$

where the sum is taken over all pairs (X, Γ) as above. Then I splits:

LEMMA 5. *The submotive I_K is stable under the μ_m-action. Moreover, There is a decomposition*

$$H^3(W) = I \oplus M$$

defined over $\mathbb{Q}$. This decomposition is also stable under the μ_m-action over K.

PROOF. Let $\Gamma_m \in Z^3(W \times_K W)$ be the graph of σ. Take any pair (X, Γ) where X is a smooth projective curve over $\mathbb{Q}$ and $\Gamma \in Z^2(X \times_{\mathbb{Q}} W)$ is an algebraic correspondence. Then $X_K := X \times K$ is also a smooth projective curve over $\mathbb{Q}$ and $\Gamma \circ \Gamma_m \in Z^2(X \times_K W) = Z^2(X_K \times_{\mathbb{Q}} W)$. Hence the first part follows.

For the second part, from Theorem 4-1, Proposition 4-3 and Theorem 5-6 of [**Kl**] it follows that the ring $A := (Z^3(W \times_K W)/Z^3(W \times_K W)_{\mathrm{hom}}) \otimes \mathbb{Q}$ is a semi simple $\mathbb{Q}$-algebra. In the Theorem 5-6 of [**Kl**] the field of definition is not specified, but the proof goes similarly if rationality is taken into consideration. Therefore there is a decomposition

$$H^3(W_K) = I_K \oplus M_K$$

as an A-module. Note that M_K is stable under the action of μ_m. Since I is a $\mathbb{Q}$ rational submotive of $H^3(W)$ this decomposition is also defined over $\mathbb{Q}$. □

We will consider W over $\mathbb{Q}$ and also over K. We get the following

LEMMA 6. *M_K is an R module.*

PROOF. First note that

$$\sum_{i=0}^{m-1} \sigma^k = \pi^* \pi_* \in \operatorname{End} M_K.$$

From the conjecture $\theta(S)$ in the notation of [**Kl**] it follows that $\pi^* H^3(S \times \mathbb{P}^1) \subset I$. □

Hence there gives rise to the decomposition:

$$M = \bigoplus_{n|m} M_n$$

where $M_n := K_n M$.

In this situation, the conjecture of Beilinson and Bloch asserts the validity of the equalities:

$$\operatorname{ord}_{s=2} L(M_{n\mathbb{Q}}, s) = \dim_{\mathbb{Q}} G_n^{\operatorname{Gal}(K/\mathbb{Q})}$$

and, respectively,

$$\operatorname{ord}_{s=2} L(M_{nK}, s) = \dim_{\mathbb{Q}} G_n$$

for all $n|m$.

Let $M_m = K_m M = KM$. Let l be a prime which is inert in K. Then the etale realization $M_{ml} = M_m \otimes_{\mathbb{Q}} \mathbb{Q}_l$ is a K_l vector space and we have the decomposition

$$M_{ml} \otimes_{\mathbb{Q}_l} K_l = \sum_{\delta \in \operatorname{Gal}(K/\mathbb{Q})} V_\delta$$

where V_δ is the subspace on which σ acts as ζ^δ. Since the action of K_l on M_{ml} is semilinear, for an element $\gamma \in \operatorname{Gal}(\bar{\mathbb{Q}}/\mathbb{Q})$ we have

$$\gamma(V_\delta) = V_{\overline{\gamma}^{-1}\delta}$$

where $\overline{\gamma}$ is the image of γ to $\operatorname{Gal}(K/\mathbb{Q})$. From this fact we have the following

LEMMA 7. *As a Galois module,*

$$M_{ml} \otimes_{\mathbb{Q}_l} K_l = \operatorname{Ind}_{\operatorname{Gal}(\bar{K}/K)}^{\operatorname{Gal}(\bar{\mathbb{Q}}/\mathbb{Q})} V_{id}$$

where $id \in \operatorname{Gal}(K/\mathbb{Q})$ is the identity element.

Therefore, it is not difficult to see that

$$L(M_{mK}, s) = L(M_{m\mathbb{Q}}, s)^{[K:\mathbb{Q}]}$$

up to the bad factors. Hence the theorem provides a "relative" evidence for the validity of this conjecture.

Acknowledgment. The author is grateful to Chad Schoen for discussions and also to Hiroshi Saito for valuable comments.

References

[Be] A. A. Beilinson, *Height pairing between algebraic cycles*, Current Trends in Arithmetical Algebraic Geometry (Arcata, CA., 1985), Contemp. Math., vol. 67, Amer. Math. Soc., Providence, RI, 1987, pp. 1–24.

[Bl] S. Bloch, *Algebraic cycles and values of L-functions*, J. Reine Angew. Math. **350** (1984), 94–108.

[De] P. Deligne, *Cohomologie étale*, Lecture Notes in Math. vol. 569, Springer-Verlag, Berlin-New York, 1977.

[Kl] S. L. Kleiman, *The standard conjectures*, Motives (Seattle, WA, 1991), Proc. Sympos. Pure Math. vol. 55, Part 1, Amer. Math. Soc., Providence, RI, 1994, pp. 3–20.

[KR] N. Koblitz and D. Rohrlich, *Simple factors in the Jacobian of a Fermat curve*, Canad. J. Math. **30** (1978), 1183–1205.

INSTITUTE OF MATHEMATICS, UNIVERSITY OF TSUKUBA, TSUKUBA, IBARAKI, 305 JAPAN
E-mail address: `kimurak@math.tsukuba.ac.jp`

Centre de Recherches Mathématiques
CRM Proceedings and Lecture Notes
Volume **24**, 2000

p-adic Abel-Jacobi maps and p-adic heights

Jan Nekovář

This is a slightly expanded version of the author's lecture at the 1998 Banff Conference "The Arithmetic and Geometry of Algebraic Cycles." Its aim is to give a survey of foundational results on p-adic Abel-Jacobi maps and p-adic height pairings for algebraic cycles on varieties over number fields. Some of the results are new; their proofs will be given elsewhere.

1. p-adic Cycle Class Maps

1.1. Let X be a proper and smooth variety over a field K. Fix a separable closure K^{sep} of K and put $\overline{X} = X \times_K K^{\text{sep}}$, $G_K = \text{Gal}(K^{\text{sep}}/K)$. We also fix a prime number $p \neq \text{char}(K)$. Throughout this article, all cohomology groups with $\mathbb{Q}_p$-coefficients refer to continuous cohomology in the sense of [**Ja1**]].

1.2. The fundamental object of our interest is the p-adic cycle class map

$$cl_X\colon \text{CH}^n(X) \to H^{2n}(X_{\text{et}}, \mathbb{Q}_p(n))$$

with values in arithmetic étale cohomology of X. The link to geometric étale cohomology $V^j = H^j(\overline{X}_{\text{et}}, \mathbb{Q}_p)$ is provided by the Hochschild-Serre spectral sequence

$$(1.1) \qquad {}^{\text{et}}E_2^{i,j} = H^i(G_K, V^j(n)) \implies H^{i+j}(X_{\text{et}}, \mathbb{Q}_p(n)),$$

which can also be viewed as a Leray spectral sequence for the structure morphism $X \to \text{Spec}(K)$.

1.3. It is a non-trivial fact that (1.1) degenerates at E_2, i.e., that ${}^{\text{et}}E_2^{i,j} = {}^{\text{et}}E_\infty^{i,j}$. For X smooth and projective over K, this follows from Deligne's Thesis [**De1**] and his proof of the Hard Lefschetz Theorem [**De2**] (together with the existence of an appropriate formalism of derived categories for continuous cohomology [**Ek**]). The general case reduces to the previous one by using de Jong's theorem on alterations [**deJ**].

As a consequence of this degeneration, the filtration $F^*H^*(X_{\text{et}}, \mathbb{Q}_p(n))$ defined by (1.1) satisfies

$$gr_F^j H^m(X_{\text{et}}, \mathbb{Q}_p(n)) \xrightarrow{\sim} H^j(G_K, V^{m-j}(n)).$$

1991 *Mathematics Subject Classification.* 14F30, 11G09, 11F85, 14C25.
Received by the editors July 21, 1998; Accepted October 20, 1998.
This is the last form of the paper.

One obtains, therefore, a sequence of induced maps

$$\begin{aligned}
\mathrm{cl}_{0,X}\colon & \mathrm{CH}^n(X) \to H^0(G_K, V^{2n}(n)) \\
\mathrm{cl}_{1,X}\colon & \mathrm{Ker}(\mathrm{cl}_{0,X}) \to H^1(G_K, V^{2n-1}(n)) \\
\mathrm{cl}_{j,X}\colon & \mathrm{Ker}(\mathrm{cl}_{j-1,X}) \to H^j(G_K, V^{2n-j}(n)) \\
\mathrm{cl}_{j,X} \otimes \mathbb{Q}_p\colon & \mathrm{Ker}(\mathrm{cl}_{j-1,X}) \otimes \mathbb{Q}_p \to H^j(G_K, V^{2n-j}(n))
\end{aligned}$$

1.4. The subgroup

$$\mathrm{CH}^n(X)_{\mathrm{hom}\sim 0} := \mathrm{Ker}(\mathrm{cl}_{0,X}) \subseteq \mathrm{CH}^n(X)$$

is the subgroup of *homologically trivial cycles* (modulo torsion) and $\mathrm{cl}_{1,X}$ is the p-adic Abel-Jacobi map.

If $\mathrm{char}(K) = 0$, then one can assume without loss of generality that there is an embedding $\sigma\colon K^{\mathrm{sep}} \hookrightarrow \mathbb{C}$. Putting $X_{\mathbb{C}} = \overline{X} \times_{K^{\mathrm{sep}},\sigma} \mathbb{C}$ and using standard comparison isomorphisms between étale and singular cohomology over $\mathbb{C}$

$$\begin{array}{ccccc}
\mathrm{CH}^n(\overline{X}) & \longrightarrow & H^{2n}(\overline{X}_{\mathrm{et}}, \mathbb{Q}_p(n)) & \xrightarrow[\sigma_*]{\sim} & H^{2n}((X_{\mathbb{C}})_{\mathrm{et}}, \mathbb{Q}_p(n)) \\
\downarrow{\scriptstyle \sigma_*} & & & & \downarrow{\scriptstyle \simeq} \\
\mathrm{CH}^n(X_{\mathbb{C}}) & & \longrightarrow & & H^{2n}(X_{\mathbb{C}}^{\mathrm{an}}, \mathbb{Q}(n)) \otimes_{\mathbb{Q}} \mathbb{Q}_p
\end{array}$$

we see that $\mathrm{Ker}(\mathrm{cl}_{0,X})$ is independent on the choice of p.

If $\mathrm{char}(K) > 0$, then the independence of $\mathrm{Ker}(\mathrm{cl}_{0,X})$ on p is unknown in general.

2. Varieties Over Number Fields

2.1. From now on, K will be a finite extension of $\mathbb{Q}$ (hence $H^i(G_K, V^j(n)) = 0$ for $i > 2$). In this case the cycle class maps are expected to behave reasonably well.

CONJECTURE 2.1. (1) $\mathrm{Im}\,(\mathrm{cl}_{0,X} \otimes \mathbb{Q}_p) \overset{?}{=} H^0(G_K, V^{2n}(n))$ *(a special case of Tate's conjectures* [**Ta**]).

(2) $\mathrm{Ker}(\mathrm{cl}_{1,X} \otimes \mathbb{Q}_p) \overset{?}{=} 0$ *and* $\mathrm{Im}\,(\mathrm{cl}_{1,X} \otimes \mathbb{Q}_p) \overset{?}{=} H^1_f(G_K, V^{2n-1}(n))$ *(a special case of conjectures of Bloch-Kato* [**BK**]).

2.2. Let us recall the definition of Bloch-Kato style Selmer groups [**BK**] appearing in Conjecture2.1(2). Let V be a $\mathbb{Q}_p$-representation of G_K, i.e., a finite dimensional vector space over $\mathbb{Q}_p$ with a continuous $\mathbb{Q}_p$-linear action of G_K (e.g., $V = V^j(n)$). The subspace $H^1_f(G_K, V) \subset H^1(G_K, V)$ is defined by the Cartesian square

$$\begin{array}{ccc}
H^1_f(G_K, V) & \hookrightarrow & H^1(G_K, V) \\
\downarrow & \blacksquare & \downarrow \\
\prod_v H^1_f(G_v, V) & \hookrightarrow & \prod_v H^1(G_v, V),
\end{array}$$

in which v runs through all non-Archimedean primes of K, $G_v = \mathrm{Gal}(\overline{K}_v/K_v)$ and

$$H^1_f(G_v, V) = \begin{cases} H^1_{\mathrm{ur}}(G_v, V) = H^1(G_v/I_v, V^{I_v}), & \text{if } v \nmid p \\ \mathrm{Ker}(H^1(G_v, V) \to H^1(G_v, V \otimes_{\mathbb{Q}_p} B_{\mathrm{cris}})), & \text{if } v \mid p \end{cases}$$

Replacing $B_{\rm cris}$ by $B_{\rm st}$ (resp. by $B_{\rm dR}$), one obtains—for $v \mid p$—groups $H^1_{\rm st}(G_v, V)$ (resp. $H^1_g(G_v, V)$) satisfying

$$H^1_f(G_v, V) \subseteq H^1_{\rm st}(G_v, V) \subseteq H^1_g(G_v, V) \subseteq H^1(G_v, V).$$

We refer to [**Fo1**] and [**Fo2**] for more details on Fontaine's rings $B_{\rm cris} \subset B_{\rm st} \subset B_{\rm dR}$.

EXAMPLES 2.2. (1) $n = 1$. In this case $V^1(1) = V_p(A) = T_p(A) \otimes_{\mathbb{Z}_p} \mathbb{Q}_p$, where $A = {\rm Pic}^0_{X/K}$ is the connected Picard variety of X. Descent theory for the Abelian variety A gives maps into classical Selmer groups

$$A(K) \otimes \mathbb{Z}/p^n\mathbb{Z} \hookrightarrow {\rm Sel}(A/K, p^n) \subset H^1(G_K, A_{p^n}).$$

Passing to the projective limit and tensoring with $\mathbb{Q}_p$ one obtains the cycle class map for homologically trivial divisors

$${\rm cl}_{1,X} \otimes \mathbb{Q}_p \colon A(K) \otimes \mathbb{Q}_p \hookrightarrow H^1_f(G_K, V_p(A)) \subset H^1(G_K, V_p(A))$$

(as $A(K)$ is finitely generated by the Mordell-Weil theorem, and

$$\left(\varprojlim_n {\rm Sel}(A/K, p^n)\right) \otimes_{\mathbb{Z}_p} \mathbb{Q}_p = H^1_f(G_K, V_p(A))$$

by [**BK**, 3.11]). The conjectural equality 2.1(2)

$${\rm Im}\,({\rm cl}_{1,X} \otimes \mathbb{Q}_p) \stackrel{?}{=} H^1_f(G_K, V_p(A))$$

is equivalent to the finiteness of $\text{Ш}(A/K)_{p^\infty}$, the p-primary part of the Tate-Šafarevič group of A over K.

(2) $n = d = \dim(X)$ (assuming that X is equidimensional). The Albanese variety ${\rm Alb}_{X/K}$ can be identified with the dual $\hat{A}$ of $A = {\rm Pic}^0_{X/K}$ and $V^{2d-1}(d)$ with $V_p(\hat{A})$. The cycle class map ${\rm cl}_{d,X} \otimes \mathbb{Q}_p$ is the composition of the (surjective) Albanese map

$${\rm CH}_0(X) \otimes \mathbb{Q}_p \twoheadrightarrow \hat{A}(K) \otimes \mathbb{Q}_p$$

and the (injective) descent map $\hat{A}(K) \otimes \mathbb{Q}_p \hookrightarrow H^1_f(G_K, V_p(\hat{A}))$. The injectivity of the Albanese map is a very difficult open problem. The surjectivity of the descent map is, as before, equivalent to the finiteness of $\text{Ш}(A/K)_{p^\infty}$.

2.3. As a basic check of consistency of the Conjecture 2.1.2 one would like to show that

$$\text{(2.1)} \qquad {\rm Im}\,({\rm cl}_{1,X} \otimes \mathbb{Q}_p) \stackrel{?}{\subseteq} H^1_f(G_K, V^{2n-1}(n))$$

For $n = 1$ or $n = \dim(X)$ this follows from the discussion in Examples 2.2. In general, (2.1) boils down to a local question at each non-Archimedean prime v of K. Changing perspective, we fix v and consider a proper and smooth variety X_v over K_v. The question is whether

$$\text{(2.2)} \qquad {\rm Im}\,({\rm cl}_{1,X_v} \otimes \mathbb{Q}_p) \stackrel{?}{\subseteq} H^1_f(G_v, V^{2n-1}(n))$$

(now with $V^j = H^j((\overline{X}_v)_{\rm et}, \mathbb{Q}_p))$. At primes not dividing p the situation is more or less under control.

PROPOSITION 2.3. *Let $v \nmid p$. If X_v has potentially good reduction or, more generally, if the purity conjecture for the monodromy filtration* [**Ja2, RZ**] *on V^{2n-1} holds, then $H^1(G_v, V^{2n-1}(n)) = 0$.*

PROOF. Tate's local Euler characteristic formula [**Mi**, Theorem. I.2.8] [**Se**, II.5.7 Theorem 5] gives (for $V = V^{2n-1}(n)$)

$$\dim_{\mathbb{Q}_p} H^1(G_v, V) = \dim_{\mathbb{Q}_p} H^0(G_v, V) + \dim_{\mathbb{Q}_p} H^0(G_v, V^*(1)). \tag{2.3}$$

Our assumptions, together with Weil conjectures proved by Deligne [**De2**], imply that the geometric Frobenius $\mathrm{Fr}(v) \in G_v/I_v$ acts on both V^{I_v} and $(V^*(1))^{I_v}$ with weights < 0. It follows that the R.H.S. of (2.3) is equal to zero. □

3. Varieties Over p-adic Fields

3.1. In this section, $v \mid p$ (i.e. K_v is a finite extension of $\mathbb{Q}_p$) and X_v is proper and smooth over K_v. Our goal is to verify (2.2).

THEOREM 3.1. (i) *If X_v has potentially good reduction, then* (2.2) *holds.*
(ii) *In general,*

$$\mathrm{Im}\,(\mathrm{cl}_{1,X_v} \otimes \mathbb{Q}_p) \subseteq H^1_{\mathrm{st}}(G_v, V^{2n-1}(n)) = H^1_g(G_v, V^{2n-1}(n)).$$

If the p-adic version of the purity conjecture for the monodromy filtration for V^{2n-1} holds [**Mo**, Conjecture 3.2.7], *then*

$$H^1_{\mathrm{st}}(G_v, V^{2n-1}(n)) = H^1_f(G_v, V^{2n-1}(n))$$

and (2.2) *again holds.*

3.2. The purity conjecture for the monodromy filtration is known to hold (for all V^j and all v) for Abelian varieties and for compactified Kuga-Sato varieties over modular curves [**Sa**]. The validity of (2.1) in the latter case implies that the main result of [**Ne1**] holds for all p.

3.3. We now explain an approach to the proof of Theorem 3.1 based on [**Ne3**]. The part (i) was proved independently by Nizioł[**Ni**] (she assumed X_v to be smooth and projective, but the general case follows by the same argument as in 3.7 below). A special case of (ii) is due to A. Langer [**La**]. Let us distinguish two cases.

3.4. Good Reduction Case. Assume that X_v has a model $\mathfrak{X}_v$, proper and smooth over $\mathcal{O}_v$ (the ring of integers of K_v). According to Fontaine's conjecture C_{cris}, which is now a theorem [**FM, Fa, KaM, Ts**], all V^j are *crystalline* representations of G_v ([**Fo1, Fo2**]). One defines

$$H^i_f(G_v, -) = \mathrm{Ext}^i_{\mathrm{Rep}_{\mathrm{cris}(G_v)}}(\mathbb{Q}_p, -),$$

where $\mathrm{Rep}_{\mathrm{cris}(G_v)}$—a full subcategory of $\mathrm{Rep}(G_v)$—is the category of crystalline representations of G_v (for $i = 1$ this is compatible with the definition given in 2.2). For an object V of $\mathrm{Rep}_{\mathrm{cris}(G_v)}$ one has

$$H^0_f(G_v, V) = H^0(G_v, V), \quad H^1_f(G_v, V) \subseteq H^1(G_v, V), \quad H^i_f(G_v, V) = 0 \quad (i > 1).$$

The key ingredient of the proof of Theorem 3.1 in the good reduction case is the *syntomic cohomology* $H^i((\mathfrak{X}_v)_{\mathrm{syn}}, s_{\mathbb{Q}_p}(n))$ of Fontaine-Messing [**FM**], which is a p-adic analogue of Deligne cohomology (more precisely, of Beilinson's "absolute Hodge cohomology" [**Be**]). Recall that there are two kinds of cohomology theories, geometric and arithmetic (= absolute), that come in pairs [**Ja3**, 10.5]. For example, we have:

Geometric theory	Arithmetic theory
étale cohomology of $\overline{X}$	étale cohomology of X
Betti cohomology (over $\mathbb{C}$)	absolute Hodge cohomology
crystalline cohomology of $\mathfrak{X}_v$	syntomic cohomology

Given the limited scope of this article, we have to treat syntomic cohomology as a black box. One of its important features is the existence of a commutative diagram of cycle class maps

$$\begin{array}{ccc} \mathrm{cl}^{\mathrm{syn}}_{\mathfrak{X}_v}\colon \mathrm{CH}^n(\mathfrak{X}_v) & \longrightarrow & H^{2n}((\mathfrak{X}_v)_{\mathrm{syn}}, s_{\mathbb{Q}_p}(n)) \\ \downarrow \mathrm{res} & & \downarrow \alpha_{FM} \\ \mathrm{cl}^{\mathrm{et}}_{X_v}\colon \mathrm{CH}^n(X_v) & \longrightarrow & H^{2n}((X_v)_{\mathrm{et}}, \mathbb{Q}_p(n)), \end{array} \tag{3.1}$$

where α_{FM} is the (highly non-trivial) Fontaine-Messing map (for $p > 2$ it was constructed in [**FM**]; the case of $p = 2$ is treated in [**Ts**]).

One possible definition of the syntomic cycle class $\mathrm{cl}^{\mathrm{syn}}_{\mathfrak{X}_v}$ uses the fact that $\mathrm{CH}^n(\mathfrak{X}_v) \otimes \mathbb{Q}$ is canonically isomorphic to the weight n eigenspace of $K_0(\mathfrak{X}_v) \otimes \mathbb{Q}$ with respect to Adams operations ψ^k; the usual formula $\mathrm{cl}^{\mathrm{syn}}_{\mathfrak{X}_v} = (-1)^{n-1} c_n^{\mathrm{syn}}/(n-1)!$ then makes sense. Similarly, one has $\mathrm{cl}^{\mathrm{et}}_{X_v} = (-1)^{n-1} c_n^{\mathrm{et}}/(n-1)!$. The compatibility between the syntomic c_n^{syn} and étale c_n^{et} Chern classes of vector bundles with respect to the Fontaine-Messing map reduces, by the splitting principle, to the compatibility of the two c_1's of a line bundle, which can be verified explicitly. See [**Ts**] for more details, which are also valid in the logarithmic case discussed in 3.5 below.

3.5. Semistable Reduction Case. Assume that X_v has a proper semistable model $\mathfrak{X}_v$ over $\mathcal{O}_v$ (i.e., étale locally, $\mathfrak{X}_v$ is given by

$$\mathrm{Spec}(\mathcal{O}_v[T_1, \ldots, T_n]/(T_1 \ldots T_m - \pi_v)),$$

where $m \leq n$ and π_v is a uniformizer of $\mathcal{O}_v$). In this case $\mathfrak{X}_v$ has a canonical log-structure $\mathfrak{X}_v^{\times}$ induced from X_v (see [**HK**] for a crash course on log-structures). According to Fontaine-Jannsen's conjecture C_{st}, proved in [**Ka, Ts**] all V^j are *semistable* representations of G_v [**Fo1, Fo2**]. As in 3.4, one defines

$$H^i_{\mathrm{st}}(G_v, -) = \mathrm{Ext}^i_{\mathrm{Rep}_{\mathrm{st}(G_v)}}(\mathbb{Q}_p, -),$$

where $\mathrm{Rep}_{\mathrm{st}(G_v)}$—a full subcategory of $\mathrm{Rep}(G_v)$—is the category of semistable representations of G_v (again, for $i = 1$ this is compatible with the definition given in 2.2). For an object V of $\mathrm{Rep}_{\mathrm{st}(G_v)}$ one has [**FPR**, Chapter II, Proposition 3.3.8]

$$\begin{array}{ll} H^0_{\mathrm{st}}(G_v, V) = H^0(G_v, V), & H^1_{\mathrm{st}}(G_v, V) \subseteq H^1(G_v, V), \\ H^2_{\mathrm{st}}(G_v, V) \twoheadrightarrow H^2(G_v, V), & H^i_{\mathrm{st}}(G_v, V) = 0 \quad (i > 2). \end{array}$$

More precisely, [**FPR**] gives an explicit complex computing cohomology $H^i_{\mathrm{st}}(G_v, V)$. The map

$$H^2_{\mathrm{st}}(G_v, V) \twoheadrightarrow H^2(G_v, V)$$

is surjective, being dual to the inclusion

$$H^0(G_v, V^*(1)) = D_{\mathrm{st}}(V^*(1))^{N=0,\varphi=1} \cap F^0 D_{\mathrm{dR}}(V^*(1)) \hookrightarrow D_{\mathrm{st}}(V^*(1))^{N=0,\varphi=1}.$$

Here $D_{\mathrm{st}}(V^*(1)) = H^0(G_v, V^*(1) \otimes_{\mathbb{Q}_p} B_{\mathrm{st}})$ denotes the filtered (φ, N)-module associated to $V^*(1)$ and $F^i D_{\mathrm{dR}}(V^*(1)) = H^0(G_v, V^*(1) \otimes_{\mathbb{Q}_p} F^i B_{\mathrm{dR}})$.

There is a logarithmic version of syntomic cohomology and of the diagram (3.1):

$$
\begin{array}{ccc}
\mathrm{CH}^n(\mathfrak{X}_v) & \longrightarrow & H^{2n}((\mathfrak{X}_v^\times)_{\mathrm{syn}}, s^{\log}_{\mathbb{Q}_p}(n)) \\
\Big\downarrow{\scriptstyle \mathrm{res}} & & \Big\downarrow{\scriptstyle \alpha_{FM}} \\
\mathrm{CH}^n(X_v) & \longrightarrow & H^{2n}((X_v)_{\mathrm{et}}, \mathbb{Q}_p(n)),
\end{array}
\tag{3.2}
$$

Again, the log-syntomic cycle class map can be defined using Chern classes with values in log-syntomic cohomology [**Ts**].

THEOREM 3.2. *If X_v has good (resp. semistable) reduction and if $\mathfrak{X}_v$ is a proper and smooth (resp. proper and semistable) model of X_v over $\mathcal{O}_v$, then there is a canonical spectral sequence*

$$ {}^{\mathrm{syn}}E_\infty^{i,j} = {}^{\mathrm{syn}}E_2^{i,j} = H^i_f(G_v, V^j(n)) \implies H^{i+j}((\mathfrak{X}_v)_{\mathrm{syn}}, s_{\mathbb{Q}_p}(n)) $$

(resp.

$$ {}^{\log-\mathrm{syn}}E_2^{i,j} = H^i_{\mathrm{st}}(G_v, V^j(n)) \implies H^{i+j}((\mathfrak{X}_v^\times)_{\mathrm{syn}}, s^{\log}_{\mathbb{Q}_p}(n))) $$

compatible (via α_{FM}) with the Hochschild-Serre spectral sequence

$$ {}^{\mathrm{et}}E_2^{i,j} = H^i(G_v, V^j(n)) \implies H^{i+j}((X_v)_{\mathrm{et}}, \mathbb{Q}_p(n)). $$

3.6. See [**Ne3**] for the proof in the good reduction case. The semistable reduction case will be treated in an updated version of [**Ne3**]]; a special case was proved by A. Langer [**La**]. In the good reduction case there is also a version with $\mathbb{Z}_p$-coefficients, valid for $K_v/\mathbb{Q}_p$ unramified and $p-1 > \max(i+j, n)$. These results have interesting applications to the study of the image of the cycle class map:

COROLLARY 3.3. *If X_v has potentially good reduction, then*

$$ \mathrm{Im}\left(cl_{X_v}\colon \mathrm{CH}^n(X_v) \to H^{2n}((X_v)_{\mathrm{et}}, \mathbb{Q}_p(n))\right) \cap F^2H^{2n}((X_v)_{\mathrm{et}}, \mathbb{Q}_p(n)) = 0. $$

3.7. Proof of Theorem 3.1 (Assuming Theorem 3.2). For every finite Galois extension K_v'/K_v and $* \in \{\varnothing, f, g, \mathrm{st}\}$ one has

$$ H^1_*(\overline{K}_v/K_v, V) = H^1_*(\overline{K}_v/K_v', V)^{\mathrm{Gal}(K_v'/K_v)}. $$

One is free, therefore, to replace K_v by K_v' and X_v by its base change to K_v'. This reduces (i) to the case of good reduction, in which case everything follows from the commutative diagram (3.1) and Theorem 3.2 (as the restriction map $\mathrm{CH}^n(\mathfrak{X}_v) \to \mathrm{CH}^n(X_v)$ is surjective).

(ii) According to de Jong's theorem on alterations [**deJ**] there is—possibly after a finite extension of scalars K_v'/K_v—an alteration $\pi\colon X_v' \to X_v$ such that X_v' has semistable reduction. The existence of trace maps in étale cohomology implies that the induced maps $\pi^*\colon V^j \to V'^j$ are split injective. It follows that the statement

$$ \mathrm{Im}\,(\mathrm{cl}_{1,X_v} \otimes \mathbb{Q}_p) \overset{?}{\subseteq} H^1_{\mathrm{st}}(G_v, V^{2n-1}(n)) $$

reduces to a similar statement for X_v', in which case we conclude as before by the commutative diagram (3.2) and Theorem 3.2.

The quotient $H^1_{\mathrm{st}}(G_v, V)/H^1_f(G_v, V)$ (for $V = V^{2n-1}(n)$) is dual to

$$ D_{\mathrm{st}}(V^*(1))^{N=0,\varphi=1} $$

[**Ne2**, Corollary 1.18]. The p-adic version of the purity conjecture for the monodromy filtration for V^{2n-1} ([**Mo**, Conjecture 3.27]; cf. [**Il**, Question 3.13.2], [**Fo3**,

Conjecture 2.4.3-5]), implies that φ acts on $D_{\mathrm{st}}(V^*(1))^{N=0}$ with weights < 0; thus $D_{\mathrm{st}}(V^*(1))^{N=0,\varphi=1}$ vanishes. The equality $H^1_{\mathrm{st}} = H^1_g$ for potentially semistable representations is proved in [**Hy**] (see also [**Ne2**, Proposition 1.24.3]). This concludes the proof of Theorem 3.1. □

3.8. Perhaps it is worth pointing out that in the good reduction case the weights of the action of φ on $D_{\mathrm{cris}}(V^j) = H^0(G_v, V^j \otimes_{\mathbb{Q}_p} B_{\mathrm{cris}})$ are equal to j ("crystalline Weil conjectures"). For $\mathfrak{X}_v$ smooth and projective over $\mathrm{Spec}(\mathcal{O}_v)$ this was proved by Katz and Messing [**KM**], the general case by Chiarellotto and Le Stum [**CLS**].

3.9. If $Z \subset X_v$ is a homologically trivial cycle of codimension n, the statement

$$\mathrm{cl}_{1,X_v}(Z) \overset{?}{\in} H^1_{\mathrm{st}}(G_v, V^{2n-1}(n))$$

would follow if one could prove directly that $H^{2n-1}((\overline{X_v - |Z|})_{\mathrm{et}}, \mathbb{Q}_p)$ is a semistable representation of $\mathrm{Gal}(\overline{K}_v/K'_v)$ for some finite extension K'_v/K_v (i.e., it is potentially semistable). This is because of an alternative description of $\mathrm{cl}_{1,X_v}(Z)$ in terms of the extension class of Galois representations arising from the exact cohomology sequence

$$0 \to H^{2n-1}(\overline{X}_v, \mathbb{Q}_p)(n) \to H^{2n-1}((\overline{X_v - |Z|})_{\mathrm{et}}, \mathbb{Q}_p)(n) \to H^{2n}_{|\overline{Z}|}(\overline{X}_v, \mathbb{Q}_p)(n) \to \cdots$$

(see [**Ja3**, Lemma 9.4]).

4. p-adic Height Pairings

4.1. As in Section 2, let X be a proper and smooth variety over a number field K, equidimensional of dimension $\dim(X) = d$. Fix integers $i, j \geq 1$ satisfying $i + j = d + 1$. In the notation of 1.2, put $V = V^{2i-1}(i)$. Poincaré duality

$$V^{2i-1}(i) \times V^{2j-1}(j) \to V^{2d}(d+1) \xrightarrow{\mathrm{Tr}} \mathbb{Q}_p(1)$$

then identifies $V^{2j-1}(j)$ with $\mathrm{Hom}_{\mathbb{Q}_p}(V, \mathbb{Q}_p(1)) = V^*(1)$.

4.2. We assume that the following assumption is satisfied:

(A) At each non-Archimedean prime v of K where X has bad reduction, the purity conjecture for the monodromy filtration for V^{2i-1} (or, equivalently, for V^{2j-1}) holds.

It follows from Theorem 3.1 that under the assumption (A) there are p-adic Abel-Jacobi maps

$$\mathrm{CH}^i(X)_{\mathrm{hom}\sim 0} \to H^1_f(G_K, V)$$
$$\mathrm{CH}^j(X)_{\mathrm{hom}\sim 0} \to H^1_f(G_K, V^*(1))$$

THEOREM 4.1 ([**Ne2**]). *Under the assumption* (A) *there are height pairings*

$$\langle\ ,\ \rangle \colon H^1_f(G_K, V) \times H^1_f(G_K, V^*(1)) \to \mathbb{Q}_p$$

depending on the following additional data:

(a) *A continuous homomorphism* $\ell \colon \mathbb{A}^*_K/K^* \to \mathbb{Q}_p$.

(b) $\mathbb{Q}_p$*-linear splittings of Hodge filtrations*

$$F^i H^{2i-1}_{\mathrm{dR}}(X_v/K_v) \leftrightarrows H^{2i-1}_{\mathrm{dR}}(X_v/K_v)$$

for all primes $v \mid p$ *(where* $X_v = X \times_K K_v$*).*

COROLLARY 4.2. *There are p-adic height pairings*

$$\mathrm{CH}^i(X)_{\mathrm{hom}\sim 0} \times \mathrm{CH}^j(X)_{\mathrm{hom}\sim 0} \to \mathbb{Q}_p$$

depending on the data (a), (b) *and factoring through p-adic Abel-Jacobi maps.*

REMARKS 4.3. (1) The assumptions of [**Ne2**, 2.1.2, 7.2] under which the height pairings are constructed, follow from (A), possibly after a finite extension of the field K (cf. the discussion in 3.7–3.8 above).

(2) It is still an open problem whether the real-valued height pairing factors through the complex-analytic Abel-Jacobi map (if $\min(i,j) > 1$).

(3) Sometimes the assumption (A) can be verified only for a suitable direct factor $W \subset V$ coming from a "submotive" of $h^{2i-1}(X)(i)$. In that case one gets only pairings

$$H^1_f(G_K, W) \times H^1_f(G_K, W^*(1)) \to \mathbb{Q}_p$$

depending on a similar set of data.

(4) By class field theory, ℓ can be viewed as a $\mathbb{Z}_p$-linear map $\mathrm{Gal}(K_\infty/K) \to \mathbb{Q}_p$, where K_∞ is the compositum of all $\mathbb{Z}_p$-extensions of K inside $\overline{K}$. The dependence of $\langle\ ,\ \rangle$ on ℓ is linear, which means that once the choices of splittings in (b) are fixed one obtains a "universal" height pairing with values in the finite-dimensional $\mathbb{Q}_p$-vector space $\mathrm{Gal}(K_\infty/K) \otimes_{\mathbb{Z}_p} \mathbb{Q}_p$.

(5) The dependence on ℓ can be seen already in the following simple example: let $X = \mathbb{P}^1_K$, $x \in K^* \setminus \{1\}$, $Z = (0) - (\infty)$, $Z' = (1) - (x)$. Consider first the real valued height: as $\mathrm{CH}^1(X)_{\mathrm{hom}\sim 0} = 0$, the global height $\langle Z, Z'\rangle = 0 \in \mathbb{R}$. However, the local heights $\langle Z, Z'\rangle_v = \log |x|_v$ are non-zero (unles x is a root of unity) and

$$\langle Z, Z'\rangle = \sum_v \langle Z, Z'\rangle_v = \log\left(\prod_v |x|_v\right) = 0$$

vanishes by the product formula. In the p-adic situation, ℓ defines continuous homomorphisms $\ell_v \colon K_v^* \to \mathbb{Q}_p$ for all primes v, and the local p-adic heights are equal to $\langle Z, Z'\rangle_v = \ell_v(x)$. The global height

$$\langle Z, Z'\rangle = \sum_v \langle Z, Z'\rangle_v = \sum_v \ell_v(x) = 0$$

vanishes, as $\ell|_{K^*} = 0$.

(6) In certain cases there are canonical splittings of the Hodge filtrations in (b), which then define "canonical p-adic height pairings." This happens whenever there is, for each prime $v \mid p$, an exact sequence of representations of G_v

$$0 \to V_v^+ \to V \xrightarrow{\pi} V_v^- \to 0$$

with the property that π induces an isomorphism $F^0 D_{\mathrm{dR}}(V) \xrightarrow{\sim} D_{\mathrm{dR}}(V_v^-)$. For example, for $i = 1$ this happens iff $A = \mathrm{Pic}^0_{X/K}$ has ordinary reduction at all $v \mid p$ (see [**Ne2**] for more details).

5. p-adic Height Pairings in Families

5.1. The p-adic heights discussed in Section 4 appear in the p-adic version of the conjecture of Birch and Swinnerton-Dyer (and its higher-dimensional analogue). In the formulation of [**PR1, PR2**] *both sides of the conjecture* depend on the data (a), (b) from Theorem 4.1. Traditionally, p-adic L-functions have characters of $\mathrm{Gal}(K_\infty/K)$ as arguments. The homomorphism ℓ then defines a direction in which derivatives of the p-adic L-function are taken. Perrin-Riou's p-adic L-function depends on an extra parameter, of which the splittings (b) are a special case.

5.2. Apart from Perrin-Riou's new parameter, p-adic L-functions tend to depend on additional variables. For example, Hida's theory of ordinary families of p-adic modular forms [**Hi**] produces two-variable p-adic L-functions, with weight as the second variable [**GS, Ki**].

On the algebraic side, the theory gives p-adic families of representations of G_K of the following type (for $K = \mathbb{Q}$):

(1) R is a complete local ring, finite and flat over $\Lambda = \mathcal{O}_F[[X]]$, where $\mathcal{O}_F$ is the ring of integers in a finite extension F of $\mathbb{Q}_p$.
(2) T is an R-module of finite type, flat over Λ, with an R-linear continuous action of G_K. The action is unramified outside a finite set S of primes of K. One can assume that S contains all primes dividing p and all Archimedean primes of K.
(3) For every prime $v \mid p$ in K there is an exact sequence of $R[G_v]$-modules
$$0 \to T_v^+ \to T \to T_v^- \to 0,$$
with $T_v^\pm$ flat over Λ.

5.3. In concrete terms, T constructed by Hida interpolates Galois representations associated to modular forms as follows: for every even integer $k \geq 2$ in a fixed residue class modulo $2(p-1)$ and a prime ideal $\mathfrak{p}_k \subset R$ above $(X+1-(1+p)^{k-2}) \subset \Lambda$, $V_k = (T/\mathfrak{p}_k T) \otimes_{\mathcal{O}_F} F$ is a two-dimensional representation of $G_\mathbb{Q}$ over F, associated to—up to a Tate twist—a Hecke eigenform of weight k on $\Gamma_1(Np)$ (for $p \nmid 6N$). Our normalization of the Tate twist (cf. [**Gr**]) is such that V_k is a direct factor of $H^{k-1}(\overline{X}_{\mathrm{et}}, \mathbb{Q}_p(k/2))$, where X is a compactified Kuga-Sato variety of dimension $k-1$, fibered over the modular curve $X_1(Np)$. In this case

$$\wedge_F^2 V_k \xrightarrow{\sim} F(1), \quad \mathrm{Hom}_F(V_k, F(1)) \xrightarrow{\sim} V_k. \tag{5.1}$$

The ring R is a local component of the ordinary part of the projective limit of Hecke algebras (with $\mathbb{Z}_p$-coefficients) acting on the Jacobians $J_1(Np^r)$ of modular curves $X_1(Np^r)$. Up to a twist, T is the Pontrjagin dual of the R-part of
$$\varinjlim_r J_1(Np^r)_{p^\infty}(\overline{\mathbb{Q}})^{\mathrm{ord}},$$
where the superscript "ord" refers to the subgroup on which the Hecke operator $T(p)$ is not topologically nilpotent [**Hi**].

5.4. In the abstract setup of 5.2(1)–(3), put $\omega_R = \mathrm{Hom}_\Lambda(R, \Lambda)$ and
$$T^* = \mathrm{Hom}_\Lambda(T, \Lambda) = \mathrm{Hom}_R(T, \omega_R),$$
$$(T^*)_v^\pm = \mathrm{Hom}_\Lambda(T_v^\mp, \Lambda) = \mathrm{Hom}_R(T_v^\mp, \omega_R).$$

In the situation of 5.3 there is a canonical skew-symmetric isomorphism $T^*(1) \xrightarrow{\sim} T$, which interpolates the isomorphisms (5.1).

Plater [**Pl**] defined Selmer groups

$$H^1_f(G_K,T)^0 \subset H^1_f(G_K,T) \subset H^1(G_K,T)$$

and—under certain assumptions—height pairings

$$H^1_f(G_K,T)^0 \times H^1_f(G_K,T^*(1))^0 \to R \tag{5.2}$$

that interpolate the height pairings

$$H^1_f(G_K,V_k) \times H^1_f(G_K,V_k^*(1)) \to F$$

of Theorem 4.1.

5.5. Selmer complexes. It seems that the correct setup for duality theorems and p-adic height pairings involves *Selmer complexes* [**Ne4**] rather than Selmer groups. The idea is that one should consider local conditions for complexes rather than for cohomology groups, as in 2.2. In the situation of 5.2, let K_S be the maximal extension of K (inside $\overline{K}$) unramified outside S and $G_{K,S} = \mathrm{Gal}(K_S/K)$. Denote by $C^\bullet_{\mathrm{cont}} G_{K,S}, T)$ the complex of non-homogeneous continuous cochains (with respect to the pro-finite topologies) on $G_{K,S}$ with values in T and put, for every non-Archimedean prime $v \in S$,

$$U_v^+(T) = \begin{cases} C^\bullet_{\mathrm{cont}} G_v/I_v, T^{I_v}), & \text{if } v \nmid p \\ C^\bullet_{\mathrm{cont}} G_v, T_v^+), & \text{if } v \mid p \end{cases}$$

The *Selmer complex* $\mathbb{R}\widetilde{\Gamma}_f(G_{K,S},T)$ is defined as the simple complex associated to

$$\left(\begin{array}{ccc} C^\bullet_{\mathrm{cont}} G_{K,S}, T) & \longrightarrow & \bigoplus_{v\in S_f} C^\bullet_{\mathrm{cont}} G_v, T) \\ & & \uparrow \\ & & \bigoplus_{v\in S_f} U_v^+(T) \end{array} \right)$$

(where $S_f \subset S$ is the subset of non-Archimedean primes of S). Its cohomology sits in an exact sequence

$$\cdots \to \widetilde{H}^i_f(G_{K,S},T) \to H^i(G_{K,S},T) \oplus \bigoplus_{v\in S_f} H^i(U_v^+(T))$$
$$\to \bigoplus_{v\in S_f} H^i(G_v,T) \to \widetilde{H}^{i+1}_f(G_{K,S},T) \to \cdots$$

The appropriate generalization of (5.3.1) is given by the following

THEOREM 5.1 ([**Ne4**]). *In the situation of* 5.2(1)–(3), *assume that either* $p > 2$ *or* K *is totally imaginary. Then, for every continuous homomorphism*

$$\ell\colon \mathrm{Gal}(K_\infty/K) \to R,$$

there is a "height pairing"

$$\mathbb{R}\widetilde{\Gamma}_f(G_{K,S},T) \overset{\mathbb{L}}{\otimes}_R \mathbb{R}\widetilde{\Gamma}_f(G_{K,S},T^*(1)) \to \omega_R[-2],$$

which depends R-linearly on ℓ.

5.6. Selmer complexes $\mathbb{R}\widetilde{\Gamma}_f(G_{K,S}, T)$ are better than classical Selmer groups even for elliptic curves (when R and Λ are replaced by $\mathbb{Z}_p$). For example, if $K = \mathbb{Q}$ and $T = T_p(E)$ is the Tate module of an elliptic curve E over $\mathbb{Q}$ with *split multiplicative* reduction at p, then $T_p^+ = \mathbb{Z}_p(1)$ and $T_p^- = \mathbb{Z}_p$. The group $H^0(G_p, T_p^-) = \mathbb{Z}_p$ then contributes to the kernel of $\widetilde{H}^1_f(G_{K,S}, T) \to H^1(G_{K,S}, T)$; there is an exact sequence

$$0 \to \mathbb{Q}_p \to \widetilde{H}^1_f(G_{K,S}, T_p(E)) \otimes_{\mathbb{Z}_p} \mathbb{Q}_p \to H^1(G_K, V_p(E)) \to 0.$$

This is consistent with the phenomenon discovered by Mazur, Tate and Teitelbaum [**MTT**]: the term $H^0(G_p, T_p^-)$ is responsible for a trivial zero of the p-adic L-function and

$$\widetilde{H}^1_f(G_{K,S}, V_p(E)) = \widetilde{H}^1_f(G_{K,S}, T_p(E)) \otimes_{\mathbb{Z}_p} \mathbb{Q}_p$$

is the corresponding "extended Selmer group." A variant of Theorem 5.1 then defines a height pairing

$$\widetilde{H}^1_f(G_{K,S}, T_p(E)) \times \widetilde{H}^1_f(G_{K,S}, T_p(E)) \to \mathbb{Z}_p.$$

5.7. Coleman [**Co**] constructed p-adic analytic families of non-ordinary modular forms. It would be interesting to extend previous constructions to this case.

References

[Be] A.A. Beilinson, *Notes on absolute Hodge cohomology*, Applications of Algebraic K-theory to Algebraic Geometry and Number Theory, (Boulder, CO, 1983), Contemp. Math., vol. 55, Part I, Amer. Math. Soc., Providence, RI, 1986, pp. 35–68.

[BK] S. Bloch and K. Kato, *L-functions and Tamagawa numbers of motives*, The Grothendieck Festschrift, vol. I, Progr. Math., vol. 86, Birkhäuser, Boston, 1990, pp. 333–400.

[CLS] B. Chiarellotto and B. Le Stum, *Sur la pureté de la cohomologie cristalline*, C. R. Acad. Sci. Paris Sér. I Math. **326** (1998), 961–963.

[Co] R. Coleman, *p-adic Banach spaces and families of modular forms*, Invent. Math. **127** (1997), 417–479.

[De1] P. Deligne, *Théorème de Lefschetz et critères de dégénérescence de suites spectrales*, Inst. Hautes Études Sci. Publ. Math. **35** (1968), 259–278.

[De2] P. Deligne, *La conjecture de Weil.* II, Inst. Hautes Études Sci. Publ. Math. **52** (1980), 137–252.

[Ek] T. Ekedahl, *On the adic formalism*, The Grothendieck Festschrift, Vol. II, Progr. Math., vol. 87, Birkhäuser, Boston, 1990, pp. 197–218.

[Fa] G. Faltings, *Crystalline cohomology and p-adic Galois representations*, Algebraic Analysis, Geometry, and Number Theory (Baltimore, MD, 1988), (J.-I. Igusa, ed.), Johns Hopkins University Press, Baltimore, 1990, pp. 25–79.

[Fo1] J.-M. Fontaine, *Le corps des périodes p-adiques*, Périodes p-adiques (Bures-sur-Yvette, 1988), Astérisque, vol. 223, Soc. Math. France, Paris, 1994, pp. 59–101.

[Fo2] ______, *Représentations p-adiques semi-stables*, in: Périodes p-adiques , Astérisque, vol. 223, Soc. Math. France, Paris, 1994, pp. 113–184.

[Fo3] ______, *Représentations ℓ-adiques potentiellement semi-stables*, Périodes p-adiques , Astérisque, vol. 223, Soc. Math. France, Paris, 1994, pp. 321–347.

[FI] J.-M. Fontaine and L. Illusie, *p-adic periods: a survey*, Proceedings of the Indo-French Conference on Geometry (Bombay, 1989), Hindustan Book Agency, Delhi, 1993, pp. 57–93.

[FM] J. M. Fontaine and W. Messing, *p-adic periods and p-adic étale cohomology*, Current Trends in Arithmetic Algebraic Geometry, Contemp. Math., vol. 67, Amer. Math. Soc., Providence, RI, 1987, pp. 179–207.

[FPR] J.-M. Fontaine and B. Perrin-Riou, *Autour des conjectures de Bloch et Kato: cohomologie galoisienne et valeurs de fonctions L*, Motives (Seattle, WA, 1991), 599–706, Proc. Sympos. Pure Math., vol. 55, Part 1, Amer. Math. Soc., Providence, RI, 1994, pp. 599–706.

[Gr] R. Greenberg, *Elliptic curves and p-adic deformations*, Elliptic Curves and Related Topics, CRM Proc. Lecture Notes, vol. 4, Amer. Math. Soc., Providence, RI, 1994, pp. 101–110.

[GS] R. Greenberg and G. Stevens, *p-adic L-functions and p-adic periods of modular forms*, Invent. Math. **111** (1993), 407–447.

[Hi] H. Hida, *Elementary theory of L-functions and Eisenstein series*, London Math. Soc. Stud. Texts, vol. 26, Cambridge Univ. Press, Cambridge, 1993.

[Hy] O. Hyodo, $H^1_g(K,V) = H^1_{\mathrm{st}}(K,V)$ (unpublished).

[HK] O. Hyodo and K. Kato, *Semi-stable reduction and crystalline cohomology with logarithmic poles*, Périodes p-adiques , Astérisque, vol. 223, Soc. Math. France, Paris, 1994, pp. 221–268.

[Il] L. Illusie, *Autour du théorème de monodromie locale*, Périodes p-adiques , Astérisque, vol. 223, Soc. Math. France, Paris, 1994, pp. 9–57.

[Ja1] U. Jannsen, *Continuous étale cohomology*, Math. Ann. **280** (1988), 207–245.

[Ja2] ———, *On the l-adic cohomology of varieties over number fields and its Galois cohomology*, Galois Groups over $\mathbb{Q}$ (Berkeley, CA, 1987) (Y. Ihara, K. Ribet, and J.-P. Serre, eds.), Math. Sci. Res. Inst. Publ., vol. 16,, Springer, New York-Berlin, 1989, pp. 315–360.

[Ja3] ———, *Mixed motives and algebraic K-theory*, Lecture Notes in Math., vol. 1400, Springer-Verlag, Berlin, 1990.

[deJ] A. J. de Jong, *Smoothness, semi-stability and alterations*, Inst. Hautes tudes Sci. Publ. Math. **83** (1996), 51–93.

[Ka] K. Kato, *Semi-stable reduction and p-adic étale cohomology*, Périodes p-adiques , Astérisque, vol. 223, Soc. Math. France, Paris, 1994, pp. 269–293.

[KaM] K. Kato and W. Messing, *Syntomic cohomology and p-adic étale cohomology*, Tôhoku Math. J. **44** (1992), 1–9.

[KM] N. Katz and W. Messing, *Some consequences of the Riemann hypothesis for varieties over finite fields*, Invent. Math. **23** (1974), 73–77.

[Ki] K. Kitagawa, *On standard p-adic L-functions of families of elliptic cusp forms*, p-adic Monodromy and the Birch and Swinnerton-Dyer Conjecture (Boston, MA, 1991), Contemp. Math., vol. 165, Amer. Math. Soc., Providence, RI, 1994, 81–110.

[La] A. Langer, *Local points of motives in semi-stable reduction*, Compositio Math. **116** (1999), 189–217.

[MTT] B. Mazur, J. Tate, and J. Teitelbaum, *On p-adic analogues of the conjectures of Birch and Swinnerton-Dyer*, Invent. Math. **84** (1986), 1–48.

[Mi] J. S. Milne, *Arithmetic duality theorems*, Perspect. Math., vol. 1, Academic Press, Boston, MA, 1986.

[Mo] A. Mokrane, *La suite spectrale des poids en cohomologie de Hyodo-Kato*, Duke Math. J. **72** (1993), 301–337.

[Ne1] J. Nekovář, *Kolyvagin's method for Chow groups of Kuga-Sato varieties*, Invent. Math. **107** (1992), 99–125.

[Ne2] ———, *On p-adic height pairings*, Séminaire de théorie des nombres (Paris 1990/91) (S. David, ed.), Progr. Math., vol. 108, Birkhäuser, Boston, 1993, pp. 127–202.

[Ne3] ———, *Syntomic cohomology and p-adic regulators*, 1998 (preprint; available at `http://www.dpmms.cam.ac.uk/~nekovar/papers/`)

[Ne4] ———, *Selmer complexes* (in preparation).

[Ni] W. Nizioł, *On the image of p-adic regulators*, Invent. Math. **127** (1997), 375–400.

[PR1] B. Perrin-Riou, *Fonctions L p-adiques d'une courbe elliptique et points rationelles*, Ann. Inst. Fourier (Grenoble) **43** (1993), 945–995.

[PR2] B. Perrin-Riou, *Fonctions L p-adiques des représentations p-adiques*, Astérisque, vol. 229, Soc. Math. France, Paris, 1995.

[Pl] A. Plater, *Height pairings in families of deformations*, J. Reine Angew. Math. **486** (1997), 97–127.

[RZ] M. Rapoport and T. Zink, *Über die lokale Zetafunktion von Shimuravarietäten. Monodromiefiltration und verschwindende Zyklen in ungleicher Charakteristik*, Invent. Math. **68** (1982), 21–101.

[Sa] T. Saito, *Modular forms and p-adic Hodge theory*, Invent. Math. **129** (1997), 607–620.

[Se] J.-P. Serre, *Cohomologie galoisienne*, 5th edition, Lecture Notes in Math. vol. 5, Springer-Verlag, Berlin, 1994.

[Ta] J. T. Tate, *Algebraic cycles and poles of zeta functions*, Arithmetical Algebraic Geometry (Purdue Univ., 1963) (O. F. G. Schilling, ed.), Harper & Row, New York, 1965, pp. 93–110.

[Ts] T. Tsuji, *p-adic étale cohomology and crystalline cohomology in the semi-stable reduction case*, 1996 (preprint).

Department of Pure Mathematics and Mathematical Statistics, University of Cambridge, 16 Mill Lane, Cambridge CB2 1SB, United Kingdom

E-mail address: nekovar@dpmms.cam.ac.uk

Centre de Recherches Mathématiques
CRM Proceedings and Lecture Notes
Volume **24**, 2000

Crystalline Fundamental Groups and p-adic Hodge Theory

Atsushi Shiho

Contents

1. Introduction	381
2. Definition of Fundamental Groups	383
3. Statement of the Main Theorems	386
4. The Category $C(\mathbb{Z}(\mathrm{Sch}\,/S))$	388
5. Outline of Proofs	390
Acknowledgments	396
References	396

1. Introduction

In arithmetic geometry, various cohomologies such as Betti, etale, de Rham and crystalline ones have been studied. Their definitions are quite different, but all of them satisfy some good properties (such as Poincaré duality etc.), and in some situations, there exist various comparison theorems between them: The comparison theorem between Betti cohomology and etale cohomology is proved by Artin and Grothendieck. The comparison theorem between Betti cohomology and de Rham cohomology is known as Hodge theory. The comparison theorem between de Rham cohomology and crystalline cohomology is proved by Berthelot and Ogus [**BO2**], and the comparison theorem between etale cohomology and de Rham cohomology or crystalline cohomology is known as p-adic Hodge theory, which has been developed by many people including Fontaine, Messing, Faltings, Hyodo, Kato, Tsuji and Niziol (see [**FM, KM, Fa1, Fa2, HK, Ka2, Ts1, Ts2, Ts3, Ni1, Ni2**] etc.). Grothendieck proposed the philosophy that cohomology theory is motivic (i.e., the above cohomology theories have 'deeper origin') from such facts and most mathematicians believe his philosophy today.

This philosophy is extended to (the pro-unipotent quotient of) rational fundamental group by Deligne in [**De1**]. Historically, the (topological) rational fundamental group was first considered by Quillen in [**Qui**]. The etale realization of

1991 *Mathematics Subject Classification.* 14F30, 11G09, 11F85, 14C25.
Received by the editors March 5, 1999; Accepted June 22, 1999.
This is the final version of the paper.

fundamental group was defined by Grothendieck. The de Rham realization of rational fundamental group was constructed independently by Sullivan [**Su**] and Chen [**Ch**]. The theory of mixed Hodge structure on rational fundamental group was considered by Morgan [**Mo**] and Hain [**Ha2**], and Deligne gave the construction of these fundamental groups by means of Tannakian categories and fiber functors in [**De1**]. That is, he defined Betti, l-adic, and de Rham realizations of rational fundamental groups of smooth algebraic varieties as the groups of tensor automorphisms of fiber functors from Tannakian categories to the categories of finite-dimensional vector spaces, and proved various comparison theorems between them. If we follow his philosophy, there should be a crystalline realization of rational fundamental group for a smooth variety X over a field k of characteristic $p > 0$, and if X is liftable to a field of characteristic zero, there should be the comparison theorem between de Rham rational fundamental group and crystalline rational fundamental group (which is the π_1-version of Berthelot-Ogus theorem), and there should be the comparison theorem between etale fundamental group and de Rham rational fundamental group or crystalline rational fundamental group (which is the π_1-version of p-adic Hodge theory). But in [**De1**], the crystalline realization is defined only for smooth varieties which are liftable to a field of characteristic zero, and in this case, the crystalline realization is defined by introducing the crystalline Frobenius to the de Rham realization. Since the crystalline realization should be defined for any smooth variety X over a field of positive characteristic (even if X is not liftable), and since the crystalline realization should depend only on the reduction, the construction of the crystalline rational fundamental groups in [**De1**] is not the best possible way.

In the papers [**Sh2, Sh3**], we have given various definitions (which are compatible) of crystalline rational fundamental groups for smooth varieties over a perfect field of characteristic $p > 0$ by using the theory of Tannakian categories and proved fundamental properties of them, including the comparison theorem with de Rham rational fundamental groups (the π_1-version of Berthelot-Ogus theorem). Moreover, recently we have proved a part of p-adic Hodge theory (crystalline conjecture) for fundamental groups, that is, the comparison theorem between p-adic rational fundamental groups and crystalline rational fundamental groups. In this paper, we will give a survey on the above results briefly, without proofs. Detailed proof will be given in a forthcoming paper [**Sh4**], in which more general results (for example, more about p-adic Hodge theory) will be proven.

The content of each section is as follows: In Section 2, after giving a review on Tannakian category, we will review the definitions of 'pro-unipotent version' of p-adic etale fundamental groups (which we call p-adic rational fundamental groups) and de Rham rational fundamental groups. Then we will give a definition of crystalline rational fundamental groups by using the theory of Tannakian categories. In Section 3, we will state the main theorems in this paper. The first one is a π_1-version of Berthelot-Ogus theorem, and the second one is a π_1-version of crystalline conjecture. In Section 4, we will give a definition of a category $C(\mathbb{Z}(\mathrm{Sch}/S))$ of 'complex of schemes over S' for a scheme S and define various cohomologies for objects in this category. In Section 5, we will give an outline of our proof of the main theorems. The proof consists of two steps: First, we will express various fundamental groups in terms of cohomology groups of some objects in the category $C(\mathbb{Z}(\mathrm{Sch}/S))$. Second, we will prove the crystalline conjecture for the object in

$C(\mathbb{Z}(\mathrm{Sch}/S))$ which appeared in the above expression. This is shown by reducing to the case of cohomology groups.

Notation. For a field K, denote the algebraic closure of K by $\overline{K}$. For a pro-unipotent algebraic group G over a field K of characteristic zero, we will denote the associated pro-nilpotent Lie algebra over K by $\mathrm{Lie}\, G$. We define a descending filtration $\Gamma^n \mathrm{Lie}\, G$ $(n \in \mathbb{N})$ by

$$\Gamma^0 \mathrm{Lie}\, G := \mathrm{Lie}\, G, \quad \Gamma^{n+1} \mathrm{Lie}\, G := [\Gamma^n \mathrm{Lie}\, G, \mathrm{Lie}\, G]$$

and denote the quotient $\mathrm{Lie}\, G/\Gamma^n \mathrm{Lie}\, G$ by $\mathrm{Lie}^n G$.

2. Definition of Fundamental Groups

In this section, after giving a review on Tannakian category, we will review the definitions of 'pro-unipotent version' of p-adic etale fundamental groups (which we call p-adic rational fundamental groups) and de Rham rational fundamental groups, which are essentially due to Deligne [**De1**], in the form we use later. Then we will give a definition of crystalline rational fundamental groups by using the theory of Tannakian categories. This definition was given in [**Sh2**].

First we recall the definition and basic properties on Tannakian categories.

DEFINITION 2.1. Let k be a field. A tensor category $\mathcal{C}$ is called a Tannakian category over k if it is a rigid Abelian tensor category such that $\mathrm{End}(\underline{1}) = k$ (where $\underline{1}$ is the unit object in $\mathcal{C}$) holds and there exists a fiber functor (that is, an exact faithful tensor functor) $\mathcal{C} \to \mathrm{Vec}_K$ for a field K containing k, where Vec_K denotes the category of finite-dimensional vector spaces over K. If there exists a fiber functor $\mathcal{C} \to \mathrm{Vec}_k$, then the tensor category $\mathcal{C}$ is called a neutral Tannakian category.

For a neutral Tannakian category, we have the following structure theorem ([**SR**], see also [**DM, De2**]).

THEOREM 2.2 (Saavedra Rivano). *Let k be a field and $\mathcal{C}$ be a neutral Tannakian category over k. Let ω be a fiber functor $\mathcal{C} \to \mathrm{Vec}_k$. Then the functor*

$$(k\text{-algebras}) \to (\text{groups}); \quad R \mapsto \mathrm{Aut}^{\otimes}(\mathcal{C} \xrightarrow{\omega} \mathrm{Vec}_k \to \mathrm{Mod}_R)$$

is representable by an affine pro-algebraic group over k. Here $\mathrm{Aut}^{\otimes}$ denotes the group of tensor automorphisms and Mod_R denotes the category of R-modules. We denote the representing pro-algebraic group by $G(\mathcal{C}, \omega)$ in this paper.

Moreover, the functor ω induces an equivalence of categories

$$\mathcal{C} \to \mathrm{Rep}_k(G(\mathcal{C}, \omega)),$$

where, for a pro-algebraic group G over a field k, $\mathrm{Rep}_k(G)$ denotes the category of finite-dimensional rational representations of G over k.

We are interested only in the pro-unipotent quotient of rational fundamental groups. So it is convenient to introduce the notion of nilpotence on the side of Tannakian categories as follows:

DEFINITION 2.3. Let $\mathcal{C}$ be a Tannakian category and let $\underline{1}$ be the unit object of $\mathcal{C}$. Then, an object $E \in \mathcal{C}$ is called nilpotent if there exists a filtration

$$0 = E_0 \subset E_1 \subset \cdots \subset E_n = E$$

such that $E_i/E_{i-1} \cong \underline{1}$ holds for $1 \le i \le n$. We define the nilpotent part $\mathcal{NC}$ of the Tannakian category $\mathcal{C}$ as the full Tannakian subcategory of $\mathcal{C}$ which consists of nilpotent objects. A Tannakian category $\mathcal{C}$ is said to be nilpotent if $\mathcal{C} = \mathcal{NC}$ holds.

By definition, the following corollary of Theorem 2.2 holds:

COROLLARY 2.4. *Let $\mathcal{C}$ be a nilpotent neutral Tannakian category over a field k and let $\omega\colon \mathcal{C} \to \mathrm{Vec}_k$ be a fiber functor. Then $G(\mathcal{C}, \omega)$ is a pro-unipotent algebraic group over k.*

Now we define p-adic rational fundamental groups (here p is a fixed prime number) for schemes. For a scheme X, denote the category of smooth $\mathbb{Q}_p$-sheaves by $\mathcal{C}_p(X)$. If $H^0_{\mathrm{et}}(X, \mathbb{Q}_p)$ is a field, the category $\mathcal{C}_p(X)$ is a Tannakian category. In particular, the nilpotent part $\mathcal{NC}_p(X)$ is also Tannakian. So we will define p-adic rational fundamental group as follows:

DEFINITION 2.5. Let X be a scheme and $x \to X$ be a geometric point. Assume $H^0_{\mathrm{et}}(X, \mathbb{Q}_p) = \mathbb{Q}_p$ holds. Then we define the p-adic rational fundamental group $\pi_1^p(X, x)$ of X with base point x by

$$\pi_1^p(X, x) := G(\mathcal{NC}_p(X), \omega_x).$$

Here $\omega_x : \mathcal{NC}_p(X) \to \mathrm{Vec}_{\mathbb{Q}_p}$ is the functor defined by $\omega_x(E) := E_x$ (= the stalk of E at x).

REMARK 2.6. The existence of a geometric point $x \to X$ assures that the category $\mathcal{NC}_p(X)$ is in fact a nilpotent *neutral* Tannakian category. Similar remarks hold for other rational fundamental groups, which we will define later in this section.

Let the notations be as above. Then, by Corollary 2.4, the group $\pi_1^p(X, x)$ is a pro-unipotent algebraic group over $\mathbb{Q}_p$. To see the relation between the p-adic rational fundamental group and the usual etale fundamental group of Artin-Grothendieck, let us introduce the pro-analytic topology on the group $\pi_1^p(X, x)(\mathbb{Q}_p)$ as follows.

DEFINITION 2.7. Let the notations be as above. Then the group $\pi_1^p(X, x)(\mathbb{Q}_p)$ is expressed as the projective limit

$$\pi_1^p(X, x)(\mathbb{Q}_p) = \varprojlim_G G(\mathbb{Q}_p),$$

where G runs through the quotient $\pi_1^p(X, x) \to G$ such that G is a unipotent algebraic group. Then we define the pro-analytic topology on $\pi_1^p(X, x)(\mathbb{Q}_p)$ as the projective limit of the topology on $G(\mathbb{Q}_p)$ (G is as above) which is induced by the usual topology on $\mathbb{Q}_p$.

Then, the relation between the p-adic rational fundamental group and the usual etale fundamental group is described as follows (we omit the proof, but one can check easily):

PROPOSITION 2.8. *Let X be a scheme and $x \to X$ be a geometric point. Then there exists a canonical map*

$$f\colon \pi_1^{\mathrm{et}}(X, x) \to \pi_1^p(X, x)(\mathbb{Q}_p)$$

such that f is continuous if we define the topology on the left hand side (resp. right hand side) by the profinite topology (resp. the pro-analytic topology). Moreover, the map f satisfies the following universal property: Let G be a unipotent algebraic group over $\mathbb{Q}_p$ and assume given a continuous homomorphism $g\colon \pi_1^{\mathrm{et}}(X, x) \to$

$G(\mathbb{Q}_p)$, where the topology on $G(\mathbb{Q}_p)$ is induced by that on $\mathbb{Q}_p$. Then there exists a unique homomorphism $g' : \pi_1^p(X,x) \to G$ of pro-algebraic groups over $\mathbb{Q}_p$ such that the following diagram is commutative:

$$\begin{array}{ccc} \pi_1^{\mathrm{et}}(X,x) & \xrightarrow{f} & \pi_1^p(X,x) \\ {\scriptstyle g}\downarrow & & \downarrow{\scriptstyle \mathbb{Q}_p} \\ G(\mathbb{Q}_p) & = & G(\mathbb{Q}_p). \end{array}$$

So $\pi_1^p(X,x)$ is the pro-unipotent algebraic group over $\mathbb{Q}_p$ which approximates $\pi_1^{\mathrm{et}}(X,x)$ most closely.

Next we will review the definition of de Rham rational fundamental groups. Let k be a field. For a scheme X smooth over $\operatorname{Spec} k$, let $\mathcal{C}_{\mathrm{dR}}(X/k)$ be the category of coherent sheaves with integrable connection on X over k. If $H^0_{\mathrm{dR}}(X/k)$ is a field, it is a Tannakian category. In particular, the nilpotent part $\mathcal{NC}_{\mathrm{dR}}(X/k)$ is a Tannakian category. So we define the de Rham rational fundamental group, which is a pro-unipotent algebraic group over k, as follows:

DEFINITION 2.9. Let X be a scheme over $\operatorname{Spec} k$ and let x be a k-valued point of X. Assume that $H^0_{\mathrm{dR}}(X/k) = k$ holds. Then we define the de Rham rational fundamental group $\pi_1^{\mathrm{dR}}(X/k,x)$ of X over k with base point x by

$$\pi_1^{\mathrm{dR}}(X/k,x) := G(\mathcal{NC}_{\mathrm{dR}}(X/k),\omega_x).$$

Here $\omega_x : \mathcal{NC}_{\mathrm{dR}}(X/k) \to \mathcal{NC}_{\mathrm{dR}}(x/k) \simeq \mathrm{Vec}_k$ is the functor defined by the pull-back to x.

Now we will give the definition of crystalline rational fundamental groups. We will use the category of nilpotent isocrystals (which are considered to be the p-adic analogue of smooth $\mathbb{Q}_l$-sheaves) as the Tannakian category defining crystalline rational fundamental groups. Let k be a perfect field of characteristic $p > 0$ and W be the Witt ring of k. Denote the fraction field of W by K_0. Let X be a scheme which is smooth over k. Let $\mathcal{C}_{\mathrm{cris}}(X/W)$ be the category of isocrystals on the crystalline site $(X/W)_{\mathrm{cris}}$ of X over $\operatorname{Spf} W$. If $\mathbb{Q} \otimes_{\mathbb{Z}} H^0_{\mathrm{cris}}(X/W)$ is a field, then one can show that the category $\mathcal{C}_{\mathrm{cris}}(X/W)$ is Tannakian. In particular, the nilpotent part $\mathcal{NC}_{\mathrm{cris}}(X/W)$ is also a Tannakian category. So we can define the crystalline rational fundamental group, which is a pro-unipotent algebraic group over K_0, as follows:

DEFINITION 2.10. Let X be a scheme which is smooth over k, and let x be a k-valued point of X. Assume $\mathbb{Q} \otimes_{\mathbb{Z}} H^0_{\mathrm{cris}}(X/W) = K_0$ holds. Then we define the crystalline rational fundamental group $\pi_1^{\mathrm{cris}}(X/W,x)$ of X over W with base point x by

$$\pi_1^{\mathrm{cris}}(X/W,x) := G(\mathcal{NC}_{\mathrm{cris}}(X/W),\omega_x).$$

Here $\omega_x : \mathcal{NC}_{\mathrm{cris}}(X/W) \to \mathcal{NC}_{\mathrm{cris}}(x/W) \simeq \mathrm{Vec}_{K_0}$ is the functor defined as the pull-back to $(x/W)_{\mathrm{cris}}$.

REMARK 2.11. 1. In the papers [**Sh2, Sh3**], crystalline rational fundamental groups are called as crystalline fundamental groups, for short.
2. In the papers [**Sh2, Sh3, Sh4**], the definitions of various rational fundamental groups for certain log schemes are given. (For basic definitions concerning log schemes, see [**Ka1**].) In the case of log schemes, it is no more

true that, for example, the logarithmic analogue of the category $\mathcal{C}_{\mathrm{dR}}(X/k)$ is Tannakian. But one can check that the nilpotent part $\mathcal{NC}_{\mathrm{dR}}(X/k)$ is Tannakian under certain condition.

3. There is a definition of the crystalline rational fundamental groups by using the category of nilpotent isocrystals on (log) convergent site ([**Sh2**]) or the category of nilpotent overconvergent isocrystals ([**Sh3**]). It is shown in [**Sh2, Sh3**] that they are in fact compatible in an appropriate sense.

3. Statement of the Main Theorems

In this section, we will state the main theorems in this survey paper. The first one is the rational-π_1-version of the Berthelot-Ogus theorem, which is the comparison theorem between crystalline rational fundamental groups and de Rham rational fundamental groups. The second one is the rational-π_1-version of the crystalline conjecture in p-adic Hodge theory, which is the comparison theorem between p-adic rational fundamental groups and crystalline rational fundamental groups.

REMARK 3.1. The usual crystalline conjecture (crystalline conjecture for cohomology groups) has been already proved by Faltings [**Fa1**], Tsuji [**Ts1**] and Niziol [**Ni1**] by different methods.

Throughout this section, let k be a perfect field of characteristic $p > 0$, let W be the Witt ring of k, and let V be a complete discrete valuation ring of mixed characteristic whose residue field is k. Denote the fraction field of W and V by K_0, K respectively. Now we state our first main theorem:

THEOREM 3.2 (Berthelot-Ogus theorem for rational π_1). *Consider the following diagram of schemes*

$$\begin{array}{ccccc} X & \longrightarrow & \widetilde{X} & \longleftarrow & X_0 \\ \downarrow & & \downarrow{\scriptstyle f} & & \downarrow \\ \operatorname{Spec} K & \longrightarrow & \operatorname{Spec} V & \longleftarrow & \operatorname{Spec} k, \end{array}$$

where two squares are Cartesian and f is proper smooth geometrically connected morphism. Let $\tilde{x}$ be a V-valued point of $\widetilde{X}$ and denote $\tilde{x} \otimes_V K$ and $\tilde{x} \otimes_V k$ by x, x_0 respectively. Then there exist canonical isomorphisms of nilpotent Lie algebras

$$\mathrm{Lie}^m\, \pi_1^{\mathrm{dR}}(X/K, x) \cong \mathrm{Lie}^m\, \pi_1^{\mathrm{cris}}(X_0/W, x_0) \otimes_{K_0} K,$$

which are compatible with respect to m.

Before we state the crystalline conjecture for rational fundamental groups, we will recall some properties of the rings of Fontaine B_{dR}, B_{cris} which are needed here. (For definitions of these rings, see [**Fo2**], for example.)

B_{dR} is a complete discrete valuation field with residue field $\widehat{\overline{K}}$ (:= the p-adic completion of the algebraic closure of K). In particular, it is a K-algebra and it has a descending filtration $\{\mathrm{Fil}^n B_{\mathrm{dR}}\}_{n=-\infty}^{\infty}$ induced by the discrete valuation. Moreover, B_{dR} is endowed with an action of the Galois group $\mathrm{Gal}(\overline{K}/K)$.

B_{cris} is a subring of B_{dR} and which is stable under the action of $\mathrm{Gal}(\overline{K}/K)$. Moreover, it is a K_0-algebra and it is endowed with an automorphism $\varphi\colon B_{\mathrm{cris}} \to B_{\mathrm{cris}}$ which is Frobenius-linear as an automorphism of a K_0-algebra.

Then, we can state the crystalline conjecture for rational fundamental groups as follows:

THEOREM 3.3 (Crystalline conjecture for rational π_1). *Consider the following diagram of schemes*

$$\begin{array}{ccccc} X & \longrightarrow & \widetilde{X} & \longleftarrow & X_0 \\ \downarrow & & \downarrow{\scriptstyle f} & & \downarrow \\ \operatorname{Spec} K & \longrightarrow & \operatorname{Spec} V & \longleftarrow & \operatorname{Spec} k, \end{array}$$

where two squares are Cartesian and f is a proper smooth geometrically connected morphism. Let $\tilde{x}$ be a V-valued point of $\widetilde{X}$ and denote $\tilde{x} \otimes_V K$ and $\tilde{x} \otimes_V k$ by x, x_0 respectively. Denote $X \otimes_K \overline{K}$ and $x \otimes_K \overline{K}$ by $\overline{X}, \bar{x}$ respectively. Then there exist isomorphisms of nilpotent Lie algebras

$$\operatorname{Lie}^m \pi_1^p(\overline{X}, \bar{x}) \otimes_{\mathbb{Q}_p} B_{\mathrm{cris}} \cong \operatorname{Lie}^m \pi_1^{\mathrm{cris}}(X_0/W, x_0) \otimes_{K_0} B_{\mathrm{cris}},$$

which are compatible with respect to m. Moreover, the above isomorphisms satisfy the following properties:

1. *The above isomorphisms are functorial with respect to the pair $(\widetilde{X}, \tilde{x})$.*
2. *If we define the action of the Galois group $\operatorname{Gal}(\overline{K}/K)$ on the left hand side (resp. the right hand side) by $\sigma \otimes \sigma$ (resp. $\mathrm{id} \otimes \sigma$) for $\sigma \in \operatorname{Gal}(\overline{K}/K)$, the above isomorphisms are compatible with the action of $\operatorname{Gal}(\overline{K}/K)$.*
3. *If we define the action of the Frobenius operator φ on the left hand side (resp. the right hand side) by $\mathrm{id} \otimes \varphi$ (resp. $F^* \otimes \varphi$, where F^* is the automorphism of $\operatorname{Lie}^m \pi_1^{\mathrm{cris}}(X_0/W, x_0)$ induced by the absolute Frobenius $F\colon X_0 \to X_0$), the above isomorphisms are compatible with the action of φ.*
4. *Tensoring with B_{dR} over B_{cris} and using the Berthelot-Ogus theorem for rational π_1, we obtain the isomorphisms*

   $$\operatorname{Lie}^m \pi_1^p(\overline{X}, \bar{x}) \otimes_{\mathbb{Q}_p} B_{\mathrm{dR}} \cong \operatorname{Lie}^m \pi_1^{\mathrm{dR}}(X/K, x) \otimes_K B_{\mathrm{dR}}.$$

   If we define the descending filtration Fil^n ($n \in \mathbb{Z}$) on the left hand side (resp. the right hand side) by

   $$\operatorname{Fil}^n := \operatorname{Lie}^m \pi_1^p(\overline{X}, \bar{x}) \otimes_{\mathbb{Q}_p} \operatorname{Fil}^n B_{\mathrm{dR}},$$

   (resp.

   $$\operatorname{Fil}^n := \sum_{i+j=n} \operatorname{Fil}^i \operatorname{Lie}^m \pi_1^{\mathrm{dR}}(X/K, x) \otimes_K \operatorname{Fil}^j B_{\mathrm{dR}},$$

   where $\operatorname{Fil}^i \operatorname{Lie}^m \pi_1^{\mathrm{dR}}(X/K, x)$ is the Hodge filtration defined by Wojtkowiak [**Wo**], *which we recall in Remark* 5.6), *the above isomorphism preserves the filtrations.*

REMARK 3.4. 1. In p-adic Hodge theory for cohomologies, more general result (stable conjecture) is known, which is proved by Tsuji [**Ts1**], Faltings [**Fa2**] and Niziol [**Ni2**]. We can also prove a rational-π_1-version of stable conjecture by modifying our proof of Theorem 3.3, and it will be written in [**Sh4**].

2. In the paper [**Sh2**], we proved Theorem 3.2 by comparing Tannakian categories directly. Here we will present another method of proof, which will be included in [**Sh4**].

4. The Category $C(\mathbb{Z}(\mathrm{Sch}/S))$

In this section, we will introduce the category $C(\mathbb{Z}(\mathrm{Sch}/S))$ of 'complex of schemes over S' for a scheme S and define various cohomologies for objects in this category.

A complex of schemes is a generalization of the notion of cosimplicial scheme which is considered, for example, in [**Wo**]. The important point is that, by using complexes of schemes, one can also obtain the structure of algebraic groups on rational fundamental groups from geometric morphisms, which we will see in the next section. It is analogous to the fact that the group structure of the topological fundamental group of a topological space X is induced by the structure of the H-space on the loop space of X.

On the other hand, we would like to remark that the notion of complexes of schemes over a spectrum of a field is related to the derived category of mixed motives of Hanamura [**Han**] or Voevodsky [**Vo**]: In fact, a complex of schemes defines naturally an object in the derived category of mixed motives of Hanamura or Voevodsky, and our objects are very special ones in their categories. But we would like to treat only complexes of schemes for some reasons: partly because we need to work over the spectrum of discrete valuation rings, partly because it is easy to treat (especially when we treat crystalline cohomologies), and partly because it is sufficient to treat only them for our purpose.

Let S be a Noetherian scheme, and denote the category of schemes of finite type over S by (Sch/S). First we introduce the category $\mathbb{Z}(\mathrm{Sch}/S)$ as follows:

DEFINITION 4.1. Let the notations be as above. Then we define the category $\mathbb{Z}(\mathrm{Sch}/S)$ as follows: An object is the same as that in (Sch/S). And for X, $Y \in \mathbb{Z}(\mathrm{Sch}/S)$, the set of homomorphisms is defined by

$$\mathrm{Hom}_{\mathbb{Z}(\mathrm{Sch}/S)}(X,Y) := \bigoplus_{i\in I}\Big(\bigoplus_{f\in \mathrm{Hom}_{(\mathrm{Sch}/S)}(X_i,Y)} \mathbb{Z}\cdot f\Big),$$

where $X = \coprod_{i\in I} X_i$ is the decomposition of X into the connected components.

Then, we define the category $C(\mathbb{Z}(\mathrm{Sch}/S))$ as follows:

DEFINITION 4.2. Let the notations be as above. Then the category $C(\mathbb{Z}(\mathrm{Sch}/S))$ is defined as follows: An object in this category is a diagram in the category $\mathbb{Z}(\mathrm{Sch}/S)$ of the form

$$\cdots X_{-1} \xrightarrow{f_{-1}} X_0 \xrightarrow{f_0} X_1 \xrightarrow{f_1} X_2 \xrightarrow{f_2} \cdots$$

such that $X_{-n} = \varnothing$ holds for sufficiently large n and $f_{i+1}\circ f_i = 0$ holds for all $i \in \mathbb{Z}$. We will denote the above object simply by $\{X_i, f_i\}$. A morphism between $X := \{X_i, f_i\}$ and $Y := \{Y_i, g_i\}$ is a collection $h := \{h_i\}$ of morphisms $h_i\colon X_i \to Y_i$ $(i \in \mathbb{Z})$ in the category $\mathbb{Z}(\mathrm{Sch}/S)$ such that $g_i \circ h_i = h_{i+1}\circ f_i$ holds for all $i \in \mathbb{Z}$.

We will define n-truncation and tensor product for objects in the category $C(\mathbb{Z}(\mathrm{Sch}/S))$ as follows:

DEFINITION 4.3. Let $X := \{X_i, f_i\}$ be an object in $C(\mathbb{Z}(\mathrm{Sch}/S))$. Then the n-truncation $X^{\leq n} := \{X_i^{\leq n}, f_i^{\leq n}\}$ is defined as follows:

$$X_i^{\leq n} := \begin{cases} X_i & \text{for } i \leq n, \\ \varnothing & \text{for } i > n, \end{cases} \qquad f_i^{\leq n} := \begin{cases} f_i & \text{for } i < n, \\ 0 & \text{for } i \geq n. \end{cases}$$

DEFINITION 4.4. Let $X := \{X_i, f_i\}$, $Y := \{Y_i, g_i\}$ be objects in $C(\mathbb{Z}(\mathrm{Sch}/S))$. Then we define the tensor product $X \otimes Y := \{(X \otimes Y)_i, h_i\}$ as follows:

$$(X \otimes Y)_i := \coprod_{j+k=i} X_j \times_S Y_k,$$

$$h_i := \sum_{j+k=i} f_j \times \mathrm{id} + (-1)^j \mathrm{id} \times g_k.$$

Next we define various cohomologies of sheaves on complex of schemes. Here we will consider one of the following three situations:

SITUATION 4.5.

1. Let S be a Noetherian scheme. Fix an integer n and for $X \in (\mathrm{Sch}/S)$, denote the category of complexes of sheaves of $\mathbb{Z}/n\mathbb{Z}$-modules on the etale site of X by $\mathrm{Sh}(X)$. Let $\bullet$ be the word 'etale' (or 'et' for short), and let $S_\bullet$ be the etale site of S.
2. Let S be a Noetherian scheme and for $X \in (\mathrm{Sch}/S)$, denote the category of complexes of sheaves of $\mathcal{O}_X$-modules on Zariski site of X by $\mathrm{Sh}(X)$. Let $\bullet$ be the word 'Zariski' (or 'Zar' for short), and let $S_\bullet$ be the Zariski site of S.
3. Assume given a triple $(\tilde{S}, I, \gamma)$, where $\tilde{S}$ is a Noetherian scheme, I is a sheaf of ideals on $\tilde{S}$, and γ is a PD-structure on the ideal I. Let S be the closed subscheme of $\tilde{S}$ defined by the ideal I. For $X \in (\mathrm{Sch}/S)$, denote the category of complexes of sheaves of $\mathcal{O}_{X/\tilde{S}}$-modules on crystalline site $(X/\tilde{S})_{\mathrm{cris}}$ by $\mathrm{Sh}(X)$. Let $\bullet$ be the word 'crystalline' (or 'cris' for short), and let $S_\bullet$ be the crystalline site $(S/\tilde{S})_{\mathrm{cris}}$.

Let the situation be as in one of the above situations and assume given a finite commutative diagram $D := \{X_i, f_j\}_{i \in I, j \in J}$ in the category in $\mathbb{Z}(\mathrm{Sch}/S)$, where I, J are index sets. By definition of the category $\mathbb{Z}(\mathrm{Sch}/S)$, the morphism f_j has the following expression:

$$f_j := \sum_{k \in K_j} c_k g_k,$$

where each c_k is a non-zero integer, each g_k is a morphism in (Sch/S), and K_j is an index set (depending on j). For i, $i' \in I$, let $M(i, i')$ be the set of morphisms from a connected component of X_i to $X_{i'}$ in (Sch/S) which are realized by a composite of g_k's, and let M be the set $\coprod_{i,i' \in I} M(i, i')$. Then we define the notion of complex of sheaves on D as follows:

DEFINITION 4.6. Let the notations be as above. Then a complex of sheaves F on D is the following set of data:

1. For each $i \in I$, an element F_{X_i} in $\mathrm{Sh}(X_i)$.
2. For each $f\colon Y \to X_{i'}$ in M (where Y is a connected component of X_i), a morphism $F(f)\colon F_{X_{i'}} \to f_*(F_{X_i}|_Y)$, which satisfies the cocycle condition for composable morphisms in M.

Here we give some examples.

EXAMPLE 4.7. Let $X := \{X_i, f_i\}$ be an object in $C(\mathbb{Z}(\mathrm{Sch}/S))$ which is bounded above (that is, $X^{\leq m} = X$ holds for some $m \in \mathbb{N}$). Then we can regard X naturally as a finite diagram in $\mathbb{Z}(\mathrm{Sch}/S)$.

1. Let N be a $\mathbb{Z}/n\mathbb{Z}$-module and let the situation be as in 1. in Situation 4.5. For $i \in \mathbb{N}$, define F_{X_i} by $F_{X_i} := N$ (constant sheaf which is sitting in degree 0). Then, F_{X_i}'s form a complex of sheaves on X, with the natural transition maps. We will denote this complex of sheaves on X simply by N.
2. Let the situation be as in 2. in Situation 4.5. For $i \in \mathbb{N}$, define F_{X_i} by $F_{X_i} := \Omega^{\cdot}_{X_i/S}$. Then, F_{X_i}'s form a complex of sheaves on X, with the natural transition maps. We will denote this complex of sheaves on X by $\Omega^{\cdot}_{X/S}$.
3. Let the situation be as in 3. in Situation 4.5. For $i \in \mathbb{N}$, define F_{X_i} by $F_{X_i} := \mathcal{O}_{X_i/\tilde{S}}$ (sitting in degree 0). Then, F_{X_i}'s form a complex of sheaves on X, with the natural transition maps. We will denote this complex of sheaves on X by $\mathcal{O}_{X/\tilde{S}}$.

Let the situation be as in Situation 4.5. Let $X := \{X_i, f_i\}$ be a complex of schemes which is bounded above, and let $F := \{\{F_{X_i}\}_i, \{F(f)\}_{f \in M}\}$ be a complex of sheaves on X, where M is as in the paragraph before Definition 4.6. Now we define the $\bullet$-cohomology of X with coefficient F.

First, one can show (by using the results in [**SD**, §1]) that there exists a homomorphism of complexes of sheaves on X

$$\varphi\colon F \to I := \{\{I_{X_i}\}_i, \{I(f)\}_{f \in M}\}$$

such that the induced homomorphisms $F_{X_i} \to I_{X_i}$ are quasi-isomorphisms for all i and I_{X_i} consists of flasque sheaves on X_i. Let $p_i\colon X_i \to S$ be the structure morphism. For $g \in M(i, i+1)$, denote the morphism $p_{i+1,*}I_{X_{i+1}} \to p_{i,*}I_{X_i}$, which is obtained by applying $p_{i+1,*}$ to $I(g)$, by $p_*I(g)$. For $g \in M(i, i+1)$, let $c_g \in \mathbb{Z}$ be the coefficient of g in f_i, and define

$$p_*I(f_i)\colon p_{i+1,*}I_{X_{i+1}} \to p_{i,*}I_{X_i},$$

by $p_*I(f_i) := \sum_{g \in M(i,i+1)} c_g p_*I(g)$. Then $p_*I := (p_{i,*}I_{X_i}, p_*I(f_i))$ defines a complex in the category $\mathrm{Sh}(S)$, that is, a double complex of sheaves on S. Then we define the $\bullet$-cohomology of X with coefficient F by

$$H_{\bullet}(X, F) := H^m(R\Gamma(S_{\bullet}, \mathrm{tot}(p_*I))).$$

One can show that this definition is independent of the choice of the homomorphism $\varphi\colon F \to I$.

Next, let the situation be as in Situation 4.5 and let $X := \{X_i, f_i\}$ and $Y := \{Y_i, g_i\}$ be complexes of schemes which are bounded above and let $h := \{h_i\}$ be a morphism from X to Y. Then we can regard $D := \{X_i, Y_i, f_i, g_i, h_i\}$ as a diagram in $\mathbb{Z}(\mathrm{Sch}/S)$. Let F be a complex of sheaves on the diagram D. Then one can show that there exists the morphism of $\bullet$-cohomology

$$h^*\colon H^m_{\bullet}(Y, F) \to H^m_{\bullet}(X, F),$$

which is naturally induced by h. Details will be written in [**Sh4**].

5. Outline of Proofs

In this section, we will give an outline of the main theorems stated in Section 3. The proof consists of two steps: First, we will express various fundamental groups in terms of cohomology groups of an object (which plays a role of 'loop space') in the category $C(\mathbb{Z}(\mathrm{Sch}/S))$. Second, we will prove the Berthelot-Ogus theorem and the crystalline conjecture for the object in $C(\mathbb{Z}(\mathrm{Sch}/S))$ which appeared in the above expression.

First, we will introduce an object in the category $C(\mathbb{Z}(\mathrm{Sch}/S))$ which plays a role of 'loop space'.

DEFINITION 5.1. Let X be an object in (Sch/S) and let x be an element of $X(S)$. Then we define a complex of schemes $\Omega_x X := \{(\Omega_x X)_n, d_n\}$ over S as follows: $(\Omega_x X)_n$ is the n-fold fiber product of X over S (which we will denote by X^n), and $d_n : (\Omega_x X)_n \to (\Omega_x X)_{n+1}$ is defined by $d_n := \sum_{i=0}^{n+1}(-1)^i d_n^i$, where d_n^i is defined by the following formula:

$$\begin{aligned} d_n^0(x_1,\dots,x_n) &= (x,x_1,\dots,x_n),\\ d_n^i(x_1,\dots,x_n) &= (x_1,\dots,x_i,x_i,\dots,x_n) \quad (1\le i\le n),\\ d_n^{n+1}(x_1,\dots,x_n) &:= (x_1,\dots,x_n,x). \end{aligned}$$

Next, we introduce several morphisms between $\Omega_x X$'s or their tensor products.

DEFINITION 5.2. Let X and x be as in the above definition and denote $\Omega_S S$ by the symbol $*$.

1. We define a morphism of complex of schemes
$$\Lambda := \{\Lambda_n\}_{n\ge 0} : \Omega_x X \to \Omega_x X \otimes \Omega_x X$$
as follows: For $i, j \ge 0$ such that $i+j=n$ holds, define the map $\Lambda_{i,j} : X^n \to X^i \times_S X^j$ by
$$\Lambda_{i,j} := \sum_\sigma \mathrm{sgn}(\sigma) f_\sigma,$$
where σ runs through (i,j)-shuffles and the map f_σ is defined by
$$f_\sigma((x_1,\dots,x_n)) := ((x_{\sigma(1)},\dots,x_{\sigma(i)}),(x_{\sigma(i+1)},\dots,x_{\sigma(n)})).$$
Then the map
$$\Lambda_n : X^n \to \coprod_{i+j=n} X^i \times_S X^j$$
is defined by $\Lambda_n := \sum_{i+j=n} \iota_{i,j}\circ\Lambda_{i,j}$, where $\iota_{i,j}$ is the natural inclusion $X^i\times_S X^j \hookrightarrow \coprod_{i+j=n} X^i\times_S X^j$.
2. We define a morphism of complex of schemes
$$\Delta := \{\Delta_n\}_{n\ge 0} \colon \Omega_x X\otimes\Omega_x X \to \Omega_x X$$
as follows: For $i, j\ge 0$ such that $i+j=n$ holds, define the map
$$\Delta_{i,j}\colon X^i\times_S X^j \to X^n$$
by
$$\Delta_{i,j}((x_1,\dots,x_i),(x_{i+1},\dots,x_n)) := (x_1,\dots,x_n)$$
and define Δ_n by $\Delta_n := \sum_{i+j=n}\Delta_{i,j}$.
3. We define a morphism of complex of schemes
$$e := \{e_n\}_{n\ge 0}\colon * \to \Omega_x X$$
by defining e_n by
$$e_n := (x,x,\dots,x)\colon S^n = S \to X^n.$$

One can check, by direct calculation, that the above morphisms are well-defined, that is, they define a morphism in the category $C(\mathbb{Z}(\mathrm{Sch}/S))$.

We define p-adic, de Rham and crystalline cohomology of 'loop spaces' as follows:

DEFINITION 5.3. 1 (p-adic version). Let X be a proper scheme over $\operatorname{Spec} k$, where k is an algebraically closed field and let x be a k-rational point of X. Then we define $H^*_p(\Omega_x X)$ by

$$H^*_p(\Omega_x X) := \varinjlim_n (\mathbb{Q} \otimes_{\mathbb{Z}} \varprojlim_m H^*_{\mathrm{et}}(\Omega_x X^{\leq n}, \mathbb{Z}/p^m\mathbb{Z})).$$

2 (de Rham version). Let k be a field of characteristic zero and let X be a scheme smooth over k and let x be a k-valued point of X. Then we define $H^*_{\mathrm{dR}}(\Omega_x X)$ by

$$H^*_{\mathrm{dR}}(\Omega_x X) := \varinjlim_n H^*_{\mathrm{Zar}}(\Omega_x X^{\leq n}, \Omega^{\cdot}_{\Omega_x X^{\leq n}/k}).$$

3 (Crystalline version). Let k be a perfect field of characteristic $p > 0$ and let W be the Witt ring of k. Denote the fraction field of W by K_0. Let X be a scheme which is proper smooth over k and x be a k-valued point of X. Then we define $H^*_{\mathrm{cris}}(\Omega_x X)$ by

$$H^*_{\mathrm{cris}}(\Omega_x X) := \varinjlim_n (\mathbb{Q} \otimes_{\mathbb{Z}} H^*_{\mathrm{cris}}(\Omega_x X^{\leq n}/W, \mathcal{O}_{\Omega_x X^{\leq n}/W})).$$

Let $\bullet$ be one of the words $\{p, \mathrm{dR}, \mathrm{cris}\}$, and when $\bullet = p$ (resp. $\bullet = \mathrm{dR}$, $\bullet = \mathrm{cris}$) holds, assume we are in the situation of 1. (resp. 2., 3.) in Definition 5.3. Then, the condition in Definition 5.3 assures that there exists a canonical isomorphism of graded vector spaces

$$H^*_\bullet(\Omega_x X) \otimes H^*_\bullet(\Omega_x X) \cong H^*_\bullet(\Omega_x X \otimes \Omega_x X)$$

which is induced by Künneth morphism. Then, the morphisms Λ, Δ and e induce the homomorphisms

$$\Lambda^* \colon H^0_\bullet(\Omega_x X) \otimes H^0_\bullet(\Omega_x X) \to H^0_\bullet(\Omega_x X),$$
$$\Delta^* \colon H^0_\bullet(\Omega_x X) \to H^0_\bullet(\Omega_x X) \otimes H^0_\bullet(\Omega_x X),$$
$$e^* \colon H^0_\bullet(\Omega_x X) \to H^0_\bullet(*) = k_\bullet,$$

where $k_\bullet$ is defined by

$$k_p := \mathbb{Q}_p, \quad k_{\mathrm{dR}} := k, \quad k_{\mathrm{cris}} := K_0.$$

Let

$$\Delta' \colon H^0_\bullet(\Omega_x X) \to H^0_\bullet(\Omega_x X) \otimes H^0_\bullet(\Omega_x X)$$

be the morphism $2^{-1}(\Delta - \tau \circ \Delta)$, where $\tau(x \otimes y) = y \otimes x$ $(x, y \in H^0_\bullet(\Omega_x X))$, and define $QH^0_\bullet(\Omega_x X)$ by

$$QH^0_\bullet(\Omega_x X) := \operatorname{Ker}(e^*)/\Lambda^*(\operatorname{Ker}(e^*) \otimes \operatorname{Ker}(e^*)).$$

Then one can check, by direct calculation, the following proposition:

PROPOSITION 5.4. *Let the situations be as above. Then Δ' induces the morphism*

$$\Delta' \colon QH^0_\bullet(\Omega_x X) \to \Lambda^2 QH^0_\bullet(\Omega_x X),$$

and $QH^0_\bullet(\Omega_x X)$ is endowed a structure of (ind-nilpotent) coLie algebra over $k_\bullet$ by the morphism Δ'.

Then, the first step for the proof of the main theorem is described as follows:

THEOREM 5.5. 1 (p-adic version). *Let X be a scheme which is proper over* $\operatorname{Spec} k$*, where k is an algebraically closed field and let x be a k-rational point of X. Assume moreover that $H^0_{\text{et}}(X, \mathbb{Q}_p) = \mathbb{Q}_p$ holds. Then we have the canonical isomorphism*

$$\operatorname{Lie} \pi_1^p(X, x) \cong (QH^0_p(\Omega_x X))^*$$

of pro-nilpotent Lie algebras over $\mathbb{Q}_p$.

2 (de Rham version). *Let k be a field of characteristic zero. Let X be a scheme which is smooth over k and let x be a k-valued point of X. Assume moreover that $H^0_{\text{dR}}(X/k) = k$ holds. Then we have the canonical isomorphism*

$$\operatorname{Lie} \pi_1^{\text{dR}}(X/k, x) \cong (QH^0_{\text{dR}}(\Omega_x X))^*$$

of pro-nilpotent Lie algebras over k.

1. (Crystalline version). *Let k be a perfect field of characteristic $p > 0$ and let W be the Witt ring of k. Denote the fraction field of W by K_0. Let X be a scheme which is proper smooth over k and let x be a k-valued point of X. Assume moreover that $\mathbb{Q} \otimes_{\mathbb{Z}} H^0_{\text{cris}}(X/W) = K_0$ holds. Then we have the canonical isomorphism*

$$\operatorname{Lie} \pi_1^{\text{cris}}(X/W, x) \cong (QH^0_{\text{cris}}(\Omega_x X))^*$$

of pro-nilpotent Lie algebras over K_0.

We will not give a proof of the above theorem here, since we need some tools such as bar constructions. Details will be written in [**Sh4**].

REMARK 5.6. Here we remark on the Hodge filtration on the Lie algebra of the de Rham fundamental group $\operatorname{Lie}^m \pi_1^{\text{dR}}(X/k, x)$. By the above theorem, it suffices to define the Hodge filtration on the group

$$H^*_{\text{dR}}(\Omega_x X) := \varinjlim_n H^*_{\text{Zar}}(\Omega_x X^{\leq n}, \Omega^{\cdot}_{\Omega_x X^{\leq n}/k}).$$

For $l \in \mathbb{N}$, let $\operatorname{Fil}^l \Omega^{\cdot}_{\Omega_x X^{\leq n}/k}$ be the stupid filtration of $\Omega^{\cdot}_{\Omega_x X^{\leq n}/k}$, that is,

$$\operatorname{Fil}^l \Omega^i_{\Omega_x X^{\leq n}/k} := \begin{cases} 0 & \text{for } i < l, \\ \Omega^i_{\Omega_x X^{\leq n}/k} & \text{for } i \geq l. \end{cases}$$

Then the Hodge filtration $\operatorname{Fil}^l H^*_{\text{dR}}(\Omega_x X)$ on $H^*_{\text{dR}}(\Omega_x X)$ is defined by

$$\operatorname{Fil}^l H^*_{\text{dR}}(\Omega_x X) := \operatorname{Im}(\varinjlim_n H^*_{\text{Zar}}(\Omega_x X^{\leq n}, \operatorname{Fil}^l \Omega^{\cdot}_{\Omega_x X^{\leq n}/k}) \to H^*_{\text{dR}}(\Omega_x X)).$$

Since we have expressed various rational fundamental groups in terms of cohomology groups, what we have to show is the following theorems, which is the second step for the proof of the main theorems:

THEOREM 5.7. *Let the notations be as in Theorem 3.2. Then we have the canonical isomorphisms*

$$H^0_{\text{dR}}(\Omega_x X) \cong H^0_{\text{cris}}(\Omega_{x_0} X_0) \otimes_{K_0} K$$

which are compatible with homomorphisms Λ^*, Δ^* and e^*.

THEOREM 5.8. *Let the notations be as in Theorem 3.3. Then we have the canonical isomorphisms*

$$H_p^0(\Omega_{\bar{x}}\overline{X}) \otimes_{\mathbb{Q}_p} B_{\mathrm{cris}} \cong H^0_{\mathrm{cris}}(\Omega_{x_0}X_0) \otimes_{K_0} B_{\mathrm{cris}},$$

which are compatible with homomorphisms Λ^*, Δ^* and e^* and satisfy the following properties:

1. *The above isomorphism is functorial with respect to the pair $(\widetilde{X}, \tilde{x})$.*
2. *If we define the action of the Galois group $\mathrm{Gal}(\overline{K}/K)$ on the left hand side (resp. right hand side) by $\sigma \otimes \sigma$ (resp. $\mathrm{id} \otimes \sigma$) for $\sigma \in \mathrm{Gal}(\overline{K}/K)$, the above isomorphism is compatible with the action of $\mathrm{Gal}(\overline{K}/K)$.*
3. *If we define the action of the Frobenius operator φ on the left hand side (resp. right hand side) by $\mathrm{id} \otimes \varphi$ (resp. $F^* \otimes \varphi$, where F^* is the automorphism of $H^0_{\mathrm{cris}}(\Omega_{x_0}X_0)$ induced by the absolute Frobenius $F\colon X_0 \to X_0$), the above isomorphism is compatible with the action of φ.*
4. *Tensoring with B_{dR} over B_{cris} and using Theorem 5.7, we obtain the isomorphism*

$$H_p^0(\Omega_{\bar{x}}\overline{X}) \otimes_{\mathbb{Q}_p} B_{\mathrm{dR}} \cong H^0_{\mathrm{cris}}(\Omega_{x_0}X_0) \otimes_K B_{\mathrm{dR}}.$$

If we define the descending filtration Fil^n $(n \in \mathbb{Z})$ of the left hand side (resp. right hand side) by

$$\mathrm{Fil}^n := H_p^0(\Omega_{\bar{x}}\overline{X}) \otimes_{\mathbb{Q}_p} Fil^n B_{\mathrm{dR}},$$

(resp.

$$\mathrm{Fil}^n := \sum_{i+j=n} \mathrm{Fil}^i\, H^0_{\mathrm{dR}}(\Omega_x X) \otimes_K \mathrm{Fil}^j\, B_{\mathrm{dR}},$$

where $\mathrm{Fil}^i\, H^0_{\mathrm{dR}}(\Omega_x X)$ is the Hodge filtration defined in Remark 5.6), the above isomorphism preserves the filtration.

We will give an outline of the above theorems briefly. First let us consider Theorem 5.7. By definition of cohomologies, we have the spectral sequences

$$E_1^{s,t} = H^t_{\mathrm{dR}}(X^{-s}/K) \implies H^{s+t}_{\mathrm{dR}}(\Omega_x X),$$
$$E_1^{s,t} = H^t_{\mathrm{cris}}(X_0^{-s}/W) \otimes \mathbb{Q} \implies H^{s+t}_{\mathrm{cris}}(\Omega_{x_0} X_0),$$

where, for a negative integer s, X^s and X_0^s is defined to be empty. By following the method of Berthelot-Ogus in [**BO2**], we can construct the homomorphism

$$\Phi_n\colon H^n_{\mathrm{dR}}(\Omega_x X) \to H^n_{\mathrm{cris}}(\Omega_{x_0}X_0) \otimes_{K_0} K$$

and we can show that there exists the following morphism of spectral sequences

$$\begin{array}{ccc} H^t_{\mathrm{dR}}(X^{-s}/K) & \Longrightarrow & H^{s+t}_{\mathrm{dR}}(\Omega_x X) \\ {\scriptstyle \Phi_{s,t}}\downarrow & & \downarrow{\scriptstyle \Phi_{s+t}} \\ H^t_{\mathrm{cris}}(X_0^{-s}/W) \otimes_W K @ & \Longrightarrow & H^{s+t}_{\mathrm{cris}}(\Omega_{x_0}X_0) \otimes_{K_0} K, \end{array}$$

where $\Phi_{s,t} : H^t_{\mathrm{dR}}(X^{-s}/K) \to H^t_{\mathrm{cris}}(X_0^{-s}/W) \otimes_W K$ is the isomorphism of the usual Berthelot-Ogus theorem. Since $\Phi_{s,t}$ are isomorphisms, Φ_n is isomorphic by the above diagram. In particular, Φ_0 is an isomorphism, as desired. The compatibility with Λ^*, Δ^* and e^* follows from the functoriality of the construction of Φ_n.

Next let us consider Theorem 5.8. Fix an integer n, and let us denote the group $\mathbb{Q}\otimes\varprojlim_m H^*_{\mathrm{et}}(\Omega_{\bar{x}}\overline{X}^{\leq n},\mathbb{Z}/p^m\mathbb{Z}(r))$ simply by $H^*_p(\Omega_{\bar{x}}\overline{X}^{\leq n})(r)$ and the group $\mathbb{Q}\otimes H^*_{\mathrm{cris}}(\Omega_{x_0}X_0^{\leq n}/W)$ simply by $H^*_{\mathrm{cris}}(\Omega_{x_0}X_0^{\leq n})$. Then we have spectral sequences

$$E_1^{s,t} := H^t_{\mathrm{et}}(\overline{X}^{-a(s)},\mathbb{Q}_p(r)) \implies H^{s+t}_p(\Omega_{\bar{x}}\overline{X}^{\leq n})(r),$$
$$E_1^{s,t} := \mathbb{Q}\otimes H^t_{\mathrm{cris}}(X_0^{-a(s)}/W) implies H^{s+t}_{\mathrm{cris}}(\Omega_{x_0}X_0^{\leq n}),$$

where $a(s)$ is defined by

$$a(s) := \begin{cases} s & \text{for } s\geq -n,\\ 1 & \text{for } s< -n,\end{cases}$$

and for a negative integer s, $\overline{X}^s$ and X_0^s are defined to be empty. Moreover, by following the method use by Tsuji [**Ts1**], we can define the syntomic cohomology $H^*(\Omega_{\bar{x}}\overline{X}^{\leq n},\mathcal{S}^{\sim}_{\mathbb{Q}_p}(r))$ of $\Omega_{\bar{x}}\overline{X}^{\leq n}$ and one can show that there exists the following spectral sequence for large r:

$$E_1^{s,t} := H^t(\overline{X}^{-a(s)},\mathcal{S}^{\sim}_{\mathbb{Q}_p}(r)) \implies H^{s+t}(\Omega_{\bar{x}}\overline{X}^{\leq n},\mathcal{S}^{\sim}_{\mathbb{Q}_p}(r)).$$

Again by following the method of Tsuji, we can define the homomorphisms

$$\tilde{\alpha}_{m,r}\colon H^m(\Omega_{\bar{x}}\overline{X}^{\leq n},\mathcal{S}^{\sim}_{\mathbb{Q}_p}(r))\otimes_{\mathbb{Q}_p}B_{\mathrm{cris}}\to H^m_{\mathrm{cris}}(\Omega_{x_0}X_0^{\leq n})\otimes_{K_0}B_{\mathrm{cris}},$$
$$\tilde{\beta}_{m.r}\colon H^m(\Omega_{\bar{x}}\overline{X}^{\leq n},\mathcal{S}^{\sim}_{\mathbb{Q}_p}(r))\otimes_{\mathbb{Q}_p}B_{\mathrm{cris}}\to H^m_p(\Omega_{\bar{x}}\overline{X}^{\leq n})\otimes_{\mathbb{Q}_p}B_{\mathrm{cris}},$$

such that they fit into the following morphisms of spectral sequences:

(5.1)
$$\begin{array}{ccc} H^t(\overline{X}^{-a(s)},\mathcal{S}^{\sim}_{\mathbb{Q}_p}(r))\otimes_{\mathbb{Q}_p}B_{\mathrm{cris}} & \Longrightarrow & H^{s+t}(\Omega_{\bar{x}}\overline{X}^{\leq n},\mathcal{S}^{\sim}_{\mathbb{Q}_p}(r))\otimes_{\mathbb{Q}_p}B_{\mathrm{cris}}\\ \downarrow{\scriptstyle \alpha'_{s,t,r}} & & \downarrow{\scriptstyle \tilde{\alpha}_{s+t,r}}\\ H^t_{\mathrm{cris}}(X_0^{-a(s)}/W)\otimes_W B_{\mathrm{cris}} & \Longrightarrow & H^{s+t}_{\mathrm{cris}}(\Omega_{x_0}X_0^{\leq n})\otimes_{K_0}B_{\mathrm{cris}},\end{array}$$

(5.2)
$$\begin{array}{ccc} H^t(\overline{X}^{-a(s)},\mathcal{S}^{\sim}_{\mathbb{Q}_p}(r))\otimes_{\mathbb{Q}_p}B_{\mathrm{cris}} & \Longrightarrow & H^{s+t}(\Omega_{\bar{x}}\overline{X}^{\leq n},\mathcal{S}^{\sim}_{\mathbb{Q}_p}(r))\otimes_{\mathbb{Q}_p}B_{\mathrm{cris}}\\ \downarrow{\scriptstyle \beta'_{s,t,r}} & & \downarrow{\scriptstyle \tilde{\beta}_{s+t,r}}\\ H^t_{\mathrm{et}}(\overline{X}^{-a(s)},\mathbb{Q}_p(r))\otimes_{\mathbb{Q}_p}B_{\mathrm{cris}} & \Longrightarrow & H^{s+t}_{\mathrm{et}}(\Omega_{\bar{x}}\overline{X}^{\leq n},\mathbb{Q}_p(r))\otimes_{\mathbb{Q}_p}B_{\mathrm{cris}},\end{array}$$

where $\alpha'_{s,t,r}$ (resp. $\beta'_{s,t,r}$) is the homomorphism naturally induced by the homomorphism $\alpha_{t,r}$ (resp. $\beta_{t,r}$) for $\widetilde{X}^{-a(s)}$ defined in [**Ts1**, 4.8] (resp. [**Ts1**, 4.9]). By [**Ts1**, 3.4.4], $\beta'_{s,t,r}$ is isomorphic if $t\leq r$ holds. By this and the diagram (5.2), one can check that $\tilde{\beta}_{0,r}$ is isomorphic for large r. Let us fix sufficiently large r and an element $0\neq t\in\mathbb{Q}_p(1)$, and define the homomorphism

$$\tilde{\gamma}\colon H^0_p(\Omega_{\bar{x}}\overline{X}^{\leq n})\otimes_{\mathbb{Q}_p}B_{\mathrm{cris}}\to H^0_{\mathrm{cris}}(\Omega_{x_0}X_0^{\leq n})\otimes_{K_0}B_{\mathrm{cris}}$$

by the composite

$$H^0_p(\Omega_{\bar{x}}\overline{X}^{\leq n})\otimes_{\mathbb{Q}_p}B_{\mathrm{cris}}\xrightarrow{t^r\otimes\mathrm{id}}H^0_p(\Omega_{\bar{x}}\overline{X}^{\leq n})(r)\otimes_{\mathbb{Q}_p}B_{\mathrm{cris}}$$
$$\xrightarrow{\tilde{\alpha}_{0,r}\circ\tilde{\beta}_{0,r}^{-1}}H^0_{\mathrm{cris}}(\Omega_{x_0}X_0^{\leq n})\otimes_{K_0}B_{\mathrm{cris}}\xrightarrow{\mathrm{id}\otimes t^{-r}}H^0_{\mathrm{cris}}(\Omega_{x_0}X_0^{\leq n})\otimes_{K_0}B_{\mathrm{cris}}.$$

One can check that $\tilde{\gamma}$ is independent of the choice of r and t. Then one can show that $\tilde{\gamma}$ is isomorphic, by using the diagrams (5.1) and (5.2) and reducing to the crystalline conjecture for cohomology groups. By taking the inductive limit with respect to n, we obtain the desired isomorphism

$$H_p^0(\Omega_{\bar{x}}\overline{X}) \otimes_{\mathbb{Q}_p} B_{\text{cris}} \to H_{\text{cris}}^0(\Omega_{x_0}X_0) \otimes_{K_0} B_{\text{cris}}.$$

The properties described in Theorem 5.8 is shown by the method similar to Hyodo-Kato-Tsuji ([**Ka2, Ts1**]). Details will be given in [**Sh4**].

Acknowledgments

The author would like to express his profound gratitude to Professor Takeshi Saito for valuable advice. He also would like to express his thanks to Professors Yukiyoshi Nakajima, Takeshi Tsuji and Nobuo Tsuzuki for valuable comments on his work. He would like to express his thanks to Takako Fukaya and Professor Kazuya Kato for encouragement while he was working on this subject, and to Professor Takeshi Kajiwara for encouragement while he was writing this survey paper. He would like to express his thanks to Professors Shuji Saito and Noriko Yui for giving him an opportunity to give a talk at the Banff Conference and an opportunity to write this survey paper.

The author was supported by JSPS Research Fellowship for Young Scientists in 1996-97, when [**Sh2**] and a part of [**Sh3**] were written. He was partially supported by Kawai Foundation when he attended to the Banff Conference. Currently, he is supported in part by the Grants-in-Aid for the Encouragement of Young Scientists, the Ministry of Education, Science, Sports and Culture, Japan.

References

[Be1] P. Berthelot, *Cohomologie cristalline des schémas de caractéristique $p > 0$*, Lecture Notes in Math., vol. 407, Springer-Verlag, Berlin-New York.

[Be2] ———, *Cohomologie rigide et cohomologie rigide à supports propres* (preprint).

[BO1] P. Berthelot and A. Ogus, *Notes on crystalline cohomology*, Princeton Univ. Press, Princeton, NJ, 1978.

[BO2] ———, *F-isocrystals and de Rham cohomology.* I, Invent. Math. **72** (1983), 159–199.

[BG] A. K. Bousfield and V. K. A. M. Gugenheim, *On* PL *de Rham theory and rational homotopy type*, Mem. Amer. Math. Soc., vol. 8, no. 179, Amer. Math. Soc., Providence, RI, 1976.

[Ch] K.-T. Chen, *Iterated path integrals*, Bull. Amer. Math. Soc. **83** (1977), 831–879.

[De1] P. Deligne, *Le groupe fondamental de la droite projective moins trois points*, Galois Groups over $\mathbb{Q}$ (Berkeley, CA, 1987), Math. Sci. Res. Inst. Publ., vol. 16, Springer, New York-Berlin, 1989, pp. 79–297.

[De2] ———, *Catégories tannakiennes*, The Grothendieck Festschrift. II, Progr. Math., vol. 87, Birkhäuser, Boston, 1990, pp. 111–195.

[DM] P. Deligne and J. S. Milne, *Tannakian categories*, Hodge Cycles, Motives, and Shimura Varieties, Lecture Note in Math., vol. 900, Springer Verlag, 1982, pp. 101–228.

[Fa1] G. Faltings, *Crystalline cohomology and p-adic etale cohomology*, Algebraic Analysis, Geometry and Number Theory (Baltimore, MD, 1988) (J-I. Igusa, ed.), Johns Hopkins Univ. Press, Baltimore, MD, 1989, pp. 25–80.

[Fa2] ———, *Almost etale extensions*, preprint no. MPI1998-3, Max-Planck-Institut für Mathematik, Bonn, 1998.

[Fo1] J.-M. Fontaine, *Sur certain types de représentations p-adiques du groupe de Galois d'un corps locaux; construction d'un anneau de Barsotti-Tate*, Ann. of Math. (2) **115** (1982), 529–577.

[Fo2] ———, *Le corps des périodes p-adiques*, Périodes p-adiques (Bures-sur-Yvette, 1988), Astérisque, vol. 223, Soc. Math France, Paris, 1994, pp. 59–111.

[FM] J.-M. Fontaine and W. Messing, *p-adic periods and p-adic étale cohomology*, Current Trends in Arithmetical Algebraic Geometry (Arcata, CA, 1985), Contemp. Math., vol. 67, Amer. Math. Soc., Providence, RI, 1987, pp. 179–207.

[Gr] sga1 A. Grotendieck (ed), *Revêtements étales et groupe fondamental*, Lecture Notes in Math., vol. 224, Springer-Verlag, Berlin-New York, 1971.

[GM] P. A. Griffiths and J. W. Morgan, *Rational homotopy theory and differential forms*, Progr. Math., vol. 16, Birkhäuser, Boston, MA, 1981.

[Ha1] R. M. Hain, *The geometry of the mixed Hodge structure on the fundamental group*, Algebraic Geometry, Bowdoin, 1985 (Brunswick, MN, 1985), Proc. Sympos. Pure Math., vol. 46, Part 2, Amer. Math. Soc., Providence, RI, 1987, pp. 247–282.

[Ha2] ———, *The de Rham homotopy theory of complex algebraic variety*. I, K-theory **1** (1987), 271–324.

[Han] M. Hanamura, *Mixed motives and algebraic cycles*. II, preprint no. MPI1997-106, Max-Planck-Institut für Mathematik, 1997.

[HK] O. Hyodo and K. Kato, *Semi-stable reduction and crystalline cohomology with logarithmic poles*, Périodes p-adiques (Bures-sur-Yvette, 1988), Astérisque, vol. 223, Soc. Math France, Paris, 1994, pp. 221–268.

[Ka1] K. Kato, *Logarithmic structures of Fontaine-Illusie*, Algebraic Analysis, Geometry and Number Theory (Baltimore, MD, 1988) (J-I. Igusa, ed.), Johns Hopkins Univ. Press, Baltimore, MD, 1989, pp. 191–224.

[Ka2] ———, *Semi-stable reduction and p-adic etale cohomology*, Périodes p-adiques (Bures-sur-Yvette, 1988), Astérisque, vol. 223, Soc. Math France, Paris, 1994, pp. 269–293.

[KM] K. Kato and W. Messing, *Syntomic cohomology and p-adic étale cohomology*, Tôhoku Math. J. **44** (1992), 1–9.

[Mo] J. W. Morgan, *The algebraic topology of smooth algebraic varieties*, Inst. Hautes Études Sci. Publ. Math. **48** (1978), 137–204.

[NA] V. Navarro Aznar, *Sur la theorie de Hodge-Deligne*, Invent. Math. **90** (1987), 11–76.

[Ni1] W. Niziol, *Crystalline conjecture via K-theory* (preprint).

[Ni2] ———, *Semi-stable conjecture for vertical log-smooth families* (preprint).

[O1] A. Ogus, *F-isocrystals and de Rham cohomology*. II. *Convergent isocrystals*, Duke Math. J. **51** (1984), 765–850.

[O2] ———, *The convergent topos in characteristic p*, The Grothendieck Festschrift. III, Progr. Math., vol. 88, Birkhäuser, Boston, 1990, pp. 133-162..

[Qui] D. Quillen, *Rational homotopy theory*, Ann. of Math. (2) **90** (1969), 205–295.

[SR] N. Saavedra Rivano, *Catégories tannakiennes*, Lecture Notes in Math., vol. 265, Springer-Verlag, Berlin-New York, 1972.

[SD] B. Saint-Donat, *Techniques de descente cohomologique*, Théorie des topos et cohomologie étale des schémas, Lecture Notes in Math., vol. 270, Springer-Verlag, Berlin-New York, 1972, pp. 83–162.

[Sh1] A. Shiho, *Tangential map and log scheme*, Hokkaido Univ. Tech. Rep. Ser. in Math **47** (1996), 28–36 (Japanese).

[Sh2] ———, *Crystalline fundamental groups* I. *Isocrystals on log crystalline site and log convergent site*, 1997 (preprint).

[Sh3] ———, *Crystalline fundamental groups*. II. *Overconvergent isocrystals*, 1997 (preprint).

[Sh4] ———, *Fundamental groups, homotopy groups and p-adic Hodge theory* (in preparation).

[Su] D. Sullivan, *Infinitesimal computation in topology*, Inst. Hautes tudes Sci. Publ. Math. **47** (1977), 269–331.

[Ts1] T. Tsuji, *p-adic étale cohomology and crystalline cohomology in the semi-stable reduction case* (preprint).

[Ts2] ———, *On p-adic Hodge theory*, Sūrikaisekikenkyūsho Kōkyūroku **1026** (1998) 63–75 (Japanese).

[Ts3] ———, *p-adic Hodge theory in the semi-stable reduction case*, Proceedings of the International Congress of Mathematicians, Vol. II (Berlin, 1998) Doc. Math. (1998), Extra Vol. II, 207–216.

[Vo] V. Voevodsky, *Triangulated categories of motives over a field* (preprint).

[Wo] Z. Wojtkowiak, *Cosimplicial objects in algebraic geometry*, Algebraic K-Theory and Algebraic Topology (Lake Louise, AB, 1991) (P. G. Goerss and J. F. Jardine, eds.), NATO

Adv. Sci. Inst. Ser. C Math. Phys. Sci., vol. 407, Kluwer Acad. Publ., Dordrecht, 1993, pp. 287–327.

MATHEMATICAL INSTITUTE, TÔHOKU UNIVERSITY, AOBA, SENDAI 980-8578 JAPAN
E-mail address: shiho@math.tohoku.ac.jp

Centre de Recherches Mathématiques
CRM Proceedings and Lecture Notes
Volume **24**, 2000

Thompson Series, and the Mirror Maps of Pencils of $K3$ Surfaces

Helena Verrill and Noriko Yui

ABSTRACT. The mirror maps of pencils of certain $K3$ surfaces are shown to be related in a mysterious way to Thompson series arising from some genus zero subgroups of $\mathrm{PSL}_2(\mathbb{R})$. The pencils of $K3$ surfaces discussed here are one-(or multi-)parameter deformations of weighted orbifold Calabi-Yau hypersurfaces of dimension 2, and those obtained via toric construction from root lattices such as A_1^3, A_3 (and possibly C_3).

This note is intended to highlight the mysterious relationship, the so-called *Lian-Yau mirror moonshine phenomenon* between mirror maps and Thompson series (and ultimately the Monster and Monstrous Moonshine) with a number of explicit examples.

Some open problems and generalizations will also be discussed. This article is meant to give a brief exposition of a monograph by Verrill and Yui [**VY**]. It is also hoped to complement the article by Doran [**Dor**] in this volume, where he demystifies the Lian-Yau mirror-moonshine phenomenon from the point of view of Picard-Fuchs differential equations.

CONTENTS

1. Introduction 400
2. The Monster and Thompson series 402
3. Lattice Polarized $K3$ Surfaces 406
4. Moduli Spaces of Lattice Polarized $K3$ Surfaces 409
5. Picard-Fuchs Differential Equations and Monodromy Groups 412
6. The Mirror Maps and Thompson Series. I 415
7. The Mirror Maps and Thompson Series. II 417
8. Remarks and Open Problems on Mirror Maps 421
9. The Mirror $K3$ Surfaces, and Mirror Symmetry 424
10. Generalizations 426
Acknowledgments 430
References 430

1991 *Mathematics Subject Classification.* 14D05, 11F03, 14J28.
Received by the editors October 16, 1998; Accepted February 18, 1999.
This is the final form of the paper.

1. Introduction

Let $\Gamma = \mathrm{PSL}(2,\mathbb{Z})$ and $\mathfrak{H} := \{z \in \mathbb{C} \mid \mathrm{Im}(z) > 0\}$ be the upper half complex plane. It is a well known classical fact that the elliptic modular j-invariant is a uniformizer of level 1, that is, $j\colon \mathfrak{H}^*/\Gamma \to \mathbb{P}^1(\mathbb{C})$ gives a complex analytic isomorphism, and that it has a Fourier expansion at ∞ of the form:

$$j(q) = \frac{1}{q} + 744 + \sum_{n\geq 1} c_j(n)q^n \quad \text{with } c_j(n) \in \mathbb{Z} \text{ all } n \geq 1$$

where $q := e^{2\pi i z}$, $z \in \mathfrak{H}$. Its reciprocal $1/j(q)$ is a holomorphic function with q-expansion of the form:

$$\frac{1}{j(q)} = q - 744q^2 + 356652q^3 - 140361152q^4 + O(q^5).$$

The power series $j(q)$ also arises from the representation theory of the Monster group $\mathbb{M}$, in the following way. Put $J(q) = j(q) - 744$. Let $V = V_{-1} + \sum_{n\geq 1} V_n$ be a graded $\mathbb{Q}$-vector space with $\dim(V_{-1}) = 1$ and $\dim(V_n) = c_j(n)$ for each $n \geq 1$. There is a reducible character $\rho_n\colon \mathbb{M} \to \mathrm{GL}(V_n)$ which sends $g \in \mathbb{M}$ to $\rho_n(g) := \mathrm{Tr}(g \mid V_n)$. This gives rise to the power series, called a *Thompson series*:

$$T_g(q) := q^{-1} + \sum_{n\geq 1} \mathrm{Tr}(g \mid V_n)q^n.$$

In particular, if $g = \mathrm{Id}_{\mathbb{M}}$, the identity element of $\mathbb{M}$, the corresponding Thompson series is given by $T_{\mathrm{Id}_{\mathbb{M}}} = J(q) = j(q) - 744$.

The above construction may be extended in the following way: Γ may be replaced by certain *genus zero congruence subgroup G of level $N > 1$* of Γ, and j by the corresponding (normalized, or specified) *uniformizer*. The Monstrous Moonshine Conjecture of Conway and Norton [**CN**] asserts that *the Thomson series $T_g(q)$ (for any element $g \in \mathbb{M}$) should coincide with the Fourier expansion of a (normalized, or specified) uniformizer for some genus zero subgroup G of level ≥ 1 (depending on the order of g) of* Γ. The Monstrous Moonshine Conjecture has been proved in the affirmative by Borcherds [**Bo**], using generalized Kac-Moody algebras and the no-ghost theorem from string theory.

Recently, Lian and Yau [**LY1, LY2**] have noticed the so-called the *mirror-moonshine phenomenon* that Thompson series appear as the mirror maps of pencils of certain lattice polarized $K3$ surfaces with generic Picard number 19. By a $K3$ surface, we mean a simply connected compact complex manifold of dimension 2 with a nowhere vanishing holomorphic 2-form. Any $K3$ surface X is equipped with a unique lattice structure in the sense that $H_2(X,\mathbb{Z})$ is a lattice isomorphic to the so-called $K3$-*lattice* $\mathcal{L} := U_2^3 \perp (-E_8)^2$. (Here U_2 is the hyperbolic lattice of rank 2 and E_8 is the even unimodular negative definite lattice of rank 8.) Let $\mathcal{M} \subset \mathcal{L}$ be a sublattice of rank $t+1$ and signature $(1,t)$ with $t \leq 19$. The orthogonal complement $\mathcal{M}^\perp$ in $\mathcal{L}$ is a lattice of signature $(2, 20-(t+1))$. An embedding $\mathcal{M} \hookrightarrow \mathcal{L}$ is said to be *primitive* if $\mathcal{L}/\mathcal{M}$ is free. As a consequence of the surjectivity of the period map (see Morrison [**Mo**, Corollary 1.9]), for any primitive sublattice $\mathcal{M} \hookrightarrow \mathcal{L}$ of signature $(1,t)$ (resp. $\mathcal{T} \hookrightarrow \mathcal{L}$ of signature $(2, 20-(t+1))$ with $t \leq 19$, there exists an algebraic $K3$ surface X and an isometry (as lattices) $\mathrm{Pic}(X) \simeq \mathcal{M}$ (resp. $\mathrm{Pic}(X)^\perp \simeq \mathcal{T}$). Here $\mathrm{Pic}(X)$ is the group of algebraic cycles on X; it is a free finitely generated Abelian group of rank $\rho(X) \leq 20$, called the Picard group of X. $\mathrm{Pic}(X)^\perp$ denotes the orthogonal complement of $\mathrm{Pic}(X)$ in $H_2(X,\mathbb{Z})$. (A $K3$ surface is called *algebraic* if its function field has the transcendental degree 2.)

Let $\mathcal{M} \subset \mathcal{L}$ be an even sublattice of rank $t+1$ and signature $(1,t)$ which has a unique embedding into $\mathcal{L}$. An *$\mathcal{M}$-polarized $K3$ surface* is a $K3$ surface X with a fixed primitive embedding $\mathcal{M}$ into the Picard group $\mathrm{Pic}(X)$. We further assume that the fixed primitive embedding $\mathcal{M} \hookrightarrow \mathrm{Pic}(X)$ contains a pseudo-ample or ample divisor class. (An pseudo-ample class is a class D such that $D^2 > 0$ and $D \cdot C \geq 0$ for all curves C on the $K3$ surface X.) For instance, if $\mathcal{M}$ is hyperbolic (i.e., contains U_2), then this assumption is fulfilled. Then one can construct the moduli space of those $\mathcal{M}$-polarized $K3$ surfaces via the period map and the period domain. It has dimension $19-t$, and in some cases, it may be identified with well-understood spaces (e.g., the quotient of the symmetric domain of type IV on $\mathcal{T} = \mathcal{M}^\perp$ by some arithmetic group). For detailed discussion on moduli spaces of lattice polarized $K3$ surfaces and their structures, the reader is referred to the excellent article by Dolgachev [**Dol**].

We will confine ourselves to some specific lattices of rank $t+1 = 19$, e.g.,

$$\mathcal{M} = U_2 \perp (-E_8)^2 \perp \langle -2N \rangle \quad \text{for some integer } N \geq 1.$$

We will construct pencils of $\mathcal{M}$-polarized $K3$ surfaces with generic Picard number 19. The moduli space of such $\mathcal{M}$-polarized $K3$ surfaces has dimension 1, and furthermore, is identified with the space $\mathfrak{H}/\Gamma_0(N)^+$ for some genus zero subgroup $\Gamma_0(N)^+$. The group $\Gamma_0(N)^+$ (or its subgroups of finite index) may be identified with the monodromy group of the Picard-Fuchs differential equation of pencils of $\mathcal{M}$-polarized $K3$ surfaces.

The mirror map of a pencil of $\mathcal{M}$-polarized $K3$ surfaces X_z with parameter $z \in \mathbb{P}^1$ is naively defined to be the inverse of the period map. More precisely, the Picard-Fuchs differential equation of a pencil of $\mathcal{M}$-polarized $K3$ surfaces X_z is of order 3, and near a maximally degenerating point (which may be assumed to be the point $z=0$), it has a unique holomorphic solution

$$w_0(z) = 1 + \sum_{n\geq 1} c_n z^n$$

and also another solution of the form

$$w_1(z) = \log w_0(z) + \sum_{n\geq 1} d_n z^n.$$

Introduce the new variable

$$2\pi i t := \frac{w_1(z)}{w_0(z)}.$$

Then $q = e^{2\pi i t}$ gives an invertible analytic map from a disc $|z| < S$ to some disc $|q| < R$. The inverse, $z = z(q)$ with $z(0) = 1$ is called the *mirror map* of the family X_z. By the definition, the mirror maps are holomorphic in the discs of definition. Lian and Yau [**LY2**] have proved that the reciprocal of the mirror maps have q-expansions with integer coefficients, and have derived congruences for the coefficients modulo p, which in turn used to prove the divisibility of the instanton numbers by certain integers. Further they have observed that the reciprocal of the mirror maps plus some constant happen to be Thompson series $T_g(q)$ ($g \in \mathbb{M}$) corresponding to some genus zero subgroup of Γ. Based on their observation, they have formulated the so-called *mirror-moonshine conjecture*, that the mirror maps of pencils of certain lattice polarized $K3$ surfaces are *modular*, and this is the subject of Doran's paper [**Dor**] in this volume.

In this paper, we will discuss several examples of pencils of lattice polarized $K3$ surfaces with Picard number 19 whose mirror maps are not always modular, but commensurable with Thompson series. (Here two holomorphic q-series $f(q)$ and $g(q)$ are said to be *commensurable* if there is a polynomial $P(x,y) \in \mathbb{Q}[x,y]$ such that $P(f(q),g(q)) = 0$.) The mirror maps which are commensurable with Thompson series take transcendental values for any algebraic number in $\mathfrak{H}$ not belonging to an imaginary quadratic field.

We will also include possible generalizations of our discussions on Picard number 19 $K3$ surfaces to those $K3$ surfaces with smaller Picard numbers, e.g., ≤ 18. As an illustration, we construct pencils of $\mathcal{M}$-polarized $K3$ surfaces with

$$\mathcal{M} = U_2 \perp (-E_8)^2 \quad \text{so } \mathcal{M}^{\perp} = U_2^2.$$

In this case, the moduli space has dimension 2, and the period domain may be identified with certain Hilbert modular spaces. We extend the definition of the mirror maps, and then compute them. We observe that the mirror maps in our examples are commensurable not only with Thompson series, but also with modular functions associated to higher genus modular groups.

This project has been somewhat motivated by the mirror symmetry conjecture for $K3$ surfaces. This is mainly based on observation (or experience) that the period integrals of $K3$ surfaces appear to be more tractable in computation on mirror partners, or it is easier to construct lattice polarized $K3$ surfaces with large Picard numbers if one passes onto mirror partners. However, due to the lack of time and space, we will not go into detailed discussions on mirror symmetry for $K3$ surfaces here; the readers interested on these developments are referred to the vast bodies of literatures on the subject. There seems to be several variants of "mirror symmetry" for $K3$ surfaces. Some of the origins of mirror symmetry phenomenon might be traced back to the Arnold strange duality [**Ar**]. Our source of references include Dolgachev [**Dol**], Kobayashi [**Ko**], Batyrev [**Ba**], Gritsenko and Nikulin [**GN1, GN2, GN3**], Kyoji Saito [**Sa**], and W. Ebeling [**Eb**] (and references therein), among others.

2. The Monster and Thompson series

2.1. The Monster group. Let $\mathbb{M}$ denote the Monster group. It is the largest sporadic finite simple group with order, $\#\mathbb{M} \sim 8 \times 10^{53}$. We denote by $\mathcal{S}$ the set of all primes dividing $\#\mathbb{M}$. Then

$$\mathcal{S} = \{2,3,5,7,11,13,17,19,23,29,31,41,47,59,71\}.$$

Incidentally, Thompson has proved that $\mathbb{M}$ is realizable as a Galois group over $\mathbb{Q}$ using the fact that $\mathbb{M}$ is a homomorphic image of $\hat{\pi}_1^{\mathrm{alg}}(\mathbb{P}^1 - \{0,1,\infty\})$, and the Riemann Existence Theorem. The constructive realization is, however, still open.

2.2. Genus Zero Subgroups of $\mathbf{PSL_2(\mathbb{R})}$. Let

$$\Gamma = \mathrm{PSL}_2(\mathbb{Z}) = \left\{ \begin{pmatrix} a & b \\ c & d \end{pmatrix} \,\middle|\, a,b,c,d \in \mathbb{Z}, ad-bc = 1 \right\} \Big/ \{\pm I_2\}.$$

For any integer $N \geq 2$, let

$$\Gamma_0(N) := \left\{ \begin{pmatrix} a & b \\ c & d \end{pmatrix} \in \mathrm{PSL}_2(\mathbb{Z}) \,\middle|\, c \equiv 0 \pmod{N} \right\},$$

and

$$\Gamma_0(N)+ := \text{the normalizer of } \Gamma_0(N) \text{ in } \mathrm{PSL}_2(\mathbb{R}).$$

Let

$$\mathfrak{H} := \{z = x + iy \in \mathbb{C} \mid y > 0\}$$

be the upper half complex plane. Then Γ acts on $\mathfrak{H}$ by the linear fractional transformation. The compactified quotient space

$$\mathfrak{H}^*/\Gamma := \mathfrak{H}/\Gamma \cup \{i\infty\}$$

is a compact Riemann surface of genus zero, and hence there is a uniformizer

$$j\colon \mathfrak{H}^*/\Gamma \simeq \mathbb{P}^1(\mathbb{C}), \quad z \mapsto j(z)$$

where j is the elliptic modular j-function. If we set $q = e^{2\pi i z}$ with $z \in \mathfrak{H}$, it has the Fourier expansion at ∞:

$$j(q) = q^{-1} + 744 + 196884q + 21493760q^2 + \cdots = q^{-1} + 744 + \sum_{n=1}^{\infty} c_j(n)q^n$$

with $c_j(n) \in \mathbb{Z}$ for every n. The j function is called the *level* 1 *uniformizer*.

Observe that the reciprocal of $j(q)$ has the Fourier expansion also with integer coefficients whose first six terms are given as follows:

$$\begin{aligned}\frac{1}{j(q)} &= \frac{q}{(1 + 744q + \sum_{n=1}^{\infty} c_j(n)q^{n+1})} \\ &= q - 744q^2 + 356652q^3 - 140361152q^4 + 49336682190q^5 \\ &\qquad - 16114625669088q^6 + O(q^7).\end{aligned}$$

The function $1/j(q)$ will have an interpretation as the mirror map of a certain family of $K3$ surfaces.

More generally, we may consider certain discrete subgroups of $PSL_2(\mathbb{R})$. Let G be a discrete subgroup of $PSL_2(\mathbb{R})$ such that $\Gamma_0(N) \subseteq G \subseteq \Gamma_0(N)+$ for some integer $N \geq 1$. Then we can define the quotient space $\mathfrak{H}^*/G = \mathfrak{H}/G \cup \{\text{cusps of } G\}$, which is a compact Riemann surface. We are particularly interested in those groups G which give rise to genus zero compact Riemann surfaces. We will call them *genus zero congruence subgroups of level N of $PSL_2(\mathbb{R})$.*

THEOREM 2.1 (Fricke, Ogg [**Og**] and Thompson [**Th**]). (a) *For $G = \Gamma_0(N) \subset \Gamma$, $\mathfrak{H}^*/\Gamma_0(N)$ is genus zero only for*

$$N = 1, 2\text{–}10, 12, 13, 16, 18, 25.$$

(b) *For $G = \Gamma_0(N)+ \subset \mathrm{PSL}_2(\mathbb{R})$ with N prime, $\mathfrak{H}^*/\Gamma_0(N)+$ is genus zero if and only if*

$$N = 2, 3, 5, 7, 11, 13, 17, 19, 23, 29, 31, 41, 47, 59, 71.$$

(b′) *For $G = \Gamma_0(N)+ \subset \mathrm{PSL}_2(\mathbb{R})$ with N non-prime, there are only finitely many conjugacy classes of genus zero congruence subgroups of* $\mathrm{PSL}_2(\mathbb{R})$ *which are commensurable with Γ.*

The assertions (a) and (b) are due to Fricke and Ogg, while the finiteness in (b′) is due to Thompson [**Th**].

Note that the values of prime N occurring in (b) are exactly the same primes in the set $\mathcal{S}$! If N is not prime, the list of genus zero congruence subgroups is much larger; confer the list of Conway and Norton [**CN**].

2.3. A Uniformizer or a Hauptmodul. Associated to any genus zero subgroup G such that $\Gamma_0(N) \subseteq G \subseteq \Gamma_0(N)+$ in Theorem 2.1, there is an analytic complex isomorphism between the two genus zero compact Riemann surfaces, called a *uniformizer*, or a *Hauptmodul*

$$f\colon \mathfrak{H}^*/G \to \mathbb{P}^1(\mathbb{C}), \quad z \mapsto f(z)$$

which has a pole at ∞ with residue 1, so that its Fourier expansion at ∞ is of the form:

$$f(q) = q^{-1} + \sum_{n=0}^{\infty} c_f(n)q^n \quad \text{with } c_f(n) \in \mathbb{C} \text{ for every } n \geq 0.$$

REMARKS 2.2. (a) From the definition, a uniformizer f is defined up to linear fractional transformation, that is, one can have the choice of the values of three distinct points on $\mathbb{P}^1$, so that we may impose that f has a pole at ∞ with a residue 1. Then its q-expansion start with $1/q + c_f(0) + \cdots$. After performing finite translation to f, f may still have a pole at ∞ with a residue 1, but with a different constant term. This raises a question: *Is there any canonical choice for the constant term of f?*

The constant term has no relevance in the Moonshine at all, and usually taken to be 0. Outside the Moonshine, the constant term seems to have a "canonical" choice, which was explained to us by A. Sebbar, involving the Rademacher constant.

(b) The coefficients $c_f(n)$ are complex numbers. For which genus zero subgroups G, are $c_f(n)$ algebraic integers (or a rational integers)? These questions have not been answered yet. According to R. Borcherds (unpublished), if G contains the Fricke involution, then the coefficients of $f(q) = 1/q + \cdots$ are positive (see Chen and Yui [**CY**, p. 258 Remarks (1)]). Asai, Kaneko and Ninomiya [**AKN**] have determined the signs of the coefficients of $1/j(q)$ that they strictly alternating.

EXAMPLE 2.3. We shall list several examples of genus zero congruence subgroups of $\mathrm{PSL}(2,\mathbb{R})$ whose Hauptmoduln have Fourier expansion (at ∞) with $c_f(n) \in \mathbb{Z}$ for every $n \geq 1$. We identify them with Thompson series in the table of Conway and Norton [**CN**]. Let $\eta(z)$ denote the eta-function:

$$\eta(z) = e^{2\pi i/12} \prod_{n=1}^{\infty} (1 - e^{2\pi i n z}) = q^{1/24} \prod_{n=1}^{\infty} (1 - q^n) \quad \text{where } z \in \mathfrak{H}.$$

Here $[a_{-1}, 0, a_1, a_2, a_3, \ldots]$ represents a power series $a_{-1}/q + 0 + a_1 q + a_2 q^2 + a_3 q^3 + \cdots$.

(a) Let $G = \Gamma_0(5)$. Then $j = (f^2 + f + 19)^3/f$ and

$$f(q) + 1 = [1, 0, 9, 10, -30, 6, -25, 96, 60, -250, 45, -150, \cdots].$$

This is the Thompson series 5B in [**CN**].

(b) Now we take the group $G = \Gamma_0(6)+3$. This is a subgroup of the normalizer of $\Gamma_0(6)$ in $\mathrm{PSL}_2(\mathbb{R})$ obtained by adjoining Atkin-Lehner involutions of the form

$\left(\begin{smallmatrix} 3a & b \\ 6c & 3d \end{smallmatrix}\right)$ with $a, b, c, d \in \mathbb{Z}$ to $\Gamma_0(6)$. Then $f(q) = (\eta(q)\eta(q^3)/[\eta(q^2)\eta(q^6)])^6$ and

$$f(q) + 6 = [1, 0, 15, -32, 87, -192, 343, -672, 1290, -2176, 3705, -6336, \ldots].$$

This is the Thompson series 6C in [**CN**].

(c) Let $G = \Gamma_0(13)+$. Then $f(q) = \eta(q)^2/\eta(q^{13})^2 + 13\eta(q^{13})^2/\eta(q)^2$ and

$$f(q) - 12 = [1, 0, 12, 28, 66, 132, 258, 468, 843, 1428, 2406, 3900, 6253, 9780, 15144, \ldots].$$

This is the Thompson series 13A in [**CN**].

(d) Let $G = \Gamma_0(71)+$. The Thompson series is not expressible in terms of the η-functions alone; it has the expression

$$f(q) = \frac{\theta(4,2,8) - \theta(6,2,12)}{2\eta(q)\eta(q^{71})} - 1.$$

Here θ is the theta function defined by $\theta(a,b,c) = \sum q^{(ax^2+bxy+cy^2)/2}$ where the sum runs over pairs $(x,y) \in \mathbb{Z}^2$. This is the Thompson series 71A in [**CN**], and the first few terms of the q-expansion are:

$$f(q) - \frac{1}{3} = [1, 0, 1, 1, 1, 1, 2, 2, 3, 3, 4, 4, 6, 6, 7, 8, 10, 11, 13, 14, 17, 19, 22, 24, \ldots].$$

Now we will define Thompson series from the representation theory of the Monster group.

THEOREM-DEFINITION 2.4. *There exists a graded vector space over* $\mathbb{Q}$:

$$V = V_{-1} + \sum_{n=1}^{\infty} V_n$$

with

$$\dim V_{-1} = 1, \quad \dim V_n = c_j(n) = \text{the } n\text{th coefficient of } j(q).$$

The Monster $\mathbb{M}$ *acts on this vector space respecting the grading.*

(a) *There exists a reducible character*

$$\rho_n \colon \mathbb{M} \to \mathrm{GL}(V_n) \colon g \mapsto \rho_n(g) := \mathrm{Tr}(g \mid V_n)$$

such that

$$\rho_n(g) \in \mathbb{Z} \quad \text{for every } n \geq 1 \text{ and } \rho_{-1}(g) = 1.$$

(b) *For each* $g \in \mathbb{M}$*, a* Thompson series *is defined by*

$$T_g(q) := q^{-1} + \sum_{n=1}^{\infty} \mathrm{Tr}(g \mid V_n) q^n \quad \text{with } q = e^{2\pi i z}, z \in \mathfrak{H}.$$

If $g = \mathrm{Id}_{\mathbb{M}}$*, then the Thompson series is*

$$T_1(q) = j(q) - 744.$$

Conway and Norton [**CN**] have formulated the famous conjecture on Thompson series, called the *Monstrous Moonshine*, which has been proved by Borcherds [**Bo**].

THEOREM 2.5 (Monstrous Moonshine Conjecture; Theorem of Borcherds). *A Thompson series* $T_g(q)$ *is a Hauptmodul for a genus zero subgroup* G, $\Gamma_0(N) \subseteq G \subseteq \Gamma_0(N)+$*, of level* N*, where* N *is divisible by* $o(g)$*, the order of* g.

REMARKS 2.6. (a) We will be dealing only with the *fundamental* Thompson series $T_g(q)$ (for the precise definition, confer Conway and Norton [**CN**]). In this case, $N = o(g)$, and in particular, G is the subgroup $\Gamma_0(N)$ extended by some Atkin-Lehner involutions of $\Gamma_0(N)$. In fact, if N is a prime in the set $\mathcal{S}$, then $G = \Gamma_0(N) + w_N$ where $w_N = \left(\begin{smallmatrix} 0 & 1 \\ -N & 0 \end{smallmatrix}\right) \in \mathrm{PGL}_2(\mathbb{Q})^+$.

(b) Borcherds' proof uses the Lie superalgebra with $\mathbb{Z} \times \mathbb{Z}$ grading, that is, the Frenkel-Lepowsky-Meurmann vertex operator algebra. The monster $\mathbb{M}$ acts on this Lie superalgebra respecting the grading, and one has the representation V_n of $\mathbb{M}$. His proof, however, does not explain conceptually why the formal power series $T_g(q)$ are Hauptmoduln for some genus zero subgroups of Γ. Recently Cummins and Gannon [**CG**] addressed this problem.

(c) The theorem of Borcherds implies that the coefficients $c_f(n)$ of the Fourier expansion $f(q)$ for genus zero congruence subgroups in Theorem 2.1 are rational integers for every $n \geq 1$, which answers the question in Remark2.2(b) in this case. However, the Monstrous Moonshine Conjecture does not give any information about the constant term $c_f(0)$.

3. Lattice Polarized $K3$ Surfaces

The general reference for this section is the paper by Dolgachev [**Dol**]. All surfaces are defined over $\mathbb{C}$.

DEFINITIONS 3.1. A $K3$ *surface* is a smooth simply connected surface $X \subset \mathbb{P}^3$ with trivial canonical bundle. In other words, a $K3$ surface X has the following properties:

(1) $H^1(X, \mathcal{O}_X) = 0$, and

(2) $K_X \equiv 0$, that is, the canonical bundle is trivial.

A $K3$ surface X is said to be *algebraic* if it admits an ample line bundle which defines a projective embedding of X into a projective space. Algebraic $K3$ surfaces may be characterized by the transcendental degree of the function field $\mathbb{C}(X)$ of X thus: a $K3$ surface is algebraic if and only if the transcendental degree of $\mathbb{C}(X)$ is 2.

All $K3$ surfaces discussed in this paper are assumed to be algebraic.

3.1. Numerical Characters of a $K3$ Surface. (a) The Hodge numbers of X. The (i,j)th Hodge number of X is

$$h^{i,j}(X) := \dim H^j(X, \Omega_X^i).$$

By the duality of the Hodge cohomology groups, $h^{i,j}(X) = h^{j,i}(X)$. Condition (1) implies that $h^{1,0}(X) = 0$. Condition (2) implies that the geometric genus of X, $p_g(X) := h^{0,2}(X) = \dim H^0(X, \Omega_X^2) = \dim H^0(X, \mathcal{O}_X) = 1$. The Hodge numbers of X form the Hodge diamond:

$$\begin{array}{ccccc} & & 1 & & \\ & 0 & & 0 & \\ 1 & & 20 & & 1 \\ & 0 & & 0 & \\ & & 1 & & \end{array}$$

(b) The Betti numbers of X. The i-th Betti number of X is

$$b_i(X) := \dim_{\mathbb{Q}_\ell} H^i(\overline{X}, \mathbb{Q}_\ell) = \dim_{\mathbb{C}} H^i(X, \mathbb{C}).$$

By the duality of the cohomology groups with coefficients in $\mathbb{Q}_\ell$ or $\mathbb{C}$, $b_i(X) = b_{4-i}(X)$ for i, $0 \le i \le 4$. There is the Hodge decomposition

$$b_i(X) = \sum_{j=0}^{i} h^{j,2-j}(X).$$

Therefore, $b_0(X) = 1 = b_4(X)$, $b_1(X) = 0 = b_3(X)$, and $b_2(X) = 22$.

(c) The Euler characteristic of X is

$$E(X) := \sum_{i=0}^{4} (-1)^i b_i(X) = 24.$$

EXAMPLES 3.2. (a) Any smooth quartic surface X in $\mathbb{P}^3$ is a $K3$ surface. For instance, let X be defined by a homogeneous equation of degree 4:

$$\mathcal{F}(X_0, X_1, X_2, X_3) = 0 \subset \mathbb{P}^3 \quad \text{with } \deg(\mathcal{F}) = 4,$$

e.g.,

$$\mathcal{F}(X_0, X_1, X_2, X_3) = X_0^4 + X_1^4 + X_2^4 + X_3^4 + 4\lambda X_0 X_1 X_2 X_3 = 0.$$

with some parameter λ such that $\lambda^4 \neq 1$.

(b) Consider the Fermat surface of degree m defined by the equation

$$V\colon X_0^m + X_1^m + X_2^m + X_3^m = 0 \subset \mathbb{P}^3.$$

Then V is a smooth projective surface with geometric genus $p_g(V) = \binom{m-1}{3}$. Therefore, V is a $K3$ surface if and only if $m = 4$. We wish to construct more examples of $K3$ surfaces. For this, we introduce a *weight* relative to the Fermat surface V of degree m. Let $Q = (q_0, q_1, q_2, q_3) \in (\mathbb{Z}^\times)^4$ be a quadruple of integers having the following properties:

(i) $q_i \ge 1$ and $q_i \mid m$ for every i, and

(ii) $\gcd(q_i, q_j, q_k) = 1$ for any distinct triple i, j, k.

Put $\boldsymbol{\mu}_Q = \mu_{q_0} \times \mu_{q_1} \times \mu_{q_2} \times \mu_{q_3}$ where μ_{q_i} denote the cyclic group generated by q_i-th roots of unity, ζ_{q_i}. Then $\boldsymbol{\mu}_Q$ acts on V with the action $(X_0, X_1, X_2, X_3) \mapsto (\zeta_{q_0}^{e_0} X_0, \zeta_{q_1}^{e_1} X_1, \zeta_{q_2}^{e_2} X_2, \zeta_{q_3}^{e_3} X_3)$ with $e_i \in \mathbb{Z}/q_i\mathbb{Z}$ for i, $0 \le i \le 3$. We may define the quotient $Y := V/\boldsymbol{\mu}_Q$. Then Y is defined by the equation

$$Y\colon Y_0^{m/q_0} + Y_1^{m/q_1} + Y_2^{m/q_2} + Y_3^{m/q_3} = 0 \subset \mathbb{P}^3(Q)$$

in the weighted projective 3-space $\mathbb{P}^3(Q)$ where $\deg(Y_i) = q_i$ for i, $0 \le i \le 3$. From the construction, Y is a singular surface with at most Abelian (cyclic) quotient singularities. The minimal resolution, X, of Y exists.

The main result of Goto [**Go**] (cf. Yonemura [**Yos**] where he classified all Gorenstein weighted hypersurfaces which are $K3$ surfaces) is that there are exactly 14 pairs $\langle m, Q\rangle$ for which the minimal resolutions X of $Y = V/\boldsymbol{\mu}_Q$ are $K3$ surfaces. In fact, such a pair must satisfy the additional condition

(iii) $m = q_0 + q_1 + q_2 + q_3$.

All the 14 pairs are tabulated as follows.

m	Q	Re	m	Q	Re
4	$(1,1,1,1)$		12	$(1,2,3,6)$	
6	$(1,1,1,3)$		12	$(1,1,4,6)$	
6	$(1,1,2,2)$		18	$(1,2,6,9)$	
8	$(1,1,2,4)$		20	$(1,4,5,10)$	W_{12}
10	$(1,2,2,5)$		24	$(1,3,8,12)$	E_{14}
12	$(2,3,3,4)$		30	$(2,3,10,15)$	
12	$(1,3,4,4)$	U_{12}	42	$(1,6,14,21)$	E_{12}

Under the column "Re", we list the corresponding names of the 14 exceptional unimodal hypersurface singularities of Arnold (see e.g., Dolgachev [**Dol**] or Ebeling [**Eb**] for the entire list).

(c) We may consider orbifolds of one-or multi-parameter deformations of the $K3$ surfaces in the above table.

For instance,

$$Y_0^{m/q_0} + Y_1^{m/q_1} + Y_2^{m/q_2} + Y_3^{m/q_3} + m\,zY_0Y_1Y_2Y_3 = 0$$

where z is a parameter such that $z^m \neq 1$ in the orbifold ambient space $\mathbb{P}^3(Q)/G$ with a suitable finite group G.

We may also consider two-parameter deformations, for instance, corresponding to the pairs $\langle m, Q\rangle = \langle 6,(1,1,2,2)\rangle$, $\langle 8,(1,1,2,4)\rangle$, $\langle 12,(1,1,4,6)\rangle$, we may define the two-parameter deformation:

$$Y_0^m + Y_1^m + Y_2^{m/q_2} + Y_3^{m/q_3} + z\,Y_0Y_1Y_2Y_3 + w\,Y_0^{m/2}Y_1^{m/2} = 0$$

in the orbifold ambient spaces $\mathbb{P}^3(Q)/G$ with suitable finite groups G, where z, w are parameters.

(d) Let A be an Abelian variety of dimension 2, e.g., $A = E \times E'$ a product of two elliptic curves. Let $\iota\colon A \to A$ be an involution such that $\iota(a) = -a$. The quotient $A/\langle\iota\rangle$ is a singular surface with 16 nodes. Let $\mathrm{Kum}(A)$ be a minimal resolution of $A/\langle\iota\rangle$. Then $\mathrm{Kum}(A)$ is a $K3$ surface, called the *Kummer surface*.

3.2. The $K3$ Lattice of X. For any $K3$ surface X, there is a unique holomorphic 2-form ω up to scalar multiplication (as $p_g(X) = 1$). We know that $H_2(X,\mathbb{Z}) \simeq \mathbb{Z}^{22}$. Let $\{\gamma_1, \gamma_2, \dots, \gamma_{22}\}$ be a basis for $H_2(X,\mathbb{Z})$. The cup-product (= the intersection pairing)

$$H_2(X,\mathbb{Z}) \times H_2(X,\mathbb{Z}) \to \mathbb{Z}\colon (\gamma_1,\gamma_2) \mapsto \int_X \gamma_1\cdot\gamma_2$$

yields a unimodular symmetric bilinear form, which is even of rank 22 and with signature $(3,19)$. This implies, by the classification theorem of such lattices, that

$$H_2(X,\mathbb{Z}) \simeq U_2^3 \perp (-E_8)^2 =: \mathcal{L}.$$

Here $U_2 = \left(\begin{smallmatrix}0&1\\1&0\end{smallmatrix}\right)$ is the standard hyperbolic plane, and E_8 is the even unimodular negative definite lattice of rank 8. We call $\mathcal{L}$ the $K3$ *lattice*.

3.3. The Picard Group of X. Let X be a $K3$ surface. Let $\mathrm{Pic}(X)$ denote the Picard group of X, that is, the group of divisors on X modulo linear equivalence. Then $\mathrm{Pic}(X) \subset H^2(X,\mathbb{Z})(=\mathrm{Hom}(H_2(X,\mathbb{Z}),\mathbb{Z}))$, and is parametrized by algebraic cycles on X. Pic(X) has no torsion, and indeed, is a free finitely generated Abelian group. Let $\rho(X)$ be the rank of Pic(X). Then

$$\mathrm{Pic}(X) \simeq \mathbb{Z}^{\rho(X)} \quad \text{with } 1 \leq \rho(X) \leq 20.$$

By the Hodge Index Theorem, Pic(X) is a lattice with signature $(1, \rho(X) - 1)$. We may call Pic(X) the Picard lattice.

3.4. The Group of Transcendental Cycles. Let

$$T(X) := \mathrm{Pic}(X)^{\perp}$$

be the orthogonal complement of Pic(X) in $H^2(X, \mathbb{Z})$. Then $T(X)$ forms a group, called the group of transcendental cycles on X. One sees immediately from the construction that $T(X)$ is endowed with a lattice structure with rank $22 - \rho(X)$, and signature $(2, 19 - (\rho(X) - 1))$. We may call $T(X)$ the transcendental lattice.

4. Moduli Spaces of Lattice Polarized $K3$ Surfaces

Let X be a $K3$ surface with the $K3$ lattice $\mathcal{L}$. Let $\alpha\colon H_2(X, \mathbb{Z}) \to \mathcal{L} = U_2^3 \perp (-E_8)^2$ be an isomorphism, which is called a *marking*. We fix α once and for all, and call (X, α) a marked $K3$ surface.

Now we will define lattice polarized $K3$ surfaces. Let $\mathcal{M}$ be a sublattice of $\mathcal{L}$. It is not in general true that any sublattice $\mathcal{M} \subset \mathcal{L}$ of rank $t+1$ and signature $(1, t)$ with $t \leq 19$ occur as the Picard lattice of some algebraic $K3$ surfaces. Some extra structure is required. We ought to introduce *primitive* sublattices: A sublattice $\mathcal{M} \subset \mathcal{L}$ is said to be *primitive* if $\mathcal{L}/\mathcal{M}$ is free. Let $\mathcal{M} \hookrightarrow \mathcal{L}$ be a primitive sublattice of signature $(1, t)$, $t \leq 19$. (So $\mathcal{M}$ may contain at most 1 copy of U_2 as a sublattice.)

We have the following results which can be found, e.g., in Morrison [**Mo**, Corollary 1.9].

THEOREM 4.1. (a) *For any primitive sublattice $\mathcal{M} \hookrightarrow \mathcal{L}$ of signature $(1, t)$ with $t + 1 \leq 20$, there exists an algebraic $K3$ surface X and an isometry* Pic$(X) \simeq \mathcal{M}$.

(b) *For any primitive sublattice $\mathcal{T} \hookrightarrow \mathcal{L}$ of signature $(2, 20 - (t+1))$ with $t+1 \leq 20$, there exists an algebraic $K3$ surface X and an isometry $T(X) \simeq \mathcal{T}$.*

(In fact, (b) is deduced from (a): Take $\mathcal{M} = \mathcal{T}^{\perp}$ in $\mathcal{L}$ and apply (a) to get $T(X) = \mathrm{Pic}(X)^{\perp} \simeq \mathcal{M}^{\perp} = \mathcal{T}$.)

Nikulin [**N1**] (see also Morrison [**Mo**]) found some condition on the existence of unique primitive embeddings of even lattices into $\mathcal{L}$, which impose strong consequences for the structure of the Picard lattices and transcendental lattices of $K3$ surfaces.

THEOREM 4.2. (a) *If $t+1 \leq 10$, then every even lattice $\mathcal{M}$ of signature $(1, t)$ occurs as the Picard group of some algebraic $K3$ surface, and the primitive embedding $\mathcal{M} \hookrightarrow \mathcal{L}$ is unique. The transcendental lattice of an algebraic $K3$ surface X with $\rho(X) \leq 10$ is uniquely determined by its signature and discriminant-form.*

(b) *If $12 \leq t + 1 \leq 20$, then every even lattice $\mathcal{T}$ of signature $(2, 20 - (t + 1))$ occurs as the transcendental lattice of some algebraic $K3$ surface, and the primitive embedding $\mathcal{T} \hookrightarrow \mathcal{L}$ is unique.*

The Picard group of an algebraic $K3$ surface X with $12 \le \rho(X) \le 20$ is uniquely determined by its signature and discriminant-form.

(c) *If $t+1 = 11$, every even lattice of signature $(1, 10)$ (resp. $(2, 9)$) can occur as the Picard group (resp. the group of transcendental cycles) on some algebraic $K3$ surface. (However, no uniqueness for the primitive embedding.)*

In any event, we will be looking at primitive sublattices of $\mathcal{L}$ which occur as either the Picard lattices or the transcendental lattices of some algebraic $K3$ surfaces.

DEFINITION 4.3. Let $\mathcal{M}$ be a primitive sublattice of $\mathcal{L}$ of rank $1+t$ and signature $(1, t)$. An (marked) *$\mathcal{M}$-polarized $K3$ surface* is a pair $(X_{\mathcal{M}}, \varphi_\alpha)$ where $X_{\mathcal{M}}$ is an algebraic $K3$ surface, and $\varphi_\alpha := \alpha^{-1}|_{\mathcal{M}} \colon \mathcal{M} \to \mathrm{Pic}(X_{\mathcal{M}})$ is a primitive lattice embedding (i.e., isometry of lattices).

(If $t+1 \le 10$, Theorem 4.2(a) guarantees the existence of such a pair. If $t+1 \ge 12$, take $\mathcal{M} = \mathcal{T}^\perp$ in $\mathcal{L}$ in Theorem 4.2(b) and apply (a) for the existence of such a pair.)

An (marked) $\mathcal{M}$-polarized $K3$ surface $(X_{\mathcal{M}}, \varphi_\alpha)$ is called *pseudo-ample* if $\varphi_\alpha(\mathcal{M})$ contains a pseudo-ample divisor class.

Next we wish to determine the space of isomorphism classes of pseudo-ample (marked) $\mathcal{M}$-polarized $K3$ surfaces. For this, we need to introduce *periods* of $K3$ surfaces.

4.1. The Period Map and the Period Domain. Let (X, α) be a marked $K3$ surface. Fix a $\mathbb{Z}$-basis for $H_2(X, \mathbb{Z})$, say, $\{\gamma_1, \gamma_2, \dots, \gamma_{22}\}$. Let ω be a holomorphic differential 2-form on X. For each $\gamma_i \in H_2(X, \mathbb{Z})$, the integral $\int_{\gamma_i} \omega$ is called a *period* of X. The *period map* of X is the map which assigns to a marked $K3$ surface (X, α), the vector:

$$P((X, \alpha)) := \left[\int_{\gamma_1} \omega : \int_{\gamma_2} \omega : \cdots : \int_{\gamma_{22}} \omega\right] \in \mathbb{P}^{21},$$

called the *period point* of X. The images of the period map is called the *period domain* and denoted by Ω.

The integral $\int_\gamma \omega$ is equal to the cup product of ω which is in $H^{2,0}$, and the Poincaré dual of γ, which, when γ is an algebraic cycle, is in $H^{1,1}$. Hence the cup product is in $H^{3,1}$, but this group is zero, and so for $\gamma \in \mathrm{Pic}(X)$, we have $\int_\gamma \omega = 0$. (Here $H^{i,j}$ stands for $H^j(X, \Omega_X^i)$.)

The period map (and period domain) classify the marked $K3$ surfaces up to isomorphism. The next theorem is known as Torelli's theorem.

THEOREM 4.4. *Let (X, α) and (X', α') be marked $K3$ surfaces. If $P((X, \alpha)) = P((X', \alpha'))$, then there is an isomorphism of algebraic surfaces $f \colon X \to X'$ such that $\alpha = \alpha' \circ f_*$ where f_* denotes the induced isomorphism on the second cohomology groups.*

Now we can describe the moduli space of $\mathcal{M}$-polarized $K3$ surfaces and its structure. We shall follow the exposition of Dolgachev [**Dol**].

THEOREM 4.5. *Let $\mathcal{M}$ be a primitive sublattice of $\mathcal{L}$ of rank $t+1$ and signature $(1, t)$ with $10 \le t \le 19$, and let $\mathcal{T} = \mathcal{M}^\perp$ in $\mathcal{L}$, a lattice of rank $22 - (t+1)$ and signature $(2, 19-t)$. Assume that $(X_{\mathcal{M}}, \varphi_\alpha)$ is pseudo-ample, that is, the embedding $\varphi_\alpha \colon \mathcal{M} \to \mathrm{Pic}(X_{\mathcal{M}})$ contains a pseudo-ample divisor class. Then the coarse moduli* *space of marked pseudo-ample $\mathcal{M}$-polarized $K3$ surfaces exists, and is denoted by $\mathfrak{M}^{\alpha}_{K3,\mathcal{M}}$.*

The space $\mathfrak{M}^{\alpha}_{K3,\mathcal{M}}$ may be identified with an open subset of the quadric $\mathcal{Q} \subset \mathbb{P}(\mathcal{T} \otimes \mathbb{C}) \simeq \mathbb{P}^{19-t}$ defined by the quadratic form $b(x,y)$ corresponding to $\mathcal{T}$, and by the (Riemann) inequality $b([\xi],[\bar{\xi}]) > 0$.

We may get rid of the marking by introducing the group of isometries. Let

$$\Gamma_{\mathcal{M}} := \{\delta : \mathcal{L} \to \mathcal{L} \mid \delta|_{\mathcal{M}} = id|_{\mathcal{M}}\}$$

be the group of isometries, and let

$$\mathfrak{M}_{K3,\mathcal{M}} := \mathfrak{M}^{\alpha}_{K3,\mathcal{M}}/\Gamma_{\mathcal{M}}.$$

Then $\mathfrak{M}_{K3,\mathcal{M}}$ is independent of markings, and defines a quasi-projective algebraic variety, and it has dimension $\dim \mathfrak{M}_{K3,\mathcal{M}} = 19 - t$. *Furthermore, it is isomorphic to a certain bounded symmetric homogeneous space of type* IV.

We may regard $\mathfrak{M}_{K3,\mathcal{M}}$ as the moduli space of $\mathcal{M}$-polarized $K3$ surfaces.

For detailed discussions on the statement in Theorem 4.5, the reader is referred to the paper by Dolgachev [**Dol**].

Now we consider some specific sublattices. The moduli space of a primitive sublattice $\mathcal{M} \subset \mathcal{L}$ of rank 20 ($t = 19$) is of dimension zero. This is the isomorphism classes of of singular $K3$ surfaces. It was shown by Shioda and Inose [**SI**, Theorem 4] that there is a natural $1-1$ correspondence from the set of singular $K3$ surfaces to the set of positive definite even integral binary quadratic forms, modulo the $\mathrm{SL}_2(\mathbb{Z})$-equivalence. The latter set is countably finite. (See Shioda and Inose [**Sa**]] for detail).

We consider specific primitive sublattices of $\mathcal{M} \subset \mathcal{L}$ of rank 19 ($t = 18$).

THEOREM 4.6 (Dolgachev [**Dol**]). *Let $\mathcal{M} \simeq U_2 \perp (-E_8)^2 \perp \langle -2N \rangle$ and consider $\mathcal{M}$-polarized $K3$ surfaces. Then the moduli space is given by*:

$$\mathfrak{M}_{K3,\mathcal{M}} = \Omega/\Gamma_{\mathcal{M}} \simeq \mathfrak{H}/\Gamma_0(N)^* \quad \textit{for some integer } N.$$

Here $\Gamma_0(N)^* := \Gamma_0(N) + w_N$ denote the Fricke double cover *of $\Gamma_0(N)$ obtained by adjoining to $\Gamma_0(N)$ the Fricke involution $w_N = \begin{pmatrix} 0 & -1/\sqrt{N} \\ \sqrt{N} & 0 \end{pmatrix} \in \mathrm{PSL}_2(\mathbb{R})$.*

REMARK 4.7. Recall that $\mathfrak{H}/\Gamma_0(N)^*$ is the moduli space of pairs (E, C_N) of elliptic curves E and cyclic isogenies of order N, C_N, up to the action of the Fricke involution $w_N((E,C_N)) = (E/C_N, E[N]/C_N)$. Theorem 4.6 of Dolgachev asserts that $\mathfrak{H}/\Gamma_0(N)^*$ is also the moduli space of $\mathcal{M}$-polarized $K3$ surfaces. So the two moduli spaces are the same. This gives rise to a connection between $\mathcal{M}$-polarized $K3$ surfaces with Picard number 19 and elliptic curves.

4.2. The Shioda-Inose Structure, and a Theorem of Morrison. Morrison [**Mo**] has shown that any $K3$ surface X with Picard number 19 always admits a Shioda-Inose structure, which we shall discuss here. For details, the reader should consult Morrison's article [**Mo**].

Let X be an algebraic $K3$ surface. An involution ι on X is called a *Nikulin involution* if $\iota^*(\omega) = \omega$ for every $\omega \in H^{2,0}(X)$.

We say that X *admits a Shioda-Inose structure* if there is a Nikulin involution ι on X with rational quotient map $\pi : X \to Y$ such that Y is a Kummer surface, and π_* induces an isometry between the transcendental lattices $T(X)(2)$ and $T(Y)$. In this case, there is a complex torus Z whose Kummer surface is Y and an isometry $T(X) \simeq T(Z)$.

The following conditions are equivalent.

(i) X admits a Shioda-Inose structure.
(ii) There exists an Abelian surface A such that $Y = \mathrm{Kum}(A)$ and an isometry $T(X) \simeq T(A)$.
(iii) There is a primitive embedding $T(X) \hookrightarrow U_2^3$.
(iv) There is an embedding $(-E_8)^2 \hookrightarrow \mathrm{Pic}(X)$.

Note that the Picard number of any Abelian surface is at most 6. We can reformulate the above results in terms of the Picard numbers $\rho(X)$:

(a) If $\rho(X) = 19$ or 20, then X admits a Shioda-Inose structure.
(b) If $\rho(X) = 18$, X admits a Shioda-Inose structure if and only if $T(X) \simeq U_2 \perp T'$ for some even lattice T'.
(c) If $\rho(X) = 17$, X admits a Shioda-Inose structure if and only if $T(X) \simeq U_2^2 \perp T'$ for some even lattice T'.
(d) If $\rho(X) < 17$, X admits no Shioda-Inose structure.

We now apply Morrison's results to $\mathcal{M}$-lattice polarized $K3$ surfaces in Theorem 4.6.

PROPOSITION 4.8. *The notation and the hypothesis of Theorem* 4.6 *remain in force. Let X_z be an $\mathcal{M}$-polarized $K3$ surface. Then X_z admits a Shioda-Inose structure. More precisely, there exist an Abelian surface $A := E_\tau \times E_\tau/C_N$, a Kummer surface $Y = \mathrm{Kum}(A/\pm 1)$ and a canonical involution ι on X_z such that $X/\langle\iota\rangle$ is birationally isomorphic to the Kummer surface* $\mathrm{Kum}(A/\pm 1)$.

5. Picard-Fuchs Differential Equations and Monodromy Groups

Let X_z be a 1-parameter family of $K3$ surfaces parametrized by $z \in B$ where $B := \mathbb{P}^1 \setminus \{z \mid X_z \text{ singular}\}$. There is, up to scalar multiplication, a unique holomorphic differential 2-form $\omega_z \in H^2(X_z, \mathbb{C})$.

DEFINITION 5.1. For a fixed $z_0 \in B$, let $\pi(B, z_0)$ denote the fundamental group. Then we have the monodromy representation:

$$\pi(B, z_0) \to \mathrm{Aut}(\mathbb{P}(H_2(X_z, \mathbb{Z})).$$

The image of this map, denoted, G, is the *monodromy group* of the family $\{X_z\}_{z\in B}$.

Let $\{\gamma_{1_z}, \gamma_{2_z}, \ldots, \gamma_{22_z}\}$ be a $\mathbb{Z}$-basis for $H_2(X_z, \mathbb{Z})$, which is flat with respect to z. Then the period map is the map:

$$B \to \mathbb{P}^{21}/G\colon z \mapsto \left[\int_{\gamma_{1_z}} \omega_z : \cdots : \int_{\gamma_{22_z}} \omega_z\right]$$

where each function $\int_{\gamma_{i_z}} \omega_z$ is a period.

The Picard-Fuchs differential equation is a differential equation satisfied by the periods with the same monodromy group.

PROPOSITION 5.2. *Let X_z be a* 1*-parameter family of $K3$ surfaces parametrized by $z \in B$. Suppose that $\rho(X_z) = r$ for generic z, and put $k := 22 - r$. Then the period of X_z satisfies a Picard-Fuchs differential equation of order k.*

PROOF. For a $K3$ surface X_z, the second de Rham cohomology group $H^2_{\rm dR}(X_z)$ has dimension 22. If $\rho(X_z) = r$, then $\dim \mathrm{Pic}(X_z) = r$, so $H^2_{\rm dR}/\mathrm{Pic}(X_z)$ has dimension $k = 22 - r$. Hence there must be a linear relation satisfied by the classes:

$$[\omega_z], \left[\frac{\partial \omega_z}{\partial z}\right], \ldots, \left[\frac{\partial^i \omega_z}{\partial^i z}\right], \ldots, \left[\frac{\partial^k \omega_z}{\partial^k z}\right]$$

in $H^2_{\rm dR}/\mathrm{Pic}(X_z) \otimes \mathbb{C}$. Hence there are $g_0, g_1, \ldots, g_i, \ldots, g_k \in \mathbb{C}$ such that

$$\mathcal{G} := g_0\omega_z + g_1\frac{\partial \omega_z}{\partial z} + \cdots + g_i\frac{\partial^i \omega_z}{\partial^i z} + \cdots + g_k\frac{\partial^k \omega_z}{\partial^k z} \in \mathrm{Pic}(X_z),$$

which means that $\int_\gamma \mathcal{G} = 0$ for any $\gamma \in H_2(X_z, \mathbb{Z})$. Note that for fixed z, the g_i are elements of $\mathbb{C}$; as z varies, they are $\mathbb{C}^\infty$ functions of z. Since integrating around a cycle $\gamma \in H_2(X_z, \mathbb{Z})$ commutes with differentiation with respect to z, in the sense that

$$\int_\gamma g_i \frac{\partial^i \omega_z}{\partial z^i} = g_i \frac{d^i}{dz^i}\int_\gamma \omega_z,$$

the linear relation becomes a differential equation for $\int_\gamma \omega_z$ upon changing the order of integration and differentiation. □

The fact that a family of $K3$ surfaces with generic Picard number 19 admits a Shioda-Inose structure is reinterpreted in terms of Picard-Fuchs differential equations as follows.

5.1. Digression. Given a family of elliptic curves, we may obtain a pencil of $K3$ surfaces by taking a certain fiberwise product of the elliptic curves, and taking the quotient to obtain Kummer surfaces. Then the Picard-Fuchs differential equation of the resulting family of $K3$ surfaces is the symmetric square of the family of elliptic curves. This is described in Peters [**Pe**]. Conversely, for a $K3$ surface, we have the relation $\Omega \wedge \Omega = 0$ (the Riemann equality), which implies that there is a quadratic relation satisfied by the solutions of the Picard-Fuchs equation. When the Picard lattice has dimension 19, this implies that the Picard-Fuchs differential equation is the symmetric square of some second order differential equation. By Morrison [**Mo**], a pencil of $K3$ surfaces of Picard number 19 is related to a pencil of elliptic curves by a 2 : 2 map. The second order differential equation turns out to be the Picard-Fuchs differential equation for this family of elliptic curves.

REMARK 5.3. Doran [**Dor**] (in particular, Section 5.2) has described in some detail the symmetric square structure of the Picard-Fuchs differential equations of $\mathcal{M}$-polarized $K3$ surfaces, from points of view of differential equations.

5.2. Examples of Lian and Yau. We follow the exposition due to Lian and Yau [**LY2**]. (More examples can be found in Lian and Yau [**LY1**].) We consider 1-parameter deformation of the following two $K3$ surfaces: (a) the quartic surfaces, and (b) the double planes branched along a sextic, in weighted projective spaces. Further, we pass onto suitable orbifold ambient spaces, by taking the quotients of these pencils by the symmetry groups preserving the 2-form. This will lead us into the realm of mirror partners of $K3$ surfaces (which we will address in the section 8 below).

These two pencils of $K3$ surfaces have the following defining equations in the orbifold ambient spaces, respectively:

(a) $X_0^4+X_1^4+X_2^4+X_3^4+z^{-1/4}X_0X_1X_2X_3=0\subset\mathbb{P}^3((1,1,1,1)/G$;
1. (b)] $X_0^6+X_1^6+X_2^6+X_3^2+z^{-1/6}X_0X_1X_2X_3=0\subset\mathbb{P}^3((1,1,1,3))/G$,

where $G=(\mathbb{Z}/4\mathbb{Z})\times(\mathbb{Z}/4\mathbb{Z})$ for (a) and $(\mathbb{Z}/6\mathbb{Z})\times(\mathbb{Z}/6\mathbb{Z})\times(\mathbb{Z}/2\mathbb{Z})$ for (b), and z denotes a parameter in both cases.

We denote by X_z the 1-parameter family of smooth $K3$ surfaces obtained from the surfaces defined above by resolving singularities. Then X_z is a lattice polarized with the Picard number $\rho(X_z)=19$ for generic z.

The period integral of X_z satisfies the Picard-Fuchs differential equation of order 3, which were determined by Lian and Yau [**LY2**]. In fact, they determined the Picard-Fuchs differential equations for the 1-parameter families of $K3$ surfaces (a) and (b) in the weighted projective spaces $\mathbb{P}^3((1,1,1,1))$ and $\mathbb{P}^3((1,1,1,3))$, respectively.

(Lian has explained to us that the Picard-Fuchs differential equations for the quotients X_z (the mirror partners) coincide with those of the 1-parameter families in the orbifold ambient spaces. This is a general phenomenon whenever the mirror manifolds come from the orbifold construction of the original manifolds. This is because the desingularization process gives rise only to exceptional divisors and they are all algebraic, and hence it does not change the transcendental part which governs the Picard-Fuchs differential equation.)

We start with the following third order Picard-Fuchs equation of the form

$$\left(\Theta^3-z\left(\sum_{i=0}^{3}r_i\Theta^i\right)\right)f(z)=0$$

where z is the variable, and $\Theta=z\frac{d}{dz}$. The constants r_i are integers subject to the relation $r_0=(2r_1-r_3)/4$, $r_2=3r_3/2$, $r_3\neq 0$. After simplification, we obtain an equation of the form

$$\left(\Theta^3-r_3z\left(\Theta+\frac{1}{2}\right)\left(\Theta^2+\Theta+\frac{r_1}{r_3}-\frac{1}{2}\right)\right)f(z)=0.$$

For (a), the Picard-Fuchs differential equation is given by specifying $(r_3,r_1)=(256,176)$ and for (b), $(r_3,r_1)=(1728,1104)$.

Now we shall actually determine the monodromy groups for families of $K3$ surfaces with generic Picard number 19.

PROPOSITION 5.4. *Let X_z be a pencil of $K3$ surfaces with Picard number $\rho(X_z)=19$ for generic z. Let $T:=T(X_z)=Pic(X_z)^{\perp}$ denote the lattice of transcendental cycles on X_z of rank* 3, *and* $\mathrm{disc}(T)$ *the intersection matrix of T. Put*

$$\mathrm{SO}(T):=\{M\in\mathrm{PSL}_3(\mathbb{R})\mid M^T\,\mathrm{disc}(T)M=\mathrm{disc}(T)\}$$

and

$$\mathrm{SO}(T,\mathbb{Z}):=\mathrm{SO}(T)\cap\mathrm{PSL}_3(\mathbb{Z}).$$

Then the monodromy group of X_z is isomorphic to a subgroup of $SO(T,\mathbb{Z})$.

PROOF. It follows from Section 5.1 that the Picard-Fuchs differential equation of X_z is of order 3, and is the symmetric square of an order 2 differential equation (associated to the underlying elliptic curve). The space of local solutions V_{z_0} near a point z_0, of the Picard-Fuchs differential equation has dimension 3, and the monodromy group, which describes how elements of $\pi_1(B,z_0)$ act on $\mathbb{P}(V_{z_0})$, is isomorphic to a subgroup of $\mathrm{PGL}_3(\mathbb{C})$. If we consider the action as an action on twisted cycles, i.e., V_{z_0} as the stalk of $H_2(X, \mathbb{Z}) \otimes H^{0,2}$ at z_0, then because there is only one holomorphic differential 2-form up to multiplication by scalars, and because $H_2(X_z, \mathbb{Z})$ is a lattice, and the monodromy group preserves this lattice, we may express elements of the monodromy group in terms of a basis of the lattice $H_2(X_z, \mathbb{Z})$. So the coefficients are in $\mathbb{Z}$ and the determinant is 1. Also, the action of $\pi_1(B, z_0)$ must preserve the intersection matrix of $H_2(X_z, \mathbb{Z})$. This implies that the monodromy group is a subgroup of $\mathrm{SO}(T, \mathbb{Z})$. □

PROPOSITION 5.5. *Let*

$$\mathcal{M} \simeq U_2 \perp (-E_8)^2 \perp \langle -2N \rangle$$

and let X_z be a pencil of $\mathcal{M}$-polarized $K3$ surfaces. Then $T(X_z) \simeq U_2 \perp \langle 2N \rangle$. The intersection matrix of $T(X_z)$ is given by

$$\begin{pmatrix} 0 & 1 & 0 \\ 1 & 0 & 0 \\ 0 & 0 & 2N \end{pmatrix}.$$

The monodromy group of X_z is isomorphic to a congruence subgroup of $\mathrm{PSL}_2(\mathbb{R})$ *of level N.*

PROOF. By Proposition 5.4, the monodromy group is a subgroup of $\mathrm{SO}(T, \mathbb{Z})$. Choose a basis (a, b, c) for $T(X_z)$ such that the intersection matrix $\mathrm{disc}(T(X_z))$ is given by the above 3×3 matrix, i.e., $ab + Nc^2 = 0$. Because of this relation, we may replace (a, b, c) by (ξ, η). This relation also implies that the differential equation for X_z is the symmetric square of an order 2 differential equation. The monodromy group is then isomorphic to a subgroup of $\mathrm{PSL}_2(\mathbb{R})$. The singularities of the Picard-Fuchs differential equation will provide further necessary information to determine cusps and elliptic points of this subgroup, and finally the level of this group and group itself in question. (For details, confer Stienstra-Beukers [**SB**], Peters and Stienstra [**PS**] and Verrill [**V**].) □

6. The Mirror Maps and Thompson Series. I

Let X_z be a pencil of $K3$ surfaces with Picard number $\rho(X_z) = 19$ for generic z.

Now we are interested in the actual "local" solution of the Picard-Fuchs differential equation of a pencil of $K3$ surfaces. This will lead to the definition of the mirror map.

DEFINITION 6.1. The Picard-Fuchs differential equation of X_z of order 3 has a unique solution

$$\omega_0(z) = \sum_{n \geq 0} c_n z^n$$

which is holomorphic near $z = 0$ with $\omega_0(0) = 1$. The existence of such a solution is guaranteed by the maximal unipotency assumption. There is also a unique solution of the form

$$\omega_1(z) = \log(z)\omega_0(z) + \sum_{n \geq 1} d_n z^n.$$

The coefficients c_n, d_n are polynomials in the parameters r_2 and r_3. Let $t = \omega_1(z)/(2\pi i\omega_0(z))$ be a parameter. Then

$$q := e^{2\pi it} = z\exp\left(\frac{\sum d_n z^n}{\sum c_n z^n}\right)$$

gives an invertible analytic map from a disc $|z| < S$ to some disc $|q| < R$. The inverse, denoted by $z(q)$ with $z(0) = 1$, is the *mirror map* of the family of $K3$ surfaces in question. Observe that the mirror maps are holomorphic functions in the respective discs. The mirror map is viewed as the inverse to the (truncated) period map.

6.1. The Existence of Maximally Degenerating Points. For Picard-Fuchs differential equations of pencils of algebraic $K3$ surfaces arising from orbifold construction, the existence of maximally degenerating points has been established by Hosono, Lian and Yau [**HLY**]. The local solutions to the Picard-Fuchs differential equations are well-defined in neighborhoods of the points of maximally degeneracy. (In many examples in [**HLY**], $z = 0$ may be taken as a maximally degenerating point.) For the complementary discussion of this topic, the reader is referred to the paper by Doran [**Dor**].

Lian and Yau [**LY2**] have proved that the reciprocal of the mirror maps have q-expansions with integer coefficients, and have derived modulo p congruences for these integral coefficients. These fact are used to show the divisibility of the instanton numbers by certain integers. Further they have observed that the reciprocal of the mirror maps are sometimes Hauptmodul of genus zero subgroups of $\mathrm{PSL}_2(\mathbb{R})$. Based on their observation, they have formulated the following "conjecture".

6.2. The Mirror-Moonshine Conjecture (Lian and Yau [**LY2**]; cf. Doran [**Dor**]). Let X_z be a 1-parameter family of lattice polarized algebraic $K3$ surfaces with Picard number 19 from orbifold construction. Let $z(q)$ be the mirror map for X_z. Then, for some constant $c \in \mathbb{Z}$, the q-series $\frac{1}{z(q)} + c$ is a Thompson series $T_g(q)$ for some element g in the Monster group. When this is the case, we may say that the mirror map is *modular.*

The following examples of Lian and Yau [**LY2**] give evidence to the mirror moonshine conjecture.

THEOREM 6.2 (Lian and Yau [**LY2**]; see also [**LY1**] for more examples).

(a) *The mirror map for the one-parameter family of the diagonal quartics in Section* 5.2(a) *is given by the series:*

$$z(q) = q - 104q^2 + 6444q^3 - 311744q^4 + 13018830q^5 - 493025760q^6 + \cdots$$

and taking $c = -96$,

$$\frac{1}{z(q)} - 96 = \frac{1}{q} + 8 + 4372q + 96256q^2 + 1240002q^3 + \cdots$$

is the Thompson series 2A *for* $\Gamma_0(2)+$ *in Conway and Norton list* [**CN**].

(b) *The mirror map for the one-parameter family of the sextic hypersurfaces in* $\mathbb{P}^3((1,1,1,3))$ *in Section 5.2(b) is given by the series:*

$$z(q) = q - 744q^2 + 356652q^3 - 140361152q^4 + 49336682190q^5 - \cdots$$

and taking $c = 0$,

$$\frac{1}{z(q)} = j(q) = \textit{Thompson series of level } 1.$$

(c) *The monodromy group of* X_z *is isomorphic to* $\Gamma_0(2)+$ *for* (a), *and* Γ *for* (b).

Therefore, in both cases, the mirror maps are modular.

7. The Mirror Maps and Thompson Series. II

In this section we will discuss construction of lattice polarized $K3$ surfaces from root lattices, and then compute their mirror maps.

(In this section, due to conflict of the notations, we will use λ in place of z for parameters.)

7.1. A Pencil of Varieties Associated to a root lattice. Let R be a root system (of rank n), and let M_R be the root lattice generated by R, and let N_R be its dual lattice. The Weyl chambers in $N_R \otimes \mathbb{Q}$ is defined as follows: For each $r \in R$, let $H_r := \{s \in N_R \otimes \mathbb{Q} \mid \langle s, r \rangle = 0\}$. A Weyl chamber is the closure of any connected component of $N_R \otimes Q \setminus \cup_{r \in R} H_r$. Let Σ_R be the fan in $N_R \otimes \mathbb{Q}$ consisting of the Weyl chambers, together with all their sub-faces. Denote by $X(\Sigma_R)$ the toric variety associated to the fan Σ_R. Let Δ_R be the polyhedron in $M_R \otimes \mathbb{Q}$ with vertices given by $r \in R$, and let $L(\Delta_R)$ be the space of Laurent polynomials with support in Δ_R. We write $\mathbf{x}^r$ for the monomial in $L(\Delta_R)$. Define the polynomial $\chi_R := \sum_{r \in R} \mathbf{x}^r$ (which is the character of the adjoint representation of R). Then $\chi_R \in L(\Delta_R)$, and we have a rational function $\chi_R : X(\Sigma_R) \to \mathbb{P}^1$. For $\lambda \in \mathbb{P}^1$, X_λ is defined to be the closure in the ambient toric variety $X(\Sigma_R)$ of the inverse image of λ under χ_R, i.e., set $X_\lambda := \overline{\chi_R^{-1}(\lambda)} \subset X(\Sigma_R)$. Let $\pi : X_R \to X(\Sigma_R)$ be the blow up of the base locus. We denote by $\mathcal{X}_R$ the family $\bar{\chi}_R : X_R \to \mathbb{P}^1$ with fibers $X_\lambda := \overline{\pi^{-1}(\bar{\chi}_R = \lambda)}$. The family $\mathcal{X}_R$ is the object that we are interested in what follow.

For fuller discussion on Calabi-Yau hypersurfaces in toric varieties, the reader is referred to the article of Batyrev [**Ba**].

EXAMPLES 7.1. Let $\mathbb{R}^n$ be the vector space with the standard basis $\{e_i \mid i = 1, 2, \dots, n\}$, and denote x^{e_i} by X_i.

(a) Let $R = A_1 \times A_1 \times A_1$, which we may write simply A_1^3. Let $\{e_i \mid i = 1, 2, 3\}$ be the standard basis of $\mathbb{R}^3$. Then M_R is a sublattice of $\mathbb{R}^3$ of rank 3 with simple roots $\pm e_i$ with $i = 1, 2, 3$. The character of the adjoint representation of A_1^3 is given by $\chi_{A_1^3} = \sum_{i=1}^3 (X_i + X_i^{-1})$. For $\lambda \in \mathbb{P}^1$, the fibers X_λ are locally defined by

$$X_1 + X_1^{-1} + X_2 + X_2^{-1} + X_3 + X_3^{-1} = \lambda + \lambda^{-1}.$$

(b) Let $R = A_3$. The M_R is a sublattice of $\mathbb{R}^4$ of rank 3 with roots $\{e_i - e_j \mid 1 \le i, j \le 4, i \neq j\}$. The character of the adjoint representation of A_3 is given by $\chi_{A_3} = \sum_{1 \le i,j \le 4, i \neq j} X_i X_j^{-1}$. For $\lambda \in \mathbb{P}^1$, the fibers X_λ are locally defined by

$$(X_1 + X_2 + X_3 + X_4)(X_1^{-1} + X_2^{-1} + X_3^{-1} + X_4^{-1}) = \lambda + 4.$$

We shall discuss examples of pencils of $K3$ surfaces constructed from the root lattices A_1^3, and A_3. Picard-Fuchs differential equations, monodromy groups and mirror maps may be determined explicitly for such families of $K3$ surfaces, as we will see in what follow. All families of $K3$ surfaces considered here are $\mathcal{M}$-polarized $K3$ surfaces where

$$\mathcal{M} = U_2 \perp (-E_8)^2 \perp \langle -2N \rangle$$

with some positive integer $N > 1$.

7.2. Example of Beukers, Peters, and Stienstra. The first example 7.1(a) was constructed by Beukers and Peters [**BP**] interpreting the Apéry recurrence relation for $\zeta(3)$ in terms of arithmetic of a pencil of $K3$ surfaces. Subsequently, Peters and Stienstra [**PS**] considered the Fermi-threefold corresponding to the root lattice A_1^3. It is defined by the affine equation:

$$\lambda + \lambda^{-1} + X_1 + X_1^{-1} + X_2 + X_2^{-1} + X_3 + X_3^{-1} = 0$$

It fibers via the projection to $\mathbb{P}(\lambda)$ into surfaces whose closure are $K3$ surfaces. The generic Fermi-fiber, X_λ, is a pencil of $K3$ surfaces with Picard number 19 and

$$\mathrm{Pic}(X_\lambda) = U_2 \perp (-E_8)^2 \perp \langle -12 \rangle \quad \text{and} T(X_\lambda) = U_2 \perp \langle 12 \rangle.$$

The holomorphic differential 2-form is given by

$$\omega(\lambda) = \frac{1}{X_3 - X_3^{-1}} \frac{dX_1}{X_1} \wedge \frac{dX_2}{X_2}.$$

The period is a function of λ and satisfies a third order Picard-Fuchs differential equation:

$$(\lambda^2 - 4)(\lambda^2 - 36)\frac{d^3F}{d\lambda^3} + 6\lambda(\lambda^2 - 20)\frac{d^2F}{d\lambda^2} + (7\lambda^2 - 48)\frac{dF}{d\lambda} + \lambda F = 0.$$

The fact that this differential equation has power series solutions of the form

$$F(\lambda) = \sum_{n \geq 0} v_n \lambda^{2n+1} \quad \text{where } v_n = \sum_k \binom{n}{k}^2 \binom{n+k}{k}^2$$

is equivalent to the assertion that v_n is a solution to the Apéry recurrence relation for $\zeta(3)$ (which was used in the proof of the irrationality of $\zeta(3)$).

In [**PS**], the monodromy group is explicitly determined, and given as a genus zero subgroup $P\Gamma_1(6,2)^* := \Gamma_1(6,2)^*/\langle w_6^2 \rangle$ of $PSL_2(\mathbb{R})$. It is defined as follows:

$$\Gamma_1(6) = \left\{ \begin{pmatrix} a & b \\ c & d \end{pmatrix} \middle| \; a \equiv d \equiv 1 \pmod 6,\ c \equiv 0 \pmod 6 \right\},$$

$$\Gamma_1(6,2) = \left\{ \begin{pmatrix} a & b \\ c & d \end{pmatrix} \middle| \; c \equiv 6b \pmod{12} \right\},$$

$$\Gamma_1(6)^* = \text{group generated by } \Gamma_1(6) \text{ and } w_6 := \begin{pmatrix} 0 & 1 \\ -6 & 0 \end{pmatrix}$$

and

$$\Gamma_1(6,2)^* = \text{group generated by } \Gamma_1(6,2) \text{ and } w_6.$$

The group $P\Gamma_1(6,2)^*$ is an index 2 subgroup of $\Gamma_0(6)+6$ (which comes from the condition $c \equiv 6b \pmod{12}$) and a representative for the second coset is $\left(\begin{smallmatrix} 1 & 0 \\ -6 & 1 \end{smallmatrix}\right)$.

The function

$$\lambda(\tau) = \left(\frac{\eta(\tau)\eta(6\tau)}{\eta(2\tau)\eta(3\tau)}\right)^6$$

generates the field of modular functions for $\Gamma_1(6,2)$. It has the Fourier expansion of the form:

$$\begin{aligned}\lambda(q) = q^{1/2}(1 &- 6q + 15q^2 - 20q^3 + 15q^4 - 12q^5 + 37q^6 - 96q^7 + 156q^8 - 180q^9 \\ &+ 171q^{10} - 192q^{11} + 333q^{12} - 618q^{13} + 936q^{14} - 1100q^{15} \\ &+ 1092q^{16} - 1224q^{17} + 1885q^{18} - 3084q^{19} + O(q^{20})).\end{aligned}$$

This function should give rise to the mirror map of the pencil of $K3$ surfaces. However, the factor $q^{1/2}$ is not desirable. To get rid of this factor, we may take the square $\lambda(q)^2$. Then the function $\lambda(\tau)$ is related to Thompson series 6B, $6+6$ in the following sense:

$$\left(\frac{1}{\lambda(\tau)}\right)^2 = \left(\frac{\eta(2\tau)\eta(3\tau)}{\eta(\tau)\eta(6\tau)}\right)^{12}$$

is the Thompson series in the Monster class 6B with symbol $6+6$. (In fact, $\lambda(q)$ is transformed under the element $\left(\begin{smallmatrix} 1 & 0 \\ -6 & 1\end{smallmatrix}\right)$ to $-z(q)$, and hence $\lambda(q)^2$ is a modular function for $\Gamma_0(6)+6$.)

We summarize the above discussion into the following theorem.

THEOREM 7.2. *The map $1/\lambda(q)$ itself is not a Thompson series, though its square $(1/\lambda(q))^2$ is identified with the Thompson series $T_{6\mathrm{B}}$ with symbol $6+6$, associated to the subgroup $\Gamma_0(6)+6$.*

Now we will reconsider the example of Beukers, Peters and Stienstra from a different angle which is more suitable in the context of mirror moonshine. For instance, we do not want the $q^{1/2}$ factor in the mirror map. This may removed by changing variable q to q^2. We will discuss this point of view in the following section.

7.3. The Example of Beukers, Peters, and Stienstra Revisited. In this section, we will give a reinterpretation of the results of Peters and Stienstra [**PS**] and the discussions in 7.2, which is more suitable in the context of mirror-moonshine. We retain the notation in 7.2. One can show that the monodromy group $P\Gamma_1(6,2)^*$ is isomorphic to the group $\Gamma_0(12|2)+6$, in the notation of Conway and Norton [**CN**]. In fact, $\Gamma_0(12|2)$ is the group of matrices of of the form $\left(\begin{smallmatrix} a & b/2 \\ 12c & d\end{smallmatrix}\right)$, a, b, c, $d \in \mathbb{Z}$ with determinant 1, and $\Gamma_0(12|2)+6$ is the group generated by $\Gamma_0(12|2)$ and the involution w_6.

Now we we take $\lambda = 1/T_{12\mathrm{F}}$ where $T_{12\mathrm{F}}$ is the Thompson series for this group; it is given by

$$T_{12\mathrm{F}}(q) = \left(\frac{\eta(q^4)\eta(q^6)}{\eta(q^2)\eta(q^{12})}\right)^6 = [1,0,6,0,21,0,56,0,126,0,258,0,498,0,924,0,\ldots].$$

Then a solution of the Picard-Fuchs equation is given by

$$F(\tau) = \eta(2\tau)\eta(4\tau)\eta(6\tau)\eta(12\tau).$$

It can be numerically verified that locally, in terms of q-expansions, we have an equality:

$$F(\lambda) = \sum_{n\geq 0} v_n\lambda^{2n+1} = q - q^3 - 2q^5 + q^9 + 4q^{11} - 2q^{13} + 2q^{15} + O(q^{17})$$

where v_n are as in Section 7.2. The function

$$\lambda(\tau) = \left(\frac{\eta(2\tau)\eta(12\tau)}{\eta(4\tau)\eta(6\tau)}\right)^6$$

generates the field of modular functions for $\Gamma_0(12|2)+6$. It has the Fourier expansion of the form:

$$\begin{aligned}\lambda(q) = q - 6q^3 + 15q^5 - 20q^7 + 15q^9 - 12q^{11} + 37q^{13} - 96q^{15} + 156q^{17} \\ - 180q^{19} + 171q^{21} - 192q^{23} + 333q^{25} - 618q^{27} + 936q^{29} - 1100q^{31} \\ + 1092q^{33} - 1224q^{35} + 1885q^{37} - 3084q^{39} + O(q^{40}).\end{aligned}$$

This function should be the mirror map of the pencil of $K3$ surfaces.

We summarize the above discussion into the following theorem.

THEOREM 7.3. *The mirror map $1/\lambda(q)$ is the Thompson series $T_{12\mathrm{F}}$ with symbol $12 \mid 2+6$, associated to the subgroup $\Gamma_0(12 \mid 2)+6$.*

7.4. Problem. The above two calculations of the mirror map for the example of Beukers, Peters and Stienstra poses a question.

Is there a canonical choice for the mirror map of a pencil of $K3$ surfaces?

Now we will give a brief discussion on the example constructed by Verrill [**V**] corresponding to the root lattice A_3. The construction is done along the same line as in Peters and Stienstra [**PS**].

7.5. The Example of Verrill. Verrill [**V**] considered the variety Z_λ in $\mathbb{P}^3$ defined by

$$(X_1+X_2+X_3+X_4)(X_1^{-1}+X_2^{-1}+X_3^{-1}+X_4^{-1}) = \lambda+4$$

where λ is a parameter. This variety corresponds to the root lattice A_3. Let $B_1\mathbb{P}^3$ be $\mathbb{P}^3$ blown up in the points

$$(0:0:0:1),\ (0:0:1:0),\ (0:1:0:),\ (1:0:0:0).$$

Let $B_2\mathbb{P}^3$ be $B_1\mathbb{P}^3$ blown up along the transforms of the six lines

$$L_{ij} : X_i = X_j = 0 \quad \text{for } i,j = 1,2,3,4.$$

Let X_λ denote the closure of Z_λ in $B_2\mathbb{P}^3$. X_λ is locally given by

$$(1+x+xy+xyz)(1+z+zy+zyx) = (\lambda+4)xyz,$$

after putting

$$x = X_1/X_2,\ y = X_2/X_3,\ z = X_3/X_4.$$

Then X_λ defines a pencil of $K3$ surfaces with Picard number 19 for generic λ, and

$$\mathrm{Pic}(X_\lambda) = U_2 \perp (-E_8)^2 \perp \langle -6\rangle \quad \text{and} \quad T(X_\lambda) = U_2 \perp \rangle 6 \langle ..$$

The period satisfies a Picard-Fuchs differential equation of the form:

$$\lambda(\lambda+4)(\lambda-12)\frac{d^3y}{d\lambda^3} + 6(\lambda^2-7\lambda-12)\frac{d^2y}{d\lambda^2} + \frac{(7\lambda^2-12\lambda-96)}{\lambda+4}\frac{dy}{d\lambda} + \frac{\lambda}{\lambda+4}y = 0,$$

where λ is the variable.

The monodromy group of X_λ is shown to be isomorphic to a genus zero subgroup $\Gamma_0(6)+3$, and the corresponding Thompson series is

$$-\lambda(q) = \left(\frac{\eta(q)\eta(q^3)}{\eta(q^2)\eta(q^6)}\right)^6 + 4$$

with the Fourier expansion

$$\frac{1}{q} - 2 + 15q - 32q^2 + 87q^3 - 192q^4 + 343q^5 - 672q^6 + O(q^7).$$

This is the Thompson series 6C in the Conway and Norton list [**CN**].

Now the mirror map for this family of $K3$ surfaces is given explicitly by

$$-\frac{1}{\lambda(q)} = q + 2q^2 - 11q^3 - 20q^4 + 102q^5 + 170q^6 - 832q^7 - 1336q^8 + 6361q^9 + 9996q^{10} + O(q^{11})$$

We summarize the above discussion into the following

THEOREM 7.4. *The mirror map* $-1/\lambda(q)$ *is the Thompson series* $T_{6\mathrm{C}}$, *associated to the genus zero subgroup* $\Gamma_0(6) + 3$.

8. Remarks and Open Problems on Mirror Maps

In this section, we collect some concluding remarks, open problems, and discuss possible generalizations of the mirror moonshine phenomenon to lattice polarized $K3$ surfaces with smaller Picard numbers.

8.1. Commensurability. The mirror map calculated in Section 7.2 for the pencil of $K3$-surfaces constructed by Beukers, Peters and Stienstra gives a counter-example to the mirror-moonshine conjecture 6.2, that is, the reciprocal of the mirror map itself is not a Hauptmodul, but it is algebraically (polynomially) related to a Hauptmodul. This suggests the following definition.

DEFINITION 8.1. Let f and g be two holomorphic functions in some open disc $|z| < R$. Suppose that f, g have power series expansions $f(q)$, $g(q)$ at $q = 0$ ($q = e^{2\pi i z}$). We say that $f(q)$ and $g(q)$ are *commensurable* if there is a polynomial $P(x, y) \in \mathbb{Q}[x, y]$ such that $P(f(q), g(q)) = 0$.

Then the mirror map in Section 7.2 is commensurable with the Thompson series $T_{6\mathrm{B}}$, in fact, if we put $f(q) = T_{6\mathrm{B}}$ and $g(q) = 1/\lambda(q)$, then $P(x, y) = x - y^2$.

The question of which mirror maps are commensurable to a Thompson series was answered by Doran [**Dor**, §§5.3–5.5].

8.2. Transcendence of Mirror Maps. The mirror maps which are commensurable with Thompson series assume transcendental values at algebraic numbers $z \in \mathfrak{H}$ not belonging to imaginary quadratic fields. This follows from a result of Chen and Yui in [**CY**] on singular moduli of Thompson series (or rather the converse to that) that Thompson series $T_g(z)$, $g \in \mathbb{M}$ is transcendental for any algebraic number $z \in \mathfrak{H}$ not belonging to an imaginary quadratic field.

8.3. More Examples of $K3$ Surfaces Arising From Root Lattices. There should be more examples of lattice polarized $K3$ surfaces arising from root lattices.

a) The root lattice $A_1 \times A_2$ gives rise to a lattice polarized $K3$ surface. The character of the adjoint representation gives a polynomial

$$\begin{aligned}\chi_{A_1 \times A_2} &= X_1 + X_2 + X_1 X_2 + X_1^{-1} + X_2^{1} + X_1^{-1} X_2^{-1} + X_3 + X_3^{-1} \\ &= \frac{(1 + X_1 + X_1 X_2)(1 + X_2 + X_2 X_1)}{X_1 X_2} - 3 + X_3 + X_3^{-1}.\end{aligned}$$

We obtain a pencil of $K3$ surfaces:

$$\frac{(1+X_1+X_1X_2)(1+X_2+X_2X_1)}{X_1X_2}=\lambda+3+X_3+X_3^{-1}.$$

The generic $K3$ surface X_λ of this family has Picard-lattice of rank 17, and so we expect the Picard-Fuchs differential equation to be of order 5. The root lattice A_2 gives rise to a pencil of elliptic curves: $(X+Y+Z)(XY+YZ+ZX)=(\lambda+3)XYZ$, as described in Verrill [**V**]. This family was studied by Stienstra and Beukers [**SB**]. Thus the root lattice $A_1\times A_2$ gives rise a family of elliptic $K3$ surfaces, since each $K3$ fiber is given by fixing λ, and as X_3 varies, we get a double cover of the A_2 pencil of elliptic curves.

(b) Stienstra and Dolgachev have pointed out to us that since the C_3 lattice (whose roots will be described below) sits inside the A_1^3 lattice, there should be a pencil of $K3$ surfaces associated to the root system of the C_3 lattice, which will be related to the A_1^3 family by the pull back of a double cover of the base space.

Recall from Example 7.1(a) that the lattice A_1^3 is isomorphic to $\mathbb{Z}^3$, with roots given by $\pm e_i$, where e_1,e_2,e_3 are the standard basis of $\mathbb{R}^3$. The lattice C_3 has roots given by

$$\{\pm 2e_i, e_i\pm e_j \mid 1\le i,j\le 3, i\ne j\}.$$

So, we have corresponding monomials X_i^2, X_iX_j and $X_iX_j^{-1}$. Although the character of the adjoint representation of the lattice C_3 will not define a family of $K3$ surfaces, if we take an alternative choice of weight (i.e, choice of coefficients for the monomials), we obtain

$$\sum X_i^2+\sum X_i^{-2}+2\sum X_iX_j+2\sum X_iX_j^{-1}.$$

Factor this and, set equal to u^2-6, to obtain the following equation:

$$(X_1+X_1^{-1}+X_2+X_2^{-1}+X_3+X_3^{-1})^2-6=u^2-6.$$

This can be rewritten as

$$X_1+X_1^{-1}+X_2+X_2^{-1}+X_3+X_3^{-1}=u,$$

which is the same as one of the families of $K3$ surfaces appearing in Beukers and Peters [**BP**].

The Picard-Fuchs equation for the A_1^3 case can be transformed to the C_3 case by a substitution, replacing the variable λ by the variable u, where $\lambda+\lambda^{-1}=u$. Then the solution to the Picard-Fuchs equation will be the same as the solution in the A_1^3 case, but in terms of a different local variable. Now the local variable u is given by the Hauptmodul, $u=T_{6\mathrm{A}}$. In the notation of Conway and Norton [**CN**], the monodromy group of this $K3$ pencil is $\Gamma_0(6)+$ and the mirror map is the reciprocal of the Thompson series 6A.

In terms of

$$\lambda=T_{12F}^{-1}=\left(\frac{\eta(2\tau)\eta(12\tau)}{\eta(4\tau)\eta(6\tau)}\right)^6,$$

a solution of the Picard-Fuchs equation is

$$\eta(2\tau)\eta(4\tau)\eta(6\tau)\eta(12\tau).$$

It can be verified from relations in Table 3 of Conway and Norton [**CN**] that

$$T_{12C} = T_{12F} + \frac{1}{T_{12F}}$$

and that

$$T_{6\mathrm{A}}(q^2) = (T_{12\mathrm{C}}(q))^2.$$

So, for the C_3 case, if the local variable u is given by $u = T_{6\mathrm{A}}$, a solution to the Picard-Fuchs equation is $\eta(\tau)\eta(2\tau)\eta(3\tau)\eta(6\tau)$.

The details will appear elsewhere.

8.4. From Thompson Series to Mirror Maps: the Inverse Problem. We have discussed several examples of pencils of lattice polarized $K3$ surfaces with Picard number 19 whose mirror maps have turned out to be commensurable with Thompson series of genus zero congruence subgroups of $\mathrm{PSL}_2(\mathbb{R})$ of level N.

Dolgachev has asked the inverse problem: *Given Thompson series and genus zero subgroup $G \subseteq \Gamma_0(N)+$, can one construct a pencil of lattice polarized $K3$ surfaces which admits G as its monodromy group, and its mirror map is commensurable with the prescribed Thompson series*?

For instance, consider the primes N in Theorem 2.1(b). For each prime N, we have $\Gamma_0(N)+ = \Gamma_0(N)^*$. Then the inverse problem is to construct a pencil of $K3$ surfaces with Picard lattice $U_2 \perp (-E_8)^2 \perp \langle -2N \rangle$, and having $\Gamma_0(N)+$ as monodromy group.

8.5. The Inverse Problem: Some Attempts. Given a congruence subgroup $G \subset \mathrm{PSL}_2(\mathbb{Z})$, Shioda [**Sh**] has described the construction of the modular surface, B_G, attached to G, which is a family of elliptic curves parametrized by $\mathfrak{H}/G$.

In [**SB**], under the following assumptions (which are described in [**SB**, (14.4)]):

1. G has no elliptic elements,
2. all parabolic elements have trace 2,
3. the index $[\mathrm{PSL}_2(\mathbb{Z}) : G]$ is divisible by 12, and
4. $\mathfrak{H}/G$ is genus zero,

it is shown that the Picard-Fuchs equation for B_G has a solution in terms of modular forms for the group G (see [**SB**, Theorem 14.9]).

If we take a fiber product over the base of such B_G, then we have a family of Abelian surfaces. If for B_G, the elliptic curve over $\tau \in \mathfrak{H}$ is E_τ, then in the fiber product, the fiber over τ is $E_\tau \times E_\tau$. The base of the space will still be $\mathfrak{H}/G$. Alternatively, we can pair up isogenous curves, under an involution $\sigma \in \mathrm{PSL}_2(\mathbb{R})$, such that $\sigma^2 \in G$. Then we can obtain a family of Abelian surfaces over $\mathfrak{H}/\langle G, \sigma \rangle$, as described in [**Pe**, §5]. In [**Pe**], the description of the Picard-Fuchs equation for these products is given. In the first case, the mirror map will be the reciprocal of a uniformizing parameter for $\mathfrak{H}/G$, and in the second case for $\mathfrak{H}/\langle G, \sigma \rangle$. By making a change of variable, we should be able to write the mirror map as a Thompson series.

So far, this is a family of Abelian surfaces. In the case where we can find a global zero section, and a globally well defined involution on the fibers, we can quotient out by the involution, and, after resolution of singularities, obtain a family of $K3$ surfaces, which will have the same mirror map.

So, we should go through all the groups in the Conway and Norton list in [**CN**], and check which satisfy the conditions of Stienstra and Beukers (14.4) in [**SB**], and for which the quotient by the involution is properly defined on fibers, so that we obtain a family of $K3$ surfaces with mirror map a Thompson series for G, or $\langle G, \sigma >$, for appropriate σ. In the Conway and Norton list in [**CN**], over half the groups have "index" given as divisible by 12. However, these groups are mostly not subgroups of $\mathrm{PSL}_2(\mathbb{Z})$, so the results of Shioda [**Sh**] and [**SB**] would have to be extended, and may apply in slightly different ways. One would expect this would still lead to a large collection of examples of pencils of $K3$ surfaces with mirror map given by a Thompson series.

Confer also Doran [**Dor**, Remark 5.21] where he describes a method for the inverse problem.

9. The Mirror $K3$ Surfaces, and Mirror Symmetry

There are several variants of mirror symmetry for $K3$ surfaces. The origin of mirror symmetry for $K3$ surfaces might be traced back to the strange duality of Arnold [**Ar**], where he found 14 exceptional unimodal hypersurface singularities, and strange dualities among them. Recently, Arnold's strange duality has been recaptured in the context of "mirror symmetry" for $K3$ surfaces.

9.1. Mirror Symmetry for $K3$ Surfaces Due to Dolgachev[Dol]. Dolgachev's starting point was to describe the strange duality of Arnold as a two-dimensional analogue of the mirror symmetry conjecture for Calabi-Yau threefolds. The readers are referred to the paper of Dolgachev [**Dol**] where he discusses the construction of mirror pairs for pseudo-ample lattice polarized $K3$ surfaces.

For pseudo-ample lattice polarized $K3$ surfaces, the mirror symmetry conjecture is formulated as follows: Let $\mathcal{M} \subset \mathcal{L}$ be a primitive sublattice which has a unique embedding into $\mathcal{L}$ and let $\mathcal{M}^{\perp}$ be the orthogonal complement of $\mathcal{M}$ in $\mathcal{L}$. Consider a certain sublattice $\mathcal{N} \subset \mathcal{M}^{\perp}$ admitting an orthogonal splitting $\mathcal{M}^{\perp} = U_2 \perp \mathcal{N}$. Then there exists a pair of pseudo-ample lattice polarized $K3$ surfaces $X_{\mathcal{M}}$ and $X_{\mathcal{N}}$ such that

$$\rho(X_{\mathcal{M}}) + \rho(X_{\mathcal{N}}) = 20.$$

In other words, a mirror partner of $X_{\mathcal{M}}$ exists if and only if the lattice U_2 splits from $\mathrm{Pic}(X_{\mathcal{M}})^{\perp}$.

For instance, if $\mathcal{M} = U_2 \perp (-E_8)^2 \perp < -2N >$ for some integer N, then $\mathcal{M}^{\perp} = U_2 \perp \langle 2N \rangle$, and take the sublattice $\mathcal{N} = \langle 2N \rangle$. Then $(X_{\mathcal{M}}, X_{\mathcal{N}})$ is a mirror pair and it satisfies the equality: $\rho(X_{\mathcal{M}}) + \rho(X_{\mathcal{N}}) = 19 + 1 = 20$.

If $\mathcal{M} = U_2 \perp (-E_8)$, then $\mathcal{M}^{\perp} = U_2 \perp \mathcal{N}$ with $\mathcal{N} = U_2 \perp (-E_8)$. So $(X_{\mathcal{M}}, X_{\mathcal{N}})$ is a mirror pair.

M. Reid (unpublished) and Yonemura [**Yos**] listed all 95 families of weighted projective Gorenstein $K3$ hypersurfaces in terms of degree and weight. Recently, S-M. Belcastro [**Be**] has computed, using toric geometry, lattice theory and elliptic surface techniques, the Picard lattices for the Reid's 95 families. Further she has shown that the Reid's list is not closed under the mirror symmetry, that is, there are families of $K3$ surfaces in the Reid's list whose mirror families are not on his list.

The actual construction of lattice polarized $K3$ surfaces with prescribed Picard lattice, and its mirror partner remains an open problem.

9.2. Variants of Mirror Symmetry for $K3$ Surfaces. There are other dualities due to several authors which may be regarded as mirror symmetry for $K3$ surfaces. Due to lack of time and space, we are not going into detailed discussions on these variants; the interested readers are referred to the following articles.

(a) Kobayashi's duality of weight systems [**Ko**]. A *weight system* is a quadruple $W = (w_1, w_2, w_3; N)$ of positive integers with $N = c_1w_1 + c_2w_2 + c_3w_3$ ($c_i \in \mathbb{N}$); w_i are called weights and N the degree. W is *reduced* if $\gcd(w_1, w_2, w_3) = 1$. Two reduced weight systems $W = (w_1, w_2, w_3; N)$ and $W' = (w_1', w_2', w_3'; N')$ are said to be *dual* if there exists a 3×3 matrix Q with non-negative entries such that

$$(w_1, w_2, w_3)Q = (N, N, N) \quad \text{and} \quad Q(w_1', w_2', w_3')^t = (N', N', N')^t.$$

He proved that for the weight system of one of the 14 exceptional unimodal singularities, there exists a unique dual weight system, which belongs to the dual singularity in the sense of Arnold.

(b) Batyrev's polar duality between certain special convex polyhedra [**Ba**]. Let M be a free Abelian group of rank r, $N := \mathrm{Hom}(M, \mathbb{Z})$ the dual group, and $\langle\ ,\ \rangle$ the natural non-degenerating pairing. Denote by $M_{\mathbb{R}}$ (resp. $N_{\mathbb{R}}$) the scalar extension $M \otimes_{\mathbb{Z}} \mathbb{R}$ (resp. $N \otimes_{\mathbb{Z}} \mathbb{R}$). A convex r-dimensional polyhedron $\Delta \subset M_{\mathbb{R}}$ is called *reflexive* if

(i) all vertices of Δ belong to the lattice $M \subset M_{\mathbb{R}}$,
(ii) the zero vector belong to the interior of Δ, and
(iii) all vertices of the polar polyhedron $\Delta^* := \{b \in N_r \mid \langle a, b\rangle \geq -1, a \in \Delta\}$ belong to the dual lattice $N \subset N_{\mathbb{R}}$.

If $\Delta \subset M_{\mathbb{R}}$ is a reflexive polyhedron, then $\Delta^* \subset N_{\mathbb{R}}$ is again a reflexive polyhedron, and $(\Delta^*)^* = \Delta$. So this gives rise to a natural involution $\Delta \leftrightarrow \Delta^*$, and *polar duality*. Toric geometry is the mathematical tool to understand polar dualities of convex polyhedra. Let $\mathbb{P}_\Delta$ denote the toric variety associated to reflexive polyhedra Δ. Let $T_M := \mathrm{Spec}\,\mathbb{C}[M]$ be the algebraic torus with lattice of characters M, and let $Z_f \subset T_M$ be the affine hypersurface in T_M defined by the equation

$$f(x_1, x_2, \ldots, c_r) = \sum_{m \in M \cap \Delta} a_m x^m = 0$$

where the set $\{a_m\}_{m \in M \cap \Delta}$ consists of generically chosen complex numbers. Then the projective closure of Z_f in the toric variety $\mathbb{P}_\Delta$ is a normal irreducible variety $\overline{Z}_f$ having trivial canonical class. One may repeat the construction to the polar reflexive polyhedron Δ^* and the dual torus $T_N := \mathrm{Spec}\,\mathbb{C}[N]$, to obtain another affine hypersurface $Z_g \subset T_N$ defined by the equation

$$g(y_1, y_2, \ldots, y_r) = \sum_{n \in N \cap \Delta^*} b_n y^n = 0.$$

Denote by $\overline{Z}_g \subset \mathbb{P}_{\Delta^*}$ the projective compactification of Z_g in $\mathbb{P}_{\Delta^*}$. Then the pair $(\overline{Z}_f, \overline{Z}_g)$ is conjectured to be *mirror symmetric.*

In Section 7.1, we discussed several examples of toric varieties arising from root lattices. However, we did not construct mirror partners of those varieties.

(c) The notion of arithmetic mirror symmetry of Gritsenko and Nikulin: they have developed the general theory for reflective lattices T with 2 negative squares and reflective automorphic forms on homogeneous domains of type IV defined by T, which they have regarded as mirror symmetric to the theory of elliptic and parabolic hyperbolic reflection groups and corresponding hyperbolic root systems.

There are vast amount of works by the same authors, and we do not have space nor time to discuss their work here in detail; see series of papers by the same authors on automorphic forms related to the Lorentzian Kac-Moody algebras and mirror symmetry, for instance, [**GN1, GN2, GN3**].

(d) Kyoji Saito's dualities of characteristic polynomials of the monodromy operators of the 14 exceptional unimodal singularities, and mysterious connection with characteristic polynomials of the automorphisms of the Leech lattice [**Sa**]. Let h be a positive integer (so-called the Coxeter number), and let $\phi(\lambda)$ and $\phi^*(\lambda)$ be the cyclotomic polynomials whose roots are h-th roots of unity. We can decompose $\phi(\lambda)$ and $\phi^*(\lambda)$ as follows:

$$\phi(\lambda) = \prod_{m|h} (\lambda^m - 1)^{\chi_m}, \quad \phi^*(\lambda) = \prod_{m|h} (\lambda^m - 1)^{\chi^*_m}$$

where the exponents χ_m and χ^*_m are integers subject to the relation $\chi_m + \chi^*_{h/m} = 0$. The polynomial $\phi(\lambda)$ is considered as the characteristic polynomial of the monodromy for one of the 14 exceptional unimodal singularities. Then one can associate a modular function (a product of the η-function) by defining

$$\eta_\phi(z) := \prod_{m|h} \eta(mz)^{\chi_m}.$$

Similarly, we have

$$\eta_{\phi^*}(z) := \prod_{m|h} \eta(mz)^{\chi^*_m}.$$

Saito [**Sa**] proved that the relation between the exponents $\chi(m)$ and $\chi^*_{h/m}$ is equivalent to the identity

$$\eta_\phi(-\frac{1}{hz})\eta_{\phi^*}(z)\sqrt{d} = 1$$

where $d := \prod m^{\chi_m}$ is called the discriminant. (This is a generalization of the classical identity $\eta(-\frac{1}{z}) = \sqrt{-iz}\eta(z)$.) It follows from this identity that $\eta_{\phi\phi^*}$ is a modular function for the group $\Gamma_0(N)$.

Kyoji Saito's work is purely of arithmetic nature, and does not seem to address geometry behind his duality. This gives rise to

QUESTION. *Are there any pencils of (lattice polarized) $K3$ surfaces whose mirror maps have any relation to $\eta_{\pi\pi^*}$?*

(e) Ebeling's paper [**Eb**] on the dualities of Dolgachev, Kobayashi, and Saito, and their generalizations and extensions (especially new examples of mirror symmetric families of $K3$ surfaces). He assigns to a weight system $(w_1, w_2, w_3; N)$ defining a weighted $K3$ hypersurface a certain product of η-functions, generalizing the works of Kobayashi and K. Saito mentioned above.

10. Generalizations

10.1. Generalizations. Let $\mathcal{M} \hookrightarrow \mathcal{L}$ and let $X_\mathcal{M}$ be an $\mathcal{M}$-polarized $K3$ surface with a fixed primitive embedding $\mathcal{M} \to \mathrm{Pic}(X_\mathcal{M})$ such that the image of $\mathcal{M}$ contains a pseudo-ample divisor class. Let $\mathcal{T} = \mathcal{M}^\perp$ and let $\mathcal{T}'$ be the dual of $\mathcal{T}$.

Nikulin [**N2**] (see also a series of papers by Gritsenko and Nikulin, e.g., [**GN1, GN2, GN3**]), and Borcherds *et al.* [**BKPS**] have addressed the question of the existence of automorphic forms on the period space of marked $\mathcal{M}$-polarized $K3$ surfaces with Picard number $t+1 \leq 20$.

Nikulin [**N2**] conjectured that there are only finite number of lattices $\mathcal{M}$ such that there is an automorphic form vanishing only on $\mathcal{M}$-polarized $K3$ surfaces which have a norm -2 vector in $\mathcal{T}'$. Several examples of lattices are provided by Nikulin [**N2**]. Gritsenko and Nikulin (in a series of papers, e.g., [**GN1, GN2, GN3**]) and also Borcherds *et. al.* [**BKPS**] have constructed those automorphic forms of weight 12 on the moduli space of $\mathcal{M}$-polarized $K3$ surfaces. These automorphic vanish on divisors $v^\perp$ of vectors v in $\mathcal{T}'$ with $0 > (v,v) \geq -2$.

Are there any relation of these automorphic forms to mirror maps of lattice polarized $K3$ surfaces?

Borcherds *et al.*, and Gritsenko and Nikulin have addressed this question via Kac-Moody algebras. Nikulin [**N2**] proved that this is not true in general. Further, Gritsenko and Nikulin in a series of papers, e.g., [**GN2, GN3**] have given partial answer to this question that reflective automorphic forms are very exceptional; they have listed possible lattices which give rise to reflective automorphic forms. (Also they gave almost classification of such lattices of rank 17.)

We are not going into detailed discussions on this question; we merely illustrate the general phenomena by a couple of examples.

We consider pencils of lattice polarized $K3$ surfaces with Picard number 18, and compute their mirror maps. These examples give rise to "bi-modular" moduli spaces and the mirror maps are identified with Thompson series in some cases.

EXAMPLE 10.1. Let $\mathcal{M} \subset \mathcal{L}$ be a primitive sublattice of rank 18 given by

$$\mathcal{M} = U_2 \perp (-E_8)^2, \quad \text{so } \mathcal{M}^\perp = U_2^2.$$

Consider $\mathcal{M}$-polarized $K3$ surfaces. By the results in Section 4.2, these $K3$ surfaces admit Shioda-Inose structures. The moduli space $\mathfrak{M}_{K3,\mathcal{M}}$ is of dimension 2. We have several examples of $\mathcal{M}$-polarized $K3$ surfaces where the period maps take values in $\mathfrak{H} \times \mathfrak{H}$ and the mirror maps are expressed in terms of Thompson series. We will give a brief exposition of this connection; the details will be discussed in [**VY**].

Consider the two-parameter deformation of the weighted diagonal $K3$ surface with $m=8$ and $Q=(q_0,q_1,q_2,q_3)=(1,1,2,4)$. It is defined by the equation

$$X_0^8 + X_1^8 + X_2^4 + X_3^2 + z\,X_0X_1X_2X_3 + w\,X_0^4X_1^4 = 0$$

in the orbifold ambient space $\mathbb{P}^3((1,1,2,4))/G$ where $G=(\mathbb{Z}/2\mathbb{Z})\times(\mathbb{Z}/4\mathbb{Z})$. It is shown that the minimal resolution of this surface is an $\mathcal{M}$-polarized $K3$ surface. Now introduce redundant deformation parameters, $\mathbf{c}=(c,c_0,c_1,c_2,c_3,c_4)$, and let

$$f(\mathbf{c}) = cX_0X_1X_2X_3 + c_0X_0^8 + c_1X_1^8 + c_2X_2^4 + c_3X_3^2 + c_4X_0^4X_1^4 = 0.$$

then the period integral is defined by

$$\omega_0(x,y) = \int \frac{c}{f(\mathbf{c})}\,d\mu$$

where

$$d\mu = \sum (-1)^i q_i dX_0 \wedge \ldots \widehat{dX_i} \wedge \cdots \wedge dX_3$$

and the local coordinates (for the deformation space) are given by $x=(c_2c_3^2c_4)/c^4$ and $y=c_0c_1/c^2$. The Picard-Fuchs differential equation may be determined using the GKZ hypergeometric system (cf. Hosono-Lian and S. T. Yau [**HLY**]). There are four solutions altogether to the Picard-Fuchs differential equation. The fundamental solution about $x = y = 0$ is given by

$$\omega_0(x,y) = \sum_{n,m} c(n,m)x^n y^m \quad \text{where } c(n,m) = \frac{(4n)!}{(2n!)n!(m!)^2(n-2m)!}.$$

The other two solutions are obtained as follows: Replace n, m by $n+\rho_1$, $m+\rho_2$ in $\omega_0(x,y)$. Then we have

$$\omega_1 = \partial_{\rho_1}\omega_0|_{\rho_1=0}, quad\omega_2 = \partial_{\rho_2}\omega_0|_{\rho_2=0}$$

where ∂_{ρ_k} 's represent the derivatives with respect to the indices introduced in the standard way in $\omega_0(x,y)$.

Now put

$$q_1 := e^{\omega_1/2\pi i\omega_0}, \quad q_2 := e^{\omega_2/2\pi i\omega_0}$$

and express the variables x, y in terms of the new variables q_1, q_2. The mirror maps are the inverses and given by

$$x(q_1,q_2) = \frac{1}{T_{2\mathrm{B}}(q_1)} + \frac{1}{T_{2\mathrm{B}}(q_1q_2)} - 128\frac{1}{T_{2\mathrm{B}}(q_1)T_{2\mathrm{B}}(q_2)}$$

and

$$y(q_1,q_2) = \frac{1}{T_{2A}(q_1)T_{2A}(q_1q_2)}\frac{1}{x(q_1q_2)^2}.$$

Therefore, the mirror maps are commensurable with Thompson series $T_{2\mathrm{A}}$ and $T_{2\mathrm{B}}$ associated to the groups $\Gamma_0(2)^*$ and $\Gamma_0(2)$, respectively. The period domain is explicitly given by $\mathfrak{H} \times \mathfrak{H}/\Gamma_0(2) \times \Gamma_0(2)^* \simeq \mathfrak{H}/\Gamma_0(2) \times \mathfrak{H}/\Gamma_0(2)^*$, which is a "bimodular" space of genus zero (or a Hilbert modular space).

EXAMPLES 10.2 (These are also "bimodular" examples.). (a) Carrying out the similar analysis as in Example 10.1 for a pencil of $K3$ surfaces with $\langle m, Q\rangle = \langle 6, (1,1,2,2)\rangle$, we can compute the mirror maps. They are commensurable with Thompson series $T_{3\mathrm{A}}$, $T_{3\mathrm{B}}$ associated to the genus zero groups $\Gamma_0(3)^*$, $\Gamma_0(3)$. The moduli space is again a product of two one-dimensional spaces: $\mathfrak{H}/\Gamma_0(3) \times \mathfrak{H}/\Gamma_0(3)^*$.

(b) Consider the $K3$ surface with $\langle m, Q\rangle = \langle 12, (1,1,4,6)\rangle$. In this case, we can also compute the mirror maps. However, they no longer correspond to genus zero subgroups. In fact, the mirror maps are given by $T_{1\mathrm{A}}, \widetilde{T}(q)$ where $T_{1\mathrm{A}}$ is the Thompson series of level 1, but $\widetilde{T}(q)$ is the modular function associated to the group $G := \langle \left(\begin{smallmatrix} 0 & 1 \\ -1 & 1 \end{smallmatrix}\right), \left(\begin{smallmatrix} 1 & 2 \\ 0 & 1 \end{smallmatrix}\right)\rangle$, which is a genus one modular group. In this case, the moduli space is isomorphic to $\mathfrak{H}/\Gamma \times \mathfrak{H}/G$.

EXAMPLE 10.3. Matsumoto [**Ma**] computed the inverse of the period map of the 4-dimensional family of $K3$ surfaces, the double cover of $\mathbb{P}^2$ branching along 6 lines. This family was discussed in detail in the book of Yoshida [**Yos**]. Let

$$L_j = \{(t_1,t_2,t_3) \in \mathbb{P}^2 \mid x_{1j}t_1 + x_{2j}t_2 + x_{3j}t_3 = 0\} \quad (1 \le j \le 6)$$

be six lines in general position on $\mathbb{P}^2$ with homogeneous coordinates (t_1, t_2, t_3), and let $\mathbf{L}$ be the system of ordered six lines. Let $S'(\mathbf{L})$ be the double cover of $\mathbb{P}^2$ branching along the union of six lines, and let $S(\mathbf{L})$ be the smooth surface obtained from $S'(\mathbf{L})$ by replacing $\binom{6}{2} = 15$ singular points by 15 smooth rational curves. Then $S(\mathbf{L})$ is a $K3$ surface with a nowhere vanishing holomorphic 2-form:

$$\omega(\mathbf{L}) = \prod_{j=1}^{6}(x_{1j}t_1 + x_{2j}t_2 + x_{3j}t_3)^{-1/2}dt$$

where $dt = t^1dt^2 \wedge dt^3 + t^3dt^1 \wedge dt^2 + t^2dt^3 \wedge dt^1$. This family of $K3$ surfaces has 16 algebraic cycles, namely, 15 algebraic curves arising from the desingularization and the pull back of the generic line in $\mathbb{P}^2$. The Picard lattice has rank 16, and the transcendental lattice is a rank 6 lattice of the form $U_2(2) \perp U_2(2) \perp (-I_2)(2)$. (Here $U_2(2) = \left(\begin{smallmatrix} 0 & 2 \\ 2 & 0 \end{smallmatrix}\right)$, and $I_2(2)$ is the 2 times of the 2×2 identity matrix.) The Picard lattice is then $(-E_8)^2$. The moduli space is a bounded symmetric domain of type IV of dimension 4. The period map is defined as follows. Choose a system of cycles $c_j(\mathbf{L})$ $(1 \le j \le 6)$. Then the periods are defined by

$$z_j(\mathbf{L}) := \int_{c_j(\mathbf{L})} \omega(\mathbf{L}) \quad (1 \le j \le 6).$$

Let $\mathbf{X}(3,6)$ denote the configuration space of six lines in $\mathbb{P}^2$. Then the period map is computed as a map $\mathbf{X}(3,6) \to \mathbb{P}^5 : \mathbf{L} \to [z_1(\mathbf{L}), \dots, z_6(\mathbf{L})]$. The period domain is given as follows. Let $\mathbf{H}_2$ denote a bounded symmetric domain of type $I_{2,2}$:

$$\mathbf{H}_2 := \left\{ W \in \mathrm{Mat}(2,2,\mathbb{C}) \;\middle|\; \frac{W - W^*}{2i} > 0 \right\}$$

where $\mathrm{Mat}(2,2,\mathbb{C})$ is the set of 2×2 complex matrices, $W^* :=^t \overline{W}$ and $i = \sqrt{-1}$. On $\mathbf{H}_2$, the transposition T and the group $U(2,2) := \{g \in \mathrm{GL}(4,\mathbb{C}) \mid g^*Jg = J\}$ with $J := \left(\begin{smallmatrix} 0 & I_2 \\ -I_2 & 0 \end{smallmatrix}\right)$ act according to the following rules:

$$T \cdot W = {}^tW, : g \cdot W = (AW + B)(CW + D)^{-1}$$

for $g \in \mathbf{H}_2$ and $g = \left(\begin{smallmatrix} A & B \\ C & D \end{smallmatrix}\right) \in U(2,2)$. ($\mathbf{H}_2$ looks like a generalized Siegel upper half plane.) The moduli space is then described explicitly. First the monodromy group of the period map is $\Gamma_M(1+i)$ which is the index 2 subgroup of the semi-direct product

$$\Gamma_T(1+i) := \Gamma(1+i) \rtimes \langle T \rangle$$

where $\Gamma(1+i)$ is the principal congruence modular group of level $(1+i)$ defined by

$$\Gamma(1+i) := \{g \in \mathrm{GL}(4,\mathbb{Z}[i]) \mid g^*Jg = J, g \equiv I_4 \bmod (1+i)\}$$

and $\langle T \rangle$ denotes the group generated by the transposition T. The moduli space is given by $\mathbf{H}_2/\Gamma_M(1+i)$. Next the theta functions Θ are defined on $\mathbf{H}_2$ with characters $a, b \in \{0, (1+i)/2\}^2$, $a^*b \in \mathbb{Z}$ as follows:

$$\Theta \begin{bmatrix} a \\ b \end{bmatrix} (W) := \sum_{n \in \mathbb{Z}[i]^2} \mathbf{e}[(n+a)^*W(n+a) + 2\,\mathrm{Re}\{b^*n\}].$$

Their squares Θ^2 are holomorphic modular forms of weight 2 with respect to the group $\Gamma_T(1+i)$ with character det. There are 10 linearly independent theta functions and this gives rise to the map $F : \mathbf{H}_2/\Gamma_T(1+i) \to \mathbb{P}^{10-1}$ by

$$W \to [\dots, \Theta \begin{bmatrix} a \\ b \end{bmatrix} (W)^2, \dots].$$

This is the inverse of the period map $\mathbf{X}(3,6)/\heartsuit \to \mathbf{H}_2/\Gamma_T(1+i)$. (Here $\heartsuit$ is an involution on $\mathbf{X}(3,6)$ which is equivalent under the period map, to the action of T on $\mathbf{H}_2$.) The final step is to lift this to the inverse of the period map $\mathbf{X}(3,6) \to \mathbf{H}_2/\Gamma_M(1+i)$. This is done by making Θ into a holomorphic modular form of weight 4 with respect to the group $\Gamma_T := \{g \in \mathrm{PGL}(4,\mathbb{Z}[i]) \mid g^*Jg = J\} \rtimes \langle T\rangle$, which has simple zeros on the branching locus of the covering $\mathbf{H}_2/\Gamma_M(1+i) \to \mathbf{H}_2/\Gamma_T(1+i)$.

However, the mirror map does not seem to have any interpretation in terms of Thompson series.

This is a brute force construction of the mirror map for this 4-dimensional family of $K3$ surfaces. One natural question is : *Can this family of $K3$ surfaces be realized via orbifold or toric construction*?

Acknowledgments

We would like to thank I. Dolgachev, J. Stienstra and S.-T. Yau for their help, interests in this paper (and a monograph in preparation) and encouragements during its preparation.

We are also indebted to S. Hosono for his explaining physics around the mirror symmetry conjecture and mirror maps, Masahiko Saito for answering many geometric questions, C. Doran and B. Lian for their help in clarifying many questions concerning Picard-Fuchs differential equations, J. McKay and A. Sebbar for answering questions on Moonshine and Hauptmoduln (and Thompson series), and I. Chen and Y. Goto for reading through and pointing out several improvements to the earlier version. We thank also K. Yoshikawa for bringing the paper by Matsumoto to our attention.

Special thanks are due to D. Zagier for his constructive criticisms which are very helpful clarifying many ambiguous points in the paper, and also for making available his manuscript [**Za**] in preparation, where he derives a simple way of computing the mirror maps of some of our examples (e.g., Examples 7.1 and in Section 7.2 as an application of relations among derivatives of modular forms.

During the preparation of this note, N. Yui was partially supported by a research grant from NSERC. H. Verrill was a postdoctoral fellow at Queen's University supported by N. Yui's NSERC research grant for the period of two years from 1996 April. Subsequently Verrill held a postdoctoral fellowship for a year at Max-Planck Institut für Mathematik, Bonn. After May 1999, Verrill will take up a postdoctoral fellowship at University of Copenhagen.

The paper was completed at Max-Planck-Institut für Mathematik, Bonn in the spring of 1999 where both Verrill and Yui were members of the institute.

References

[Ar] V. I. Arnold, *Critical points of smooth functions and their normal forms*, Uspehi Mat. Nauk. **30** (1975), 3–65 (Russian); English transl., Russian Math. Surveys **30** (1975), 1–75.

[AKN] T. Asai, M. Kaneko, and H. Ninomiya, *Zeros of certain modular functions and an application*, Comment. Math. Univ. St. Pauli **46** (1997), 93–101.

[Ba] V. Batyrev, *Dual polyhedra and mirror symmetry for Calabi-Yau hypersurfaces in toric varieties*, J. Algebraic Geom. **3** (1994), 493–535.

[Be] S.-M. Belcastro, *Picard lattices of families of $K3$ surfaces*, math.AG/9809008, 1998.

[BP] F. Beukers and C. Peters, *A family of $K3$ surfaces and $\zeta(3)$*, J. Reine Angew. Math. **351** (1984), 42–54.

[Bo] R. E. Borcherds,*Monstrous moonshine and monstrous Lie superalgebras*, Invent. Math. **109** (1992), 405–444.

[BKPS] R. Borcherds, L. Katzarkov, T. Pantev, and N. Shepherd-Barron, *Families of $K3$ surfaces*, J. Algebraic Geom. **7** (1998), 183–193.

[CN] J Conway and S. Norton, *Monstrous moonshine*, Bull. London Math. Soc. **11** (1979), 308–339.

[CG] C. J. Cummins and T. Gannon, *Modular equations and the genus zero property of moonshine functions*, Invent. Math. **129** (1997), 413–443.

[CY] I. Chen and N. Yui, *Singular values of Thompson series*, Groups, Difference Sets, and the Monster (Columbus, OH, 1993), Ohio State Univ. Math. Res. Inst. Publ., vol. 4, de Gruyter, Berlin, 1996, pp. 255–326.

[Dol] I. Dolgachev, *Mirror symmetry for lattice polarized $K3$ surfaces*, J. Math. Sci. **81** (1996), 2599–2630.

[Dor] C. Doran, *Picard-Fuchs uniformization: Modularity of the mirror maps and mirror-moonshine*, in this volume.

[Eb] W. Ebeling, *Strange duality, mirror symmetry, and the Leech lattice*, alg-geom/9612010, 1996.

[Go] Y. Goto, *Arithmetic of certain weighted diagonal surfaces over finite fields*, J. Number Theory **59** (1996), 37–81.

[GN1] V. A. Gritsenko and V. V. Nikulin, *$K3$ surfaces, Lorentzian Kac-Moody algebras and mirror symmetry*, Math. Res. Lett. **3** (1996), 211–229.

[GN2] ———, *Automorphic forms and Lorentzian Kac-Moody algebras*. I, Internat. J. Math. **9** (1998), 153–199.

[GN3] ———, *Automorphic forms and Lorentzian Kac-Moody algebras*. II, Internat. J. Math. **9** (1998), 201–275.

[HLY] S. Hosono, B. H. Lian, and S.-T. Yau, *GKZ-generalized hypergeometric systems in mirror symmetry of Calabi-Yau hypersurfaces*, Comm. Math. Phys. **182** (1996), 535–577.

[Ko] M. Kobayashi, *Duality of weights, mirror symmetry and Arnold's strange duality*, 1994 (preprint).

[LY1] B. H. Lian and S.-T. Yau., *Mirror maps, modular relations and hypergeometric series*, XIth International Congress of Mathematical Physics (Paris, 1994), International Press, Cambridge, MA, 1995, pp. 163–184.

[LY2] ———, *Arithmetic properties of mirror maps and quantum coupling*, Comm. Math. Phys. **176** (1996), 163–191.

[Ma] K. Matsumoto, *Theta functions on the bounded symmetric domain of type $I_{2,2}$ and the period map of a 4-parameter family of $K3$ surfaces*, Math. Ann. **295** (1993), 383–409.

[Mo] D. Morrison, *On $K3$ surfaces with large Picard number*, Invent. Math. **75** (1984), 105–121.

[N1] V. V. Nikulin, *Integer symmetric bilinear forms and some of their applications*, Izv. Akad. Nauk. SSSR **43** (1979), 111–177 (Russian).

[N2] ———, *A remark on discriminants of $K3$ surfaces moduli as sets of zeros of automorphic forms*, J. Math. Sci. **81** (1996), 2738–2743.

[Og] A. P. Ogg, *Hyperelliptic modular curves*, Bull. Soc. Math. France **102** (1974), 449–462.

[Pe] C. Peters, *Monodromy and Picard-Fuchs equations for families of $K3$ surfaces and elliptic curves*, Ann. Sci. École Norm. Sup. (4) **19** (1986), 583–607.

[PS] C. Peters and J. Stienstra, *A pencil of $K3$ surfaces related to Apéry's recurrence for $\zeta(3)$ and Fermi surfaces for potential zero*, Arithmetic of Complex Manifolds (Erlangen, 1988) (W.-P. Barth and H. Lange, eds.), Lecture Notes in Math., vol. 1399, Springer, Berlin 1989, pp. 110–127.

[Sa] K. Saito, *On a duality of characteristic polynomials for regular systems of weights*, preprint no. RIMS-993, Kyoto University, 1994; *Duality for regular systems of weights*, Asian J. Math. **2** (1998), 983–1048.

[Sh] T. Shioda, *On elliptic modular surfaces*, J. Math. Soc. Japan **24** (1972), 20–59.

[SI] T. Shioda and K. Inose, *On sngular $K3$ surfaces*, Complex Analysis and Algebraic Geometry (W. Baily and T. Shioda, eds.) Kinokuniya Shoten, Tokyo, 1977, pp. 119–136.

[SB] J. Stienstra and F. Beukers, *On the Picard-Fuchs equation and the formal Brauer group of certain elliptic $K3$ surfaces*, Math. Ann. **271** (1985), 269–304.

[Th] J. G. Thompson, *A finiteness theorem for subgroups of* $\mathrm{PSL}(2,\mathbb{R})$ *which are commensurable with* $\mathrm{PSL}(2,\mathbb{Z})$, The Santa Cruz Conference on Finite Groups (Santa Cruz, CA,

1979), Proc. Symp. Pure Math., vol. 37, Amer. Math. Soc. Providence, RI, 1980, pp. 533–555

[V] H. Verrill, *Root lattices and pencils of varieties*, J. Math. Kyoto Univ. **36** (1966), 423–446.

[VY] H. Verrill and N. Yui, *The mirror maps of certain pencils of $K3$ surfaces and Thompson series* (in preparation).

[Yon] T. Yonemura, *Hypersurface simple $K3$ singularities*, Tôhoku J. Math. (2) **42** (1990), 351–380.

[Yos] M. Yoshida, *Hypergeometric functions, my love*, Aspects Math. vol. E32, Friedr. Vieweg & Sohn, Braunsweig, 1997.

[Za] D. Zagier, *Derivatives of modular forms and their applications* (in preparation).

Max-Planck-Institut für Mathematik, Vivatsgasse 7, 53111 Bonn, Germany

Current address: Matematisk Institut Københavns Universitet, Universitetsparken 5, DK-2100 København Ø, Denmark

E-mail address: verrill@math.ku.dk

Department of Mathematics and Statistics, Queen's University Kingston, Ontario K7L 3N6, Canada

E-mail address: yui@mast.queensu.ca